New Zealand, with its long isolation from other lands, and latitudes extending from subtropical to subantarctic, has a unique flora and highly diverse vegetation. This book is a comprehensive description of the vegetation ranging from its origins to the major communities that the plants exist within. The text, supported by over 300 photographs, maps and diagrams, will make an outstanding contribution to the understanding of the biology of these islands.

VEGETATION OF NEW ZEALAND

VEGETATION OF NEW ZEALAND

Peter Wardle

Land Resources Division
Department of Scientific and Industrial Research

THE BLACKBURN PRESS

Vegetation of New Zealand

ISBN-10: 1-930665-58-X
ISBN-13: 978-1-930665-58-3

Library of Congress Control Number: 2002106382

THE BLACKBURN PRESS
P. O. Box 287
Caldwell, New Jersey   07006 U.S.A.
973-228-7077
www.BlackburnPress.com

Dedicated to
Professor Geoff Baylis
whose inspiration enabled the *Vegetation
of New Zealand* to be commenced; and
Margaret Wardle
whose patience enabled it to be completed.

PREFACE
to the 2002 Reprinting

The twelve years since the publication by Cambridge University Press of "Vegetation of New Zealand" have seen advances in vegetation science in New Zealand, and continuing changes in the vegetation itself. The references in this preface provide a lead into the recent New Zealand literature, and they are largely selected for their reviews rather than their content of original research. Many papers published in the *New Zealand Journal of Botany* and the *New Zealand Journal of Ecology* report progress in topics of traditional concern, such as regeneration patterns in forest, the effects of introduced plants and herbivores, the ecology of introduced plants perceived as conservation weeds, carbon assimilation and nutrient cycling, and the declining indigenous content of short tussock grasslands. The last phenomenon is especially apparent in the spectacular increase in *Hieracium* spp., and considerable efforts have been made to understand the complex reasons for this (1). There also have been comparisons of the structure and composition of New Zealand vegetation with other parts of the world, especially tropical pine and temperate South America, in view of shared physiognomic and floristic features (2-4).

The publication of a modern grass flora has been of major benefit to ecologists (5). By 2000, vascular plant species with validly published names totalled nearly 2,090 native and 1,755 fully naturalised (6). New names for native species include some previously recognised by tag names, but there have also been reductions to synonymy, especially within *Carmichaelia*, which now has 23 species, even after subsuming three small genera previously regarded as endemic to New Zealand (7). Since the previous edition of "Vegetation of New Zealand" was written, 75 more introduced plant species have become fully naturalised and a further 119 have been reported as occasionally growing wild; all but eight of these had been introduced through horticulture (8, 9).

Application of molecular biology and cladistics is improving understanding of the flora and its relationships. In particular, it has become clearer that the large number of species in genera such as *Ranunculus*, *Myosotis* and *Hebe* results from arrival of ancestral taxa by transoceanic dispersal in the late Tertiary, followed by rapid diversification (10, 11). This may also apply to the complex of gnaphalioid taxa traditionally assigned to *Helichrysum, Gnaphalium, Raoulia, Leucogenes* and *Ewartia* (12). However, that New Zealand has itself been a centre for dispersal is also supported by recent work in, for example, *Myosotis* (11) and *Metrosideros* (13).

A recent paper has reasserted the view that the New Zealand forest flora is basically Gondwanan, albeit greatly modified by mid-Tertiary extinctions and subsequent immigration from the north-west via former island "stepping-stones" (14). Other authors have presented arguments for more recent origins, including suggestions that the forest flora may be entirely derived from ancestors that reached New Zealand, mainly from Australia, long after the Tasman Sea formed (15,16). Estimations of the times of separation of New Zealand species from their nearest overseas relatives are providing increasing support for the latter viewpoint. Even some extant genera represented by Cretaceous or early Tertiary fossils seem to have disappeared from the record and then reappeared, perhaps through reintroduction. *Nothofagus menziesii* in subgenus *Lophozonia* seems critical in this respect. On the one hand, its present distribution within New Zealand, indicating an ability to disperse only over a few km, and the presence of the *Lophozonia* pollen type in the New Zealand record from the upper Cretaceous suggest that its ancestors were in New Zealand before the Tasman Sea opened. On the other hand, molecular evidence indicates that it diverged from its Australian relatives in subgenus *Lophozonia* as recently as the Miocene (17).

Knowledge of the quaternary history of New Zealand vegetation continues to be extended. During the last glacial, maximum forest remained extensive north of lat. 37°S, but southwards it became increasingly limited, with the extent and location of refugia still being unclear. From the middle of the Holocene, *Nothofagus* increased and expanded into areas previously occupied by conifer/broad-leaf forests, and it seems likely that this was a response to a shift to less equable climates, characterised by disturbance events that favoured the establishment of *Nothofagus* seedlings (18, 19). Opinions remain divided concerning the present regional "beech gaps", though they may not be altogether incompatible. Leathwick suggests that explanations lie in historical events rather than present environments (20), whereas Haase considers that outliers of *Nothofagus fusca* near the northern boundary of the Westland beech gap have become established beyond the usual environmental limits where they can compete with the dominants of conifer/broad-leaf forest, through occupying sites disturbed by earthquakes 250-350 years ago (21).

Recent archeological and palynological evidence suggests that Polynesian colonisation of New Zealand, leading to direct and indirect influences on vegetation, was as recent as 1200-1400 AD (22), though still-contentious dating on bones suggests that *Rattus exulans* may have been introduced several centuries earlier (around 2000 BP) as a result of landfalls by earlier voyagers (23).

Continuing botanical exploration˙is furthering understanding of the relationships between plant communities, species and environments, for example in geologically diverse north-west Nelson, the Otago plateau mountains, and far-southern Campbell

Island (24-26). Some new species have also been discovered, especially in rugged parts of Northland and limestone terrain in the South Island (27). Further populations of local or threatened species are also turning up, e.g. *Pittosporum obcordatum, Hebe armstrongii* and *Coprosma pedicillata,* although this is offset by losses or reductions in other localities or of other species. Some threatened shrubs that occupy recent alluvial or colluvial surfaces, such as *Muehlenbeckia astonii, Olearia hectorii* and *Hebe cupressiodes,* survive as adults in scattered localities whereas seedlings are scarce or absent, apparently because they succumb to browsing in grazed areas and to competition from introduced grasses where grazing is excluded (28, 29). Continued survival may depend on intensive management or cultivation in gardens. Many lowland herbs, for example, *Urtica linearifolia, Chenopodium detestans, Ischnocarpus novae-zelandiae* and *Lepidium kirkii,* are in similar circumstances. Several rare native plants, including *Scutellaria novae-zelandiae* (30) and *Pittosporum obcordatum* (31), have been the subjects of autecological studies aimed at identifying factors underlying their rarity and the management needed to ensure their survival. A New Zealand icon, *Cordyline australis,* has also been extensively studied, because of recent "sudden decline", probably attributable to a phytoplasma (32).

Explanations of the origin of the characteristic New Zealand growth form known as *divaricate* or, as I prefer, *filiramulate,* continue to be polarised between browsing by extinct ratite birds and response to past and present climates. Currently, proponents are testing their hypotheses experimentally, on the one hand, by using emus or ostriches as surrogate moas (33), and on the other, by manipulating the environment and observing the physiological effects (34).

The variable flowering of New Zealand plants has shown to be synchronous in at least 17 different tree and herb species in *Chionochloa, Nothofagus, Phormium,* and *Elaeocarpus* at widely separated sites in step with climate cues associated with La Niña weather patterns (35). The snow tussocks (*Chionochloa* spp.) have the most extreme variable flowering patterns documented worldwide (36).

Recently it has been shown that the three species of native honey eaters (Meliphagidae) derive nectar, not only from larger, more colourful flowers, but also from species with small, aggregated flowers considered primarily entomophilous, for example, *Pseudopanax arboreus, Weinmannia racemosa* and *Pittosporum* spp. (37). Flowers of *Peraxilla* and *Alepis* (Loranthaceae) are ready to be pollinated while tepals still form a closed tube; the tepals separate (explosively in the case of *Peraxilla)* when they are manipulated by native bees or meliphagous birds in order to reach the nectar (38, 39). Declines in abundance of honey eaters are a likely cause of suboptimal seed set in ornithophilous species, with potential effects on regeneration (40). The root parasite *Dactylanthus taylori* has been added to the list pollinated by the bat *Mystacina tuberculata* (41). There have also been studies on seed production (42), seed rain and soil seed banks (43-45), germination (46), and the distribution of fleshy fruit by native and introduced birds (47) or, in the case of filiramulate shrubs, by lizards (48).

Interest continues in the role of disturbance in regeneration of New Zealand forests. Severe droughts as well as earthquakes are reflected in the demography of *Nothofagus* forest in the Buller district (49). Evidence from tree ring patterns, ages of tree cohorts, and carbon-14 dating of buried wood has revealed that landslides in the Southern Alps, triggered by earthquakes along the Alpine Fault in 1826, 1717 and *ca.* 1630 and 1460 AD, led to establishment of even-aged cohorts in conifers and *Nothofagus,* not only on the mountain slopes but also on the flood plains below as a result of aggradation (50). A severe earthquake in 1994 caused destruction in *Nothofagus* forests directly through landslides and through deaths of damaged trees over a longer period (51).

Among papers dealing with ecological concepts may be mentioned the proposal that positive-feedback mechanisms can lead to "switches" resulting in abrupt temporal or spatial boundaries between communities (52). Studies of soil-plant relationships include facilitation of primary succession to forest by the nitrogen-fixing shrub *Carmichaelia odorata* (53), the effects of burrowing seabirds and tuatara (*Sphenodon*) on soil fertility and tree-seedling density (54), and the ways in which above- and below-ground components of ecosystems in native forest are affected by introduced herbivores. The latter have generally negative effects, not only on the diversity of understoreys, but also on habitat diversity in the litter and on populations of most groups of litter-dwelling meso- and macrofauna (55). The richness and diversity of the epiphytic guild in lowland rain forest was confirmed in a census on three trees of different species in south Westland, that supported 61 vascular and 94 non-vascular species distributed among 15 epiphytic communities (56). Alpine plants communities are usually considered to have low rates of turnover, through individual plants being long-lived and reproducing vegetatively rather than by seed. However, a population of the rare, localised *Myosotis oreophila* fluctuated by 40% during five years, and presumably experienced total turnover during this period (57).

With a few notable exceptions, such as the five-fold increase of *Hieracium lepidulum* in the understoreys of *Nothofagus solandri* forest between 1970 and 1993 (58), intact, continuous native vegetation resists invasion by exotic species (59, 60). Disturbance, whether human-related or natural, provides most of the immediate opportunities for invaders. Nevertheless, the vigour of naturalised species representing functional types that are poorly represented in or absent from the native flora, such as trees adapted to very cold winters, semi-arid climates and recent alluvial substrates, poses problems for the long-term survival of much native non-forest vegetation (4, 61).

Problems of displacement of native plants by introduced species are most acute in successional vegetation where adventive pioneers are usually present, and often dominant, even in remote localities. A spectacular example is the recent occupation of coastal cliffs in Taranaki by the Chilean giant herb *Gunnera tinctoria*, which totally suppresses the small native plants, including *Gunnera* spp., that formerly occupied this habitat. Once established, introduced grasses and shrubs can strongly impede secondary succession back to native forest (62, 63).

The low-altitude, settled parts of New Zealand have suffered the greatest reduction of native vegetation, and accordingly, remnants have disproportionately high conservation value. Forest remnants, as well as being subjected to damage by wild and domestic animals, are also especially prone to invasion by weeds, mostly garden escapes (64, 65). To those mentioned in the first edition may be added *Cotoneaster* spp., *Jasminum humile, Berberis glaucocarpa* and *Tropaeolum speciosum*. In some small forest reserves, native trees form a canopy over understoreys that are almost completely adventive. Restoration of these requires intensive, coordinated effort; and to an encouraging extent this has been forthcoming, especially through work by voluntary groups.

Management of native vegetation has changed fundamentally as a result of state sector reorganisation beginning in the 1980s. The Department of Conservation now controls the national parks, scenic reserves and nearly all indigenous vegetation administered by the former Forest Service. There are also reserves administered by local authorities, and an increasing number of "conservation covenants", i.e., private-owned land where owners have contracted to physically and legally protect native vegetation. State-owned forestry companies retain ownership of some areas of native forest with a potential to yield timber, but recent plans to do so were opposed by conservation organisations, and eventually cancelled by the government. This has been a contentious decision, even among ecologists, some of whom consider that forests that regenerate in response to natural disturbance should be able to sustainably yield timber without loss of natural values. At present, timber production from native forests is permitted only on private land where owners have approved plans for sustainable yield, and on Maori tribal land. Anomalously, private-owned forest and other native vegetation can still be cleared for agriculture, subject to approval by local authorities.

Much non-forested mountainous land in the South Island has been grazed under long-term leases from the government. Currently, many lessees are arranging to buy the parts of the leaseholds with the greatest farming value, in return for relinquishing areas of conservation value to the Department of Conservation. This is expected to add about 1 000 000 ha to the conservation estate, mainly at high altitudes.

Although nature conservation is now the primary objective for about a third of New Zealand's land area, the problems of protecting the native vegetation and fauna from introduced plants and animals remain. Australian possums are still regarded as a major cause of destruction of forest canopies, and they also prey on bird nests and are a vector for bovine tuberculosis. Trapping for fur can provide local control; but for control over extensive areas, it is deemed necessary to spread sodium monofluoroacetate (1080) poison from the air, a practice that is highly controversial (66). Threats to native fauna from introduced omnivores and carnivores, especially rats and mustelids, pose even more urgent problems, with the survival of several native birds, including kiwi (*Apteryx* spp) and kaka (*Nestor meridionalis*) becoming problematical on the main islands of New Zealand (67). The resources available to the Department of Conservation fall short of ideal, resulting in a perceived neglect of parts of the conservation estate that attracts criticism both from conservationists and from those who object to so much land being unavailable for extraction of resources. At present, the department concentrates much of its effort within limited areas, such as offshore and outlying islands, and in mainland areas known to have high levels of indigenous biodiversity. Within these areas, there have been encouraging successes in eliminating introduced herbivores and predators, with corresponding benefits to native animals and plants (68, 69).

REFERENCES

1. A.B. Rose, L.R. Basher, S.K. Wiser, K.H. Platt, I.H. Lynn (1998). *New Zealand Journal of Ecology* 22: 121-140.
2. S.R.P. Halloy, A.F. Mark. (1996). *Journal of the Royal Society of New Zealand* 26: 41-78.
3. A.F. Mark, K.J.M. Dickinson, J. Allen, R. Smith, C.J. West. (2001). *Austral Ecology* 26: 423-440.
4. P. Wardle, C. Ezcurra, C. Ramírez, S. Wagstaff (2001). *New Zealand Journal of Botany* 39: 69-108.
5. E. Edgar, H.E. Connor (2000). *Flora of New Zealand. V. Grasses.* Manaaki Whenua Press, Lincoln, New Zealand.
6. A. D. Wilton, I. Breitwieser (2000). *New Zealand Journal of Botany* 38: 537-549.
7. P.B. Heenan (1996). *New Zealand Journal of Botany* 34: 157-178.
8. P.A. Williams *pers. comm.*
9. I.A.E. Atkinson, E.K. Cameron (1993). *Trends in Ecology and Evolution* 8: 447-451.
10. S.J. Wagstaff, M.J. Bayly, P.J. Garnock-Jones, D.C. Albach (2002 in press). *Annals of the Missouri Botanical Garden* 83.
11. R.C. Winkworth, S.J. Wagstaff, D. Glenny, P.J. Lockhart (in press). *Trends in Ecology and Evolution.*
12. J.M. Ward, I. Breitwieser (1998). *New Zealand Journal of Botany* 36:165-172.
13. S.D. Wright, C.G. Yong, J.W. Dawson, D.J. Whittaker, R.C. Gardner (2000). *Proceedings of the National Academy of Sciences USA* 97: 4118-4123.
14. D.E. Lee, W.G Lee, N. Mortimer (2001). *Australian Journal of Botany* 49: 341-356.
15. M.S. McGlone, R.P. Duncan, P.B.Heenan (2001). *Journal of Biogeography* 28: 199-216.
16. M.S. Pole (1994). *Journal of Biogeography* 21: 625-35.
17. P.G. Martin, J.M. Dowd (1993). *Australian Journal of Systematic Botany* 6: 441-448.
18. M.S. McGlone, D.C. Mildenhall, M.S. Pole (1996). Pp. 83-130 in T.T. Veblen, R.S. Hill, J. Read (eds.). *The Ecology and Biogeography of Nothofagus forest.* Yale University Press, New Haven.
19. M.S. McGlone, M.J. Salinger, N.T. Moar (1993). Pp. 294-317 in H. E. Wright, J.E. Kutzbach, T. Webb III, W.F. Ruddiman, F.A. Street-Perrrott, P.J. Bartlein (eds). *Global Climates since the Last Glacial Maximum.* University of Minnesota Press, Minneapolis.
20. J.R. Leathwick (1995). *Journal of Vegetation Science* 6: 237-248.
21. P. Haase (1999). *Journal of Biogeography* 26: 1091-1099.
22. M.S. McGlone, J.M. Wilmshurst (1999). *Quaternary International* 59: 17-26.
23. R.N. Holdaway (1996). *Nature* 384: 225-226.
24. P.A. Williams (1993). *New Zealand Journal of Botany* 31: 65-90.
25. K.J.M. Dickinson, A.F. Mark, B.I.P. Barrett, B.H. Patrick. (1998). *Journal of the Royal Society of New Zealand* 28: 83-156.

26. C.D. Meurk, N.M. Foggo, J.B. Wilson (1994). *New Zealand Journal of Ecology* 18: 123-168.
27. B.P.J. Molloy, E. Edgar, P.B. Heenan, P.J. de Lange (1999). *New Zealand Journal of Botany* 37: 41-50.
28. G.M. Rogers (1996). *New Zealand Journal of Botany* 34: 227-240.
29. D. Widyatmoko, D.A. Norton (1997). *Biological Conservation* 82: 193-201.
30. P.A. Williams (1992). *New Zealand Journal of Ecology* 16: 127-135.
31. B.D. Clarkson, B.R. Clarkson (1994). *New Zealand Journal of Botany* 32: 155-168.
32. P. Simpson (2000). *Dancing Leaves. The Story of New Zealand's Cabbage tree, Ti Kouka.* Canterbury University Press, Christchurch, New Zealand.
33. W.G. Lee, W.J. Bond, G. Wells, J. Wass, D. Harder, J. Craine (2001). Abstract in programme for New Zealand Ecological Society 50th Jubilee Conference.
34. C.J. Howell, D. Kelly, M.H. Turnbull (2002 in press). *Functional Ecology* 16:
35. E.M. Schauber, D. Kelly, P. Turchin, C. Simon, W.G. Lee, R.B. Allen, I.J. Payton, P.R. Wilson, P.E. Cowan, R.E. Brockie (in press). *Ecology.*
36. D. Kelly, A.L. Harrison, W.G. Lee, I.J. Payton, P.R. Wilson, E.M. Schrauber (2000). *Oikos* 90: 477-488.
37. I. Castro, A.W. Roberston (1997). *New Zealand Journal of Ecology* 21: 169-180.
38. E.J. Godley (1979). *New Zealand Journal of Botany* 17: 441-466.
39. J.J. Ladley, D. Kelly (1995). *Nature* 378: 766.
40. A.W. Robertson, D. Kelly, J.J. Ladley, A.D. Sparrow (1999). *Conservation Biology* 13: 499-508.
41. C.E. Ecroyd (1996). *New Zealand Journal of Ecology* 20: 81-100.
42. N.J. Enright, A.D. Watson (1992). *New Zealand Journal of Botany* 30: 29-44.
43. G. Sem, N.J. Enright (1996). *New Zealand Journal of Botany* 34: 215-226.
44. A.T. Moles, D.R. Drake (1999). *New Zealand Journal of Botany* 37: 679-85.
45. A.T. Moles, D.W. Hodson, C.J. Webb (2000) *Oikos* 89: 541-45.
46. C.J. Burrows (1999). *New Zealand Journal of Botany* 37: 277-288.
47. P.A. Williams, B.J. Karl (1996). *New Zealand Journal of Ecology* 20: 127-145.
48. J.M. Lord, J. Marshall (2001). *New Zealand Journal of Botany* 39: 567-588.
49. P. Vittoz, G.H. Stewart, R.P. Duncan (2001). *Journal of Vegetation Science* 12: 417-26.
50. A. Wells, R.P. Duncan, G.H. Stewart (in press) *Journal of Ecology.*
51. R.B. Allen, P.J. Bellingham, S.K. Wiser (1999). *Ecology* 80: 708-14.
52. J. B. Wilson, A.D.Q. Agnew (1992). *Advances in Ecological Research* 23: 263-336.
53. P.J. Bellingham, L.R. Walker, D.A. Wardle (2001). *Journal of Ecology* 89: 861-875.
54. S.W. Mulder, S.N. Keall (2001). *Oecologia* 127: 350-60.
55. D.A. Wardle, G.M. Barker, G.W. Yeates, K.I. Bonner, A. Ghani (2001). *Ecological Monographs* 71: 587-614.
56. R.J. Stanley, K.J.M. Dickinson, A.F. Mark (2001). *Journal of Biogeography* 28: 1033-1049.
57. R.G.M. Hofstede, K.J.M. Dickinson, A.F. Mark (2000). *Arctic and Alpine Research* 30: 227-240.
58. S.K. Wiser, R.B. Allen, P.W. Clinton, K.H. Platt (1998). *Ecology* 79: 2071-2081.
59. L.K. Jesson, D. Kelly, A.D. Sparrow (2000). *New Zealand Journal of Botany* 38: 451-468.
60. R.B. Allen, W.G. Lee (1989). *New Zealand Journal of Botany* 27: 491-98.
61. W.G. Lee (1998). *Royal Society of New Zealand Miscellaneous Series* 48: 91-101.
62. T.R. Partridge (1992). *Journal of Applied Ecology* 29: 85-91.
63. S.K. Wiser, R.B. Allen, K.H. Platt (1997). *New Zealand Journal of Botany* 35: 505-515.
64. M.C. Smale, R.O. Gardner (2000). *Pacific Conservation Biology* 5: 83-93.
65. R.J. Standish, A.W. Robertson, P.A. Williams (2001). *Journal of Applied Ecology* 38:

1253-1263.

66. J. Innes, G. Barker (1999). *New Zealand Journal of Ecology* 23: 111-127.
67. P.R. Wilson, B.J. Karl, R.J. Toft, J.R. Beggs, R.H. Taylor (1998). *Biological Conservation* 83: 175-185.
68. R.H. Taylor, B.W. Thomas (1993). *Biological Conservation* 65: 191-198.
69. R.A. Empson, C.M. Miskelly (1999). *New Zealand Journal of Ecology* 23: 241-254.

CONTENTS

Introduction   xviii

1 THE PHYSICAL AND BIOLOGICAL ENVIRONMENT   1
Topography and geology   1
Weather and climate   3
Flora   5
Fauna   6

2 ORIGINS AND HISTORY OF THE FLORA AND   8
   VEGETATION
The fossil record   8
The elements of the native flora   11
Plant distributions and Quaternary history of the vegetation   14
The human period   15

3 PLANT FORM IN RELATION TO HABITAT   20
Growth forms of woody plants   20
    Tall trees   20
    Broad-leaved small trees and shrubs of lowland forest and bush   22
    Broad-leaved small trees and shrubs of upland, southern and infertile habitats   24
    Small podocarps   26
    Epacrid and ericad trees and shrubs   26
    Other small-leaved native shrubs   27
    Dwarf shrubs   27
Herbaceous growth forms   28
    Dicot herbs   28
    Tussocks and other monocot growth forms   29
    Ferns and fern-allies   31
Special and dependent growth forms   31
    The filiramulate habit   31
    Mat and cushion plants   33
    Lianes   34
    Vascular epiphytes   35

Vascular parasites and saprophytes   36
Insectivorous plants   36
Leaf dimensions   36
Woody plants   36
Herbaceous dicots   38
Herbaceous monocots   38
Ferns   39
Deciduous, summer-green and annual plants   39
Bud structure and shoot growth in woody plants   40
Sclerophylly and xeromorphy   40
Roots   43

4   REPRODUCTIVE ASPECTS, SEEDLING FORM AND   52
     LONGEVITY
Floral characters   52
Phenology   55
Variable flowering and fruiting   56
Fruit characters and dispersal modes   58
Ecological implications of dispersal mode   59
Seed mass; quantity of viable seed; germination   60
Seedling and juvenile forms   64
Longevity   64

5   DESCRIPTION, NOMENCLATURE AND CLASSIFICATION   71
     OF VEGETATION, ENVIRONMENT AND ECOLOGICAL
     PROCESSES
Vegetation classes   71
Allocation of material among chapters   71
Arrangement within chapters   73
Defining and naming plant communities   74
Divisions of vegetation and habitat based on temperature gradients   74
Altitudinal belts   74
Latitudinal zones   78
The coastal–inland gradient   80
Zonation due to temperature inversion   81
Compression of altitudinal zones on summits   82
The moisture factor   83
Rocks and soils   85
Rock types   85
Soil horizons   86
Nutrient ions   87
Soil developmental sequences   88
Soil fertility classes   90
Vegetation processes   91

6   BOTANICAL PROVINCES   92

7 FOREST 111
Conifer/broad-leaved forests of the lower altitudes 111
  Structure and gradients 111
  Mixed forests of Northland and Auckland 114
  Volcanic Plateau 120
  Taranaki to the Marlborough Sounds 123
  Western South Island 126
  Southern districts 132
  Eastern South Island 132
High-altitude conifer/broad-leaved forests 134
  Central Westland 134
  Southern districts 136
  Eastern South Island 136
  Southern North Island 136
  Northern zone 139
Coastal forests 140
Beech forests 140
  Beech forest species 141
  Beech forests of the northern South Island 143
  Central and southern South Island 147
  North Island axial ranges 152
  Other beech stands of the North Island 156
Exotic forests 157

8 BUSH, HEATH, SCRUB AND FERNLAND 161
Temperate bush 161
  Structure and composition 161
  Northern coastal bush 165
  Coastal bush of central and southern New Zealand 167
  Bush of steep, unstable slopes 169
  Griselinia littoralis bush 170
  Bush on fertile alluvial flats and fans 172
Primary tree- and shrub-heaths and subalpine bush 173
  Structure and gradients 173
  Primary tall heaths and subalpine bush in Central Westland 176
  Western South Island below lat. 43°40′ 179
  Stewart Island 180
  Western Nelson 184
  North Island 184
  Eastern South Island 188
  Tree-heaths of southern coasts 190
Penalpine shrub-heath 191
Shrub-heaths of mires and frost flats 193
  Volcanic Plateau 193
  South Island 194
Kanuka–manuka heaths 195
  Ecology of manuka and kanuka 195
  Other species of kanuka–manuka heaths 197
  Manuka heath of the gumlands 198
  Manuka scrub established on pasture 201
  Kanuka tree-heath 202
  Kanuka and manuka heath on dry sites 203

Subalpine secondary heaths   205
Grey scrub   207
    Lower altitudes in the South Island   207
    Volcanic Plateau   208
    Other primary grey scrub communities   209
    Secondary grey scrub   209
Subalpine *Hebe* communities   210
Shrubland of *Cassinia* (tauhinu) and similar plants   211
Naturalised shrubs other than *Erica* and *Hakea* species   212
Fernland   213

9   GRASSLAND AND HERBFIELD   216
Chionochloa grasslands and related vegetation   216
    The major grasses   217
    Important composites   219
    Other plants of the *Chionochloa* grasslands   220
Chionochloa communities   223
    Mountains of Western Nelson   223
    Westland   226
    Fiordland   230
    Stewart Island   232
    Eastern South Island south of the Rakaia River   234
    Northern Canterbury and Marlborough   241
    High country in Sounds-Nelson province   242
    Tararua and Rimutaka Ranges   243
    Northern axial ranges   243
    The high volcanoes   244
Short-tussock and related grasslands of inland districts   244
    History   244
    The main grasses, sedges and rushes   247
    Dicot and monocot forbs   248
    Cryptogams   249
    Short-tussock and related grasslands of southern Canterbury   256
    Short-tussock and related grasslands in other regions   258
    Adventive-dominated grasslands and herbfields of inland districts   259
Maritime grasslands   263
    History   263
    The grass species   264
    Sedges and rushes   265
    Other native plants   266
    Adventive dicots   266
Maritime grassland communities   267
    Canterbury   267
    Western South Island   272
    Manawatu   272
    Hawkes Bay   275
    Northland   276
Patterns and processes affecting pasture composition   276
    Ecology of pasture management   280
Lawn and turf communities   281
Coastal tussock grassland of Rakiura province   281

10 WETLAND VEGETATION 284

Physical characteristics of wetlands 284

Adaptations to aquatic habitats 286

The history of wetlands 286

Saline wetlands 287

The nature of salt-tolerance 288
Plants of saline habitats 288

The salt-marsh communities 291

Northern zone 292
Central zone 294
Southern coasts 297
West coast of the South Island 298
Inland saline habitats 299

Lakes, ponds and streams 300

Aquatic plants 300
Manapouri: a near-pristine lake with a mountainous catchment 303
Rotorua Lakes 303
A eutrophic lake 305
Upland tarns and pools 306
Thermal water 307

Eutrophic lowland swamps 309

Monocot herbs 310
Trees, shrubs, dicot herbs, ferns and mosses 311
Northern zone 312
Southern North Island 314
Western South Island 314
Eastern South Island 316
Inland South Island 318
Ditches and water-races 320

Oligotrophic lowland mires and wet heaths 321

The species 321

Oligotrophic mire communities 323

Northland 323
Waikato Basin 324
Volcanic Plateau 324
Western South Island 326
Eastern Fiordland 331
Southland 331

Wet heath communities 334

Moraines and fluvioglacial terraces in Westland and Western Nelson 334
Fiordland and Stewart Island 336
Plateaus of Western Nelson 336
Volcanic Plateau 337

Mountain wetlands 337

Cushion and mat plants 338
Plants with other growth forms 338
Mountain wetlands of Stewart Island 339
Cushion bogs on the southern borders of the Otago plateaus 340
Soligenous wetlands of the Central Otago mountains 341
Canterbury and Marlborough 341
Wetlands on western ranges of the South Island 343
North Island mountain wetlands 348

11   OPEN OR PATCHY VEGETATION ON PRIMARY SURFACES   350
     AND DEPLETED LANDS
     Coastal sand and gravel   350
       The environment   350
     The plants of unconsolidated coastal deposits   351
       Woody plants   351
       Sand binders   352
       Succulents   353
       Other herbs   353
       Non-vascular plants   354
     Communities on coastal sand and gravel   355
       A selection of North Island dune systems   355
       South Island dunes   356
       Shingle beaches   359
     Inland flood-plains, dunes, moraines, tephra and depleted   360
     lands
       The land-forms   360
     Native pioneers   362
       Dicot herbs   362
       Monocot herbs   362
       Woody plants   363
       Cryptogams   363
     Adventive pioneers   364
       Herbaceous species   364
       Woody plants   365
     Pioneering vegetation   366
       Lowland flood-plains of Central Westland   366
       Dry river flats and severely depleted lands in inland regions of the South Island   367
       Unconsolidated tephra   368
       Moist river flats dominated by adventive herbs   372
     Debris and boulder fields   372
       Slope debris   372
       Boulder fields   374
     Cliffs, bluffs and rock outcrops   375
       The rupestral habitat   375
     The rupestral flora   376
       Trees, shrubs and lianes   376
       Succulent dicots   378
       Other dicot herbs   379
       Monocot herbs   381
       Ferns and fern-allies   382
       Non-vascular plants   382
     The vegetation of cliffs and similar land forms   383
       Eastern coasts and islands in the northern zone   383
       A North Island gorge   384
       Hawkes Bay   384
       Eastern Wairarapa   384
       Vicinity of Cook Strait   385
       Kaikoura Ecological Region   385
       Spenser Ecological Region   388
       Western Nelson   389

Westland and the western Alps   391
Fiordland   393
Inland Canterbury   394
Volcanic cliffs of the eastern South Island coast   395
Rakiura province   396
Ultramafic surfaces   396
Surville Cliffs near North Cape   397
Sounds–Nelson   397
South Westland   400
Inland Southland   400
Vegetation of thermal areas   400

12   THE ALPINE AND NIVAL BELTS   405
Species of high altitudes   405
Scree plants   406
Cushion and mat plants   409
Other herbaceous dicots   412
Other grasses, sedges and rushes   414
Shrubs   414
Ferns and lycopods   415
Bryophytes and lichens   415
Alpine communities   416
Eastern Alps   416
Western Alps   418
Central Otago   419
Waimakariri catchment, northern Canterbury   425
Marlborough   427
Western Nelson   429
Northern mountains   430

13   OUTLYING ISLANDS   432
Kermadec Islands   432
History and flora   432
Communities   435
Chatham Islands   436
Environment and history   436
Species of bush and tree-heath   437
Tall woody communities   438
Shrubland and fernland   442
Grassland   444
Wetlands   444
Sand dunes and cliffs   445
Auckland Islands   446
Environment and history   446
Far-southern plants present on the Auckland Islands   447
Vegetation gradients   451
Forest and shrubland   451
Grassland, herbfield and fen of the lower altitudes   453
Chionochloa grassland, low scrub and bogs 454
Vegetation of the hill tops   455
Coastal cliffs   455

Campbell Island   456
   Environment, history and flora   456
   Tall heath and scrub   457
   Herbaceous vegetation   458
   Offshore islets   460
Antipodes Islands   461
Bounty Islands   463
Macquarie Island   463
   Environment, history and flora   463
   Vegetation   463

14   BIOMASS, GROWTH, NUTRITION AND TOLERANCES   466
Biomass and related parameters   467
   Biomass   467
   Standing volume   467
   Basal area   469
   Stand height and complexity   471
Production and growth   471
   Dry matter production   471
   Litter fall   475
   Volume increment   475
   Diameter increment   475
   Shoot growth and height increment   477
Carbon assimilation   480
Nutrient element contents of plant material   485
   Foliar nutrient concentrations   485
   Nutrient elements on a land area basis   489
   Seasonal variations in nutrient content   490
Mycorrhizas and phosphorus uptake   492
Nitrogen in ecosystems   495
Water relations among woody plants   497
Temperature tolerances   499
Conclusions   505

15   SUCCESSION, RETROGRESSION AND INVASION   507
   What is succession?   507
   Arrangement of material   508
Primary succession   509
   Low-altitude successions in central Westland   509
   Montane and subalpine successions in central Westland   511
   Beech forest successions in the western South Island   514
   Sequences below tree-limit in the eastern South Island   516
   Successions on alluvial flats dominated by adventive plants   521
   Dune sequences   521
   Successions on volcanic parent materials   524
   Ultramafic surfaces   527
   Penalpine and alpine successions   527
Secondary succession at low altitudes   531
   Succession on northern off-shore islands   532
   Succession to lowland forest on the northern mainland   533

Forest successions near Wellington   538
Secondary succession through legumes in the east of the South Island   540
Prevalent stages in secondary succession towards lowland forest   540
Secondary successions at higher altitudes   547
Mt Cook National Park   547
Arthurs Pass, Hawdon Ecological Region   547
Tararua Range   549
Other localities   551
Invasion   553
Conclusions   554
Some successional concepts in the New Zealand context   557

16   DISTURBANCE, REGENERATION AND TRENDS IN   559
      NATIVE FOREST
Stand structure in relation to life history of species   559
Requirements of native tree seedlings   560
The role of vegetative reproduction   562
Damage to forest canopies and understoreys   562
Die-back attributable to damaged root systems   563
Die-back attributable to pathogens attacking stems   564
Agents of defoliation   564
Die-back related to drought   567
Canopy destruction by wind and snow   567
Modification of lower tiers by introduced mammals   573
Other animal influences   576
Forest collapse through multiple causes   579
Population structure and regeneration   583
Beech forests   583
Forests dominated by broad-leaved trees   586
Kaikawaka   590
Tall podocarps in the west of the South Island   594
Tall podocarps in the eastern South Island and North Island   599
Kauri   601
Conclusions   602

Epilogue   606

Appendixes

1   Abbreviations   611

2   Common names and Latin equivalents   612

References   615

Index of plant genera and species   637

General index   666

INTRODUCTION

The natural vegetation of New Zealand is unique and diverse, in keeping with its isolation from other lands (Fig. 0.1), latitudes extending from subtropical to subantarctic, a rugged and varied landscape, and habitats ranging from thermal pools to perpetual snows. The floristic relationships and prevalent growth forms in these south-temperate islands contrast remarkably with those of the north-temperate vegetation familiar to most botanists. Lowland forests, with succulent-fruited conifers, evergreen broad-leaved trees, rampant lianes, epiphytes and tree ferns closely resemble those of the New Guinea highlands, 4000 km to the north-west, whereas in cooler districts there are extensive stands of small-leaved, evergreen beeches, comparable with those in temperate Chile, 7500 km to the east. Native grasslands above the forest limit and in the drier districts are mostly dominated by evergreen tussocks, thereby resembling those of tropical high mountains rather than northern temperate grasslands. There are also distinctive shrublands, herbfields, wetlands and cushion plant communities.

Very many New Zealanders, laymen as well as scientists, are deeply interested in the native flora and vegetation, partly because these are so often associated with areas of the country frequented for relaxation and recreation, partly because people realise that native plants impart a unique aspect to our country and, not least, because of a nostalgia that becomes more intense as the landscape in the populated parts of the country comes to bear ever less of its original character. Although human settlement began only with the arrival of Polynesians about 1000 years ago, it has led to profound changes, especially since European colonisation began early in the nineteenth century. Introduced plants now dominate the settled districts, and there are some even in habitats that have been scarcely exposed to human activity.

A comprehensive account of New Zealand vegetation was last provided in the 1928 edition of *The Vegetation of New Zealand*, written by Leonard Cockayne for the *Die Vegetation der Erde* series. This remains a valuable source of information, although many localities described by Cockayne no longer support native vegetation, and others that were remote or even unexplored in Cockayne's time have since been visited and

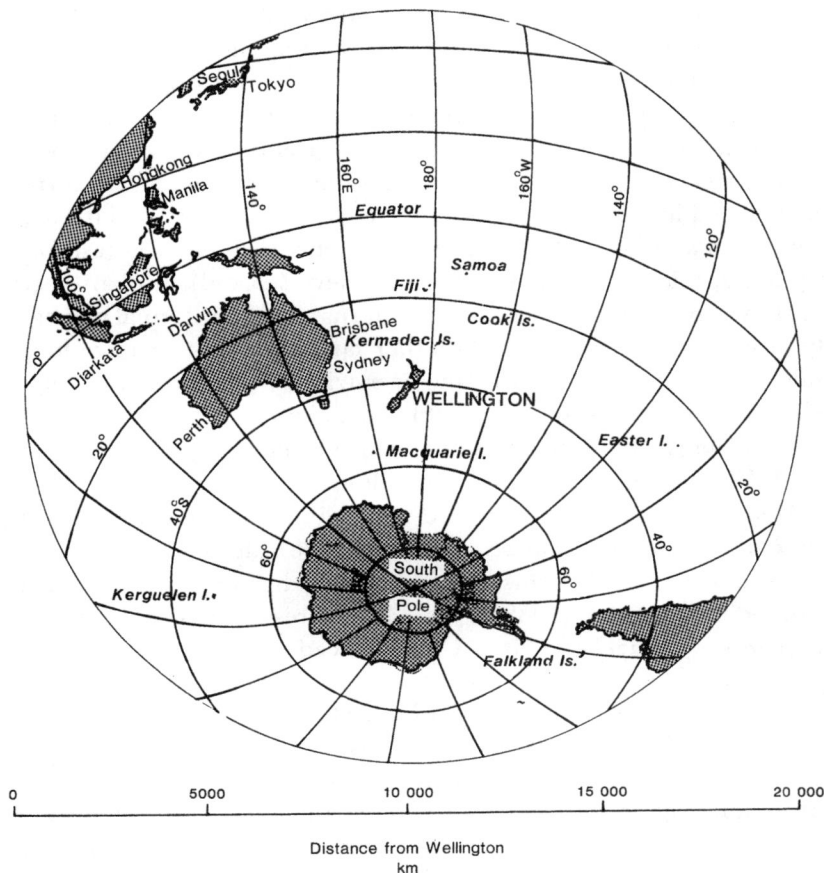

Fig. 0.1. The hemisphere centred on New Zealand (from Wards 1976).

described botanically. Cockayne's ecological work, which he began only in his fortieth year in 1895, represents almost the total plant-ecological endeavour in New Zealand up to 1928, the value of observations by other botanists, surveyors and explorers notwithstanding. Since 1928, and particularly since 1945, there has been a great deal of ecological research, which provides a basis for a deeper understanding of our vegetation than was available to Cockayne. The present volume attempts to co-ordinate these studies into a single treatise that devotes most space to the unique indigenous vegetation. The briefer treatment of adventive-dominated vegetation reflects the limited amount of ecological information and does not include ephemeral communities such as those of crops and urban wasteland, although the latter have recently been described for Auckland city (Esler 1988b, and earlier papers in the series).

Chapters 1–4 describe the New Zealand environment, flora and fauna, and discuss the origins, relationships, life forms and reproductive aspects of the indigenous vegetation. Chapter 5 is a synopsis of vegetation types, habitat classes and environmental processes, and serves also to define the terms in which these are described. It is followed by an outline of the geographic divisions of the country (Chapter 6). Chapters 7–13 contain expanded descriptions of plant communities, preceded where appropriate by information on their structure and characteristic species and genera. Chapters 14–16 discuss ecological functions and processes.

For brevity, well-established common plant names are used, but corresponding binomials are listed in Appendix 2 and repeated at the first mention in each chapter. Where there is no ambiguity, the generic name only may be used (as, for example, where only one species of the genus occurs in New Zealand). A plant-species index lists currently accepted binomials as well as alternative and informal names, indicates the family of each genus, and codes whether taxa are native or introduced. In the general index, bold numerals indicate major entries and definitions of terms.

The reference list covers significant publications since 1928, but the book draws on a far wider range of material. Sources not cited specifically include the *Flora of New Zealand* (C.J. Webb *et al.* 1988 and earlier volumes) and *New Zealand Alpine Plants* (Mark & Adams 1973), which provided much information on distributions and habitats. *Forest Vines to Snow Tussocks* (Dawson 1988), although written for a general readership, discusses growth forms and origins of indigenous plants in greater detail than attempted here. Esler (1988*a*) provides corresponding information for the adventive plants of Auckland city. Unreferenced information includes what is currently accepted as common knowledge, as well as my own unpublished data.

Unpublished material at Botany Division, DSIR, Lincoln, has been invaluable. It consists of collectors' data on herbarium labels, reports on vegetation, and species lists; A.P. Druce has been an indefatigable compiler of the last.

For permission to use copyright material, I am indebted to the following: Antarctic Division (Department of the Arts, Sport, Environment, Tourism and Territories, Tasmania); Auckland University Press; Blackwell Scientific Publications (Australia); British Ecological Society; Cambridge University Press; Department of Scientific and Industrial Research, New Zealand (including Division of Land and Soil Science); Elsevier Publishers (Amsterdam); Forest Research Institute (Ministry of Forestry, New Zealand); Gebrüder Borntraeger (Berlin); Government Printing Office, Wellington; Institute of Arctic and Alpine Research (University of Colorado); Institute of Terrestrial Ecology (Grange-over-Sands, England); Kluwer Academic Publishers (Amsterdam); National Archives of New Zealand; New Zealand Ecological Society; New Zealand Institute of Foresters; Plant Science Department (University of Canterbury, Christchurch); Royal Society of New Zealand; Dr Graeme Stevens (Wellington); Springer Verlag (Berlin); University of Chicago Press; University of Hawaii Press.

Finally, it is a pleasure to acknowledge the assistance and information provided by colleagues, including botanists, ecologists, and many others who love New Zealand's plants, animals and landscapes. Special thanks are due to Rowan Buxton, Sherryn Kelly and illustrators Pat Brooke and Jonathan Poff, who assisted directly with preparing this book.

THE PHYSICAL AND BIOLOGICAL ENVIRONMENT

Topography and geology
(Fig. 1.1)

The New Zealand archipelago is centred on three main islands which, with adjacent smaller islands, cover 263 830 km², and extend over a distance of 1500 km from latitude 34°09' to 47°17'S; these will be referred to as *mainland* New Zealand. Several outlying groups, with a total area of 1876 km², lie far to the north (Kermadec Islands), east (Chatham Islands) and south (Snares, Bounty, Antipodes, Auckland and Campbell Islands, and Macquarie Island, which is administratively Australian). These extend the latitudinal range from 29°15' to 54°30'. Despite distances of up to 1000 km from the main islands, they are connected with the latter by submarine platforms and ridges, and share most of their plant species with them.

The main islands are mountainous. Mesozoic greywackes form an axis extending from the north-east to the south of the North Island, and across Cook Strait form the high ranges of the north-east of the South Island. They continue southwards as the eastern flanks of the Southern Alps, which reach 3764 m at Mt Cook, but westwards they grade into schists, which in turn are sharply bounded by a major tectonic feature, the Alpine Fault. Further south, the schists extend eastwards as high plateaus. Granites, gneisses and Palaeozoic sedimentary rocks are extensive in the north-west and south-west of the South Island. There are also ultramafic outcrops in the South Island, and a small one in the far north of the North Island.

These Mesozoic and Palaeozoic rocks are also exposed in Chatham, Bounty and Campbell Islands, and underlie the whole New Zealand archipelago including the submarine platforms. However, surface rocks in the lower-lying parts of New Zealand are mainly Tertiary and Quaternary in age, and sedimentary and volcanic in origin. Volcanic deposits are especially prominent in the centre and west of the North Island, where there are several active and dormant volcanoes. Quaternary glaciers sculptured valleys throughout the central and western mountains of the South Island, and rivers spread glacial debris over the plains; the North Island was far less affected. Holocene processes have formed alluvial valleys, estuaries and inlets, and provided most of the weathered veneer in which soils have formed.

Fig. 1.1. Geology of the main islands.

Legend:

p | (Precambrian)- Paleozoic sedimentary and associated volcanic rocks.

m | (Permian)- Triassic-Jurassic (lower Cretaceous) greywacke and associated volcanic and sedimentary rocks.

t | Tertiary (upper Cretaceous) sedimentary rocks.

q | Quaternary sediments

■ | Upper Cretaceous-Recent volcanic rocks

g | Granite and other plutonic rocks

s | Schist

x | Rocks include ultramatic outcrops

0   50   200 km

Table 1.1. *Precipitation (mm) and screen temperatures (°C) for nine climatic stations*

| | latitude | | alt. | precip. (mean | air temperature mean daily | | Extreme |
	°	′	(m)	annual; mm)	Jan.	July	minima
Raoul Id	29	15	38	1538	21.7	16.2	7.4
Kaitaia	35	04	80	1418	19.3	11.7	− 0.5
Waiouru	39	28	823	1048	13.8	4.0	− 9.0
Wellington	41	17	126	1240	16.4	8.2	− 1.9
Hokitika	42	43	39	2783	15.3	7.2	− 3.2
Christchurch	43	32	7	666	16.6	5.9	− 7.1
Alexandra	45	16	141	343	17.0	2.6	− 11.7
Invercargill	46	25	0	1037	13.7	5.1	− 7.4
Campbell Id	52	33	15	1361	9.3	4.7	− 6.7

Source: New Zealand Meteorological Service 1982.

Weather and climate
(Table 1.1; Fig. 1.2)

New Zealand lies entirely within the zone of mid-latitude westerlies, except for the Kermadec Islands, which are more influenced by south-easterly trade winds. Weather is therefore determined by successive east-moving anticyclones and depressions, and associated fronts. Often, a regular pattern prevails, beginning with the passage of an anticyclone over a few days, that gives generally light winds and clear skies; then the approach of a depression, with skies becoming cloudier and north-westerly winds strengthening; and finally, the passage of one or several cold fronts from the south-west, accompanied by rain but presaging the arrival of the next anticyclone as pressures rise. The departures from regularity are provided by warm and occluded fronts, and cyclones originating in the tropics to the north-west, which strike the North Island especially, bringing torrential rain. Droughts can occur at any time of year, but are most severe during the warmer months when evaporation is greatest.

The translation of atmospheric systems into weather and climate is profoundly influenced by topography. In the South Island, the unbroken axial chain of mountains sharply differentiates climatic regions. Rainfall is highest in the west, with mean annual totals exceeding 10 000 mm on the westernmost ranges. The rate of precipitation can also be very high, 120 mm in 24 hours being not uncommon, and there tends to be a seasonal maximum during October, November and December.

When pressures are very low, westerly storms can carry the rain as far as the eastern foothills, but usually rain and cloud cease abruptly a few kilometres east of the island's Main Divide. East of here, clear skies, low humidities and strong, warm, gusty, often gale-force north-westerly winds constitute föhn conditions. Annual rainfall along the eastern coast is generally between 500 and 750 mm, and is mostly brought by southerly fronts. It increases on seaward slopes of the eastern ranges, but intermontane basins lie in rain shadows from both east and west, and their mean annual precipitation can be as low as 350 mm.

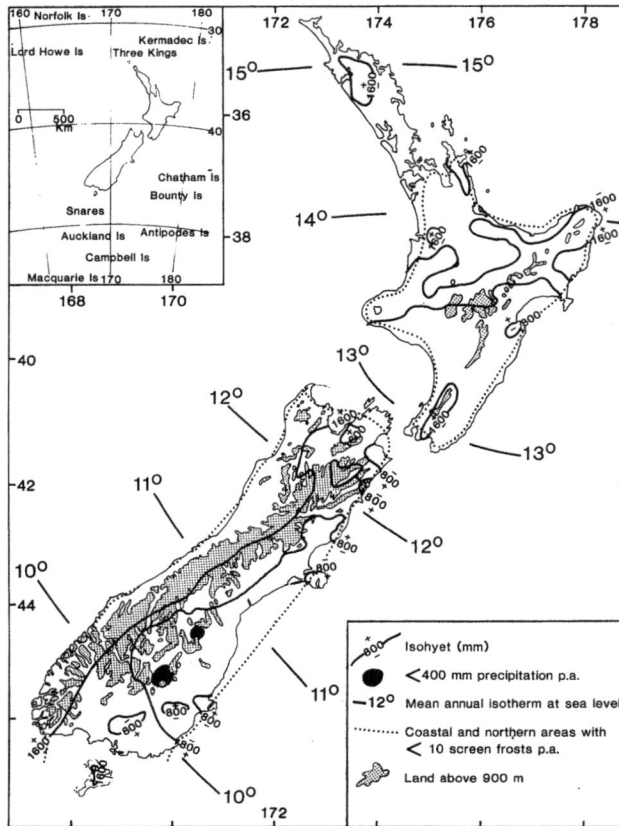

Fig. 1.2. Relief and climate of New Zealand.

Conditions become very dry in the east when anticyclones and dry north-westerlies persist for long periods. Practically rainless periods of two or more months, and runs of years with rainfall well below average, are recorded in some districts. In the west, rainless periods of up to a month result from sequences of anticyclones and easterly weather.

Sunshine reaches 2400 hours per annum in localities east of the Divide, and even on western coasts values approaching 2000 h reflect concentration of rain in torrential falls. Western mountains, however, are subject to afternoon cloud frequencies exceeding 90% of observations. Sunshine hours on eastern coastal hills are also reduced by orographic cloud.

Diurnal temperature range narrows during windy, cloudy weather, but widens during clear, still, anticyclonic conditions. Then, hot summer days can produce convectional cloud and thunderstorms, and winter nights bring frosts that are severe in inland valleys and frequently extend to coastal plains. The extremes of 42 °C and − 20 °C were measured in eastern South Island districts that are subject to hot föhn winds and severe frosts.

Along and west of the Main Divide, the permanent snow line lies at a height of about 2000 m at latitude 44° S and there are extensive névés and large glaciers. Eastwards, the snow line rises in response to lower annual precipitation and warmer summers; the Kaikoura Ranges in the north-east lack permanent snow although they exceed 2800 m. Despite this, low altitudes are more subject to snow-falls in the east of the South Island than in the west, as they are more exposed to southerly storms. It is only in the south and south-east of the island that snow lies at sea level during most winters, although only for a few days at a time.

In the North Island the climatic pattern is less strongly differentiated. It is generally warmer and although areas east of the axial ranges are relatively dry and experience föhn winds, no localities show such extremes of temperature and rainfall as in the South Island. There is a pronounced winter rainfall maximum, and droughts during the warmer seasons can be quite severe. Snow at sea level is almost unknown. The greatest climatic contrast is between the central plateau, which is subject to heavy snow-falls and severe frosts, and the Northland peninsula, which is practically frost-free near the sea.

The climates of the outlying islands, from the subtropical Kermadecs to subantarctic Macquarie, are truly oceanic, with daily variations of temperatures exceeding the seasonal range. In the far-southern islands, seasonal phenology is probably controlled as much by day length as by temperature.

Flora
(sources of statistics as in plant species index, p. 637)

The indigenous flora contains some 2300 vascular species, about 85% being endemic. The largest numbers are in the Asteraceae (*c.* 340 species), Poaceae (200), Cyperaceae (180), Scrophulariaceae (130), Orchidaceae (100) and Apiaceae (90). There are 20 species of native conifer, all but three belonging to the Podocarpaceae, a family with fleshy, few-seeded fruit instead of typical cones. Several important families are poorly represented, notably Fabaceae (27–50 species, only two being herbaceous), Brassicaceae (*c.* 30), Solanaceae (2–3) and Lamiaceae (2). The small numbers of Myrtaceae (19 species), Rutaceae (3) and Proteaceae (2), and absence of Mimosoideae contrast greatly with neighbouring Australia. No families are endemic.

The largest genera include the pan-temperate *Carex* (73 species), *Epilobium* (*c.* 40), *Ranunculus* (40), *Poa* (35), *Myosotis* (> 30) and *Gentiana* s.l. (*c.* 25); the mainly Australian *Olearia* (> 30) and *Rytidosperma* (18) and the mainly New Zealand *Hebe* (*c.* 80), *Celmisia* (> 50), *Aciphylla* (*c.* 40), *Carmichaelia* (15–38), *Dracophyllum* (*c.* 30), *Leptinella* (*c.* 25), *Brachyglottis* (*c.* 25) and *Chionochloa* (> 21). These, like the vascular flora as a whole, are best represented in shrubby and herbaceous vegetation. *Coprosma* (*c.* 50 species), *Uncinia* (30) and *Astelia* (> 13), which are also mainly New Zealand genera, are well represented in habitats ranging from high-altitude grasslands to the undergrowth of lowland forests. Forest trees belong to genera with few New Zealand species, the most important being *Metrosideros* (12 species, 5 of which are trees), *Nothofagus* (4), *Beilschmiedia* (2), *Weinmannia* (1–2), *Laurelia* (1), *Agathis* (1), *Libocedrus* (2), *Phyllocladus* (3), *Podocarpus* (4, 2 being shrubby), *Prumnopitys* (2),

Dacrycarpus (1) and *Dacrydium* (1). *Leptospermum* (1 species), *Kunzea* (2) and *Discaria* (1) dominate large areas of seral or open woody vegetation.

About 35 endemic genera are currently recognised. Other than *Raoulia* with *c.* 24 species and *Hoheria* with 4–5, all of these have only 1–3 species, and most are closely related to larger native genera. Four endemic genera belong to the Brassicaceae and three to the Fabaceae, despite the poor showing of these families overall.

Native ferns number 173 species, including eight tall tree ferns; nearly half of the species but only one genus (*Loxsoma*) are endemic. Fern-allies comprise *Lycopodium* (10 species), *Phylloglossum* (1), *Tmesipteris* (4) and *Psilotum* (1).

About 140 native vascular species are listed as vulnerable or endangered (Wilson & Given 1989), and nine are probably extinct, including the monotypic, endemic mistletoe *Trilepidea adamsii*, which was last seen in the Waikato region in 1954. *Pittosporum obcordatum* var. *kaitaiaensis* is now known only in cultivation. *Logania depressa* has not been seen since it was originally collected in the Moawhango district (Volcanic Plateau).

The introduced vascular flora contains 1860 species, the largest families being Asteraceae (215 species), Poaceae (220), Fabaceae (121), Brassiceae (74), Caryophyllaceae (54), Lamiaceae (54), Scrophulariaceae (52) and Solanaceae (50). The largest genera are *Juncus* (31 species), *Trifolium* (25), *Rubus* (24), *Carex* (22), *Eucalyptus* (20), *Solanum* (19), *Salix* (16), *Senecio* (16), *Veronica* (16), *Oxalis* (15), *Euphorbia* (14) and *Rosa* (14).

There are over 500 species of moss, over 500 hepatics and over 1000 macro-lichens; these, like the large fungal flora, have been largely neglected by ecologists.

Fauna

Because New Zealand had been isolated from the rest of the world for some 60–80 million years, its primeval fauna was unusual and of limited variety. Although beetles, moths and flies had speciated profusely, other insect groups of world-wide importance, such as butterflies, were poorly represented, and there were no social, long-tongued bees. There were about 25 species of freshwater fish, as well as primitive frogs (*Leiopelma*), skinks, geckos and the tuatara (*Sphenodon punctata*), the only survivor of a reptilian group that flourished during the early Mesozoic. Mammals were represented only by seals, cetaceans, and two species of bat, one of which constitutes an endemic family. Birds of the sea coast and wetlands were abundant, and generally comparable to the avifauna of similar habitats in Australia and further afield. Forest birds were more distinctive. Those with the power of flight, which included several endemic genera and two endemic families, were predominantly insectivorous, frugivorous or nectar-seeking. Only parakeets (*Cyanoramphus*) seem to have been largely seed-eating. Two birds lived exclusively in high-mountain terrain, these being a so-called wren (*Xenicus gilviventris*) and the parrot *Nestor notabilis*.

A remarkable assemblage of flightless birds included rails (*Gallirallus*, *Notornis* and *Aptornis*), a goose (*Chemiornis*), a parrot (*Strigops*) and above all, those two very different ratite groups, kiwis (*Apteryx*) and the extinct moas (Dinornithiformes). Kiwis mainly seek earthworms and soil-dwelling larvae. Moas were herbivores, and existed as about 12 species that ranged in height from one to three metres. They were solidly built and, if they were as abundant as subfossil bones and gizzard stones suggest, must have had a large influence on the vegetation. They probably grazed mainly in forest and its margins, filling niches usually occupied by large mammals. The largest invertebrate herbivores are weta, which are nocturnal, flightless Orthoptera that, in *Deinacrida heteracantha*, can attain a body length of 82 mm and weight of 71 g. There are also earthworms to 1.4 m long, carnivorous snails with shells to 11 cm across, and the primitive arthropod *Peripatus*.

The arrival of Polynesians was a disaster for the native fauna. Many birds became extinct, including moas and the giant eagle (*Harpagornis*) that probably preyed on them (Holdaway 1989). Animals such as the tuatara became restricted to off-shore islands, whereas *Strigops* and *Notornis* persisted with greatly reduced ranges in the main islands. People played a direct part in these extinctions, through hunting the larger creatures and greatly increasing the incidence of fire. Feral dogs probably created havoc, but the Polynesian rat or kiore (*Rattus exulans*) was the more damaging introduction. Doubtless swarming at times in plague proportions – as it still did in the earlier days of European settlement – it overwhelmed many of the large invertebrates, as well as reptiles, frogs and ground-nesting birds. Immense colonies of sea birds and seals continued to thrive on distant islands, but it is likely that on the main islands, there was a decline towards the generally small, scattered, inaccessible populations of the present day.

Deliberate and accidental introductions of animals during European colonisation marked a second phase of faunal transformation. Domestic sheep, pigs, goats, horses and cats all became wild as, briefly, did dogs. Inevitably Norway rats, black rats and mice also invaded, largely displacing kiore and exploiting even more niches. European passerine birds were introduced to remedy the paucity of native birds in man-made habitats and other birds were brought in for game. Salmonid fish and two Australian frogs are abundantly naturalised. Deer of several species, chamois, thar, Australian possums and wallabies, hares, rabbits, and mustelids to control rabbits were also introduced, despite protests from people less imbued with wilful ignorance.

Probably, most introduced insects are confined to adventive vegetation. The spread of honey and bumble bees throughout New Zealand, however, may have brought new forces to bear on floral evolution (p. 54). The arrival of omnivorous wasps (*Vespa germanica* in the 1940s, *V. vulgaris* in the 1980s), and their subsequent population explosions may correlate with population collapse in several species of native bird that had weathered earlier threats; it is likely that other native animals are also vulnerable.

2

ORIGINS AND HISTORY OF THE FLORA AND VEGETATION

The floristic history of New Zealand centres on four events: breakup of the Gondwana super-continent (Fig. 2.1) including the separation of New Zealand from Australia in the late Cretaceous 60–80 million years ago; the Rangitata orogeny of the early Cretaceous, followed by relative quiescence that, by the Oligocene, led to New Zealand being reduced to small, warm, low-lying islands; renewed tectonic activity culminating in the Kaikoura orogeny of the Pliocene and Quaternary, accompanied by cooling, periodic glaciation and volcanism (Fig. 2.2); and, finally, the arrival of humans during the past millennium.

The fossil record
(from Mildenhall 1980)

The fossil record, consisting mostly of spores, pollen, wood and leaf impressions, is selective and incomplete. For instance, *Beilschmiedia* must have had a long history in New Zealand, yet its pollen is rarely found, even in recent deposits in regions where it is a forest dominant. A taxon is likely to have been present in a region long before the oldest detected fossil; and with increasing age taxonomic affinities of fossils become less certain.

The ancestry of the present-day forests lies firmly in Gondwanaland. Late Cretaceous forest floras in New Zealand scarcely differed from those of southern Australia. Conifers, some with lineages extending back to still earlier times, included the extant *Agathis, Dacrydium* s.l. and *Podocarpus* s.l., as well as *Microstrobus, Microcachrys* and, probably, *Athrotaxis*, which now survive only in Australia. Ferns included *Dicksonia, Cyathea, Gleichenia* and possibly *Hymenophyllum* and *Blechnum*. Angiosperms made their first New Zealand appearance, and included *Nothofagus, Ascarina, Gunnera*, Winteraceae (= *Pseudowintera?*) and Proteaceae.

Throughout the Tertiary there was accretion of further modern genera. Most of these have affinities with Melanesia, Queensland and Malaysia, examples being *Metrosideros* (possibly present as early as the Palaeocene), *Dysoxylum, Elytranthe* s.l., *Freycinetia* and *Rhopalostylis* (Eocene), *Myrsine, Elaeocarpus* and *Weinmannia*

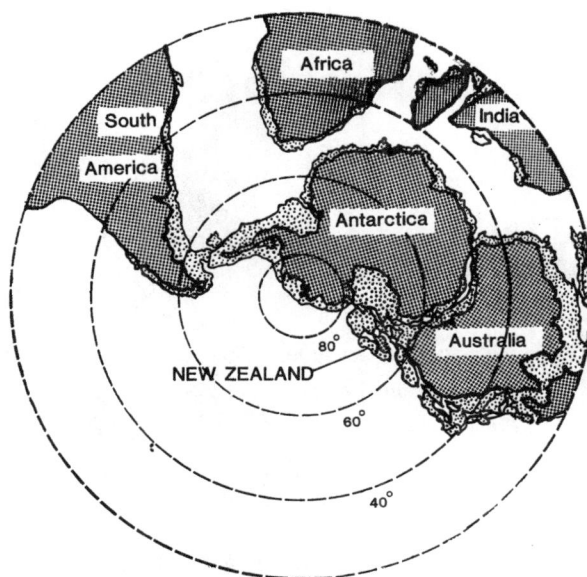

Fig. 2.1. Gondwanaland configuration in the late Cretaceous, after commencement of rifting (from Stevens 1980).

(Oligocene), *Nephrolepis, Alectryon, Macropiper* and *Cordyline* (Miocene), and *Eugenia* (= *Syzygium?*), *Loranthus* (= *Ileostylus?*) and *Arthropodium* (Pliocene). Several lowland forest genera, including *Tmesipteris, Ackama, Streblus, Nestegis* and *Pennantia*, are first detected in the Quaternary. These genera must have either dispersed across wide ocean gaps, or been already present in New Zealand but undetected in the fossil record, or spread from regions or habitats as yet unknown palaeobotanically, perhaps situated on the now-submerged ridges that link New Zealand with Melanesia.

Even more perplexing are 'primitive' genera confined to fragments of the Gondwanan continent, such as *Laurelia* (Oligocene) and, especially, *Phyllocladus* (Eocene) as its copious pollen output makes it unlikely to have escaped notice had it been present in earlier times. *Griselinia* and *Aristotelia* are further genera with southern distributions that are first reported from the Miocene, as is *Coriaria*, which has an interrupted modern distribution in both hemispheres. Endemic and New Zealand-based genera also progressively appear: *Phormium* (Eocene), *Pseudowintera* and *Coprosma* (Oligocene), *Melicytus* and *Ixerba* (Miocene), and *Geniostoma* (Quaternary).

Cocos, Bombax and *Cupania* belong to a tropical element that was present in our flora only during the Tertiary. The section of *Nothofagus* that produces pollen of the *N. brassii* type, *Casuarina* and *Ephedra* were present throughout the Tertiary, but diminished during the Pliocene and had all but disappeared by the Pleistocene. *Ilex* ranged from Eocene to early Miocene, capsules and leaves indistinguishable from those of *Eucalyptus* are known from Miocene deposits (Pole 1989), and *Racosperma* extended from the Miocene to late Pleistocene. As these warmth-demanding and

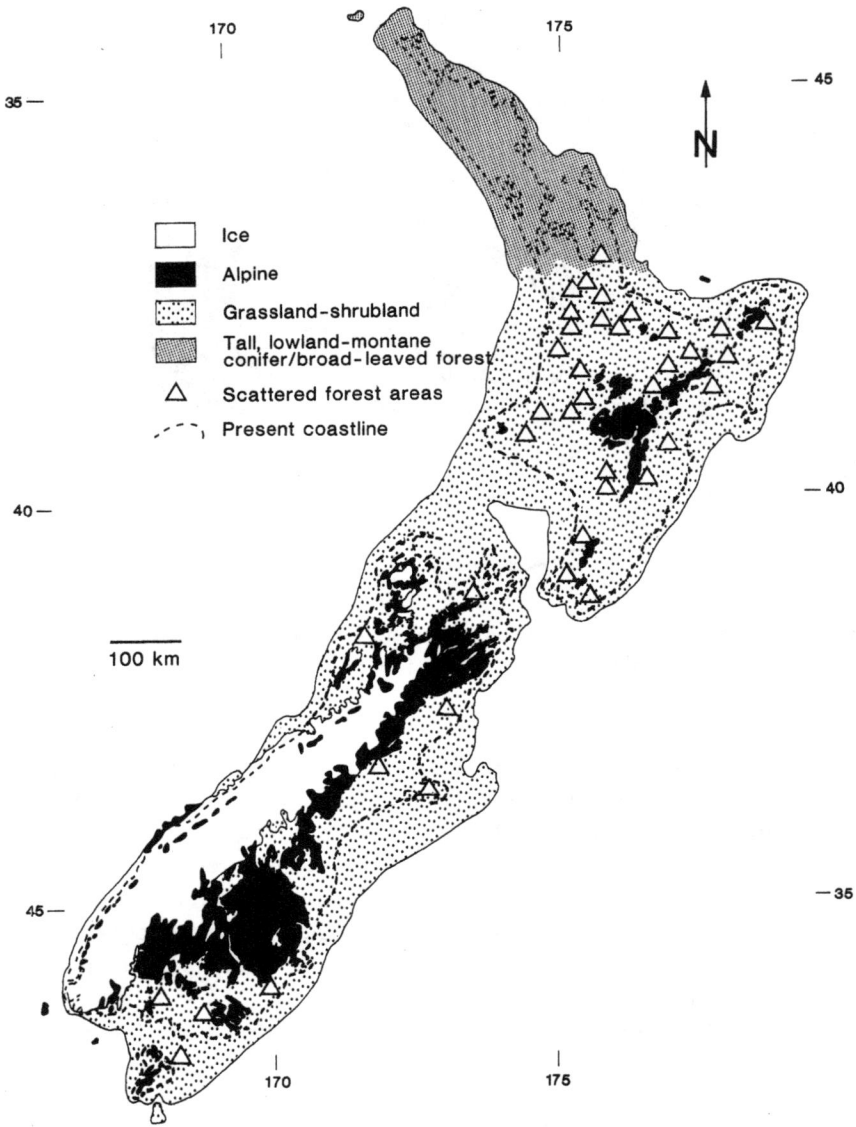

Fig. 2.2. New Zealand at the time of maximum late Pleistocene glaciation (from McGlone 1988).

Australian elements diminished, herbaceous and shrubby genera adapted to cool climates entered the fossil record. *Pimelea, Gentiana, Hebe* and *Colobanthus* are first reported from the Pliocene, whereas *Euphrasia, Drapetes, Forstera, Myosotis, Rumex, Oreostylidium, Psychrophila, Ranunculus* and *Notothlaspi* appear only during the Quaternary; *Epilobium*, however, is identified as early as the Oligocene. The proportions of grasses, composites, crucifers and umbellifers also increased as climates cooled.

Table 2.1. *Distribution of native vascular species among major habitat classes*

vegetation	belt	terrestrial	aquatic	% total
forest	Warm-temperate	358		20
	Cool-temperate	61		3
coastal	Warm-temperate	71	17	5
	Cool-temperate	30	2	2
shrubby & herbaceous	Warm-temperate	338	92	24
	Cool-temperate	371	37	22
	Inland-temperate	121	1	7
	Penalpine & alpine	227		13
restricted to outlying islands		91	6	5

Note:
The table is based on 1823 named species. Altitudinal belts are defined in Chapter 5. Species that characteristically occur in more than one habitat-class are assigned to the one that has the taller vegetation or warmer climate. 'Aquatic' includes plants in which at least the base of the stem is usually submersed.

The elements of the native flora

Although forests covered some 75% of the New Zealand mainland before human impact (Newsome 1987), they contain only 23% of the native vascular species, these being largely residual from richer Tertiary forests (Table 2.1). Nearly half of the species are more or less confined to habitats that are cool-temperate or colder, or to dry, frosty, inland localities; their origins pose one of the most intriguing questions in plant geography, in view of the isolation and presumed mid-Tertiary warmth and reduction of the New Zealand archipelago. Since early fossil records of cool-climate taxa are so incomplete, the problem will be resolved only through better understanding of taxonomic and evolutionary relationships. Nevertheless, the available facts provide useful insights.

Four ecologically relevant floristic assemblages may be recognised. The first comprises species known or inferred to have Miocene or earlier roots in New Zealand, and contains most forest trees of the lowlands, as well as several that are abundant in cooler upland forests. It is improbable that the latter have evolved their cool-climate tolerances since the Miocene, especially species such as *Nothofagus menziesii* and the podocarps *Phyllocladus alpinus* and *Lagarostrobos colensoi*, which have ecologically comparable relatives in Tasmania. Endemic, monotypic genera mainly confined to warmer parts of New Zealand are probably also survivors from the Tertiary; examples are *Knightia, Ixerba, Rhabdothamnus, Entelea* and *Teucridium* on the main islands, *Elingamita* on the Three Kings Islands, and *Myosotidium* and *Embergeria* on the Chatham Islands. *Tecomanthe* and *Xeronema*, which are Melanesian genera each with a single species on off-shore islands of northern New Zealand, may be in a similar category.

The second assemblage contains cold-tolerant plants belonging to small, taxonomically isolated, seemingly ancient genera. They occupy wet, infertile soils and most occur on other fragments of Gondwanaland, especially Tasmania or Fuegia. The best examples are *Lepidothamnus, Donatia, Liparophyllum, Gaimardia, Oreobolus, Carpha* and the endemic *Halocarpus, Tetrachondra* and *Oreostylidium*. Possibly, this assemblage and cool-climate taxa in the preceding assemblage survived the period of mid-Tertiary warmth in habitats similar to those that permit survival at low altitudes in the west of the South Island today, such as peneplains with soils too leached to support continuous forest, cool ravines, and ridges shrouded in orographic cloud.

Such refugia, however, cannot account for *Hectorella*, a monotypic genus confined to very high altitudes in the South Island, and one of the two members of its family, the other being *Lyallia* in Kerguelen. Possibly it reached New Zealand from a locality from which it has since vanished; if so, plants of apparent Gondwanan affinity may have greater potential for dispersal than is usually conceded, despite the failure of most to reach small, isolated islands such as those south of New Zealand. The composite *Haastia* is another small, high-altitude genus of uncertain affinity.

The third assemblage is of species in larger, more widely-distributed genera, that are likely to be late, transoceanic immigrants in that they, or very close relatives, occur also in other lands. The Australian element is the most important, and includes *Discaria, Rytidosperma, Dichelachne, Baumea*, and many epacrids, ferns and orchids; *Myoporum debile, Pomaderris apetala, Epilobium gunnianum, Sprengelia incarnata, Cryptostylis subulata* and *Scirpus polystachys* are such recent arrivals that they still have restricted ranges near western coasts. This element accounts for much of the flora of low-altitude grasslands and heathlands. The latter, together with related wetlands, seem merely depauperate derivatives of the rich Australian heath and sclerophyll vegetation. Yet they may not be entirely of recent origin, as they support the endemic, monotypic restiad *Sporadanthus*; one of the dominant genera, *Leptospermum*, although represented only by a single species that also grows in Australia, has a fossil pollen history that possibly extends from the Palaeocene; and the fossil records of *Casuarina, Eucalyptus* and *Racosperma* suggest greater diversity in the past. Species shared with other regions must have dispersed over great distances, especially the north-temperate *Carex pyrenaica, Deschampsia caespitosa, Trisetum spicatum* and the native form of *Festuca rubra*.

The fourth assemblage comprises the bulk of the New Zealand shrubby and herbaceous flora, that has evolved in response to the new environments of the Quaternary era. It can be subdivided into five groups:

1 Genera cosmopolitan in cool regions, but with a secondary centre of evolution in New Zealand. In *Ranunculus* and *Epilobium*, lineages concentrated in New Zealand have outlying representatives in Australia. The New Zealand species of *Gentiana*, together with a few Australian species, seem to fall in the same complex as South American species. Other examples include *Euphrasia, Myosotis, Poa* and *Luzula*.

2 Genera concentrated in other southern lands, but with distinctive species-clusters in New Zealand; e.g. *Helichrysum* (Australia and Africa), *Olearia* (Australia), *Parahebe* (Australia and New Guinea), and *Ourisia* (South America, but one species in Tasmania).

3 Genera with a large proportion of their species in the New Zealand lowlands, that can be reasonably assumed to have radiated into mountain environments as these became available; e.g. *Carmichaelia*, *Dracophyllum*, *Coprosma*, *Brachyglottis*, *Hebe* and *Astelia*.

4 Genera with their main development at high altitudes in New Zealand, but with a few species reaching or confined to the lowlands; e.g. *Aciphylla*, *Anisotome*, *Celmisia* and *Chionochloa*.

5 Small genera of higher altitudes or latitudes that are satellites of larger genera, e.g. *Pachycladon* and *Notothlaspi* (allied to *Lepidium*), *Cheesemania* (*Cardamine*), *Corallospartium* (*Carmichaelia*), *Lignocarpa* (*Anisotome*), *Leucogenes* (*Helichrysum–Raoulia* complex), *Dolichoglottis* (*Senecio–Brachyglottis* complex), *Damnamenia* and *Pleurophyllum* (*Olearia–Celmisia* complex), and *Chionohebe* (*Hebe–Parahebe* complex). *Notospartium* and *Chordospartium* (*Carmichaelia*) and *Pachystegia* (*Olearia*), lower-altitude genera almost confined to the north-east of the South Island, are of similar status. All these genera are endemic, excepting *Cheesemania* and *Chionohebe* which have 1–2 species in Australia.

Apart from *Raoulia* (perhaps present in New Guinea), *Hebe* (depending on the taxonomic treatment of the Tasmanian *Veronica formosa*) and *Carmichaelia* (which is on Lord Howe Island), every genus listed in groups 1–4 is also represented in Australia. In genera such as *Melicytus*, *Coprosma*, *Celmisia* and *Chionochloa*, or subgeneric clusters such as the 'alpine' ranunculi, where the species and morphological diversity are concentrated in New Zealand, the few species endemic in Australia are usually thought to result from westwards dispersal. However, in the *Anisotome–Aciphylla* complex and *Leptinella*, which also have most of their species in New Zealand, the Australian outliers are the most primitive (Webb 1986; Lloyd 1972). Similar evidence from *Epilobium* led Raven (1973) to propose that the New Zealand mountain flora is predominantly of Australian and, ultimately, Eurasian origin. Alternatively, origins may be diverse, with New Zealand itself acting as a centre of dispersal (Wardle 1978*b*). Among lowland and coastal plants, *Sophora microphylla* has reached Chile and Tristan da Cunha, two species of *Hebe* have reached Chile, and solitary species of *Hebe*, *Fuchsia* and *Chionochloa* on Rapa, Tahiti and Lord Howe Island respectively are presumably derived from New Zealand ancestors. Among upland plants, the presence of *Epilobium brunnescens*, *E. tasmanicum* and *Gingidia montana* in south-east Australia, and *Chionohebe ciliolata* in Tasmania, is best explained through westwards dispersal across the Tasman Sea, despite the contrary prevailing winds. The meagre flora of Macquarie Island (p. 463) also mainly results from dispersal of New Zealand species against prevailing winds.

Continuing evolution is suggested by numerous clusters of closely related species, and by the prevalence of hybrids, which often involve parents of very dissimilar appearance (e.g. shrubby and herbaceous species of *Brachyglottis*) or species assigned to different genera (*Gaultheria* × *Pernettya*, *Celmisia* × *Olearia*, *Leucogenes* × *Raoulia*, *Anaphalis* × *Helichrysum*, *Forstera* × *Phyllachne*). More isolated taxa possibly result from earlier bursts of evolution; examples may include the small genera in Group 5, and distinctive species within larger genera, e.g. *Hebe cupressoides, Coprosma talbrockiei, Brachyglottis bifistulosa* and *Celmisia philocremma.*

Plant distributions and Quaternary history of the vegetation

Even by a generous estimate, less than 25% of indigenous vascular species are distributed as widely through the New Zealand mainland as their apparent ecological tolerances would suggest. This decreases to less than 20% with exclusion of orchids and pteridophytes, for which corresponding estimates are 50% and 60% respectively. That is, the great majority of species appear to occupy only portions of their potential range, with many existing as widely disjunct populations.

The northern part of the North Island and the northern and southern parts of the South Island are recognisable as floristic 'centres', and the intervening areas as 'gaps' (Fig. 2.3). McGlone (1985) has explained the centres as being older and geologically more stable than the gaps; the south of the North Island was under sea from the Pliocene to the lower Pleistocene, and the middle portion of the South Island has been interpolated where the northern and southern parts of the island were separated through a lateral displacement of 450 km along the Alpine Fault which, according to some views, has been achieved since the upper Miocene. This has been accompanied by vertical displacement forming the Southern Alps, which is places are still rising by as much as 12 mm annually.

Although regional environmental differences established through tectonic events must have considerable bearing on plant geography, many details of plant range seem better explained through more recent events, especially the disruption of interglacial vegetation during glaciations. During the last glaciation, from 120 000 to 10 000 years ago, forest in South Island localities was, at times, so reduced in extent that tree pollen contributed scarcely 1% of the pollen rain, although higher values were reached during warmer stadials (Moar & Suggate 1979). Even in the Waikato basin around 38° S, grassland or scrub were the main vegetation between 20 000 and 14 000 years ago (Newnham *et al.* 1989). Reduction in mean temperatures was probably about 4.5 °C, equivalent to lowering altitudinal zones by 830–850 m, but the effects were probably intensified through severe temperature inversions, outbreaks of polar air, droughts and resulting fires.

Temperate forest remained extensive, at least in the far north of the northern floristic centre (Dodson *et al.* 1988). In the north of the South Island, great diversity of habitat may have allowed considerable survival of species. Centres of endemism in eastern Marlborough and Fiordland may be related to steeply plunging coastlines, which reduce migration distances between coastal refugia and present habitats.

Forests of conifers and broad-leaved trees spread throughout New Zealand as climates warmed, with the expansion taking place some 3500 years earlier in northern regions than in southern parts of the South Island, where it was not complete until about 9500 years ago. Pollen evidence indicates that the forests on the lowland strip west of the Southern Alps have scarcely changed since, but after the mid-Holocene in the west of the North Island, rimu (*Dacrydium cupressinum*) decreased in favour of matai (*Prumnopitys taxifolia*) and other species that tolerate drought better. Contemporaneously in the south of the South Island, rimu increased at the expense of

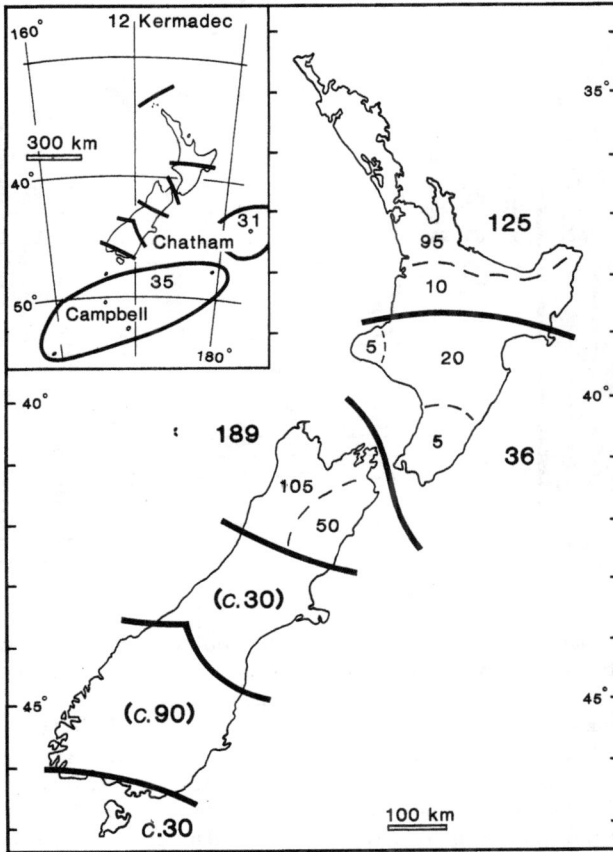

Fig. 2.3. Numbers of endemic species in different parts of New Zealand (from Table 13.1, McGlone 1985 and Wilson 1987, except for data in brackets, which refer only to dicots).

matai (Fig. 2.4). About 2500 years ago, fire permanently destroyed much of the forest in Central Otago (McGlone 1988). The frost-sensitive small trees *Ascarina lucida* and *Dodonaea viscosa* also decreased during the upper Holocene, and glaciers in the Southern Alps made periodic resurgences.

Spread of beeches during the Holocene has been much slower, with large tracts of apparently suitable habitat being still unoccupied (p. 554). Their rise to dominance occurred at different times in different localities (Fig. 2.5), but the pace seems to have quickened since the mid-Holocene.

The human period

After AD 900, Maori fires swept most of the forest from the eastern half of the South Island and from much of the North Island as well. In drier and higher regions, it was replaced by tussock grasslands whereas in milder, moister parts of New Zealand

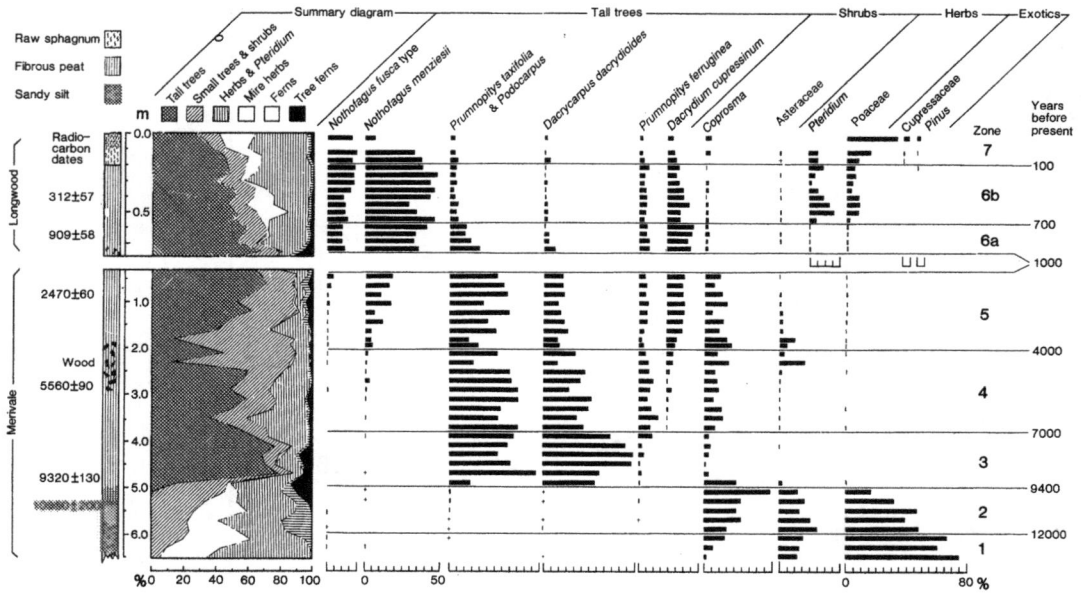

Fig. 2.4. Pollen diagrams from two localities in the Longwood Range, Te Wae Wae ER., showing variations in main life forms and key taxa (from McGlone & Bathgate 1983). The lower locality is at 60 m a.s.l.; the upper locality is at 670 m and still surrounded by *Nothofagus menziesii* forest. Pollen percentages are based on all dryland plants except pteridophytes. The six pollen zones show (1) dominance of grasses at the end of the last glaciation; (2) increase of shrubs, small trees and, finally, tree ferns with warming; (3) rapid spread of podocarp-dominated forest; (4) decrease of kahikatea (*Dacrycarpus*), and appearance of rimu and beech; (5) increase of rimu and beeches; (6) destruction of forest by Maori fires; (7) European settlement.

bracken fern (*Pteridium esculentum*) prevailed. Repeated burning maintained and extended grassland and fernland.

The Maori were Neolithic agriculturalists and eventually cultivated considerable areas of fertile lowland soil, but at first the tropical crop plants that they brought with them could be grown only in the warmest parts of the North Island. Eventually, however, they developed techniques that enabled them to grow sweet potatoes (*Ipomoea batatas*) at least as far south as Christchurch. Bracken rhizomes were dug up as a source of starch, so that the best bracken ground became semi-cultivated. Karaka (*Corynocarpus laevigatus*) probably had its natural range greatly extended by planting around coastal settlements, as the kernels were an important article of diet once they had been detoxified by steaming. The Maori also gathered wild plant material for food, shelter, medicine, fibre and timber.

Plants that arrived with the Maori from tropical Polynesia include a few garden escapes that persist in the north (*Cordyline terminalis, Colocasia esculenta*, sweet potato, and possibly *Hibiscus diversifolius*). The northern weeds *Alternanthera sessilis* and *Bidens pilosa* and the more widespread *Solanum americanum* may also be Maori introductions. Maori traditions that karaka and the small tree *Pomaderris apetala*,

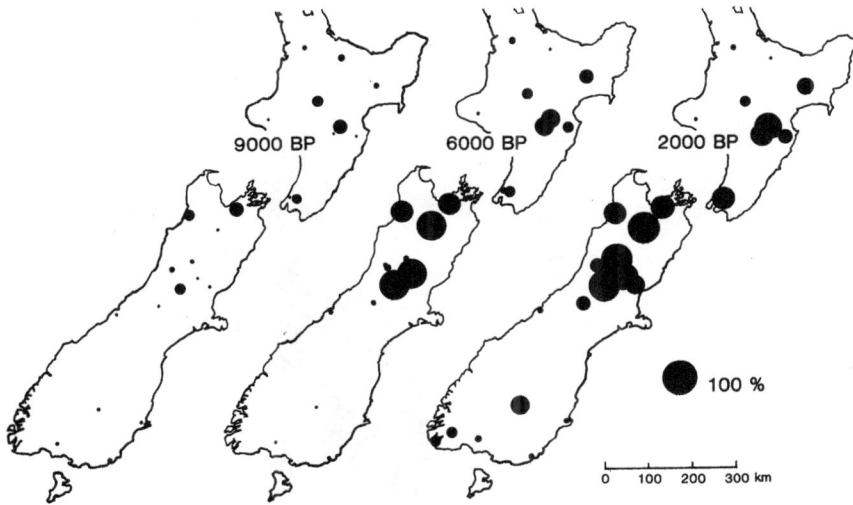

Fig. 2.5. Increase of *Nothofagus* at sites with a complete Holocene record, represented as percent of total tree pollen (McGlone 1988).

which grows on a stretch of western coast about latitude 38° 30′ S, were introduced by their forebears seem improbable, as the former is endemic and the latter is shared only with Tasmania.

Europeans appeared in the seventeenth and eighteenth centuries as voyagers, and then, after 1790, in search of seals, whales, kauri timber and *Phormium* fibre. From 1830 onwards a growing influx of permanent settlers began. Whereas the first 900 years of Maori occupation led to a population that was probably less than 200 000, from 1800 to the present the population increased to three million, acquiring ever more powerful technology as it did so. Crops, pastures of introduced grasses, and *Pinus* forests have now displaced native vegetation from most of the lowlands and accessible hill country, and areas have been completely cleared for mining, roads and urban development. Timber has been extracted from native forests. Grazing of sheep and cattle on tussock grassland and fernland has entailed burning of unpalatable growth, broadcasting of seed of palatable grasses and clovers, and application of fertilisers. Introduced birds are mostly found in man-made or modified habitats, but feral mammals are contained by no barriers except those of impassable terrain; their influence on the natural vegetation may not be as obvious as where man plays a direct role, but is profound.

Massive naturalisation of plants began with European colonisation. Many cultivated for food, fodder, shelter, timber or ornament became wild, and many others, including some universal followers of man, were brought unintentionally. Nearly every region of the globe has supplied introductions, although most can be traced to temperate Europe, North America or Australia. The native versus naturalised status is often uncertain, especially for cosmopolitan plants such as *Cotula coronopifolia* and *Calystegia sepium*, and many grasses and weeds shared with Australia. The success of the naturalised flora is largely in habitats created by man, but

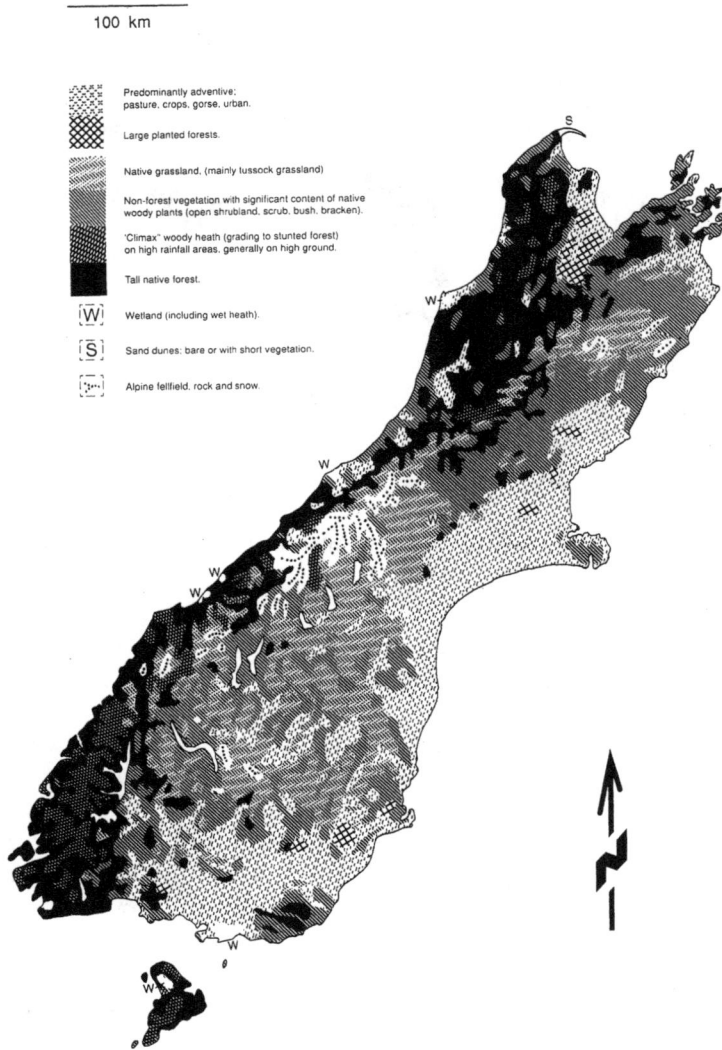

Fig. 2.6. Distribution of the main types of native and adventive plant cover in New Zealand (from Newsome 1987). Types consisting largely of small, isolated areas (e.g. alpine and wetland) are under-represented.

also reflects low competitiveness of native species in some natural habitats, such as flood plains, dunes and lakes.

Although the degradation and loss of so much native vegetation is regrettable, the cultural landscapes are often of great scenic appeal and contain much of botanical interest. Moreover, native vegetation, albeit often greatly modified, still covers about 60% of New Zealand (Figs 2.6 and 2.7); more than a quarter of it is protected in national parks and other reserves.

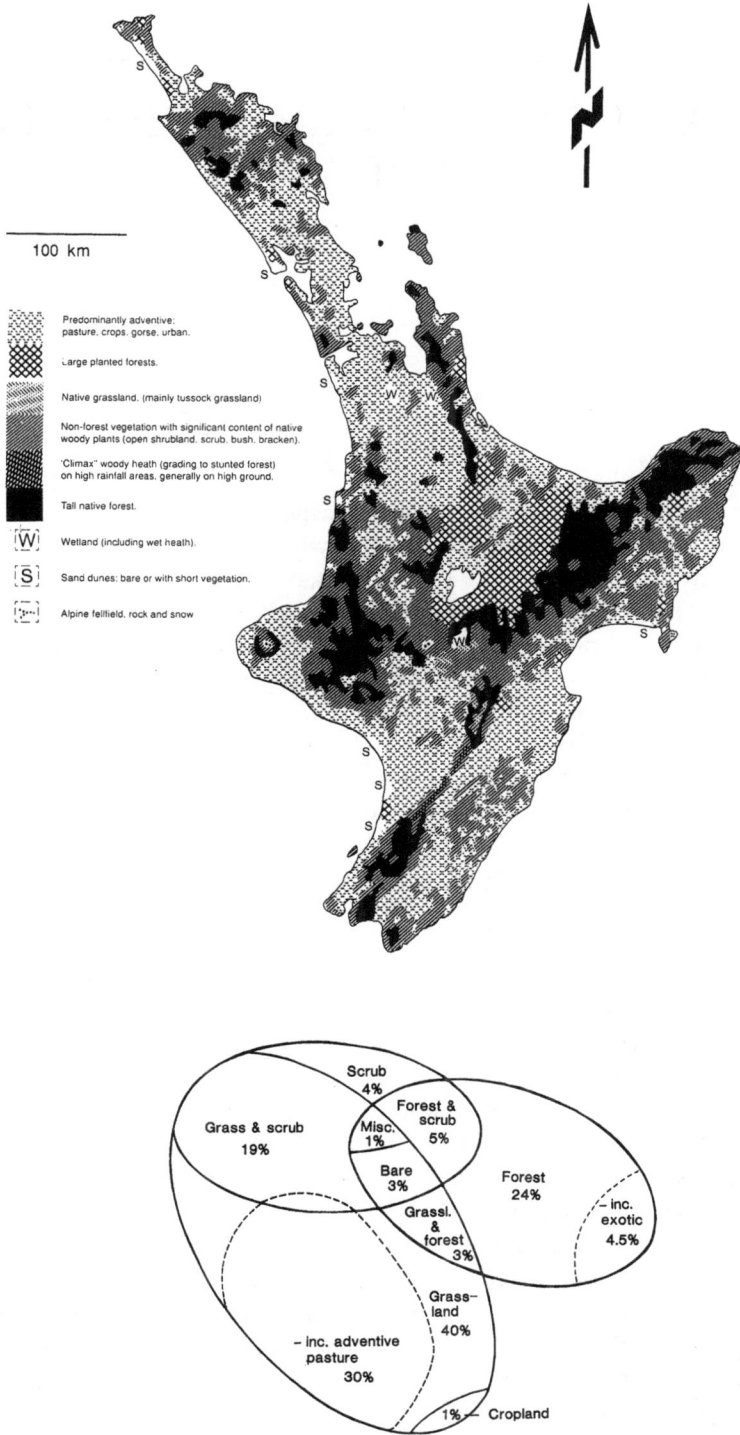

Fig. 2.7. Proportions of plant cover in New Zealand (from Newsome 1987).

3

PLANT FORM IN RELATION TO HABITAT

This chapter describes ecologically significant aspects of the vegetative morphology of representative, mainly native species. To test whether plant form can be objectively related to environment, Allan (1937) calculated biological spectra, as devised by Raunkiaer (1934), for the indigenous and adventive floras and major vegetation types. According to these, the native flora consists predominantly of hemicryptophytes and chamaephytes (i.e. with overwintering buds situated just below and just above the soil surface, respectively), but there is also a strong phanerophyte (tree and shrub) element. The adventive flora contains a higher proportion of therophytes (annuals). Allan (1937) concluded that biological spectra 'do not give a more accurate picture of plant environments than would careful summaries of the plant formations'.

Growth forms of woody plants

Tall trees (normally > 15 m)

Kauri (*Agathis australis*) is the giant of the New Zealand forest. Young trees, or *rickers*, have tall, slender trunks and deep, narrow crowns with short lateral branches, but once they exceed 35–40 cm diameter at breast height (d.b.h.) they shed lower branches and a stout, columnar bole develops, which divides into several limbs at a height that is commonly around 12 m. These ascending limbs bear the massive, more or less flat-topped crown. Large kauri reach 60 m, with trunks up to 7 m in diameter and canopy spreads exceeding 30 m (Ecroyd 1982).

Podocarps form the tallest tier where kauri is absent; indeed, rimu (*Dacrydium cupressinum*) and kahikatea (*Dacrycarpus dacrydioides*) can exceed 60 m. Matai (*Prumnopitys taxifolia*) and lowland totara (*Podocarpus totara*) reach diameters of 2 m. Miro (*Prumnopitys ferruginea*), mountain totara (*Podocarpus hallii*) and *Halocarpus kirkii* seldom exceed 25 m in height. Silver pine (*Lagarostrobos colensoi*) occasionally attains the canopy in tall forest. In closed stands, trunks of these podocarps are long and tapered, whereas the crowns are small. Widely spaced trees bear much wider crowns on heavy limbs. Mature trees tend to be stag-headed, with a disproportionately small amount of live foliage. Kahikatea develops large buttresses, in keeping with its often waterlogged habitats.

Phyllocladus trichomanoides is an erect tree with a narrow, conical crown of whorled branches; *P. glaucus* is shorter, but with larger cladodes and a wider crown. The two species of *Libocedrus* also maintain deep, conical crowns until they became stag-headed in old age, although exposed *L. bidwillii* (kaikawaka) trees can develop a 'flag' form, with branches only on the leeward side of the trunk.

Saplings and young trees of the tall conifers are straight and slender, those of rimu having bronze-green, weeping branchlets. In most species, severely deformed trees are infrequent. Totara are the exception, in that young trees in the open are often bushy and multi-stemmed, and trees on exposed coasts and mountains can be reduced to procumbent bushes.

Tall dicot trees form a tier beneath the conifers, or the main canopy where the latter are sparse or absent. Species of *Beilschmiedia, Nothofagus* and *Elaeocarpus*, as well as *Ixerba brexioides, Nestegis* spp. and *Litsea calicaris* at the lower height limit of the tall-tree class, have single, straight trunks, although butt-sweep often develops on steep slopes. Crowns are generally quite narrow in maturity; for example, in red beech (*Nothofagus fusca*) canopy spread scarcely exceeds 16 m even in the open. *Knightia excelsa* has a fastigiately branched, columnar crown, and *Vitex lucens* has a relatively short trunk that bears a canopy up to 25 m across (Burstall & Sale 1984). Mature red beech and *Laurelia novae-zelandiae* have large buttresses.

Most of these dicot trees are excluded from extreme sites, so that grossly misshapen specimens are infrequent. However, under the influence of wind or drifting snow at the upper forest limit, *Nothofagus solandri* var. *cliffortioides* and *N. menziesii* often die back from the top and coppice from the base, leading to stunted plants that tend to have procumbent, adventitiously rooting main stems (Fig. 3.1). The distinctive juvenile (p. 68) of *Elaeocarpus hookerianus* can develop wind-sculptured 'flags' when growing on exposed heathland.

In tawa (*Beilschmiedia tawa*), *Quintinia* and *Weinmannia* spp., multi-stemmed saplings may develop into trees with several trunks, although competition often reduces these to one (p. 562). *Weinmannia*, moreover, usually begins life as an epiphyte when growing in closed forest, and develops a 'false' trunk through the fusion of descending roots. These eventually encase the host, most often a tree fern. Both true and false trunks augment radial growth by producing and fusing with adventitious roots, and become irregular and gnarled.

When established in the open on gentle terrain, the three main tree species of *Metrosideros* (*M. excelsa, M. robusta* and *M. umbellata*) have the form of giant, rounded bushes that are multi-stemmed from the base; *M. excelsa* can attain a canopy spread of 52 m (Burstall & Sale 1984). In closed vegetation, a single, crooked main trunk develops. This can be procumbent or extend horizontally from cliffs, and its direction of growth seems to respond more to the best light source than negatively to gravity. Moreover, *M. robusta* usually and *M. umbellata* often have epiphytic beginnings and, as in *Weinmannia*, develop false trunks that encase the host tree. In *M. robusta*, these trunks can exceed 4 m in diameter. All species, especially *M. excelsa*, also produce hanging aerial roots from their limbs.

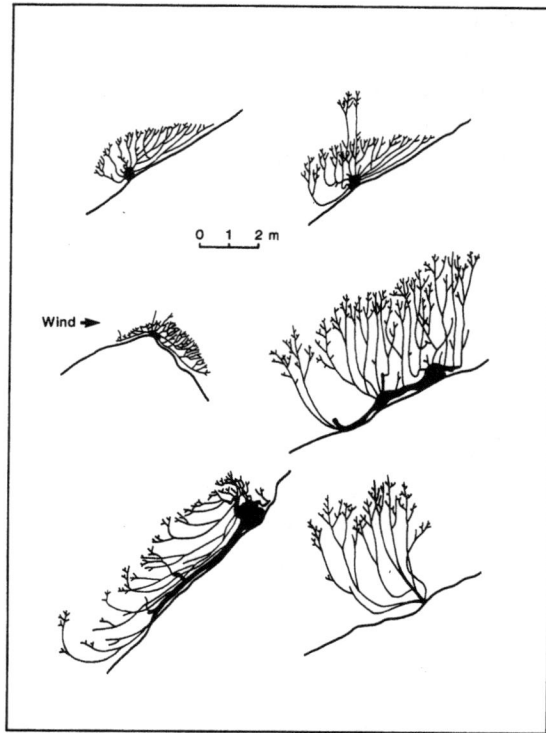

Fig. 3.1. Profiles of mountain beech growth forms at the upper forest limit in the Craigieburn Range (Norton & Schoenenberger 1984).

Broad-leaved small trees and shrubs of lowland forest and bush

Small broad-leaved trees, some scarcely more than shrubs, form a subcanopy tier in tall forest, and the canopy in shorter vegetation that will be referred to as 'bush' (Chapter 5). The upper height limit is usually about 10 m, although it is convenient to include also karaka (*Corynocarpus laevigatus*), kohekohe (*Dysoxylum spectabile*) and *Plagianthus regius*, which are commonly up to 15 m tall. Four groups can be recognised, as follows.

(1) Long-lived trees with stout trunks and mainly glossy leaves

Karaka, kohekohe, and the northern *Pisonia brunoniana* and *Planchonella costata* normally have straight, erect trunks and are confined to mild coastal districts. The mainly coastal ngaio (*Myoporum laetum*) has crooked trunks that are often prostrate. *Griselinia littoralis*, which is widespread in cooler districts, begins life perched on trees, mossy boulders or logs, or on the ground where undergrowth is sparse. The trunk, whether true or formed of fused descending roots, is thick-barked, gnarled, often over 1 m in diameter, and freely produces adventitious shoots, enabling the tree to resprout after fire. *G. lucida*, although usually epiphytic, can become a small tree on coastal cliffs.

Fuchsia excorticata appears to grow towards the direction of most light, its trunk being crooked, leaning, and often prostrate near its base. Diameters are up to 50 cm, and the bark is thin, flaking and underlain by chlorenchyma. In mahoe (*Melicytus ramiflorus*), individual stems are erect and fairly short-lived, but replacement stems are produced from a broad lignotuber. *M. macrophyllus* and *M. lanceolatus* are slender relatives, of northern and upland forests respectively.

(2) Slender, fast-growing, short-lived trees and large shrubs

The best examples are the northern, coastal whau (*Entelea arborescens*), and wineberry (*Aristotelia serrata*) which grows throughout the mainland. *Pseudopanax* includes the lowland lancewoods *P. crassifolius* and the much rarer *P. ferox*, which have thick leaves that are long and narrow in the unbranched saplings. The lowland *P. edgerleyi* and the upland *P. simplex* have broader, thinner leaves. *P. arboreus* (five-finger), the mainly upland and southern *P. colensoi*, and the northern *P. laetus* and *P. lessonii*, like *Schefflera digitata*, all have palmate leaves. The larger-leaved pittosporums include two common lowland trees, *Pittosporum eugenioides* and the smaller *P. tenuifolium*. The latter has several close relatives including the southern and upland *P. colensoi*, and the northern coastal *P. crassifolium*. The pepperwoods, named for their hot-tasting leaves, are *Pseudowintera axillaris* in lowland forests of central and northern New Zealand, and *P. colorata* in upland and southern forests.

Myrsine australis is abundant throughout lowland forest and bush, whereas *M. salicina* grows mainly in the North Island and mild western districts of the South Island. *Carpodetus serratus* is most abundant on moist slips and stream banks, *Hedycarya arborea* is widespread in lowland forest except in south-eastern districts of the South Island, and *Ascarina lucida* is abundant along the western seaboard of the South Island but very local in the North Island. Further coastal trees are akeake (*Dodonaea viscosa*), which is mainly seral in northern and central New Zealand; *Melicytus novae-zelandiae*, which is northern except for outlying populations on islands in Cook Strait, where it meets the smaller-leaved *M. obovatus*; and the northern island endemic *Meryta sinclairii*.

Large-leaved coprosmas include the seral *Coprosma robusta*, and *C. lucida* and *C. grandifolia*, which are more common in forest. The coastal taupata (*C. repens*), which has leaves with a mirror-like gloss, can become a small tree or thrive as prostrate shrubs on exposed rocks.

Olearia rani is abundant on dryish forested slopes southwards to latitude 41° 31' S. *O. paniculata* grows on North Island and eastern South Island coasts, extending inland in rocky gorges. *O. arborescens* is along stream banks and in mesic scrub; satellite species include *O. furfuracea*, which is a pioneer on dry hills throughout the North Island. *O. avicenniifolia* is common throughout the South Island, as a pioneer and in rocky places. *O. fragrantissima* grows in scattered localities between Banks Peninsula, Lakes and Catlins Ecological Regions (ERs).

Rangiora (*Brachyglottis repanda*) is abundant in moist lowland seral vegetation through northern and central New Zealand. *B. hectorii* is confined to limestone in the north-west of the South Island, whereas *B. kirkii* is abundant in the shrub tier of northern forests and as an epiphyte extends to cooler North Island districts.

Alseuosmia is a genus of 0.5–2 m tall, sparingly branched shrubs largely confined to moist northern forest, but two species reach western districts of the South Island.

Tall hebes are often prominent in disturbed areas. *Hebe salicifolia* is abundant throughout the South Island and *H. stricta* through the North Island; they overlap in the Marlborough Sounds. *H. parviflora* var. *arborea*, at 7 m one of the tallest hebes, is especially common near Wellington. *Coriaria arborea* is a toxic shrub, that in spring produces succulent, unbranched shoots up to 4.5 m long that later develop leafy laterals bearing tassels of dark, shining berries. It colonises stream banks, moist gravel fans and slips and often persists in seral bush, becoming a small, irregular tree.

(3) Trees with small-leaved, divaricating juvenile forms (Table 3.1)
Most species are widely distributed on relatively fertile sites such as river flats and talus, but *Pittosporum turneri* is a rare tree confined to the Volcanic Plateau.

(4) The palm-form
Rhopalostylis, a true palm, has pinnate leaves, and its stems develop their final girth through enlargement of the apical meristem before height growth commences. Tree ferns have finely divided fronds, and the trunks consist of a thick cortex of adventitious roots and leaf traces surrounding the central stem. In *Cordyline*, leaves are simple and strap-like, and trunks thicken through multiplication of vascular bundles. Since the cambium is regenerated from stem parenchyma, *Cordyline* is seldom killed by fire. The wide-leaved, upland *C. indivisa* and the short-stemmed, northern *C. pumilio* usually remain unbranched, but the common cabbage tree (*C. australis*) ultimately develops a large, many-headed crown. *C. banksii* grows as loose clumps, mainly branching from the base.

Broad-leaved small trees and shrubs of upland, southern and infertile habitats
Woody composites with leathery, tomentose leaves are a distinctive element in cool-climate vegetation. In the open, most grow as rounded bushes, but in dense stands they develop crooked main trunks with lower portions that are often procumbent and adventitiously rooting. Bark in many is thin, papery and flaking.

Olearia avicenniifolia and *O. arborescens* ascend to the limit of tall woody vegetation, where the latter hybridises with the aptly named mountain holly *O. ilicifolia* and *O. lacunosa*. *Olearia moschata* is a much smaller shrub growing at still higher altitudes.

The macrocephalous olearias, so-called because of their large capitula, have thick, finely serrate leaves. In *O. colensoi*, a mountain species, the lower branches sweep down-hill when growing on steep slopes and take root so that nearly impenetrable thickets develop. In the south-west *O. colensoi* descends to the coast, and on Stewart Island meets *O. lyallii*, which becomes a tree up to 10 m tall, despite retaining prostrate lower trunks. *O. oporina* grows on coasts and islets around Foveaux Strait and along the outer coast of Fiordland. There are also two macrocephalous olearias on the Chatham Islands.

Taxa currently recognised in a complex of broad-leaved *Brachyglottis* species are *B. elaeagnifolia* and *B. buchananii* on the ranges of the North and South Island respectively, with the latter merging into *B. rotundifolia* on the coasts of Fiordland and

Table 3.1. *Woody plants with filiramulate habit (left) and nearest larger-leaved relatives (right)*

Trees (only juveniles filiramulate)

Prumnopitys taxifolia	sd	*P. andina* in South America; *P. ferruginea*
Pittosporum turneri, virgatum	sd	Possibly *P. patulum*
Elaeocarpus hookerianus	sd	*E. dentatus*
Plagianthus regius	sd	Chatham Island juveniles wf
Hoheria sexstylosa	sf	*H. populnea* juveniles wf
Hoheria angustifolia	sd	*H. sexstylosa*
Hoheria lyallii, glabrata	sf	
Carpodetus serratus	wd	
Sophora microphylla	sd	Chatham Island & some mainland varieties wf; also *S. tetraptera*
Nothofagus fusca, solandri	wd	
Streblus heterophyllus	sd	*S. banksii* (northern & central zone coasts)
Pennantia corymbosa	sd	*P. baylisiana* (Three Kings Islands)
Coprosma arborea	wf	

Shrubs

Muehlenbeckia astonii	sf	Lianoid *Muehlenbeckia* spp.
Pittosporum divaricatum, P. rigidum & several other spp.	sd	Possibly *P. patulum*
Melicytus micranthus	sd	Hybridises with *M. ramiflorus*
Melicytus alpinus (branchlets become thick spines)	sd	Seems to grade to large-leaved spp. *via* a series including *M. crassifolius*
Myrsine divaricata	sd	Hybridises with *M. salicina*
Lophomyrtus obcordata	sd	Hybridises with *L. bullata*
Neomyrtus pedunculata	sf	
Aristotelia fruticosa	sd	Hybridises with *A. serrata*
Plagianthus divaricatus	sd	Rare hybrids with *P. regius*
Corokia cotoneaster	sd	Hybridises with *C. buddleoides*
Sophora prostrata	sd	Hybridises with *S. microphylla*
Discaria toumatou (spinous dwarf shoots)	sd	Closest relatives also sd, in Australia & Patagonia
Alseuosmia banksii var. *banksii*	wf	Hybridises with larger-leaved species
Pseudopanax anomalus	sd	Hybridises with *P. simplex*
Melicope simplex	sd	Hybridises with *M. ternata*
Coprosma propinqua	sd	Hybridises with *C. robusta*
Coprosma spathulata	sf	*C. arborea* (wf juvenile)
Coprosma rigida & > 15 other species	sd	No interfertile large-leaved species
Olearia capillaris	wd	*O. arborescens*
Olearia fragrantissima	wf	
Olearia odorata	sd	*O. hectorii* wf
Olearia virgata	sd	*O. lineata* sf; possibly *O. traversii*
Rhabdothamnus solandri	wf	Larger-leaved plants on off-shore islands
Teucridium parvifolium	wd	

Lianes

Clematis marata and related spp.	sf	*C. foetida* and other larger leaved species
Muehlenbeckia complexa	sf	*M. australis*
Scandia geniculata	sf	*S. rosifolia*
Parsonsia capsularis forms with linear adult leaves	sf	Other *Parsonsia* forms, sf only in juveniles
Brachyglottis sciadophila	sf	Shrubby *Brachyglottis* spp.

Note:
f, Filiramulate only; d, filiramulate and divaricating; s,w, habit strongly or weakly exhibited.

Stewart Island. They all have rayless capitula, unlike the showy *B. stewartiae*, a tree confined to islands west of Stewart Island. *Traversia baccharoides*, a monotypic segregate from *Senecio*, is a shrub with brittle, sticky twigs and glabrous leaves that grows on steep, rocky places in northern parts of the South Island.

The araliad *Pseudopanax colensoi*, in 3- and 5-foliolate subspecies, is abundant except where eliminated by introduced mammals. *P. linearis*, a high-altitude equivalent of the lancewoods, grows on western mountains of the South Island. *Pseudowintera traversii* is endemic to infertile uplands in the north-west of the South Island.

In striking contrast to other native woody vegetation, mountain ribbonwoods (*Hoheria glabrata* and *H. lyallii*) have soft, deciduous leaves and white, apple-like blossom; they grow on relatively fertile soils in South Island mountains. In the Tararua and Rimutaka ranges, an evergreen ribbonwood in the *H. sexstylosa* complex occupies similar habitats.

Small podocarps

Silver pine, pink pine (*Halocarpus biformis*), celery pine (*Phyllocladus alpinus*) and *Podocarpus acutifolius* can grow either as erect trees over 10 m tall, or form low bushes that spread through root suckers in silver pine, and by layering branches in the other species. Yellow-silver pine (*Lepidothamnus intermedius*) is a small tree that forms colonies of leaning, crooked stems. It hybridises with the mat-forming *L. laxifolius*, reputedly the world's smallest conifer. *Halocarpus bidwillii* can reach 3 m, but is usually an erect, bushy shrub. *Podocarpus nivalis* (snow totara) has procumbent to prostrate or buried stems that can form layered colonies extending as much as 16 m downhill from the point of establishment.

Epacrid and ericad trees and shrubs

Cyathodes juniperina and *C. fasciculata* are shrubs of open habit that are often abundant in forest or shrubland on dry or infertile soils. The other 6–7 New Zealand species are procumbent (*C. empetrifolia*) or dwarf shrubs (p. 27). In wet, high-altitude forest in the South Island, *Archeria traversii*, a shrub with a stout, gnarled trunk, is usually abundant. *A. racemosa* occurs instead on summits of the Barrier islands, Coromandel Range and Raukumara Range. *Epacris pauciflora* and *Sprengelia incarnata* are spindly shrubs of wet heathland; at high altitudes the low-growing *Epacris alpina* replaces *E. pauciflora*.

Species in *Dracophyllum* subgenus *Oreothamnus* have needle-like leaves, and encompass a variation in stature and leaf-length that is commensurate with their ecological range. *D. longifolium* is a major dominant of upland heaths through most of the South Island, and varieties or closely related species extend from the Coromandel Peninsula to Campbell Island. It can be 12 m tall, although the main stem is often leaning or horizontal, especially on steep slopes. *D. lessonianum* and *D. subulatum* are smaller erect shrubs, abundant on northern heaths and central North Island tephra soils respectively.

D. uniflorum is a shrub of high altitudes that usually forms dense clumps, 1–2 m tall, with ascending stems that layer in their lower portions. On mountains north of the

Manawatu Gorge it is replaced by *D. recurvum*, which is lower and more spreading; this and *D. strictum* are the only species in the section with terminal inflorescences.

Subgenus *Dracophyllum* is unique among native dicots in its strap-like, parallel-veined leaves held in terminal tufts. Young plants are single-stemmed but branching patterns diverge when flowering begins. In the upland *D. traversii* and the northern *D. latifolium* the inflorescence is large and terminal. Further vegetative growth is therefore from lateral buds, and a many-headed tree results. Other species in the section have small, lateral inflorescences. *D. fiordense* forms clumps of stout, erect, unbranched stems up to 3 m tall, whereas *D. menziesii* is a procumbent, sparingly branched shrub; both are plants of south-western mountains. *D. townsonii* is an erect shrub confined to mountains in the north-west of the South Island.

The only native members of the Ericaceae are in *Gaultheria* and *Pernettya*. These have broad, reticulate-veined leaves, but occupy typical heath habitats, and some are also rupestral. They range from erect shrubs up to 2 m tall, such as *G. rupestris* and *G. antipoda*, to high-altitude mat-plants.

Other small-leaved native shrubs

Manuka (*Leptospermum scoparium*) is the most abundant New Zealand shrub, especially in seral vegetation. In the open under favourable conditions, it is a fast-growing, more-or-less fastigiate bush, but in closed stands it develops single, spindly, naked stems up to 8 m tall. In exposed sites, manuka becomes prostrate, and on wet, sterile soils can be creeping and matted. Strains vary in leaf form, flower size and colour, and architecture (Yin *et al.* 1984). Kanuka (*Kunzea ericoides*) often codominates with manuka in seral shrubland but becomes a tree up to 20 m tall. There are also prostrate ecotypes and, on Great Barrier Island, a shrub separated as *K. sinclairii*.

The New Zealand cassinias, a group of closely related, 1–2 m tall shrubs, are often abundant in secondary shrublands and on coastal sand. In foliage they are remarkably similar to *Olearia solandri*, *Pomaderris phylicifolia* and *Brachyglottis cassinioides*. All these plants, as well as *Olearia nummulariifolia*, *O. cymbifolia*, and various small shrubs in the genera *Hebe*, *Helichrysum* and *Pimelea*, have closely spaced leaves (p. 38), whereas in *Coprosma pseudocuneata*, an abundant shrub in the undergrowth of high-altitude forest and tall scrub, leaves are more distant. In most small-leaved coprosmas, and in several convergent species in other genera, leaves or leaf-clusters are separated by disproportionately long internodes, and branchlets diverge at wide angles. This habit is discussed on p. 31.

Dwarf shrubs

Woody plants under 20 cm tall are important in native grasslands and other open vegetation from the coast to the highest altitudes. Most of the species have very small leaves and belong to genera that contain taller relatives; *Pimelea*, *Gaultheria*, *Cyathodes*, *Dracophyllum*, *Coprosma* and *Hebe* provide examples. Only three have leaves that much exceed the leptophyll category (p. 37), these being the rare, northern-coastal *Fuchsia procumbens*, *Rubus parvus*, which grows on river terraces in the west of the South Island, and *Coprosma serrulata*, which is an understorey species in high-altitude shrubland.

Basal parts of the stems are usually prostrate and adventitiously rooting, but *Lepidothamnus laxifolius, Cyathodes fraseri, C. pumila, Pentachondra pumila, Gaultheria depressa, Pernettya* spp. and mat-forming coprosmas are rhizomatous. The semi-rhizomatous *Rubus parvus* and *Muehlenbeckia axillaris* and the trailing *M. ephedroides* have lianoid ancestry. Many other dwarf shrubs are cushion plants (p. 33).

Herbaceous growth forms

Dicot herbs

Dicot herbs can be conceptually derived from shrubby prototypes, through reduction of longevity, stature and woodiness. The 2 m tall poroporo (*Solanum aviculare* and *S. laciniatum*) and the 1 m tall *Haloragis erecta* are bushes of disturbed lowland habitats, that might equally be regarded as soft-wooded shrubs or large herbs. Smaller plants in *Parahebe* and section of Lignosae of *Celmisia* develop wood, but this remains narrow through the length of the stem. *Parietaria debilis, Pratia physaloides, Urtica australis, Sarcocornia quinqueflora* and willow-herbs such as *Epilobium melanocaulon* are bushy, erect, tap-rooted, perennial herbs with leaves spaced along the branchlets.

Trailing stems are common among plants of saline habitats, such as *Tetragonia tetragonioides, Spergularia media* and species of *Atriplex* and *Chenopodium*, but usually prostrate herbs root at the nodes, as in *Nertera, Epilobium brunnescens* and *Ourisia caespitosa*.

One reduction of the 'archetypal' bushy herb is to an annual herb with leaves and inflorescences distributed along one or a few main stems. Many adventive weeds fit this description, e.g. *Solanum nigrum, Fumaria muralis* and *Linum catharticum*. The vegetative phase may be further reduced to a single tap-rooted rosette; this is followed by the reproductive phase, and then by death of the plant. Usually the inflorescence is much larger than the leaf rosette, and may include cauline leaves. This category has many adventive herbs, including most umbellifers, crucifers and many composites. Species range from small annuals such as *Cardamine hirsuta* to tall biennials such as *Verbascum* spp., hemlock (*Conium maculatum*) and *Beta vulgaris. Gentiana grisebachii*, the very local wetland plant *Brachyscome linearis* (p. 108) and *Myosurus minimus*, a species of semi-arid habitats, are among the few native annuals. Monocarpic native biennials or perennials are mostly plants of high altitudes, such as *Gentiana divisa, G. corymbifera, Notothlaspi rosulatum, Cheesemania fastigiata* and, probably, *Myosotis macrantha*.

Alternatively, the rosette phase may be of indeterminate longevity, and produce flowers year after year, as in most native dicot herbs. In a few native and introduced plants, e.g., *Taraxacum* and *Kirkianella*, the rosette tends to remain solitary, as inflorescences are axillary. More usually, the stock develops a number of crowded leaf rosettes, still served by the original roots, as the original stem apex yields dominance after damage or flowering. This is normal in dandelion-like plants such as *Microseris* and *Hypochoeris*, as well as in most species of *Aciphylla*, the larger species of *Anisotome* (e.g. *A. pilifera, A. haastii* and *A. latifolia*), *Geranium sessiliflorum* and *Lepidium sisymbrioides*.

In many species, the branches of the stock elongate and produce adventitious roots, so that clumps expand indefinitely; examples are *Anisotome aromatica, Plantago lanigera* and *Ranunculus multiscapus*. Massive branching stocks are produced by some ourisias, *Ranunculus lyallii* and *Myosotidium*. Stolons carry daughter rosettes away from the parent in *Hieracium pilosella, Ranunculus repens, R. membranifolius, Gunnera* spp. and *Eryngium vesiculosum*.

Rhizomes mostly have long internodes, and nodes that give rise to roots and leafy dwarf shoots. They have originated in several ways. *Calystegia soldanella* and *Dichondra* are derived from convolvulaceous lianes. *Acaena* grades from the sprawling *A. glabra*, to species such as *A. anserinifolia* that weave through low vegetation but take root where they touch the soil, to fully rhizomatous species such as *A. inermis*. The subscandent *Wahlenbergia gracilis* may be compared with the rhizomatous *W. albomarginata*, or the suberect *Leptinella featherstonii* of the Chatham Islands and the sprawling *L. plumosa* of the far-southern islands with rhizomatous mainland species.

Alternatively, rhizomes could develop through further elongation of the rootstocks of clumped plants. For example, *Aciphylla congesta* forms wide, low clumps and *A. pinnatifida* is fully rhizomatous. Stolons can become rhizomes through burial, as in *Gunnera dentata*.

These morphological series are not necessarily evolutionary series. In *Celmisia*, for example, broad clumps as in *C. spectabilis* might plausibly be derived from the rhizomatous habit represented by *C. glandulosa*, or from species such as *C. ramulosa* that have an open, shrubby habit with leaves spread along elongated stems, or from the many species like *C. semicordata* that have leaves crowded on a single rootstock.

Tussocks and other monocot growth forms

Luzuriaga parviflora and the epiphytic orchid *Dendrobium cunninghamii* have branching, leafy, aerial shoots of more or less determinate length. The leafy shoots of *Earina* are also determinate, but branch only from the base. In *Bulbophyllum*, leaves are borne singly on bulb-like internodes that interrupt thread-like rhizomes. In many grasses and some sedges and rushes, leafy stems trail on the surface (e.g. *Agrostis stolonifera, Schoenus maschalinus, Juncus articulatus*) or scramble through vegetation (e.g. *Microlaena polynoda, Pennisetum clandestinum*), taking root where they touch the soil. Nearly all other herbaceous monocots in New Zealand, except those that are fully aquatic (p. 301), have linear or reduced leaves tufted at ground level on short, erect shoots that usually terminate in an inflorescence. However, the inflorescence axis may bear more distant leaves or leafy bracts.

In densely tufted monocots, new shoots grow up beside existing shoots, often being contained by the sheaths of closely packed, long, narrow, evergreen leaves; the habit seems associated with longevity, and probably confers the advantage that the plants are not readily invaded and displaced by competitors. In New Zealand it is seen in *Phormium, Astelia* and many grasses, sedges and rushes, and is common from the coast to the high mountains, and from forests and swamps to open grasslands. Usually, dead leaves and scapes collapse around the perimeter, sometimes accumulating as a pile of litter, but the plant itself remains green and leafy.

In tussock grasses, however, dead sheaths, leaf bases and culms remain trapped. As plants grow older and larger, accumulating dead material forces the growing points of tillers upwards and outwards. This appears to set a size limit beyond which tussocks either collapse and die, or break up into a ring of daughter tussocks. The swamp sedges *Carex secta* and *C. sectoides* continue upward growth until the leafy crown becomes supported on a 'trunk', up to 1.5 m tall, of interwoven living and dead roots and tillers. *Chionochloa acicularis* and *Poa litorosa* develop similar trunks on wet, peaty ground in Fiordland and the far-southern islands respectively.

Most of the 'tall tussocks' or chionochloas, and the even taller cortaderias, have relatively broad leaves that are abscised when they die, so that plants appear more or less green throughout the year. In contrast, in the 'short tussocks' (mostly species of *Poa* and *Festuca*) living leaves are very narrow, in-rolled and strict, and fail to overtop persisting dead leaves and culms, so that the plants remain unchanging shades of pale brown. This evergreen or ever-brown appearance reflects milder winters than prevail in north-temperate grasslands; similar tussocks dominate grasslands on the mountains of south-east Australia and on tropical high mountains.

In monocots of more open habit, new shoots diverge widely, either bursting through the leaf sheaths or arising behind them. They may then grow upwards, to form broad, open tufts, as in many species of *Rytidosperma*, *Luzula* or *Carex*. Alternatively, some shoots may grow indefinitely as rhizomes; such plants are especially characteristic of loose, sparsely vegetated sand, e.g. *Carex pumila*, *Spinifex sericeus* and *Pyrrhanthera exigua*.

No firm line separates the tufted from the rhizomatous habit; single genera, especially *Carex* and *Poa*, may include the whole spectrum. Even a single genotype can cover the range. For instance, *P. cockayneana* on stable flats forms tussocks scarcely distinguishable from those of silver tussock (*P. cita*), whereas in other circumstances, upper internodes of some tillers elongate and give rise to daughter tillers; this leads to tufts connected by rhizomes on unstable river shingle, and stoloniferous swards on steep, moist slopes. Usually tufted species of *Libertia* can also develop stolons and rhizomes, with *L. peregrinans* being fully rhizomatous.

Wetland monocots often have massive rhizomes that ramify below the surface of the soil or submerged mud. In many *Juncus* species, the rhizome system is contracted, and the aerial shoots or leaves are crowded into dense tussocks. Rhizomes are far-spreading and the aerial shoots distant in *Typha* and *Schoenoplectus*. Three restiad genera, each with a single New Zealand species, have extensive 'rafts' of closely branched rhizomes; rush-like stems that rise densely from the rhizomes are unbranched in *Leptocarpus* and sparingly branched in *Empodisma* and *Sporadanthus*.

Most ground orchids and *Bulbinella* die back after flowering to a perennating swollen root, whereas the inconspicuous grassland herbs *Iphigenia novae-zelandiae* and *Hypoxis hookeri* die back to corms. Several adventive monocots that perennate and disperse through bulbs or corms have become well-established, notably *Allium triquetrum* in urban wasteland, *Crocosmia ×crocosmiiflora* along roadsides and downstream from points of introduction, *Watsonia bulbillifera* along northern stream margins, and *Sisyrinchium* spp. in open grassland.

Ferns and fern-allies

All *Lycopodium* species have leafy aerial stems, which are scrambling in *L. volubile*, creeping and rooting in *L. scariosum*, connected by rhizomes in *L. fastigiatum*, and tufted in *L. varium*. The two psilophyte genera also have tufted stems, which are rootless in *Tmesipteris* and leafless in *Psilotum*. In the minute lycopod *Phylloglossum drummondii*, leaves are tufted on a tuber.

All true ferns, excepting aquatic species, have fronds inserted directly on root-bearing rhizomes. In many species fronds are inserted singly on a creeping rhizome. Some of these, including bracken (*Pteridium esculentum*), *Histopteris incisa*, and species of *Hypolepis*, are robust ferns with extensive, buried rhizomes. Others, with superficial rhizomes, are mostly epiphytic or rupestral (p. 35), but *Hymenophyllum demissum* and *H. bivalve* are usually terrestrial, and *H. multifidum* is often so.

In the *Blechnum 'capense'* complex and the much smaller *B. penna-marina*, fronds are crowded towards the ends of shallow, branching rhizomes and densely cover considerable areas. In other ferns, fronds form terminal rosettes or tufts, which are solitary in *Lastreopsis hispida*, *L. glabella* and most species of *Asplenium* and *Polystichum*, whereas in *Cyathea colensoi* and the usual form of *Dicksonia lanata*, they are borne on shallow, branching rhizomes.

Some tufted ferns bear fronds on a trunk-like caudex. Solitary species, in which each caudex represents an independent plant, include *Marattia salicina*, *Polystichum vestitum* and *Leptopteris superba*. In *Blechnum discolor* and a northern form of *Dicksonia lanata*, caudices are stout, matted with aerial roots and dead frond bases, usually less than 50 cm tall, and connected by rhizomes to form extensive colonies, whereas caudices in *Blechnum fraseri* colonies are up to 1 m tall and very slender. Tree ferns are an extension of this category; all except *Dicksonia squarrosa* are solitary.

Special and dependent growth forms

The filiramulate habit

Many New Zealand shrubs in diverse genera and families have minute apical growing points, and slender, interlacing twigs. Leaves are distant or in distant clusters on dwarf shoots, usually spreading, thin and short-lived, and less than 1 cm long (Figs 3.2 and 3.3). Several tree juveniles, again in diverse families, and some lianes, exhibit the same form. The habit is referred to as 'divaricating', but because this term emphasises wide-angle branching, which is absent in some of the plants under consideration, yet frequent in plants of quite different appearance, I propose the term *filiramulate* instead, to emphasise the slender, wiry twigs which may be divaricating, zig-zagging, or merely flexuous. In exposed habitats or under severe browsing the filiramulate habit leads to dense, rounded shrubs in which most of the leaves, flowers, fruit and growing apices are protected within a mesh of interlocking, springy branches. These are regarded as 'typical divaricating shrubs', and there has been much debate as to whether they have evolved in response to browsing by now-extinct avifauna or in

Fig. 3.2. Two interfertile *Coprosma* spp., the filiramulate *C. propinqua* and the large-leaved *C. robusta*. Lamina lengths *c.* 5 and 80 mm respectively.

response to climatic adversity (Atkinson & Greenwood 1989; McGlone & Webb 1981).

Most filiramulate species have clearly evolved from larger-leaved forest species within New Zealand, and many hybridise with non-filiramulate relatives (see Table 3.1). Some tree juveniles, e.g. *Carpodetus*, and forest shrubs such as *Rhabdothamnus* have weaker filiramulate tendencies. It may be that lax filiramulate forms evolved as mid-Tertiary forests adapted to cooler climates, a rejuvenated landscape and an evolving 'megafauna', and that closely divaricating forms mostly represent phenotypic adaptation to more extreme situations. The remarkable convergence among unrelated genera must reflect New Zealand's isolation, which prevented immigration of lineages already tolerant of browsing or climatic adversity.

Filiramulate plants are found in salt-marshes, swamps, forests, successional scrub, and on dry or frosty terraces, with many species showing a wide amplitude, none more so than *Coprosma propinqua*. However, the number of species and density of individuals are greatest in forest margins, bush and tall scrub on relatively fertile soils. Small-leaved shrubs with divaricating branching are conspicuous in various semi-arid regions, notably eastern Patagonia, but in contrast to New Zealand plants they are mostly thorny (D.R. McQueen, personal communication).

Fig. 3.3. Branching habit of *Myrsine divaricata*.

Mat and cushion plants

Mat plants are creeping or rhizomatous herbs and dwarf shrubs in which the foliage presents a continuous flat surface. Leafy mats are developed in various species, including most native cudweeds (*Gnaphalium*) and the adventive hawkweed *Hieracium pilosella*, but many New Zealand mat plants have minute, close-set leaves on short, crowded shoots; some are moss-like.

Mat plants growing among other vegetation have diffuse margins, but when colonising bare ground most develop circular outlines. Mats remain low because they continue to grow radially while central shoots scarcely grow at all, whereas in cushion plants central and marginal shoots grow at similar, albeit very slow rates. Some cushion plants are extremely compact, with a shell of close-set leaves and shoots, and an interior packed with stems and dead material.

Mat and cushion plants have evolved convergently from many genetic stocks, and through reduction of several other growth forms. A morphological series leading to herbaceous mat plants is demonstrated by the wiry, layering *Helichrysum bellidioides*, the smaller, more compact, rhizomatous *H. filicaule*, the usually compact *Raoulia monroi*, and the very compact *R. australis*. *Coprosma* provides woody series, from large to small-leaved erect shrubs, through sprawling, adventitiously rooting dwarf shrubs such as *C. cheesemanii*, to mat plants such as *C. perpusilla* and *C. petriei*. Reduction can also be achieved phenotypically, as in *Myriophyllum propinquum*, which develops buoyant stems up to 1 m long in permanent ponds and streams, or forms short, close turf on muddy ground that is only periodically submerged. Manuka becomes a rhizomatous mat on wet, infertile soils.

Mat plants may change to cushion forms in difficult environments where radial spread is restricted. For example, *Raoulia australis* and *Coprosma atropurpurea* grow as circular mats on bare, flat ground, but become hard, rounded cushions if their shoots trap fine loess while intervening surfaces are deflated through frost-heave and wind. In other cushion plants the form is inherent. *Pimelea pulvinaris* is a woody cushion plant with layering branches, which grows in very dry places and has evolved from a dwarf shrub of more open habit, such as *P. oreophila*. Woody raoulias are tap-rooted cushions, whereas the herbaceous *Phyllachne colensoi* has tap-roots and also sends adventitious roots into the soil and its own peaty interior.

Anisotome shows a sequence from loosely clumped rosettes on a branched, adventitiously rooting stock, as in *A. aromatica*, through small, dense mats as in *A. flexuosa* to the alpine, cushion-forming *A. imbricata*. In most species of *Colobanthus*, rounded cushions of tightly packed rosettes arise from a single tap-rooted stock. Adventitiously rooted cushions are found in *Poa maniototo*, *Agrostis muscosa* and *Luzula ulophylla* of very dry habitats and the alpine *Poa pygmaea*, *Luzula colensoi* and *L. pumila*. *Colobanthus* and these grasses and luzulas are successful on bare soil subject to needle ice, as they remain intact even when partly frost-heaved.

Cushion plants are most abundant and diverse at high latitudes and altitudes in the Southern Hemisphere. In New Zealand they grow in a wide range of severe habitats from the permanent snowline to the seashore, and from bogs to dry gravel terraces (Gibson & Kirkpatrick 1985). In cold environments they are probably an evolutionary response to shoot activity being favoured only in the relatively calm, diurnally heated layer near the ground, and uptake of water and nutrients being limited by cold, raw soils. Uptake is also impaired in acidic, saturated soils. In dry, open, sandy habitats, the compact habit may protect the plant from excessive evaporation and abrasion. Mat plants occupy most of these environments, for the same reason. Some are effective pioneers, or persist where taller plants are destroyed by grazing. Others are important in short turf where water levels fluctuate along margins of lakes, tarns and estuaries (Chapter 10).

Lianes

Lianes are common in most lowland vegetation except tidal and submersed wetlands. They can be classified according to mode of climbing.

(1) Woody lianes
  Tendrils: Native and adventive *Clematis* and *Passiflora* spp.
  Adventitious roots: *Metrosideros* spp., *Freycinetia baueriana*.
  Twining: *Ripogonum scandens*, *Parsonsia* spp., *Muehlenbeckia* spp.
  Prickles: Native and adventive *Rubus* spp.
  Scrambling: *Fuchsia perscandens*, *Brachyglottis sciadophila*.
(2) Semi-woody lianes
  Tendrils: *Sicyos australis*.
  Twining: Native and adventive *Calystegia* spp.
  Scrambling: *Scandia geniculata*, *Tetragonia trigyna*, *Senecio mikanioides* (adventive).

(3) Ferns and lycopods
   Twining rachis: *Lygodium articulatum.*
   Adventitious roots: *Blechnum filiforme.*
   Scrambling: *Lycopodium volubile.*

Most indigenous woody species are forest lianes that reach the canopy. *Blechnum filiforme* attains the lower limbs, whereas *Lygodium* and the lycopod form tangles in the understorey. *Metrosideros* spp. and *Blechnum filiforme* also carpet the forest floor where the leaves, like those on the lower climbing stems, are much smaller than those on the fertile shoots in the canopy. Much of the mass of *Freycinetia* also exists as dense, 1–2 m tall thickets in the understorey, supported by stilt-like roots.

The semi-woody and adventive lianes mainly grow in forest margins, bush and modified vegetation. Many weak-stemmed herbs and dwarf shrubs also become lianoid when trailing through dense vegetation; *Microlaena polynoda* and *Einadia* spp., which grow in coastal bush and scrub respectively, and the heathland sundew *Drosera peltata*, are obligate scramblers.

Vascular epiphytes

In very moist forests, most species can occur as epiphytes, at least as seedlings. The following appear especially well adapted to the epiphytic mode, and mostly also thrive on rocks where fissures or pockets of humus allow initial establishment; those marked † grow terrestrially as often as epiphytically.

Juveniles of trees (pp. 21–2): *Metrosideros robusta, M. umbellata*†, *Griselinia littoralis*†, *Pseudopanax arboreus*†, *P. colensoi*†.
Shrubs: *Griselinia lucida, Pittosporum cornifolium, P. kirkii, Brachyglottis kirkii*†.
Creeping and rooting angiosperm herbs: *Peperomia* 2 spp. †, *Bulbophyllum* 2 spp.
Tufted monocot herbs: *Astelia solandri, Collospermum* 2 spp., the orchids *Dendrobium cunninghamii, Earina* 2 spp., *Drymoanthus adversus.*
Tufted ferns and fern-allies: *Asplenium polyodon, A. flaccidum, Hymenophyllum pulcherrimum, Ctenopteris heterophylla*, most *Grammitis* spp., *Tmesipteris* spp., *Lycopodium varium*†.
Creeping and rooting ferns: most epiphytic filmy ferns including the thicker-leaved kidney fern (*Trichomanes reniforme*), *Phymatosorus* 3 spp., *Pyrrosia serpens, Anarthropteris lanceolata, Rumohra adiantiformis.*

Juvenile trees and *Griselinia lucida* eventually send roots to the ground, whereas other shrubs and all the herbaceous species can thrive indefinitely without contacting the soil. Epiphytes must maintain a viable water balance through dry periods, and most probably achieve this through physiological means. However, *Bulbophyllum* has swollen internodes subtending the leaves, and *Collospermum* has leaf-bases shaped to hold rain water. Filmy ferns, like bryophytes, shrivel during dry weather and recover on wetting.

Vascular epiphytes mainly grow where humus collects in forks and on large horizontal branches and leaning trunks; decaying bryophytes and bark are the main sources. *Bulbophyllum* and *Pyrrosia* tend to be mostly on branches high in the canopy, whereas *Grammitis billardierei* is usually near the bases of trunks. *Drymoanthus*, like some filmy ferns, clings to vertical trunks. *Tmesipteris* spp. and *Trichomanes venosum* are almost confined to stems of tree ferns and *Hymenophyllum malingii* is usually on living or dead trunks of *Libocedrus bidwillii*, but no other vascular epiphytes are host-specific.

Vascular parasites and saprophytes

Dactylanthus taylorii is the only native vascular plant that is fully and directly parasitic. It forms a gall-like woody structure on tree roots, and is best known from the central North Island. The adventive dodder (*Cuscuta epithymum*) is common in low-producing grassland, and also parasitises *Raoulia tenuicaulis* mats. Broomrape (*Orobanche minor*) is found mainly among introduced shrubs.

Non-green, almost leafless *Gastrodia* orchids occur widely but sparingly. *G. minor* and *G. sesamoides* can share root-inhabiting fungi with *Leptospermum scoparium* and *Racosperma melanoxylon* respectively, and *G. cunninghamii* associates with *Nothofagus* via *Armillaria mellea*; they presumably draw nutrients from the trees via the fungi (Campbell 1962, 1963). *Thismia rodwayi* and the orchid *Yoania australis* are small, rarely seen, non-green monocots of mainly northern forests.

Photosynthetic semi-parasites are more important ecologically. Those attached to the roots of their hosts through haustoria include the 15 herbs and semi-woody shrubs in the genus *Euphrasia*, the adventive *Parentucellia viscosa*, and two members of the Santalaceae, i.e. the small northern tree *Mida salicifolia* and *Exocarpus bidwillii*, a practically leafless shrub of high-altitude shrublands (Fineran 1962). The tiny, leafless mistletoe *Korthalsella* represents the Viscaceae. *K. salicornioides* on *Leptospermum* and *Kunzea*, and *K. lindsayi* and *K. clavata* on filiramulate shrubs, convincingly mimic the shoots of their hosts. Leafy mistletoes are assigned to five small genera of the Loranthaceae. *Peraxilla* spp. and *Alepis flavida* are showy plants, with red and yellow flowers respectively, which grow almost exclusively on *Nothofagus*. They do not show the mimicry which is so striking among Australian loranths and probably for this reason have been severely depleted by possums. *Ileostylus micranthus*, despite a much wider distribution and host range, also seems much depleted, and *Trilepidea adamsii* is probably extinct. In contrast, *Tupeia antarctica* has secured its survival through copiously parasitising introduced plants, especially *Chamaecytisus palmensis*.

Cassytha paniculata, a slender, rampant liane that attaches itself to shrubs by haustoria, is an Australian plant that extends to semi-coastal heath in Northland.

Insectivorous plants

Only two genera are represented. *Drosera* grows on bogs and heaths that are at least seasonally wet, whereas *Utricularia* grows in acidic pools.

Leaf dimensions

Woody plants

Leaf shape and size impart much of the appearance of vegetation, and are also significant in its functioning. A selection of relatively broad (length less than five times width), net-veined leaves, classified according to Raunkiaer's (1934) scheme, are discussed first.

1. In the native vegetation, only the compound leaves of *Rhopalostylis* and tree ferns qualify as **megaphylls**, i.e. >1640 cm² (corresponding to oblong leaves with dimensions of *c*. 80 × 25 cm).

2. The **macrophyll** category, i.e. >180 cm² (*c*. 20 × 10 cm) includes *Meryta sinclairii*, the palmately compound leaves of *Schefflera digitata* and some *Pseudopanax* spp., and the pinnately compound leaves of *Dysoxylum spectabile* and, barely, *Alectryon excelsus*.

3. **Mesophylls** lie in the range 20–180 cm², i.e. from *c*. 6 × 4 cm. They predominate in mixed forest (p. 71) and bush, and among woody composites in upland and southern coastal vegetation. Species are listed below in approximate order of diminishing size (†indicates the leaflet of a compound leaf).

(*a*) ≤50 cm²: *Pisonia brunoniana, Entelea arborescens, Brachyglottis repanda, Beilschmiedia tarairi, Dysoxylum†, Coprosma grandifolia, Macropiper excelsum, Pseudopanax arboreus†, Schefflera†.*

(*b*) <50 cm²: *Corynocarpus laevigatus, Aristotelia serrata, Melicytus ramiflorus, Alseuosmia macrophylla, Weinmannia racemosa, Olearia colensoi, Nestegis cunninghamii, Griselinia littoralis, Hedycarya arborea, Pittosporum eugenioides, Alectryon†, Olearia arborescens, Coprosma robusta, Pseudowintera axillaris, Myrsine salicina, Olearia rani, Fuchsia excorticata.*

Pseudopanax crassifolius, P. ferox, P. linearis, Ixerba brexioides, Knightia excelsa, Elaeocarpus dentatus and *Olearia lacunosa* have unbranched saplings bearing long, stiff, narrow linear leaves; in *P. crassifolius* dimensions are commonly 100 × 2 cm. Adult leaves are shorter, wider mesophylls (Fig. 4.6).

4. **Microphylls** are in the range 2.25–20 cm², i.e. from *c*. 2 × 1.5 cm. They include important dominants of mixed forest: *Metrosideros* spp., *Agathis australis, Beilschmiedia tawa, Elaeocarpus hookerianus* and *Phyllocladus* spp. (in respect of pinnate cladodes). They also include *Nothofagus fusca, N. truncata* and some smaller trees and shrubs, e.g. *Planchonella costata, Brachyglottis buchananii, Coriaria arborea, Myoporum laetum, Geniostoma rupestre, Dodonaea viscosa, Carpodetus serratus, Hebe salicifolia, Pennantia corymbosa, Toronia toru, Ascarina lucida, Pittosporum tenuifolium, Olearia traversii* and *Plagianthus regius.*

5. **Nanophylls** (25–225 mm², or from *c*. 4 × 8 mm) include *Nothofagus solandri, N. menziesii*, a few species of lowland bush, such as *Hoheria angustifolia, Streblus heterophyllus* and *Coprosma rotundifolia*, and upland shrubs such as *Hebe odora, Olearia moschata* and *Gaultheria crassa.*

Other leaves, including those in Raunkiaer's **leptophyll** category (<25 mm²) are best classified morphologically as follows.

1. Strap-like, parallel-veined leaves borne as dense terminal tufts on sparingly branched stems. Species total only 11 but are distributed from northern coasts to the upper forest limits, except in cold, dry districts. Leaves are about 150 cm long in *Freycinetia*, and in *Cordyline* range from 30–60 cm in *C. pumilio* to 200 cm in *C. indivisa*. In *Dracophyllum* subgenus *Dracophyllum* they range from 15 cm long in *D. menziesii* to 70 cm in *D. fiordense.*

2. Leaves of filiramulate shrubs and juvenile trees (p. 31).

3. Narrow, firm, usually sharp-tipped needle leaves. Spreading needles occur in *Podocarpus, Prumnopitys* and *Cyathodes juniperina*. In *Dracophyllum* subgenus

Oreothamnus laminas are usually nearly erect, and vary in length from only 2 mm in the high-altitude cushion plant *D. muscoides* to as much as 20 cm in *D. longifolium*; in the taller species, juvenile leaves are larger, strap-like and reflexed.

4. Ericoid leaves generally < 1 cm long, closely spaced on slender shoots, more or less spreading, and usually firm-textured. Species include manuka, kanuka, cassinias and similar plants (p. 27), all pimeleas other than *P. longifolia* and *P. gnidia*, *Olearia nummulariifolia*, *O. cymbifolia*, and many epacrids and dwarf shrubs. They are most characteristic of heathlands, but also grow in forest understories, in grassland and at high altitudes.

5. The cupressoid habit distinguished by scale leaves, which are usually closely appressed to the stem. It occurs in *Dacrydium*, *Dacrycarpus*, *Lagarostrobos*, *Lepidothamnus*, *Halocarpus*, *Libocedrus* and sections of *Helichrysum* and *Hebe*.

6. The final reduction is to leaflessness, which is discussed under xeromorphy (p. 40).

Herbaceous dicots

The usual dicot leaf is simple with pinnate main veins, elliptic to oblong outline, and entire or toothed margins. However, linear leaves with subparallel veins feature in *Oreostylidium subulatum*, *Dolichoglottis* spp., most of the densely tufted species of *Celmisia*, and *Aciphylla* in respect of leaf segments. Pinnately lobed leaves are found mainly among composites such as *Senecio, Lagenifera* and the Chicorioideae. In *Acaena* and *Potentilla anserinoides* leaves are pinnate, and in *Leptinella* they range from linear through lobed to bipinnate. Nearly all apioid umbellifers have 1–2-pinnate leaves; exceptions are simple, linear leaves in *Lilaeopsis* and a few species of *Aciphylla*.

Lobed leaves of radial design are found in *Geranium*, hydrocotyloid umbellifers, *Gunnera* and *Stilbocarpa*. Other Hydrocotyloideae and *Oxalis* have trifoliolate leaves. In *Ranunculus*, leaves range from peltate (*R. lyallii*), through 1–2-ternate, to bipinnate (*R. gracilipes*).

The following selection of species is listed in order of diminishing leaf area.

Macrophylls: *Anisotome latifolia, Pleurophyllum speciosum, Stilbocarpa polaris, Aciphylla scott-thomsonii, Myosotidium hortensia, Embergeria grandifolia*; all these except the *Aciphylla* are confined to outlying islands.

Mesophylls: *Ranunculus lyallii, Celmisia semicordata, Gingidia montana, Elatostema rugosum, Ourisia macrophylla*.

Microphylls: *Dolichoglottis scorzoneroides, Senecio minimus, Anisotome aromatica, Geum parviflorum, Gentiana corymbifera, Brachyglottis bellidioides, Urtica incisa, Celmisia viscosa, Plantago raoulii, Aciphylla monroi, Gnaphalium audax*.

Nanophylls: *Acaena anserinifolia, Celmisia gracilenta, C. glandulosa, Leptinella squalida, Geranium sessiliflorum, Nertera* aff. *dichondrifolia, Epilobium glabellum, E. brunnescens*.

Leptophylls include many species of suboptimal habitats, such as *Sarcocornia quinqueflora* and *Crassula* spp. (coastal), *Nertera depressa* and *Australina pusilla* (forest floor), the epiphyte *Bulbophyllum pygmaeum* and nearly all mat and cushion plants.

Herbaceous monocots

Leaves of all native herbaceous monocots have parallel veins, and nearly all are narrow-linear. Exceptions are the narrow-lanceolate leaves of *Arthropodium cirratum* and various orchids, some of which are heart-shaped (e.g. *Adenochilus gracilis*), orbicular or broadly oblong (*Corybas* spp.). In cross-section, most leaves are flat (e.g. *Libertia*), folded (*Phormium* and most grasses), double-folded (species of *Astelia* and *Carex*), incurved (other *Carex* spp., *Desmoschoenus spiralis*) or involute (many narrow-leaved grasses). Leaves are tubular in the orchids *Microtis* and *Prasophyllum*. In some rushes (e.g. *Juncus maritimus*) and rush-like sedges they are scarcely indistinguishable from the culms, and in many others are reduced to basal scales. Restiads are distinguished by culms that bear whorls of scale leaves at regular intervals.

Only *Phormium tenax* leaves, with dimensions up to 300×12 cm, qualify as megaphylls. Macrophylls include *P. cookianum*, large astelias such as *A. fragrans*, *Arthropodium cirratum* and *Cortaderia* spp.

Mesophylls include *Chionochloa conspicua* and *Gahnia xanthocarpa*. Most of the important tussock grasses, including *Chionochloa pallens* and *Festuca novae-zelandiae*, are microphylls, as are sedges such as *Carex coriacea*. The smaller grasses, such as *Poa colensoi* and most *Rytidosperma* spp., and *Astelia linearis* are nanophylls and cushion plants such as *Oreobolus* spp. are leptophylls.

Ferns

Fern fronds typically show high orders of pinnate dissection. The main exceptions are:

1-pinnate: *Nephrolepis, Doodia, Pellaea, Ctenopteris*, the larger aspleniums such as *A. polyodon* and *A. oblongifolium*, the smallest filmy ferns, and *Blechnum* spp. other than *B. fraseri* (2-pinnate) and the two species below.
Pinnately lobed to entire: *Blechnum colensoi, B. nigrum, Phymatosorus*.
Entire: *Grammitis, Pyrrosia, Anarthropteris, Ophioglossum, Hymenophyllum armstrongii, Trichomanes reniforme*, and the aquatic ferns.

In outline, fronds of *Marattia salicina*, tree ferns, *Cyathea colensoi, Dicksonia lanata* and *Blechnum* 'black spot' qualify as megaphylls, and many other abundant ferns as macrophylls, e.g. *Polystichum vestitum, Blechnum discolor* and *Leptopteris superba*. At the other extreme, fronds of *Hymenophyllum minimum* on coastal rocks, *H. armstrongii* on tree trunks and *Grammitis poeppigiana* on alpine ledges are less than 25 mm^2 in area.

Deciduous, summer-green and annual plants

Indigenous vegetation, from northern coasts to the alpine limits, is overwhelmingly evergreen, although in some mountain plants, especially *Dracophyllum uniflorum*, red coloration in the leaves intensifies during winter. The only fully deciduous trees and shrubs are *Fuchsia excorticata, Plagianthus regius, Hoheria glabrata, H. lyallii, Discaria toumatou* and filiramulate olearias, and even in these, deciduousness reflects short-lived leaves and winter-cessation of growth, rather than deep winter dormancy. By the end of winter, *Aristotelia* spp., *Brachyglottis hectorii, Urtica ferox*, *Muehlenbeckia* vines and some races of *Sophora microphylla* are almost leafless in colder districts. *Nothofagus fusca* saplings may also lose all but their youngest leaves, but no New Zealand beech is fully deciduous.

Native herbaceous plants are likewise mostly evergreen perennials, but in fertile wetlands (Chapter 10) the species of *Bolboschoenus* and *Schoenoplectus*, several rhizomatous *Carex* species and *Typha orientalis* die to the ground each winter. The rhizomatous ferns *Histiopteris incisa* and *Hypolepis millefolium* are summer-green even in sheltered habitats, as are rhizomatous, semi-woody species of *Coriaria*. However, fronds of New Zealand bracken (*Pteridium esculentum*) usually survive through winter, in contrast to bracken in northern lands. Most ground orchids, *Bulbinella* and *Iphigenia* are summer-green; *Hypoxis* is the only summer-dormant geophyte, but may not be native. The coastal *Eryngium vesiculosum* and the high-mountain plants *Dolichoglottis, Stellaria roughii*, several *Ranunculus* spp. and the tufted fern *Polystichum cystostegia* are also summer-green. Annuals in the native flora are few and inconspicuous (p. 28). That the native flora contains so few deciduous, summer-green and annual plants reflects isolation as much as environment, for such plants form a large proportion of the naturalised flora.

Bud structure and shoot growth in woody plants
(Wardle 1963a, 1978c; Haase 1987 and earlier papers)

Woody plants range from those with 'specialised' resting buds enclosed by protective scales, to others with 'unspecialised' buds invested by foliage leaves that have not yet expanded. The latter include the largest buds, such as those of *Cordyline* and *Dracophyllum* section *Dracophyllum*, as well as minute buds of filiramulate species. Specialised buds typically produce a determinate number of leaves and internodes in a single, rapid flush early in the growing season; one of the more striking phenological events in New Zealand is the wave of light green that moves up a beech-clad mountainside in early summer, contrasting with the dark green of mature evergreen foliage above and below.

In the lowlands, fast-growing, broad-leaved trees such as *Pittosporum* spp. probably flush more than once during the growing season, but this needs to be verified. In *Podocarpus totara* and *P. hallii* flushes are remarkably unco-ordinated, varying from shoot to shoot within a tree, with some shoots flushing twice and others less than once a year.

Species with unspecialised buds have the anatomical potential to grow continuously, and although some may do so in the lowlands, most are limited by constraints of physiology and climate. Examples occur at high altitudes, e.g. species of *Halocarpus, Dracophyllum* and *Hebe*, as well as low altitudes (*Dacrydium cupressinum, Melicytus* spp. and many others). Even the deciduous *Fuchsia excorticata* has unspecialised resting buds. In contrast to the determinate growth in beech trees, rapidly growing beech saplings produce shoots of indeterminate length ending in unprotected buds that often fail to survive the winter.

Sclerophylly and xeromorphy

Most New Zealand trees and shrubs and many important herbaceous plants, such as *Phormium, Chionochloa* and most *Celmisia* species, have leathery or firm-textured leaves, since the evergreen habit requires stronger structures than the deciduous habit. Further, leaves of most native plants are microphylls or smaller, and some remain green and attached for many years (over 50 years in *Halocarpus biformis*). Although reduced leaves can be an adaptation to dry conditions (e.g. in *Stellaria gracilenta* and low-altitude species of *Raoulia* and *Colobanthus*), New Zealand examples are more often related to inherently slow growth (as in podocarps), infertile, phosphorus-deficient soils (epacrids and manuka), or cold climates (e.g. 'whipcord' hebes).

In a number of woody plants of dry habitats, photosynthesis is carried out mainly by green twigs. Those of *Muehlenbeckia ephedroides* remain terete, but in *Carmichaelia* and its satellite genera they range from slightly flattened to strap-like and > 1 cm wide in *C. williamsii*. In the lianes *Clematis afoliata* and *Rubus squarrosus*, petioles are fully developed and function as climbing organs through twining and prickles respectively. *Discaria* has green, spine-tipped dwarf shoots in the axils of thin, 1 cm long, obovate leaves. Seedlings and shaded parts of all these plants are leafier; *Rubus squarrosus* is often fully leafy. Leaves are also reduced in the non-xeric genus *Phyllocladus*, the semi-parasites *Cassytha* and *Exocarpus*, and most rushes and rush-like plants, where the photosynthetic surfaces are leaf-like cladodes, slender twigs, and culms respectively.

Plants with linear leaves usually have stomata arranged in longitudinal bands separated by strips of parenchyma. In xeromorphic species, these become deep grooves and ridges respectively. The grooves may be further protected by hairs as in many grasses, or by wax as in *Aciphylla*. In xeromorphic grasses, including most tussock grasses, the grooves close over during dry conditions, through inrolling or curving of the leaves. The leafless twigs of *Muehlenbeckia ephedroides*, *Chordospartium stevensonii*, *C. muritai* and *Corallospartium crassicaule* also have deep stomatal grooves which, in the last two species, are conspicuously tomentose.

Pale hairs that obscure or completely cover leaves and twigs are confined to plants of open, stony habitats, and seem adapted to intense radiation rather than drought *per se*. Among the best New Zealand examples, *Haastia recurva* and the cushion plants *H. pulvinaris*, *Raoulia eximia* and *Chionohebe ciliolata* grow at very high altitudes. *Craspedia incana* grows on debris at mid-altitudes, and *Pseudognaphalium luteoalbum* and the adventive *Verbascum thapsus* are commonest at low altitudes. Twigs of the cupressoid species of *Helichrysum* are enveloped in tomentum except for the abaxial leaf surfaces, which lack stomata but allow light to reach the mesophyll.

In very many New Zealand plants leaves are glossy on their upper surfaces (at least when mature), and tomentose only on lower surfaces. In the coastal trees *Metrosideros excelsa* and *Pittosporum crassifolium* this morphology contrasts with the glabrous leaves of more inland relatives. Leaves are thinly tomentose in most woody composites of lowland forests, such as *Olearia rani*, *Brachyglottis repanda* and *B. hectorii*, although *B. kirkii* is glabrous. In *Pachystegia*, a shrub that roots in rock crevices in eastern Marlborough, leaves are very glossy above and woolly below.

Adaptation to drought does not readily explain similar leaves in upland or alpine species of *Olearia*, *Celmisia*, *Brachyglottis* and *Astelia*, which are as prevalent on wet, cloudy mountains as on the driest and sunniest. Measurements support an alternative hypothesis, that the smooth, green upper surface absorbs radiant energy, while the tomentum below inhibits heat loss through convection and evapotranspiration. This raises leaf temperatures considerably, especially during calm, sunny intervals (Fig. 3.4).

The differences between *Celmisia petriei* and *C. lyallii*, vicarious species with linear

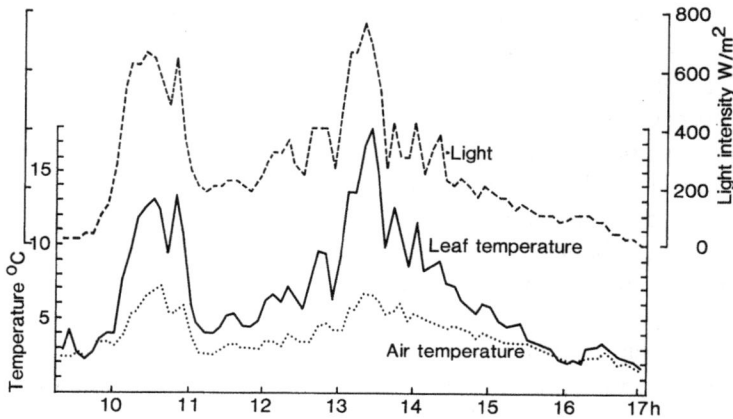

Fig. 3.4. Microclimate and leaf temperature in *Olearia colensoi*, measured on 12 April 1983 at 1120 m a.s.l. in the Paparoa Range, Western Nelson, on a partly cloudy day with light wind.

leaves 20–30 cm long, are clearly related to climate. The first grows west of the South Island Divide, and has coriaceous leaves up to 3 cm wide and 0.4 mm thick. In *C. lyallii*, which grows east of the Divide, leaves are rigid, almost spiny-tipped, <1 cm wide, and 1 mm thick. In *C. petriei* the stomata are on the flat, lower leaf surface beneath a thick layer of tomentum, whereas in *C. lyallii* both stomata and tomentum are confined to deep grooves that are almost closed tubes.

The 10 mm long leaves of the eastern South Island *Olearia cymbifolia* are strongly revolute, in contrast to the rounder, flatter leaves of its close relative, the widespread *O. nummulariifolia*. Yet *O. crosby-smithiana* of Fiordland, which has narrow-linear leaves revolute almost to the midrib, grows in as wet a climate as the related *O. lacunosa*, which has broader, flatter leaves. In *Chionochloa*, species of wet ground, such as *C. rubra*, have the narrowest, most involute leaves.

Although spines or prickles are well-developed in many naturalised plants, they occur in only 7–8 native genera. Among these, *Rubus, Discaria, Cyathodes, Eryngium* and the probably native *Hibiscus diversifolius* evolved prickles or spines outside New Zealand. Most *Aciphylla* species have extremely sharp, rigid spines terminating narrow leaflets, but a few, including the two Australian representatives (p. 13), have quite soft leaves. Abortion of minute apical buds, followed by secondary thickening, produces blunt spines in *Carmichaelia petriei* and *Melicytus alpinus*. Spiny and prickly plants are most evident in open, droughty, grazed habitats, although this does not apply to the forest liane *Rubus cissoides*, nor to large aciphyllas, such as *A. horrida*, that are confined to high altitudes on the wettest mountains.

The small, sparse, short-lived leaves of most filiramulate shrubs seem designed to be readily shed during dry spells, at little cost to the plant. Two small plants of dry, stony sites, *Crassula sieberiana* and the adventive *Sedum acre*, have succulent leaves, but otherwise succulence is characteristic of saline habitats (Chapter 10).

Table 3.2. *Root : shoot ratios of seedlings from good and poor sites on Te Hunga,*
Kaimai Range

	Quintinia acutifolia	*Ixerba brexioides*	*Nothofagus menziesii*	*Weinmannia racemosa*
Poor site 850 m	0.36	0.04	0.09	0.05
Good site 750 m	0.47	0.24	0.36	0.48

Note:
Values are the mean of 10 seedlings 15–30 cm in height. Shoot = leaf + stem; root = tap root + fine roots.
Source: Jane & Green 1986.

Roots

Ready production of adventitious roots by procumbent and rhizomatous stems is widespread in New Zealand vegetation, and descending aerial roots of woody epiphytes and some trees are conspicuous in moist, lowland forest (p. 35). Ascending aerial roots, or pneumatophores, are developed by mangrove (*Avicennia resinifera*) and *Syzygium maire*, a tree of swampy ground. However, there is little information as to how roots are distributed through the soil profile, perhaps because patterns are so influenced by soil conditions. In forest trees, Hinds & Reid (1957) distinguish vertically descending tap-roots, laterals arising from the tap-root, and peg roots descending vertically from the laterals. They describe three major patterns.

In tap-rooted species, the tap-root remains prominent. Laterals and pegs may also be well developed, the former usually spreading widely and shallowly (kahikatea, kauri, *Litsea*). In plate-rooted species, most of the root system consists of shallow, spreading laterals; the tap-root is superseded and peg roots are usually weakly developed (*Nothofagus*, kaikawaka, rimu, miro, matai, tawa). In heart-rooted species, a number of major roots descend steeply from the base of the bole (*Elaeocarpus dentatus*, silver pine, *Beilschmiedia tarairi*, *Larix*, *Pinus radiata*).

Major lateral roots and fine roots are usually concentrated in the upper 10–20 cm of soil, especially the humus layer. On the summit of the Kaimai Range, erosion of humus has exposed a net of *Ixerba* and *Nothofagus menziesii* roots suspended through sporadic contacts with the poorly-drained, stony clay loam beneath, whereas kamahi (*Weinmannia racemosa*), *Myrsine salicina* and *Quintinia* are rooted more deeply, and kaikawaka roots penetrate to bed-rock at 80 cm, even in wet soil. Root:shoot ratios of seedlings are depressed on poor sites, especially in kamahi and *Ixerba* (Table 3.2).

Fig. 3.5 contrasts root development in tawa and rimu growing on deep pumice sand. Tawa initially has a strong tap-root, which becomes superseded by laterals that are mostly contained in a melanised layer underlying the thin humus. They extend only 3–5 m radially, i.e. about half as far as the crown. There is much grafting between the roots of the same and adjacent trees.

In rimu, as in all native conifers except *Libocedrus*, rootlets bear lateral nodules (p. 494). Seedlings are initially shallow-rooted, especially when growing in shade, and

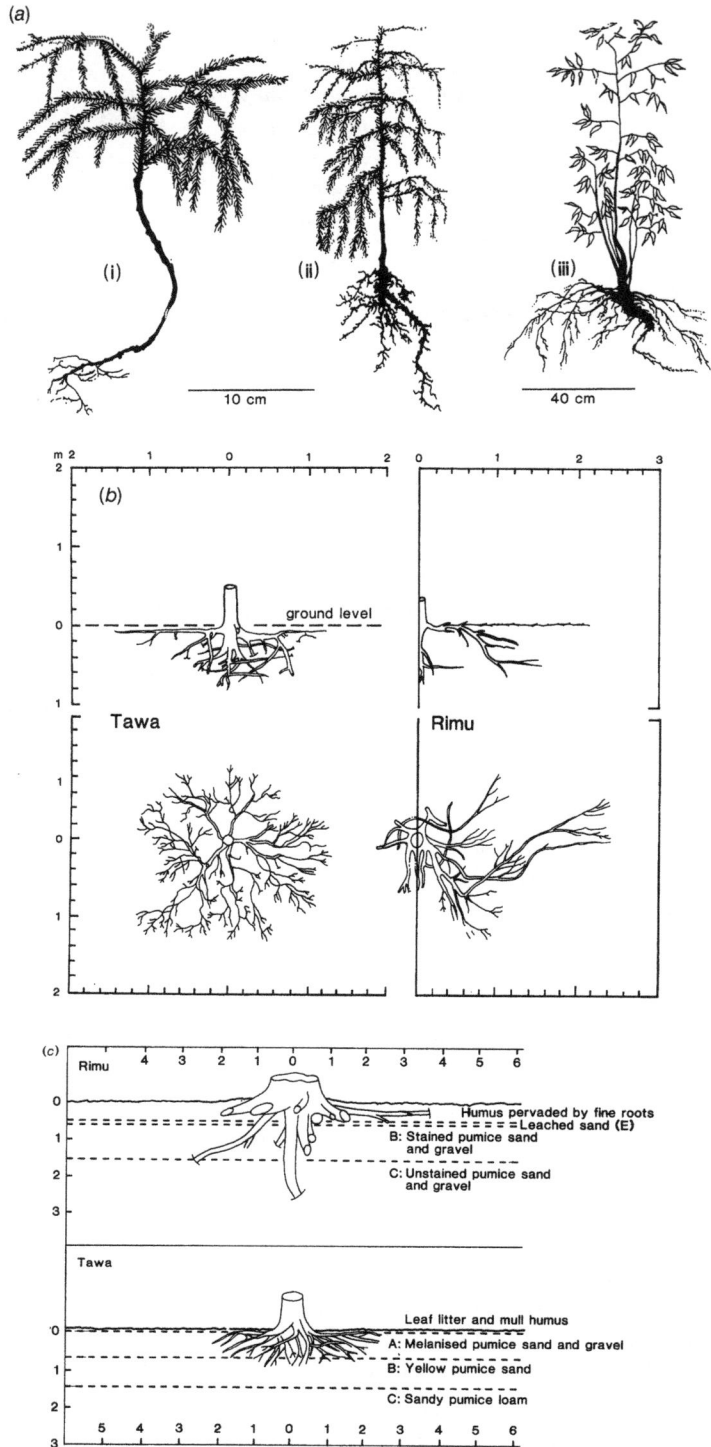

Fig. 3.5. Root development in rimu (*Dacrydium cupressinum*) and tawa (*Beilschmiedia tawa*) in pumice soil on the Volcanic Plateau (Cameron 1963). (*a*) Seedlings of (i) rimu in dense forest (ii) rimu in a successional *Kunzea–Weinmannia* stand (iii) tawa under dense tawa canopy (scales approximate). (*b*) Profile and plan of 15 cm d.b.h. saplings. (*c*) Mature root systems.

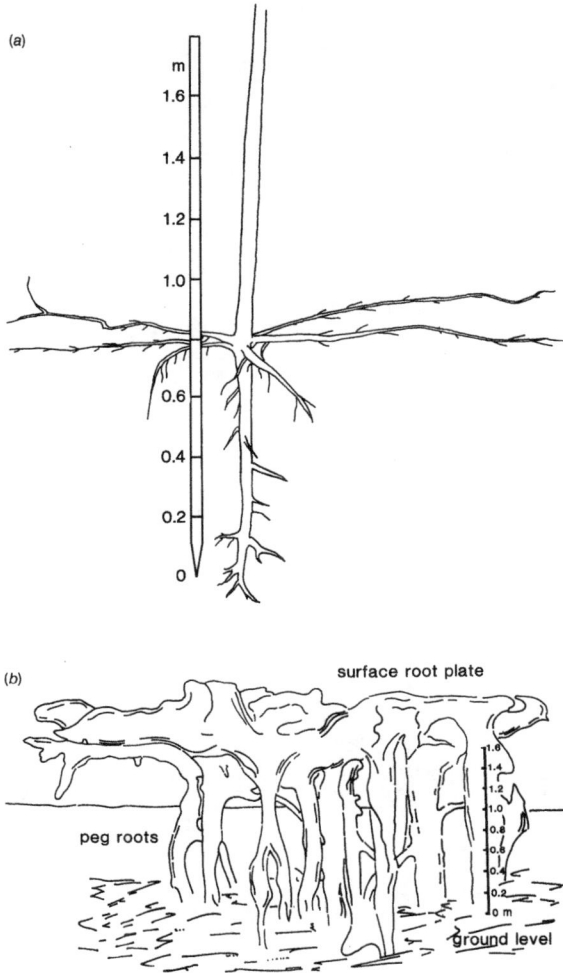

Fig. 3.6. Kauri (*Agathis australis*) roots from Northland (R.C. Lloyd, unpublished). (*a*) Tap root of sapling. (*b*) Root system of a mature tree exposed by deflation of a swampy surface.

saplings develop spreading lateral roots. In the adult tree, these form a massive plate contained largely in the deep litter and humus around the base of the tree. Shallow roots extend >9 m from the plate, far beyond the area covered by the crown. Large 'peg roots' descend from the plate, to depths exceeding 3 m.

Allan (1926) found lateral roots of a matai tree extending 19.5 m before disappearing into the ground. In former kauri land, deflation of swampy ground exposes massive root plates that are almost solid through fusion. From these, pegs descend deeply through saturated and indurated horizons (Fig. 3.6). *Nothofagus* species have most of their roots concentrated into a plate <60 cm thick, and grafting is common. Peg roots are generally weak, and absent in shallow or waterlogged soils, as

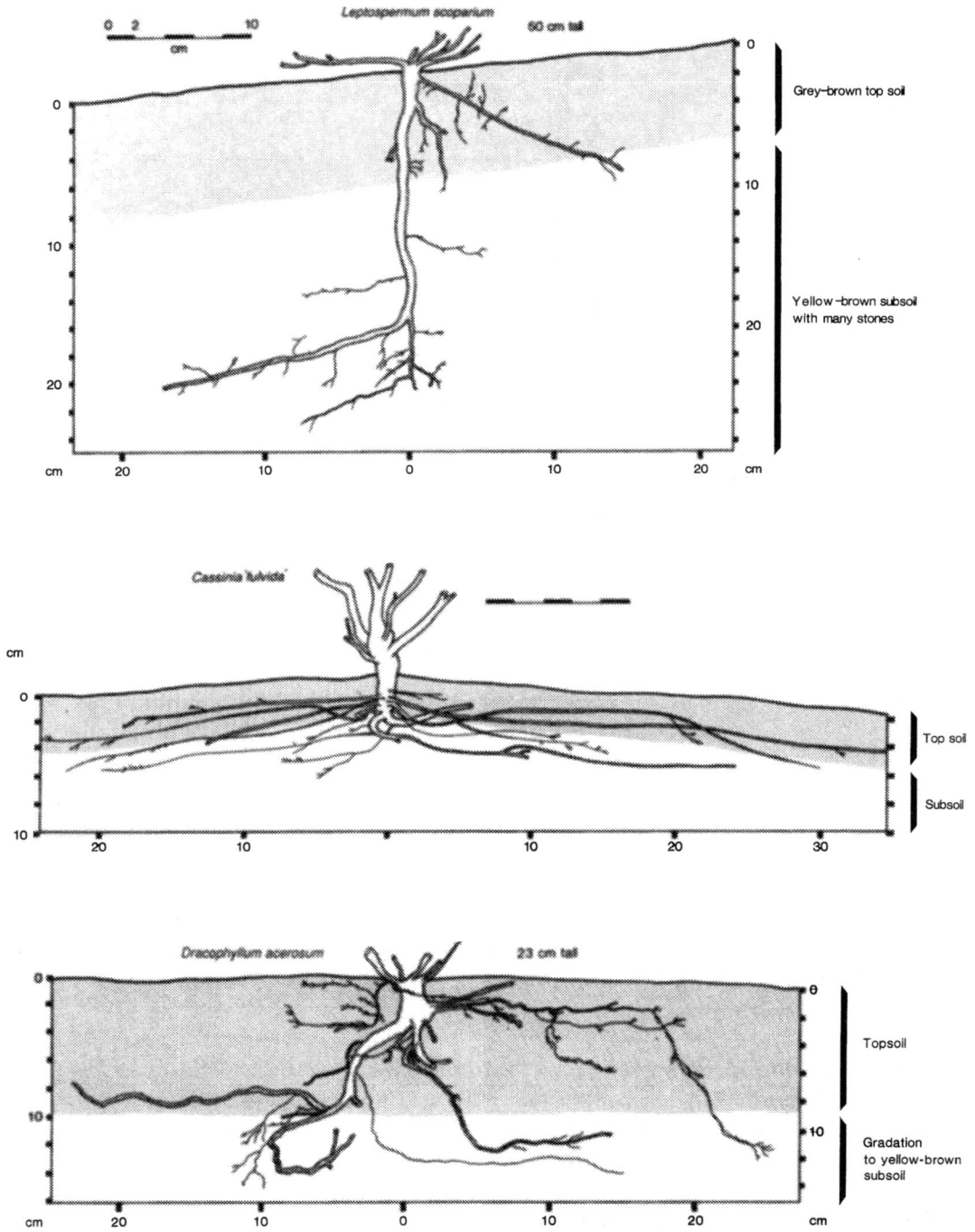

Fig. 3.7. Root systems of shrubs growing on yellow-brown earth, in open montane shrubland; Puketeraki ER (J.A. Wardle, unpublished).

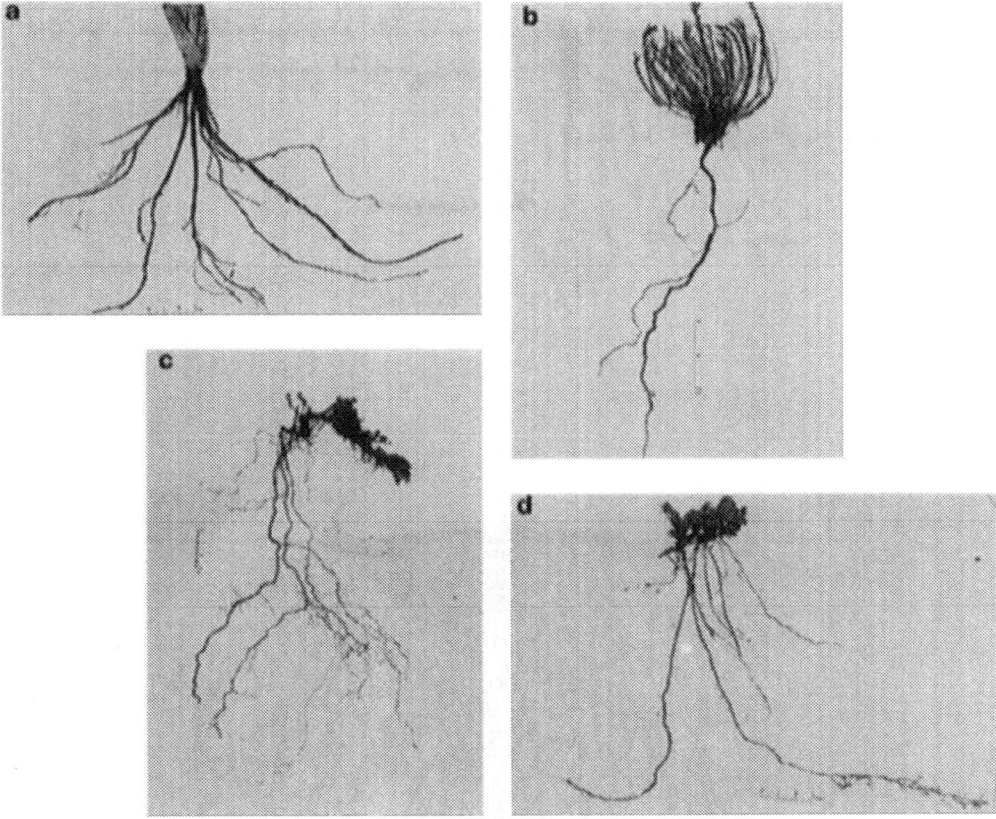

Fig. 3.8. Root systems of dicot herbs in subalpine grassland on yellow-brown earth (*a,c*) and in recent alluvial gravel (*b,d*); Puketeraki ER. (*a*) Seedling *Aciphylla aurea*; (*b*) *Epilobium melanocaulon*; (*c*) *Raoulia subsericea*; (*d*) *Raoulia tenuicaulis*.

roots cannot survive in horizons subjected to more than 50 days' continuous flooding (see Fig. 16.2). Nevertheless, hard beech (*N. truncata*) roots were found at 1.6 m depth in clayey subsoil.

The prevalence of shallow rooting in forest soils probably reflects dependence on nutrient cycling through litter fall. However, it may be that deeply descending roots function disproportionately to their mass in bringing up water and nutrients from lower horizons.

In shrubland and grassland on leached soils, roots of most species are concentrated in the humus-enriched upper horizon (Figs 3.7 and 3.8; Table 3.3). Species of relatively fertile, recent soils, such as the tree *Sophora microphylla* and the shrubs *Discaria toumatou* and *Coprosma propinqua*, have deeply descending tap roots. Roots

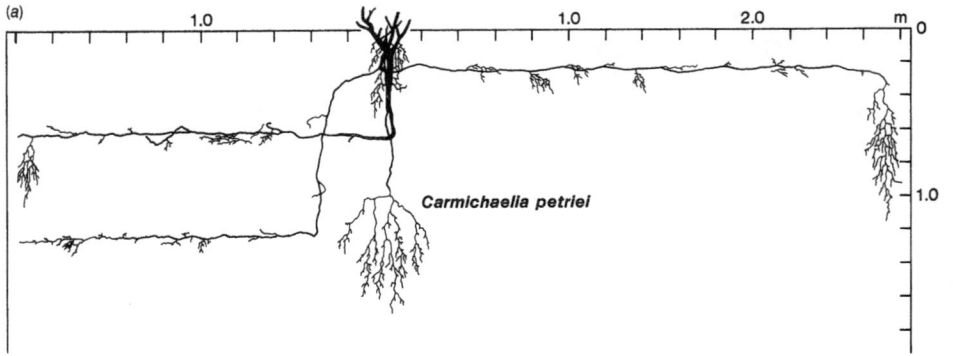

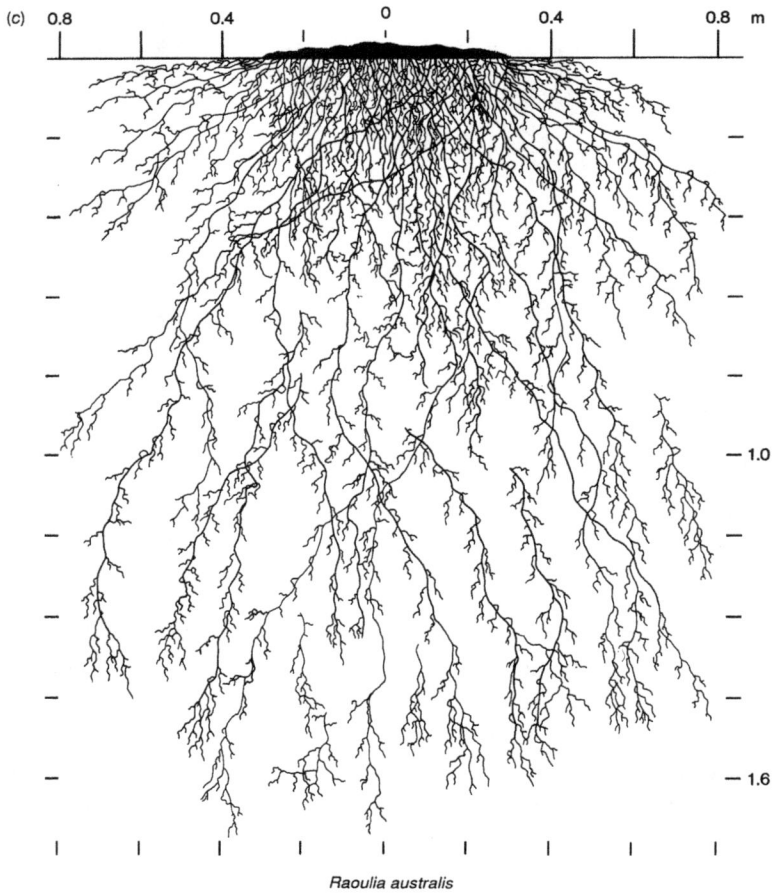

Fig. 3.9. Root systems of two shrubs (*a,b*) and two herbs (*c,d*), growing in sand or gravel in semi-arid Central Otago (McIndoe 1932).

Melicytus alpinus

Lepidium sisymbrioides

Table 3.3. *Decrease in roots (percentage of total) with increasing soil depth in Chionochloa grassland, Paddle Hill Creek, Heron ER*

Depth (cm)	C. rigida		C. macra	
	a	b	a	b
0–5	50	43	33	31
5–10	22	25	22	18
10–15	10	12	14	16
15–20	5	6	10	12
20–25	4	4	7	9
25–30	3	3	5	6
30–35	2	3	5	5
35–40	1	2	4	3
40–50	3	2		
Total*	3382	2692	2814	2010

Notes:

a, Beneath tussocks; b, between tussocks.

* g root/m² soil.

Source: Williams 1977.

descend deepest in coarse gravels (Figs 3.8 and 11.5) and under semi-arid conditions; the deeply rooting tussock of *Poa cita* in Fig. 3.10 contrasts with the same species on stony clay loam, where few roots descend below 10 cm.

The morphology of roots, especially the finest rootlets, has received little attention except in respect of mycotrophy (p. 492), but could reveal much of ecological significance. For example, in mid-altitude tussock grasslands, roots grade from sparse and relatively coarse to abundant and very finely divided in the sequence *Chionochloa* spp.→*Festuca novae-zelandiae*→*Poa cita*→*Agrostis capillaris* (Fig. 3.10). This parallels an ecological sequence from large, slow-growing tussocks characteristic of undisturbed vegetation to small, fast-growing grasses that increase in response to disturbance. Especially fine roots bearing copious root hairs form dense brushes in *Knightia excelsa* as in many other members of the Proteaceae (Webb *et al.* 1990), and form a close felt in *Empodisma minus* in the Restionaceae (p. 321). Both the families are characteristic of poor soils in the southern hemisphere.

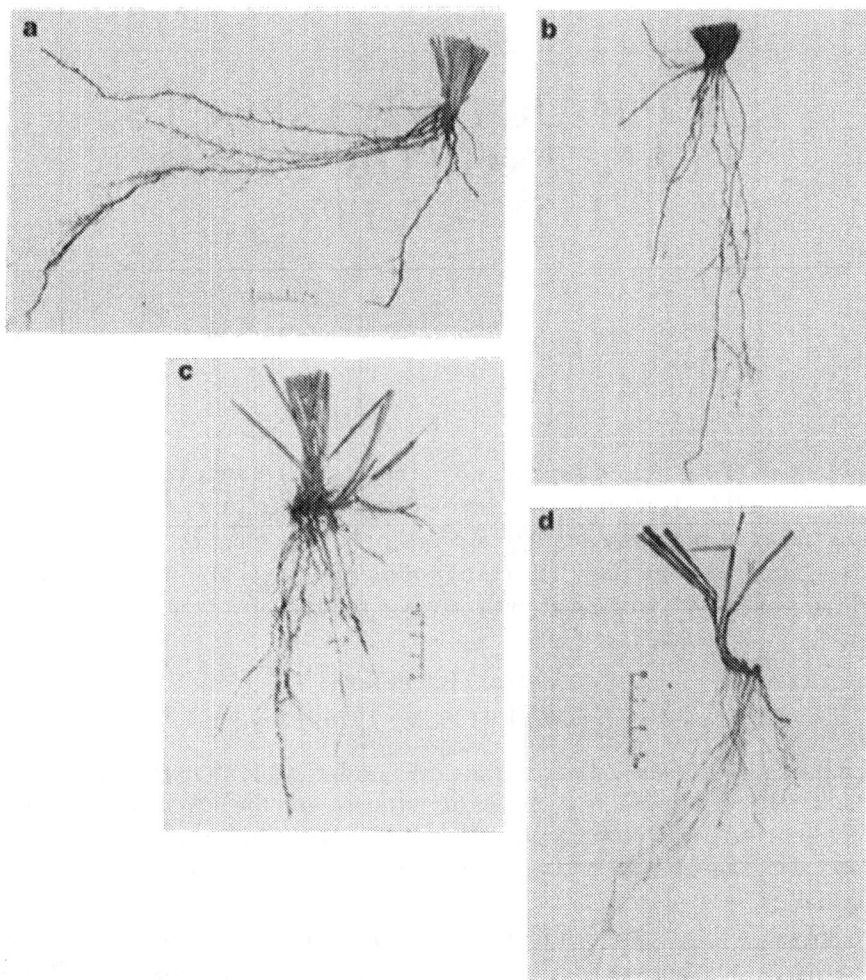

Fig. 3.10. Roots of grasses in subalpine grassland on yellow-brown earth; Puketeraki ER. (*a*) *Chionochloa rigida*; (*b*) *Festuca novae-zelandiae*; (*c*) *Poa cita*; (*d*) *Agrostis capillaris*.

4

REPRODUCTIVE ASPECTS, SEEDLING FORM AND LONGEVITY

The composition of vegetation depends on the reproductive success of the species, and on how long individuals monopolise their sites. These, in turn, are functions of the successive stages of plant life-history. Accordingly, this chapter follows a sequence from floral characters, especially aspects of pollination, through production, dispersal and germination of seed, and the growth forms of seedlings and saplings, to longevity.

Floral characters
(based on reviews by Godley 1979; Lloyd 1985)

The main pollen vectors in the native flora are wind, moths, short-tongued bees, flies and birds. Beetles, the four native species of butterfly, and the now rare native bats also transfer pollen. Anemophilous groups include conifers, beeches, *Ascarina*, *Macropiper*, *Coprosma*, *Coriaria*, all grasses, sedges, rushes and restiads, and other monocots of similar habit.

Many flowers that are presumably entomophilous are small, greenish and inconspicuous to the human eye; this includes the flowers of most umbellifers and araliads, some species of *Melicytus* including *M. ramiflorus*, *Pseudowintera*, *Hedycarya*, *Laurelia*, *Beilschmiedia*, *Muehlenbeckia*, *Ileostylus*, *Dodonaea*, *Melicope*, *Geniostoma*, *Nestegis* spp., *Astelia*, *Collospermum*, and many orchids such as the very common *Microtis unifolia* and *Prasophyllum colensoi*. Some orchids have larger green flowers, those of *Pterostylis* being up to 5 cm long.

Cockayne (1928) estimated that among 'showy' flowers, 61% overall and 78% of mountain species are white. The more conspicuous white, near-white or mostly white flowers include those of *Weinmannia*, *Ixerba*, *Carpodetus*, *Pennantia*, *Pimelea*, *Leptospermum*, *Kunzea*, four lianoid *Metrosideros* spp., *Libertia*, *Cordyline*, *Clematis paniculata*, *Ranunculus lyallii*, *R. buchananii*, most Scrophulariaceae, especially in the genera *Hebe*, *Parahebe*, *Chionohebe*, *Ourisia* and *Euphrasia*, most gentians, and ray florets of most composites, especially in the genera *Olearia*, *Celmisia*, *Pachystegia*, *Raoulia*, *Cassinia* and *Traversia*, as well as *Brachyglottis kirkii*, *B. hectorii* and *Dolichoglottis scorzoneroides*.

52

The showiest yellow flowers are those of *Sophora, Corokia, Bulbinella, Taraxacum* and its relatives, most *Ranunculus* spp., and the rays of *Dolichoglottis lyallii*, of most *Brachyglottis* spp., and of the few species of *Senecio*, such as *S. lautus*, that have them. In most *Clematis* spp. flowers are greenish shades of yellow.

In *Phormium tenax, Clianthus puniceus, Peraxilla* spp., *Alseuosmia* and the monotypic *Rhabdothamnus solandri*, flowers are 1–8 cm long and usually shades of red; *Xeronema callistemon, Knightia excelsa* and most species of *Metrosideros* have generally smaller red flowers that are aggregated into conspicuous clusters. Together with *Sophora* and *Fuchsia* these are also the plants most visited by birds, which include three species of Meliphagidae with specialised brush-like tongues as well as several more casual visitors. In *Fuchsia excorticata*, nectar-bearing flowers that attract bell-birds (*Anthornis melanura*) are green; they turn red when they are no longer rewarding for birds (Delph & Lively 1989). Flowers of most *Pittosporum* spp. are deep-red and night-scented, but some are yellow (e.g. *P. eugenoides, P. kirkii*) or white (*P. dallii*). The flowers of *Metrosideros* and *Knightia*, as well as the massed but inconspicuous flowers of *Astelia, Collospermum* and *Freycinetia*, are also visited by the endemic bat *Mysticina*.

Flowers of most native brooms (*Carmichaelia* and its satellite genera) are shades of blue or pink, and have a close relationship with native bees. *Pratia physaloides* flowers are also blue, and those of *Wahlenbergia* species grade from blue to white. In macrocephalous olearias (p. 24) disc florets are usually purplish, whereas rays are white, purplish or absent.

Many flowers that are predominantly white have localised areas of colour, e.g. bluish lines in parahebes, pink lines in some gentians, and yellow throats in most ourisias. Within genera or species that are predominantly white in New Zealand, some taxa may show more colour than the majority, e.g. reddish flowers in *Epilobium billardiereanum*, pinkish in *E. wilsonii*, yellow in *Euphrasia cockayneana* and *Myosotis australis*, reddish yellow in *M. macrantha*, red in *Parsonsia capsularis* var. *rosea* and pink in some geraniums. In far-northern plants of *Leptospermum scoparium*, petals are usually pink towards the base, and red forms have been propagated for horticulture.

In the large genus *Hebe*, most species have small white flowers aggregated into spikes and racemes, but the blue or purple tints that characterise much of the *Veronica* complex in other floras are retained to a degree in most lowland and coastal hebes, and are intense in some, e.g. *Hebe hulkeana* and *H. speciosa*. The dense, deep purple racemes of the latter, like the relatively large, yellow flowers of *Carmichaelia williamsii*, are possibly bird-pollinated.

The high incidence of coloured flowers, mostly blue or pink, on the far-southern islands contrasts with comparable mainland environments. In the wide-ranging genera *Epilobium* and *Myosotis*, as well as in the endemic Chatham Islands genus *Myosotidium*, the colours approximate to those that prevail in the same or related genera in other lands; in *Epilobium*, as in *Hebe benthamii, Anisotome* spp. and gentians, they also represent deepening of hues present in some New Zealand mainland species. The far-southern endemic genus *Pleurophyllum*, like the related macrocephalous olearias, has purplish discs with rays, where present, being pink. Likewise there is a hint of the blue of *Damnamenia* flowers in a few celmisias, especially Australian species and the Southland endemic *C. thomsonii* (Mark & McSweeney 1987). Populations of some of these far-southern species also contain variable proportions of plants with white flowers (p. 450), so perhaps selection pressures for flower colour or the lack of it are weak. The islands have extremely depauperate faunas of pollinating insects.

White or bright-coloured flowers are generally well displayed, often in dense inflorescences. In araliads and most umbellifers, the inconspicuousness of individual flowers is offset through massing into compound umbels. In the tall species of *Aciphylla*, the umbels are quite small but borne laterally on erect scapes which, in *A. scott-thomsonii*, grow up to 4 m tall, forming the most spectacular inflorescence in the flora (see Fig. 9.4). Often, however, inconspicuous flowers are borne singly or in small clusters among the foliage. The small flowers of *Melicytus* and *Myrsine* mostly occur on the upper branchlets below the leaves, and the larger, bird-pollinated flowers of *Fuchsia excorticata* and *Dysoxylum spectabile* can be on major limbs.

At 12–13%, the proportion of dioecy is high and although many of the dioecious species belong to genera that exhibit dioecy in other parts of the world, other New Zealand examples are unusual in their family (*Aciphylla, Anisotome, Melicytus*) or genus (species of *Aristotelia, Rubus, Passiflora, Clematis* and probably others). Among species with hermaphrodite flowers or flowers of different sexes on the same plant, self-fertility has been demonstrated for many or most species of *Ranunculus, Epilobium, Cardamine, Parahebe* and leafless species of *Juncus*, whereas self-incompatibility may be more common in larger or woodier plants, including *Pseudowintera, Corokia* (C.J. Webb, personal communication), *Discaria, Cordyline, Phormium* and, probably, *Nothofagus*. In *Phormium tenax*, larger and far more numerous seeds result from outcrossing (Craig & Stewart 1988).

Native pollinating insects are generally unspecialised and pollination mostly promiscuous; hence the paucity of striking colours, large sizes or complex designs. Most flowers are of the shallow 'dish' type, with exposed pollen and nectar. Tubular flowers are also common, but generally short except in *Euphrasia disperma* where the narrow corolla tube is up to 6 cm long. The absence of long-tongued bees, which are credited with greater ability to discriminate among flowers of different colour and shape than other insects, may largely account for the lack of specialisation.

Conversely, the characteristics of the flora largely result from selective immigration and establishment of species able to succeed in the absence of specialised pollinators. For example, the small, shallow, white flowers of *Leptospermum scoparium* and *Discaria toumatou* occur in the same or related species in Australia, and probably evolved there. For the majority of New Zealand species with dimorphic flowers, the trait is equally evident in relatives elsewhere, although it appears to have developed within New Zealand in *Leptinella, Hebe, Lepidium, Gentiana* and *Bulbinella*. The number of taxa with unspecialised floral characters has been further increased through evolutionary radiation into the new habitats that arose during orogeny and cooling. The naturalisation of long-tongued bees is a factor in the success of some adventive plants, an example being *Trifolium pratense*, which set seed poorly in New Zealand before bumble bees (*Bombus* spp.) were introduced in 1885.

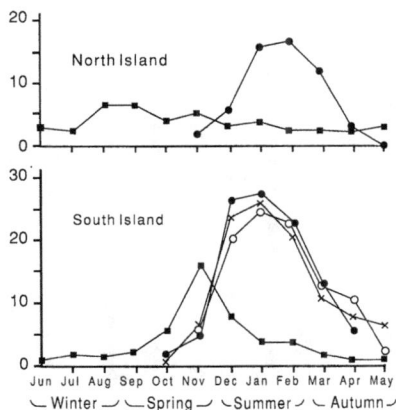

Fig. 4.1. No. of native species flowering in each month at four localities (Godley 1978). North Island: ■, Little Barrier Island, Auckland ER (18 species); ●, Mt Ruapehu 1280–1930 m (30 species). South Island: ■, Port Hills, Banks Peninsula E.R. (26 species); ×, ○, ●, successive years at Cupola Basin, Spenser ER, >1200 m (31 species).

Phenology

(based on reviews by Godley 1979; Wardle 1978c)

In the northern lowlands there are native species in flower throughout the year, although more flower in early spring than at other times. With increasing latitude, flowering time becomes more concentrated and the peak moves towards late spring. Above the upper forest limit, flowering is confined to the warmer half of the year and peaks in full summer (Fig. 4.1).

Seasonal peaks would probably appear sharper were intensity of flowering to be measured, but the season is nevertheless attenuated, especially at low altitudes. This is partly because some species flower outside the main season, e.g. *Metrosideros fulgens* from late summer to early winter, and *Pseudopanax arboreus* through winter to early spring. Some species have protracted flowering, e.g. *Hebe salicifolia* year-round but with a summer peak, *Beilschmiedia tawa* from January to June, and *Avicennia resinifera* from mid-summer to spring with an autumn peak. *Melicytus ramiflorus* flowers in several discrete episodes between October and May (Powlesland *et al.* 1985). Each *Metrosideros umbellata* tree flowers in a concentrated burst, but flowering time among trees varies from July to March. *Sophora microphylla* has geographic races with different flowering times that span from May to December. At high altitudes members of the Ranunculaceae are among the first to flower, yet in *Ranunculus sericophyllus, R. pachyrrhizus* and *Psychrophila obtusa* flowering may follow the edges of retreating snow banks into late summer.

Fruiting at low altitudes also occurs year-round, although it is concentrated between late summer and early winter. Small seeds produced in small capsules, as in *Weinmannia* and *Hebe*, ripen quickly after fertilisation, whereas *Dacrydium cupressi-*

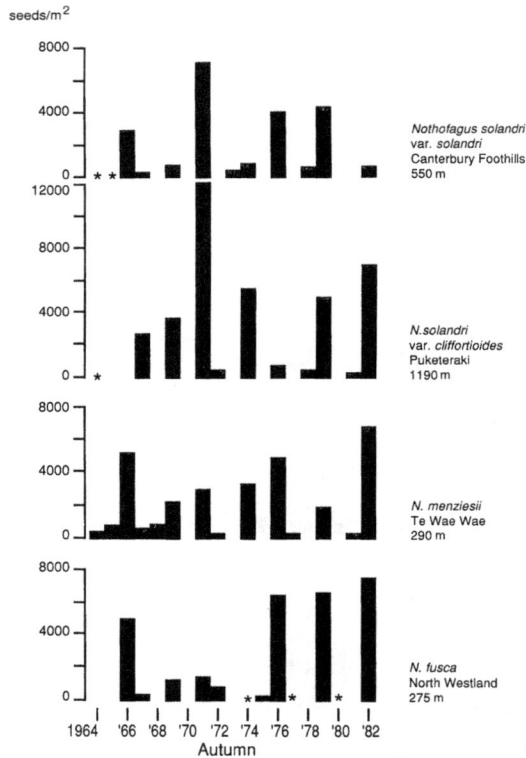

Fig. 4.2. Annual production of *Nothofagus* seed between 1964 and 1982 (Wardle 1984);
*, no record.

num, Coprosma lucida and *Metrosideros umbellata* take over a year to ripen ovules, drupes and woody capsules respectively.

Variable flowering and fruiting

Quasi-triennial variations in flowering intensity are evident in *Nothofagus, Cordyline, Phormium*, and robust, long-lived mountain herbs. Compilation of partial records (especially Connor 1966; Campbell 1981; Brockie 1986; Haase 1986*a*) shows that in recent decades flowering was concentrated in the following summers: 1953–4, 56–7, 59–60, 62–3, 65–6, 68–9, 70–1, 73–4, 75–6, 78–9, 81–2, 85–6, and 88–9. In several summers, notably 1970–1, 73–4, 78–9 and 81–2, flowering was especially prolific, involving most of New Zealand and most 'periodic' species; conversely, flowering was especially sparse in 1969–70, 76–7 and 79–80. Other flowering and non-flowering years involve fewer species or are less universal. The best records for the pattern are from species of *Nothofagus* (Fig. 4.2) and *Chionochloa. Phormium tenax* invariably coincides with the same years, but sometimes it also flowers well in the preceding year, as in 1977–8 and 84–5. Paired flowering years seem to be the rule in *P. cookianum*, at

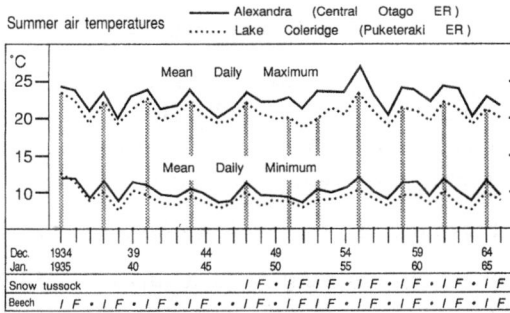

Fig. 4.3. Flowering (F) and inductive (I) seasons in relation to summer temperatures in *Chionochloa* and fusca-group beeches (Connor 1966).

least near Wellington. The large *Aciphylla* spp., which are conspicuously periodic, sometimes coincide with the general flowering year (as in 1973–4) and sometimes with the preceding year (as in 1977–8 and 84–5).

In *Nothofagus, Chionochloa* and most alpine forbs (Mark 1970), floral primordia are induced during the summer or autumn of the year preceding flowering, and for *Nothofagus* and *Chionochloa* there is correlation with summer warmth (Fig. 4.3). For lowland *Phormium*, Brockie (1986) has shown strong correlation with monthly maximum temperatures for the preceding April, May and June. Conversely, poor flowering may reflect insufficient temperature stimulus for induction, insufficient build-up of reserves during a cool growing season, or depletion of reserves during a preceding flowering season.

Some of the larger *Celmisia* species (but not *C. spectabilis*) show wide year-to-year variation in flowering. As flowering years tend to coincide with those of other periodic plants, previous summer warmth may have an influence, but Mark (1970) showed that abortion of floral initials during winter is more important. This may explain why, in a population of the alpine *C. viscosa*, 13 years elapsed between the flowering years of 1965–6 and 78–9, with five consecutive years in which no flowers at all were noted.

Beilschmiedia tawa and some lowland podocarps also produce variable quantities of seed. During the span of 35 years under discussion, good seed crops in *Dacrycarpus dacrydioides* coincided with at least four general flowering years. Seed years in rimu (*Dacrydium cupressinum*) tend to be paired, with quantities of seed being correlated positively with temperatures during the summer of ripening and negatively with seed production and summer temperatures two years previously, when ovules are fertilised (Norton *et al.* 1988). In *Metrosideros umbellata* good flowering years seem to coincide with weak flowering of other periodic plants, but since individual trees flower at different times over several months, the record may be unreliable. Further periodic species include *Dracophyllum traversii, Olearia colensoi, Brachyglottis buchananii* and, at least in upland habitats, species of *Pseudopanax*. Some large far-southern herbs, including species of *Pleurophyllum, Chionochloa* and *Anisotome*, show periodicity that approximately coincides with mainland New Zealand (C.D. Meurk, unpublished).

Fruit characters and dispersal modes

Representative terrestrial species are classified below according to primary dispersal adaptations. There may also be secondary modes; for instance, large disseminules can be transported by gravity, small or hairy disseminules are likely to be transported incidentally by animals, and most disseminules can be carried by water.

1 Succulent fruit (groups and species in approximate order of decreasing size).

   (i)   Drupe (i.e. containing a single seed surrounded by a bony integument) 30–20 mm long: *Corynocarpus > Planchonella, Beilschmiedia, Dysoxylum > Prumnopitys ferruginea, Vitex, Coprosma macrocarpa*.

   (ii) Berry (i.e. containing several or many seeds) 30–20 mm long: *Solanum aviculare, S. laciniatum*.

   (iii) Drupe 20–10 mm long: *Elaeocarpus > Hedycarya, Toronia > Nestegis cunninghamii > N. lanceolata, Syzygium, Rhopalostylis > Ripogonum* (the last two scarcely fleshy).

   (iv) Berry 20–10 mm long: *Rubus parvus, Fuchsia procumbens > Dianella > Astelia linearis > A. fragrans*.

   (v)   Berry or drupe 10–2 mm long.
         Trees, shrubs and lianes: other *Fuchsia* spp., *Corokia macrocarpa, Coprosma lucida, Alseuosmia > Nestegis montana*, other *Rubus* spp., *Coprosma robusta, C. perpusilla, Myrsine salicina, Prumnopitys taxifolia, Pennantia, Pseudopanax, Lophomyrtus, Myoporum*, other *Corokia* spp. *> Griselinia, Coprosma propinqua, Pseudowintera, Pentachondra, Neomyrtus, Melicytus, Pernettya, Carpodetus > Coprosma rotundifolia, Aristotelia, Streblus > Cordyline, Coprosma rhamnoides, Schefflera, Cyathodes > Macropiper, Myrsine australis, Ascarina, Pimelea prostrata*. Herbs: most *Astelia* spp. *> Tetragonia trigyna > Collospermum > Nertera, Einadia triandra > Gunnera*.

2 Small seed or nut seated on or only partly enclosed by a fleshy structure. *Muehlenbeckia, Gaultheria, Coriaria*, all podocarps except *Prumnopitys*.

3 Disseminules dispersing through adhesion.
   (i) Seeds or fruits viscid: most *Pittosporum* spp., *Pisonia*.
   (ii) Barbs, hooks, awns and spines: most *Acaena* spp. (barbed processes on cupule), *Nothofagus menziesii* (glandular hooks on cupule), *Uncinia* spp. (hooked spines), many grasses (scabrid awns which may also help to bury the seed).

4 Fruit or seed with structures for wind transport.
   (i)   Winged: *Agathis, Libocedrus, Knightia, Dodonaea, Hoheria* (obscure in *H. lyallii* and *H. glabrata*), *Nothofagus, Gingidia montana, Phormium*.
   (ii) Plumes, tufted hairs or pappus: *Weinmannia, Epilobium, Clematis, Laurelia, Parsonsia*, most Asteraceae.
   (iii) Mobile inflorescence: *Spinifex* (inflorescence a 30 cm diameter, spherical cluster of long-awned spikelets), *Lachnagrostis* (open panicle).

5 Small seeds mostly shed from capsules, or one-seeded fruit such as achenes; special structures for dispersal minute or absent.

Trees and shrubs: *Leptospermum, Kunzea, Metrosideros, Quintinia, Pittosporum eugenioides, Geniostoma, Plagianthus, Pomaderris, Discaria, Entelea, Carmichaelia* and related genera, *Dracophyllum, Sprengelia, Archeria, Hebe.*

Herbs: *Ranunculus, Gentiana, Lagenifera, Euphrasia, Libertia, Bulbinella,* many umbellifers and crucifers, and all Caryophyllaceae, orchids, rushes and restiads.

6 Spores.

Most cryptogams.

Ecological implications of dispersal mode

Some 14% of indigenous seed-plant species and 60% of the woody genera have succulent fruit, and they occur in all tiers of the forest from the canopy to ground level (*Nertera*), and from the coast almost to the limits of vegetation (*Coprosma perpusilla*). Large and medium-sized drupes are confined to forest trees, except for those of the tall liane *Ripogonum*, and the only extant bird to ingest the largest is the pigeon *Hemiphaga novae-seelandiae* (Clout & Hay 1989). The berries of the low, dense shrub *Melicytus alpinus* are almost inaccessible to birds, being on the undersides of the branches, but they are eaten by geckos. Lizards also possibly disperse the minute, cryptic drupes of *Gunnera monoica* (Whitaker 1987). Seeds that are not securely enclosed within a succulent structure are likely to fall off when the latter is eaten, at least in podocarps (Beveridge 1964).

Lee *et al.* (1988) point out that the basic fruit colour in *Coprosma* is red, which attracts birds through contrasting with a green background. It prevails in the large-leaved lowland species as well as in low-growing alpine species. In the filiramulate species, however, the reduced leaf area is associated with a wide variety of drupe colours, and it is suggested that they attracted an array of dispersers, many now rare or extinct, that included nectar-eating birds and lizards.

Adaptations to cling to fur or feather were important in primeval New Zealand, to judge by the number of species in *Acaena* (> 14 including four without barbs) and *Uncinia* (> 30). Mammals effectively transport their seeds at present, and presumably ground-dwelling birds did so in the past.

Other than *Phormium* and some umbellifers, only tree species have wings visible to the unaided eye, whereas species with hairs for dispersal range from lowland trees to alpine herbs. Minuteness is no less effective for wind-dispersal, as shown by the Australian origin of many of our orchids. Seeds can be thrown some distance when dry capsules are shaken by the wind, but explosive dispersal, so characteristic of gorse (*Ulex europaeus*), common broom (*Cytisus scoparius*) and the garden weed *Cardamine hirsuta*, is reported only for matagouri (*Discaria toumatou*) among native plants (Daly 1969). For some wide-ranging taxa with small seeds, chance transport by birds is likely; this probably applies to wetland plants such as rushes and almost certainly to *Hebe salicifolia* and *H. elliptica*, which New Zealand shares with the Chilean coast.

The preponderance of disseminules adapted for dispersal by birds or wind, or through their small size by either, probably reflects the transoceanic ancestry of much of the flora, and the available dispersal agents within New Zealand. For example, the absence of large nuts reflects an absence of mammals to cache them. Only *Leptospermum scoparium* has capsules that release seeds *en masse* after fire, which probably reflects the infrequency of fire in prehuman times.

Few non-aquatic species seem specifically adapted for water dispersal, although water is important as an incidental agent as shown, for instance, by spread of *Nothofagus* along river banks. Many introduced plants also spread downstream, whether by seed or, as in willows, by vegetative fragments. The utricle surrounding the nut in *Carex* may provide flotation. *Avicennia* seeds germinate as soon as they fall to float in the tide (p. 291).

The long, slightly winged pods of *Sophora microphylla* can be blown across bare ground, but spread upslope is probably through birds ingesting the hard-coated, 5–8 mm long, yellow seeds. These seeds also float and resist salt water, allowing the species to disperse far to the east (p. 13) (Godley 1968). *Tetragonia tetragonioides*, which has a bony endocarp, and other strand plants presumably also resist salt water and abrasion by sand.

The hard, shiny, black, brown or red nuts of *Gahnia*, which are suspended from anther filaments when ripe, seem likely to attract seed-eating birds. In the grass *Elymus tenuis* and several dryland sedges, slender, elongating culms carry seeds away from the parent, to as far as 3 m in *Carex testacea*. Elongating peduncles in the water-plant *Ruppia* may have a similar function.

Although many native plants, both woody and herbaceous, spread and multiply through horizontal shoots that take root or through roots that produce suckering shoots, with eventual separation from the parent, few show specialised forms of vegetative reproduction. *Asplenium bulbiferum* fronds produce plantlets that fall off and become independent. Deciduous buds, or gemmae, are a mode of propagation in some bryophytes, especially species of *Marchantia* and *Bryum*; soredia in lichens have the same function.

Seed mass; quantity of viable seed; germination

[Sources of information additional to those cited in the text or Table 4.3 are: *Celmisia*, Scott 1975; *Gentiana*, Simpson & Webb 1980; *Metrosideros umbellata*, Wardle 1971.]

The nutrient content and calorific value of storage tissue in seeds correlates with the size of newly germinated seedlings and their ability to survive early adversity. No data are available for the New Zealand flora, but the seed masses in Table 4.1 indicate how species are likely to rank, despite including much other tissue, especially in seeds enclosed within thick, bony structures (Table 4.2). Clearly, most seeds are very small, and even the largest do not attain the acorn dry mass of the common European oak *Quercus robur*, i.e. 3500 mg with its shell and 3000 mg without.

Sexually dimorphic species tend to have high levels of fruit set on female plants (e.g. many native apioid umbellifers), whereas in species with all individuals functionally both maternal and paternal, only a proportion of flowers (as low as 4–7% in

Table 4.1. *Mean dry mass (mg) of sound seeds in samples of native woody species*

Corynocarpus laevigatus	2700	Myoporum laetum	1.8
Beilschmiedia tawa	1140	Celmisia spectabilis	1.7
Rhopalostylis sapida (1)	548;453	Libocedrus plumosa	1.6
Prumnopitys ferruginea	480;196	Chionochloa macra	1.3
Elaeocarpus dentatus	358	Pentachondra pumila	1.2
Rhopalostylis sapida (2)	262;226	Chionochloa pallens	1.0
Hedycarya arborea	171	Brachyglottis buchananii†	0.8*
Ripogonum scandens	141;124	Melicytus ramiflorus	0.7
Prumnopitys taxifolia	111	Schefflera digitata	0.6
Corokia macrocarpa	61	Carpodetus serratus	0.4
Sophora microphylla	58	Coriaria arborea	0.4
Myrsine salicina	29	Geniostoma rupestre	0.3
Griselinia littoralis	26	Olearia ilicifolia†	0.3*
Knightia excelsa	19	Leptospermum scoparium	0.2
Phyllocladus trichomanoides	16	Weinmannia racemosa	0.1
Pennantia corymbosa	14	Fuchsia excorticata	0.08
Coprosma robusta	14	Dracophyllum traversii†	0.07*
Lepidothamnus intermedius	14	Quintinia acutifolia	0.07
Halocarpus kirkii	12	Metrosideros robusta	0.06
Laurelia novae-zelandiae	12	Dracophyllum longifolium	0.05
Pittosporum tenuifolium	11	Hebe salicifolia	0.05*
Dacrycarpus dacrydioides	10	Archeria traversii	0.03*
Cyathodes fraseri	10	Epacris pauciflora	0.02
Dodonaea viscosa	9.5	Gaultheria crassa	0.02
Agathis australis	9.5	Gaultheria depressa	0.02
Pseudopanax arboreus	9.0	Dracophyllum pronum	0.02
Myrsine divaricata	7.8		
Nothofagus truncata	7.7		
Lophomyrtus obcordata	7.0		
Nothofagus fusca	6.8		
Hoheria populnea	6.6		
Coprosma propinqua	6.5		
Aciphylla aurea	5.2		
Pittosporum eugenioides	5.1		
Dacrydium cupressinum	4.8		
Lepidothamnus laxifolius	4.8		
Melicytus micranthus	4.2		
Plagianthus regius	4.2		
Cordyline australis	4.1		
Pseudowintera colorata	3.7		
Nothofagus menziesii	3.6		
Nothofagus solandri	3.4		
Coprosma rhamnoides	2.7		
Hoheria angustifolia	2.6		
Cyathodes empetrifolia	2.4		
Entelea arborescens	2.0		
Lagarostrobos colensoi	1.8		

Notes:
Some values include an envelope that is not strictly part of the seed, e.g. the nut of *Nothofagus*.
* Air-dry, i.e. *c.* 10% heavier than oven-dry
† Haase 1986
(1) Chatham Islands
(2) Mainland New Zealand
Source: R.P. Buxton, unpublished.

Table 4.2. *Ratio of hard outer layers to total dry mass (%) and mass of contents (mg) for some of the larger seeds in Table 4.1*

	%	mg
Corynocarpus laevigatus	46	1250
Beilschmiedia tawa	16	960
Rhopalostylis sapida (1)	27	330
Rhopalostylis sapida (2)	28	176
Ripogonum scandens	3	120
Hedycarya arborea	33	116
Prumnopitys ferruginea	91	43
Elaeocarpus dentatus	90	37
Myrsine salicina	48	15
Corokia macrocarpa	93	5

Source: R.P. Buxton, unpublished.

Leptospermum scoparium) may develop ovaries; and only a proportion of ovaries may develop into fruit (e.g. 9–13% in *Discaria toumatou* (Lloyd *et al.* 1980)).

In many species only a low percentage of seed is sound, e.g. 34% in *Olearia ilicifolia* and < 4% in some samples of *Celmisia* species. In *Nothofagus*, the percentage is much higher in good mast years (54–68%) than in poor years (3–11%). Many-seeded fruits consistently have low proportions of filled seed, e.g. 13% in a sample of *Metrosideros umbellata*. In such species, the optimal number of sound seed is probably far fewer than the number of ovules. Insect larvae further reduce the quantity of sound seed, notably in high-altitude species of *Chionochloa* (White 1975*b*) and *Celmisia*. Nevertheless, in good years the amount of viable seed produced can be prodigious (Table 4.3).

Apparently sound seeds of native species usually prove to have high viability, often approaching 100%, but in *Celmisia* only 9–12% may be viable. Some reports of poor or zero germinability, as in some gentians and *Phyllocladus alpinus*, may indicate failure to provide appropriate conditions.

In the laboratory, most native species germinate fastest under warm, moist conditions; 20 °C is usually effective. Species of cool habitats can germinate at much lower temperatures, e.g. < 5 °C in some celmisias. In warm lowlands, autumn germination is usual in kauri (*Agathis australis*) (Ecroyd 1982), is reported for *Beilschmiedia tawa* and beeches, and is probably widespread among other species. With increasing altitude, significant germination becomes restricted to spring and early summer. Beeches, gentians, *Dracophyllum traversii* and matagouri (Daly 1969) benefit from moist, cool stratification, which probably discourages premature germination. Light periods assist or are essential for germination in lowland hebes (Simpson 1976), *Olearia ilicifolia*, *Brachyglottis buchananii*, *Dracophyllum traversii*, *Celmisia* spp., *Metrosideros umbellata* and *Leptospermum scoparium*.

Most small-seeded species germinate within 30 days, although there may be sporadic emergence after the main germination, e.g. beyond 115 days in *Olearia*

Table 4.3. *Numbers of sound seed produced in one year
by tree and shrub species*

	no. of seeds	reference
Seeds/m²		
Dacrydium cupressinum	8–5000	1
Beilschmiedia tawa	10–>100	2
Nothofagus menziesii	11–3100	3
N. solandri v. *cliffortioides*	3–5100	3
Olearia colensoi	1500–2000	4
Seeds/tree		
Prumnopitys ferruginea	32 000	5
Dacrydium cupressinum	200 000	5
Dacrycarpus dacrydioides	4 500 000	5
Phyllocladus alpinus	6000–20 000	6
Weinmannia racemosa	6 000 000	7
Leptospermum scoparium	2 500 000	8
Dracophyllum traversii	65 000–750 000	9
Brachyglottis buchananii	43–30 000	9
Olearia ilicifolia	17 600–243 000	9

References:
1 Franklin 1968; Norton *et al.* 1988
2 Knowles & Beveridge 1982
3 Wardle 1984
4 Wardle *et al.* 1971
5 Beveridge 1964
6 Wardle 1969 (clones 2–3 m in diameter; >90% of seed
   appears sound, but no germination achieved
   experimentally)
7 Wardle & MacRae 1966
8 Grant 1967 (10 yr old shrub; figure may include non-viable
   seed)
9 Haase 1986*c,e, f*

ilicifolia. Some gentians show protracted sporadic germination. Delayed germination is to be expected in seed with thick or impervious integuments; this includes most species with medium-sized or large drupes, and also many with small drupes, notably coprosmas. Further examples are *Vitex lucens*, in which 2–4 seeds are enclosed in a woody endocarp (Godley 1979), and *Sophora* spp. which, like many adventive legumes, have impervious seed coats. In *Vitex, Coprosma, Corokia* and the water plant *Myriophyllum* (Simpson & Burrows 1978) a defined portion of the integument opens or falls out to release the embryo. In other species the integument must decay or abrade to allow germination.

Miro and matai (*Prumnopitys* spp.) seed may lie buried up to four years before they germinate (Beveridge 1973). On Banks Peninsula, seedlings of kaikawaka (*Libocedrus bidwillii*) appeared several years after all adult trees had died, to judge from growth ring counts (Wardle 1978*a*). Following fencing of the Hapupu Reserve on the

Chatham Islands, *Melicytus chathamicus* seedlings appeared in great profusion although few live trees were found. There is also a report of rimu seed deeply buried in peaty soil germinating on exposure (J.L. Bathgate, unpublished).

Enright & Cameron (1988) differentiated germinable seed beneath a remnant of kauri forest according to whether it had lain less or more than two years. The recent seed fall was dominated by four small native trees; in one of these, *Kunzea ericoides*, 20% of the seed was older. Most of the > 2 yr seed consisted of *Cordyline australis*, the understorey shrub *Geniostoma rupestre*, the native weedy herb *Solanum americanum* and lesser quantities of adventive weeds of open habitats. Seed beneath regenerating bush in Canterbury consists largely of adventive species that have been carried in or, like gorse and common broom, have persisted from earlier vegetation; among native dominants, only *Sophora microphylla* contributes to the portion of the seed bank lying below 2.5 cm (Partridge 1989).

Nothofagus is probably typical of most indigenous species in its unremarkable retention of viability. Although the seed can be stored up to eight years under optimal conditions, few remain viable on the forest floor 18 months after seed-fall (Wardle 1984).

Seedling and juvenile forms
(based on Godley 1985)

In *Corynocarpus, Beilschmiedia* and *Sophora*, cotyledons remain below ground as storage organs and seedlings are relatively robust. In most other native plants, cotyledons are raised above the soil to function as green leaves; in *Agathis, Griselinia, Euphrasia disperma, Elaeocarpus* and Loranthaceae, they are already green within the seed. *Pisonia* cotyledons measure 4.5 × 4 cm, but most seedlings are small and delicate, in keeping with the prevalence of small seeds (Figs 4.4 and 4.5).

Young trees usually pass through varied juvenile stages (Fig. 4.6). Typically, sapling leaves are larger than those of either the seedling or mature phase, and strikingly so in the group that includes *Pseudopanax crassifolius*. In trees with filiramulate juveniles, on the other hand, sapling leaves are smaller than those of the adult. Sapling leaves are also often more intricate than those of either the seedling or adult, especially in the araliad *Pseudopanax simplex*.

Trees and shrubs in which adult leaves are greatly reduced in size as in cupressoid species, sparse, or virtually absent as in *Phyllocladus alpinus*, invariably have leafy seedlings or juveniles. In 'whipcord' hebes, spreading leaves can also be induced by shading. The transition from needle leaves to scales is very abrupt in *Halocarpus* species, which also develop scattered 'reversion' shoots. In herbaceous plants, juvenile leaves and those borne on inflorescence axes are smaller and less complex than those of the fully developed vegetative phase.

Longevity

Except in very open vegetation, opportunities for successful reproduction lie largely in the death of established plants. Every plant established from seed or vegetatively has

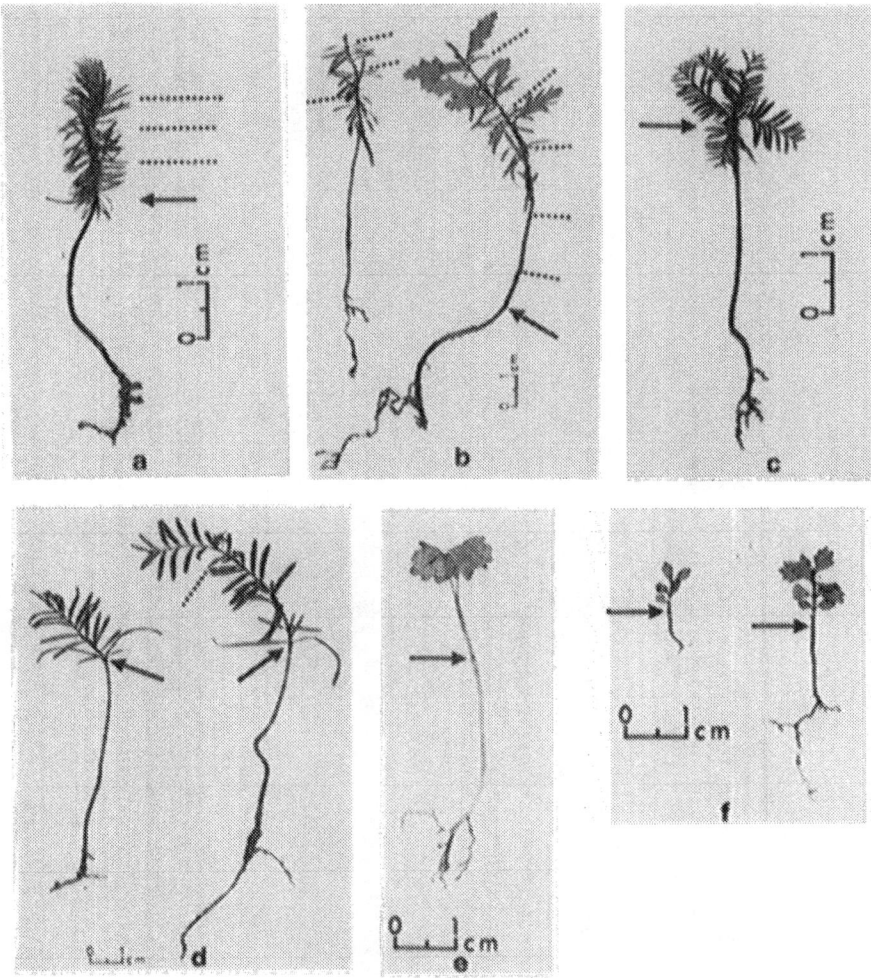

Fig. 4.4. A selection of tree seedlings. Arrows indicate the insertion of cotyledons; dotted lines show the points of cessation of growth, presumably at the end of the growing season. (*a*) *Dacrydium cupressinum*; (*b*) *Phyllocladus alpinus*; (*c*) *Dacrycarpus dacrydioides*; (*d*) *Prumnopitys ferruginea*; (*e*) *Hoheria glabrata*; (*f*) *Weinmannia racemosa*; (*g*) *Quintinia acutifolia*; (*h*) *Nothofagus solandri* var. *cliffortioides*; (*i*) *Coprosma* aff. *ciliata*; (*j*) *Myrsine divaricata*; (*k*) *Griselinia littoralis*; (*l*) *Pseudopanax simplex*; (*m*) *Pseudopanax crassifolius*; (*n*) *Elaeocarpus dentatus*; (*o*) *Hedycarya arborea*; (*p*) *Aristotelia serrata*; (*q*) *Coprosma propinqua* × *robusta*; (*r*) *Sophora microphylla*. Species collected in spring, all in forest except (*p–r*), which are open-grown.

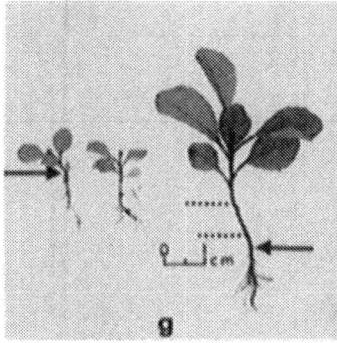

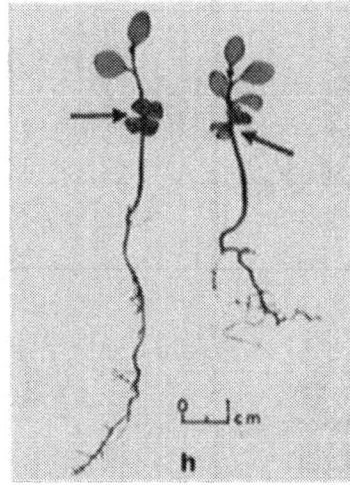

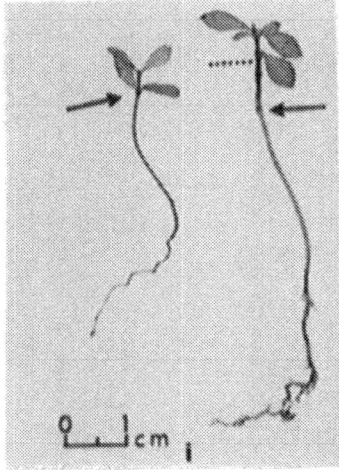

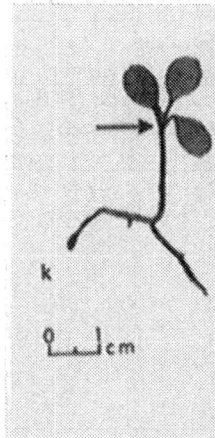

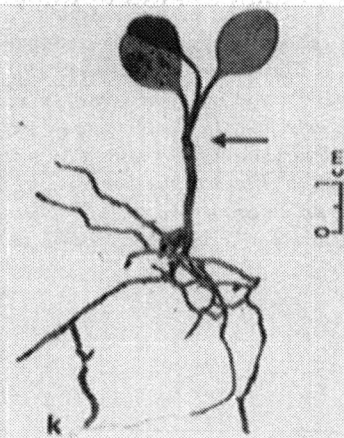

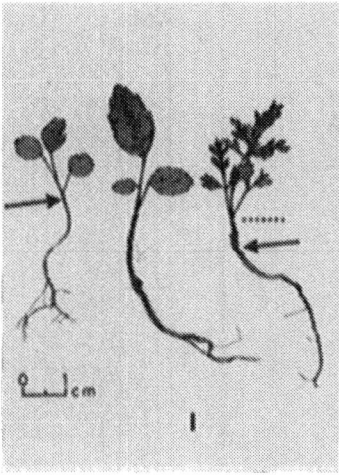
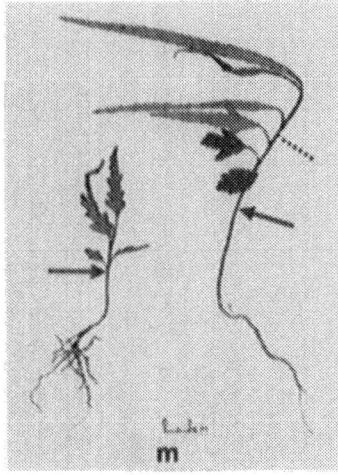

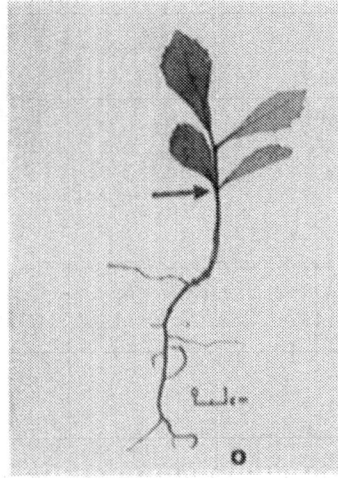

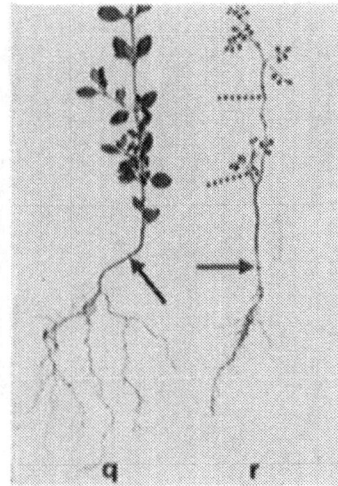

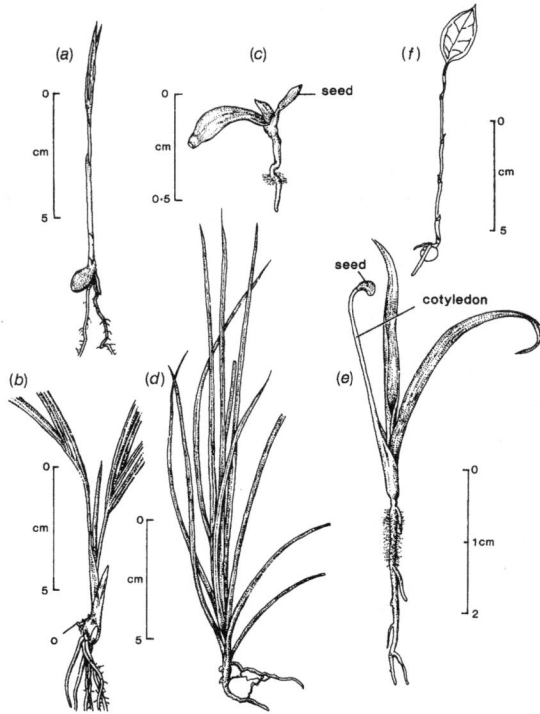

Fig. 4.5. Seedlings of woody monocots. (*a,b*) *Rhopalostylis sapida* (O, aborted original apex); (*c,d*) *Freycinetia baueriana*; (*e*) *Cordyline australis*; (*f*) *Ripogonum scandens*. (Parts (*a–e*) from Tomlinson & Esler 1973; (*f*) from Macmillan 1972.)

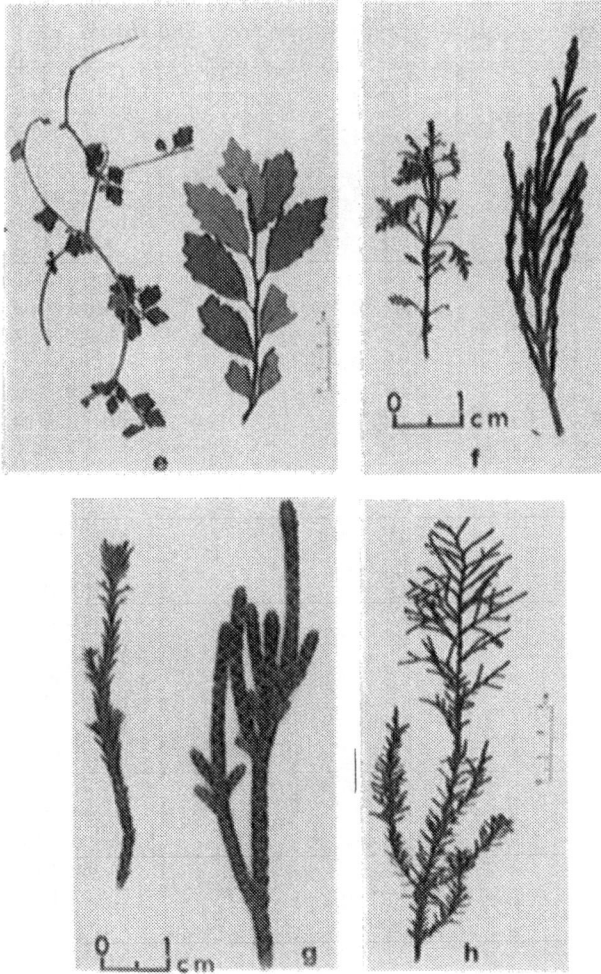

Fig. 4.6. Juvenile (left) and adult (right) stages of woody plants (*a*) *Pseudopanax crassifolius*; (*b*) *Elaeocarpus dentatus*; (*c*) *Elaeocarpus hookerianus*; (*d*) *Pseudopanax simplex*; (*e*) *Pennantia corymbosa*; (*f*) *Hebe cupressoides*; (*g*) *Hebe lycopodioides*; (*h*) *Halocarpus bidwillii*.

Table 4.4. *Life span of native trees and shrubs*

Species	Years	Source
Agathis australis	600–1700	Ahmed & Ogden 1987
Prumnopitys taxifolia	1400	P. Wardle, unpublished
Dacrydium cupressinum	600–1200	Norton *et al.* 1988
Lagarostrobos colensoi	573*	Dunwiddie 1979
Halocarpus biformis	1000	Wardle 1963*a*
Phyllocladus glaucus	441*	Dunwiddie 1979
Phyllocladus alpinus	260*†	Dunwiddie 1979
Libocedrus bidwillii	720*	Dunwiddie 1979
Beilschmiedia tawa	200–400	Knowles & Beveridge 1982
Metrosideros umbellata	400–500	Wardle 1971
Metrosideros umbellata	>1084*	Smith *et al.* 1985
Weinmannia racemosa	450	P. Wardle, unpublished
Nothofagus sol. v. cliffort.	300–360	Wardle 1984
Nothofagus truncata	400–500	Wardle 1984
Nothofagus fusca	450–600	Wardle 1984
Nothofagus menziesii	600	Wardle 1984
Kunzea ericoides	160	P. Wardle, unpublished
Cyathodes fasciculata	36	Bray 1989
Aristotelia serrata	30	Bray 1989
Schefflera digitata	24	Bray 1989
Coprosma grandifolia	24	Bray 1989
Entelea arborescens	≥10	Milliner 1947
Solanum aviculare	7	Bray 1989
Dracophyllum traversii	500–600	Haase 1986*c*
Dracophyllum longifolium	220*	P. Wardle, unpublished
Olearia lacunosa	300	Wardle 1963*a*
Olearia ilicifolia	240	Haase 1986*f*
Hoheria glabrata	>150†	Haase 1987
Olearia lyallii	>140	Lee *et al.* 1983*a*
Olearia colensoi	>117†	Wardle *et al.* 1971

Notes:
* Largest number of growth rings reported.
† Coppicing or layering prolongs life of individuals.

an inherent maximum longevity that is a function of growth rate and ultimate size, although the realised life span is usually reduced by external factors such as competition, predation and catastrophe. Among trees, conifers live longer than broad-leaved species of comparable stature, and small, seral species have the shortest life-span (Table 4.4).

Vegetative renewal, especially layering and coppicing, increases the longevity of many trees and shrubs (p. 562). This is even more prevalent among indigenous herbs, few of which are annuals or monocarpic perennials (p. 28). Vegetatively spreading plants are potentially immortal, but space imposes a constraint. There is also an inherent limit to the size of densely clumped herbs such as tussock grasses; nevertheless some, including the larger chionochloas, live decades if not centuries.

DESCRIPTION, NOMENCLATURE AND CLASSIFICATION OF VEGETATION, ENVIRONMENT AND ECOLOGICAL PROCESSES

This chapter summarises the major vegetation types and the environmental gradients and processes that determine them, while defining the terms in which these are described. Such terms are mostly informal and interpretative, and are not intended to substitute for precise nomenclature based on floristics or structure, such as is necessary for mapping or inventory of vegetation.

Vegetation classes

Allocation of material among chapters

Vegetation is divided among Chapters 7–12 primarily according to structural criteria. Chapter 7 discusses forest dominated by trees > 10 m tall. At lower altitudes, indigenous forests are dominated by tall conifers, by broad-leaved evergreen trees that are usually less tall, or by mixtures in which the conifers (mostly podocarps) form an open overstorey to the main canopy of broad-leaved trees. These are generally referred to as *conifer/broad-leaved* or '*mixed*' forests, but can also be described as podocarp, broad-leaved, podocarp/broad-leaved, etc., according to the dominant component. Northern mixed forests, including those dominated by the massive kauri (*Agathis australis*), are described first and in some detail, as they can be regarded as an ensemble from which other mixed forests have segregated, becoming simpler in structure and poorer in species as latitude and altitude increase.

For descriptive convenience, the term 'broad-leaved' excludes the beeches (*Nothofagus*). In the colder or drier localities, beeches dominate extensive forests of relatively simple structure to the exclusion of other canopy trees. Smaller stands of beech occur among other forest, usually on harsher sites such as ridge crests, and there are also forests in which beeches mingle with conifers and broad-leaved trees; terms such as kauri/beech or (podocarp)/beech are used for these. The chapter concludes with a brief account of exotic forests.

Chapter 8 covers varied communities dominated by shrubs or trees < 10 m tall. One set has canopies of small trees or large shrubs, mostly of the same species that form lower tiers in tall forest. Its communities are prominent near the coast, in gullies, on unstable slopes, in successions, and where a former canopy of tall trees has been lost, usually by logging, windthrow or disease. Although such vegetation has ecological and floristic affinities with forest, its height range of 3–10 m overlaps those of forest and scrub, and it is generally too dense to call woodland. It therefore seems appropriate to apply the vernacular term *bush* in the sense that it was first used by early European settlers to distinguish dense secondary woody vegetation near populated areas from the tall forest in the hinterland (Johnston 1981); although, in New Zealand, 'bush' soon came to include tall forests as well. Most of the trees are broad-leaved evergreens, but the few native deciduous trees, the palm-like *Cordyline*, the palm *Rhopalostylis sapida* and tree ferns can also be important. Where severe environments, such as prevail on exposed coasts and high ridges, reduce normally tall trees to low stature, it is more convenient to refer to stunted forest.

Much of the woody vegetation on very infertile soils and at high altitudes is referred to as *heath*, because it is dominated by shrubs or small trees with close-set ericoid, needle-like or cupressoid leaves. These are usually malleable as to stature; manuka (*Leptospermum scoparium*) can vary from creeping mats to small trees, according to site. Heath is qualified as tree-, shrub-, or dwarf-, according to whether the dominants are > 3 m, 0.2–3 m, or < 0.2 m tall.

Primary heath mostly consists of slow-growing epacrids, small podocarps and manuka. At high altitudes and on southern coasts woody composites with broad, tomentose leaves in the genera *Olearia* and *Brachyglottis* can be abundant or dominant. Upland and coastal heath can alternate with or merge into bush occupying more fertile soils.

Secondary heath, which usually develops after fire, can be a stage of succession towards primary heath or forest. It has fewer species than primary heath, occurs over a wider range of soil fertility, and the dominants are usually faster-growing. Manuka is the most widespread species, but species of *Dracophyllum* and the introduced *Hakea* can also be abundant or dominant. Stands of kanuka (*Kunzea ericoides*) are conveniently discussed with secondary heaths, even though they can attain a stature of 20 m and usually occupy better soils.

Many heathland soils are wet, because rock or impervious pans impede drainage. Because of recurrent fires, dominance tends to alternate between rush-like sedges (especially species of *Baumea* and *Lepidosperma*) and *Gleichenia* fern, and manuka and other shrubs. *Gumlands* are wet heathlands occupying former kauri land in the north of the North Island, in which the shrub phase predominates. *Pakihi* is similar heathland confined to the west of the South Island, but as the herbaceous phase predominates, it is described with wetland in Chapter 10.

General terms applicable to shrubby vegetation are scrub (if dense) or shrubland (if open). They include shrub-heaths, and many other communities of native or introduced shrubs. Filiramulate shrubs are prominent in seral communities in frosty localities. Because their leaves are small and mostly obscured by twigs, stands have a dark appearance and are referred to as *grey scrub*. Filiramulate lianes, nearly leafless native brooms (*Carmichaelia*) and, in dry South Island localities, the spiny matagouri (*Discaria toumatou*) are also important in grey scrub. Bracken (*Pteridium esculentum*) fernland is included in Chapter 8, because of its close successional relationship with woody vegetation.

Grasslands are assigned to three main categories (Chapter 9). *Chionochloa* grasslands mainly grow at high altitudes, although in the recent past they were also extensive on cool, southern lowlands, following Polynesian deforestation. Most species of the dominant genus are tall 'snow tussocks'; locally, dominance may be yielded to or shared with large forbs to form *herbfield* or *tussock-herbfield* respectively. *Short-tussock* grasslands of dryish inland districts are visually dominated by species of *Poa* and *Festuca* although smaller plants, now mostly introduced species, usually form most of the cover. In the driest districts, some of these grasslands have been depleted to the extent that moss-like raoulias and short-lived adventive grasses and forbs dominate. In maritime regions (p. 80) grasslands consist mostly of introduced cool-temperate grasses and forbs, except where native plants, especially *Rytidosperma* species, remain important on poor soils. They have been developed at the expense of forest, shrubland, wetland, fernland and native grasslands. In northern districts, adventive species of warm-temperate origin are prominent.

Wetland vegetation is permanently or seasonally waterlogged, partly covered in shallow water, or completely submerged, i.e. aquatic. Chapter 10 concentrates on herbaceous wetland vegetation, but pays only passing attention to wet grasslands since these, as well as tree- and shrub-dominated wetlands, are discussed in earlier chapters. Sections deal respectively with saline communities that are all coastal apart from tiny inland areas, mostly in Central Otago; the vegetation of lakes, ponds and streams (planktonic and marine vegetation lie outside the scope of the book); *fertile lowland swamps* dominated by vigorous plants such as *Carex* sedges, *Typha* and *Phormium*; *infertile lowland mires* generally dominated by rush-like sedges, restiads or *Gleichenia* fern; *wet heaths* with vegetation similar to that of infertile mires, but with little or no peat; and the various mountain wetlands.

Chapter 11 deals with the plants of recent, unweathered, mineral surfaces provided by coastal sand and gravel, inland river-beds, severely depleted inland grasslands that have come to botanically resemble river-beds, slope debris and boulder fields, cliffs and rock outcrops, ultramafic rocks, and thermal areas. The low-growing vegetation of high altitudes is reserved for Chapter 12. Although the vegetation discussed in these two chapters is generally sparse, it is diverse and includes most of the local endemics in the flora. Chapter 13 covers the outlying islands, where floristic and environmental differences from the mainland give rise to distinctive vegetation.

Arrangement within chapters

Within Chapters 7–9, each important vegetation type is introduced with general comment on its structure, composition and ecological status, followed by such autecological information on individual species or genera as has not been adequately covered in Chapters 3–4. Next, the vegetation type is described in some detail, usually in respect of the region where it is most extensive or best developed. This is followed by briefer descriptions of other examples, concluding with those that are most distant or least typical. Discussion of processes within vegetation types and successional trends that link different types are mostly reserved for Chapters 14–16.

Chapter 10 contains more complete information on growth forms, autecology and vegetation processes that are largely unique to the wetland environment. Chapters 11–13 are also largely self-contained.

Defining and naming plant communities

Within plant communities, tiers are described in sequence from the uppermost down, followed by listing of lianes, epiphytes and vascular parasites. Communities are designated according to the dominant species, genus or family, e.g. *Chionochloa pallens* grassland, *Dracophyllum* heath or podocarp forest.

Where more precise naming is required, I separate tiers by a solidus (/) and codominant species in tiers by a dash (–). Significant non-dominant species, or species that form a very open tier, are enclosed by brackets. For example, (rimu)/hard beech/ *Blechnum discolor* forest indicates an open overstorey of *Dacrydium cupressinum*, dominance by *Nothofagus truncata* in the main canopy, and a well-developed tall fern tier of *Blechnum discolor*. (In a system developed by Atkinson (1985), / and – indicate the structure and floristic composition of the canopy alone.)

Although naming and classification in this book is 'intuitive', it draws on more objective studies that employ contrasting approaches. One emphasises the proportions of dominant species, an example being McKelvey's (1984) classification of South Island forests. The other approach is floristic analysis, usually through clustering and ordination techniques that also explore relationships of species and communities to environmental gradients, as in Partridge & Wilson's (1988) study of salt marshes. Species that differentiate communities need not be prominent ecologically, as Reif & Allen (1988) show in complex forest vegetation. Further, stands that are developmentally related but of quite different structure may be grouped together through floristic similarities.

Hybrid techniques use floristic comparisons, in which species are weighted according to sociological criteria such as abundance or dominance. Examples are Connor's (1964, 1965) accounts of tussock grasslands, J.A. Wardle's (1970) classification of *Nothofagus solandri*-dominated communities, and Leathwick's (1987) account of vegetation near Pureora on the Volcanic Plateau.

Divisions of vegetation and habitat based on temperature gradients

Altitudinal belts

The most regular zonation in New Zealand vegetation is defined by altitude (Fig. 5.1) and can be correlated with broad parameters such as mean annual temperatures. Underlying factors are more complex, the most important probably being the temperature regimes necessary for growth and reproduction, and the limits of cold-tolerance at various phases in the life history and seasonal growth cycle. Altitudinal belts are defined by horizontal upper boundaries that correspond to the highest altitudes reached by certain growth forms and species (Fig. 5.2). The following describes the belts on forested coastal mountains that are exposed to neither strong prevailing winds nor temperature inversion.

(1) In the *warm-temperate* belt, forests are luxuriant with lianes, epiphytes, tree ferns,

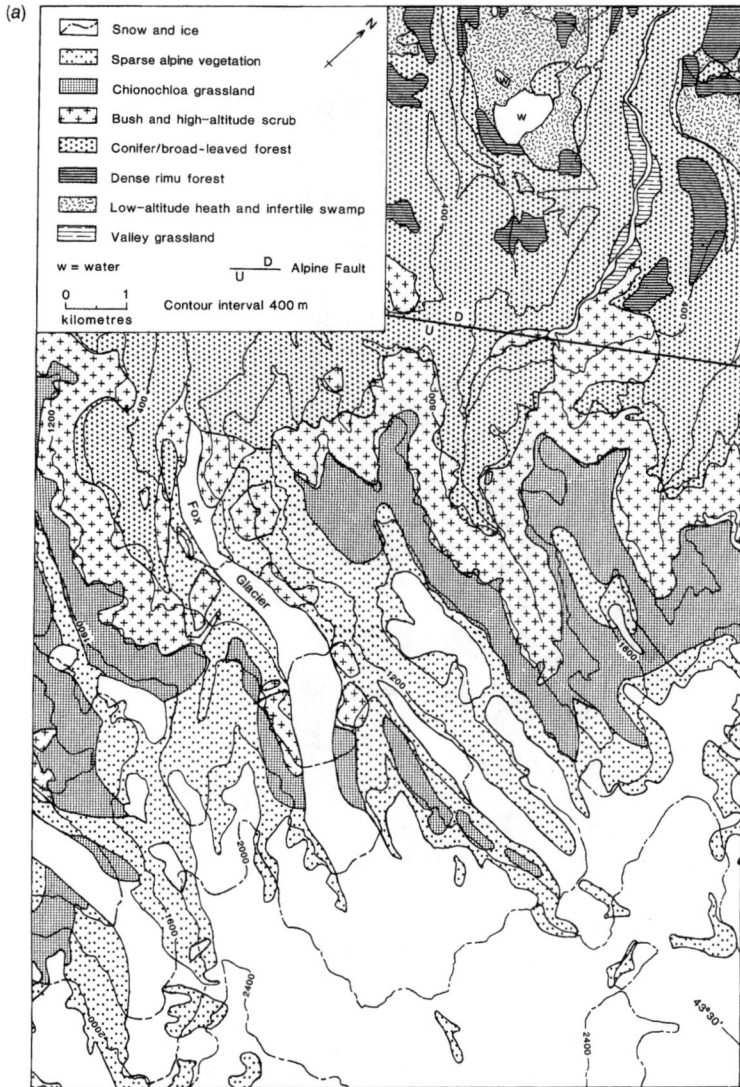

(a)

Snow and ice	
Sparse alpine vegetation	
Chionochloa grassland	
Bush and high-altitude scrub	
Conifer/broad-leaved forest	
Dense rimu forest	
Low-altitude heath and infertile swamp	
Valley grassland	

w = water

0   1
kilometres

Contour interval 400 m

D
— Alpine Fault
U

Fig. 5.1. Vegetation belts on four mountains. (a) Western slopes of Southern Alps at Fox Glacier, Central Westland (Department of Lands and Survey 1982). (b) Siberia valley, South-western Alps near Lake Wanaka (Mark 1977). (c) Mt Pureora, an extinct volcano showing depressed belts characteristic of isolated summits (Leathwick et al. 1988). (d) South-western sector of Mt Ruapehu, an active volcano (Atkinson 1981).

Fig. 5.1 (cont.)

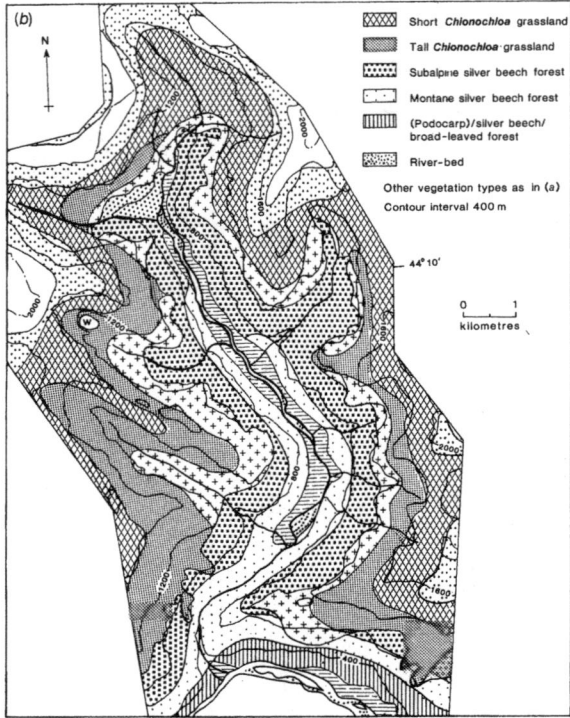

N

Short *Chionochloa* grassland

Tall *Chionochloa* grassland

Subalpine silver beech forest

Montane silver beech forest

(Podocarp)/silver beech/
broad-leaved forest

River-bed

Other vegetation types as in (a)

Contour interval 400 m

44°10'

0   1
kilometres

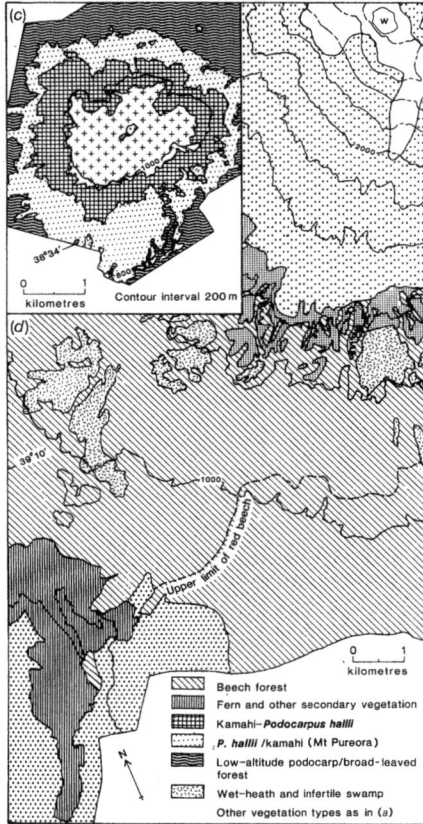

Contour interval 200 m

0   1
kilometres

38°34'

39°10'

Upper limit of red beech

N

0   1
kilometres

Beech forest

Fern and other secondary vegetation

Kamahi–*Podocarpus hallii*

P. hallii /kamahi (Mt Pureora)

Low-altitude podocarp/broad-leaved
forest

Wet-heath and infertile swamp

Other vegetation types as in (a)

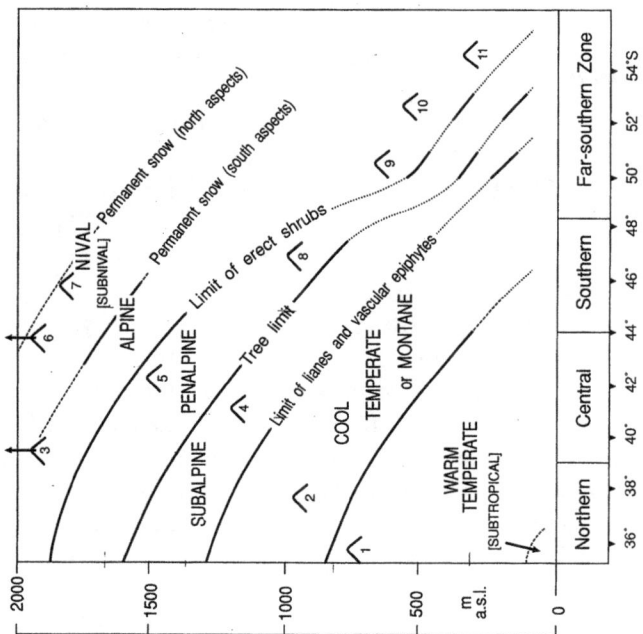

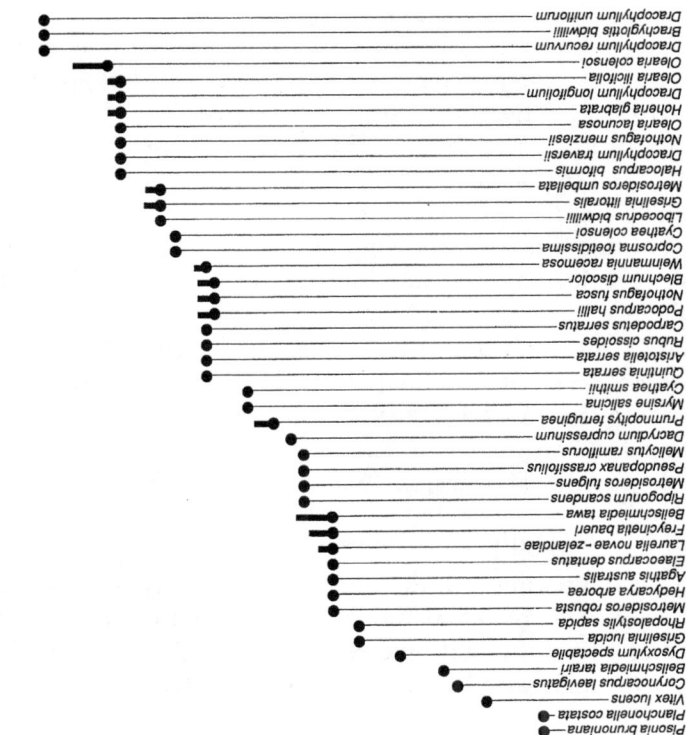

Fig. 5.2. Relative upper limits of common plants, mainly woody, on maritime slopes near western coasts (left). These limits define belts that decrease in altitude with increasing latitude (right). Numbers represent the following summits.

1. Mataraua Forest   716 m
2. Mt Te Aroha   953 m
3. Mt Taranaki   2518 m
4. Mt Stokes   1204 m
5. Paparoa Range   1501 m
6. Mt Tasman   3497 m
7. Mt Soaker   1853 m
8. Mt Anglem   676 m
9. Auckland Island   668 m
10. Campbell Island   569 m
11. Macquarie Island   434 m

and species of generally Melanesian affinities. Important constituents such as *Beilschmiedia*, kauri, kiekie vine (*Freycinetia baueriana*,) nikau palm (*Rhopalostylis sapida*) and the epiphyte *Collospermum hastatum* are confined thereto, as are many other genera and species. The upper limit is irregular, as most of the species also require moderately fertile soil and tolerate only light frosts.

(2) The *cool-temperate* or *montane* belt supports less luxuriant mixed forests and most forests in which beeches are prominent. The upper limits coincide with those of lianes, vascular epiphytes, and the important trees kamahi (*Weinmannia racemosa*) and red beech (*Nothofagus fusca*).

(3) The *subalpine* belt is occupied by almost pure stands of silver or mountain beech (*Nothofagus menziesii* or *N. solandri* var. *cliffortioides*). Where these are absent either locally or over large areas, there is dense tree-heath with epacrids, woody composites and conifers; this grades into shrub-heath on higher and more exposed sites, and into subalpine bush on lower, more sheltered sites with deeper soils. The upper boundary of the subalpine belt corresponds to the upper limit of indigenous trees and shrubs more than 1–2 m tall; this boundary is attained, but seldom exceeded, by many tall woody species. It is correlated with a mean January temperature around 11 °C (McCracken 1980). However, as trees reach identical limits on 'sunny' (i.e. north and west) and 'shady' (south and east) aspects, the causal relationship is probably with nocturnal minima (p. 504).

(4) The *penalpine* (i.e. 'almost alpine') belt is characterised by *Chionochloa* tussock grasses and robust species of *Astelia, Celmisia, Aciphylla* and *Ranunculus*. Shrubs < 1 m tall, especially *Podocarpus nivalis* and species of *Coprosma* and *Dracophyllum*, may dominate on steep spurs and stony ground. Because native trees are absent, most ecologists have regarded this as the lower part of the alpine belt, but spontaneous spread and experimental establishment of exotic trees, especially north-temperate conifers, indicates that this is invalid. Although mid-summer temperatures are lower then they are in the uppermost north-temperate forest belts, growing seasons are longer (Bliss & Mark 1974; Wardle 1985). Snow lies continuously for 2–5 months, at least on shady aspects, but this is not long enough to restrict the growing season.

(5) The *alpine* belt lies above the limits of extensive grassland. Its vegetation is short, patchy, and greatly influenced by aspect and duration of snow cover, and there are large expanses of bare, rocky ground. On the highest mountains, a *nival* belt extends above the usual lower limit of perpetual ice and snow, with cryptogams and infrequent vascular plants occupying the limited areas that are not permanently snow-covered. Especially on the drier mountains, there can be an extensive *subnival* belt that is free of snow for at least part of the summer and autumn, but lacks any continuous vegetation.

Latitudinal zones

Temperatures decrease with increasing latitude, and differences between summer and winter day length increase. Warmth-demanding species progressively drop out towards the south (Fig. 5.3), and altitudinal belts descend (Fig. 5.2). Latitudinal zones are defined according to which altitudinal belt provides the original lowland vegetation.

(1) The *subtropical* zone is fully developed only on the Kermadec Islands and even

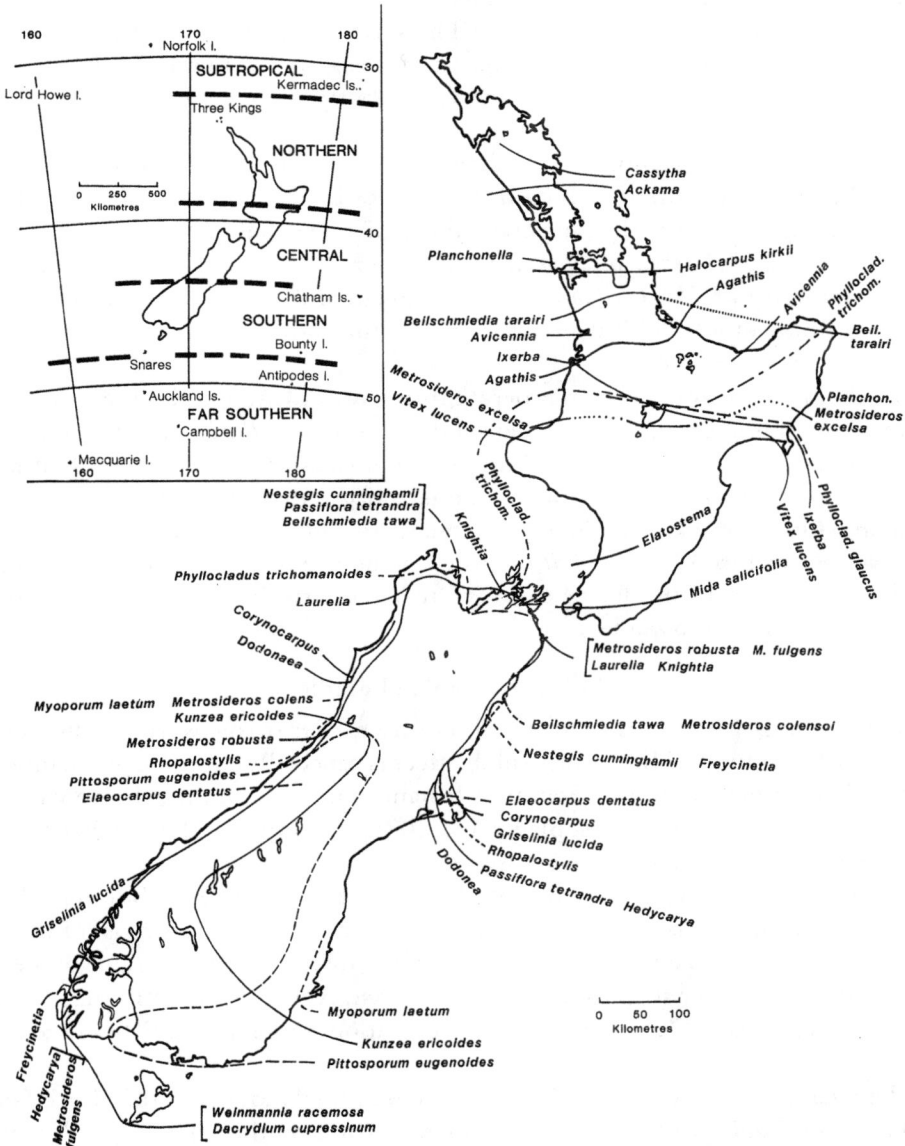

Fig. 5.3. Latitudinal zones (inset) and natural southern limits of some common woody plants.

here the vegetation appears more temperate than subtropical, reflecting its derivation from the New Zealand mainland. Some subtropical species reach the northern part of the North Island, but of these only mangrove (*Avicennia resinifera*) is important.

(2) The *northern* zone extends from the Three Kings Islands to latitude 39° S. It has the fullest development of warm-temperate forests, and contains the limits of such important species as kauri, pohutukawa (*Metrosideros excelsa*) and puriri (*Vitex lucens*).

(3) The *central* zone extends south to latitude 44°, the limit of warm-temperate forest and bush. In the east, the zone boundary includes Banks Peninsula and the Chatham Islands, coinciding with the southern limits of nikau palm, karaka (*Corynocarpus laevigatus*), hinau (*Elaeocarpus dentatus*) and other prominent species. On the western coast, northern rata (*Metrosideros robusta*), nikau and hard beech reach their main southern limits at 42° 30′ but, together with other northern species, have sporadic occurrences further south.

(4) In the *southern* zone, cool-temperate forest prevails to sea level, although a few warm-temperate species, such as kiekie and *Cyathea medullaris*, persist along western coasts. The southern boundary includes the Snares Islands just beyond latitude 48°.

(5) The *far-southern* zone comprises the island groups south of the New Zealand mainland. On the Auckland Islands, stunted forest with some cool-temperate species, such as the tree fern *Cyathea smithii*, grows along sheltered harbours, but subalpine heath covers most of the lower slopes. On Macquarie Island, tussock grasses and robust herbs prevail to sea level.

The coastal–inland gradient

With increasing distance from the sea, there are harder frosts, warmer summer days, and lower humidities. This 'continental' effect is especially pronounced in intermontane basins. Impact on the vegetation is enhanced because of the relatively low cold-tolerance of most indigenous plants (p. 502). To express the gradient, coastal, maritime and interior zones are distinguished.

In the *coastal* zone, plants benefit from mild temperatures but may have to withstand strong winds and salt. The *maritime* zone encompasses most of the North Island, all of Stewart Island, a strip 40–50 km deep along the western side of the South Island, and a narrower strip along its eastern coast. At the lowest altitudes, July mean daily minima exceed 0 °C and the natural vegetation is mostly multi-tiered temperate forest.

The *interior* zone covers inland parts of the South Island, but in the North Island is well expressed only where the Volcanic Plateau abuts on the axial ranges. Temperate forests of this zone generally lack vascular epiphytes, lianes such as *Metrosideros* spp., *Ripogonum* and kiekie, tree-ferns, and important tier-dominants such as kamahi and *Blechnum discolor*. The upper limit of the interior-temperate belt is therefore defined by fewer species, which include red beech, *Coprosma linariifolia* and remaining lianes, especially *Rubus* spp.

Interior-subalpine forests are mostly mountain beech stands of very simple composition, at least as far as vascular plants are concerned, but a higher forest limit on inland ranges than on coastal ranges (1500 m versus 1200 m at latitude 42°) reflects

Fig. 5.4. Subalpine *Nothofagus solandri* var. *cliffortioides* forest separating penalpine from valley-floor grasslands. The deforested area in the left foreground has grassland and *Dracophyllum* shrubs. The glaciated mountain in upper left is Barth (2430 m). Ahuriri valley, Lakes ER.

longer or warmer growing seasons. Where heath and bush occupy the subalpine belt, they too are floristically poorer than in maritime regions; for instance, large-leaved dracophyllums and *Olearia colensoi* are absent. Maritime–interior differences are less evident in penalpine and alpine vegetation, being expressed in floristics rather than physiognomy. Since maritime climates are less equable along eastern than western coasts, most frost-sensitive species reach higher latitudes in western districts, whereas species more dependent on summer warmth tend to grow further south in eastern districts (Fig. 5.3).

Zonation due to temperature inversion

On still, clear nights, the air near the ground becomes cold and dense, and tends to flow down slopes and collect over valley floors and basins. Habitats subject to such inversions are often as much as 8 °C colder than the slopes above, and may support plants otherwise characteristic of higher or more inland areas. The phenomenon is most marked in successional, low, or open vegetation, because forest reduces the intensity of temperature inversion.

The situation is indicated by prefixing *valley*. Thus, valley-cool-temperate forest can occupy valley floors below slopes that support warm-temperate forest; and valley-penalpine grassland is common on flats in mountain valleys where slopes support subalpine forest (Fig. 5.4). In the high mountains, alpine and nival plants can descend well below their usual altitudes on cold, shaded sites or where avalanche snow persists.

Fig. 5.5. Depressed upper limit of *Nothofagus menziesii* forest in a basin occupied by subalpine grassland. Turks Head Range, North-west Nelson.

The converse of these effects is found in valleys where interior-temperate vegetation prevails, except in a narrow belt above the influence of deep temperature inversions. Such *mid-slope* vegetation can include species such as kamahi or *Cyathea smithii* growing inland of their usual range.

Effects of temperature inversion are enhanced in valley heads that have been widened by glacial erosion. Here, cold air flowing down high, steep cirque walls gathers on the floors, and its drainage is constricted by narrow gorges down-valley. As a result, subalpine beech forests fail to enter cirques or peter out as mid-slope wedges (Figs 5.5 and 5.6). Similar forest patterns occur in subalpine passes (Wardle 1985).

Compression of altitudinal zones on summits

On summits of hills and mountains, vegetation is somewhat stunted, and species characteristic of the altitude can mingle with others more typical of higher altitudes. In the Coromandel Range, for instance, kauri and rewarewa (*Knightia excelsa*) grow with southern rata (*Metrosideros umbellata*) and *Griselinia littoralis*. Adverse factors include

Fig. 5.6. Upper limit of *Nothofagus menziesii* forest in the Hunter valley, Lakes ER. In the foreground, the limit is depressed towards a valley head, in favour of shrub-heath with mainly *Phyllocladus alpinus*, *Podocarpus nivalis*, *Dracophyllum longifolium* and *D. uniflorum*.

shallow, rocky soils, strong winds, and orographic cloud that deflects sunlight. On rounded, wooded summits the cloud cap, through suppressing evapotranspiration, contributes to waterlogging of soils, yet the vegetation is drought-susceptible (p. 579).

The moisture factor

Precipitation is a major determinant of vegetation, although its effectiveness is greatly influenced by other factors, especially temperature, water-holding capacity of the soil, and wind. Decreasing temperatures with increasing altitude and latitude cause decreasing evaporation, thereby increasing the amount of available water, but this may be offset by an increasing wind-run and decreasing ability of roots to take up water from cold soils. Short vegetation such as grassland is usually densest on shady aspects, because less water is lost through evapotranspiration. In contrast, sunny aspects support more vegetation at very high altitudes, and at lower altitudes favour some plants that are close to their upper limits; concentration of rimu trees on north-facing slopes of mountain valleys is an example.

Deep soils hold more water than shallow soils. Fine-textured soils hold more water than coarse-textured soils although it can be less available to plants. More water is lost as run-off from compact soils on steep slopes than from porous soils on flat surfaces. Wind increases evaporation where humidity is low, but where humid on-shore winds are forced to rise over coastal ranges, the air cools below the dew-point so that orographic cloud forms.

Part of the precipitation falling on vegetation is intercepted by the canopy and lost through evaporation. The rest reaches the ground either as through-fall or stem-flow, but a portion of this evaporates from the soil surface or is taken up by plants and transpired. Only the surplus is available to recharge aquifers and streams. Both gorse (*Ulex europaeus*) and native forest can completely intercept showers of less than 0.5–2 mm, and interception losses in native forest are 26–39% of annual precipitation (Aldridge 1968; Pearce *et al.* 1982). Furthermore, stem-flow can direct as much as 25% of precipitation toward tree bases. Understorey plants may therefore be subjected to severe drought in dry localities or during dry periods. Precipitation is also greatly concentrated where tall plants intercept driving rain, fog or snow; over a 5-month period on Mt Cargill near Dunedin, *Chionochloa rigida* tussocks trapped five times as much water as nearby rain gauges (Rowley 1970).

Despite these complexities, it is useful to distinguish four moisture zones. *Humid* climates with precipitation between 1000 and 2500 mm are the most prevalent, but western parts of the South Island and some high ground in the North Island are *superhumid*. In both zones, structurally complex forests are the original vegetation on maritime lowlands, and grade into simpler beech or conifer/broad-leaved communities in cooler or more inland districts. However, forests in the superhumid zone have the greatest abundance of epiphytic and terrestrial bryophytes, and because of soil leaching tend to be dominated by less demanding species such as rimu and kamahi, which in turn are replaced by heath communities on the least fertile soils.

Subhumid conditions occur in the east of the two main islands. The transition from humid lies near the 1000 mm isohyet but may exceed 1200 mm in areas east of the Southern Alps subject to föhn winds. Although forest is the original vegetation, fire has long since led to its replacement by grassland and scrub except for remnants in moist, sheltered habitats. Dominant trees include matai (*Prumnopitys taxifolia*) and totara (*Podocarpus totara*) in lowland remnants, and mountain totara (*P. hallii*), *Griselinia littoralis* and mountain beech at higher altitudes.

Only intermontane basins in Central Otago, southern Canterbury and Marlborough are *semi-arid*, with rainfall < 500 mm. The primitive vegetation was probably mostly grass- and shrub-dominated, and has been severely degraded, especially on sunny aspects and dry terraces. Even here, charcoal, logs and remnant trees are evidence of former woodlands of *Phyllocladus alpinus*, kanuka and other small trees.

Precipitation increases dramatically with rising altitude. On the western flanks of the Southern Alps, where it exceeds 10 m annually, torrential rain combines with tectonic activity to make erosion a major disruptive factor, whereas fire is most important on drier mountains. Snow avalanches occur on high mountains throughout, but most frequently along and west of the alpine Divide. However, within intact vegetation west–east differences are less obvious than at lower altitudes, as even in the driest districts penalpine and alpine soils are nearly always moist.

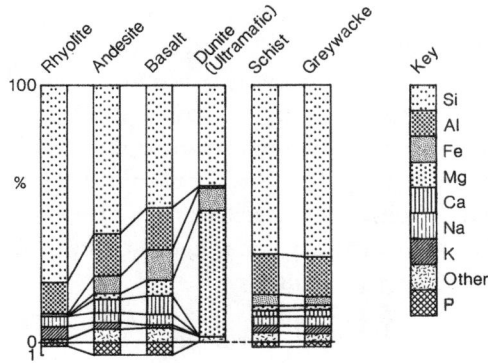

Fig. 5.7. Composition of selected rocks, as percent weight of oxides. 'Other' includes phosphorus, which is also shown on a wider scale below the horizontal axis (from Challis 1971; Williamson 1939).

Rocks and soils
(fuller discussion in Molloy 1988)

Rock types

Primary, or igneous, rocks form a sequence, beginning with *acidic* rocks in which silicon is dominant, and progressing towards *mafic* rocks as the content of iron, magnesium and calcium increases (Fig. 5.7). Granite and rhyolite, diorite and andesite, and gabbro and basalt are plutonic and volcanic pairs in this sequence. Beyond this, there are *ultramafic* rocks rich in iron and magnesium, as well as containing elements such as nickel, chromium and cobalt.

Because plutonic rocks have cooled slowly, their elements are combined into minerals that are wholly crystalline; and soil fertility is influenced, not only by composition of these minerals, but by the readiness with which they weather. In acidic rocks, most of the silica is in the form of quartz crystals, which are almost devoid of nutrients and highly resistant to weathering, although silica is dissolved very slowly by water, especially in warmer climates. Feldspars are aluminium silicates that contain potassium and sodium (especially in acidic rocks) and calcium (especially in intermediate and mafic rocks), and weather more readily than quartz. Mafic rocks have silicates of magnesium, iron, calcium and aluminium, whereas ultramafic rocks consist largely of olivine, a silicate of iron and magnesium that weathers readily. Micas, which are aluminosilicates of various metals, are common in most igneous rocks, and also weather readily. Apatite supplies phosphorus (p. 87), and minor minerals supply trace elements.

Volcanic rocks originate as lavas, as ignimbrites transported in extremely hot gaseous avalanches or pyroclastic flows and which are often welded, and as air-fall 'ash' and pumice, i.e. *tephra*, which is unconsolidated. Because they have cooled quickly, volcanic rocks consist largely of amorphous, undifferentiated glass; in andesite and rhyolite, glass predominates.

Sedimentary rocks consist of fragments and mineral particles derived from older rocks, or of biological detritus as in limestone and coal. Weathering of rocks such as calcareous mudstones provides fertile soils, in contrast to soils derived from quartz sands. The indurated sandstones known as greywacke contain abundant feldspar and have average compositions approximating to those of granite and diorite. Schists have a similar chemical composition but during metamorphosis chlorite and mica crystals are formed; as these weather readily, the fertility of soils derived from schist is initially greater than those derived from greywacke. Both schist and greywacke formations contain strata derived from sedimentary and igneous rocks other than sandstones, and these give rise to soils of differing fertility.

Soil horizons

Transformation of rocks into soil is accompanied by incorporation of organic matter, mostly resulting from death and decay of plants. *Litter* falling on the surface is converted by soil organisms into humus; litter and humus constitute the *O horizon*. *Mull* humus is produced by litter rich in nutrients, mainly in climates that are warm and not excessively wet, under broad-leaved trees characteristic of richer soils, such as puriri and mahoe (*Melicytus ramiflorus*), and under low-altitude grasslands. It breaks down rapidly and the residue is incorporated into the upper or *A horizon* of the mineral soil, chiefly by earthworms, to form granular or nut-like aggregates. Mull grades through *moder* into *mor*, which derives from litter low in nutrients, characterises infertile soils under cool, wet climates, and accumulates as a discrete surface layer. Mor humus is formed most copiously by trees that grow rapidly in poor soils, such as native beeches and exotic pines. Cones of litter and mor humus over 1 m thick build up around the boles of large kauri trees.

The rate of breakdown of humus is influenced by physical and chemical composition of the litter, moisture confent, aeration, and temperature. Its decay produces humic colloids and acids that have considerable effects on soil development. Especially under waterlogged conditions, decay may be so slow and incomplete that humus accumulates as peat, which can become so thick that no plant roots penetrate to the mineral soil beneath.

Weathering of rocks is effected by water, oxygen, carbon dioxide and humic acids. Physical comminution produces sand and silt. Chemical alteration, especially of mica, chlorite and feldspar, produces clays such as vermiculite, kaolin and allophane. Iron released by weathering forms complex precipitates that impart the yellow-brown hues to many subsoils or *B horizons*. Clays retain cations and anions on charged loci, and exchange them with plants via the soil water. Ions not taken up by plants may be displaced from the clay lattices by hydrogen ions and lost through leaching. However, phosphate ions may become strongly bonded to clay, i.e. *occluded*. Allophane is so fine-textured as to seem amorphous, and has especially strong affinity for phosphorus; it predominates among clays formed from volcanic glass, and can also be abundant in soils derived from greywacke.

Under the influence of high rainfall, forest cover and acidic humus, clays release iron and aluminium ions as pH falls below 4.8–4.5. Together with colloidal humus, these ions are leached out of the *E* (eluvial) *horizon*, which develops between the A and B horizons. E horizons are dominated by quartz and clays such as kaolin and biedellite,

P
(ppm)

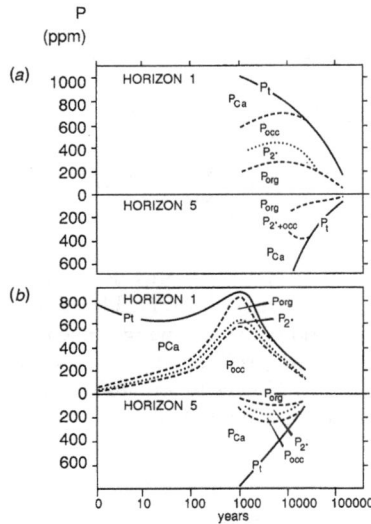

Fig. 5.8. Changes in forms of phosphorus in soils of increasing age and profile differentiation, in two horizons of unspecified depth (from Walker & Syers 1976). (a) A flight of alluvial terraces near Reefton (North Westland). (b) Outwash surfaces down-valley from Franz Josef Glacier (Central Westland). Phosphorus fractions: P_t, total; P_{Ca}, calcium-bound; P_{occ}, occluded; P_{2°, secondary (i.e. Al- and Fe-bound); P_{org}, organic.

have reduced capacity to retain nutrients, are dense and impermeable, and are subject to waterlogging and poor aeration (Park 1972). Some soils formed under kauri have thick, rock-hard pans dominated by silica.

Much of the colloidal ferric iron and humus is precipitated in the B horizon, where they can constitute iron-humus pans that impede drainage, so that the horizons above become *gleyed*, i.e. subject to alternating reducing and oxidising conditions. Further precipitation of iron occurs in the weathering parent material, or *C horizon*, which can also become cemented.

Weathering mantles or *regoliths* several metres thick have developed on very old, stable land surfaces in New Zealand, which can date back to the Tertiary. The soils, which are characterised by high concentrations of kaolin and oxides of iron and aluminium relative to silica, are most extensive in Northland but occur as far south as Stewart Island.

Nutrient ions

Phosphorus is the most universally limiting nutrient for plant growth, and sequential changes in vegetation are often related to its availability (Fig. 5.8). Calcium-bound phosphorus released from apatite during rock weathering is soluble. Some is taken up directly by plants, and some is converted to less soluble aluminium- and iron-bonded phosphorus. There is steady loss through leaching, and an increasing proportion of the remainder becomes occluded and thereby inaccessible to plants. Cations and sulphate ions released through weathering are also leached unless taken up by plants, except in

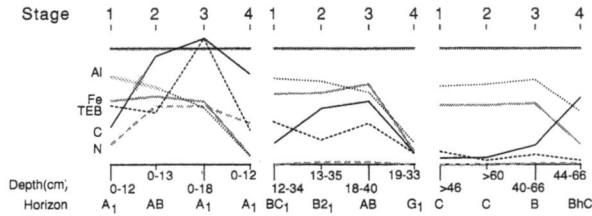

Fig. 5.9. Changes in soil chemistry at three depths, across a flight of alluvial terraces near Reefton, North Westland (from Ross *et al.* 1977). The lower horizontal line shows zero concentrations, and the upper shows the following concentrations: C, 10%; N, 1%; Al, 10%; Fe, 5% (all as g/100 g oven-dry soil); TEB (total exchangeable bases; 5 meq/100 g soil). Stages: (1) Recent; (2), (3) Holocene and Last Glaciation, both yellow-brown earths; (4) Penultimate Glaciation, gley podzol.

semi-arid regions, where evaporation may draw solutes up to the surface, to reach concentrations that are toxic to most plants. However, potassium and sodium may be replenished from the sea, *via* precipitation. The ultimate source of nitrogen is the atmosphere, but most higher plants can absorb it only as soluble ammonium, nitrate or nitrite ions. These are synthesised during electrical storms but the main source is through bacterial decomposition of organic matter and excreta, and ultimately, the atmospheric nitrogen fixed by blue-green algae and anaerobic bacteria. Where phosphorus is readily available, as in recent soils and fertilised pastures, legumes and certain other higher plants can fix large quantities of nitrogen through microbial symbiosis (p. 496).

In superhumid climates, nutrients held in the soil organic matter become increasingly important, as those associated with the mineral fractions become leached or occluded. However, they become available to plants only as breakdown of organic matter leads to mineralisation, i.e. release into the soil solution as inorganic ions. Where fire consumes vegetation and the litter layer, rapid mineralisation of nutrients may temporarily stimulate plant growth, but the ultimate effect on podzols and other infertile soils is loss of the nutrient pool and degraded vegetation, as in pakihi and gumland.

Soil developmental sequences

Recent soils become more fertile and retentive of water and nutrients as weathering and incorporation of humus proceed (Fig. 5.9). The rate of improvement depends on the parent material (e.g., soft rocks > hard, calcareous rocks > siliceous, schist > greywacke), and can be rapid in mild, humid localities where legumes or other nitrogen-fixing plants are vigorous.

Yellow-brown earths, characterised by the colour of the well-developed B horizon, are the next stage where humid climates or low water-retaining capacity promote leaching. These soils already show loss of soluble bases and occlusion of phosphorus. Under moderate to high rainfall, podzolised yellow-brown earths develop as the bleached E horizon becomes differentiated. Eventually, formation of a cemented iron–

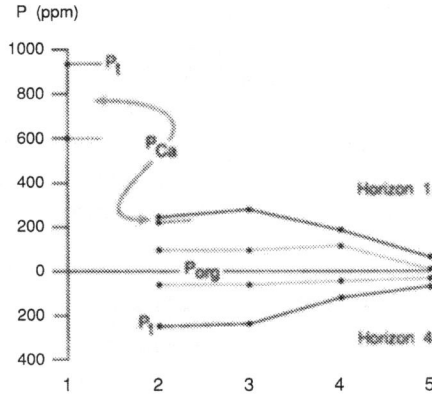

Fig. 5.10. Changes in phosphorus concentration along a catena at Kaiteriteri, North-west Nelson (Walker & Syers 1976). (1) Parent granite at 6 m; (2) steepland; (3) hill; (4) ridge top; (5) deeply weathered ridge top. P fractions as in Fig. 5.8. Horizon depths not specified.

humus pan in the B horizon gives rise to gleyed podzols, which are extensive on high gravel terraces in the west of the South Island.

The corresponding sequence in the central North Island is raw volcanic material→yellow-brown pumice soils→yellow-brown loams→brown granular loams or brown granular clays. In yellow-grey earths, developed under subhumid climates, there is less loss of bases from the upper horizons, the yellow-brown Fe–Al–humus complexes are not formed, and wetting and drying lead to dense B horizons. In brown-grey earths, developed under semi-arid conditions, bases are not leached, and clay pans form.

In localities where annual precipitation exceeds 10 m even recent soils may be extremely low in nutrients (Basher et $al.$ 1985). These soils nevertheless support species characteristic of relatively fertile soils under lower rainfall, which suggests that the markedly different heath vegetation on the least fertile gley podzols may be due, not to low nutrient status per se, but to ionic imbalance, such as excess H^+ or Al^{3+} ions. Herein also may lie the clue as to why gley podzols and ultramafic soils support such similar vegetation, despite their very different pH and cation status.

The more mature phases of soil development are best seen on flat or gently sloping ground with good vertical drainage. On hilly country, podzols may occupy crests of ridges and spurs, whereas on slopes and in gullies lateral seepage, rejuvenation of profiles by erosion, or downslope drift of soil material maintain immature profiles and higher fertility levels (Fig. 5.10). Such topographic soil gradients, or $catenas$, are reflected in the vegetation; for example, tall, slow-growing conifers may occupy ridges, whereas small, fast-growing, broad-leaved trees may occupy intervening depressions.

Much New Zealand terrain is exceedingly broken and unstable, and landslides are frequent and extensive. These provide two main kinds of surface, i.e. the bedrock or sole exposed in the upper part of the slip and the debris that piles up below. Except on the most massive landslides, the sole usually consists of shattered rock that has already undergone considerable weathering. Likewise, landslide debris contains a proportion of weathered rock as well as dislodged soil, plant remains and, often, rafts of living vegetation.

The vigour and completeness with which vegetation colonises fresh surfaces decreases with increasing altitude and exposure, so that increasing areas are occupied by rocky surfaces, such as scree, which are at early stages of soil development and yet may be thousands of years old (p. 406).

Soil fertility classes

It is scarcely possible at present to correlate natural vegetation with direct measures of soil fertility, except in special situations such as saline soils. Correlations with soil morphology are better, but seldom unique; for instance, pakihi communities of almost constant species content can be found on soils of quite different morphology, but which share low nutrient status, high or perched water tables, and shallow rooting zones (Mew 1983). The composition and vigour of natural vegetation are themselves good indicators of fertility, and on this basis, soils are ranked into broad fertility classes as follows.

1. Naturally *fertile* soils are of limited extent, examples being recent, deep, fine-textured alluvial soils, and colluvial soils derived from basalt or calcareous rocks. They can support arable farming and highly productive pastures, but native species tend to be excluded by adventive competitors adapted to a vigorous regime of phosphorus uptake, nitrogen fixation and nutrient recycling. Through pastoral management this regime has been extended over large areas of lower inherent fertility.

2. *Semi-fertile* soils include most of the younger or less-leached soils and support the more demanding kinds of native vegetation, such as lowland forests dominated by matai and totara, red beech stands, seral bush, dense bracken, and grasslands of silver tussock (*Poa cita*) or *Chionochloa* 'robust'. Less demanding pasture grasses such as cocksfoot (*Dactylis glomerata*) and Yorkshire fog (*Holcus lanatus*) thrive on these soils.

3. *Infertile* soils are strongly leached. Examples of native vegetation are forests of kauri, rimu and hard beech, manuka shrubland, and *Chionochloa flavescens* grassland. Until fertilisers are applied, pastures are lean and dominated by low-producing grasses, especially danthonias (*Rytidosperma* spp.) and browntop (*Agrostis capillaris*).

4. *Ultra-infertile* soils are intensely leached, effectively shallow because of impervious pans or underlying rock, and support heath communities. Over large areas, infertile soils have been reduced to ultra-infertile soils through burning the original forest and organic horizon.

5. Soils derived from *ultramafic* rocks contain large proportions of magnesium and unusual metallic ions, including nickel and chromium, which can be toxic. These soils support heathlands very similar to those on ultra-infertile soils, but there are unique taxa.

6. *Saline* soils occur in coastal and semi-arid localities. Plants show features such as succulence to cope with excess of sodium chloride.

7. *Enriched* soils occur where high concentrations of animals, e.g. in sheep camps or bird colonies, result in levels of nutrients, especially nitrogen, that are toxic to most

plants, but allow vigorous growth of a tolerant few, such as nettles (*Urtica*) or *Lepidium oleraceum*.

8. *Raw* soils consist of unweathered parent material such as screes and fresh alluvium. They are initially barren, but fertility rises as they weather and plant succession proceeds.

9. *Peats* range from swamp peats, which occur in low-lying areas throughout New Zealand, to the blanket peats that cover most terrain in the Chatham and far-southern islands. They vary from fertile to ultra-infertile, according to whether they are irrigated by nutrient-rich ground water, or nourished purely by rain. Southern coastal peats can have their fertility enhanced by salt spray; near-saturation probably offsets the effects of NaCl.

Vegetation processes

The vegetation at any given place also represents a point of time in the interactions among available biota, physical and biological disturbances that vary in severity, frequency and duration, vagaries of weather, and secular changes in climate. *Succession* is the process of revegetation after disturbance; complete successions begin on bare surfaces. Usually, the pioneering species form open communities that, except in the harshest environments, are soon replaced by denser, taller communities dominated by different sets of species. In the absence of further massive disturbances or environmental changes, steady-state or *climax* vegetation eventually ensues, which may change locally in time and space but remains uniform overall. Especially under high rainfall and prolonged soil leaching, the climax may ultimately *retrogress* to low, open, slow-growing vegetation. Successions are *primary* if they begin on unweathered parent material, and *secondary* if they begin on soil developed under preceding vegetation. *Regeneration* is the process whereby species maintain their position in a community; the death of individual plants is compensated by the establishment of seedlings or by vegetative reproduction. These concepts are discussed more fully in Chapters 15–16.

6

BOTANICAL PROVINCES

Primarily with the aim of identifying nature conservation needs, New Zealand has been divided into 85 ecological regions (ERs) (Fig. 6.1) and 268 ecological districts (EDs) on the basis of land form, geology, climate and biological content. Larger botanical units can also be defined, which Cockayne (1928) formalised as *provinces*; I have adopted these, with some modification of names and boundaries to accommodate the ecological regions.

1. *Northland* province has rocks ranging from ultramafic to siliceous, from igneous to sedimentary, and from Permian to Recent. Excepting the Kermadec Islands, it is the warmest part of New Zealand. Most of the province is a hilly peninsula indented by wide, shallow inlets, those on the west being almost land-locked by high dunes, whereas those in the east are open to the sea. The primitive vegetation was nearly all warm-temperate forest, with kauri (*Agathis australis*) dominating on the uplands. Although some large blocks of kauri forest remain, most has been cleared to gumland scrub, and that in turn to pasture (Fig. 6.2). The more fertile basins are now mostly given to dairy farming. Bush and low vegetation were extensive on coastal headlands, and still are on off-shore islands.

Several lowland species, including *Ackama rosifolia* and *Cassytha paniculata*, are widespread in the province but do not occur further south. Endemics restricted to off-shore islands include *Tecomanthe speciosa* on the Three Kings Islands, *Meryta sinclairii* on Three Kings and Taranga Island and *Xeronema callistemon* on the Poor Knights and Hen Islands, these being the sole New Zealand representatives of their genera. Other species endemic to the Three Kings are *Carex elingamita, Myrsine oliveri, Pennantia baylisiana, Coprosma macrocarpa, Hebe insularis* and *Elingamita johnsonii*, a tree which represents a monotypic genus confined to West Island. Off-shore islands also support large-leaved forms, some recognised as distinct taxa, e.g. *Streblus smithii, Alectryon grandis* and *Brachyglottis repanda* var. *arborescens* of the Three Kings, and the more widespread *Macropiper excelsum* var. *psittacorum*.

Te Paki ER has the few known trees of the white-flowered rata *Metrosideros bartlettii* and several taxa confined to the ultramafic outcrop at North Cape (p. 397).

Chionochloa bromoides grows on coastal cliffs of eastern Northland. Although the highest hills of western Northland do not exceed 776 m, they support several southern trees (p. 140), as well as the endemic shrub *Coprosma waima* (Druce 1989).

2. *Auckland* province contains steeply rolling country north of the Auckland isthmus, and the gently rolling Waikato basin to the south. Blocks of rugged hills, from the Waitakere Range in the north to the Herangi Range in the south, abut on western coasts. In the east, steep ranges of the Coromandel region form a striking boundary rising to 963 m in Mt Te Aroha. There are also many islands, Great Barrier being the largest.

Most of the higher ground consists of ancient volcanic ranges, whereas Quaternary sediments fill the depressions. Peat is extensive in Waikato ER, and limestone in Tainui ER. Well-preserved late Tertiary and Quaternary volcanic cones are scattered throughout; Rangitoto Island is only a few centuries old. The primitive vegetation was mainly warm-temperate forest, but bracken (*Pteridium esculentum*) and wetlands predominated when European settlement began. Many warm-temperate species reach southern limits in the province, including mangrove (*Avicennia resinifera*) at Raglan Harbour and kauri near Kawhia (Fig. 5.3). Conversely, summit vegetation includes silver (*Nothofagus menziesii*) and red (*N. fusca*) beeches on Mt Te Aroha, *Phyllocladus alpinus* and kaikawaka (*Libocedrus bidwillii*) on Te Moehau (892 m), and many other southern plants.

Trilepidea adamsii is the only notable endemic (p. 6). *Pseudopanax laetus* is characteristic of the province, but its range extends a little further south and east. *Brachyglottis myrianthos* is almost restricted to the Coromandel Peninsula, where it is abundant in broken forested terrain, but there are outlying populations, one as distant as Puketi Forest in Northland.

3. *Volcanic Plateau* province is thickly blanketed with Quaternary volcanic deposits. It consists of plateaus that exceed 1000 m in the south, intervening depressions, and the Bay of Plenty lowlands. Tongariro ER contains three active volcanoes, including Ruapehu, at 2797 m the highest mountain in the North Island. Central Volcanic Plateau includes Lake Taupo, which is the largest lake in New Zealand and the centre of cataclysmic eruptions. Further north, Rotorua ED abounds in hot pools and mud craters (pp. 307 and 400) and was the scene of the devastating Tarawera eruption of 1886. Tephra plains abut on axial ranges in the east, and Jurassic greywackes crop out in the Hauhungaroa Range in the west. There are also off-shore volcanic islands, including the active White Island.

The original vegetation along the Bay of Plenty was much the same as on the Auckland lowlands, with taraire (*Beilschmiedia tarairi*), broad-leaved tawa (*B. tawaroa*) and mangrove among species at or near their southern limits. Inland regions are frosty, and such characteristic warm-temperate plants as nikau palm (*Rhopalostylis sapida*), *Freycinetia baueriana* and pukatea (*Laurelia novae-zelandiae*) are absent, except in favoured localities. Magnificent podocarp forests were widespread into the twentieth century, although there were also extensive bracken and scrub as a result of eruptions and Maori fires. Until recently, the provincially endemic shrub *Dracophyllum subulatum* covered extensive areas on frosty plains.

In the south and east, tussock grassland and associated wetlands covered large areas.

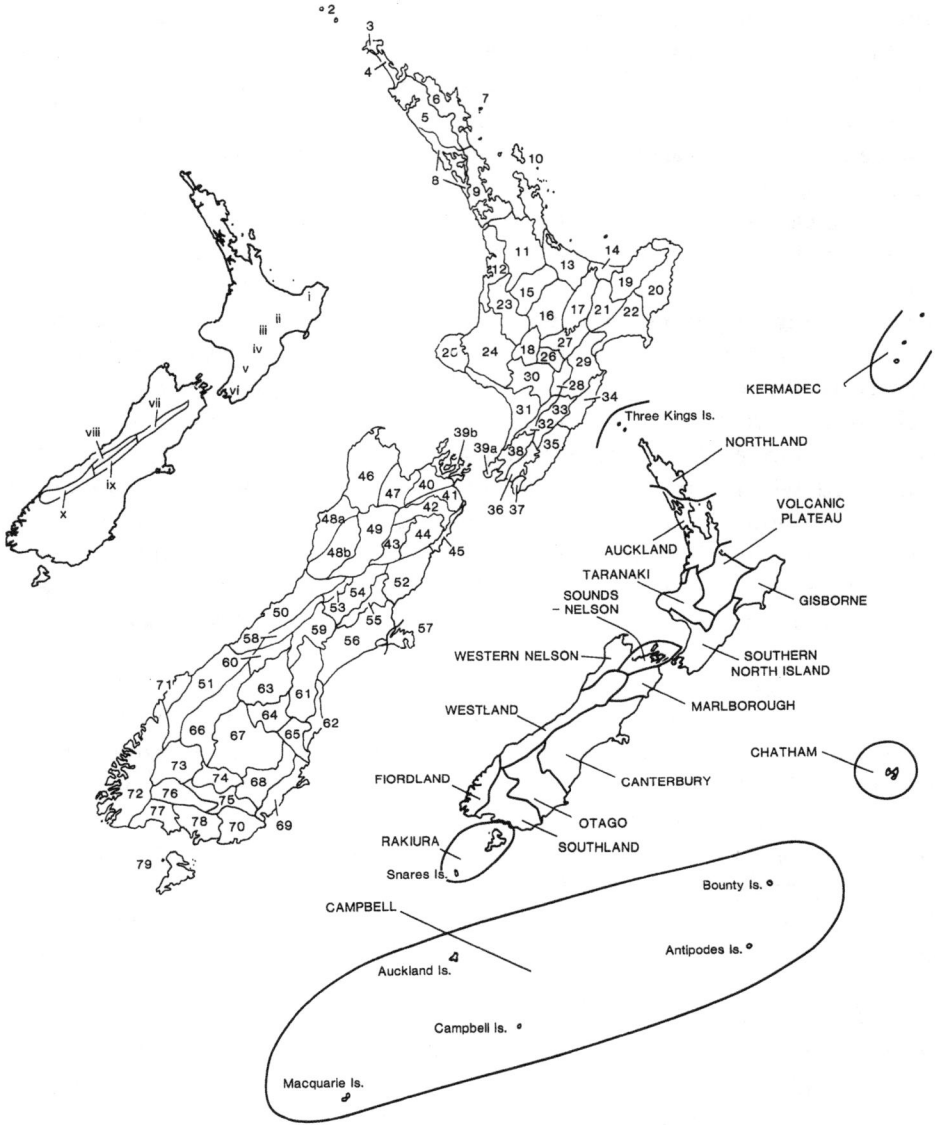

Fig. 6.1. Botanical provinces and regions. Provincial boundaries (right). Axial ranges (left), numbered as follows: i, Raukumara; ii, Huiarau; iii, Kaimanawa and Kaweka; iv, Ruahine; v, Tararua; vi, Rimutaka; vii, North-eastern Alps; viii, Western Alps; ix, Eastern Alps; x, South-western Alps.
Ecological regions of mainland New Zealand (centre) (MacEwan 1987):
2, Three Kings; 3, Te Paki; 4, Aupouri; 5, Western Northland; 6, Eastern Northland; 7, Poor Knights; 8, Kaipara; 9, Auckland; 10, Coromandel; 11, Waikato; 12, Tainui; 13, Northern Volcanic Plateau; 14, Whakatane; 15, Western Volcanic Plateau; 16, Central Volcanic Plateau; 17, Eastern Volcanic Plateau; 18, Tongariro; 19, Raukumara; 20, East Cape; 21, Urewera; 22, Wairoa; 23, King Country; 24, Eastern Taranaki; 25, Western Taranaki (Egmont); 26, Moawhango; 27, Kaimanawa; 28, Ruahine; 29, Hawkes Bay;

Fig. 6.2. Pasture with patches of native bush, including a kauri tree in foreground. Near Warkworth, Northland (photo A.D. Thomson).

They contain many species otherwise found only in the South Island; 17 of these, e.g. *Ranunculus recens*, *Euphrasia disperma* and *Tetrachondra hamiltonii*, are known in the North Island only from Moawhango ER (Rogers 1989). In contrast, the vegetation on the upper slopes of the high volcanoes is sparse and floristically poor (Fig. 6.3).

Today, the province contains the country's largest conifer plantations, but since the 1930s when the cobalt deficiency problem was solved, pastoral farming has increased to cover the greatest area. Beginning with the 'kiwi fruit' (*Actinidia*) boom of the 1970s, horticulture has become important in the Bay of Plenty.

4. *Gisborne* province consists of steep ranges and narrow, deep valleys (Fig. 6.4) with a few pockets of gentle terrain along the coast. Tephra erupted from Lake Taupo covers large areas. Raukumara and Urewera ERs west of the axial ranges are underlain by Mesozoic greywackes, whereas East Cape and Wairoa consist mainly of indurated

---

30, Rangitikei; 31, Manawatu; 32, Manawatu Gorge; 33, Pahiatua; 34, Eastern Hawkes Bay; 35, Eastern Wairarapa; 36, Wairarapa Plains; 37, Aorangi; 38, Tararua; 39a, Wellington; 39b, Sounds; 40, Richmond; 41, Wairau; 42, Inland Marlborough; 43, Molesworth; 44, Clarence; 45, Kaikoura; 46, North-west Nelson; 47, Nelson; 48, North Westland; 49, Spenser; 50, Central Westland (Whataroa); 51, Aspiring; 52, Lowry; 53, Hawdon; 54, Puketeraki; 55, Canterbury Foothills; 56, Canterbury Plains; 57, Banks; 58, D'Archiac; 59, Heron; 60, Tasman; 61, Pareora; 62, Wainono; 63, Mackenzie; 64, Waitaki; 65, Kakanui; 66, Lakes; 67, Central Otago; 68, Lammerlaw; 69, Otago Coast; 70, Catlins; 71, Olivine; 72, Fiord; 73, Mavora; 74, Waikaia; 75, Gore; 76, Southland Hills; 77, Te Wae Wae; 78, Makarewa; 79, Rakiura.

Fig. 6.3. The volcano Ngauruhoe (2287 m). In the foreground at 1350 m, bands of red tussock/*Rytidosperma setifolium* on either side of a sand flat with *Raoulia albosericea*. The lava flow at the base of the cone predates arrival of Europeans, but on the flanks flows are as recent as 1954.

Tertiary marine sediments, including slip-prone mudstones. Inland, warm-temperate mixed forests rise to montane beech forests. Nearer the coast, most of the land has been cleared to pasture, which has been savaged by erosion during cyclonic storms.

Endemic species include *Coriaria pottsiana*, the coastal *Brachyglottis perdicioides* and, by virtue of the demise of Northland populations, *Clianthus puniceus*. The ranges of *Chionochloa flavicans* and *Jovellana sinclairii* are centred on the province. *Ixerba brexioides* and *Phyllocladus glaucus* find their southern limits in the Gisborne ranges, whereas pohutukawa (*Metrosideros excelsa*) and puriri (*Vitex lucens*) extend a little further south on the Taranaki coast.

5. *Taranaki* province contains two very different landscapes. The western region contains the beautifully symmetrical volcano Mt Taranaki (2518 m), its ring-plain, and the Pouakai and Kaitake Ranges to the north, which are stumps of earlier volcanoes. Originally clothed in dense conifer/broad-leaved forest up to the tree limit, except for Maori clearings along the coasts and valleys, it now supports intensive dairy farming up to 450 m above sea level, the lower boundary of Egmont National Park.

The regions to the east include 'papa', or soft Tertiary mudstone country of deep, sub-parallel valleys and precipitous slopes. The natural vegetation is mostly warm-

Fig. 6.4. Bush and successional vegetation on steep walls, and warm-temperate forest on ridges in the Motu gorge, Gisborne province.

temperate forest, with patches of black (*Nothofagus solandri*) or hard (*N. truncata*) beech on sandstone ridges. Maori settlements had penetrated the larger valleys, and the inherently fertile soils have led to pasture establishment on relatively broad valleys and ridges of the King Country and Rangitikei ERs. The region between is characterised by deep gorges and narrow ridges that rise to 500–700 m; early twentieth-century attempts at pastoral development have largely been abandoned except around its margins (Fig. 6.5).

Papa gorges support *Brachyglottis turneri*, the province's most distinctive endemic. An uncommon rata vine, *Metrosideros carminea*, reaches its southern limit on the northern Taranaki coast.

6. *Southern North Island* is bisected by the axial ranges. These intercept heavy rainfall and are still largely clad in mixed forests at lower altitudes, and beech at higher altitudes except on either side of the Manawatu Gorge. In the rain shadow to the east, most of the forest had been replaced by fern and grass by the time of European settlement. Hawkes Bay ER is rolling country, except for small plains bordering the sea. In Eastern Hawkes Bay and Eastern Wairarapa, hills are generally steeper, and composed of diverse Tertiary rocks. Further south, are the now-agricultural Wairarapa Plain and the still-forested greywackes of the Aorangi Range.

Manawatu ER consists of the recent alluvial plains of the Manawatu and smaller rivers, the largest expanses of sand dune in New Zealand, and lower Quaternary marine strata separated from those of Hawkes Bay by the early Quaternary uplift of the axial ranges. The vegetation was a mosaic of swamp, forest and grassland at the time of

Fig. 6.5. Forest remnants and regrowth on steep valley sides, high terraces and ridges cleared to pasture, bracken in the foreground. Manganui-a-te-ao valley, Taranaki ER.

European settlement, but is now mainly rich dairy pasture. Wellington ED takes in the south-western spurs of the axial ranges, and its remaining native vegetation is mainly mixed forest.

The few species endemic to Southern North Island province include *Chionochloa beddiei* and *Brachyglottis compacta* on coastal cliffs in the south-east, and *Hebe evenosa* in the Tararua Range. *Elatostema rugosum* and *Mida salicifolia* are among the few species with southern limits in the province.

7. Crests of the *axial ranges* that rise above the subalpine forest limit, from Mt Hikurangi in the north to the Rimutaka Range in the south, have more in common with each other than the provinces they lie in. The prevailing vegetation is grassland dominated by *Chionochloa pallens* and scrub of *Olearia colensoi* and *Dracophyllum filifolium*. The Tararua Range supports *Dracophyllum uniflorum*, *Brachyglottis adamsii* and *Raoulia rubra*, which are otherwise South Island mountain plants, whereas *Dracophyllum recurvum* grows only north of the Manawatu Gorge. *Ourisia vulcanica* and *Parahebe spathulata* extend from the central volcanoes to adjacent axial ranges. The depth and extent of tephra increases towards the Volcanic Plateau, and in the Kaimanawa Range penalpine grasslands merge into those of the high tephra plains.

8. *Sounds–Nelson* province supports extensive beech forests, mixed forests being mainly confined to lower-altitude, moderately fertile sites, especially near the coast. Sounds ED is a complex of drowned valleys, narrow ridges and islands carved in greywacke and schist. Much of its forest was cleared by Maori; European settlers

Fig. 6.6. Barren ultramafic terrain of Red Hills in the foreground; behind, Permian argillites of the Gordon Range with partly eroding penalpine grassland (mainly *Chionochloa australis*) above mountain beech (*Nothofagus solandri* var. *cliffortioides*) forest. Sounds–Nelson province.

cleared still more, but as farming has been only marginally economic, large areas are reverting to scrub and bush. Cook Strait ED encompasses the islands beyond the shelter of the Sounds, including Kapiti and Mana. Richmond ER mainly consists of the Richmond Range, which rises to penalpine ridges. It is structurally continuous with the Sounds, but separated from Marlborough province by the Wairau Fault. The beech forests of the slopes and valleys are still largely intact; maritime influences are less evident than in the Sounds, especially on the southern flanks where hard beech and tawa (*Beilschmiedia tawa*) are lacking. The penalpine crest of the Richmond Range supports the locally endemic *Celmisia macmahonii*, *C. rutlandii*, *C. cordatifolia* and *Hebe gibbsii*; the first two celmisias occur also on Mt Stokes (1204 m), the highest point in the Sounds.

Nelson ER mainly comprises the low country bounded by Tasman Bay in the north and by mountains in the west, south and east, but it is convenient to include the eastern ranges that contain the Nelson 'Mineral Belt' with its distinctive ultramafic vegetation and endemic taxa (Fig. 6.6). Otherwise, the region is largely covered by Lower Pleistocene gravels that rise to 1015 m in the south, and slope northwards as rolling ridges separated by alluvial valleys. The forests, chiefly of beech, have been receding southwards and upwards since Maori settlement, and the trend continues today. Now the valleys are mainly given over to horticulture, and the gravel hills support ever-expanding pine forests.

9. *Western Nelson* province, consisting of North-west Nelson ER and part of North

Fig. 6.7. Penalpine karst landscape of Palaeozoic marble, Mt Owen, North-west Nelson; *Hoheria lyallii* trees on the debris slope below the cliffs.

Westland, is a landscape of deep, narrow valleys and gorges and broad, rolling summits that generally do not exceed 1500 m. Although granites and related metamorphic and igneous rocks predominate, sedimentary rocks range from Cambrian to Recent, and include calcareous and siliceous strata. Pleistocene glaciation was localised, but periglacial regolith is extensive.

Arthur ED in the north-east includes Ordovician marble, which forms the highest summits of the province (Mt Owen 1875 m, Mt Arthur 1777 m) and is penetrated by the deepest caves known in the Southern Hemisphere (Fig. 6.7). Wangapeka ED forms the rugged core of North-west Nelson, whereas Heaphy and Ngakawau EDs consist largely of high coastal peneplains. The Paparoa Range has an axis of granite and gneiss flanked by sedimentary rocks that include limestones and coal measures. The province is terminated in the north by Farewell Spit.

Although beech forests are the main vegetation, luxuriant warm-temperate forests clothe the lower slopes, and enclaves of montane and subalpine forest without beech occur in the south. There are also extensive heaths on ancient terraces and peneplains, and penalpine grasslands. Significant areas of farmland exist only within the Golden Bay, Karamea (Fig. 6.8), and Foulwind districts; sawmilling and mining for coal, gold and other minerals have been the main economic activities.

Western Nelson is floristically the richest part of New Zealand, partly on account of numerous endemics such as *Pseudowintera traversii, Clematis marmoraria, Aciphylla hookeri, Brachyglottis hectorii, Celmisia dallii* and *Astelia skottsbergii*, but also because

Fig. 6.8. Nikau persisting in a coastal pasture near Karamea, North-west Nelson.

a considerable number of taxa have disjunct distributions that lie partly in this province (some extending to adjoining provinces) and partly in regions much further north or south. The most ecologically important of these are the beeches. Further examples with gaps across Southern North Island include *Sticherus flabellatus*, *Dracophyllum traversii*, the orchid *Yoania australis*, and several lowland forest species (listed on p. 132). Examples of gaps across the central South Island are mostly high-altitude, e.g. *Celmisia petriei*, *Microlaena thomsonii*, and the vicariant pairs *Brachyglottis adamsii* (Arthur ED to Tararua Range) – *B. revoluta* (southern zone), and *Dracophyllum townsonii* – *D. menziesii*. *Astelia subulata* and *Mitrasacme montana* otherwise occur only south of Foveaux Strait and in Tasmania respectively.

10. *Westland* province is bounded by the sea and the Southern Alps. The terrain east of the Paparoa Range consists of blocks of early Palaeozoic and Precambrian igneous and sedimentary rocks, separated by Quaternary moraines, fluvioglacial gravels and alluvium transported across the Alpine Fault from the Alps. The Maruia valley and valleys of Rotoroa ED (Spenser ER) are confined by broad ridges of the ancient rocks, whereas further south-west a continuous system of fluvioglacial and alluvial terraces occupies the Grey–Inangahua tectonic depression. Beech forests still cover most of the area up to the tree limit, except for recent flats once clad in mixed forests and now mostly cleared to pasture. In Rotoroa ED, where rainfall can be as low as 1500 mm, there are still remnants of short-tussock grassland and matagouri (*Discaria toumatou*) scrub in the wider valleys. In North Westland ER, from the Totara Flat district to the Taramakau River, the terraces contain the sequence from beech dominance in the north to the mixed forests of Central Westland.

Central Westland ER is mantled with debris from large glaciers that flowed from the Alpine divide to beyond the present coastline; ancient granites and greywackes

Fig. 6.9. Silver beech forest on slopes, mainly kahikatea (*Dacrycarpus dacrydioides*) forest enclosing mires on the terrace in the centre. Paringa Valley, South Westland.

crop out only as isolated hills. The piedmont west of the Alpine Fault and the lower slopes of the Southern Alps are clothed in dense mixed forests; however, recent alluvial soils are mostly cleared to pasture, and outwash terraces in the north that have been logged and burnt have developed pakihi. In Paringa ED (Aspiring ER) Palaeozoic rocks again prevail, and there are also late Cretaceous and Tertiary strata along the coast. The northern Paringa hills are mainly in mixed forest, which grades into silver beech forest on the southern hills and adjoining slopes of the Southern Alps (Fig. 6.9). Forest on the alluvial plain of Haast ED is mostly mixed, but in Olivine ER (Fig. 6.10) beech forest is almost ubiquitous, except for coastal bush, and on ultramafic mountains east of the Alpine Fault and moraine plateaus to the west that contain a large proportion of ultramafic boulders, where heathland prevails.

No provincial endemics are reported. Warm-temperate species drop out one by one towards the south, and beeches and a few other species (e.g. *Lindsaea linearis, Hebe gracillima*) are disjunct across Central Westland.

11. *Fiordland* province consists almost entirely of gneiss and granite mountains, with accordant summits that decrease in elevation from 2000 m in the north to 1000 m in the south-west. These suggest a former peneplain, now deeply dissected by hanging valleys and fiords. The original vegetation remains intact, and mostly consists of beech forests, seral communities on precipitous fiord walls (Fig. 6.11), heathlands on shallow soils on weathering-resistant rocks, and subalpine grasslands.

Darran ED in the north has a coastal area of ultramafic rocks supporting heathland. Preservation ED in the south-west is bordered by extensive marine terraces, rising to 500 m above sea level, that also support stunted vegetation including unusual low-altitude tussock grasslands. At the coast this district receives about 1800 mm of rain

Fig. 6.10. Penalpine grassland above subalpine silver beech forest in glaciated landscape of the South-western Alps and Olivine ER.

annually, only a third of that falling on the mountainous hinterland, but it is much windier. Te Anau ED covers the large lakes that feed the Waiau River and the ranges that separate their western fiords, and has outcrops of Tertiary sedimentary rocks, including limestone. Rainfall decreases to 2000 mm on the eastern shore of Lake Te Anau, and the proportion of red and mountain beeches to silver beech is greater than in other districts of Fiordland.

The province contains many penalpine and alpine endemics, including *Pachycladon crenata, Aciphylla crosby-smithii, A. pinnatifida, A. takahea, Celmisia inaccessa, Myosotis* sp., *Hebe pauciflora* and *Chionochloa ovata.*

12. *Marlborough* province lies south of the extension of the Alpine Fault that follows the Wairau Valley. Inland parts consist of high ranges, wide valleys and intermontane basins that lie as much as 1000 m above sea-level, but glaciated topography is well developed only in the west. Rocks are mostly Jurassic greywacke but the Kaikoura Ranges are a Cretaceous complex. Except for remnants of mountain beech forest in the west, mixed forest in the Kaikoura Ranges and extensive rock outcrops and screes, the vegetation now is mostly native grassland. As in all the eastern South Island high country, growing wool is the dominant land use, apart from cattle grazing in the valleys.

In the north and north-east, the high country merges into hills of mainly Tertiary sediments, including much limestone. Generally, rainfall is low (600 mm annually at Lake Grassmere), and this is reflected in short-tussock grasslands and dry pastures

Fig. 6.11. Beech forest interrupted by slips and seral vegetation on Mitre Peak (1694 m), Milford Sound (photo P.N. Johnson).

that have replaced them. However, favoured habitats such as sheltered limestone gorges support mesic vegetation, including many species endemic to the province (p. 385). Mild, moist influences prevail within a narrow strip between the sea and the steep rampart of the Seaward Kaikoura Range. Here, remnants of mixed forest include northern trees such as tawa and *Nestegis cunninghamii*.

13. *Canterbury* province has a hinterland of greywacke mountains separated by glacially carved valleys descending from the Alps, and falling abruptly to eastern hills and plains. Tussock grassland predominates in the hinterland, and there are large screes. In Puketeraki and Hawdon ERs north of the Rakaia River, much of the original beech forest remains, especially in the wetter, western valleys. To the south, however, most of the forest was conifer/broad-leaved, which had been reduced to remnants by early Maori fires. Mackenzie and Waitaki ERs comprise most of the catchment of the Waitaki River, including the wide, high Mackenzie Plains and large glacial lakes in the tributary valleys. They are botanically transitional to Central Otago.

Fig. 6.12. Downlands with sown pasture in the foreground. The slope has an adventive sward, relict *Poa cita* and cabbage trees (*Cordyline australis*), gorse scrub and planted *Pinus*. Wainono E.R., Canterbury.

Lowry ER consists of coastal hills of Mesozoic greywackes and varied Tertiary rocks, and several dry inland basins. Except for arable land, some exotic forestry and rare remnants of the original forest, the vegetation is native grassland and scrub, most of it greatly modified. Some Marlborough endemics extend into this region, and maritime conditions influence the coastal fringe, especially in Hundalee ED where nikau and kiekie occur. The Canterbury Plains consist of intersecting alluvial fans, forming the largest area of flat land in New Zealand. The tussock grasslands, swamps and rare remnants of forest that covered the area at the time of European settlement are now almost entirely displaced by agriculture. The plains either abut directly on the high mountains of the hinterland, or are separated by a belt of hills which still support native grassland and scrub. As rainfall is higher in the bordering steeplands they still carry blocks of native forest, dominated by beech north of the Rangitata River and by conifer/broad-leaved elements to the south.

In the south-east of the province, the ranges and basins of Mackenzie and Waitaki ERs give way to lower hills of Tertiary and Quaternary rocks that include basalt and limestone, which in turn merge into loess-covered downs and strips of coastal plain that constitute the Wainono ER (Fig 6.12). Pasture and arable farming have largely replaced earlier vegetation of tussock grassland and remnants of mixed forest.

Banks Peninsula is an isolated block of steep hills, based on the calderas of two late Tertiary basaltic volcanoes. Except where exposed to north-west winds, it is moister and milder than the adjoining plains, and the original vegetation was mostly luxuriant mixed forest that included several species at their southern limit. e.g. *Alectryon*

Fig. 6.13. West-facing slopes with a fire-induced altitudinal sequence from *Kunzea–Leptospermum* scrub, through bracken on mid-slopes, to subalpine short-tussock grassland grading to penalpine *Chionochloa* grassland; remnants of beech forest in the gullies. The river flats support grassland largely of adventive species. Hunter valley, Lakes ER.

excelsus and *Macropiper excelsum*. Most of this was destroyed in the nineteenth century and replaced by pasture and scrub.

This large province contains relatively few endemic species, the most unusual being the scrambling shrub *Helichrysum dimorphum* in the gorges of the Waimakariri river. Others are *Hebe lavaudiana, H. strictissima, Celmisia mackaui* and *Leptinella minor* on Banks Peninsula, and *Hebe pareora* and *Helichrysum plumeum* on rocky slopes in the south-east.

14. *Otago* province is diverse. The valleys at the heads of lakes Hawea, Wanaka and Wakatipu rise in the Alps and, in contrast to most valleys in Mackenzie ER, have sides clothed in beech forests that are still largely intact. On the slopes surrounding the lakes, however, bracken prevails (Fig. 6.13).

Central Otago consists of high schist plateaus and broad semi-arid basins. Other than the Mackenzie plains and parts of Marlborough, this is the only place in New Zealand where forest was precluded because of low rainfall, but even here treelessness is partly a result of fire (p. 84). The native grasslands are greatly depleted, but productive pastures and stone-fruit orchards have been developed through irrigation.

Fig. 6.14. Salt-pan on the Maniototo Plain, Central Otago, at 400 m; in the distance, schist terrain of Rough Ridge rises to 1040 m.

Patches of halophytes occupy pans where salts leached from Tertiary marine sediments have concentrated (Fig. 6.14).

The semi-arid landscape, despite its uniqueness within New Zealand, supports only a few endemic species. These include the halophyte *Lepidium kirkii*, the former grassland dominant *Elymus apricus*, and *Carmichaelia compacta, Myosotis albosericea* and *Stipa petriei*, which grow on steep rocky slopes, the last occurring also in Waitaki ER. Precipitation rises to almost 2500 cm annually on the plateaus, which support more continuous grasslands and extensive areas of wetland, cushion- and fellfield, and have at least 18 endemic vascular species centred on them, e.g. *Parahebe trifida, Chionohebe myosotoides, Poa pygmaea, Luzula crenulata* and *Anisotome lanuginosa*.

Lammerlaw ER consists of rolling schist hills, mostly less than 1000 m high. The *Chionochloa* grasslands that prevailed until about 1950 are now being rapidly replaced by pasture. Otago Coast ER is a strip of Tertiary hills that includes the extinct volcanoes of the Dunedin district, the largest area of flat land being the flood-prone Taieri Plain. Mixed forest prevailed originally but, despite moderate rainfall and a cool climate, most had been replaced by grassland, fern or scrub before the nineteenth century; the largest remnants are in Dunedin ED.

15. *Southland* province comprises the area south of the Otago schist country and east of Fiordland. In Catlins ER, Jurassic hills are still largely forested in the south-east and in the west supported Maori-induced grassland that has, in turn, been mostly displaced by pasture. The regions bordering Foveaux Strait consist of plains and gently rolling terrain on upper Quaternary sediments, but there are also hills of older rocks, including greywacke, volcanic and igneous rocks and limestone. Their original mixed forests have been reduced to small remnants, but there is still much beech forest in the west, especially on the old marine terraces of Te Wae Wae ER. Otherwise, rich

pastures are the prevailing vegetation. Peat domes, some still holding native vegetation, are a feature.

Inland regions comprise sediment-filled basins separated by ranges of hills. In the west, there are large blocks of beech forest and some of the largest remaining tracts of red tussock grassland; the latter, however, are being steadily converted to improved pasture. Similar grasslands, together with remnants of mainly podocarp/broad-leaved forest, also prevailed in the east, but conversion to pasture and agriculture is more complete, especially in Gore ER.

The high ranges in western Southland consist of greywacke, igneous rocks, and a continuation of the South Westland ultramafic belt. The alpine flora of the Eyre mountains has the distinctive local endemics *Celmisia philocremma* (on exposed rock ridges), *C. thomsonii, C. spedenii* (on ultramafic substrate), *Ranunculus scrithalis* (on scree), *Cheesemania wallii* and *Aciphylla lecomtei* (both extending to the Garvie Range), and *A. spedenii*, which has an outlying occurrence in central Fiordland. Southland ranges also have a vegetable sheep, which may be *Raoulia eximia*, and a disjunct occurrence of *Stellaria roughii*. Wetland turf at low altitudes in the Waiau catchment contains two local endemics, *Brachyscome linearis* and *Iti lacustris* (Garnock-Jones & Johnson 1987). *Tetrachondra* occurs more widely in Southland, as well as Moawhango ED (p. 95).

16. The high-mountain backbone of the South Island separates lowland provinces with very different climate and vegetation, but differences are less apparent above the forest limit. It is therefore convenient to recognise an *Alps* province, comprising terrain above the potential limit of native forest and superimposed on the divisions adopted for lower altitudes. West of the Main Divide, subalpine beech forests meet penalpine grasslands, except in the central portion where beech is absent and the upper subalpine vegetation is dense scrub. The corresponding sequences east of the Divide have been largely destroyed, so that fire-induced grassland on lower slopes usually merges imperceptibly into penalpine vegetation.

There are four sectors in the Alps, differing floristically more than in respect of vegetation. The North-Eastern Alps carry little permanent snow but were formerly heavily glaciated. They rise to 2338 m in Mt Travers, but between Lewis and Arthurs passes no summits exceed 2000 m. Floristically they are linked with the Kaikoura ranges to the east. Many species are centred on these northern ranges, the most conspicuous being *Haastia* vegetable sheep north and east of Lewis Pass and carpet grass (*Chionochloa australis*) which, like *Parahebe cheesemanii*, reaches Arthurs Pass.

In their central portion, the Alps increase in elevation from north to south, from 2271 m in Mt Rolleston to 3764 m in Mt Cook. They carry glaciers throughout their length, and almost continuously so for 130 km southwards from Whitcombe Pass. Uplift along the Alpine Fault is as much as 12 mm annually, but decreases towards the east. As in the north-eastern Alps, there is an eastwards gradient from schists to less metamorphosed greywackes. The Main Divide separates the Western Alps from the Eastern Alps.

The western sector experiences the highest precipitation and cloud frequency in New Zealand, resulting in lower altitudinal belts than in the east. There is marked floristic poverty, with a number of penalpine and alpine species of adjoining sectors being absent or extremely rare; examples with more northern or more southern distributions are cited above and below. *Anemone tenuicaulis, Geum uniflorum, Celmisia laricifolia, C. petriei* and *Carex lachenalii* are present both north and south of the Western Alps, and form part of the pattern of disjunction that includes *Nothofagus* at lower altitudes. *Ranunculus gracilipes, Epilobium tasmanicum, E. pycnostachyum, Gingidia decipiens, Dracophyllum pronum, Celmisia angustifolia, C. viscosa, Haastia sinclairii, Hebe epacridea, H. haastii, Brachyscome* spp., woody raoulias and whipcord hebes are widely distributed in the Alps excepting the western sector. The only typically western species that are rare east of the Divide appear to be *Colobanthus canaliculatus, Celmisia vespertina,* and *Chionohebe ciliolata* which grades eastwards into *C. pulvinaris. Ranunculus godleyanus* is confined to the central part of the Alps, growing on both sides of the Divide.

In the Eastern Alps and the ranges that diverge from them towards the Canterbury Plains, *Ranunculus grahamii* is possibly the only endemic high-altitude species, but northern and southern elements overlap more than in the west. For instance, *Dracophyllum muscoides, Anisotome capillifolia* and *Aciphylla dobsonii* extend from Otago well into Mackenzie ER; *Raoulia hectorii* and *Chionochloa rigida* cross the Rangitata River to meet the northern *Notothlaspi rosulatum, Haastia recurva* and *Raoulia mammillaris*; and the mountains between the Waimakariri and Rakaia rivers have the northern limits of *Schizeilema hydrocotyloides, Raoulia youngii* and *Leptinella dendyi* and the southern limits of *L. atrata* and *Brachyglottis bidwillii.*

The South-Western Alps consist mostly of schist ranges, that tend to form discrete blocks separated by deep valleys and low passes; they are not as high as the western and eastern sectors, although Mt Aspiring attains 3027 m and there are impressive ice-fields. They also include the Olivine ultramafic belt, and west of this, the granite and diorite of the Darran Range, culminating in Mt Tutuko (2746 m). *Epilobium purpuratum, Celmisia markii, C. bonplandii, Chionohebe armstrongii* and *Hebe fruticeti* are endemic to the South-western Alps. *Pachycladon novae-zelandiae, Aciphylla kirkii, A. hectorii, Dracophyllum muscoides, Myosotis pulvinaris, Abrotanella inconspi-cua* and *Chionohebe thomsonii* are shared with the Central Otago plateaus and mostly do not extend west of the Divide. Species that range widely in the South-western Alps and Fiordland, without extending further north, include *Ranunculus buchananii, Aciphylla congesta, Dracophyllum politum, Celmisia hectorii, Raoulia buchananii, Brachyglottis revoluta, Hebe petriei* and *Parahebe plano-petiolata.*

17. *Rakiura* province forms the southernmost portion of the New Zealand mainland. It is characterised by granite and other plutonic rocks, a cool, equable, moist climate, and a cover of mixed forest and heath; beeches are absent. Stewart Island together with Codfish and smaller offshore islands comprises most of the area. Its mountains form two massifs, separated by a sedimentary plain, and because of exposure penalpine plants are present on their summits, although the highest, Mt Anglem, is only 980 m. Foveaux ED includes the islands in Foveaux Strait, together with Bluff Hill on the South Island mainland. The Solander Islands, remnants of an early Pleistocene volcano, are 40 km south of Fiordland. The Snares Islands (see Table 13.1) are usually considered far-southern but their flora is almost the same as that of the Solander Islands.

Some distinctive plants are endemic to the coasts and small islands of the province (e.g. *Brachyglottis stewartiae, Stilbocarpa robusta*) or extend to adjacent South Island coasts (e.g. *Olearia oporina, Myosotis rakiura*). The higher summits of Stewart Island also have endemic species, e.g. *Aciphylla traillii* confined to the northern massif, and *A. stannensis* and *Celmisia polyvena* to the southern.

18. The *outlying islands* constitute three distinct provinces, which will be discussed in Chapter 13.

7

FOREST

Native forests occur from the sea coast to the subalpine tree limit and on the main islands contain some 48 species of tall and medium-sized trees, and over 70 tree species that do not normally exceed 10 m. Since these trees and other forest plants vary widely in their ecological tolerances, there is a kaleidoscopic variety of intergrading communities. Those on settled lowlands have been reduced to vulnerable remnants, and even on hilly country, extensive forest tracts are confined to rugged, remote areas. Especially on the drier mountains, deforestation has extended to the upper forest limits. Only the western side of the South Island still has forests extending unbroken from the coast to the subalpine belt.

Warm-temperate conifer/broad-leaved or 'mixed' forests are the tallest and most complex. Beginning with the kauri (*Agathis australis*) stands of Northland, they are described in a southwards sequence of decreasing complexity. Shorter, floristically poorer upper-montane equivalents are best developed in Central Westland, although smaller areas of comparable vegetation are scattered from the Coromandel Range to Auckland Island. Most high-altitude forests and most of those remaining in drier parts of New Zealand are dominated by evergreen beeches (*Nothofagus*); their extent and diversity is greatest in the north of the South Island. The expanding exotic forest estate is centred on the Volcanic Plateau.

Conifer/broad-leaved forests of the lower altitudes

Structure and gradients

The canopy of warm-temperate mixed forests is usually at least two-tiered. Conifers, i.e. kauri, *Libocedrus* and dioecious, succulent-fruited podocarps, form an upper tier that may be continuous, scattered or absent; a lower tier of tall broad-leaved trees varies conversely in density. A tier of small trees is largely suppressed where the tall canopy is continuous, but can locally provide the effective canopy, especially in tree-fall gaps and gullies; tree ferns and nikau palm (*Rhopalostylis sapida*) also belong to this

111

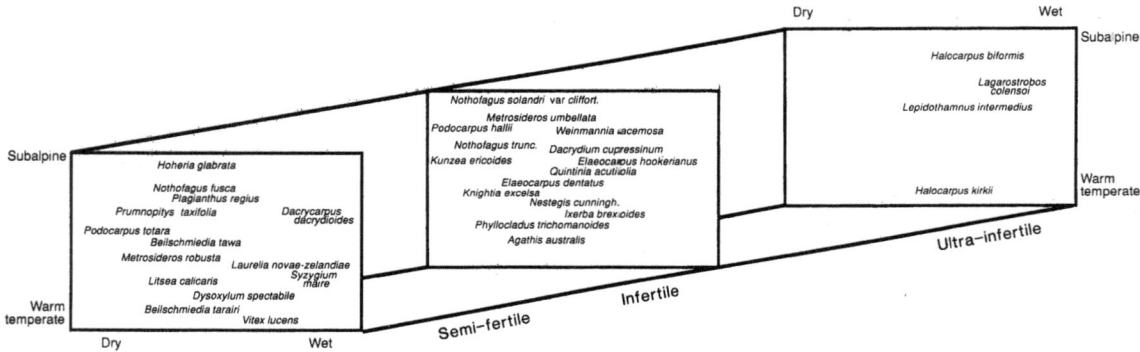

Fig. 7.1. Relative modal distributions of 33 native tree species along gradients of fertility, temperature and moisture.

tier, which merges into a tier of saplings and shrubs that may include filiramulate forms.

A tall herb tier consists mainly of large ferns, but monocot genera, especially *Astelia, Gahnia* and *Uncinia*, can be important. The ground tier, generally less than 10 cm tall, consists mainly of filmy ferns, mosses and liverworts, but there are angiosperms including small uncinias, orchids, *Nertera* and, on fertile soils, *Australina pusilla*. Lianes are abundant, with kiekie (*Freycinetia baueriana*), *Metrosideros* spp. and *Blechnum filiforme* sprawling over the forest floor as well. Obligate and incidental vascular epiphytes include seedlings of several tree species. The wetter mixed forests are especially rich in mosses, leafy and thallose liverworts and lichens; the ground-dwelling *Dawsonia superba* (to 50 cm tall), *Plagiochila stephensonii* (25 cm tall), *Monoclea forsteri* (thalli 20 cm long), and the epiphyte *Pseudocyphellaria coronata* (to 30 cm diameter) are remarkably large plants in each group, and all are widespread.

In each tier, composition follows climatic and soil fertility gradients. Prior to extensive deforestation, species demanding the highest levels of soil fertility were most abundant in regions of moderate rainfall, whereas in regions with high rainfall and leaching rates, species tolerating low fertility have the widest distribution, with the more demanding species being confined to recent alluvium and debris and soils derived from limestone. Fig. 7.1 illustrates this for several tree species, and Fig. 7.2 illustrates the climatic requirements of two species more precisely.

Among the conifers kauri and two much smaller trees, manoao (*Halocarpus kirkii*) and silver pine (*Lagarostrobos colensoi*), grow on infertile to ultra-infertile soils; only silver pine extends south of the northern zone, albeit discontinuously. Rimu (*Dacrydium cupressinum*) and miro (*Prumnopitys ferruginea*) are almost ubiquitous on such soils, but extend to semi-fertile soils where moisture is adequate. Lowland totara (*Podocarpus totara*), matai (*Prumnopitys taxifolia*) and kahikatea (*Dacrycarpus dacrydioides*) are almost confined to semi-fertile soils, the first two extending to low-rainfall regions and kahikatea being characteristic of wet gound.

Mountain totara (*Podocarpus hallii*) is a low-fertility species in warm-temperate climates, but edaphically wide-ranging in cooler climates, where it is often

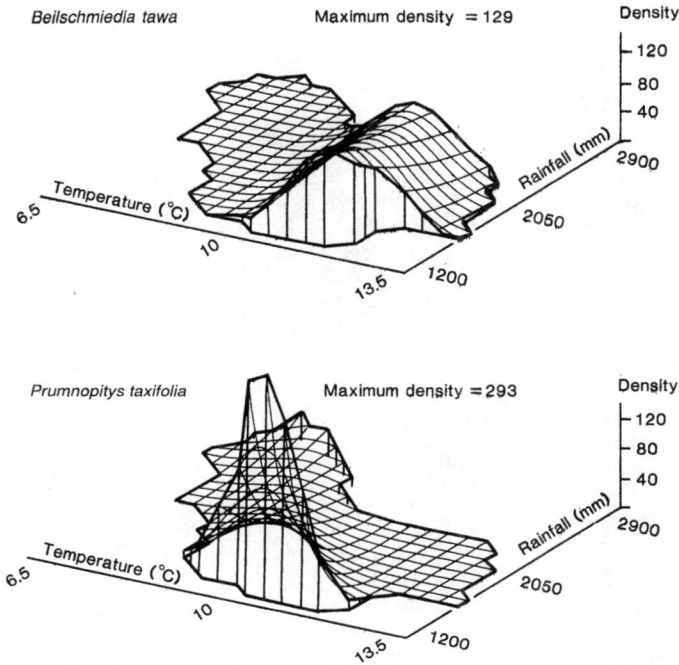

Fig. 7.2. Generalised linear model, fitted by using a Poisson error distribution to National Forest Survey data, to predict the density of tawa and matai in relation to two climatic parameters in the central North Island (J.R. Leathwick & N. Mitchell, unpublished). Density is measured as the number of stems per hectare $\geqslant 30.5$ cm d.b.h.

accompanied by kaikawaka (*Libocedrus bidwillii*). Tanekaha (*Phyllocladus trichoma-noides*) grows on infertile soils in the northern zone and, locally, in northernmost parts of the South Island; kawaka (*Libocedrus plumosa*) has a similar but more sporadic range. Toatoa (*Phyllocladus glaucus*) tends to replace tanekaha at higher altitudes in the northern zone, and a tall form of *P. alpinus* grows on ultra-infertile soils in the west of the South Island.

Among tall broad-leaved dominants in the northern zone, decreasing fertility is expressed in the sequence pukatea (*Laurelia novae-zelandiae*; on wet ground) and puriri (*Vitex lucens*)→taraire (*Beilschmiedia tarairi*; at low altitudes) and tawa (*B. tawa*)→towai (*Weinmannia silvicola*). None of these species reach the southern zone where, in humid regions, kamahi (*W. racemosa*) and southern rata (*Metrosideros umbellata*) are the only abundant tall broad-leaved trees, the former occupying a wide range of soils and the latter growing mainly on ridge crests and coastally, apart from trees of epiphytic origin. In mixed forests south of latitude 42° 20′, a tall, evergreen, broad-leaved tier is scarcely developed on recent soils and in dry eastern localities.

Among lower tiers, the fertility catena is indicated best by the larger ground ferns. Representative species are *Asplenium bulbiferum* (semi-fertile), *Blechnum* 'black spot' (semi-fertile to infertile), *B. discolor* (infertile), and *B. procerum* and *Sticherus*

Fig. 7.3. Mature kauri trees at Waipoua Forest, Northland (photo New Zealand Forest Service; National Archives of New Zealand M11876).

cunninghamii (infertile to ultra-infertile). Tree ferns mainly indicate climate; *Cyathea medullaris* and *C. cunninghamii* are warm-temperate, *C. smithii* extends highest and furthest south, *C. dealbata* is absent from inland and western regions of the South Island, *Dicksonia squarrosa* is wide-ranging, and *D. fibrosa* is abundant only on alluvial and tephra soils in districts with warm summers.

Mixed forests of Northland and Auckland

Northern forest communities occupy a gradient from leached clays and podzolised sands to more fertile soils developed over basalt, in alluvium or on colluvial slopes. Kauri occupies infertile soils, and today is mostly confined to spurs, ridges and high plateaus, although it was once widespread on lower, rolling terrain; sound logs are also recovered from swamps. Kauri rickers were being cut for ship's masts and spars before 1820. Logging reached a peak about 1905, and despite their massive size, logs were

Fig. 7.4. Dense ridge-crest kauri stand, Omahuta Forest, Northland.

extracted from very rugged country, especially in the Coromandel Peninsula. This was often achieved through 'kauri dams': logs were hauled down from ridges into water courses, and the water was then impounded behind wooden dams, to be released as a flood that swept the logs downstream. This wasted much timber, and standing trees were damaged through bleeding for resin. Only small areas of forest now retain a continuous kauri canopy, but the trees are so massive and visually dominant that any vegetation including them is referred to as kauri forest (Figs 7.3–7.5). Table 7.1 compares the canopy composition of dense and open kauri stands with conifer/broad-leaved stands that lack kauri.

On rolling terrain with moderately leached soils kauri trees tend to be clumped and some, like Tane Mahuta of Waipoua Forest, are giants. Smaller podocarp trees, mainly rimu, miro and mountain totara, are scattered among the kauri, above a broad-leaved tier mainly of tawa, taraire and towai. The shrub tier is quite open, but the tall

Fig. 7.5. Forest interior, contrasting large kauri with small stems of dicot trees. Waipoua Forest (photo W.B. Silvester).

herb tier dominated by *Gahnia xanthocarpa*, *Astelia trinervia* and kiekie is usually dense. These stands can be regarded as modal in that they have the most 'kauri' species, the commoner additional ones being:

(1) Tall broad-leaved trees: mangeao (*Litsea calicaris*), hinau (*Elaeocarpus dentatus*), northern rata (*Metrosideros robusta*), *Quintinia serrata*, *Ixerba brexioides*, *Nestegis cunninghamii* and rewarewa (*Knightia excelsa*).

(2) Small trees and shrubs: toro (*Myrsine salicina*), *M. australis*, *Hedycarya arborea*, *Coprosma lucida*, *C. grandifolia*, five-finger (*Pseudopanax arboreus*), lancewood (*P. crassifolius*), *Melicytus macrophyllus*, *Geniostoma rupestre*, *Alseuosmia macrophylla*, *Olearia rani*, *Brachyglottis kirkii*, and the epacrids *Dracophyllum latifolium* and *Cyathodes fasciculata*.

(3) Palm and tree ferns: nikau, *Cyathea medullaris*, *C. dealbata*, *Dicksonia squarrosa* and *D. lanata* (with a 2 m tall trunk in northern districts).

(4) Herb tier: *Blechnum* 'black spot'; thick litter usually suppresses bryophytes and other low-growing plants.

(5) Lianes: *Metrosideros albiflora*, *M. perforata*, *Ripogonum scandens*, and the ferns *Lygodium articulatum* and *Blechnum filiforme*.

Table 7.1. *Main canopy species in kauri stands, as mean no. of trees ≥ 30.5 cm (12 inches) d.b.h./ha*

	Forest type		
	dense kauri	kauri/podocarp/broad-leaved	
		ridges etc.	valley sides etc.
Agathis australis	111	7	
Dacrydium cupressinum	7	2	2
Prumnopitys ferruginea	17	10	4
Podocarpus hallii	17	12	4
Phyllocladus trichomanoides	2	5	2
Metrosideros robusta	+	+	5
Beilschmiedia tawa	5	+	7
Beilschmiedia tarairi	5	12	37
Weinmannia silvicola	7	25	25
Ixerba brexioides	7		
Dysoxylum spectabile	+	+	12
Vitex lucens			7
Knightia excelsa	+	+	5
Elaeocarpus dentatus	+	+	2
Corynocarpus laevigatus			2
Laurelia novae-zelandiae			7*

Notes:
+ = < 1 tree/ha
* = in gullies
Source: McKelvey & Nicholls 1959

(6) The common epiphytes are *Trichomanes reniforme* and other Hymenophyllaceae, *Asplenium flaccidum, A. polyodon, Lycopodium varium, Pittosporum cornifolium, Griselinia lucida, Astelia solandri, Collospermum hastatum, Dendrobium cunninghamii* and *Earina* spp.

Plants that occur widely but less abundantly in modal stands include *Tmesipteris* spp., *Schizaea dichotoma, Phymatosorus diversifolius, Asplenium oblongifolium, Rubus australis, Nestegis lanceolata, Metrosideros fulgens, M. diffusa, Bulbophyllum pygmaeum* and *Uncinia uncinata*.

On well-drained ridges and spurs with shallow, stony clay soils, kauri trees of modest diameter (1–2 m) can form an almost continuous upper canopy in virgin stands. A subcanopy is formed by smaller kauri and suppressed podocarp and broad-leaved trees. The tall herb tier is usually more open, and tanekaha (*Phyllocladus trichomanoides*), *Ixerba, Nestegis montana, Cyathodes fasciculata* and *Gahnia pauciflora* are more common than in modal stands, whereas tawa, taraire, mangeao, *Nestegis cunninghamii*, toro, *Alseuosmia macrophylla*, nikau, *Ripogonum, Cyathea medullaris, C. dealbata* and *Blechnum fraseri* are less common or absent. *Phebalium nudum*, a tall shrub that occurs mainly in secondary tall heath, grows sparingly on forested ridges.

Fig. 7.6. Kauri (left) and hard beech (right), Omahuta Forest, Northland. *Astelia trinervia* and *Gahnia xanthocarpa* form most of the understorey.

A kauri/hard beech (*Nothofagus truncata*) community grows locally on ridge-crests from sea level to 600 m on shallow clay soils (Fig. 7.6). In virgin stands, there are about 50 canopy trees of kauri per hectare in the height range 24–30 m, and similar numbers of hard beech form most of the subcanopy (McKelvey & Nicholls 1959).

In depressions with saturated, leached, often peaty silt more than 1 m deep, rimu can outnumber kauri; the community, which occurs only as small patches, is probably equivalent to 'terrace rimu' forests in the west of the South Island. Silver pine and manoao are rather uncommon podocarps ranging from these sites to wetter modal sites. Species excluded by wetness include hinau, rewarewa, *Olearia rani*, *Blechnum discolor*, nikau, *Alseuosmia macrophylla* and *Cyathea* spp. Conversely, *Cyathodes fasciculata* and *Ixerba* are more abundant here than on modal sites and the *Gahnia xanthocarpa–Astelia trinervia* tier can be at its densest.

Kauri communities of gullies and lower slopes with good drainage are transitional to non-kauri communities of semi-fertile soils; indeed, the presence of kauri and other 'kauri species' here probably depends on irregular terrain that provides local 'kauri sites', and on seedlings establishing on large logs and open slips where they escape vigorous competition. Species that occur in these but rarely in other kauri communities include kahikatea, pukatea, mahoe (*Melicytus ramiflorus*), kohekohe (*Dysoxylum spectabile*), karaka (*Corynocarpus laevigatus*), *Syzygium maire*, *Schefflera*

Fig. 7.7. Profile of a ridge-crest transect at Maungataniwha, Northland, including a patch of smaller trees that have grown up in a canopy gap (Dawson & Sneddon 1969). Abbreviations for profile diagrams in conifer/broad-leaved forest: *Ap, Asplenium polyodon*; *Ar, Ackama rosifolia*; *As, Astelia solandri*; *Bc, Blechnum 'capense'*; *Bd, Blechnum discolor*; *Br, Beilschmiedia tarairi*; *Bt, Beilschmiedia tawa*; *C, Coprosma* spp.; *Cc, Cyathea colensoi*; *Cci, Coprosma ciliata*; *Cd, Cyathea dealbata*, *Cde, Coprosma depressa*; *Cf, Coprosma foetidissima*; *Cg, Coprosma grandifolia*, *Ch, Collospermum hastatum*; *Cj, Cyathodes juniperina*; *Cl, Coprosma lucida*; *Cm, Cyathea smithii*; *Cp, Coprosma pseudocuneata*; *Cs, Carpodetus serratus*; *Ct, Coprosma tenuifolia*; *Dc, Dacrydium cupressinum*; *Dd, Dacrycarpus dacrydioides*; *Dq, Dicksonia squarrosa*; *Ds, Dysoxylum spectabile*; *E, Elaeocarpus dentatus*; *F, Freycinetia baueriana*; *G, Geniostoma rupestre*; *Gi, Griselinia littoralis*; *Gu, Griselinia lucida*; *H, Hedycarya arborea*; *Lb, Libocedrus bidwillii*; *Ln, Laurelia novae-zelandiae*; *Ma, Myrsine australis*; *Mav, Microlaena avenacea*; *Mb, Metrosideros robusta*; *Md, Myrsine divaricata*; *Mm, Mida salicifolia* var. *myrtifolia*; *Ms, M. salicifolia* var. *salicifolia*; *Mp, Metrosideros perforata*; *Mr, Melicytus ramiflorus*; *Mu, Metrosideros umbellata*; *N, Nestegis lanceolata*; *Or, Olearia rani*; *Pa, Pseudopanax arboreus* (Fig. 7.7), *Phyllocladus alpinus* (Fig. 7.14); *Pc, Pseudowintera colorata*; *Pcr, Pseudopanax crassifolius*; *Pf, Prumnopitys ferruginea*; *Ph, Podocarpus hallii*; *Pm, Streblus heterophyllus*; *Ps, Pseudopanax simplex*; *Pt, Prumnopitys taxifolia*; *Q, Quintinia acutifolia*; *Ra, Rubus australis*; *Rs, Rhopalostylis sapida*; *S, Schefflera digitata*; *V, Vitex lucens*; *Wr, Weinmannia racemosa*; *Ws, Weinmannia silvicola*.

digitata, Microlaena avenacea, Cyathea smithii, C. cunninghamii (in gullies), *Elatostema rugosum* (on stream banks) and *Blechnum filiforme*.

Extensive forests without kauri grow on rolling clay uplands (Fig. 7.7). Usually podocarps are only scattered, often because of logging; these are mainly rimu and miro. Canopies are formed mainly by towai or taraire, with other important species being northern rata, kohekohe, hinau and tawa. Rewarewa and nikau are more frequent than in most kauri forests. Pukatea and kahikatea grow mainly along drainage lines. Widespread kauri forest species that are usually scarce in forests without kauri are *Quintinia serrata*, *Ixerba*, *Cyathodes fasciculata*, *Gahnia xanthocarpa*, *Astelia trinervia* and *Dicksonia lanata*. These forests would have once merged into valley-floor stands characterised by lowland totara, matai, puriri, titoki (*Alectryon excelsus*) and, on wet ground, kahikatea and pukatea, but only modified vestiges and secondary patches remain.

Volcanic Plateau

At the time of European settlement, river flats on the Bay of Plenty lowlands supported magnificent stands of kahikatea, but little remains. In a reserved remnant described by Smale (1984), kahikatea trees up to 40 m tall grow scattered or as clumps, over the main canopy of tawa and scattered pukatea at 18 m. A subcanopy is formed of mahoe, young tawa and some *Hedycarya*. Well-developed lower tiers include tree seedlings, shrubs, foliage of the lianes kiekie, *Metrosideros colensoi* and *Blechnum filiforme*, other ferns, and the grass *Microlaena avenacea*. A total of 118 native vascular species was recorded on 4.5 ha of flood-plain; fewer than 10 of these appear to reflect disturbance at the forest margin.

Forest has also vanished from the lower hills except for modified remnants, but covers the sides of deep valleys emerging from the ranges and steeplands in the east and west of the district. According to McKelvey (1973), forest below 430 m in Urewera ER is characterised by large rimu and northern rata above a dense canopy of tawa, accompanied by kamahi, hinau, rewarewa, kohekohe, nikau, and on damp ground, pukatea. Smaller trees and large shrubs include *Pseudopanax* spp., *Olearia rani*, *Coprosma* spp., mahoe, *Schefflera*, rangiora (*Brachyglottis repanda*), *Pseudowintera axillaris*, toro and tree ferns.

The lowlands rise to plateaus, which are unique in the North Island in that significant areas of forest remain on easy terrain. Most of these forests grow on late Quaternary tephra and pyroclastic flows. West of Lake Taupo (McKelvey 1963), stands characterised by 'scattered large rimu and northern rata emergent over dense lower storeys dominated by tawa and containing pukatea, but lacking kohekohe and nikau' occur between 240 and 490 m. With increasing altitude, the cold-sensitive pukatea, kiekie, and then tawa decrease whereas species adapted to cooler and moister climates increase, especially rimu and kamahi (*Weinmannia racemosa*).

Table 7.2 selects three upland forest types to demonstrate decreasing podocarp densities and increasing broad-leaved dominance with increasing distance from the Lake Taupo eruptive centre. This phenomenon and the paucity of young podocarps shown in the table are discussed on pp. 525 and 599. Dense stands (L1) are dominated mainly by rimu and matai, together with miro and totara (lowland, mountain or hybrids), growing as straight, generally small-canopied trees 0.4–1 m in diameter and 30–40 m tall. The proportion of rimu is greater on slopes and rolling terrain where pumice is thinner and frosts less severe than in depressions where matai is most abundant. Other conifers include kahikatea and, locally, tanekaha and toatoa. The hardwood subcanopy consists of slender trees, mainly kamahi. Tawa is locally present in rimu-dominated stands, mainly as saplings. Lower tiers are also fairly open (Fig. 7.8).

Table 7.2. *Mean number of stems/ha in podocarp/broad-leaved forest on the Volcanic Plateau*

		L1				M1				D1			
		Tr	Pl	Sp	Sd	Tr	Pl	Sp	Sd	Tr	Pl	Sp	Sd
Dacrydium cupressinum	R	24	9	15	37	11	2	11	33	6	2	+	33
Prumnopitys taxifolia	M	39	5	8	17	12	3	+	17	+	+	+	7
Podocarpus totara/hall.	M	13	9	26	34	5	11	27	45	1	3	7	20
Prumnopitys ferruginea	M	17	7	28	59	5	5	18	57	1	3	7	53
Dacrycarpus dacrydioid.	M	1	1	+	25	+	+	+	6	+	+	+	7
Phyllocladus trichoman.	R	2	3	18	4	+	3	+	2	+	+	+	0
Weinmannia racemosa	RM	11	58	f	f	14	117	f	0	2	34	f	r
Nestegis spp.	M	2	12	45	41	3	10	37	51	+	3	8	10
Elaeocarpus dentatus	M	1	2	7	12	2	4	9	12	12	5	+	17
Beilschmiedia tawa	R	+	12	57	26	+	14	43	25	32	148	298	90
Elaeocarpus hookerian.	M	1	1	18	29	1	2	4	12	+	+	+	0
Knightia excelsa	R	+	+	+	5	+	2	1	5	1	3	14	43
Metrosideros robusta										3	+		
Small-tree species	RM	13	257	f	f	25	367	a	a	10	229	a	a

Notes:
All types lie between 450 and 900 m, and are shown in order of increasing distance from Lake Taupo.
L1 Dense mixed podocarps on deep pumice tephra
   R: mainly in rimu-dominated stands
   M: mainly in matai-dominated stands
M1 Podocarp/kamahi/small broad-leaved species, on deep tephra
D1 Rimu–northern rata/tawa, where Taupo tephra is thin or absent
   Tr = Trees > 30.5 cm d.b.h.
   Pl = Poles 10–30.5 cm d.b.h.
   Sp = Saplings 2.5–10 cm d.b.h.
   Sd = Seedlings > 15 cm tall and < 2.5 cm d.b.h. (% frequency in 0.02 ha plots)
   a, f, r: abundant, frequent, rare; + : < 1 stem/ha
Source: McKelvey 1963

In type M1 podocarps are present at only a third of the density, but as larger-diameter, mature trees (Fig. 7.9). The main canopy is formed by broad-leaved trees, and is itself differentiated into an upper tier with kamahi, *Nestegis* spp., *Elaeocarpus* spp. and local tawa, and a lower tier with *Griselinia littoralis*, toro, fuchsia (*Fuchsia excorticata*), wineberry (*Aristotelia serrata*), tree ferns, etc. This type includes pockets of dense podocarp forest around swampy depressions with deeper pumice.

Type D1 has an even lower density of podocarps, these being large trees that are frequently host to large northern rata trees, many or most of which have died during recent decades. Tawa rather than kamahi dominates the tall broad-leaved canopy (Fig. 7.10). Where terrain is dissected, in both this and the previous type, podocarps and tall broad-leaved trees tend to be concentrated on ridges and upper slopes, whereas lower slopes and gullies are dominated by small broad-leaved species and tree ferns, and

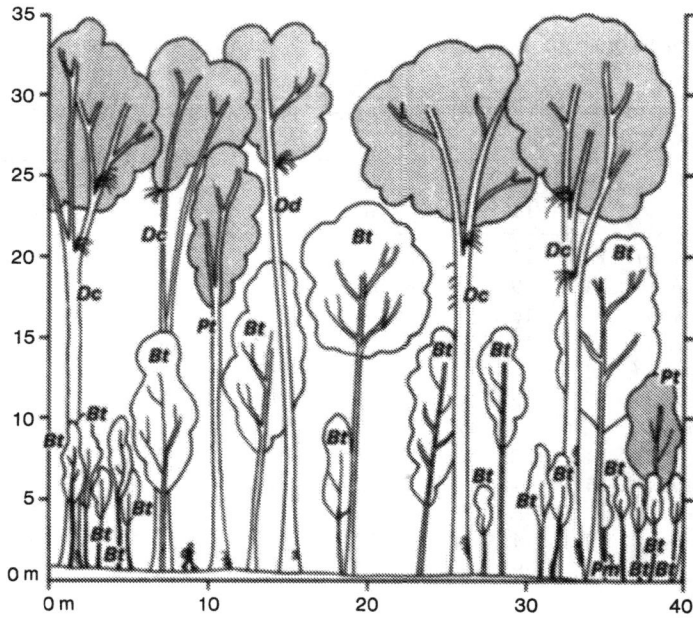

Fig. 7.8. Profile of a densely stocked podocarp stand near Te Whaiti, eastern Volcanic Plateau. (Shrubs below 3 m are omitted in Figs 7.8, 7.9 and 7.12.) (From Robbins 1962.) Abbreviations as for Fig. 7.7.

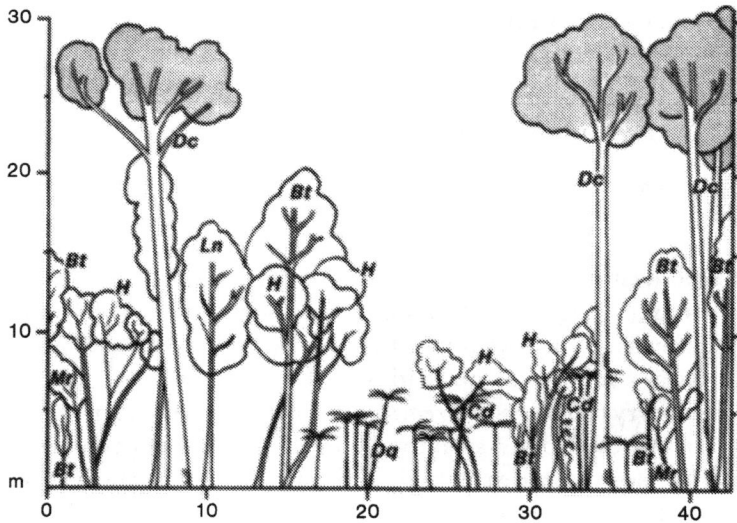

Fig. 7.9. Profile diagram in an open podocarp stand with broad-leaved trees and tree ferns forming most of canopy. Near Rotorua (Robbins 1962). Abbreviations as for Fig. 7.7.

Fig. 7.10. Tawa grove, Pureora Forest, Western Volcanic Plateau.

there is a denser fern understorey. The pumice veneer is thin or lacking and the trees root in soil derived from older volcanic or sedimentary rocks.

Fig. 7.11 relates nine vegetation types lying between 400 m and 620 m above sea level. Types 1 and 3 have a dense upper canopy of podocarps, with the larger content of matai in Type 1 reflecting proximity to flat, frosty depressions. Types 5 and 7 have an open overstorey of large podocarps; 5 is on sloping terraces and faces, whereas 7 is on dissected topography, which is reflected in the greater importance of small broad-leaved trees, *Ripogonum* and tree ferns. Type 4 grows on poorly drained terraces and is also open in the podocarp tier, but the rimu trees have comparatively small crowns and tall, slender trunks; kahikatea is locally abundant, and tawa is generally unthrifty.

Types 6 and 9, characterised by rewarewa, are on slopes and steep faces respectively; they seem seral to podocarp-dominated forest. Types 2 and 8 are gully vegetation of small broad-leaved trees, lianes, and tree ferns. They grow at higher and lower altitudes respectively, and are distinguished by the greater abundance of species such as *Dicksonia fibrosa* and *Pseudowintera colorata* in 2 and *Geniostoma* and rangiora in 8.

Taranaki to the Marlborough Sounds

In Taranaki, kohekohe-dominated semi-coastal bush ascending to 240 m and the next belt dominated by tawa survive on the Kaitake Range (Clarkson 1985). Corresponding altitudes on Mt Taranaki have been cleared to farmland, but kamahi with an overstorey of rimu and northern rata clothe slopes above 400 m (Fig. 7.12). Among the

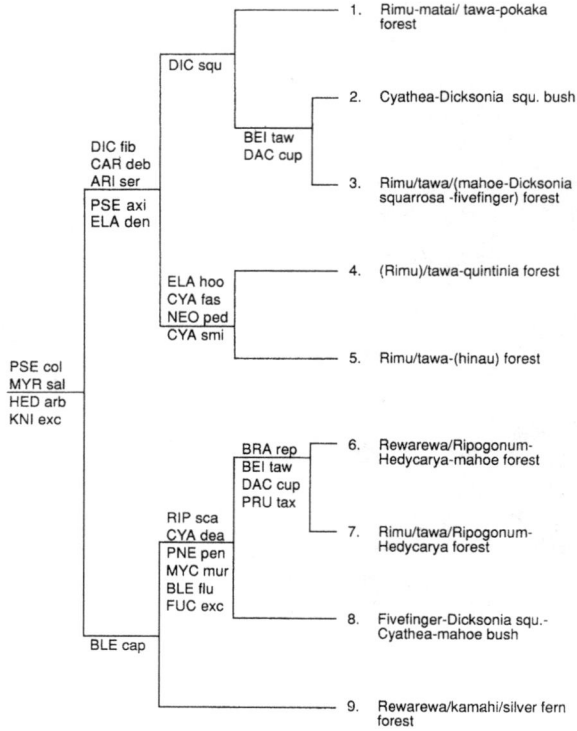

Fig. 7.11. Community relationships, derived by classification of plots and species weighted according to tier. Pureora Forest (from Leathwick 1987). Indicator species are designated by the first three letters of genus and species.

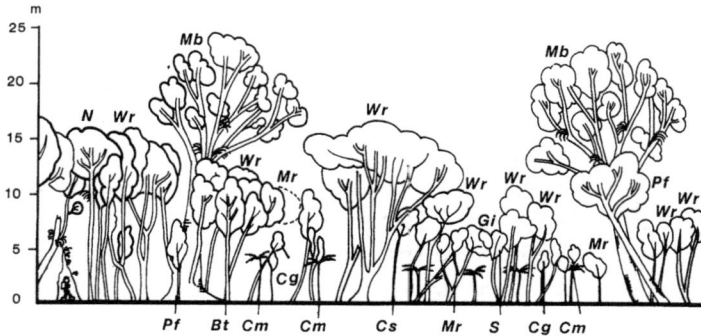

Fig. 7.12. Profile diagram in forest dominated by *Weinmannia racemosa* with an overstory of *Metrosideros robusta*. Mt Taranaki, 550 m (Robbins 1962). Abbreviations as for Fig. 7.7.

Fig. 7.13. Understorey of mixed forest near Wellington. Species include kiekie (foreground), *Cyathea dealbata* (left) and *C. smithii* (right) (photo, New Zealand Forest Service; M2759 New Zealand National Archives).

ridges and gorges of Eastern Taranaki ER, mixed forests usually have an overstorey of large podocarps. Tawa forms most of the broad-leaved canopy on sheltered, relatively fertile sites. Kamahi is widespread and, with northern rata, dominates on most ridges, especially above 350 m. *Quintinia* grows in the north of the region, but nowhere else in Taranaki province (Nicholls 1956).

For Hutt Valley in Wellington ED, Druce & Atkinson (1958) outline the sequence rimu–northern rata/tawa/kohekohe (30–240 m); rimu–northern rata/hinau/tawa (30–430 m) (Fig. 7.13); rimu–northern rata/hinau/kamahi (150–430 m); miro–rimu/kamahi (430–610 m). Where such sequences continue upwards, beeches usually become increasingly dominant, but on Mt Taranaki and parts of the Tararua Range the sequence is towards cool- or summit-temperate mixed forests. On infertile soils around Wellington and in the Marlborough Sounds beech forest descends to low altitudes; mixed forest is restricted to pockets on fertile colluvium, and its podocarps have generally been logged.

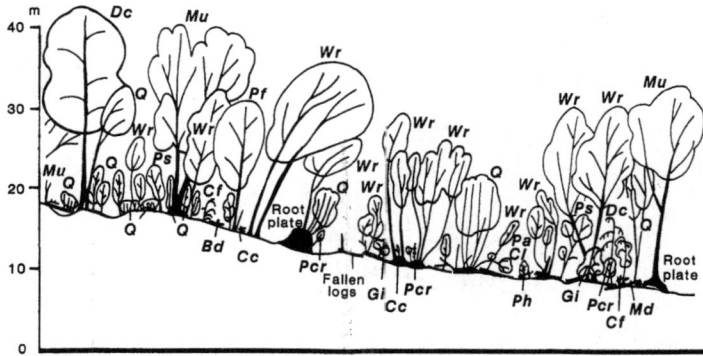

Fig. 7.14. Profile of mixed forest at 470 m on the Hohonu Range, North Westland (Reif & Allen 1988). Abbreviations as for Fig. 7.7.

Below 100 m a.s.l. in the outer reaches of Queen Charlotte Sound, tall broad-leaved forest on gentle depressions and moist flats is dominated by pukatea about 30 m tall, tawa to 20 m tall and kohekohe to 15 m tall. Hinau is also present, but there are few podocarps, these being mostly large kahikatea. The small tree and shrub tiers include *Macropiper excelsum*, mahoe, *Coprosma grandifolia*, *Hedycarya*, nikau and saplings of canopy species. Lianes and epiphytes are abundant, and include *Blechnum filiforme*, *Phymatosorus scandens*, *Metrosideros fulgens*, *Ripogonum*, *Collospermum hastatum* and *Griselinia lucida*. *Cyathea medullaris*, rangiora and *Coprosma grandifolia* increase towards gullies with less stable soil. Towards the spurs, first pukatea and then tawa and kohekohe drop out, and there is some development of a hinau–kamahi community with an overstorey of rimu and an understorey of *Cyathea dealbata*, but hard beech is usually also present and increases to dominance on the spur crests.

In valleys draining into the heads of the Sounds, mixed forest lacks the semi-coastal element, but podocarps seem more important. In the Pelorus Bridge Scenic Reserve, concave slopes have a main canopy of tawa; large rimu, matai, lowland totara, miro and kahikatea form an overstorey, and *Cyathea dealbata* forms most of the understorey. Above 330 m kamahi replaces tawa as the canopy dominant. Beech species enter on spurs and alluvial terraces. Forest on the terraces tends to be open, allowing small trees and shrubs typical of semi-fertile soil to become abundant, especially *Dicksonia fibrosa*, *Coprosma rotundifolia*, *Carpodetus serratus*, *Pennantia corymbosa* and mahoe.

Western South Island

Mild temperatures and heavy rainfall in the maritime west of the South Island lead to rapid soil leaching, so that mixed forests vary along catenas of drainage and fertility (Table 7.3). Reif & Allen's (1988) classification for Westland steeplands separates low-altitude tall forests from high-altitude equivalents, and also from shrub, bush, and heath communities, and classifies them into a '*Dacrycarpus dacrydioides* community' on recent, poorly drained sites, a '*Dacrydium cupressinum* community' on mature soils (Fig. 7.14), and a '*Prumnopitys ferruginea–Coprosma lucida* community' on immature

Table 7.3. *Mean number of trees ≥ 30.5 cm d.b.h./ha in a selection of South Island conifer/broad-leaved forest types, that are based on densities of trees*

type altitudinal range (m):	1 0–50	2 0–300	3 0–100	4 0–200	5 100–200	6 0–300	7 0–500	8 100–600
Dacrycarpus dacrydioides	31	6	224	1			2	
Dacrydium cupressinum		8	28	156	28	11	17	14
Prumnopitys ferruginea		9	2	29	6		18	11
Prumnopitys taxifolia		9	2					
Podocarpus hallii		9*		4	23	8		12
Lagarostrobos colensoi			5	3	8	50		
Libocedrus bidwillii		1			12	13		
Laurelia novae-zelandiae	9							
Metrosideros robusta	1							
Weinmannia racemosa	1	64	17	18	5	2	59	22
Metrosideros umbellata		37		8	5	4	4	28
Quintinia serrata				3	2		7	7
Elaeocarpus dentatus		2					3	
Elaeocarpus hookerianus					4			
Nothofagus menziesii			1					
Griselinia littoralis		2						
Phyllocladus alpinus					6	2		
Halocarpus biformis						12		
Leptospermum scoparium						2		

Notes:
* includes *Podocarpus totara* var. *waihoensis*
Forest types:
1 Poorly drained low-lying ground near the coast in North-west Nelson.
2 Well-drained recent alluvial sites in Westland.
3 Wet recent alluvial flats in Westland.
4 Flat to undulating older terraces in Westland.
5 Infertile soils on gentle terrain in Westland.
6 Poorly drained, ultra-infertile sites in Westland.
7 Gentle to steep slopes in Westland.
8 Gentle to steep slopes in Westland foothills with species indicating cooler conditions than 7.
Source: McKelvey 1984

soils, where plots range from former landslides dominated by broad-leaved trees to older sites with a podocarp overstorey.

The following account, which broadly agrees with Reif & Allen's (1988) treatment, is based on field descriptions of representative communities at latitude 43° 30' in Central Westland, where the largest tracts of podocarp/broad-leaved forests remaining in New Zealand extend from the Tasman Sea to the Southern Alps (Wardle 1977). The most widespread type is rimu/kamahi forest on yellow-brown earths on slopes steep enough to escape podzolisation, and on depositional surfaces not yet old enough for podzolisation to be advanced. Scattered large rimu and miro form an overstorey, over a main canopy dominated by kamahi of epiphytic origin. Mountain totara and southern rata (*Metrosideros umbellata*) are often also present.

A small-tree tier includes *Hedycarya, Pseudopanax simplex* and *Myrsine australis*, and the tree ferns *Cyathea smithii* and *Dicksonia squarrosa*; the latter is also abundant in the shrub tier, together with *Coprosma foetidissima*. *Blechnum discolor* dominates the fern tier except where there are colonies of *B.* 'black spot'. There are also copious lianes including *Ripogonum, Metrosideros fulgens* and *M. diffusa*; juvenile foliage of rata vines also covers much of the forest floor. Near the coast, stands can be overrun by a jungle of lianes, especially *Freycinetia* and *Ripogonum*. Epiphytes include filmy ferns, *Trichomanes reniforme, Asplenium polyodon, Ctenopteris heterophylla, Grammitis billardierei* (on tree bases), *Lycopodium varium, Tmesipteris tannensis* (on tree-fern trunks), *Griselinia lucida, Astelia solandri, Luzuriaga parviflora* (near the ground), *Dendrobium cunninghamii* and *Earina* spp.

Asplenium bulbiferum, Leptopteris superba, Schefflera digitata and *Microlaena avenacea* grow along water courses and in hollows. Other species include *Lastreopsis hispida, Rumohra adiantiformis, Blechnum procerum, Ascarina lucida, Pseudowintera colorata, Metrosideros perforata, Neomyrtus pedunculata, Quintinia acutifolia, Rubus cissoides, Elaeocarpus hookerianus, Griselinia littoralis, Pseudopanax colensoi, P. edgerleyi, P. crassifolius, Myrsine divaricata, Alseuosmia pusilla, Coprosma lucida, C. rhamnoides, C. colensoi, C.* aff. *ciliata, Nertera* aff. *dichondrifolia, N. depressa, Astelia fragrans* and *Libertia pulchella*.

Characteristically, rimu/kamahi stands occupy the crests of spurs and ridges, whereas broad-leaved bush grows on intervening slopes and gullies, subject to soil creep and slipping. Dense stands of young kamahi and rata trees develop on slip soles and recent moraine crests.

On rolling crests of late-Pleistocene moraines, where yellow-brown earths have developed cemented B horizons, *Quintinia* co-dominates with kamahi in the main canopy. In lower tiers *Phyllocladus alpinus, Neomyrtus pedunculata, Blechnum procerum* and *Sticherus cunninghamii* are important. *Metrosideros fulgens* and the small ferns *Lindsaea trichomanoides* and *Trichomanes reniforme* are characteristic species, and *Dicksonia lanata* occurs locally.

This kind of forest merges into dense rimu stands (Fig. 7.15) as moraines merge into poorly-drained fluvioglacial terraces with thick iron-humus pans overlain by leached silt loam or silty peat. These 'terrace rimu forests' are the largest remaining stands of native softwood timber. Rimu trees, 30 m or more tall, but very slow-growing and generally not more than 40 cm diameter, provide most of the canopy. Kamahi and *Quintinia* are abundant, mostly as slender, coppicing growth. On very wet ground, where pools lie on the surface for much of the year, rimu trees are generally smaller in diameter and mixed with silver pine, in an association transitional to tree-heath; bryophytes, including sphagnum, usually cover much of the ground.

Recent alluvial flats can support dense podocarp stands or, alternatively, large podocarps are widely scattered with crowns spreading above low broad-leaved canopies (Fig. 7.16), forming vegetation that can often be best described as bush with a podocarp overstorey. On well-drained sites, all the large podocarps can be present: rimu, lowland totara, matai, miro, kahikatea and, on frosty sites, mountain totara together with kaikawaka (*Libocedrus bidwillii*). In North Westland lowland totara is represented by hybrids between *Podocarpus totara* and *P. acutifolius*, which usually

Fig. 7.15. Dense 'terrace' rimu. Wanganui Forest, Central Westland (photo, New Zealand Forest Service; M9304 National Archives of New Zealand).

grows as colonies of layering shrubs. In Central Westland the more uniform segregate *P. totara* var. *waihoensis* prevails, and south of Haast there is only mountain totara. These totara can dominate on stony soils, but most stands existing today are pioneer or regrowth. On young soils subject to flooding and silt deposition, however, kahikatea is usually overwhelmingly dominant (Fig. 7.17).

The number of vascular species in these alluvial stands exceeds 70, and indicate relatively high fertility. They are the same as occur in bush on similar sites, and the more important are listed on p. 173.

Mixed podocarp stands on recent flats grade into rimu/kamahi forest on well-drained older terraces; and where drainage is poor, there are dense rimu–kahikatea stands on terraces intermediate in age and fertility between those dominated by rimu or kahikatea alone. Although kahikatea trees tolerate prolonged flooding, kahikatea forest is interrupted by dense colonies of *Astelia grandis* in wet depressions, and peters

Fig. 7.16. Scattered rimu (left), kahikatea and matai above main canopy of kamahi. Near Whataroa, Central Westland.

out into herbaceous swamp as the increasing extent of standing water limits the sites where kahikatea and other tree seedlings can establish (Fig. 7.18). Gradations can be towards either fertile swamps dominated by *Phormium tenax*, or infertile swamps dominated by manuka, *Empodisma minus, Baumea* spp., etc.; in the latter transitions, silver pine and *Phyllocladus alpinus* are common. Near the coast, kahikatea–swamp transitions can have almost impenetrable understories of *Gahnia xanthocarpa* and kiekie.

Northwards, along the western seaboard, further warm-temperate species enter the mixed forests: hinau and toro about latitude 43° 20′, northern rata at 42° 40′, nikau at 42° 30′ with outliers to latitude 43°, *Collospermum hastatum, Pittosporum cornifolium* and *Metrosideros colensoi* at 42° 10′, *Alseuosmia macrophylla* and *Astelia trinervia* at

Fig. 7.17. Kahikatea forest, with trees dying from silting; *Phormium tenax* in foreground. Lake Wahapo, Central Westland.

Fig. 7.18. Kahikatea invading a shrubby swamp; rimu (*Dacrydium cupressinum*) forest behind. Near Lake Kaniere, Central Westland.

41' 30', and *Libocedrus plumosa* and tanekaha at 40° 40'. The last four species, together with *Quintinia acutifolia*, are absent from the southern half of the North Island, and only tanekaha extends to Sounds–Nelson province. On the other hand, the Western Nelson mixed forests lack tawa, rewarewa, *Dysoxylum* and *Nestegis* spp., trees which reach northern coasts of the South Island. Nevertheless, they are as luxuriant as North Island lowland forests, although largely confined to recent alluvial plains (Fig. 7.19) and other fertile soils near the coast because of the prevalence of beech.

South of Central Westland, warm-temperate species become confined to sheltered coastal places and then drop out: *Quintinia* reaches latitude 43° 40', *Alseuosmia pusilla* 44°, *Gahnia xanthocarpa* 44° 20', *Metrosideros perforata* 44° 40', and *M. fulgens, Freycinetia, Ascarina* and *Hedycarya* about latitude 46°. Beeches again become prominent from latitude 43° 40', and in Fiordland descend to sea level, being absent only from exposed lower slopes facing the Tasman Sea. Nevertheless, there are areas of mixed forest, especially on flats near the coast, to as far south as the Hollyford River.

Southern districts

In Te Wae Wae ER, rimu forest is extensive on ancient marine terraces bordering Foveaux Strait, but usually has beech among it. Remnants of forest on and around the Southland plains and near Dunedin, and the still extensive forests of Catlins ER, are nearly all podocarp/broad-leaved, with rimu and kamahi prevailing on the hills. Stewart Island is completely beech-free, and again rimu/kamahi forest prevails (Fig. 7.20) with only a token presence of matai, kahikatea and the small broad-leaved trees characteristic of semi-fertile soils (Wilson 1987). Forest composition reflects exposure to strong westerly winds, infertile soils and relative floristic poverty. Southern rata is more important than in most South Island forests, and on the upper slopes and some coasts rimu is confined to aspects sheltered from the west and south-west. On poor soils, near the coast, and in exposed localities transitions toward tree-heath are extensive (p. 181); rimu forms an overstorey, but rata and kamahi shrink into the low canopy.

Eastern South Island

Forests in the rain shadow of the high mountains were mostly burnt by the Maori, and what remained was reduced to scattered fragments after European colonisation. At low altitudes towards the coast, these are characterised by matai and totara, together with kahikatea on alluvial flats and concave slopes. Towards the ranges, rimu is the predominant podocarp and the extent of beech forest increases, reflecting higher rainfall and less fertile soils.

As on the western seaboard, species drop out towards the south. The last outposts of tawa, *Nestegis, Melicope ternata, Hedycarya* and the lianes *Freycinetia* and *Metrosideros colensoi* are near Kaikoura; hinau reaches Christchurch; karaka (probably as Maori plantings), nikau, titoki, *Passiflora tetrandra, Macropiper excelsum, Dodonaea viscosa* and *Griselinia lucida* grow in sheltered localities on Banks Peninsula. Moreover, several important drought- or frost-sensitive species are rare or absent in the east of the South Island, notably kamahi, southern rata, rata vines other than *Metrosideros diffusa*, and the large tree ferns *Cyathea medullaris* and *C. cunninghamii*. In

Fig. 7.19. Kahikatea forest remnant, Aorere valley, North-west Nelson. Kiekie and *Ripogonum* (left) on lower trunks, *Collospermum hastatum* higher up (photo, New Zealand Forest Service; M2603 New Zealand National Archives).

consequence, broad-leaved canopies in inland regions and south of Banks Peninsula are comprised almost entirely of low-growing trees, such as *Griselinia littoralis*, mahoe and fuchsia, and tree ferns are largely confined to sheltered gullies.

On the other hand, forest remnants and bush in the eastern South Island contain many species that are common through much of the North Island but scarce in or absent from the forests of Westland. These species benefit from summer warmth and relatively fertile soils, and tolerate drought. They include *Cyathea dealbata, Leptopteris hymenophylloides, Pellaea rotundifolia, Polystichum richardii, Hoheria angustifolia, Streblus heterophyllus, Pittosporum eugenioides, Sophora microphylla, Parsonsia capsularis,* the tall grass *Anemanthele lessoniana,* and an array of filiramulate shrubs: *Lophomyrtus obcordata, Melicope simplex, Melicytus micranthus, Coprosma areolata, C. crassifolia, C. rubra, C. virescens,* etc.

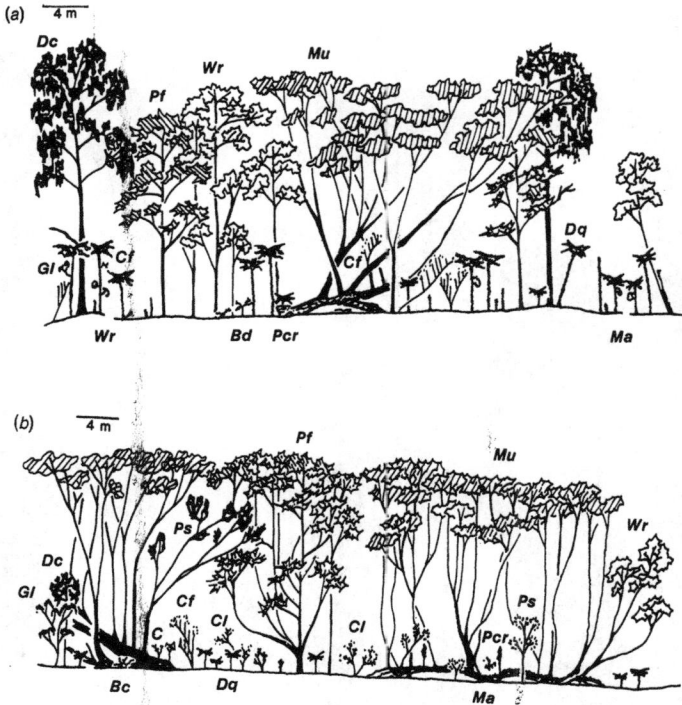

Fig. 7.20. Profiles of coastal mixed forest on Stewart Island (*a*) and Bench Island (*b*) (Veblen & Stewart 1980). Gl, *Griselinia littoralis*; other abbreviations as for Fig. 7.7.

High-altitude conifer/broad-leaved forests

In districts lacking beech, the montane–subalpine transition is occupied by mixed forests that are short and floristically poorer than the low-altitude equivalents, and the canopy is often a dense wind-roof. Similar forest grows in the heads of some glaciated valleys that are otherwise occupied by beech forest. Scattered remnants also survive in the deforested eastern regions of the South Island. This section also discusses low-statured forests on exposed summits in the northern zone and forests in shaded, frosty mountain valleys; in both situations, high-altitude species mingle with those of the lowlands.

Central Westland

Montane mixed forest is developed most extensively in the Westland 'beech gap'. On stable slopes and ridges with well-developed yellow-brown earths at latitude 43°, there is a gradual decrease in stature, with rimu dropping out by 500 m and then miro, leaving mountain totara as the only podocarp. Southern rata and kamahi dominate the main canopy, which is often dense and wind-smoothed. *Blechnum discolor* accompanied by *Astelia nervosa* forms most of the understorey.

Where soils are more leached, species such as *Phyllocladus alpinus, Gahnia procera*

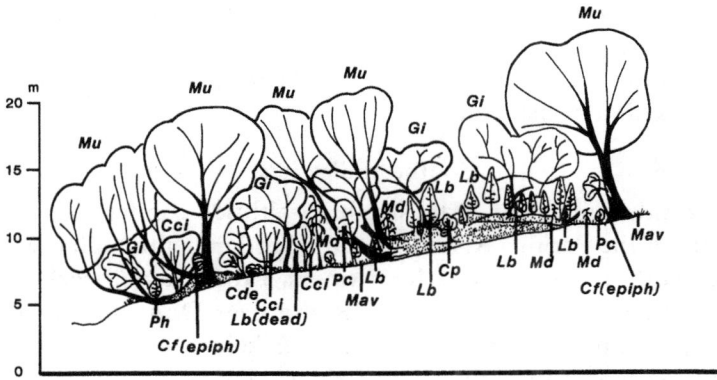

Fig. 7.21. Profile diagram in upland mixed forest at 825 m, Wilberg Range, Central Westland (Reif & Allen 1988). Abbreviations as for Fig. 7.7.

and *Sticherus cunninghamii* enter, and rimu ascends nearly 100 m higher. Pink pine (*Halocarpus biformis*) occurs on podzolised ridges, but rarely south of the Hokitika catchment. Faces and fans with more fertile soil, on the other hand, support understorey species such as *Leptopteris superba*, *Asplenium bulbiferum*, *Cyathea colensoi*, *Pseudowintera colorata* and *Schefflera*, together with the tree ferns *Cyathea smithii* and *Dicksonia squarrosa*, and the lianes *Rubus cissoides*, *Metrosideros diffusa* and, on seaward aspects, *M. fulgens*. *Archeria traversii* is common on rocky spurs and bluffs.

On seaward aspects above 650 m, the forest becomes subalpine in character and dominated by rata (Fig. 7.21), with a scattered overstorey of kaikawaka; in inland valleys this change can occur as low as 550 m. On ridges, rata forms a dense wind-roof, supported on spreading, often semi-prostrate trunks, whereas on faces taller, more open rata trees have often begun as epiphytes on kaikawaka. Mountain totara can codominate. Other characteristic species are *Pseudopanax simplex*, *P. colensoi* var. *ternatus*, *P. linearis*, *Griselinia littoralis*, *Coprosma foetidissima*, *C. pseudocuneata*, the filiramulate *C. ciliata* and *C. cuneata*, *Pseudowintera colorata*, *Astelia nervosa*, and the ferns *Cyathea colensoi*, *Blechnum* 'mountain', and on immature soils, *Leptopteris superba* and *Polystichum vestitum*. Largely as a result of trampling and browsing by deer, *Microlaena avenacea* covers extensive areas of more or less disturbed soil. On stable ground there are colonies of the giant moss *Dendroligotrichum dendroides*. *Rubus cissoides* ascends to 700 m on semi-fertile soils. The only largish vascular epiphytes are *Asplenium flaccidum* and *Lycopodium varium*, but bryophytes and filmy ferns (especially *Hymenophyllum multifidum*) thickly mantle trunks and branches, *Weymouthia* and *Papillaria* festoon twigs in the lower stories, and *Usnea* grows densely on the upper canopy twigs. Above 900 m, this forest merges into bush and heath.

On floors and lower slopes of mountain valleys subject to hard frost and shading by ridges, upland plants descend to low levels, leading to unusual combinations; for example, kaikawaka and mountain totara may grow with lowland podocarps (p. 128). Other upland species that descend include *Olearia ilicifolia*, *O. colensoi*, *Hoheria glabrata, Aristotelia fruticosa* and *Cyathea colensoi.* Conversely, lowland plants may be absent from upper valley reaches that are still below their usual altitudinal limits, or they may be restricted to sunny slopes; rimu, kamahi and lianes other than *Rubus cissoides* show this kind of distribution.

Southern districts

Like its low-altitude equivalent, extensive high-altitude conifer/broad-leaved forest ceases just north of the Paringa River in South Westland, although there are isolated patches as far south as the Arawata headwaters. From there, beech forests extend unbroken to Foveaux Strait. In Catlins ER and near Dunedin, however, lowland mixed forests pass up into upland forests with kaikawaka and mountain totara, together with pink pine on poorly drained ridges. On Stewart Island, kaikawaka is absent, and because of severe exposure forest does not extend above the limit of rimu.

Eastern South Island

A thousand years ago, high-altitude mixed forests probably clothed most mid-mountain slopes east of the Divide and the crests of higher hills, except where beech forests prevailed and on north to west aspects in the driest inland basins. Like their low-altitude equivalents, these rain-shadow forests have been reduced to scattered, depleted remnants. That they are still vulnerable was shown by the accidental burning of mountain totara forest in Mt Cook National Park in 1970 (p. 547). An impression of the original extent and composition can be gained by piecing together evidence from remnants, and from widely distributed soil charcoals and durable logs, mainly of mountain totara (Fig. 7.22).

Kaikawaka and mountain totara grow together on Banks Peninsula, at Mt Peel in South Canterbury, on the Seaward Kaikoura Range and, often with pink pine, close to the Main Divide from the Waimakariri headwaters to Mt Cook. Only mountain totara occurs in the remnants in drier districts, such as the Inland Kaikoura Range (Williams 1989) (Fig. 7.23). The broad-leaved canopy lacks both kamahi and southern rata, except for very local occurrences of stunted trees. Instead, smaller trees prevail, especially *Griselinia littoralis* and *Phyllocladus alpinus*: except in the few places where tall conifers are dominant, surviving stands are better regarded as woodland or bush than true forest, and fuller description is reserved for Chapter 8.

Southern North Island

Beech forests prevail throughout the mountains in the north-west and north of the South Island, and on the axial ranges of the North Island. However, there are minor areas in the Tararua and Ruahine Ranges where beech is absent at high altitudes, and a partial 'beech gap' centred on the Manawatu Gorge extends for 100 km, with continuity from lowland mixed forests to high-altitude equivalents. In these areas, kamahi dominates from the upper limit of rimu to about 900 m, and is associated with miro and mountain totara at the lower altitudes and mountain totara alone at the higher. Lower-tier species are much as in Westland, with the addition of toro and of *Collospermum microspermum* as a characteristic epiphyte. On the upper ridges and

Fig. 7.22. *Podocarpus hallii* logs in short-tussock grassland, resulting from Maori fires. Haldon Hills, Wairau ER.

spurs in the Tararua Range, kamahi stands develop wind-roofs with heights as low as 2 m.

In the Ruahine Range between *c.* 900 m and 1100 m, but especially in the western part, stands of kaikawaka and pink pine form discrete stands and transitions to beech forest. Kaikawaka, which also occurs in the adjacent part of the Kaimanawa Range to the north-west (Elder 1962), seems favoured by limestone and other Tertiary sedimentary rocks, recent tephra and Maori fires, whereas pink pine dominates on flatter ridges which are often peaty.

High-altitude mixed forest around the active central volcanoes, mostly dominated by mountain totara (Atkinson 1981), has been fragmented through recurrent eruptions and a millennium of fires. Older, subdued mountains on the Volcanic Plateau support more intact stands but, being relatively low and isolated, summit effects are pronounced. On Mt Pureora, an old volcanic dome, forests with rimu, matai, *Quintinia, Pseudopanax edgerleyi*, toro, etc., ascend to 850–900 m (Fig. 5.1). Above this, the mountain is encircled by communities in which kamahi and mountain totara form a dense wind-roof at 7–12 m. *Pseudowintera colorata* dominates the understorey. Other species include *Leptopteris superba, Cyathea smithii, Histiopteris incisa* (in canopy gaps), *Griselinia littoralis, Quintinia, Rubus cissoides, Uncinia rupestris* and mosses, especially *Hypnodendron* (Section *Mniodendron*). Above 1000 m, this merges into bush (p. 187).

Fig. 7.23. *Podocarpus hallii* trees scattered among *Festuca matthewsii* and rocks, 1330 m, Inland Kaikoura Range. *P. nivalis, P. hallii × nivalis* and *Phyllocladus alpinus* in foreground (photo P.A. Williams).

High-altitude mixed forests are well developed on Mt Taranaki, since beeches are absent. Kamahi prevails in the cloud forests that extend up from 760 m. With rising altitude stature decreases to less than 10 m, and mountain totara increases to co-dominance. Just above the altitudinal limit of kamahi at 1100 m, mountain totara, kaikawaka and *Griselinia littoralis* form the canopy before they gradually give way to tall scrub. The main lower-tier species of the cloud forest are toro, *Pseudowintera colorata, Astelia fragrans, Blechnum procerum* and *B. fluviatile*. Others include *Pseudopanax simplex, P. colensoi* and, reflecting eruption of tephra as recently as the eighteenth century (Druce 1966), *Schefflera, Carpodetus, Coprosma tenuifolia, Cyathea smithii* (below 930 m), *Leptopteris superba, Rubus cissoides* and *Microlaena avenacea*.

Fig. 7.24. Cloud forest at 910 m a.s.l., Mt Taranaki (photos B.R. Clarkson). (a) Crooked boles are kamahi; the erect bole at left is mountain totara. (b) Roots of *Pseudopanax simplex* descending a mountain totara trunk with dense filmy ferns.

Northern zone

Some northern mountains reach into the cool-temperate belt, but few exceed the upper limits of rimu and miro. Nevertheless, their isolation leads to pronounced summit effects, including low, dense canopies and presence of high-altitude species. Between 660 m and 780 m on Te Moehau in the Coromandel Range, Cranwell & Moore (1936) describe frequent kauri trees in forest dominated by rimu and kamahi; *Ixerba* is also abundant. The forest is low-statured, and on ridges even rimu may be only 6–9 m tall. Other important trees are *Phyllocladus glaucus*, miro, mountain totara, *Knightia*, *Quintinia serrata*, toro, southern rata, *Pseudopanax colensoi* and *Griselinia littoralis*.

Shrubs include *Pseudowintera axillaris*, *Corokia buddleioides*, *Alseuosmia macrophylla*, *Coprosma foetidissima*, *C. grandifolia*, *C. lucida* and *Brachyglottis kirkii*. The herbs *Blechnum 'capense'*, *Sticherus cunninghamii*, *Gahnia setifolia*, *Astelia trinervia* and *Libertia pulchella* are listed, and the abundance of the liane *Metrosideros albiflora* is noted. Perched in the crotches of kauri trees there are *Dendrobium cunninghamii* (the most abundant), *Brachyglottis kirkii, Lycopodium varium, Tmesipteris* and occasional *Pittosporum kirkii* and *Astelia solandri*. On trunks and large branches there are *Trichomanes reniforme, Hymenophyllum multifidum, Bulbophyllum pygmaeum*, and bryophytes including *Lepicolea scolopendra, Macromitrium longipes, Dicnemon calycinum, Holomitrium perichaetiale* and *Cladomnion ericoides*; on the barest bark there is *Trentepohlia*. Young plants of 16 vascular species grow as casual epiphytes.

On a high, flat ridge crest, forest with rimu and kauri also contains kaikawaka, *Phyllocladus alpinus, Libertia pulchella* and *Carpha alpina*. Species listed from forest without kauri on Te Moehau include *Freycinetia, Cyathea smithii* and *Microlaena avenacea*, the last increasing spectacularly where the understorey has been depleted through browsing. Other sources list *Ascarina lucida, Pseudopanax laetus* and *P. simplex*. *Archeria racemosa* and *Dracophyllum traversii* occur further south on the Coromandel Range.

Similar combinations of warm- and cool-temperate and even subalpine species occur on Little Barrier Island (Hamilton & Atkinson 1961), the Herangi Range (B.D. Clarkson, unpublished), Pirongia (Clayton-Greene 1977) and Mt Karioi (Clayton-Greene & Wilson 1985). In the mixed forests of the Kaimai Range and Raukumara and Urewera ERs *Ixerba* and kamahi increasingly prevail above 600 m, alternating with beech descending on ridge crests.

Forests on the highest summits in Northland are warm-temperate, but lack taraire and kauri above 450 m and tawa above 600 m, while supporting rare occurrences of southern rata, *Griselinia littoralis* and *Dracophyllum traversii*. At 600 m in Mataraua Forest, the canopy consists mainly of *Weinmannia silvicola* only 12–14 m tall, with a scattered overstorey of northern rata and rimu. Other main species include *Cyathea smithii, Dicksonia squarrosa, Blechnum discolor, Ackama rosifolia, Ixerba*, hinau, *Pseudopanax arboreus, P. edgerleyi, Schefflera*, toro, *Melicytus macrophyllus, Coprosma grandifolia, Freycinetia, Ripogonum*, nikau and, in hollows, *Syzygium* and pukatea.

Coastal forests

Along sheltered inlets and estuaries, especially in Fiordland and Stewart Island, tall, luxuriant, lowland forest reaches high-tide level. Forests of the same floristic composition, but with the canopy trees lowered and deformed by salt winds, can grow on quite exposed coasts. Other woody coastal vegetation in New Zealand is discussed with bush, heath or cliff communities in Chapters 8 and 11, including pohutukawa stands even though these often reach tall-tree stature.

Beech forests

(For comprehensive accounts, see J.A. Wardle 1984; Poole 1988.)

Beeches form continuous subalpine forests, but with decreasing altitude become increasingly confined to shallow soils on crests of ridges and spurs, where they can descend into the warm-temperate belt. Intervening faces and gullies support either mixed forest or bush of small broad-leaved trees. Beeches also occur at low altitudes on river banks, recent terraces and slips that were seeded with beech before competing species could dominate. There are also extensive forests in which beeches intermingle with podocarps, other broad-leaved trees, and even kauri (Table 7.4). These are best developed in the cool-temperate belt in regions of moderate to high rainfall, where they form broad transitions between forests of flats and lower slopes near the coast and pure beech forests of subalpine or inland localities.

Beeches are replaced by mixed forest and bush in some districts that, on all available evidence, contain suitable sites for beech. The major gaps are Mt Taranaki, parts of the Volcanic Plateau, both sides of the Manawatu Gorge, Westland between latitudes 42° 40′ and 43° 40′, and Stewart Island. There are also extensive gaps in the east of the South Island, although these are less apparent because of extensive deforestation (Fig. 7.25).

Beech forest species

With small, hard leaves, unisexual, wind-pollinated flowers, small winged nuts, relatively fast growth, gregarious habit and ectomycorrhizal rootlets, the beeches stand apart from other native trees. Three of the species, *Nothofagus solandri, N. fusca* and *N. truncata*, readily form hybrids, especially in disturbed areas. *N. menziesii* is more closely related to species in Australia and South America than to other New Zealand beeches.

Hard beech (*N. truncata*) has the most northern range (latitude 35° 15′ – 44° S), but also has the most interrupted. It also has the lowest altitudinal range: down to sea level at latitude 36° and occasionally up to 900 m at latitude 42° S. It is a species of spurs, ridges, and the better-drained podzols on high terraces, which suggests tolerance of infertile, dry soils, but intolerance of frost.

Red beech (*N. fusca*) is the tallest and fastest-growing, reaching heights of over 42 m. It is most abundant on moist, moderately fertile, freely drained soils on the terraces and lower slopes of inland mountain valleys. At latitude 42°, it ascends over 1000 m and descends almost to sea level, but seldom reaches the coast.

N. solandri, the only species with entire leaves, includes black beech (var. *solandri*) and mountain beech (var. *cliffortioides*). More or less oblong leaves in the former compare with ovate leaves in the latter, but the varieties differ markedly in stature and habitat. Black beech is usually a large tree, of low and mid-altitudes, that ranges southwards to about latitude 43° 30′. It grows on a wide range of soils, but in competition with red or hard beech is largely restricted to dry, well-lit sites. Mountain beech forms most of the drier subalpine forests in the two main islands, and is also widespread in cool, wet districts on shallow or boggy soils. In these stands it seldom exceeds 15 m in height, and can be reduced to shrub form at the upper forest limit.

In the north-east of the South Island, forests of *N. solandri* can extend unbroken from the plains and valley floors to the forest limit, and grade imperceptibly from var. *solandri* to var. *cliffortioides*. Elsewhere, there are ecotypes that cannot be referred readily to either of the named varieties (Wilcox & Ledgard 1983). One of these grows

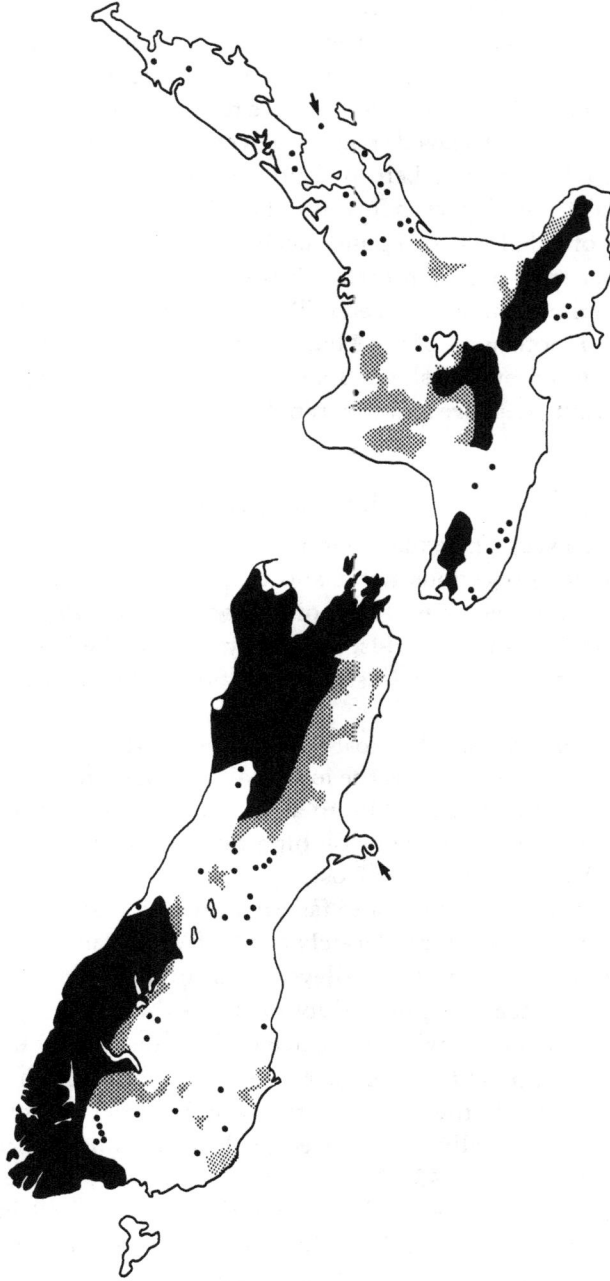

Fig. 7.25. Distribution of *Nothofagus* spp., based on presence in 9140 m (10 000 yard) grid squares (from Wardle 1984). Dots indicate isolated occurrences; stippling, discontinuous patches; and solid black, continuous beech forest.

on swampy ground and along river banks in North Westland. Another, that grows at low altitudes in western parts of Otago and Southland, approaches red beech in stature, growth rate and habitat preferences.

Silver beech (*N. menziesii*) has cherry-like bark and hooked processes on the cupule valves. Like its Australian and South American relatives, it is host to the strawberry fungus (*Cyttaria*), named for the edible fructifications that arise from large stem galls. It dominates widely in subalpine forest on the wetter mountains. Where it overlaps with mountain beech, silver beech tends to occupy moister, more fertile soils and more sheltered sites; yet it forms isolated groves near the aridity limits for native forest in Central Otago. The lower limits of silver beech descend towards the south, and it is abundant at sea level in the western fiords.

In hierarchial classifications based on cluster analysis of presence–absence data, occurrence of beech has only minor bearing on the position of communities since vascular species associated with beech forests are the same as those found in mixed forests on comparable sites (P. Wardle & R.P. Buxton; C.B. Woolmore; both unpublished). However, these associated species are much less abundant, often to the point of almost complete exclusion, and perhaps only the following four are predominantly associated with beech forest.

Chionochloa cheesemanii, a fine-leaved tussock, grows in montane and lower-subalpine beech forests from North-west Nelson to the Rimutaka Range, and very locally on other axial ranges north to Lake Waikaremoana. It occupies spurs and ridges with leached soils, and can form a continuous tier some 60 cm tall. There is an occurrence, not associated with beech, on the Seaward Kaikoura Range. In some regions *Lagenifera strangulata* is confined to beech forest. *Pittosporum patulum* is confined to beech forest and adjoining subalpine tall heath, in inland Marlborough and near Lake Ohau nearly 300 km to the south-west, and *P. dallii* grows only in subalpine forest near Takaka (p. 378).

Three large mistletoes with showy red (*Peraxilla colensoi, P. tetrapetala*) or yellow (*Alepis flavida*) flowers are almost confined to beech trees, but have become rare or extinct over large areas (p. 36). Bryophytes are often abundant on the forest floor and in superhumid districts extend up the trunks, whereas drier beech forests have a profusion of corticolous lichens.

Beech forests of the northern South Island

The north of the South Island has the largest remaining tracts of indigenous forest in New Zealand, and most of this is dominated by beeches. The following account begins with the forests of the cloudy, wet ranges adjacent to the Tasman Sea, moves to the drier hills around Tasman Bay, and concludes with the inland mountain valleys in the south-east.

Western ranges

The lowest seaward slopes on the western ranges support warm-temperate mixed forest, but hard beech occurs widely, and locally dominates on narrow spurs, which it can descend almost to sea-level, especially on well-weathered early Pleistocene gravels. Red, silver, hard and (locally) black beech follow major rivers almost to the coast. Above and inland from the coastal belt, beeches prevail. Hard beech dominates on most mid-slopes below 650 m, especially on spurs. Red beech tends to dominate on moist colluvial slopes, and silver beech usually dominates on valley floors, as well as being scattered throughout. Above the limits of hard beech, silver beech expands to occupy both gullies and spurs; red beech remains largely confined to colluvial slopes up to an altitudinal limit around 1000 m, but on limestone debris it ascends warm aspects to 1150 m. With increasing altitude, mountain beech increasingly dominates on shallow, leached soils.

Tall podocarps are sparingly represented by rimu and miro up to 640 m, and throughout by mountain totara, but the last mainly as suppressed saplings. Kaikawaka is rare, as it competes poorly in tall, close-canopied beech forest. Kamahi is the main subcanopy tree but is slender and small-crowned beneath the beech canopy. Pokaka (*Elaeocarpus hookerianus*) occurs mainly on wet ground, southern rata occupies rock outcrops, and *Quintinia* enters at lower altitudes.

Griselinia littoralis, Pseudopanax simplex and, at lower altitudes, toro are the main small trees, although the very palatable *P. colensoi* var. *ternatus* may have once been more abundant than it is today. Slender plants of *Dracophyllum traversii* and layering colonies of small *Phyllocladus alpinus* are usually present. Shrubs are locally abundant, especially beneath canopy gaps, and include *Coprosma foetidissima* and *C. microcarpa*. The tree fern *Cyathea smithii* is common below 700 m, and the sessile *C. colensoi* above that. Locally at the lower altitudes, there are large colonies of *Dicksonia lanata*, which is also sessile. By far the most important large ground fern is *Blechnum discolor*, but *Polystichum vestitum* can replace it in gullies, and there are patches of the ubiquitous *Blechnum 'capense'*. Where there is thick litter of beech or fern leaves, a ground layer is nearly absent except for bryophytes and *Nertera* on rotting logs, but on broad ridge crests there are extensive patches of *Dicranoloma* and, in lesser abundance, *Leucobryum candidum*.

In gullies and on unstable slopes and windthrown areas, especially at lower altitudes, continuous beech often gives way to communities of small trees (e.g. *Carpodetus*, wineberry, *Pseudowintera colorata* and *P. axillaris*), groves of *Cyathea smithii*, tangles of *Rubus cissoides*, patches of *Microlaena avenacea* or uncinias (especially *U. uncinata*), or dense ferns (e.g. *Blechnum discolor, B. 'capense', B. fluviatile* and *Histiopteris incisa*). In contrast, on the shallow, leached soils where mountain beech prevails, *Phyllocladus alpinus* and, at lower altitudes, rimu increase, shrubs such as *Cyathodes juniperina, C. fasciculata* and even manuka (*Leptospermum scoparium*) can be present, and the ground cover is largely *Dicranoloma*.

Forests above 1000 m are mostly dominated by silver beech, which extends to the forest limit, but mountain beech grows on shallow, infertile soils on granite and other resistant rocks, in communities transitional to tree-heath (Rose 1985; Druce *et al.* 1987). On spurs and ridges, the beech canopy can be almost continuous, with little beneath except mosses, suppressed seedlings of beech, and scattered shrubs, especially *Coprosma pseudocuneata*. Along the gradient from ridges across moist faces into gullies, the lower tiers become denser, and beech trees become larger and further apart. In areas little browsed by deer, *Polystichum vestitum* and shrubs such as *Coprosma pseudocuneata, Olearia colensoi, Brachyglottis buchananii* and *Pseudopanax*

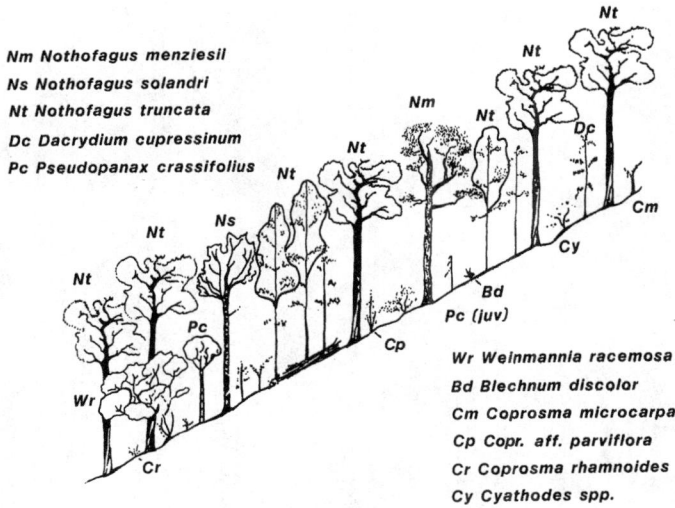

Fig. 7.26. Profile of (podocarp)/beech forest, Big Bush, Nelson ER (Benecke & Evans 1987).

colensoi are prominent, but where browsing is fairly severe the shrub tier tends to consist of small-leaved coprosmas, clumps of hedged beech saplings, and patches of *Hypolepis millefolium, Poa breviglumis* and *Uncinia* spp.

Other small trees and shrubs that can be abundant in subalpine beech forest, especially towards the heads of valleys and in transitions to tree-heath and bush, include *Dracophyllum traversii, Olearia lacunosa, O. ilicifolia* and, locally, *Hoheria glabrata*. The upper forest limit lies about 1400 m, but on mountains with low, exposed ridges and summits and in high-altitude basins it is depressed in favour of grassland and scrub.

Nelson Ecological Region
In Nelson ER, which is characterised by warm, often dry summers and infertile, podzolised yellow-brown earths, beech forests once prevailed from the coast to the upper tree-limit. Podocarp/broad-leaved stands were practically restricted to fertile alluvial flats and sheltered coastal valleys. Towards the coast, Polynesian and European clearing have reduced the forests to scattered remnants, but they are still extensive in inland parts.

At the lower altitudes, beech forests consist largely of hard beech. Black beech grows mainly on the lower ends of sharp spurs and on brows of alluvial terraces, red beech tends to be confined to moist colluvial slopes, and silver beech to frosty valley floors. Rimu and, in lesser abundance, miro are scattered through these beech forests (Fig. 7.26). Totara, matai and kahikatea grow mainly on valley floors and fertile colluvial slopes, in fingers of mixed forest rather than among beech. Lower tiers are much the same as in the western beech forests, but generally not as dense. Even *Blechnum discolor* is abundant only on moist, cool slopes, and through much of the

Fig. 7.27. Interior-temperate red beech – silver beech forest at 450 m. Sabine valley, Spenser ER.

forest, the undergrowth consists largely of small-leaved shrubs such as *Coprosma microcarpa, Cyathodes fasciculata* and *C. juniperina,* and extensive mats of *Dicranoloma* spp. and *Leucobryum*. Kanuka (*Kunzea ericoides*) can occur on sharp, dry spurs, possibly as a member of climax forest as well as a legacy of past fires, and the small tree *Coprosma linariifolia* grows on rocky slopes. On seaward slopes *Cyathea dealbata* is often abundant.

Red and silver beech continue above the limits of hard beech, and mountain beech also increases, often dominating towards tree-limit, on shallow or boggy soils, and on ridge crests. Where drainage is poor, these upland beech forests can contain kaikawaka, and heath species such as *Phyllocladus alpinus*, pink pine and *Archeria traversii*. The beech forests of Sounds ED closely resemble those of Nelson, except that mountain beech is absent.

Interior and eastern regions of the northern South Island
Hard and black beeches scarcely enter the deep inland valleys, flanked by high mountains, of Spenser ER. Probably winter frosts are too severe for the former, and summers too cool for the latter. Where rainfall is greatest, the forests from river level to around 1000 m are dominated by red and silver beeches (Fig. 7.27). Mountain beech here is mainly confined to shallow soils on glacially scoured ridges, and to the margins of stands colonising recent terraces.

On ridges, undergrowth consists largely of *Cyathodes juniperina* and *C. fasciculata*, whereas on moister slopes it consists of small-leaved coprosmas and *Blechnum discolor*. In gullies the beech canopy opens out, to admit such species as *Griselinia littoralis*, fuchsia, lancewood and *Polystichum vestitum*. Kamahi, southern rata and pokaka occur sporadically above the level where cold air ponds on valley floors, the first two mainly on rocky outcrops and edges of ravines and the last on damp ground. Podocarps are generally absent, except for mountain totara which occurs widely as seedlings and saplings but rarely as fully-grown adult trees. There is usually a well developed bryophyte ground cover with *Dicranoloma robustum* s.l., *D. plurisetum*, *Pyrrhobryum mnioides*, *P. bifarium*, *Hypnum cupressiforme*, *Ptychomnion aciculare*, *Leptotheca gaudichaudii*, *Camptochaete arbuscula* and species of *Bazzania* and *Trichocolea* (Fig. 7.28).

Mountain beech expands its edaphic range towards the east, mainly at the expense of silver beech which contracts towards moist, sheltered sites at lower altitudes. East of Lake Rotoiti, subalpine mountain beech stands extend above 1500 m, higher than any other forests in New Zealand. On spurs and ridges there may be no other vascular plants, but a few other species appear on concave slopes, notably the erect shrub *Coprosma* aff. *parviflora*, the trailing *C. depressa*, and *Hypolepis millefolium*; *Phyllocladus alpinus* enters where soils are gleyed. Even this meagre understorey has been much depleted by deer. Along upper and valley-head forest limits, the canopy tends to lower and open out, and mountain beech trees mingle with a denser understorey of subalpine shrubs, especially *Podocarpus nivalis*, *Phyllocladus alpinus* and, where precipitation is higher, *Coprosma pseudocuneata*.

These species-poor mountain beech stands extend into inland Marlborough where precipitation is less than 1200 mm annually, but fires during the past millennium have broken them into discontinuous fragments among the prevailing tussock grasslands. Red and black, but not silver or hard beeches reappear in the sporadic beech stands of Kaikoura ER (Fig. 7.29).

Central and southern South Island

Beech forests become attenuated towards the Westland 'beech gap', but the most south-western patch of beech forest in North Westland, on glacial outwash terraces near Kumara, still contains hard, red, mountain and silver beeches. In Canterbury, on the other hand, nearly all beech forest from Arthurs Pass and Lowry ER southwards consists of mountain beech (Fig. 7.30) merging into black beech where the foot-hills descend to the plains. Increasingly isolated patches extend as far as Geraldine and southern tributaries of the Rangitata River, and in these mountain beech often seems to be substituting for other beech species. For instance, at Arthurs Pass, with an annual precipitation > 5000 mm, mountain beech forest has a well-developed understorey of species normally associated with silver beech, such as *Brachyglottis buchananii*, *Pseudopanax linearis* and *Coprosma foetidissima*. In Canterbury Foothills ER, stands that are taxonomically close to black beech resemble red beech communities in having scattered rimu and other podocarps and a *Blechnum discolor* tier on moist slopes. However, remnants near Mt Grey in this ER and east of Akaroa

Fig. 7.28. Interior-temperate red beech forest with (*a*) beech seedlings and *Coprosma* spp. and (*b*) *Dicranoloma* cushions; 520 m, Maruia valley, Spenser ER.

Fig. 7.29. Mixed forest with broad-leaved trees and matai (centre) on limestone in foreground and black beech on siliceous terrain across the valley. Waima River, Kaikoura ER.

on Banks Peninsula contain both red and mountain/black beeches, and there are patches of silver beech in the Rakaia and Ashburton catchments.

Much further south, silver, mountain and red beeches coexist in a tract extending from Big Bay and Martins Bay to lakes Wakatipu and Te Anau. In the rain-drenched, western part of this tract, silver beech is overwhelmingly dominant from the valley floors to the upper forest limit at 1100 m. Mountain beech supplants it on shallow soils with stunted forest or tree-heath, whereas red beech grows on sheltered colluvial slopes and wide river terraces. An altitudinal sequence is portrayed in Fig. 7.31. Below 700 m, these beech forests contain species characteristic of temperate mixed forest such as rimu, kamahi and *Blechnum discolor*, whereas high-altitude understoreys consist mainly of small-leaved coprosmas and ferns such as *Polystichum vestitum* and *Blechnum* 'mountain' (Fig. 7.32).

On the precipitous fiord walls, the forest mantle is interrupted by numerous slips

Fig. 7.30. Mountain beech forest with seedlings in openings. Craigieburn Range, Puketeraki ER.

that expose granite bedrock at various stages of revegetation (Fig. 6.11). On steep, broken terrain, beech trees may be dispersed among species-rich shrub and herb communities, as in the kakapo habitat illustrated in Fig. 7.33.

Silver beech also prevails across the passes into the heads of east-flowing valleys, but nearer the lakes red, silver, and mountain beeches form mixed stands on lower slopes, with the last two forming the subalpine forests. The last outposts of red beech in Fiordland are near Milford Sound and Lake Monowai; beyond these, forests of silver or mountain beech prevail to the southern coast.

Most of the terrain east of the lakes was deforested during the past 1000 years, and as long ago as 2500 years in the driest parts (McGlone 1988). Previously, forest and woodland without beech had prevailed, not only on the lowlands, but also at high altitudes and bordering the semi-arid basins of Central Otago. However, surviving blocks of forest in inland Southland consist mainly of silver beech, with some mountain and red beech to as far east as the Blue Mountains (Lammerlaw ER). Silver beech also exists as tiny pockets around the fringes of Central Otago, mainly in deep ravines, and as isolated stands along the eastern hills, from Catlins ER to Burke Pass in Pureora ER.

Northwards towards the central South Island 'beech gap', beech forests also decrease in extent and variety. West of the Main Divide, first red, and then mountain beech drop out, leaving silver beech to dominate, as continuous beech forests break up into fingers and islands among the prevailing mixed forests. Hard beech grows in a few

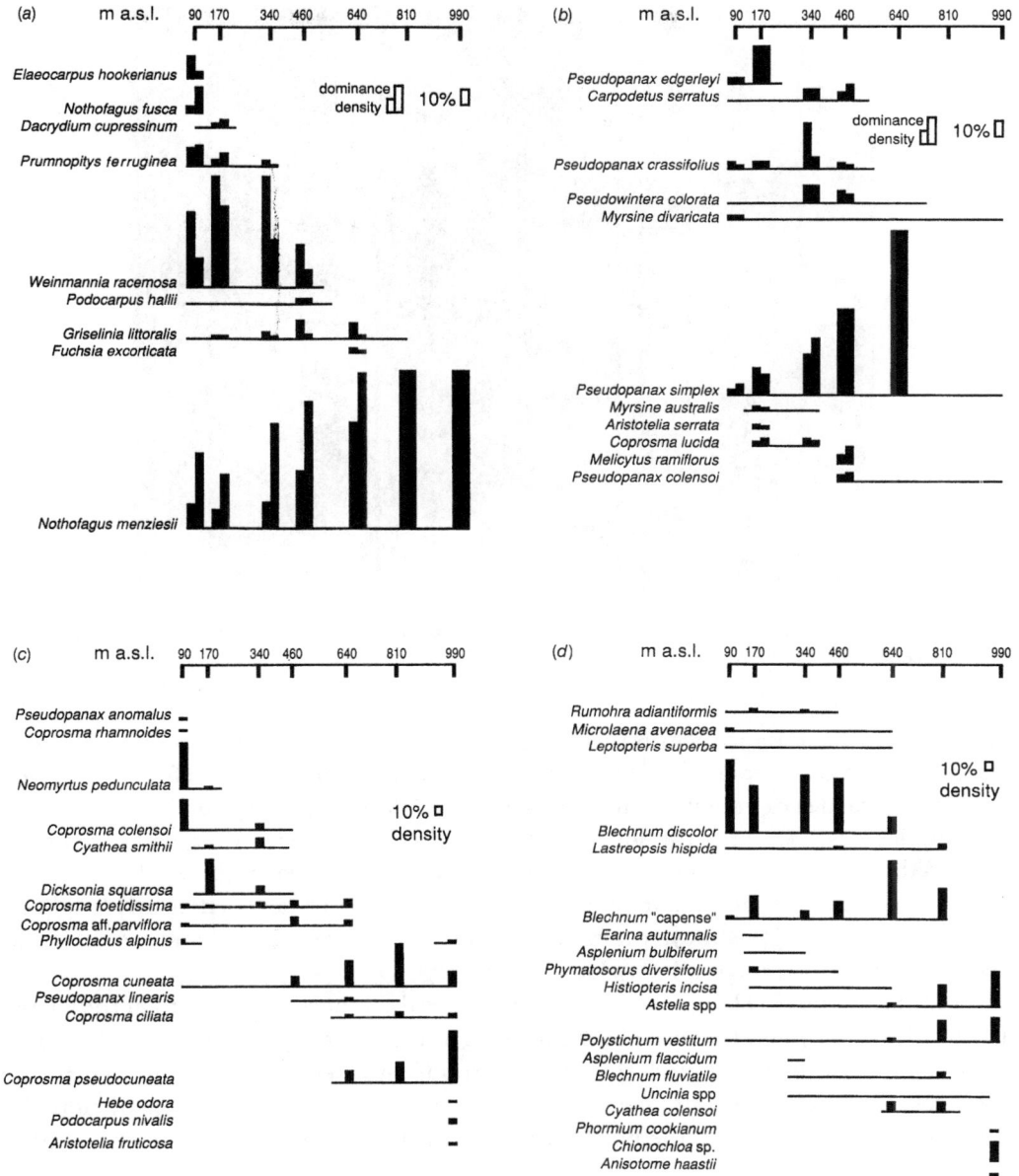

Fig. 7.31. Composition of silver beech forest along an altitudinal gradient in the Hollyford valley, northern Fiordland. Species portrayed only in their tier of greatest abundance (from Mark & Sanderson 1962). (*a*) Trees over 10.2 cm d.b.h.; (*b*) trees under 10.2 cm d.b.h., over 4.57 m tall; (*c*) shrubs 0.3–4.57 m tall; (*d*) herbs.

Fig. 7.32. *Leptopteris superba*, *Polystichum vestitum* and *Microlaena avenacea* in silver beech forest at 210 m, Gorge River, Olivine ER.

spots near Jackson Bay in South Westland, a disjunction of 250 km. East of the Divide, the northernmost outpost of red beech is at Lake Hawea. Silver beech forest occupies the upper reaches of valleys at the heads of Lakes Wanaka, Hawea and Ohau, but as rainfall decreases towards the south-east, mountain beech replaces it, and extends as scattered remnants in sheltered gullies for several kilometres beyond the lakes. Pockets of mountain beech and one of silver beech occur as far north as Mt Cook, so that among the major valleys draining eastwards from the Main Divide, only those feeding Lake Tekapo completely lack beech.

Southern-zone beech forests have fewer species than those in the north of the island (Table 7.5), the most notable absences being the shrubs *Cyathodes fasciculata* and *Coprosma microcarpa*, although the latter species is represented by *C. cuneata* instead. The coccid scale *Ultracoelostoma* and associated sooty moulds, so conspicuous in temperate beech forests in the north of the South Island, seem absent in the south.

North Island axial ranges

Beeches generally dominate the montane and subalpine belts in the North Island axial ranges, becoming restricted at lower altitudes to ridges and terraces. Although the five taxa are widely distributed, it is unusual to find all of them in the same locality. In some high-altitude forests no beeches are present (p. 136). To an extent, the distribution patterns seem related to environmental gradients, especially rainfall, but vagaries of dispersal have probably also played a role, because the mountains are relatively low and suitable beech habitat is fragmented. Eruptions may have influenced the distribution of beech in the ranges adjacent to the Volcanic Plateau.

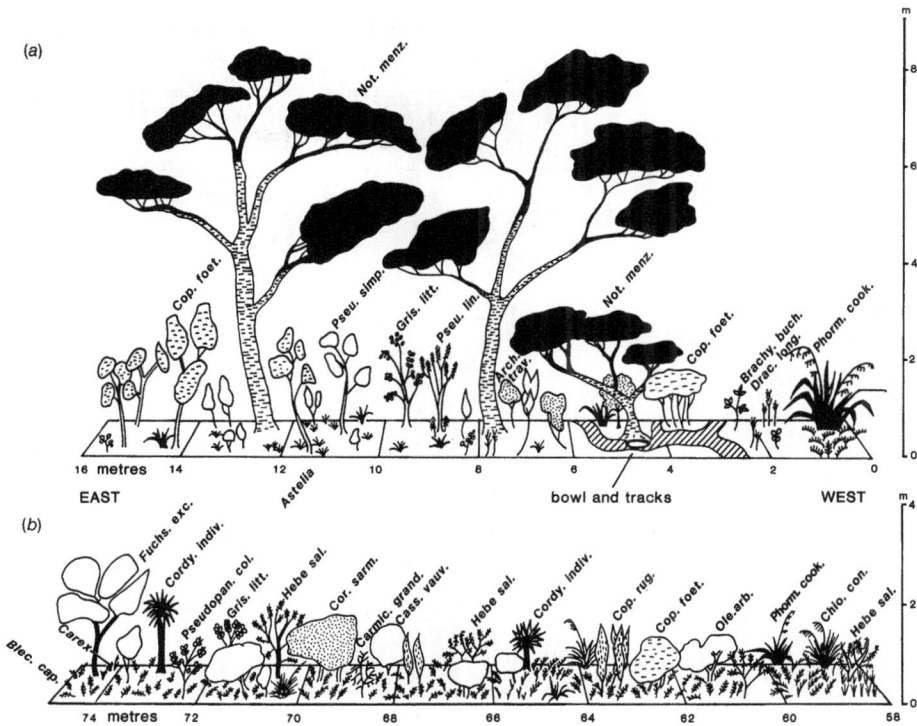

Fig. 7.33. Profiles of 1 m wide transects across silver beech forest and successional vegetation representing a territory (a) and 'garden' (b) of the almost extinct kakapo parrot (*Strigops habrotilus*); 860 m and 600 m respectively, Milford Sound. Fiordland (Johnson 1976a).

Silver beech alone forms the subalpine beech forests in the Tararua Range in the south, and is the main subalpine beech on the Raukumara, Huiarau and subsidiary ranges in the north. Mountain beech is also present on these northern ranges, especially south of Lake Waikaremoana where it can codominate with or replace silver beech. In the Kaweka and Ruahine Ranges and southern parts of the Kaimanawa Range the subalpine beech forests are largely pure mountain beech, silver beech being very local.

With descent into the montane belt, red beech joins or replaces the high-altitude beeches, except in the southern Kaimanawa Range and the adjacent Kaweka Range, where mountain beech dominates at both subalpine and lower altitudes. Black and hard beeches may enter the lower part of the montane belt, and descend on dry, infertile sites to the lowlands; on coastal slopes in Raukumara ER hard beech comes so low as to mingle with pohutukawa (*Metrosideros excelsa*). Black beech occurs almost through the full length of the axial ranges, even as pockets within the Manawatu Gorge 'beech gap', and is especially common on the flanks of the Aorangi and Rimutaka ranges, whereas hard beech is absent between latitudes 38° 50' and 40° 30'.

Table 7.4. *Mean number of trees ≥ 30.5 cm d.b.h./ha in a selection of widespread forest types ranging from podocarp/beech to pure beech stands*

type:	1	2	3	4	5	6	7	8
altitudinal range (m):	100–500	100–500	0–450	100–400	150–450	100–700	100–900	200–700
Dacrydium cupressinum	40	35	21	9	1			
Prumnopitys ferruginea	1	3	9	6		2		
Dacrycarpus dacrydioides	5		2	4	8			
Podocarpus hallii	2		4		2			
Podocarpus totara			2					
Libocedrus bidwillii	5				4			
Nothofagus truncata	8	58				60		
Nothofagus fusca				17		17	2	61
Nothofagus menziesii	6		67	64	25	11	115	32
Nothofagus soland. v. cliff.	6		6	8	62	1	1	9
Weinmannia racemosa	3	13	24	7		6	13	
Metrosideros umbellata	4	9	5			2	1	
Quintinia acutifolia		2				2		
Elaeocarpus hookerianus				+	2			
Lagarostrobos colensoi	26							
Halocarpus biformis	1							
Lepidothamnus intermedius	1				1			

Notes:
1 Sluggishly drained flat or gently sloping sites in North Westland ER.
2 Gentle to steep terrain from North Westland to Sounds ER.
3 Low-altitude forests from South Westland to Catlins ER.
4 Gentle to flat terrain in North Westland ER, soils more fertile than 1 and 5.
5 Poorly drained, generally flat, inland sites in North Westland.
6 Generally steep terrain in North Westland, North-west Nelson and Nelson.
7 Mainly high-altitude forests in South Westland and Southland.
8 Local areas in inland Southland.
Source: McKelvey 1984

The axial-range beech forests, especially those of higher altitudes, have essentially the same composition as in the South Island, although many South Island species are absent or sporadic. For example, *Hoheria glabrata* is absent throughout, high-altitude conifers other than mountain totara are rare or absent in the Tararua Range, southern rata is absent except for a few trees in the Tararua Range, and *Archeria* spp. and large-leaved dracophyllums are lacking except for *A. racemosa, D. traversii* and *D. latifolium* in Gisborne province. Expanded ecological roles for other species, notably *Olearia colensoi* and *Chionochloa conspicua*, may compensate for the absences.

The southern Kaimanawa and Kaweka Ranges are the driest in the North Island, and in structure and content their mountain beech forests resemble those in the east of the South Island. Subalpine mountain beech forests on other ranges, however, have understoreys like those of subalpine silver beech forests, indicating much wetter conditions. Low-altitude beech stands reflect the floristic content of mixed forests that

Fig. 7.34. Pocket of mountain beech at 1300 m on the northern slope of Ruapehu. Ngauruhoe in distance.

they merge into. South of latitude 39° S, there are few species that are not also frequent in podocarp/beech/broad-leaved forests of Sounds ED; *Nestegis* is an exception for, although three species extend to the South Island, they are very local there and not associated with beech forest. Northwards, the northern or disjunct *Phyllocladus trichomanoides, P. glaucus, Quintinia* and *Ixerba* appear, the last often dominating a tier beneath the beech canopy.

Other beech stands of the North Island

Except in the axial ranges, native forests of the North Island are overwhelmingly podocarp/broad-leaved or, in the north, kauri-dominated, even on high land where beech forests might be expected. Yet, with the exception of Te Paki ER, Mt Taranaki, parts of the Volcanic Plateau and extensive lowland plains, beeches are never completely absent from wide areas.

The largest of the isolated beech stands are on the side of Mt Ruapehu that was sheltered from pyroclastic flows from Lake Taupo. Red and silver beech co-dominate on the southern flank up to 1070 m, with silver beech alone extending to 1220 m, where it is replaced by mountain beech, which prevails to the tree limit of 1500 m. On the western flank, mountain beech appears at a lower altitude, and becomes the only species above 980 m. Annual precipitation exceeds 2500 mm, and the wide altitudinal extent of mountain beech and the correspondingly depressed limits of the other beeches are due to the wet, infertile soils derived from andesitic lava and still-accreting tephra. Over much of the subalpine belt, mountain beech forms tree- or shrub-heath rather than forest. Much smaller stands of red, silver, mountain and black beech are scattered elsewhere on the Volcanic Plateau (Fig. 7.34; see also p. 526), one of black beech even occupying a lake-edge cliff at Lake Taupo. Hard, silver and red beech grow together on the Mamaku Plateau north of Rotorua.

The next most extensive areas of beech forest are on the Coromandel Range, especially Mt Te Aroha (953 m). These resemble montane forests in the northernmost axial ranges, but with the addition of kauri. Silver beech caps the highest points, with red beech at mid-altitudes and hard beech descending to low altitudes on ridges and steep slopes with shallow soils. Other beech stands scattered through the North Island hill country either result from chance dispersal to suitable sites or, alternatively, are relicts from a time when climates were more favourable for beeches. Black beech caps many sandstone ridges in the southern part of Eastern Taranaki ER. In the northern part, hard beech prevails in the same position, and silver beech grows in swampy depressions on the Waitaanga plateau. There is red beech east of the axial ranges in Gisborne province. The hard beech stands in Northland (p. 118) are of special interest because of their northern location and association with kauri.

Exotic forests

Fruit- and nut-bearing trees, species of aesthetic or sentimental value and shelterbelts have been planted from the earliest days of European settlement. Planned afforestation for timber began in 1898, after it was recognised that accessible native timber was being depleted rapidly. Exotic species were used from the outset, as experience indicated that they grew faster than native trees and were easier to manage. The more extensively planted species have been *Pinus nigra, P. ponderosa, P. patula, P. contorta,* douglas fir (*Pseudotsuga menziesii*), *Racosperma melanoxylon* and species of *Larix, Chamaecyparis* and *Eucalyptus*; but commercial forestry is increasingly concentrated on a single species, *Pinus radiata*.

Some 1 200 000 ha have been planted, from northern lowland districts to cool-temperate localities in the South Island. Most of this area had been low-producing pasture, tussock grassland, scrub of both native and introduced species, fernland, sand dunes, and logged native forest after slash and residual vegetation had been burnt. Successful plantings result in a dense canopy, heavy needle-fall, thick, slowly decaying litter, and heavy demand on soil water. These conditions suppress the preceding vegetation and have been deliberately used to that end, notably in the case of nassella tussock (*Nassella trichotoma*) infestations. Nevertheless, especially in wetter, warmer districts, conifer plantations are invaded by species of native forest. In Waiotapu Forest near Rotorua, canopy closure is now around 70–80% in blocks of *Pinus nigra, P. strobus* and *P. radiata* that were planted between 1900 and 1910, probably into bracken (*Pteridium esculentum*) fernland, and tree ferns form a 4–6 m tall subcanopy that gives up to 80% cover (Fig. 7.35). *Dicksonia squarrosa* is the main species, except in moister hollows where *Cyathea medullaris* predominates. *D. fibrosa* and *C. dealbata* are also common. Bracken is still dense in windfall gaps and near roads, and sparse fronds persist throughout the forest. Five-finger, rangiora, kamahi, *Coprosma robusta, Dianella nigra, Blechnum* 'black spot', *Paesia scaberula*, and, near forest edges, *Cortaderia fulvida* are all common, and abundant in places.

Doubtless such communities will develop into native forest if left undisturbed, but

Fig. 7.35. Plantation of 75-year-old *Pinus nigra* with an understorey of *Cyathea medullaris* and *Dicksonia squarrosa*. Waiotapu Forest, northern Volcanic Plateau.

it may be centuries before pines disappear. The average density of native plants is far less than in indigenous forest developing on comparable sites, seed sources for the original large dominants are often remote, some pines have life spans of several centuries, and there are enough pine seedlings to ensure at least scattered second-generation trees. In conifer plantations in drier districts, such as Eyrewell Forest on the Canterbury Plains, undergrowth is confined to gaps and edges, and consists of pine seedlings and drought-tolerant species already present when trees were planted, such as gorse (*Ulex europaeus*), broom (*Cytisus scoparius*), kanuka (*Kunzea ericoides*) and *Rytidosperma* spp. Nevertheless, in the Canterbury foothills Rooney (1989) listed 22 native angiosperms and 25 ferns beneath plantations that are predominantly douglas fir, which casts even denser shade than pines.

Spontaneous stands of pines and other conifers, especially European larch (*Larix*

Table 7.4. *Frequencies of vascular species in seven of the most widespread beech forest types on either side of the Main Divide, in South Westland (1–4) and Lakes ER (5–7)*

forest type: mean altitude (m):	1 274	2 457	3 716	4 945	5 1000	6 950	7 900
Prumnopitys ferruginea	5	3					
Podocarpus hallii	3	5	2		2	3*	2*
Nothofagus fusca		8					
Nothofagus menziesii	9	9	10	9	10	9	
Nothofagus solandri v. cliff					3	10	10
Metrosideros umbellata	3		2				
Weinmannia racemosa	10	9	2				
Archeria traversii			2	5	2		
Aristotelia serrata		2	2				
Carpodetus serratus	6	4	2				
Dracophyllum longifolium				2	4		
Dracophyllum traversii				3			
Fuchsia excorticata	3	4					
Griselinia littoralis	9	10	10	4	3		
Hedycarya arborea	3						
Hoheria glabrata		2	5	4			
Melicytus ramiflorus	4						
Myrsine australis	2	2					
Myrsine divaricata	2	2	7	6			
Olearia ilicifolia			2	2			
Pseudopanax colensoi	3	5	4	7	2		
Pseudopanax crassifolius	8	9					
Pseudopanax simplex	5	9	9	5	6		
Pseudowintera colorata	8	9	7				
Schefflera digitata	6	3					
Podocarpus nivalis					7	3	
Phyllocladus alpinus					10	2	
Coprosma cuneata		8	10	9			
Coprosma ciliata	2	2	8	3	5	2	
Coprosma colensoi	2	2	3	2			
Coprosma foetidissima	6	10	7	5			
Coprosma pseudocuneata		2	5	10	9	4	
Coprosma rhamnoides	6	5					
Coprosma aff. parviflora					2	2	
Pseudopanax linearis			5				
Gaultheria crassa					5		
Hebe subalpina					2		
Olearia colensoi			2	2			
Brachyglottis buchananii			3	8			
Cyathea smithii	9	8	2				
Dicksonia squarrosa	4						
Asplenium bulbiferum	2	6					
Astelia nervosa	3	6	5	8			
Blechnum 'capense'	2	4					

forest type: mean altitude (m):	1 274	2 457	3 716	4 945	5 1000	6 950	7 900
Blechnum discolor	9	9					
Blechnum minus	5	7	7	3			
Hypolepis millefolium			3		2	2	
Microlaena avenacea	3	2					
Phormium cookianum				4			
Polystichum vestitum		3	8	6	6	6	
Lastreopsis hispida	3						
Leptopteris superba	2	4	3				
Coprosma depressa					4	4	
Blechnum fluviatile	4	4	7				
Blechnum penna-marina					4	4	
Hymenophyllum multifidum	2	2	6	9			
Nertera depressa	4		2				
Nertera aff. *dichondrifolia*	8	9	7				
Ranunculus reflexus			2		4	3	
Uncinia filiformis			8		3		
Uncinia rupestris			3	2			
Metrosideros diffusa	9	6					
Rubus cissoides		4	4				
Asplenium flaccidum	9	6	7	3			
Grammitis billardierei	8	9	5	3	3	3	
Hymenophyllum flabellatum	4						
Rumohra adiantiformis	4	2					
Phymatosorus diversifolius	9	5	2				
Hymenophyllum spp.					3	6	
Peraxilla tetrapetala					5	9	8
Alepis flavida						9	

Notes:

Each unit represents 10% frequency (to nearest 10%) in 'non-area' plots, with values less than 15% being excluded.

Forest types:

1 Canopy mainly silver beech and kamahi; tall fern tier mainly *Blechnum* spp.

2 Red beech dominant, with a subcanopy of silver beech, kamahi and mountain totara; dense, tall fern tier mainly *Blechnum* spp. Confined to Arawata valley.

3 Silver beech canopy; *Polystichum vestitum* dominant beneath.

4 Near the upper forest limit, with silver beech forming the canopy over high-altitude small trees and shrubs.

5 Silver beech over a dense shrubby understorey. Confined to valley heads on eastern flanks of the Main Divide.

6 Canopy of silver and mountain beech, over an open shrub tier.

7 Pure mountain beech forest, with few vascular associates other than mistletoes.

* In types 6 and 7, mountain totara is rare except as seedlings and saplings.

Further species, recorded in only one forest type with frequency < 4:

Type 1: *Dacrydium cupressinum, Neomyrtus pedunculata, Coprosma rotundifolia*; 2: *Coprosma lucida, Histiopteris incisa, Muehlenbeckia australis*; 3: *Pratia angulata*; 5: *Uncinia uncinata, Luzula picta, Lycopodium* sp.; 5: *Lagenifera* sp.; 6: *Uncinia gracilenta*.

Source: J. Wardle *et al.* 1973; Wardle & Guest 1977

decidua) and douglas fir, are developing through invading low or open vegetation. It is likely that in environments where native forest is vigorous and species-rich, these conifers will form a phase, albeit a lengthy one. On the other hand, pines and other exotic trees can invade and replace native vegetation on sites that lie near or beyond the drought- or cold-tolerance of native trees; from the viewpoint of nature conservation, the spread of species such as *Pinus contorta* at high altitudes, *P. nigra* in dry tussock grasslands, or douglas fir into dry mountain beech forests, are matters for concern.

Although there are only amenity plantings and occasional woodlots of deciduous broad-leaved exotics, most of which spread very little, sycamore (*Acer pseudoplatanus*) often invades modified native forest and has the potential to supplant it, at least in the colder districts. Willows, especially *Salix fragilis*, have spread vegetatively on river banks, flood-plains and swamps (pp. 302 and 311). Poplars planted for erosion control, especially *Populus nigra*, have also formed dense colonies through root suckers. Eucalypts are the only tall, evergreen broad-leaved trees that have spread significantly.

8

BUSH, HEATH, SCRUB AND FERNLAND

Vegetation formed by small trees, shrubs or ferns is variable in stature and composition, and usually exists as patches among other vegetation, especially forest or grassland. Some communities represent the tallest vegetation that can develop in particular environments. Stands of small broad-leaved trees or tree ferns may result from partial degradation of forest by fires, logging or wind-throw. Many other communities represent stages in primary or secondary succession, and communities dominated by the same species can be at very different stages of development.

Native shrubs still cover large areas but, increasingly, introduced shrubs are usurping their role as woody pioneers. It is difficult to define cohesive communities in shrubby successional vegetation, as shrubs usually form a temporary phase, having established among earlier vegetation and often sheltering the seedlings of their successors. The founder effect also plays a large role, especially with adventive plants which often form single-species stands or inconstant assemblages. As both adventive and native species have often invaded land intended for pasture, the varying success of attempts to destroy them through fire, herbicides and grazing has further compounded the diversity of short woody vegetation.

Chapter 8 begins with broad-leaved bush and tall shrub communities, and then describes tree- and shrub-heaths that grow mainly on infertile soils. The final sections deal with miscellaneous shrublands, including those characterised by filiramulate and adventive species.

Temperate bush

Structure and composition

On sites disturbed by either natural or human agency, tall forest is often replaced by bush in which few of the canopy species exceed 10 m. On forested steeplands as on river flats, forest forms mosaics with bush occupying the less stable or more recent soils. Scattered tall trees can also tower over a low broad-leaved canopy, blurring the distinction between bush and forest. Over extensive deforested areas in the east of the South Island, bush remnants are the tallest native vegetation.

Usually there is a large content of short-lived, seral trees, including wineberry (*Aristotelia serrata*), *Carpodetus serratus, Schefflera digitata, Pseudopanax* and *Pittosporum* species, and kanuka (*Kunzea ericoides*). Canopies of young stands may also include large-leaved coprosmas (especially *C. robusta* or *C. lucida*), *Coriaria arborea, Olearia* spp., rangiora (*Brachyglottis repanda*) or hebes. As stands mature, there is increasing prominence of trees, mostly longer-lived, that are also frequent in the subcanopy tiers of tall forest, especially *Griselinia littoralis, Myrsine australis*, lancewood (*Pseudopanax crassifolius*), mahoe (*Melicytus ramiflorus*), the gully-dwelling fuchsia (*Fuchsia excorticata*) and, in mild, coastal areas, karaka (*Corynocarpus laevigatus*) and kohekohe (*Dysoxylum spectabile*). Trees with filiramulate juveniles are important on alluvial flats throughout and on hillsides in lower-rainfall districts; *Plagianthus regius*, kaikomako (*Pennantia corymbosa*) and kowhai (*Sophora microphylla*) are the most widespread of these. Through much of the hill country east of a line joining the East Cape, Rangitikei and Eastern Wairarapa ERs, the usual kowhai is replaced by *S. tetraptera*, a larger-leaved tree without a filiramulate juvenile.

A number of broad-leaved trees and shrubs are largely confined to coastal bush. Most of these, including *Pisonia brunoniana, Planchonella costata, Entelea arborescens, Nestegis apetala* and *Pseudopanax discolor*, are endemic to the northern zone, but ngaio (*Myoporum laetum*) extends as far south as Dunedin and as much as 50 km inland in steep gorges.

Because tree ferns resprout from trunks or rhizomes after forest fires and also establish in secondary vegetation (p. 544), they are often prominent in temperate bush and can form pure stands. They are very persistent in the milder, wetter districts, but in drier, harsher districts tend to disappear from bush and forest remnants as these are reduced in size and depleted by livestock. Nikau palm (*Rhopalostylis sapida*) also survives forest clearance, to persist as groves in pasture land.

Cabbage trees (*Cordyline australis*) became abundant after the widespread deforestation that began with human settlement, growing scattered and as groves through wide expanses of open country, including swamps, fernlands and grasslands. Indeed, they were the only trees over large areas, especially on the eastern plains and downlands of the South Island. Through their longevity, they persist in secondary bush and forest. Stock eat accessible leaves, horses sometimes gnaw the trunks right through and, in the northern zone, the trees are now being decimated by disease. As a result, even this tenacious symbol of 'old New Zealand' is gradually fading from the farmscapes, while securing its position as a specimen tree in gardens, both in New Zealand and other temperate regions of the world. Both nikau and cabbage tree are extremely tolerant of coastal exposure, growing as erect trees on headlands where other vegetation is low and wind-shorn.

Small adventive trees with seeds distributed by birds have become prominent in some areas. In the northern zone, woolly nightshade (*Solanum mauritianum*), named for its large, tomentose leaves, now dominates many bush margins and gullies. Chinese privet (*Ligustrum sinense*) is vigorously invading small forest reserves, and dense, rambling thickets of *Elaeagnus* ×*reflexa* have spread from plantings. In the South Island, elder (*Sambucus nigra*) is well established on wasteland and in gullies in the east, sometimes as pure groves and sometimes in association with native trees such as mahoe. Rowan (*Sorbus aucuparia*) is invading many bush remnants, and wild cherry (*Prunus avium*) is established in high-altitude bush near Mt Cook.

Two arborescent legumes have become extensively naturalised. Brush wattle (*Paraserianthes lopantha*) is common in regenerating bush on dry slopes in Northland, and can dominate small areas; in lesser abundance, it extends southwards to Canterbury. Silver wattle (*Racosperma dealbata*) occupies similar sites, though extending to more southern and inland districts. However, most stands represent reproduction near planted trees, rather than vigorous spread.

Forest lianes, especially *Rubus cissoides*, increase greatly after tall canopy trees are destroyed by logging or wind-throw, and can also be extremely abundant in coastal bush. *R. schmidelioides*, which has much narrower leaflets than *R. cissoides*, grows in drier, more open habitats and is the main *Rubus* in seral bush on alluvial flats. *R. squarrosus* grows mainly in bush remnants and scrub in eastern maritime districts of the South Island and very locally in the North Island.

Muehlenbeckia australis and, less frequently, the smaller-leaved *M. complexa* often weigh down and smother small trees and saplings. The two *Parsonsia* spp. are also characteristic of bush and forest margins, as is *Clematis paniculata*, which has white, starry flowers up to 5 cm across (Fig. 8.1). *C. forsteri* and *C. cunninghamii*, which have smaller white flowers, and the fragrant, creamy-flowered *C. foetida* occupy similar habitats but are less common, and absent from the western side of the South Island.

Calystegia tuguriorum, a slender, soft-wooded liane with trumpet-shaped white flowers 4 cm in diameter, is common in coastal and lowland indigenous bush and scrub. The more robust, adventive *C. silvatica* replaces it in modified situations, especially in wasteland or urban fringes; the probably native *C. sepium* grows in similar habitats and also in swamps, but only in the northern and central zones. *Senecio mikanioides* and *S. angulatus* are locally abundant adventives in remnants of coastal bush and scrub. A native senecioid liane, *Brachyglottis sciadophila*, occurs very locally in bush fringes in eastern and northern lowlands of the South Island.

Traveller's joy (*Clematis vitalba*) has become a major threat to the survival of bush and forest remnants on fertile lowland sites between latitudes 39° and 44°. Like *Muehlenbeckia*, it spreads its foliage over the canopies of supporting trees but is more likely to smother and kill them. Possibly, its deciduous habit poses a hazard, in that host trees are deprived of light during the growing season but are exposed to winter cold. *C. vitalba* can be distinguished from all native *Clematis* spp. through its bipinnate leaves, and through flowering during December and January instead of spring. *Passiflora mollissima* also smothers bush remnants and forest margins in warm-temperate districts.

The undergrowth can include suppressed plants persisting from early seral stages, such as bracken (*Pteridium esculentum*). More commonly, understories are closely similar to those of tall forest, of which bush is the truncated equivalent. Reflecting the generally fertile or rejuvenated soils, however, ferns such as *Polystichum vestitum*,

Fig. 8.1. *Clematis paniculata* (photo New Zealand Forest Service; M11616 National Archives of New Zealand).

Asplenium bulbiferum, Blechnum fluviatile, B. 'capense', Hypolepis ambigua and, in warm districts, *Deparia petersenii, Lastreopsis velutina* and *Pteris macilenta* are more typical than species of less fertile soils.

Because many bush remnants and forest margins have been severely modified through human agency, especially stock grazing, they have been invaded by many shade-tolerant weeds and garden escapes. One of the most widespread of the latter is the large, tufted fern *Dryopteris filix-mas*, which tolerates drier, frostier conditions than large native ferns. A small lycopod, *Selaginella kraussiana*, is locally naturalised in native bush.

Several robust ornamental herbs form local infestations in bush remnants through vegetative spread from plant fragments. *Tradescantia fluminensis* forms thick mats that smother other understorey herbs as well as tree seedlings. In warm-temperate districts, arum lily (*Zantedeschia aethiopica*), wild ginger (*Hedychium gardnerianum*) and *Asparagus scandens* are well established.

Mycelis muralis is the most widespread adventive plant in native woody vegetation but scarcely threatens native species. *Solanum nigrum*, the possibly native *S.*

Fig. 8.2. Retreating margin of damaged coastal bush, with a taraire tree in the foreground. Maunganui Bluff, western Northland.

americanum and the probably native *Parietaria debilis* grow on disturbed ground in coastal bush. Stock-damaged bush and forest margins are the main habitats for the semi-woody *Solanum chenopodioides* and *Urtica ferox*; the sting of this 2–3 m tall, woody, native nettle has proved fatal to humans and horses. The herbaceous *U. incisa* extends inland and to the subalpine belt. *Solanum aviculare* and its close relative *S. laciniatum* are large, short-lived, soft-wooded native shrubs with conspicuous blue or white flowers, that also establish on disturbed ground, especially in coastal and maritime bush margins.

Many adventive plants that occur in damaged bush remnants are more characteristic of scrub, fernland or grassland. The more abundant include *Hypericum androsaemum*, *Rubus fruticosus*, *Conium maculatum*, *Torilis* spp., *Leycesteria formosa*, foxglove (*Digitalis purpurea*), *Verbascum thapsus* and cocksfoot (*Dactylis glomerata*).

Northern coastal bush (based on reports in *Tane* including Wright 1977)

On the main islands of New Zealand, most coastal bush has been cleared for farming or housing. Even where it has not been deliberately felled or burnt, man-made margins are exposed to salt winds that damage both canopy and undergrowth, thereby allowing grazing animals and weeds to penetrate more deeply than is usually the case in inland forests (Fig. 8.2). The most intact and floristically richest stands are on the small islands lying off eastern coasts of Northland and Coromandel and within the Hauraki

Gulf; and even here, much has been removed or modified by human activities, beginning with Maori occupation. Great variation over small distances reflects differences in exposure and soil depth, the degree of disturbance and the extent of recovery.

Kanuka establishes after fires or other clearances, and the trees can persist among the broad-leaved canopy or as an overstorey until they die of old age. Pohutukawa (*Metrosideros excelsa*) establishes in grassland and fernland, on cliffs, and on volcanic and other boulder fields that are too harsh for other pioneering species. It can form an open overstorey or continuous tall forest, or be stunted to the level of its associates. Another species, *Dodonaea viscosa*, mostly colonises dry sites such as stabilised dunes.

On sites that are more sheltered, moister and more fertile, broad-leaved species dominate from the outset. The most abundant is *Pseudopanax lessonii*, but *Entelea* can dominate slips and openings in bush and forest. Other species, especially *Myrsine australis*, can enter kanuka stands and dominate an intermediate phase.

Taupata (*Coprosma repens*), *Melicytus novae-zelandiae* and ngaio are most tolerant of exposure; the first two also occupy ground burrowed by petrels. The seedlings of taraire (*Beilschmiedia tarairi*) and broad-leaved tawa (*B. tawaroa*) are late entrants, and in very sheltered situations eventually form forest. Usually, however, they contribute only part of the canopy, in association with other broad-leaved trees. Lying between the points of extreme exposure *versus* shelter, and early successional *versus* climax, there are many combinations of species, but the following descriptions represent nodes.

(1) Bush 3–6 m tall. The most frequently dominant species are *Pseudopanax lessonii* or mahoe. Others include *Myrsine australis, Geniostoma rupestre, Coprosma robusta, C. macrocarpa, C. repens, Pittosporum crassifolium, Melicope ternata*, rangiora, five-finger (*Pseudopanax arboreus*), *Entelea, Macropiper excelsum, Hebe stricta* and *Melicytus novae-zelandiae*, as well as small plants of karaka and pohutukawa.

(2) Bush 6–12 m tall. The most frequent dominants are kohekohe, taraire, karaka, and large pohutukawa trees persisting from earlier stages. Others include puriri (*Vitex lucens*), *Hoheria populnea*, mahoe, *Streblus banksii, Pisonia, Hedycarya arborea, Litsea calicaris, Myrsine australis, Planchonella*, tawa, *Nestegis apetala, Melicope ternata* and cabbage tree. Nikau can be locally abundant. *Pittosporum crassifolium* and *Pseudopanax lessonii* may also be present as survivors from an earlier successional stage. *Meryta sinclairii* is locally abundant near the shore on the Hen and Chicken Islands.

Macropiper and *Rhabdothamnus solandri* (both as large-leaved coastal forms), *Geniostoma, Coprosma macrocarpa* and *C. arborea* are mainly in the subcanopy or shrub tier. Lower tiers are sparse in both (1) and (2) where broad-leaved canopies cast heavy shade and litter; *Doodia media, Asplenium oblongifolium, A. flaccidum* var. *haurakiense* and *Oplismenus hirtellus* are reported from most stands. Where extreme exposure, steepness or disturbance preclude a complete broad-leaved cover, they can

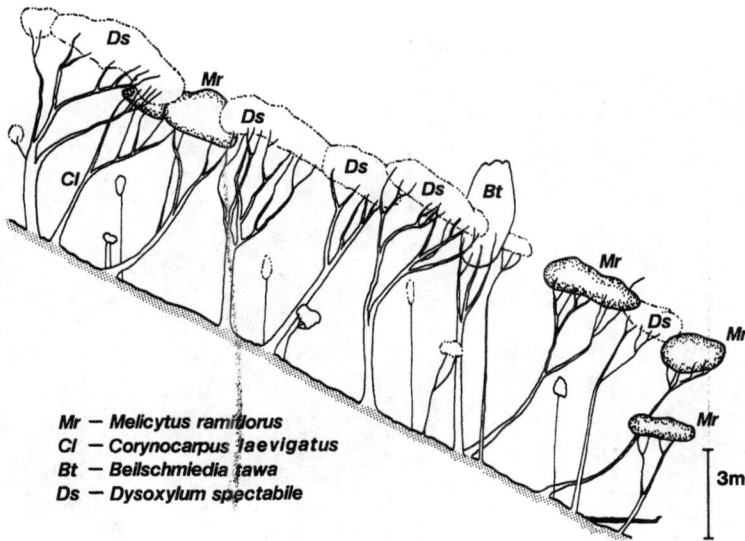

Fig. 8.3. Profile of coastal bush on Kapiti Island, Cook Strait (Esler 1967).

be accompanied by successional, rupestral and epiphytic species (p. 383). Lianoid plants can be abundant on bush margins, notably *Einadia triandra, Tetragonia trigyna, Muehlenbeckia complexa* and, locally, *Sicyos australis*.

Coastal bush of central and southern New Zealand

Remnants of coastal bush on islands in Cook Strait and Tasman Bay have proved less resilient than on northern offshore islands, because of the colder, windier climate and smaller number of species. On Kapiti Island, kohekohe bush less than 12 m tall replaces tawa forest as the climax vegetation at the lowest altitudes, particularly where there is exposure to salt winds (Fig. 8.3). Accompanying trees are tawa, mahoe (which can outnumber kohekohe on rocky talus, especially in younger stands), *Hedycarya*, lancewood, titoki (*Alectryon excelsus*), *Myrsine australis* and *Nestegis lanceolata;* *Griselinia lucida* of epiphytic origin; karaka, ngaio and *Melicope ternata* (these mainly near the sea); and five-finger, rewarewa (*Knightia excelsa*), kaikomako and *Olearia rani*, which may reflect the seral origins of the vegetation, which was partly burnt, cleared and grazed during the nineteenth century. On Trio and Stephens Islands, further species are *Olearia paniculata, Coprosma repens* and *Melicytus novae-zelandiae* (Dawson 1954).

Bush on eastern hills of the South Island consists of modified remnants that contain progressively fewer species as latitude increases (p. 132) (Fig. 8.4). Along most of the western coast, forest – albeit often stunted – usually extends almost to the shore, reflecting high rainfall and low wind speeds; bush mainly occurs on exposed headlands and cliff tops, and on debris on coastal cliffs. In Western Nelson, nikau forms groves on deep, well-drained soils and karaka is abundant on some coastal flats. Northern rata (*Metrosideros robusta*), beginning as epiphytic or rupestral seedlings, can provide a prominent overstorey. Ngaio and *Dodonaea* are also present.

Fig. 8.4. Bush remnant on hills, lower Waipara River, Lowry ER.

In Central Westland, bush on steep, coastal debris is 3–6 m tall and consists mainly of mahoe, although this can share dominance with *Hedycarya* and, in younger stands, *Coprosma lucida* (Fig. 8.5). The typical understorey species are *Carex solandri*, *Uncinia uncinata* and, where not heavily browsed by goats, *Asplenium bulbiferum* and *Astelia fragrans*. *Blechnum discolor* is usually present in places with incipient yellow-brown earths. The community is the usual habitat for *Cyathea medullaris* and for several species at or near their southern limits, e.g. *Pteris macilenta*, *Lastreopsis glabella*, *Macropiper* and, north of the Wanganui River, an isolated population of nikau. Seral equivalents are dominated by *Olearia avicenniifolia* or *Phormium tenax*, with gorse (*Ulex europaeus*) as an abundant adventive. Intermingled with pasture and gorse in the lee of dunes, there are remnants of similar bush, often with wineberry dominant and, in places, groves of *Podocarpus totara* var. *waihoensis*.

On exposed headlands and behind cliff tops, the few large rimu (*Dacrydium cupressinum*) and miro (*Prumnopitys ferruginea*) trees have wide, deformed crowns, the canopy being formed mainly by *Freycinetia baueriana*, *Ripogonum scandens* and *Metrosideros* vines, together with scattered small trees, including kamahi (*Weinmannia racemosa*), *Hedycarya* and *Coprosma lucida*. *Metrosideros perforata* forms hedges along the brows of the cliffs. *Freycinetia* thickets cover much of the Open Bay Islands (Burrows 1972).

Mahoe bush clothes coastal debris in Fiordland, but on exposed sites that are more stable, leached, or peaty there is a southwards tendency for epacrids and woody composites to increasingly dominate vegetation that is best described as coastal tree-heath (p. 191). However, in the sheltered centres of small islands in Foveaux Strait and offshore from Stewart Island, tree-heath can give way to broad-leaved bush dominated by *Myrsine chathamica*, with an undergrowth of *Asplenium obtusatum*,

Fig. 8.5. Coastal mahoe bush, Central Westland.

Phymatosorus diversifolius and, on some islands, *Stilbocarpa lyallii*. In the west of Stewart Island, *Griselinia littoralis* can dominate locally on dune hollows and coastal slopes, other prominent canopy or subcanopy species being *Carpodetus, Brachyglottis rotundifolia* and, sometimes, fuchsia.

Bush of steep, unstable slopes

On slopes of loose debris, tall trees are usually few and far between except on pockets of stable soil, and small broad-leaved trees dominate instead. Such vegetation is extensive on fault scarps in the tectonically active belt of Mesozoic sandstones extending from East Cape to Kaikoura. Along the Wairarapa Fault in the Rimutaka Range, kanuka groves form an overstorey to a main canopy of mahoe, karaka, five-finger, *Olearia paniculata* (chiefly on spurs), and rangiora, with lesser numbers of rewarewa, *Hedycarya*, lancewood, hinau (*Elaeocarpus dentatus*) and ngaio. *Cyathea*

dealbata grows mainly in the understorey. The main shrubs are *Hebe parviflora,
Coprosma propinqua, C. rhamnoides* and *Cyathodes juniperina. Microlaena avenacea,
Uncinia uncinata, U.* sp., *Gahnia pauciflora, Polystichum vestitum, Asplenium
polyodon, A. bulbiferum* subsp. *gracillimum, Phymatosorus diversifolius, Lastreopsis
glabella* and *L. velutina* form a tall herb layer that has been depleted by trampling and
browsing. Lianes include *Clematis forsteri* and *Metrosideros perforata. Collospermum
hastatum* grows on rocks and trees. On more stable slopes at higher altitude, this
community gives way to forest dominated by beeches, northern rata or kamahi.

In the Orongorongo Valley, also in the Rimutaka Range, occasional trees of miro,
hinau, rimu and rewarewa, together with logs and stumps of kamahi, mountain totara
(*Podocarpus hallii*), fuchsia, titoki and northern rata remain from once-intact tall
forest. There are also *Carpodetus, Olearia rani* and kaikomako trees, as well as *Urtica
ferox, Dicksonia squarrosa* and *Freycinetia* (Campbell 1984).

Almost everywhere, such vegetation is infested by goats and other ungulates, which
destroy palatable ferns and seedlings. To some extent these are replaced by less
palatable species, but not quickly enough to maintain cover and stability. In many
areas, scree extends even beneath surviving remnants of tall forest.

Similar vegetation is well developed along the scarp of the Alpine Fault and in the
deep, steep-sided gorges that cut across it. In Central Westland, vegetation on steep
debris slopes below 450 m is commonly dominated by kamahi (mainly as small,
coppicing trees), *Schefflera*, mahoe and *Hedycarya*. Fuchsia was abundant until
eliminated by possums. *Griselinia littoralis* and *Carpodetus* also occur in some stands.
In seral equivalents wineberry, *Hebe salicifolia* and *Coriaria arborea* co-dominate with
Schefflera. In lower tiers, *Cyathea smithii* is abundant and accompanied by *Dicksonia
squarrosa*, and *Asplenium bulbiferum* is the dominant tall herb. The liane *Ripogonum*
forms much of the canopy, and *Metrosideros diffusa* spreads over the floor. Other
consistently present species are *Lastreopsis hispida, Nertera* aff. *dichondrifolia, Rubus
cissoides, Phymatosorus diversifolius* and *Metrosideros fulgens* (Fig. 8.6).

With increasing altitude, low-altitude species such as *Ripogonum*, mahoe, *Hedycarya* and tree ferns are replaced by upland species, including *Hoheria glabrata* and
Olearia ilicifolia. Pseudowintera colorata is important in the shrub tier, and
Polystichum vestitum at first accompanies and then replaces *Asplenium bulbiferum*.
Above 750 m, this community grades into *Hoheria glabrata* bush (p. 178). On very
steep slopes, seral and subclimax bush often forms mosaics with pioneering vegetation
and fragments of tall forest with podocarps, southern rata (*Metrosideros umbellata*) and
kamahi.

Griselinia littoralis bush

In eastern South Island regions that lack beeches and are too dry for southern rata and
kamahi, large podocarps stand over a canopy of small broad-leaved trees, among
which *Griselinia littoralis* is usually the most prominent. The podocarp overstorey is
often sparse or absent, sometimes as a result of logging.

Where beech or mixed forest has been burnt, *G. littoralis* trees are likely to resprout
and assume dominance. On moist sites, seedlings establishing after fire are mostly
those of small broad-leaved trees and among these, *G. littoralis* eventually dominates

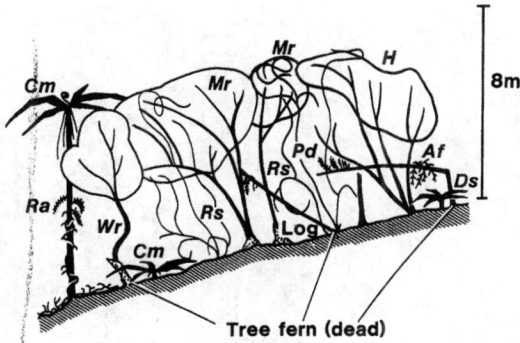

Fig. 8.6. Profile of bush on a debris slope at 320 m, Kaimata Range, North Westland (Reif & Allen 1988). Abbreviations: *Af, Asplenium flaccidum*; *Cm, Cyathea smithii*; *Ds, Dicksonia squarrosa*; *H, Hedycarya arborea*; *Mr, Melicytus ramiflorus*; *Pd, Phymatosorus diversifolius*; *Ra, Rubus australis*; *Rs, Ripogonum scandens*; *Wr, Weinmannia racemosa*.

because of its longevity; on bouldery ground it dominates from the outset. Manuka (*Leptospermum scoparium*) or kanuka may occupy drier sites, but these too are usually succeeded by *G. littoralis* and other broad-leaved species.

Conifers re-establish if there are seed trees nearby, but the process is usually slow. Nevertheless, in some mountain valleys small mountain totara trees are abundant, having established either in fire-induced grassland or among kanuka, manuka and other shrubs. *Phyllocladus alpinus, Podocarpus nivalis* and hybrids with *P. hallii* can also be present, especially on frosty terraces and at high altitudes. Such podocarp woodland is extensive on slopes in the Inland Kaikoura Range where podocarp/broad-leaved forest may have once prevailed (Williams 1989), and in the Hunter valley (Lakes ER) on slopes that once carried beech forest (Fig. 8.7). Beech recovers only from the margins of surviving beech forest and around isolated beech trees that grew up before other secondary vegetation became too dense (pp. 539 and 545). Patterns can be intricate, with tongues and patches of bush representing the original burn, but interspersed with relict and regenerated tall trees.

The floristic content of *G. littoralis* bush is as varied as the habitats that it occupies and seral stages that it represents. In addition to remnant or re-established conifers or beeches, there can be an overstorey of kanuka trees, collapsing as they yield dominance to shorter broad-leaved species. Small trees that share the canopy with *G. littoralis* and, on their favoured sites, usurp its dominant role, are *Carpodetus* (especially on small, shallow slips); fuchsia (gullies); kowhai, *Plagianthus regius, Hoheria angustifolia* and *Pittosporum eugenioides* (fertile soils); mahoe (well-drained colluvial slopes at lower altitudes, but not in inland districts); and ngaio (near the coast). Five-finger (replaced in the south and at higher altitudes by *Pseudopanax colensoi*), wineberry and *Pittosporum tenuifolium* can be prominent where there has been recent disturbance. Other small trees include lancewood, *Coprosma linariifolia, Myrsine australis* and *M. divaricata*. Subalpine trees enter at higher altitudes, especially *Phyllocladus alpinus, Dracophyllum longifolium, Olearia ilicifolia* and *Hoheria lyallii* s.l.

Fig. 8.7. Bush largely of *Griselinia littoralis* and young *Podocarpus hallii*, and expanding remnants of silver beech forest (outlined) that escaped Maori fires. Hunter valley, Lakes ER.

Rubus cissoides is often abundant. A shrub tier usually contains small-leaved coprosmas, especially *C. ciliata* and *C.* aff. *parviflora*. *Blechnum discolor* and *Cyathea smithii* can be present in maritime districts and *C. colensoi* in humid upland localities. *Polystichum vestitum* is often dense, but *Blechnum 'capense'* may replace it on more stable soils. *Uncinia uncinata* can be abundant on moist, somewhat unstable ground. Bouldery areas support small spleenworts, especially *Asplenium richardii* and *A. flabellifolium*. *Mycelis muralis* is universal. Species of surrounding grassland, such as matagouri (*Discaria toumatou*) and *Chionochloa* spp., may intermingle with the broad-leaved trees. As in other bush remnants, grazing mammals have often destroyed the indigenous seedlings and undergrowth, leading to replacement by adventive plants.

Bush on fertile alluvial flats and fans

Fertile soils can support tall podocarps, but often these are only a scattered overstorey or absent, apparently because dense, fern-dominated understoreys prevent their seedlings from becoming established; logging has further reduced the podocarp element. Composition of the broad-leaved canopy varies according to stage of succession and amount of disturbance, with the main difference from other bush communities being the prevalence of trees with straight trunks and filiramulate juveniles. Filiramulate shrubs are also well represented. The least modified examples are on river flats in Westland, where the canopy in younger stands consists of kaikomako, *Carpodetus* and *Plagianthus regius*, together with *Griselinia littoralis*,

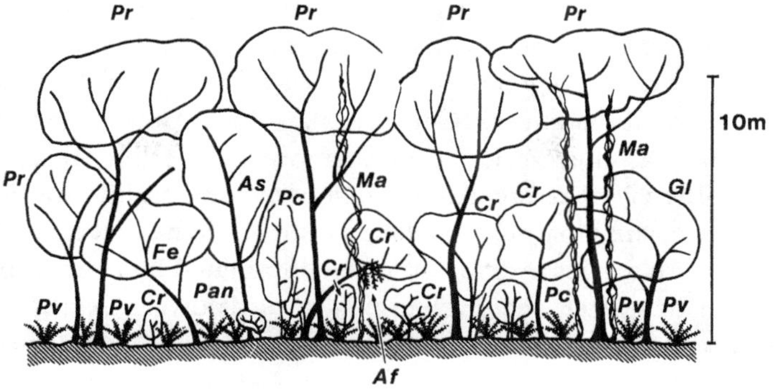

Fig. 8.8. Profile of a transect through bush on an alluvial terrace at 210 m, Karangarua valley, Central Westland (Reif & Allen 1988). Abbreviations: *Af, Asplenium flaccidum*; *As, Aristotelia serrata*; *Cr, Coprosma rhamnoides*; *Fe, Fuchsia excorticata*; *Gl, Griselinia littoralis*; *Ma, Muehlenbeckia australis*; *Pan, Pseudopanax anomalus*; *Pr, Plagianthus regius*; *Pc, Pseudopanax colensoi*; *Pv, Polystichum vestitum*.

lancewood, rare *Streblus heterophyllus* and, locally, cabbage trees (Fig. 8.8). Young trees of *Podocarpus totara* var. *waihoensis* often dominate on stony ground. In stands old enough to have a podocarp overstorey, kamahi and mahoe also contribute to the main canopy. Kowhai is common near estuaries.

In the subcanopy *Pseudowintera colorata* and *Schefflera* are usually common. *Dicksonia squarrosa* and *Cyathea smithii* can be abundant and may reach the main canopy. *D. fibrosa* also occurs in a few localities. In the shrub tier, *Coprosma rotundifolia* is abundant, and accompanied by other small-leaved coprosmas. Characteristic herbs include *Blechnum fluviatile, B. chambersii, Asplenium bulbiferum, Nertera* aff. *dichondrifolia, Microlaena avenacea, Uncinia uncinata, U. egmontiana*, other uncinias and *Astelia grandis. Leptolepia novae-zelandiae* is a characteristic species, although local. Lianes can be profuse: *Metrosideros diffusa* and *Ripogonum* in older stands, *Rubus schmidelioides, Muehlenbeckia australis* and *Parsonsia heterophylla* in younger stands and margins, and *Freycinetia* near the coast. *Phymatosorus diversifolius* and, locally, *P. scandens* are also common. *Rubus australis* is confined to forest and bush on wet recent soils in Westland, in contrast to its wider edaphic range in the northern zone.

In mountain valleys, temperature inversion brings montane species into these communities (p. 135). *Olearia ilicifolia, Hoheria glabrata* and *Myrsine divaricata* join the broad-leaved canopy, and can dominate in cold valleys to altitudes as low as 450 m. *Polystichum vestitum* forms a dense understorey, except where deer have destroyed it. *Astelia nervosa, Aristotelia fruticosa* and *Pseudopanax anomalus* can also be common.

Primary tree- and shrub-heaths and subalpine bush

Structure and gradients

In the west of the South Island and on Stewart Island, the original temperate vegetation on thin, leached soils over siliceous and ultramafic rocks, and on well-differentiated gley podzols on fluvioglacial terraces, is dominated by stunted, slow-growing, small-leaved trees and shrubs. Dwarf podocarps, manuka and epacrids are especially prominent, in vegetation that varies greatly in height, density and composition over small distances. Rimu trees are often present, and tend to form a scattered overstorey as they resist stunting more than other trees. Southern rata, beeches and kamahi may be reduced to the height of surrounding species; kamahi may even become a semi-rhizomatous, subcanopy shrub. Where soil conditions are less restricting, tree-heath grades into forest, often dominated by rimu or beeches. Where soils are shallower or wetter, the gradation is towards dwarf-heath, bog and pakihi.

Similar vegetation occurs in the subalpine belt, especially on ridges with shallow or leached soils. The most conspicuous difference between temperate tall heaths and their subalpine equivalents is the importance of broad-leaved woody composites in the latter. *Olearia colensoi*, especially, is widely dominant on wet, foggy mountains west of the South Island Main Divide, but is absent from Mt Taranaki and the Volcanic Plateau. Species in the *Brachyglottis rotundifolia* complex have a similar distribution, although they extend to less leached soils, and *B. elaeagnifolia* is abundant on Mt Taranaki. On the western mountains of the South Island and the highest summits of the northern zone, species of *Dracophyllum* with terminal tufts of recurved, strap-like leaves impart a distinctive aspect (Fig. 8.9).

On gradients towards deeper, less leached soils such as occur on colluvium and alluvium, and also on lower, more sheltered sites, the vegetation becomes taller and denser. Where both canopy and understorey include species characteristic of semi-fertile soils, such as *Griselinia littoralis*, deciduous hoherias and *Polystichum vestitum*, the vegetation is best regarded as subalpine bush, which usually forms a narrow belt on stable soils at the upper forest limit. This can grade into extensive communities of small broad-leaved trees and tall shrubs on debris, which in turn grade downwards into temperate bush of species such as *Schefflera*, wineberry and fuchsia. In the lower part of the subalpine belt there are often scattered trees and groves of kaikawaka (*Libocedrus bidwillii*), which stand taller than the surrounding bush and heath. Mountain totara, on the other hand, may be reduced to the height of the main canopy, especially where it exists as hybrids with *Podocarpus nivalis*.

Subalpine heaths also may include stunted mountain (*Nothofagus solandri* var. *cliffortioides*) or silver (*N. menziesii*) beech, and be replaced by beech forest where conditions are better. However, beeches are absent from very large areas such as Mt Taranaki, Stewart Island and much of Westland, and from local areas such as cirques and old landslides, and in these areas heath and bush occupy most of the subalpine belt. In Fiordland tree-heaths also grow at the coast, and on western slopes in Stewart Island can prevail from the shore to the highest summits.

Subalpine bush and tall, dense heaths contain few species that are not found in equivalent forest communities. Other primary heaths tend to be mosaics, in which the taller, denser parts contain the same understorey species as related forest communities, whereas the more open parts contain lower-storey plants characterstic of bog, dwarf-heath, grassland and rupestral vegetation. However, *Lepidothamnus intermedius* × *laxifolius*, the sundew *Drosera stenopetala*, *Cyathodes empetrifolia*, *Astelia linearis* and in the subalpine belt, the dwarf shrub *Myrsine nummularia* and the

Fig. 8.9. *Dracophyllum traversii*, 1050 m, Mt Arthur Track, North-west Nelson (photo, New Zealand Forest Service; M13045 National Archives of New Zealand).

trailing, large-leaved *Coprosma serrulata* reach their greatest abundances in shrub-heaths. Suppressed *Phormium cookianum* plants are nearly always present in subalpine heath, and openings provide refuges for palatable alpine herbs, notably *Anisotome haastii* and *Ranunculus lyallii*. The tussocky sedge *Gahnia procera* is abundant in tall heath on leached soils. Temperate heaths support a well developed bryophyte tier.

Lianes are rare in primary heath and the only frequent vascular epiphytes are *Lycopodium varium, Asplenium flaccidum* and some filmy ferns. The root-parasitising shrub *Exocarpus bidwillii* is common in open montane and subalpine primary heaths in the northern part of the South Island.

A large proportion of temperate heaths have been burnt, and converted to secondary vegetation dominated by manuka. In the North Island, primary communities exist only as small stands in remote localities, any more extensive stands that might have once existed (e.g. on the Northland gumlands) having long since been destroyed. Subalpine heaths also burn readily; little has escaped east of the South Island Main Divide, or through much of the Volcanic Plateau and the ranges to the east. Some of the burnt areas have regenerated to secondary heath, whereas others now support grassland or are deeply eroded.

Primary tall heaths and subalpine bush in Central Westland

The most extensive and varied primary temperate heaths in New Zealand are in Westland, on moraine plateaus dating from late Pleistocene glaciations; and the most extensive subalpine bush and heath are on the flanks of the western Alps. Temperate tree- and shrub-heaths have three main facies as follows.

(1) On poorly drained ground, an uneven canopy up to 10–12 m tall is formed by silver pine (*Lagarostrobos colensoi*), pink pine (*Halocarpus biformis*), *Phyllocladus alpinus* and manuka. In the usually dense cover of bryophytes, *Dicranoloma* is the most important and *Sphagnum* is present. *Empodisma minus* and *Gleichenia microphylla* can be abundant in transitions to bog and mire. Soils vary from deep, wet basin peats to gley podzols with less than 25 cm of peat.

(2) Yellow-silver pine (*Lepidothamnus intermedius*) can form almost pure stands on low ridges with ultra-infertile podozolised soils. Manuka is abundant, mainly as layering stems. A well-developed understorey of *Gahnia procera* forms up to 30% cover, and there is a complete, hummocky layer of moss. Associated species include silver pine, pink pine, *Cyathodes juniperina*, *Empodisma minus* and very stunted rimu trees. Near Haast, similar stands occur on peat swamps.

(3) Stunted tree-heath can occupy broad crests without an organic soil horizon. Manuka, silver pine and pink pine form the canopy as in (1) but the manuka ranges from only a few cm up to 2.5 m tall, and the podocarps are up to 3.5 m tall.

Subalpine primary heath and bush also include varied communities.

(1) Tree- and shrub-heaths on gley podzols are more frequent in the northern part of the Westland beech-free region (Fig. 8.10) than further south, where mountains are higher and more rapid tectonic uplift and recent glacial activity have prevented soil maturation. In a stand at 850 m on a 23° slope in the Hokitika catchment, scattered pink pines up to 3.5 m tall form an overstorey to a main canopy at 1–1.5 m, mainly of pink pine, *Dracophyllum longifolium*, *Archeria traversii* and *Olearia colensoi*, together with some *Pseudopanax linearis, P. simplex, Pittosporum crassicaule, Coprosma pseudocuneata* and *C. colensoi*. The understorey, developed mainly in gaps, includes *Celmisia armstrongii*, *C. walkeri*, *Schoenus pauciflorus*, *Astelia nervosa*, *Blechnum* 'mountain', *Sticherus cunninghamii* and *Lycopodium scariosum*. *Chionochloa* 'westland' grows in the widest gaps. *Podocarpus nivalis* is abundant, *Coprosma serrulata* less so. Silt loam is only 10 cm deep over schist bedrock, but similar communities grow on deeper gley podzol soils with thin iron–humus pans at depths of 20–40 cm.

(2) Dense shrub-heath on steep slopes is the most widespread woody subalpine vegetation in Central Westland (Fig. 8.11). The four most important species are *Dracophyllum longifolium*, which is concentrated on ridges and other well-drained sites, *D. traversii*, which is widely distributed, often forming an overstorey, *Olearia lacunosa*, which tends to dominate on deep, moist soils on debris cones, and *O. colensoi*,

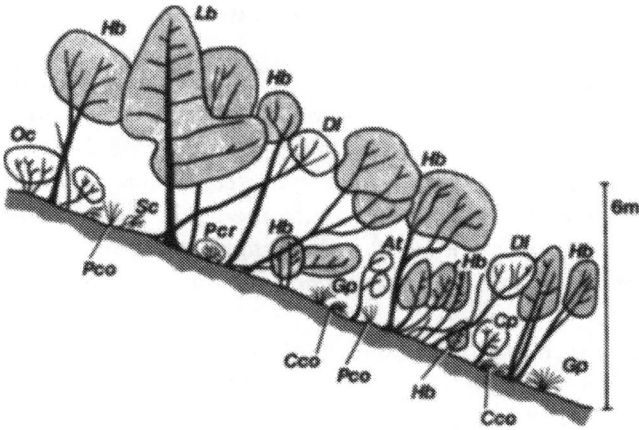

Fig. 8.10. Profile of subalpine tree-heath on a gley podzol at 1040 m, Kaimata Range, North Westland (Reif & Allen 1988). Abbreviations: *At, Archeria traversii*; *Cc, Coprosma colensoi*; *Cp, Coprosma pseudocuneata*; *Dl, Dracophyllum longifolium*; *Gp, Gahnia procera*; *Lb, Libocedrus bidwillii*; *Hb, Halocarpus biformis*; *Oc, Olearia colensoi*; *Pco, Phormium cookianum*; *Pcr, Pittosporum crassicaule*; *Sc, Sticherus cunninghamii*.

Fig. 8.11. Dense shrub-heath of *Dracophyllum longifolium* (upper left) and *Olearia colensoi*, with *Astelia nervosa* in foreground; 1050 m, Westland National Park (Wardle 1977).

which usually dominates on sites with impeded drainage, shaded southerly aspects, and on the fog-bound westernmost ranges.

Other canopy and subcanopy species include *Brachyglottis buchananii, Archeria traversii, Pseudopanax colensoi* var. *ternatus, Myrsine divaricata* and *Coprosma pseudocuneata.* At lower altitudes *Griselinia littoralis* and *Pseudopanax simplex* appear, together with kaikawaka as an overstorey. *Phyllocladus alpinus, Podocarpus nivalis* and *P. hallii* × *nivalis* are locally common close to the Main Divide. The understorey is generally sparse, consisting mainly of *Blechnum* 'mountain', *Astelia nervosa* and *Phormium cookianum. Polystichum vestitum* and *Coprosma depressa* are often also present.

(3) Tall subalpine heath grades into bush both where soils are deeper and towards the upper limit of cool-temperate forest. The transition is dominated by *Dracophyllum longifolium, D. traversii* and *Olearia lacunosa,* which are joined by *Olearia ilicifolia, Hoheria glabrata* and *Griselinia littoralis* on the more favourable sites. Often there is an overstorey with kaikawaka and, less frequently, mountain totara. The main understorey plants are *Blechnum* 'mountain' and, on the better soils, *Polystichum vestitum* and *Astelia nervosa.*

(4) On immature soils on steep, moist debris slopes between 750 m and 1050 m, *Hoheria glabrata* forms an open canopy; it may also share dominance with or yield dominance to *Griselinia littoralis, Olearia ilicifolia,* or *Pseudopanax colensoi* where deer have not reduced this species. The understorey is mostly *Polystichum vestitum,* except where this fern is locally replaced by the procumbent shrub *Coprosma depressa. Hypolepis millefolium, Ranunculus reflexus, Hydrocotyle novae-zeelandiae* var. *montana, Isolepis habra* and *Poa breviglumis* become important where the vegetation has been damaged by browsing.

This can merge into tall scrub on raw debris, with *Coprosma rugosa, Carmichaelia grandiflora, Olearia arborescens, Hebe salicifolia, H. subalpina* and young *Hoheria glabrata.* Important lower-tier plants are *Polystichum vestitum, Blechnum* 'mountain', *Coprosma depressa* and *Chionochloa conspicua.*

(5) On steep, dissected slopes with a close pattern of alternating ridges and gullies, elements of communities 1–4 often exist as a mosaic, with species as edaphically different as pink pine and *Hoheria glabrata* growing in proximity (Fig. 8.12). There are usually also enclaves of rupestral and boulder-field vegetation (Chapter 11), and of pioneering communities on debris, including patches of *Hebe* scrub (p. 210). At its upper limit, tall scrub grades into *Dracophyllum uniflorum* shrub-heath or grass-shrub mosaics. South of Franz Josef Glacier, *D. fiordense* occurs on rocky ledges at the top of the subalpine belt.

Temperate and subalpine tall heaths are generally separated by vegetation growing on the raw or rejuvenated soils of the Alpine Fault scarp and deep gorges (p. 170). There are few linking areas, the largest being on slopes at 500–900 m in the upper reaches of the Karangarua Valley, where glacial scouring has bared resistant schist rock and little soil has developed during the Holocene. The vegetation is a mosaic of stunted forest (including silver beech over part of the area), tall heath and wet heath. As with heath on the piedmont moraines, the vegetation includes silver pine, *Elaeocarpus hookerianus* and abundant manuka, but yellow-silver pine is absent

Fig. 8.12. *Dracophyllum longifolium* heath on spurs and *Hoheria glabrata* bush on debris at 1100 m, Arthurs Pass, Hawdon ER.

whereas subalpine species such as *Olearia colensoi* and *Archeria traversii* are frequent. Other transitions occur in valley bottoms subject to temperature inversions. For example, on moraines in Westland National Park, seral *Dracophyllum longifolium* heath descends as low as 800 m (p. 512).

Western South Island below latitude 43° 40′

On extensive ultramafic outcrops between Jackson Bay and Big Bay, moraines derived from them, and the hard granites of Fiordland, heaths occur from sea level to the tree-limit (Figs. 8.13 and 8.14). On other substrates, tall silver beech forests that clothe most of the subalpine belt south of Paringa restrict shorter woody vegetation to U-shaped valley heads, discontinuous patches along the tree-limit, and disturbed sites such as avalanche tracks. Even so, there are local 'beech gaps' as far south as the Arawata valley, where subalpine bush and heath prevail as in Central Westland.

Southern heaths differ from those already described mainly in the presence of stunted mountain beech. Stunted silver beech trees also occur, mainly on deeper soils. A transition from subalpine bush to beech forest can have large silver beech trees as an overstorey to a main canopy of *Hoheria glabrata*. In Fiordland, *Olearia colensoi* descends wet gullies to sea-level beneath a canopy of taller species, and the usually coastal *O. oporina* ascends exposed seaward slopes to 670 m; on the other hand, silver pine is absent south of Martins Bay, and *Phyllocladus alpinus* appears to be absent south of Dusky Sound. *Pimelea gnidia* occurs on the Cascade Plateau and in Fiordland.

Fig. 8.13. Yellow-silver pine tree-heath with *Gahnia procera* beneath, and a mountain beech tree in left foreground. On moraine containing ultramafic boulders, Gorge River, Olivine ER.

In subalpine communities, *Dracophyllum traversii* is replaced by *D. fiordense* and *D. menziesii*, and *Olearia lacunosa* may be joined or replaced by the narrower leaved *O. crosby-smithiana*.

South-west Fiordland differs from regions further north, but resembles Stewart Island, in being fully exposed to gales from the west and south-west. Consequently, blanket peat tends to form beneath tall heath near the sea, and on exposed uplands. Even at low altitudes there are exposed tablelands where trees are reduced to espalier form and, especially at higher altitudes, tall heath tends to intermingle with or be replaced by dwarf-heath and grassland (Figs 8.15 and 8.16).

Stewart Island

The lower slopes in the south of Stewart Island probably support the largest area of almost continuous primary heath in New Zealand. This reflects severe westerly exposure, a relatively subdued hilly landscape, and a leached soil mantle over weathered granite that has been neither stripped nor rejuvenated by glaciation.

Wilson (1987) recognises about nine intergrading communities. At the lower altitudes manuka and yellow-silver pine up to 9 m tall co-dominate on terraces and broad ridges, and on exposed sites with shallow soils may be stunted to only 1 m tall

Fig. 8.14. Manuka dominant in primary tree-heath at 750 m. On shallow moraine, Mt Beck, Olivine ER.

and interrupted by openings with species of dwarf-heath, such as *Cyathodes empetrifolia, Gleichenia dicarpa* and *Schizaea fistulosa*. Pink pine replaces yellow-silver pine in wet depressions. Stunted rata and kamahi are usually present, and at its best development, before it grades into forest on sheltered sites, the heath has an overstorey of rimu trees up to 18 m tall.

A shrub understorey consists mainly of *Coprosma colensoi, C. foetidissima, C. cuneata, Myrsine divaricata* and *Pseudopanax colensoi* var. *fiordense*. The herb tier,

Fig. 8.15. Manuka and *Chionochloa acicularis* in wind-exposed shrub-heath on a raised marine platform, West Cape, Fiordland.

which is usually sparse, mainly contains *Gahnia procera, Phormium cookianum* and *Blechnum procerum; Grammitis billardierei* and *Tmesipteris tannensis* are common epiphytes. Bryophyte cushions are extraordinarily well developed, and contain *Dicranoloma, Ptychomnion, Breutelia, Hypnodendron, Hypopterygium, Bazzania, Schistochila, Aneura, Riccardia, Chiloscyphus, Lepidolaena, Lophocolea, Plagiochila, Lepicolea, Lepidozia, Trichocolea, Marsupidium, Acromastigium* and *Frullania*. Pale woolly clumps of the lichen *Sphaerophorus tener* are also frequent.

Olearia colensoi mostly dominates above the 300–400 m upper limit of yellow-silver pine, in association with manuka, *Dracophyllum longifolium* and pink pine. One variant consists of *O. colensoi* with *D. longifolium* as the only other prominent tall shrub, and with mainly litter beneath instead of the usual moss hummocks. In mosaics with 4–5 m tall manuka, *O. colensoi* tends to occupy the better-drained slopes and gullies, and manuka tends to be on shallower, poorly drained soils on south or south-west aspects; beneath a manuka canopy, the understorey is mainly layering manuka and pink pine.

Subalpine terrain, as defined by the upper limit of kamahi, begins a little below 500 m and is generally clothed by dwarf-heath and grassland, which grade into *Olearia*- or manuka-dominated scrub in gullies, among rocks and in the lee of tors. On the slopes of Mt Anglem at 700–900 m, scrub of *O. colensoi* and *Dracophyllum menziesii* grows on blocky diorite.

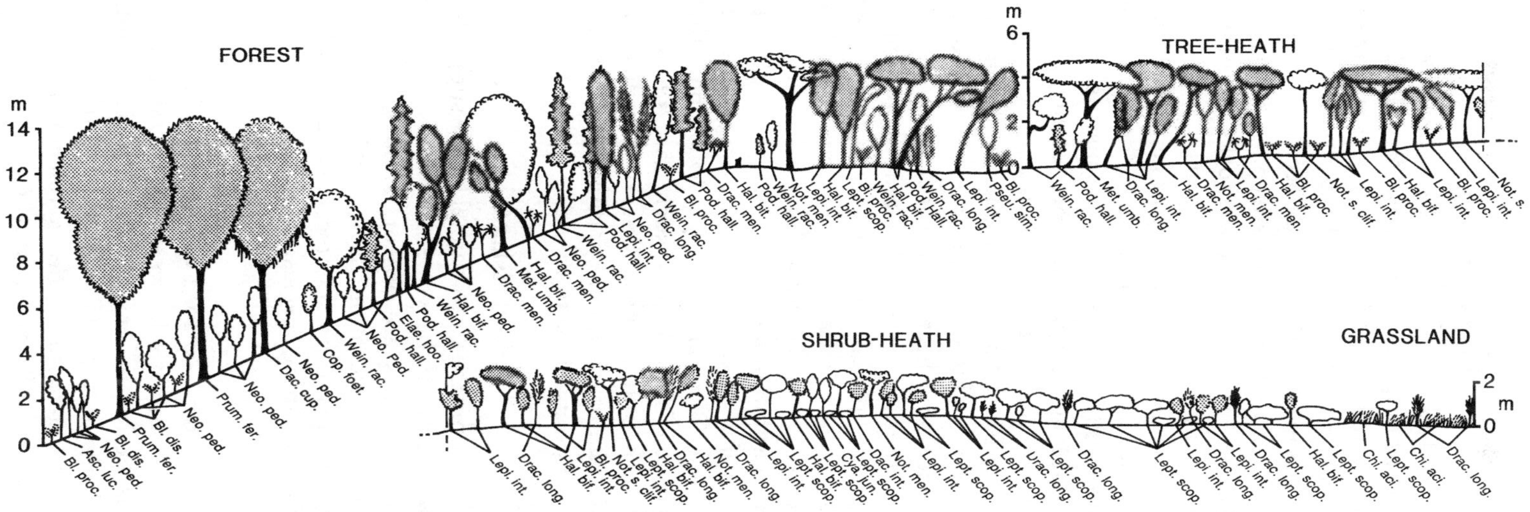

Fig. 8.16. Profile across a gradient from podocarp/broad-leaved forest in a sheltered valley, through tree- and shrub-heath to wet *Chionochloa* grassland. West Cape, Fiordland (P. Wardle *et al.* 1973).

Western Nelson

Between the Taramakau and Grey rivers, forests without beech interdigitate with beech forests, and there are isolated beech-free areas on coastal mountains as far north as the Glascow Range at latitude 41° 45'. Such areas support outliers of the high-altitude mixed forest, subalpine bush and heath typical of Central Westland. Equivalent vegetation occurs in cirque basins that beech fails to enter, although much of this was burnt during early attempts to graze sheep. *Hoheria glabrata* stands are few and small, presumably because prevailing granitic rocks and limited glacial rejuvenation of the landscape have not provided suitable soils.

Communities with beech resemble those of Fiordland. Tall heaths including small podocarps, manuka and mountain beech descend to the warm-temperate belt. They were once extensive on upper Pleistocene fluvial and marine terraces, especially south of the Mokihinui River, but logging, fire and, more recently, farm development have reduced them to remote remnants. One stand, occupying less than 1 ha of a sandstone ridge at 60 m a.s.l. near the Kohaihai River mouth, is dominated by *Dracophyllum longifolium* and kamahi and contains stunted northern rata and *Astelia trinervia*.

Upper montane and subalpine heaths are widespread on coal measures on the Denniston Plateau and southern end of the Paparoa Range (Fig. 8.17), and there are further large areas on granite peneplains north of Karamea. Because of altitude, exposure to westerly winds, soils that are often extremely shallow and infertile, and extensive fires, virgin tall heaths are patchy within a matrix of dwarf-heath and boggy grassland. Dominants such as mountain beech, pink pine, silver pine, southern rata, manuka, *Dracophyllum longifolium* and *D. traversii* range widely in altitude. Yellow-silver pine is important below 700 m, and *Olearia colensoi* above that altitude. Regional floristic richness is reflected in endemics such as *Dracophyllum townsonii* and *Pseudowintera traversii* which, unlike other species of its genus, has small thick leaves. Suckering plants of *Quintinia acutifolia* are common up to 950 m, and hard beech (*Nothofagus truncata*) enters transitions to forest.

North Island

In the North Island, perhaps only the Coromandel Range supports temperate primary tall heaths comparable with those of the western South Island. Subalpine tree-heaths are most extensive where beeches are absent, although many isolated, exposed summits of comparatively low elevation have subalpine-like heathland. Finally, because this kind of vegetation is so fragmented, there are marked floristic gaps that lead to expanded roles for remaining species.

Tararua Range

Silver beech forest does not enter U-shaped valley heads in the Tararua Range where, instead, *Griselinia littoralis* dominates between 900 or 980 m and 1160 m, on stony, moist, well-drained soils beside water courses. This is accompanied by *Olearia lacunosa*, the evergreen *Hoheria* 'tararua', and less frequently, mountain totara and *Olearia ilicifolia*. The shrub tier contains *Olearia colensoi, Brachyglottis elaeagnifolia, Myrsine divaricata, Coprosma* aff. *parviflora, C. ciliata, C. pseudocuneata, C. foetidissima, Pseudopanax simplex* and *Pseudowintera colorata*. Herbaceous tiers, as

Fig. 8.17. Stunted tree-heath with manuka, mountain beech, southern rata, *Dracophyllum longifolium*, etc.; coal-measure sandstone at 880 m on Mt Davey, Paparoa Range, North Westland.

seen in a deer-modified condition in 1959, include *Polystichum vestitum, Chionochloa conspicua, Ranunculus reflexus, Hydrocotyle* sp., *Viola filicaulis* and *Phormium cookianum.*

These stands grade laterally and upwards into shorter vegetation dominated by *Olearia colensoi.* On stable, leached soils, this is accompanied by *Dracophyllum filifolium* and the undergrowth is mostly *Blechnum* 'mountain', whereas on steep, semi-stable debris and fault breccia *Olearia colensoi* forms an open canopy about 2.5 m tall over dense *Polystichum vestitum.* The latter community extends between 1070 m and 1250 m, and is the edaphic equivalent of subalpine *Hoheria* bush in Westland. Both *Griselinia littoralis* bush and the *Olearia colensoi/Polystichum* community have forest equivalents in which silver beech trees up to 75 cm diameter and over 12 m tall form an open overstorey.

Subalpine beech forest is absent from the northern part of the range. Instead, *Olearia colensoi* and *Dracophyllum longifolium* prevail, and on gleyed soils are locally accompanied or replaced by pink pine, which is usually associated with *Phyllocladus alpinus.* Below 910–980 m, such heath grades into kamahi forest, but on broad, exposed ridges can descend below 600 m (Wardle 1962).

Northern axial ranges
Olearia colensoi heath with some pink pine extends over southern crests of the Ruahine Range. Otherwise, on the axial ranges, beech forest prevails in the subalpine belt, and

few valleys have the cirque form that excludes it. On Mt Hikurangi (1754 m) primary tree-heath extends between the summit crags and the upper limit of silver beech forest, which seems depressed through a local catabatic effect. It is probably the richest community of its kind in the North Island, with pink pine and *Olearia colensoi* sharing dominance, and *Phyllocladus alpinus, Brachyglottis elaeagnifolia* and *Dracophyllum traversii* being common. *Hebe stricta, Olearia nummulariifolia* and *O. ilicifolia* grow on the more recent soils.

Mt Ruapehu

Subalpine vegetation on this volcano is a mosaic of tall and dwarf-heath and bog, with grassland and barren terrain increasing towards and above the tree-limit, which lies at 1460–1520 m. On the south-western side, boggy mountain beech forest grades above 1040 m into tree-heath, with mountain beech, silver pine and pink pine co-dominating. Abundant associates are *Phyllocladus alpinus, Gahnia procera, Sticherus cunninghamii* and manuka; *Podocarpus nivalis*, mountain totara and kaikawaka are also present. Above 1250 m this is stunted to 3 m, and mountain beech and pink pine share dominance with *Phyllocladus alpinus, Podocarpus nivalis* and the shrubland species *Cassinia 'vauvilliersii', Dracophyllum recurvum* and *Hebe odora*; silver pine and bog pine (*Halocarpus bidwillii*) are also present, attaining much greater altitudes than elsewhere in New Zealand.

On the eastern side, subalpine shrub-heath occupies fixed dunes in barren expanses of volcanic sand and gravel. The main shrubs are *Phyllocladus alpinus* and *Dracophyllum filifolium*; others include *D. recurvum, Gaultheria crassa, Pseudopanax colensoi, Coprosma pseudocuneata, Olearia nummulariifolia, Cassinia 'vauvilliersii', Hebe odora* and *Podocarpus nivalis*. The smaller plants are mostly characteristic of dwarf-heath and grassland, e.g. red tussock (*Chionochloa rubra*), *Lepidothamnus laxifolius* and *Pentachondra pumila*. Bog pine is important on dune crests at 1050 m (Fig. 8.18).

Mt Taranaki

Like Ruapehu, this is a recent andesitic volcano, but differs in its wetter climate, steeper slopes, dense epiphytic growth of filmy ferns and bryophytes, and abundance of subalpine species at comparatively low altitudes. *Phyllocladus alpinus* and silver, pink and bog pines are all absent, probably in consequence of more fertile, better-drained soils, but the mountain has the only North Island locality for *Hoheria glabrata*.

From the upper limits of continuous forest at 1100 m to about 1150 m, there are short, wide-crowned trees of *Griselinia littoralis*, mountain totara and kaikawaka over a dense main canopy of *Brachyglottis elaeagnifolia*, accompanied by *Hebe stricta* var. *egmontiana, Pseudopanax simplex, P. colensoi* var. *colensoi, Myrsine divaricata, Coprosma pseudocuneata* and *C.* aff. *parviflora. Dracophyllum filifolium* dominates where there is poor drainage or recent slope movement. *Astelia* aff. *nervosa* and *Blechnum 'capense'* are the main herbs. The presence of *Rubus cissoides*, fuchsia and *Coprosma tenuifolia* shows that the community is really upper montane, the forest limit being depressed. Above 1220 m dense scrub merges into open shrub-heath, in which *Brachyglottis elaeagnifolia, Myrsine divaricata, Hebe odora* and *Dracophyllum filifolium* mingle with red tussock (Clarkson 1986).

Fig. 8.18. Bog pine and smaller shrubs on a steep dune in volcanic desert at 1070 m on the eastern side of Mt Ruapehu. *Rytidosperma setifolium*, *Poa cita* and a *Raoulia albosericea* mat are in the foreground.

Mt Pureora (1165 m)

Stunted forest on this extinct volcano grades above 1000 m to bush 5–8 m tall, dominated by mountain totara, *Griselinia littoralis* and *Pseudopanax simplex* (Leathwick *et al.* 1988) (Fig. 5.1). A shrub tier (40% cover) is formed mainly of *Pseudowintera colorata*, *Coprosma foetidissima* and *C.* aff. *parviflora*. Beneath, there is sparse *Blechnum* 'mountain', *Hymenophyllum multifidum* and *Dicranaloma* spp. *Rubus cissoides* is common, indicating the summit-cool-temperate status. Other species include *Polystichum vestitum*, *Asplenium flaccidum*, *Dicksonia lanata*, *Phyllocladus alpinus*, *Neomyrtus pedunculata*, *Melicytus lanceolatus*, *Olearia ilicifolia*, *Hebe stricta*, *Myrsine australis*, *Uncinia rupestris* and *Luzuriaga parviflora*. On trunks and branches there is a shroud of *Hymenophyllum multifidum* and bryophytes, including *Mastigophora flagellifera*, *Plagiochila* spp. and *Cladomnion ericoides*, with *Weymouthia* spp. pendant from twigs; these epiphytes reflect cool temperatures, moisture-laden winds and prevalent fog.

Bush gives way abruptly to shrub-heath 30 m below the gentle, wind-swept summit. This heath is 0.6–1 m tall, and in depressions consists of a dense community of bog pine, *Phyllocladus alpinus*, *Griselinia littoralis*, *Pseudopanax colensoi*, *Pseudopanax simplex*, *Coprosma foetidissima*, *C.* aff. *parviflora*, *Olearia ilicifolia*, *O. arborescens* and *Hebe stricta*. *Hymenophyllum multifidum* covers the short trunks. Exposed rises have low *O. arborescens* and bog pine among a turf with *Cyathodes empetrifolia*, *Deyeuxia avenoides*, *Oreobolus pectinatus*, *Racomitrium lanuginosum* and *Dicranoloma robustum.* Further species on the summit include *Myrsine divaricata, Gaultheria antipoda, Cassinia 'vauvilliersii', Ourisia macrophylla, Hierochloe redolens, Gahnia procera, Astelia nervosa, A. fragrans, Cordyline indivisa* and mountain totara, the last less than 1.5 m tall. A 10–30 cm veneer of Taupo airfall pumice overlies weathering andesitic boulders, and has gley-podzol characteristics. These shrub communities highlight the tendency of recent tephra to support, in close proximity, species characteristic of deep, recent soils (e.g. *Hebe stricta, Astelia fragrans*) and those normally found on shallow gley podzols (especially bog pine and *Oreobolus pectinatus*).

Coromandel Range

Table Mountain, a broad plateau lying at 820 m, is covered by wet, peaty loam. The canopy is generally dense, 10–12 m tall, and dominated by yellow-silver pine with lesser amounts of *Ixerba brexioides* and silver pine. The sparse shrub tier consists mainly of *Weinmannia silvicola, Coprosma dodonaeifolia* (closely related to *C. lucida*), *Quintinia* and *Pseudopanax discolor. Astelia* aff. *nervosa* and *Gahnia procera* form most of the tall herb tier, and there is a continuous bryophyte tier, largely of *Dicranoloma.* Where the canopy is lower and more open, manuka and *Epacris pauciflora* appear. Other species include widely scattered kauri, *Phyllocladus glaucus,* small rimu, *Pseudopanax colensoi,* southern rata, *Dracophyllum traversii, D. adamsii, Archeria racemosa, Corokia buddleioides, Toronia toru, Alseuosmia* sp., and *Gahnia ?xanthocarpa.* This isolated community resembles yellow-silver pine tree-heath in Westland, but also has links with kauri forest that contains *Halocarpus kirkii* and silver pine. Possibly, similar communities were once extensive in areas that are now gumland.

Eastern South Island

In respect of the distribution and composition of primary heath, a boundary is formed by the crest of the Southern Alps, to as far north as Lewis Pass. Further north, the boundary continues through the Victoria Range, Mt Owen and Mt Arthur to the Aorere valley; southwards, it follows the eastern flank of the Fiordland massif. With the exception of the Nelson ultramafic belt, where tall primary heath would occupy considerable areas were it not for fire, communities associated with wet, severely leached soils extend east of this boundary only as subalpine fragments, mainly where westerly rain sweeps through low passes and Foveaux and Cook Straits. For example, on some glacially scoured slopes in the headwaters of the Waimakariri River the prevailing mountain beech forest grades into stunted, open communities of mountain beech, *Phyllocladus alpinus* and pink pine. The last species also dominates, usually in association with kaikawaka, on poorly drained summits in Catlins ER and near Dunedin.

Olearia colensoi covers part of the summit of Mt Stokes (1204 m) in the Marlborough Sounds. It also occupies damp, shaded ravines close to the Main Divide and in the Takitimu mountains in Southland, and can be accompanied by other western species, such as *Dracophyllum traversii.*

Through the remainder of the eastern South Island mountains, drier variants of subalpine primary heath were once widespread but, like the forest below, have been greatly reduced by a thousand years of burning. As in the west, they were most

Fig. 8.19. Subalpine shrub-heath of *Dracophyllum longifolium* and *Phyllocladus alpinus* at 1070 m, Dobson valley, Lakes ER.

extensive in areas lacking beech forest, but elsewhere were confined to hanging valleys and other places where the beech forest limit was depressed.

Today's remnants survive in remote valley heads and places shielded by natural fire breaks (Fig. 8.19). Usually, the canopy is formed by combinations of four species. *Dracophyllum longifolium* (replaced by the related *D. acerosum* in the mid-east and *D. filifolium* in the north-east of the island) is most abundant on relatively mature, leached soils. *D. uniflorum* increases with increasing altitude, to completely replace *D. longifolium* in the penalpine belt. *Phyllocladus alpinus* is most abundant as large, layered colonies on young, rocky debris. *Podocarpus nivalis* provides an understorey to the other species but can form the canopy, especially along margins against bouldery ground and on avalanche tracks. Hybrids with *P. hallii* are frequent at altitudes approximating to the upper limit of ancient podocarp forests (Fig. 8.20). Seral shrubs can be prominent on raw soils and unstable sites. An example is *Brachyglottis monroi* scrub in the Inland Kaikoura Range, that also contains *Carmichaelia ovata, Notospartium carmichaeliae* and *Hebe traversii* (Williams 1989).

Other frequent shrubs are *Coprosma pseudocuneata, C. ciliata, Cassinia 'vauvilliersii'* and *Olearia nummulariifolia* (widespread), *Brachyglottis cassinioides* (locally abundant on debris), *B. buchananii* (chiefly on shaded ledges near the Main Divide), *Olearia cymbifolia* and *Pimelea traversii* (on eastern ranges), *Corallospartium crassicaule* (Canterbury and Central Otago), and further *Hebe* spp. (p. 211). *Myrsine nummularia, Gaultheria crassa* and *Coprosma depressa* are common dwarf shrubs. The main herbs are *Blechnum* 'mountain', *Phormium cookianum* and, on moist, well-drained soils, *Polystichum vestitum*.

Fig. 8.20. *Podocarpus hallii* (left) and *P. hallii* × *nivalis* as relics of forest destroyed by Maori fires on mid-slopes of Pisa Range, Central Otago.

Tall heath peters out into grassland on gentle slopes, especially at higher altitudes. As in the west, the transitions are marked by intermingling of shrubs with grasses and herbs, notably chionochloas, *Ranunculus lyallii* or *R. insignis*, *Celmisia semicordata*, *Anisotome haastii* and aciphyllas. These and various rupestral species (Chapter 11) are also prominent where slopes are interrupted by bluffs. At lower altitudes, wherever tall heath does not meet beech forest or seral vegetation, it merges into bush, as species such as *Griselinia littoralis*, *Olearia ilicifolia* and the deciduous *Hoheria* enter. The last forms distinctive communities on moist debris and in gullies. *H. glabrata*, which prevails near and west of the Divide, seems to grade into the hoary-leaved *H. lyallii* s.s. on eastern mountains. These *Hoheria* stands are more fragmented than their western equivalents, because lower rainfall restricts their occurrence and they have been further reduced by fire and penetrated by grazing animals. Usually, they are open and regenerating poorly, and once-dense understories of tall ferns have been largely replaced by grasses and weeds.

Chordospartium stevensonii, a beautiful, small, weeping tree with pink flowers, was known only as scattered, remnant populations in eastern Marlborough, that were being rapidly depleted as unintended victims of weed-spraying, but thriving populations have been discovered about the upper tree-limit on steep slopes in the Seaward Kaikoura Range.

Tree-heaths of southern coasts

In the cool, equable, moist, cloudy climate of southern New Zealand, exposure to frequent salt-laden westerly winds favours tall dracophyllum, macrocephalous olearias and species of *Brachyglottis*, often growing on blanket peat. This vegetation has far more in common with tall subalpine heaths than with the coastal bush of more northern districts, but it merges into bush on some islands (p. 168).

On Stewart Island, coastal tree-heath has its widest extent on western coasts and islets, being reduced on leeward shores to a discontinuous fringe at the forest edge. The most extensive community is the 'muttonbird scrub' of *Brachyglottis rotundifolia*, but on peat there is *Olearia lyallii* instead. *Dracophyllum longifolium* is usually present, and regenerates after fire to form pure stands. *Brachyglottis stewartiae* occurs on a few small islands lying off the main island. On the most exposed coasts, these species are replaced by *Olearia oporina*, which hybridises with *O. lyallii*. *Hebe elliptica* also can dominate in coastal fringes, as a seral shrub on less stable or disturbed sites such as gullies, old clearings, and places where *O. oporina* has been killed by salt.

Understoreys are usually sparse because of dense canopies and, in places, trampling or burrowing by nesting sea birds. Where one is well developed, the main species are *Phormium cookianum*, *Asplenium obtusatum*, *Blechnum durum* and, on peaty ground, *Polystichum vestitum*. The robust herb *Stilbocarpa lyallii* is important on offshore islands that are not reached by deer. Openings close to the shore contain *Carex trifida*, *Tetragonia trigyna*, *Hierochloe redolens* and *Poa* tussocks (p. 281).

The Solander Islands (Johnson 1975) are mainly covered in tree-heath, but because of their small size, isolation and extreme exposure, it includes few species. Big Solander Island shows a sequence from coastal *Hebe elliptica*, through *Brachyglottis rotundifolia*, to the summit plateau, which has *B. stewartiae* around the perimeter and *Olearia lyallii* in the centre; on Little Solander, *B. rotundifolia* and *O. lyallii* are absent. On Main Island of the Snares Group (Fineran 1964), *Hebe elliptica* dominates coastal fringes and openings. *Olearia lyallii* 8–9 m tall forms the central stands, and is accompanied by *Brachyglottis stewartiae* in the east of the island. Understoreys in both groups are like those on the islets offshore from Stewart Island. *Stilbocarpa lyallii* grows on Big Solander Island, and *S. robusta* is endemic to Little Solander and the Snares.

In the South Island, *Olearia oporina–Dracophyllum longifolium* communities form a coastal fringe on Bluff Hill and extend sporadically to Milford Sound on exposed western sea-cliffs, headlands and benches (Fig. 8.21). On the flatter surfaces they grow on blanket peat, which otherwise is little developed in Fiordland. Elements of coastal heath straggle northwards to Western Nelson where *Brachyglottis rotundifolia* and *Dracophyllum longifolium* grow on steep spurs and *Hebe elliptica* is common on the seaward fringe of woody vegetation. The last also occurs as a rare plant in Taranaki. In the east, *Dracophyllum longifolium* grows on headlands in Catlins ER and *Hebe elliptica* extends as far as Oamaru.

Penalpine shrub-heath

Tall shrubs and small trees reach a remarkably uniform limit at the same altitude as the highest beech trees in the same or similar districts. Several small shrubs continue to higher altitudes, but low temperatures and competition from grasses and herbs restrict closed shrub-heath to crests of spurs with north or west aspect, especially those with shallow, rocky soils.

Fig. 8.21. Coastal tree-heath of *Dracophyllum longifolium* with dead and sapling *Olearia oporina*, West Cape, Fiordland (P. Wardle *et al.* 1973).

Throughout the South Island and to as far as the Manawatu Gorge in the North Island, *Dracophyllum uniflorum* is the main dominant, although in high-rainfall districts it gives way to low bushes of *Olearia colensoi* on moister sites, especially steep southerly faces and ravines. *Podocarpus nivalis* is largely absent from the wettest, westernmost ranges, but towards the Main Divide it is usually co-dominant, rising to dominance on bouldery ground. East of the Divide, it forms almost pure expanses on broken greywacke. Other common shrubs are *Gaultheria crassa, Coprosma* 'penalpine' (similar to *C. pseudocuneata* but with a sprawling habit and shorter, more crowded leaves), *Brachyglottis bidwillii* north of latitude 43° 10′ and, except on the Western Alps, whipcord hebes (p. 222). Stunted plants of *Pseudopanax colensoi* var. *ternatus* are often present in the South Island. Smaller plants are those of grassland at the same altitude, including the dwarf shrubs *Myrsine nummularia* and *Coprosma cheesemanii*, and large herbs such as *Phormium cookianum, Aciphylla horrida* (Westland), *A. scott-thomsonii* (eastern South Island), *Anisotome haastii, Chionochloa* spp. (especially *C.* 'robust' in the South Island) and *Blechnum* 'mountain'. On steep bluffs, the main species of penalpine scrub intermingle with rupestral plants.

Scrub on ridge crests above tree-limit on Mt Ruapehu, the Ruahine Range and Mt Hikirangi is up to 50 cm tall, and dominated by *Podocarpus nivalis, Dracophyllum recurvum, Brachyglottis bidwillii* and the whipcord *Hebe tetragona*. *Gaultheria colensoi* is abundant on the volcanoes. As in the south, the shrubs intermingle with species representative of neighbouring *Chionochloa* grassland, bluff communities and scoria fields. On the axial ranges, *Olearia colensoi* ascends in steep, sheltered chimneys, even to the summit of Mt Hikurangi (1754 m). *Pseudopanax colensoi* var. *colensoi* also occurs sparingly to the highest limits of scrub.

Shrub-heaths of mires and frost-flats

A distinct category of heath, now greatly fragmented by fire and land-clearing, grows on valley floors and depressions subject to temperature inversions. Bog pine is the most characteristic species. On infertile mires, it tolerates wetter conditions than any other tree or shrub of comparable stature, maintaining an erect habit where even manuka becomes a creeping mat. On tephra plains and leached gravels, it is usually accompanied by *Phyllocladus alpinus*; these two podocarps show the greatest cold-tolerance measured in native woody plants (see Table 14.19). Although bog pine can regenerate after fire, more fire-tolerant shrubs and herbs have displaced it from much of its habitat.

Volcanic Plateau

Bog pine can form most of a more or less open shrub cover on frosty or boggy terraces, but becomes dense in steep-sided gullies that cut through these terraces. The other main shrubs are *Phyllocladus alpinus* (often co-dominant), silver pine (in wetter areas), *Dracophyllum subulatum* (with *D. filifolium* and hybrids at higher altitudes), and manuka. Filiramulate shrubs are abundant, especially *Coprosma propinqua* and *Olearia virgata* on the more fertile sites, and *Coprosma* aff. *parviflora* and *Myrsine divaricata* on less fertile sites. Others include *Pseudopanax anomalus, Corokia cotoneaster* and *Melicytus angustifolius. Hebe stricta* and *Carmichaelia arborea* also occur, and the locally endemic small tree *Pittosporum turneri* is confined to these communities. Swamp with *Phormium tenax* and *Carex coriacea*, bog with *Gleichenia dicarpa, Empodisma* or sphagnum, red-tussock (*Chionochloa rubra*) grassland or, on dry ground, a lichen cover of *Cladia retipora* and species of *Cladina* and *Cladonia* can form the understorey.

Dracophyllum subulatum shrub-heaths were formerly extensive on tephra plains, but most have been replaced by pasture and pine plantations, and the remainder are susceptible to invasion by *Pinus contorta* or heather (*Calluna vulgaris*). The best remaining example is on the Rangitaiki Plain at 730 m (Fig. 8.22). A dry facies mainly consists of open *D. subulatum*, with scattered, small *Poa cita* tussocks, growing on weakly welded pumice that contains iron and humus pans. Most of the ground is covered by *Racomitrium lanuginosum* and *Cladia retipora*, with lesser amounts of *Cladina leptoclada*. Other species, mainly typical of short-tussock grassland growing on dry, coarse soils, include *Lycopodium fastigiatum, Pimelea prostrata, Cyathodes fraseri, Celmisia gracilenta, Raoulia apicinigra, Hieracium pilosella*, catsear (*Hypochoeris radicata*) and low-growing manuka. After fire, shrubs are temporarily eliminated and the community becomes an open *Poa cita* grassland. In prolonged absence of fire, *D. subulatum* can become 0.5–1.8 m tall, forming *c.* 40% cover with little between except *Racomitrium* and *Cladia*. The *D. subulatum – Poa cita –* lichen mosaic seems to persist indefinitely on rises of dry, coarse pumice.

On the ring-plains surrounding the central volcanoes, there are higher-altitude *D.*

Fig. 8.22. *Dracophyllum subulatum* heath (left) and more recently burnt area with *Poa cita* (right); 730 m, Rangitaiki Plain, eastern Volcanic Plateau.

subulatum communities that merge into poorly drained red-tussock grassland. These were once seeded with heather, which now threatens to displace them, mainly through its ability to rapidly colonise burns and other bare ground.

Surface topography and soil profiles show that, prior to Maori fires, forest occurred over at least part of the area now or recently occupied by *D. subulatum* heath. On the wettest, frostiest and, perhaps, the driest sites the original vegetation may have been bog-pine heath.

The tendency for species indicating low fertility, such as bog pine and *Dracophyllum subulatum*, to mingle on tephra surfaces with species indicating semi-fertile soils, such as *Coprosma propinqua* and silver tussock (*Poa cita*), may be explained by some being able to exploit large soil volumes or to reach buried soil whereas others have their root penetration inhibited by welded or cemented horizons. On a frost-flat at Pureora, *Dracophyllum subulatum* roots are restricted by an iron–humus pan to depths of 12–20 cm whereas nearby *Coprosma propinqua* shrubs have tap roots that descend to at least 50 cm, penetrating the pan and probably reaching a topsoil buried by Taupo tephra.

South Island

Some infertile peaty swamps, wet gley–podzol soils in low-lying moraine basins, and leached terrace soils in the west of the South Island support neither forest nor tall, dense heath, though they appear never to have been burnt. The mainly rush-like vegetation, described in Chapter 10, contains scattered shrubs or thickets not more than 1–2 m tall. The main species are bog pine, *Dracophyllum palustre* and low manuka. Stunted pink pine and silver pine can be present in transitions to tree-heath, and *Olearia virgata* var. *laxiflora* grows in the drier parts of transitions to fertile swamp.

Similar mires with bog pine and manuka are scattered through eastern mountain valleys. There are also thickets of bog pine and *Phyllocladus alpinus* on well-drained, leached soils formed in the gravels of old moraines and terraces, under rainfalls

Fig. 8.23. Fringe of *Phyllocladus alpinus* and bog pine between mountain beech forest and valley-floor grassland at 800 m, Ahuriri Valley, Lakes ER.

ranging from 4000 mm per annum to as low as 600 mm (Fig. 8.23). Pollen evidence suggests that these are remnants of vegetation that was extensive when the last glaciation was drawing to a close; as taller vegetation spread, the shrub-heath became confined to the frostiest and least fertile sites and was further reduced by fire. Associated shrubs include manuka, *Cyathodes juniperina*, and filiramulate species such as *Coprosma propinqua* and *Aristotelia fruticosa* (Table 8.1). The whipcord *Hebe armstrongii*, which is almost indistinguishable from bog pine except when in flower, is known from two bog-pine stands, one wet and the other dry, in the Waimakariri catchment (Puketeraki ER). Lower-tier plants vary according to moisture and influence of grazing, from red tussock to dwarf shrubs such as *Cyathodes colensoi* and *C. fraseri*.

Kanuka–manuka heaths

Ecology of manuka and kanuka

Of all native woody plants, manuka and kanuka have responded most vigorously to disturbance. This vigour results from copious flowering that in manuka can begin when seedlings are only 5 cm tall, light seeds that are widely dispersed by wind, those of manuka being released from serotinous capsules after fire, prolific seedling establishment, unpalatability to mammals, and growth up to 40 cm annually. Kanuka is the taller and longer-lived. Manuka has wider edaphic tolerance, ranging from dry exposed ridges to swamp margins and wet heathland. Kanuka seems more competitive

Table 8.1. *Occurrence of trees, shrubs and lianes in grey scrub and related bog-pine communities*

Species	1	2	3	4	5	6	7
Carmichaelia grandiflora	f						
Rubus schmidelioides	a	a	f	f			
Muehlenbeckia complexa	f	a	f	a	+		
Sophora microphylla	ls			f	l		
Aristotelia fruticosa	f	a	o		rl	f	
Carmichaelia arborea	f					f	
Coprosma wallii	r					r	
Melicytus angustifolius	r					r	
Myrsine divaricata	f					a	
Coprosma propinqua	d	d	a	a	+	a	d
Olearia virgata or *lineata*	o		o			f	d
Coprosma aff. *parviflora*	a				+	a	f
Coprosma rigida	a					o	f
Pseudopanax anomalus	f					f	f
Melicytus alpinus		a	o				
Sophora prostrata		la		d			
Carmichaelia petriei or *robusta*		f		o			
Discaria toumatou		d	a		+		
Clematis marata or *quadribract*		o			+	o	
Cassinia 'fulvida'			f		+		
Corokia cotoneaster			f			f	a
Phyllocladus alpinus					d	f	s
Halocarpus bidwillii					d	la	lf
Leptospermum scoparium					+		f
Pittosporum turneri						lf	
Hebe stricta						f	f
Dracophyllum subulatum						f	f
Prumnopitys taxifolia						s	s
Griselinia littoralis						s	s

Notes:
1, 2, 5 and 6 are composite lists from several stands.
r, o, f, a, d, l: rare, occasional, frequent, abundant, dominant, local.
s, seedling; +, abundance not recorded.
Species listed from only one community:
1 *Coprosma* aff. *ciliata* a, *C. rugosa* ld in mountain valleys, *Muehlenbeckia australis* f, *Olearia avicenniifolia* f, *O. ilicifolia* lfs, *Hoheria glabrata* lfs, *Plagianthus regius* fs.
2 *Olearia odorata* o, *Rosa rubiginosa* a.
3 *Hebe brachysiphon* a.
4 *Calystegia turguriorum* a, *Clematis afoliata* a, *Parsonsia capsularis* a.
5 *Coprosma intertexta* +, *Pimelea traversii* o, *Cassinia 'fulvida'* +, *Corallospartium crassicaule* +.
6 *Rubus cissoides* o, *Dracophyllum longifolium* f, *Pseudowintera colorata* o, *Lagarostrobos colensoi* a, *Hebe rakaiensis* +.
7 *Erica lusitanica* f, *Pseudopanax crassifolius* f, *Pimelea tomentosa* o.
Localities:
1 Central Westland river flats below 430 m (*Sophora microphylla* only on margins of coastal lagoons).
2 Mackenzie ER; moraine and talus slopes at 750 m.
3 Cass, Puketeraki ER, in a gully notched in a fluvioglacial terrace at 640 m.
4 Marble Point, Lowry ER, on limestone debris at 270 m.
5 Bendhu, Mackenzie ER, 550 m on ancient moraine (B.P.J. Molloy, unpublished).
6 Tongariro ER on tephra at 750–800 m.
7 Pureora, Western Volcanic Plateau ER, on tephra at 500 m.

on fertile, well-drained soils. It seldom grows on waterlogged soils, but thrives on dry sites where manuka is stunted and unthrifty. Whereas manuka grows throughout the main islands and reaches the subalpine tree-limit, kanuka does not extend south of Hokitika in the west, nor west of the Clutha Valley in the south, nor into the subalpine belt.

Since the 1940s blight has killed manuka through large areas. This blight results in heavy infestation by sooty moulds, but is caused by the scale insect *Eriococcus orariensis*, which seems to have been self-introduced from Australia (Hoy 1961). Manuka blight is not important at higher altitudes or in very wet districts, and does not kill seedlings less than 50 cm tall. Moreover, its virulence has been greatly diminished through spread of a hyperparasitic fungus, *Myriangium thwaitesii*. Even so, the abundance and vigour of manuka has been reduced over wide areas, especially relative to kanuka, which is much less affected by the blight.

Although the pre-eminence of kanuka and manuka is being overtaken by adventive shrubs such as gorse and *Hakea*, they are still very prominent in secondary successions. The light-demanding seedlings establish readily in short, open, lightly-grazed pasture (Grant 1967), and both species increase when graziers are unable to maintain pasture density on hilly and infertile land. Prior to the intensive, superphosphate-dependent pasture development that began in the 1940s, manuka and kanuka covered very large areas, which earlier had supported forest or fire-induced fernland and native grassland. Both species also have roles in primary succession on river flats and slips, with manuka colonising stony surfaces and kanuka finer deposits. These primary seral stands may, in part, represent situations where rapid development of other cover had been prevented, for instance, by grazing.

Although kanuka and manuka stands are usually transient, they can also maintain themselves as more or less stable communities, for reasons that include frequent fires, harsh environments, distance from sources of forest seed, and browsing of palatable seedlings by introduced animals. Manuka is important in primary heaths, but the role of kanuka in the primitive vegetation is less certain, though it may have been an opportunist, vagile species, that also had a permanent place in woody vegetation on dry ridges and in semi-arid districts. Kanuka charcoal predominates in loess deposits formed in Canterbury during glacial times (Goh *et al.* 1977).

Other species of kanuka–manuka heaths

Erica has invaded large areas of secondary manuka heath, open bracken fernland and low-producing grassland in Sounds–Nelson province, and there are many other local infestations, from Northland to Southland. The tiny seeds are dispersed by wind and establish in large numbers on bare ground, especially after fire; the plants also resprout after fire. The species is generally *E. lusitanica*, but there are some populations of *E. arborea*.

In the northern zone, and also on hills around Tasman Bay and Golden Bay, *Hakea* species from Australia are well-established in manuka heath. Their large woody capsules release abundant seed after fire, leading to dense thickets that resist invasion by other plants. *H. sericea* has become abundant on dry ridges, although it does not displace manuka from the poorest soils. Like the less widespread *H. gibbosa*, its leaves are modified into 3–4 cm long spines. *H. salicifolia* is broad-leaved and more mesophytic, and the seedlings can establish beneath tall, open kanuka stands. Several species of *Pomaderris* occur in northern manuka heath, the most abundant being the broad-leaved, yellow-flowered *P. kumeraho*, and *P. phylicifolia* var. *ericifolia*, which occurs sparingly as far south as the Canterbury Plains.

The least fertile heathland soils, being very low in phosphorus, have escaped invasion by adventive woody legumes, but less infertile areas have been taken over by gorse and broom (*Cytisus scoparius*); brush wattle can be prominent in northern districts.

The smaller plants associated with kanuka, manuka and other invasive shrubs are mostly those of earlier vegetation that is being displaced, and the seedlings and understorey species of later phases. However, open stands contain more kinds of small, tuberous orchid than any other New Zealand vegetation although, like most of the florula of secondary heaths, they represent only a small selection from the much richer Australian heathlands. In the North Island, the scandent sundew *Drosera peltata* is common. The tiny mistletoe *Korthalsella salicornioides* and the northern parasitic liane *Cassytha paniculata* are practically confined to kanuka–manuka heath.

Manuka heath of the gumlands

Throughout Northland and much of the Coromandel Peninsula, there are tracts of infertile soils, with perched water tables, that contain kauri resin. Most are rolling uplands where leaching over tens of thousands of years has led to upper soil horizons of almost pure silica or sterile clay. Other gumlands are on Pleistocene sands, between peat swamps occupying former dune hollows. *In situ* stumps and roots of kauri and other conifers have yielded radiocarbon ages between 1390 and > 35 400 years, which suggests that the resin has accumulated over many generations of kauri trees.

Travellers' descriptions from as early as 1846 show that the prevailing gumland vegetation was manuka heath, which has since been extended and modified by human activity, especially 'gum digging'. Kauri resin has several industrial uses, and during the heyday around the turn of the century, the gumlands supported a considerable population, mainly immigrants from the Dalmatian coast of Yugoslavia. In 1905, about 10 000 tonnes of resin were recovered.

The vegetation is essentially manuka mixed with *Dracophyllum lessonianum*, rush-like sedges including *Baumea teretifolia*, *Lepidosperma australe*, *Schoenus brevifolius*, *S. tendo* and *Tetraria capillaris*, and the fern *Gleichenia dicarpa*. It is least dense where silica is exposed on rises, and thickens downslope and where soils are deeper or more fertile (Table 8.2; Fig. 8.24). The following are representative of gumland as a whole: *Lycopodium laterale*, *L. cernuum*, *L. deuterodensum*, *Blechnum 'capense'*, *Lindsaea linearis*, *Drosera peltata*, *D. pygmaea*, *Centella uniflora*, *Pimelea prostrata*, *Cyathodes fasciculata*, *C. fraseri*, *Pomaderris kumeraho*, *P. phylicifolia* var. *ericifolia*, *Gonocarpus montanus*, *Lepidosperma laterale*, *Gahnia setifolia*, *Morelotia affinis*, *Dianella nigra*, and *Rytidosperma biannulare*. In wet depressions *Empodisma*, *Cortaderia toetoe* and *Baumea rubiginosa* can be prominent, and holes made by gum diggers become filled by *Sphagnum*. *Aira* spp. and *Hypochoeris radicata* appear temporarily after fires. *Hakea sericea* has become almost universal on gumland, and can form dense colonies that

Table 8.2. *Stand structure on ridges and in depressions in gumland near Kaikohe*

	Ridge		Depression	
	dry mass	no. of stems	dry mass	no. of stems
Shrubs				
Leptospermum scoparium	61.7	399	38.4	271
Dracophyllum lessonianum	4.5	57	4.5	126
Pomaderris kumeraho	0.1	7		
Cyathodes fasciculata			0.2	3
Hakea sericea	2.8	1		
Pomaderris phylicifolia	t	1		
Epacris pauciflora			0.1	
Sedges				
Baumea teretifolia	7.7	388	30.9	1171
Tetraria capillaris			4.4	765
Schoenus brevifolius	12.2	173	9.1	138
Lepidosperma australe	6.1	146		
Schoenus tendo	4.7	20	0.3	13
Ferns and Lycopods				
Gleichenia dicarpa	8.1	6	6.9	28
Lycopodium laterale	t		4.5	17
Schizaea fistulosa	t	5	0.1	1
Lycopodium cernuum	t	2		
Others				
Dianella nigra	t	3	t	3
Gonocarpus montanus	t	4		
Litter	13.8		14.0	
Total dry mass above ground (g)	580		1128	

Notes:
Values are means of ten 1 m² plots in each habitat; dry mass as percentage of total; t, trace.
Source: Esler & Rumball 1975

exclude manuka, though mainly on the deeper soils. Gorse and bracken are generally important only on lower slopes.

Gumlands are a rich habitat for orchids. Esler & Rumball (1975) list *Acianthus fornicatus* var. *sinclairii, Bulbophyllum pygmaeum, Caladenia catenata, Chiloglottis cornuta, Corybas aconitiflorus, C. macranthus, C. oblongus, Microtis unifolia, Orthoceras strictum, Prasophyllum colensoi, P. pumilum, Pterostylis alobula, P. banksii, P. plumosa, P. graminea, P. nana, P. trullifolia, Thelymitra carnea, T. pulchella* and *T. longifolia.* They are also home to several very rare plants, notably the minute tuberous lycopod *Phylloglossum drummondii, Lycopodium serpentinum* and *Baumea complanata.*

The largest remaining area of gumland heath is on the Ahipara Plateau, at 200–300

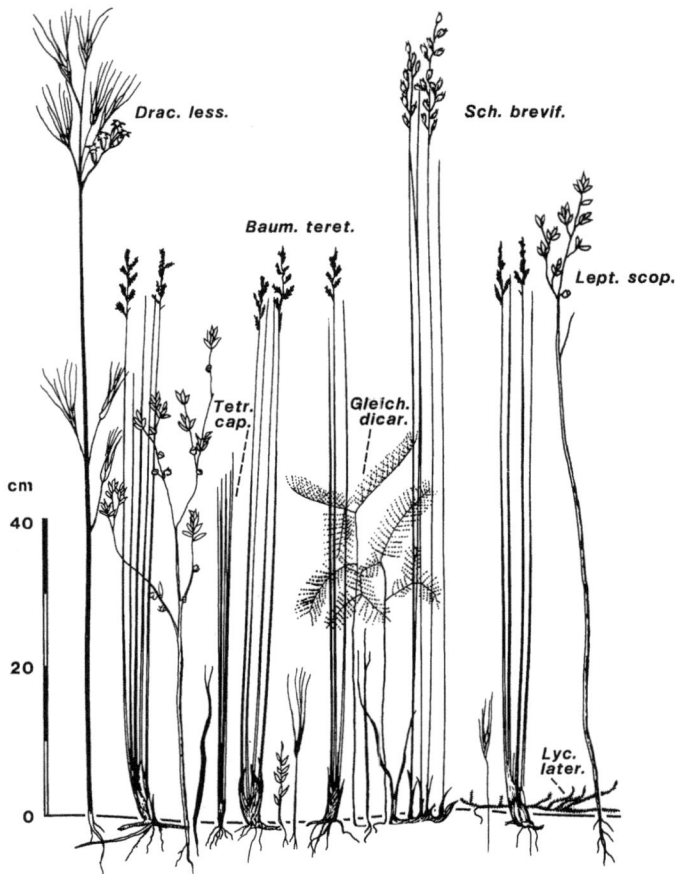

Fig. 8.24. Profile of gumland vegetation near Kaikohe, Northland (Esler & Rumball 1975).

m a.s.l. This is frequently burnt, but manuka regenerates prolifically and immediately, and grows 10–20 cm per year. Bare zones between burnt and regenerating scrub (Fig. 8.25) can form breaks against later fires. Rush-like sedges and *Gleichenia* survive fires, and in wet depressions quickly establish temporary dominance, until overtopped by manuka seedlings. Saplings of *Weinmannia silvicola* and *Coprosma lucida* occur widely in older scrub, their seed being distributed by wind and birds respectively. However, succession towards forest is vigorous only on lower slopes, with better soils, that lead down to gullies holding forest remnants.

On the North Cape plateau, clay soils derived from ultramafic rocks support low, fire-modified scrub dominated mainly by manuka, *Hebe ligustrifolia* and *Cassinia 'amoena'*. Other species, most of which also grow on the ultramafic cliffs below (p. 397), include kanuka, *Cyathodes juniperina, C. parviflora, C. fraseri, Hebe macrocarpa* var. *brevifolia, Pomaderris prunifolia* var. *edgerleyi, Cassytha, Schoenus brevifolius, Lindsaea linearis* and bracken. On soils derived from gabbro, the vegetation is more

Fig. 8.25. Gumland on Ahipara plateau, Northland. Area A has been recently burnt; the unburnt manuka in the middle (B) has responded to relaxation of competition, while competitively excluding new seedlings from its rooting zone (C).

like that of typical gumland, with abundant rush-like sedges (Thomson *et al.* 1974).

Pollen analyses show that fire has occurred and fire-tolerant heath and wetland have been present in Northland for at least 16 000 years, especially during the upper part of the Last Glaciation and since the arrival of humans (Dodson *et al.* 1988). Edaphically and floristically, gumlands have much in common with fire-induced pakihi of the western South Island. The latter have been derived from complex mosaics of forest, heath and mire; gumland probably replaced similar mosaics, which had developed as soils became impoverished under kauri and other conifers. The Ngaruka Swamp in Waipoua Forest has a central area dominated by manuka, *Gleichenia dicarpa* and *Baumea rubiginosa*, and may be a surviving example of a primitive open phase, whence fires spread into surrounding forest. On the poorest soils, the existence of kauri stands probably depends on nutrient cycling within the organic horizon. With the greatly increased frequency of fires after the arrival of humans, this horizon would have been extensively destroyed and gumland thereby entrenched.

In the past few decades, most gumland has been converted to pasture and, more recently, to *Pinus radiata* plantations through use of fertilisers and, on occasion, mechanical ripping of the silica pan. Survival of this interesting vegetation depends on establishment of reserves where fire can be accepted as a management tool.

Manuka scrub established on pasture

North of latitude 38°, the common associates of grazed manuka stands, as they grow taller and more open and admit more light, are the shrubs *Coprosma rhamnoides* and

Cyathodes fasciculata, the grasses *Oplismenus hirtellus* and *Microlaena stipoides*, and the fern *Doodia media*. *Ageratina* spp. can form an understorey in damp hollows. As there are numerous forest remnants, tree seedlings persistently invade and the less palatable eventually succeed, notably *Myrsine australis, Podocarpus totara* and *Phyllocladus trichomanoides*. Southwards, most of these species drop out at various latitudes, and in inland parts of Otago and Southland many manuka stands are distant from forest remnants and lack tree seedlings.

Kanuka tree-heath

Urewera Ecological Region

For the northern Urewera, Payton *et al.* (1984) describe stands of kanuka and kamahi that replaced mixed forest after fire. Much of the kamahi has since died (probably following drought), and deer are destroying broad-leaved seedlings. This results, 'some 50–60 years after the fire, in a relatively closed stand of kanuka, with the occasional sapling or young tree of *Carpodetus serratus* and *Pittosporum tenuifolium*, often little, if any undergrowth, and a grassy sward dominated by *Microlaena avenacea* and *Uncinia uncinata*. The sparse subcanopy is now tree fern'. Some stands have two generations of kanuka, the first being the original pioneers and the other having grown up where broad-leaved species failed to occupy canopy gaps.

Table 8.3 shows the proportions of tree and shrub species in a stand of kanuka aged 19–45 years, with diameters up to 25 cm. Only the light-demanding kanuka is represented in all size classes. Few others reach the tall seedling stage, most being held in check by deer although some may be suppressed by the kanuka canopy or herbaceous tier. The percentage frequencies (in 0.75 m² sub-plots) of the main herbs, which all tolerate or resist browsing, are *Microlaena avenacea* 92, *Uncinia uncinata* 80, *Paesia scaberula* 37, *Blechnum 'capense'* 24, *Nertera depressa* 22, *Lagenifera* sp. 20, *Rytidosperma* sp. 16, *Nertera* aff. *dichondrifolia* 14, and *Gnaphalium* sp., *Senecio jacobaea* and *Hymenophyllum* spp. each 10.

Spenser Ecological Region

Kanuka-dominated stands at 600–850 m near Lake Rotoiti are typical of those that replaced beech forest after fire over large areas in the north of the South Island (Fig. 8.26). Most of the kanuka and manuka are < 115 years old, but one kanuka tree had 139 growth rings (K.W. Briden, unpublished). The original fire may have been much earlier, as isolated beech trees are up to 280 years old, dead trees of pioneer form seem still older, and beech charcoal in the soil beneath kanuka has been radiocarbon dated at 572 ± 54 years.

Ground cover and seedlings were recorded in transects, totalling 18 m², beyond the immediate influence of beech forest margins. The canopy gives about 90% cover, and is predominantly of kanuka 6–9 m tall and up to 14 cm in diameter. A sparse shrub tier consists mainly of sapling kanuka and manuka, *Cyathodes fasciculata, C. juniperina* and *Coprosma* aff. *parviflora. Blechnum 'capense', Lycopodium volubile* and *L. scariosum* form a patchy tall herb tier. The mosses *Dicranoloma robustum, Ptychomnion aciculare* and *Leptotheca gaudichaudii* cover about 35, 20 and 5% respectively of the ground. Seedlings < 1 m tall are *Coprosma* spp. 129, *Cyathodes* spp. 18, manuka 78, and kanuka

Table 8.3. *Stand composition on a plot of 256 m², at 515 m in Urewera ER*

	class				
	A	B	C	D	E
Kunzea ericoides	4	41	8	16	20
Cyathodes fasciculata				2	
Prumnopitys taxifolia				3	
Cyathea dealbata				6	
Dicksonia squarrosa				10	
Carpodetus serratus		41	8	40	
Pittosporum tenuifolium		8		3	
Brachyglottis repanda		14			
Pseudopanax colensoi		3			
Dacrydium cupressinum		11			
Melicytus ramiflorus	35	82			
Myrsine australis	31	46			
Weinmannia racemosa	20	30			
Coprosma lucida	16	14			
Geniostoma rupestre	6	19			
Knightia excelsa	2	3			

Notes:
A Seedlings <5 cm tall; percentage frequency in 0.75 m² sub-plots
B Seedlings 6–50 cm tall; no./100 m²
C Seedlings 51–140 cm tall; no./100 m²
D Saplings 0–10 cm diameter as % of total trees and saplings
E Trees >10 cm diameter as in D
Source: Payton et al. 1984

>100. Eight beech seedlings were recorded, and only 1–3 seedlings each of *Griselinia littoralis*, lancewood and *Carpodetus*. Outside the transects, there are rare seedlings of mountain totara.

This kanuka–manuka stand, with its understorey of small-leaved shrubs, is evidently self-perpetuating. Replacement by beech forest or broad-leaved bush will be very slow, as beech seedlings are rare except within 12 m of beech trees; broad-leaved tree seedlings are even sparser, except in moist depressions. This situation may reflect past episodes of heavy browsing by deer and domestic stock, but other factors are ineffective beech regeneration, an inland climate that is too harsh for many broad-leaved trees and native conifers, and inadequate seed sources of those within their climatic range.

Kanuka and manuka heath on dry sites

From Northland to Otago, dry, exposed sites with little or no topsoil support clumps of stunted kanuka or manuka, with mainly lichen-covered ground between. On Great Mercury Island in the Hauraki Gulf, such stands have an understorey of *Cyathodes juniperina* and *C. fasciculata*, and a continuous ground cover of moss, *Cladia aggregata*, *Cladina leptoclada* and *Peltigera dolichorhiza* (Wright 1976).

Fig. 8.26. Margin between kanuka tree-heath and red beech forest at 700 m near Lake Rotoiti, Spenser ER.

On thin soils on granite ridges in Abel Tasman National Park (Western Nelson), Esler (1962) describes shrubland of 'manuka, kanuka, *Cyathodes juniperina, C. fasciculata*, with some *Pseudopanax* [presumably *P. arboreus*], stunted kamahi and *Pimelea gnidia*. Dispersed throughout the area are *Gahnia* [*setifolia, pauciflora*], *Lycopodium scariosum, L. volubile, Lepidosperma australe* and *Dianella nigra*. Three orchids, *Dendrobium cunninghamii, Earina mucronata* and *E. autumnalis*, occur as terrestrial plants'. Other species are *Cyathodes fraseri* and *Lindsaea linearis*. Lichens, including *Cladia aggregata* and *C. retipora*, form the ground cover. Although d'Urville mentioned this vegetation in 1827, it probably arose through destruction of beech forest, and much or all has been burnt since. Scattered, stunted plants of hard beech, black beech (*Nothofagus solandri* var. *solandri*), lancewood, *Coprosma lucida* and kamahi indicate that these ridges could be reoccupied by native forest, although extremely slowly because of the harsh microclimate and lack of topsoil. However, there is invasion by *Hakea salicifolia, H. sericea, Erica lusitanica, Pinus radiata* and *P. pinaster*. Gorse and bracken grow on less extreme sites.

At the outlet of Lake Rotoiti (600 m), porous, stony, sandy loam on gently rolling moraine supports open kanuka 3–5 m tall, under rainfall of 1595 mm per year. Spaces up to 15 m across between bushes and thickets are exploited by spreading kanuka roots, and support a 40% cover formed almost entirely by *Racomitrium lanuginosum*, *Stereocaulon* sp., *Cladia retipora* and *Cladina leptoclada*, whereas there is a dense mat of *Racomitrium crispulum* and *Dicranoloma robustum* s.l. beneath the bushes. Manuka is present. This vegetation extends towards beech forest fringing the Buller River, and along the boundary contains beech saplings and seedlings of *Griselinia littoralis*, lancewood and *Pittosporum tenuifolium*. However, succession to forest will be slow until a favourable soil and microclimate develop.

Kanuka stands in the Eyrewell Scientific Reserve, under rainfall of 863 mm per year, are remnants of vegetation that was once extensive on light, stony soils on the Canterbury Plains and, on the evidence of soil charcoals, existed through most of the post-glacial era. There is a mosaic of dense kanuka tree-heath, shrubland influenced by fire, and grassland, which grows on areas with the deepest, least stony soils or that are most frequently disturbed (Molloy & Ives 1972). The tall kanuka has a shrub tier of *Cyathodes juniperina*, and thick mats of *Hypnum cupressiforme* with herbs such as *Brachyglottis bellidioides*, *Celmisia gracilenta*, *Leptinella pusilla*, *Microlaena stipoides*, *Deyeuxia avenoides* and *Rytidosperma gracile*. In the shrubland, kanuka is associated with manuka, *Pomaderris phylicifolia* var. *ericifolia*, *Carmichaelia* cf. *robusta*, common broom, *Coprosma propinqua*, *C. rhamnoides*, *C. intertexta*, *Discaria*, *Cassinia 'fulvida'* and *Melicytus alpinus*. Macmillan (1976) lists the following mosses as frequent in similar vegetation in the Bankside Scientific Reserve: *Hypnum cupressiforme*, *Polytrichum juniperinum*, *Triquetrella papillata*, *Breutelia affinis*, *Hedwigia ciliata*, *Racomitrium lanuginosum*, *Bryum billardierei* and *Barbula calycina*.

In Central Otago, similar stands extend over gravel terraces and lower mountain slopes under rainfalls as low as 600 mm per year (Fig. 8.27) (Burrell 1965). Soil charcoals indicate that they replaced podocarp woodland with some kanuka after early Maori fires. Kanuka shrubs are up to 4 m tall, and are replaced by kanuka seedlings when they die. Manuka is also present. Between shrubs, there is 'a sparse turf of *Raoulia australis*, small herbs, lichens and mosses, dotted with occasional shrubs less than 0.5 m high, e.g. *Cyathodes colensoi* and *Pimelea* spp' (Burrell 1965). Above 480 m, these stands become restricted to north aspects in inverse correlation with the vigour of the native grasslands, which become sparse and depleted on north aspects and at lower altitudes.

Subalpine secondary heaths

Subalpine heaths regenerating after fire are scattered through mountainous regions, but are most extensive east of the Main Divide in the South Island. Here, subalpine grasslands have extensively replaced beech forests, mixed forests and primary heaths burnt over the last 1000 years; they have in turn been invaded by shrubs, possibly at a quickened pace since burning and grazing reduced their vigour. Resulting shrub-heaths differ from primary heaths in the following respects.

Fig. 8.27. Kanuka spreading into dry grassland, with a self-sown pine left of centre. Lower slopes of Pisa Range, Central Otago.

(1) *Chionochloa* tussocks and other grassland plants are more evident among the shrubs; on more exposed or recently burnt sites, grassland may dominate the mosaic.

(2) Dracophyllums (sect. *Oreothamnus*) are widely dominant in the older communities, but the species are not as clearly sorted ecologically as in virgin heathland. *D. uniflorum* is relatively more abundant and hybrids are frequent.

(3) Slow-growing species, notably *Phyllocladus alpinus* and *Podocarpus nivalis*, are much less prominent.

(4) Seral and filiramulate shrubs, especially hebes, cassinias and *Coprosma* aff. *parviflora*, are more abundant, and may dominate in gullies and during early stages (p. 211).

(5) *Phormium cookianum* survives fire, and in wetter districts, both in the west and along the seaward slopes of the eastern ranges, can become locally dominant. *Astelia nervosa* may also be prominent.

(6) The stands often contain manuka and/or matagouri, which may dominate on north or west aspects, especially at lower atitudes.

(7) Trees gradually reinvade where forest formerly prevailed, if seed sources are still at hand.

Grey scrub

Filiramulate shrubs attain their greatest prominence in primary-successional or semi-climax vegetation on recent fans, terraces and tephra plains subject to severe frosts. Because their leaves are small and mostly obscured by twigs, the shrubs have a dark appearance, meriting the epithet 'grey scrub'. Coprosmas are ubiquitous, but *Aristotelia fruticosa*, *Corokia cotoneaster* and others with more restricted ranges, such as *Sophora prostrata* (eastern South Island), *Muehlenbeckia astonii* (Cook Strait to Lake Ellesmere), and filiramulate species of *Pittosporum* and *Olearia* can be prominent. Lianes with filiramulate tendencies are always present, *Muehlenbeckia complexa* and *Rubus schmidelioides* being widespread whereas *Clematis* spp., forms of *Parsonsia capsularis* and the semi-woody *Scandia geniculata* are mainly in eastern districts. Near-leafless native brooms (*Carmichaelia* spp.) occur in many communities, including the almost spinous *C. petriei*, which is abundant in the Mackenzie and Central Otago ERs, and the rare, lianoid *C. kirkii*. The spinous, nitrogen-fixing matagouri is widely dominant east of the South Island divide, and usually accompanied by the also spiny, but lower-growing *Melicytus alpinus*.

Lower altitudes in the South Island

In the South Island grey scrub is distributed from rainy districts west of the Southern Alps, where it is almost confined to recent alluvial soils at low altitudes (Fig. 8.28), to dry inland basins, where it extends to steep, stony slopes and ascends to higher altitudes. There is surprising floristic uniformity, considering that rainfall varies from *c*. 10 000 mm per annum to as low as 400 mm. *Coprosma propinqua* is invariably present and usually dominant, but with decreasing rainfall and humidity, matagouri enters and dominates alone on dry, stony river flats. Almost the same assemblage of species grows on both well-drained soils and fertile swamps.

Grey scrub in moist lowland valleys is seral to forest, with many of the shrubs persisting into the forest understorey. These seral stands contain young trees, mostly with filiramulate juveniles, especially *Plagianthus regius*, *Pennantia corymbosa* and *Sophora microphylla*; west of the Main Divide, the last is restricted to the vicinity of coastal lagoons.

In the driest inland districts grey scrub on slopes seems an edaphic climax, although trees of *Sophora microphylla* are present locally. Matagouri on stony flats tends to be succeeded by tussock grassland as finer topsoil develops (p. 520). Briar (*Rosa rubiginosa*) is invariably present in dryland scrub and sometimes dominant; more localised adventives are gooseberry (*Ribes uva-crispa*), flowering currant (*R. sanguineum*) and elder. The tall whipcord *Hebe cupressoides* has local occurrences in eastern mountain valleys. As rainfall and altitude increase towards the upper reaches of mountain valleys east of the main divide, the adventives drop out, and grey scrub becomes increasingly confined to stony flats and fans (Figs. 8.29–8.31), being displaced on higher terraces and slopes by subalpine forest, bush, heath and grassland.

Despite the uniformity in the canopy, it is scarcely possible to define a characteristic assemblage of understorey and other associated species. The shrubs tend to occur as clumps and scattered bushes among herbaceous vegetation, which can consist of

Fig. 8.28. Recent alluvial flat with native and adventive grasses, and shrubs of *Myrsine divaricata* and *Carmichaelia arborea* (right). *Nothofagus menziesii* (*N. fusca*) forest occupies the older terrace in the background. Pyke River, Olivine ER.

wetland, grassland, pioneering, rupestral or ruderal plants. This is partly because the soils usually vary over short distances in respect of drainage and texture, and partly because of disturbance by human agencies, since grey scrub is usually on land required for grazing. In the moister districts, the understorey can contain species both of the preceding vegetation and of the forest that will follow; in Westland, the latter include *Polystichum vestitum, Microlaena avenacea* and *Uncinia egmontiana*. In dry localities, there may be little undergrowth beneath dense scrub.

Volcanic Plateau

Grey scrub is a significant community in frosty depressions on tephra. The filiramulate shrubs are mostly the same as in the wetter South Island districts, with *Coprosma propinqua* and *Olearia virgata* being especially prominent; and there are both swampy and well-drained facies. Like the *Dracophyllum subulatum* heaths that occupy poorer soils, they are probably fire-tolerant derivatives of bog-pine heaths, which they resemble floristically (Table 8.1). They can succeed forest, with the frost-tolerant *Phyllocladus alpinus* being the earliest and most vigorous invader. On upper, less frosty slopes, manuka tends to replace the filiramulate shrubs.

Fig. 8.29. Dense grey scrub, mainly of matagouri, *Coprosma propinqua* (most of the darker bushes) and *Hebe brachysiphon*. A mountain beech tree is in the upper left. Locality and other species as in Table 8.1, column 3.

Other primary grey scrub communities

In upper-montane and subalpine reaches of valleys, *Coprosma rugosa* is often abundant on stony flats, forming colonies through root-suckering. On fans and older terraces, *C.* aff. *parviflora* and *C. ciliata* are dominant filiramulate shrubs, but usually mingle with larger-leaved seral shrubs such as *Gaultheria rupestris* and species of *Coriaria, Hebe* and *Olearia*. There are often also young plants of species characteristic of later phases, including *Hoheria glabrata* or *H. lyallii, Podocarpus nivalis, Phyllocladus alpinus*, dracophyllums or beeches. In North-west Nelson, a *Coprosma propinqua–C. rugosa* community grows on limestone debris (Druce *et al.* 1987), and *Aristotelia fruticosa* dominates among subalpine boulders (P.A. Williams, unpublished).

Low patches of *Coprosma ciliata* or, in drier regions, *Melicytus alpinus* on penalpine boulder fields are the highest filiramulate communities. These grade into *Podocarpus nivalis* heath. At the other altitudinal pole, *Plagianthus divaricatus* and *Coprosma propinqua* in the upper reaches of salt marshes form the most salt-tolerant woody vegetation in New Zealand except for mangrove (*Avicennia resinifera*) (p. 291).

Secondary grey scrub

Filiramulate shrubs are often prominent in secondary vegetation on deforested sites, owing to their tolerance of grazing, drought, frost and wind. The communities are a mixture of open-land species such as *Coprosma propinqua* and *C.* aff. *parviflora*, and forest species such as *C. rhamnoides, C. crassifolia* and *C.* aff. *ciliata. C. propinqua* and

Fig. 8.30. Matagouri scrub at 670 m, Broken River basin, Puketeraki ER. Other shrubs include *Coprosma propinqua*, *Melicytus alpinus*, *Aristotelia fruticosa* and *Rosa rubiginosa*.

C. tenuicaulis can dominate formerly forested swamps. Wind-shorn filiramulate shrubs are often the only woody survivors where forest has been destroyed on exposed coastal slopes.

These secondary shrublands usually differ from primary stands in the presence of other forest survivors of fire or browsing, such as *Rubus cissoides* and *Pseudowintera colorata*, and in the absence of the more localised filiramulate shrubs, such as *Sophora prostrata* and *Coprosma intertexta*. They often merge with other secondary woody communities, including broad-leaved scrub, manuka and dracophyllum heaths, and *Cassinia* and *Hebe* shrublands. Much matagouri is also secondary, in so far as pastoral management, especially aerial top-dressing with superphosphate, has increased its vigour and density. Fire gives only temporary control, as matagouri resprouts from the base (Daly 1969). *Olearia odorata* has also increased in some inland localities.

Subalpine *Hebe* communities

Hebes are present in most subalpine communities on recent and semi-fertile soils excepting dense forest, and often dominate fire successions for a few decades before yielding to other woody plants or long-lived herbs and grasses. They can also dominate early phases of primary succession, and form enduring communities where soils are continually rejuvenated, e.g. by downslope drift or flushing by streams. They are advantaged through well dispersed, highly viable seed, an ability to establish on recently bared ground, and rapid growth. There are numerous species, many with rather narrow geographic and ecological ranges. Among the more widespread and

Fig. 8.31. Low matagouri scrub with some *Melicytus alpinus*, *Aristotelia fruticosa*, *Coprosma ciliata* and *Muehlenbeckia axillaris* on bouldery moraine; 820 m, Mackenzie ER.

abundant are *H. odora* (throughout New Zealand, especially on flushed ground and stream banks), *H. venustula* (North Island), *H. subalpina* (along and west of the Southern Alps), *H. glaucophylla* and *H. topiaria* (northern parts of the South Island), *H. traversii* (Marlborough and northern Canterbury), *H. brachysiphon* (Canterbury), and *H. rakaiensis* (mid-Canterbury to Southland).

Shrubland of *Cassinia* (tauhinu) and similar plants

Cassinia 'retorta' and *C. leptophylla* s.s. are coastal and maritime shrubs, the former being restricted to Northland and the latter extending southwards to Marlborough. *C. 'fulvida'* extends inland and ascends to the montane belt, but is also coastal in the South Island. *C. 'vauvilliersii'* grows mainly in subalpine shrublands, but descends to low altitudes in western and southern heathlands. These entities merge where they overlap, especially in Marlborough where plants attributable to *C. leptophylla* s.s., *C. 'fulvida'* and *C. 'vauvilliersii'* can occur in the same population.

Tauhinu has only a limited role in primary vegetation, but is a serious weed of hill-country grassland. The relevant attributes appear to be small, pappus-bearing seeds that are widely dispersed in the wind, an ability for seedlings to establish within grass swards, fairly fast growth, and, unlike most native shrubs and trees, an ability to resprout from the base after fire. On the other hand, the bushes are short-lived and in the absence of fire are succeeded by longer-living plants. Often tauhinu is the only shrub, but it also grows with manuka, dracophyllums, hebes, matagouri and filiramulate coprosmas.

Olearia solandri forms scrub with *Cassinia leptophylla*, which it closely resembles, on coastal slopes in the North Island and east of the South Island to as far as Kaikoura ER. *Brachyglottis cassinioides*, which is readily distinguished from *Cassinia* only when bearing its yellow-rayed flower heads, is subalpine in the east of the South Island, occurring sparingly in primary scrub and more abundantly after fire. Occasionally it forms hybrids with the much larger-leaved, herbaceous rosette plant *B. haastii*.

Naturalised shrubs other than *Erica* and *Hakea* species

Barberry (*Berberis darwinii*) and hawthorn (*Crataegus monogyna*) are spreading through scrubby and rocky grassland in some localities. Boxthorn (*Lycium ferocissimum*), originally planted as hedges, is now abundant on coastal slopes and cliffs in most eastern districts. Briar has spread throughout grasslands in the east of the South Island to the upper limit of the montane belt. Its range largely coincides with matagouri, the plants being either dispersed among that species or forming impenetrable thorn scrub to 3 m tall in gullies and at the foot of terraces. Rabbits once kept briar seedlings in check, but when they were brought under control in the 1950s, these seedlings were released, leading to an apparent eruption (Molloy 1976). Suckering rhizomes lead to dense clumps of canes that are practically immune to further browsing.

Flowering currant also forms clumps through vigorous basal sprouts. Seedlings escape browsing through becoming established among other shrubs, such as matagouri; once established, the bushes resist destruction through their vigour, density and height. The densest infestations are in South Canterbury (Williams 1984). On rocky slopes, gooseberry has also become established among matagouri.

Gorse is the most ubiquitous of all naturalised woody plants. In the mid-nineteenth century, it was planted as hedgerows whence it spread, suppressing pasture over large areas and denying stock access to the grazing that remained. Its biological success is due to hard-coated seeds that remain dormant for many decades when buried, but germinate quickly and profusely when brought to the surface; comparatively rapid increase in height and width of plants; the dense spiny habit that discourages penetration by animals; unpalatability of all but soft, young growth; and the ability to resprout from the base after the tops are burnt or die of old age. In New Zealand, the paucity of natural insect predators seems to be a further factor in its vigour.

The explosively dehiscing pods throw seeds only a short distance. Rapid dispersal through the country has been by other means, notably transport by machinery, animals, and fresh and salt water. Frequently, infestations begin where a road crosses a river bed, and then spread downstream to the river mouth and thence along coastal dunes and cliffs.

Gorse succeeds best where there is neither dense, continuous vegetation that prevents the seedlings from establishing, nor close, continuous grazing that eliminates them. Therefore, the most extensive stands occupy sparsely vegetated flood plains and low-producing hill pastures subject to shallow erosion. Limits to its spread are set by its cold resistance (probably about $-10\,°C$), and requirement for a moderate level of soil fertility, especially in respect of phosphorus. Once gorse is established, fire leads only to entrenchment at the expense of other plants, but if left undisturbed it can eventually be succeeded by native bush and forest (p. 540). In pasture management, the achilles heel of gorse is its susceptibility to hormonal herbicides and palatability to goats, but disturbance of the soil in areas cleared from gorse can lead to mass germination of buried seed.

Common broom shares many of the ecological characteristics of gorse, including durable seed. It differs mainly in its more rapid height growth, shorter life-span, inability to resprout from the base after fire, and tolerance of drier soils and frostier climates. Stands have covered gravel flood plains in the east of the South Island and spread up the adjacent hillsides, and there are also large areas on the Volcanic Plateau. Although a serious weed, broom makes some amends during its spectacular flowering in early summer.

Cytisus multiflorus, Teline monspessulana and *Spartium junceum* are locally well-established brooms. Tree lupin (*Lupinus arboreus*) is well established on coastal dunes as a result of planting to stabilise sand, and is also spreading on raw sand and gravel in some inland localities. Tree 'lucerne' or tagasaste (*Chamaecytisus palmensis*) is common on cliffs in some coastal localities and near Lake Taupo; recently it has attracted interest as a fodder plant. The needle-leaved *Psoralea pinnata* is locally common in Northland. *Polygala myrtifolia*, which is readily mistaken for a legume from the shape of its corolla, is abundant in coastal scrub in the same province.

Blackberry (*Rubus fruticosus* aggr., including several microspecies, the most distinctive being *R. laciniatus*) is widely abundant in secondary vegetation, especially in wetter districts. Unlike the high-climbing native species of *Rubus*, blackberry produces cane-like stems from ground level. These collapse over competing vegetation, often bearing it down. Because stems take root where they touch the ground, blackberry patches rapidly cover large areas. These are impenetrable to stock, and until the advent of herbicide sprays, grazing by goats was often the only practical means of control.

Fernland

New Zealand bracken is often regarded as a variety of the cosmopolitan *Pteridium aquilinum*, but the fronds differ from those of the north-temperate plant in that they usually remain green over winter, their texture is much firmer, pinnules are very narrow (*c.* 2 mm), and rootlets are produced on the lower part of the stipes. The extensive, fast-spreading rhizomes ramify copiously at depths down to 50 cm. Croziers emerge in spring and early summer and rapidly expand into fronds, which vary in height from 30 cm on harsh sites to over 3 m where supported by shrubs on moist, fertile soil.

On sunny aspects, bracken ascends to 800 m in the North Island and 900 m in the southern zone. Frost precludes it from the floors of interior-temperate valleys, and low rainfall restricts its distribution in semi-arid districts. As bracken requires soils that are well-drained, porous and not too compact, it grows vigorously on most alluvial, colluvial, aeolian and pumice soils and dry peats, but is generally absent from swamps, gley podzols, and soils that are shallow over solid rock. Soils supporting bracken can be stony, provided there are interstitial fines.

Bracken must have been uncommon in primeval New Zealand. The main habitats were probably coastal and inland dunes that its far-spreading rhizomes could colonise more readily than plants depending on establishment from seed. The pollen record shows temporary increases of bracken following eruptions on the Volcanic Plateau, and massive, permanent increases at the expense of forest after Polynesian settlement began about AD 900, especially in the North Island (McGlone 1983). Since bracken stands always contain large quantities of dead fronds that partially decompose into deep, dry litter, they remain fire-prone, but recover vigorously from deep rhizomes. Thus, the repeated Maori fires, which were remarked on by early European travellers from Cook onwards, maintained and extended the areas of bracken, which also invaded abandoned gardens. Harvesting of fern rhizomes for starch ceased as soon as cereal flour became available, but feral pigs, beginning with those introduced by Cook in 1769, have adopted them as their main food, and thoroughly 'plough' large patches of bracken.

European colonists assiduously destroyed forests and provided more opportunities for bracken. However, determination to establish pastures challenged and eventually overcame its supremacy. The usual practice, well described by Guthrie-Smith (1953), was to burn bracken in the spring, and broadcast seed of 'English grasses' in the ashes (p. 280). Success of pasture establishment depended greatly on whether livestock could be concentrated heavily enough to trample and break the fern croziers that emerge a few weeks after fire.

Descriptions of typical bracken communities follow; succession through bracken is discussed in Chapter 15.

Shenandoah Saddle (North Westland ER)

Virgin beech forest on a 10° slope which was burnt, probably 20–30 years ago, has been replaced by bracken averaging 1.5–2.5 m tall, with a mean cover of 90%, and laced throughout with blackberry; there are also weak-stemmed bushes of *Leycesteria*, and scattered, 3 m tall *Coprosma* aff. *parviflora* bushes, which initially established on charred, decaying beech logs. Dead bracken fronds within the perimeter of coprosma bushes are held upright, while those on the downhill side fall away to leave a gap. Other small gaps arise where fronds collapse outwards from a central point. These places receive much less fern litter, and are colonised by seedlings of coprosma, *Aristotelia serrata*, *A. fruticosa* and *Griselinia littoralis*, which have to compete with the adventive herbs *Lotus pedunculatus*, catsear, *Mycelis muralis, Crepis capillaris*, foxglove, cocksfoot and sweet vernal (*Anthoxanthum odoratum*), and a few native plants including *Epilobium chlorifolium, Stellaria parviflora, Hydrocotyle novae-zeelandiae* s.l. and *Gnaphalium* spp. Woody seedlings also continue to colonise rotting logs; within 20 m of the forest edge these include beech.

Elsewhere there is a continuous cover of living bracken fronds, above an equally continuous layer of dead fronds that are breaking down into 20 cm deep, loose litter. Over distances of many metres there are no associated plants other than blackberry and occasional *Mycelis*.

Abel Tasman National Park, Western Nelson

On a gentle coastal slope, an unbroken stand of bracken 2.5 m tall contains scattered 4 m tall gorse bushes, which presumably established when the stand was last kept open by fire or grazing. The only other species line a stream bank, a track and the foreshore, and include mahoe, manuka, *Coriaria arborea, Coprosma grandifolia, Macropiper*, blackberry and the ferns *Blechnum* 'black spot' and *Pneumatopteris pennigera*.

Orton Bradley Park, Banks Peninsula

This is typical of localities where, following destruction of forest, bracken has tended to occupy gullies and lower slopes, whereas grassland has established on the drier upper slopes. Where bracken is dense, there is virtual absence of other species, but where cover decreases to 60–80%, there is a sparse understorey including cocksfoot, *Vicia sativa, V. hirsuta, Galium aparine, Cirsium arvense* and the native *Calystegia tuguriorum*. Upslope, where bracken gives 0–40% cover, cocksfoot averages 30%, with other frequent species including *Trifolium repens, Holcus lanatus*, sweet vernal and *Lolium perenne*. Native grassland species also appear in low frequency, including silver tussock, *Elymus rectisetus* and *Wahlenbergia gracilis*.

Several other rhizomatous ferns also dominate successional vegetation. *Gleichenia microphylla* and *G. dicarpa* are important on infertile wetlands and fire-modified wet heaths (Chapter 10). In the *Blechnum 'capense'* complex, the metre-long fronds of *B.* 'black spot' form dense colonies in wet districts on vertical banks, gully sides, swamp margins, slips, road sides and damaged forest edges, and the mainly subalpine *B.* 'mountain' develops shorter colonies, especially on sides of gullies and where scrub has been burnt. The summer-green *Hypolepis millefolium* dominates montane and subalpine gullies and debris slopes on western South Island ranges, especially in places where grazing has eliminated vulnerable shrubs and grasses. *Histiopteris incisa* (also summer-green) and *Paesia scaberula* can form colonies in early stages of secondary succession.

Polystichum vestitum, which has fronds tufted on a short, stout 'trunk', is often common on shady aspects on moist, well-drained soils in cool-temperate grasslands on the east of the South Island, and can dominate, especially in gullies and on lower slopes. Originally, these sites would have supported forest with abundant *P. vestitum* in the tall-herb tier.

9

GRASSLAND AND HERBFIELD

Grasslands and mosaics of grass and scrub cover nearly 60% of New Zealand (Fig. 2.7), being distributed from the subtropical Kermadecs to far-southern Macquarie Island, and from fertile lowlands to the alpine belt. Generally they are dominated at the higher altitudes and in dry inland districts by native species of tussock form, and at the lower altitudes by sward-forming introduced species. Most have been deliberately modified for pastoral use, and almost none have escaped indirect effects of human settlement, such as grazing by feral mammals. The composition of native grasslands at the higher altitudes has been well documented, but vegetation processes are only partly understood. In contrast, factors affecting productivity in lower-altitude grasslands have been studied intensively, but little has been published about composition, except in respect of economically important species. The order of treatment in this chapter reflects the altitudinal trend in degree of modification.

Chionochloa grasslands and related vegetation

Grasslands dominated by the almost endemic genus *Chionochloa* are widespread in the mountains, and descend to sea level in southern regions. Where undisturbed, *Chionochloa* tussocks can form dense stands with thick litter beneath, but usually other grasses and herbs form lower tiers, and according to drainage and fertility these range from typical grassland assemblages to those associated with wet-heath and bog. The proportion of forbs and small shrubs increases on rocky ground, on steep, broken topography, and along watercourses; here are found many of the showiest plants of the New Zealand flora, belonging to genera such as *Hebe, Ranunculus, Celmisia* and *Ourisia*.

All *Chionochloa* grasslands below the climatic tree limit result from deforestation over the past thousand years, except on frosty or poorly drained valley floors and a few other special habitats. Being essentially seral, they tend to be invaded by tall shrubs, especially dracophyllums, cassinias, hebes, manuka (*Leptospermum scoparium*) locally and, on warm inland sites, matagouri (*Discaria toumatou*). More stable shrub–tussock mixtures form transitions to infertile wetlands and tall heaths.

One hundred and forty years of burning and grazing have depleted or eliminated *Chionochloa* tussocks from wide areas, and as their range has contracted towards moister localities and higher altitudes, they have been replaced by hitherto subordinate species, especially fescue tussocks, *Poa colensoi*, *Rytidosperma setifolium* and celmisias. Even in remote districts grazing by feral ungulates, especially red deer, has reduced the more palatable species of *Chionochloa*, thereby allowing shorter plants to prevail; until European colonisation, long-horned grasshoppers (Acrididae) had been the most conspicuous herbivores (White 1975*a*).

The major grasses

Although most *Chionochloa* species look similar and hybridise where their ranges overlap (Connor 1967), they have distinct ecological preferences. *C. conspicua*, distinguished by 1 cm wide leaves and graceful, 2 m tall inflorescences, grows on stream banks and recent slips, and also in forest and scrub, especially where disturbance has created openings. It is common in the montane and subalpine belts through most of New Zealand except Mt Taranaki and the Kaikoura ranges.

Red tussock (*C. rubra*), so named because of its pale reddish-green, narrow, inrolled leaves, is a robust grass distributed from the Huiarau Range and central volcanoes to Stewart Island, and from sea level in southern New Zealand to the lower part of the penalpine belt. Usually it grows on flat or gently rolling surfaces with deep, fine-textured soils that are acidic, slow-draining and at least seasonally waterlogged. From these modal sites, it extends on to adjacent droughty rises and moist lower slopes and stream banks. Red tussock tolerates low fertility, but responds vigorously to high fertility to form dense stands over 1.5 m tall. *C. rubra*, like other chionochloas, has become depleted on dry soils in the east of the South Island, but on damp soils large red tussocks often persist after other native grasses have given way to introduced pasture grasses. However, these readily succumb when grazing follows fire.

In some areas red tussock is replaced by other chionochloas, including three narrow-leaved species of restricted range: *C. defracta* on ultramafic soils in the north of the South Island, *C. acicularis* in western Fiordland and *C. juncea* on plateaus in Western Nelson.

Snow tussocks are so-named because they dominate penalpine grasslands. *C. pallens*, which has whitish midribs, is the most widely distributed. It is abundant on moist, well-drained, weakly weathered colluvium and alluvium on the North Island axial ranges and along and west of the South Island Main Divide, but much less common on granite mountains and eastern rain-shadow ranges. *C. pallens* has relatively soft leaves and is vulnerable to grazing.

The *Chionochloa flavescens–rigida* complex mainly comprises snow tussocks with broad, tough leaves and leaf-sheaths that fracture transversely when dry, but on infertile soils and at high altitudes it includes smaller, narrow-leaved tussocks. Some forms are adapted to moderate fertility, and others to soils almost as infertile as those tolerated by red tussock. Five taxa may be tentatively recognised. *C.* 'robust' has light-brown sheaths and leaves up to 1 cm across. It grows on crumbling bluffs, stony, coarse-textured recent colluvium and alluvium, and other sites with better than usual nutrient status, often sharing dominance with shrubs. It is distributed across the northern half of the South Island, extending through western districts into Fiordland, but its southern limits are uncertain.

C. *rigida* is the main snow tussock of eastern mountains south of the Rakaia River and also grows on the hill tops of Banks Peninsula. When European settlement began, its realm extended from the lower boundary of the alpine belt to the western edge of the Canterbury Plains and the downlands of eastern Otago; today it has all but vanished from the lower altitudes, and at mid-altitudes is mainly confined to shady aspects. In sheltered, low-altitude sites, C. *rigida* tussocks are almost as large and broad-leaved as those of C. 'robust', whereas on penalpine ridges they are smaller and slender-leaved.

Another broad-leaved tussock, C. *flavescens*, grows on leached soils in the Tararua and Rimutaka ranges. On the western Alps, the narrow-leaved C. 'westland' grows on rolling terrain with leached soils in the lower part of the penalpine belt. In Fiordland, broad-leaved snow tussocks occupy the steep, rocky sites typical of C. 'robust', but C. 'westland' extends at least to northern Fiordland, and there are also broad-leaved tussocks on wet, leached soils. I refer to this complex, together with plants on Mt Anglem in Stewart Island, as C. 'fiord'. C. *rubra* is absent over most of the ranges of C. *flavescens*, C. 'westland' and C. 'fiord', which occupy its niches, even extending on to bogs. Where these do overlap with C. *rubra*, however, the latter excludes them from the wettest sites.

C. *macra* is an eastern South Island snow tussock similar to C. *rigida*, but its leaves are narrower and less tough and the sheaths lack abscission fractures. In the southern part of its range it is largely alpine but from the Rakaia River northwards descends as low as 800 m.

On penalpine slopes in south-eastern Fiordland and the Takitimu Mountains, the main snow tussock is C. *teretifolia*, which has narrow, inrolled, pilose leaves. The similar C. *lanea* grows on the southern mountains of Stewart Island.

C. *crassiuscula* is a short tussock with leaves that curl spirally when dead and dry. It is most abundant in Westland and Fiordland, where it grows in deep, wet hollows and gullies, and on shallow, leached soils at high altitudes. The possibly conspecific C. *pungens* of Stewart Island grows on similar soils but descends almost to sea level. C. *oreophila* forms turf in late-snow hollows, especially along and west of the South Island Divide. The calcifuge C. *ovata* and the calcicole C. *spiralis*, uncommon tuft-forming species endemic to Fiordland, are more or less rupestral.

Carpet grass (C. *australis*) is very different from other chionochloas, having short, narrow, inrolled leaves crowded at the ends of prostrate tillers that extend downhill, the whole forming a thatch up to 15 cm thick. It is important on the mountains of Western Nelson and inland Marlborough, with patches scattered to northern parts of the Taramakau and Waimakariri catchments. Carpet grass mainly grows on leached, well-weathered, shallow soils on convex penalpine slopes. Although it gives an impression of advancing vigorously into snow tussock and other alpine communities, Wraight (1965) measured an extension of only 5–20 mm per year.

Poa colensoi s.l. is the commonest small grass, but other species of *Poa* (e.g. *P. kirkii*), *Hierochloe* spp. (e.g. *H. novae-zelandiae*), *Microlaena colensoi*, *Agrostis dyeri*,

Fig. 9.1. *Celmisia armstrongii* and *Chionochloa crassiuscula*; 1300 m, Arthurs Pass, North-eastern Alps.

and, indeed, most grasses of short-tussock communities also associate with snow tussocks. There are several small uncinias, the rush-like sedge *Schoenus pauciflorus* is common on wet soils, and a true rush, *Marsippospermum gracile*, appears in the alpine belt.

Important composites

Celmisia is scarcely less important in high-altitude grassland and herbfield than the snow tussocks. Subgenus *Lignosae* has sprawling, semi-woody stems, rooting towards the base. Species covering extensive areas, especially of steep, stony slopes, include *C. incana* (North Island axial ranges and north-east of the South Island), *C. hieraciifolia* (from Ruahine and Tararua ranges to Western Nelson), *C. discolor* (north of the South Island), *C. walkeri* and *C. du-rietzii* (in the west of the South Island) and *C. viscosa* (mainly alpine in the east of the South Island).

Subgenus *Pelliculatae* includes large plants with long, tufted, narrow-linear to lanceolate leaves. They survive fire because their buds and thick, short stems are protected by numerous developing leaves and, behind those, by living and dead leaf bases. This attribute and low palatability have enabled them to increase considerably, especially on grazing lands in the east of the South Island. *C. semicordata*, with silvery leaves up to 50 cm long and capitula 10 cm across, is the largest species and grows on all South Island mountains but the driest. *C. verbascifolia* has a similar range, whereas *C. traversii*, which has thick, rust-coloured tomentum, grows only on northern ranges of the South Island and in western Southland. The western *C. petriei* and *C. armstrongii* (Fig. 9.1) and the eastern *C. lyallii* have narrow-linear leaves. *C. spectabilis* (Fig. 9.2) is

Fig. 9.2. *Celmisia spectabilis*; 1100 m, Crow Valley, Eastern Alps.

the least typical member of the subgenus, being patch-forming with oblong leaves that are 1–15 cm long according to variety; nevertheless it hybridises with *C. lyallii* where they grow together. It is abundant on the axial ranges and central volcanoes of the North Island and in the north and east of the South Island, but does not reach Westland or Otago.

Most species in subgenus *Celmisia* also have firm, narrow, tufted leaves, but are smaller than in *Pelliculatae*. *C. gracilenta* is the most widespread and variable of all celmisias. The subgenus also includes alpine cushion- and mat-plants such as *C. sessiliflora*.

The endemic penalpine genus *Dolichoglottis* has tufted, linear, summergreen leaves. *D. scorzoneroides* has white-rayed capitula up to 6 cm across and grows mainly near and west of the South Island Divide (Fig. 9.3), whereas the smaller *D. lyallii* has yellow rays and grows along and east of the Divide. They hybridise, and both extend to Stewart Island. Rosette-forming species of *Brachyglottis*, especially in the *B. bellidioides* complex, are widespread except in the western Alps. *Craspedia* species, distinguished by small capitula crowded in globose inflorescences, grow in most penalpine communities.

Other plants of the *Chionochloa* grasslands

Aciphyllas are the most bizarre plants of mountain grasslands and herbfields, by virtue of their linear, rigid, spine-tipped leaf segments, and tall scapes that bear lateral umbels. *Aciphylla aurea*, with thick, yellowish pinnae up to 1 cm wide, is abundant in the east of the South Island. The narrow-leaved, glaucous *A. subflabellata* replaces it in red-tussock communities. *A. colensoi*, which differs from *A. aurea* in its conspicuous red midribs, grows on North Island axial ranges and in the north of the South Island. Along and west of the Main Divide, slender plants in the *A. lyallii–*

Fig. 9.3. *Dolichoglottis scorzoneroides*; 1500 m, Karangarua catchment, Western Alps.

crenulata complex occur instead. In Western Nelson, the somewhat similar *A. hookeri* has an alternative form so different that it was formerly regarded as two species.

Several giant aciphyllas grow in subalpine and penalpine gullies, on rocky slopes or among scrub rather than in open grassland. *A. scott-thomsonii*, which has glaucous leaves up to 1.5 m long and scapes as tall as 3 m, is widespread east of the Main Divide from Arthurs Pass southwards (Fig. 9.4). It is replaced by *A. ferox* in the north of the South Island and *A. horrida* in the west. *A. glaucescens*, which has narrow pinnae, occurs sporadically from Mt Hikurangi to Southland.

A distinct group of small aciphyllas has broad, compound umbels instead of elongated scapes, and softer, more dissected leaves. *A. monroi* is common on eastern ranges in the South Island, whereas the *A. dissecta–multisecta–divisa* complex grows on the Tararua Range and the wetter South Island ranges.

In the related genus *Anisotome*, the commonest grassland species are the small, pinnate-leaved *A. aromatica*, and *A. flexuosa*, which replaces it at high altitudes. The much larger, carrot-like *A. haastii*, which is now found mainly on rocky ground and among scrub along and west of the South Island Divide, was probably abundant in penalpine grasslands until depleted by herbivorous mammals. The same applies to the giant, summer-green, white-flowered buttercup, *Ranunculus lyallii* (Fig. 9.5). With much less grazing in the western mountains in recent years, this has made a come-back, mainly from persisting rhizomes. Smaller buttercups and the uncommon native anemone, *A. tenuicaulis*, are mainly on damp ground.

Ourisia is a genus of attractive, white-flowered herbs. Most species, including the robust *O. macrocarpa*, grow on boulder fields and steep sheltered banks, but the large-leaved *O. macrophylla* and its close relatives and the small-leaved, creeping *O. caespitosa* are common in grassland. Species of the root-parasite *Euphrasia* occur in most penalpine grasslands, *E. zelandica* being the most widespread. *Geum* is represented by white-flowered penalpine and alpine herbs with leaves borne on short,

Fig. 9.4. *Aciphylla scott-thomsonii* among *Chionochloa macra*; 1300 m, Craigieburn Range, Puketeraki ER.

thick, branching stocks, *G. parviflorum* being the most widely distributed. *Lobelia linnaeoides* is restricted to high-altitude grasslands on eastern ranges from Mt Cook southwards. *Forstera* species are creeping herbs with thick, ericoid leaves and regular, white flowers 1–2 cm across borne on slender peduncles. Most other small dicot herbs are conspecific with or closely related to plants that are also common in lower-altitude grasslands.

Penalpine grasslands, other than those of the Western Alps, are the main habitat for whipcord hebes, the commonest species being *Hebe tetragona* (North Island), *H. coarctata* (Western Nelson and North-eastern Alps), *H. lycopodioides* (eastern ranges of the South Island), and *H. hectorii* (southern zone mountains).

The mountain form of *Phormium cookianum* is often abundant in subalpine grasslands derived from forest or scrub. It is rare, however, on slopes surrounding the drier intermontane basins of the South Island. *Astelia nervosa*, growing as 1 m tall tussocks with silvery, 3 cm wide leaves, has a similar distribution and niche. *A. petriei* with broader, stiffer, light green leaves, and *A. nivicola* var. *nivicola*, which is slender and less clumped, are strictly penalpine and confined to the South Island, mainly in the west. Other than a few small orchids, among which only *Prasophyllum colensoi* is common, *Bulbinella* is the only significant summer-green bulbous geophyte at high altitudes. It has scapes of crowded, bright-yellow flowers and tufts of soft, green

Fig. 9.5. *Ranunculus lyallii*; 880 m, Hooker Valley, D'Archiac ER.

leaves, up to 50 cm long and 3 cm wide in *B. gibbsii* and *B. hookeri*, which range widely in penalpine grasslands in the North Island and north and west of the South Island. The narrow-leaved *B. angustifolia* is locally abundant in Canterbury and Otago on flushes and fertile alluvial and colluvial soils.

Lycopodium fastigiatum, Polystichum vestitum, Hypolepis millefolium, and on open, stony ground, *Blechnum penna-marina*, are important in some communities. *Polytrichum juniperinum, Racomitrium lanuginosum* and *Hypnum cupressiforme* are common penalpine mosses. Where precipitation is high these are joined by *Dicranoloma robustum* and *Psilopilum australe*. Sheltered spaces between tussocks also support moisture-demanding liverworts, including species of *Lophocolea* and *Plagiochila*.

Chionochloa communities

Discussion begins with the *Chionochloa* grasslands of Western Nelson. These are of limited extent, but include a wide range of grassland types, clearly related to the varied soil parent materials. The sequence is then southwards through climatically similar grasslands in Westland, Fiordland and Stewart Island, and then northwards through the extensive but considerably modified grasslands east of the South Island Divide. The much smaller areas in the North Island are considered last.

Mountains of Western Nelson

The parent rocks in the Matiri and Mt Owen ranges are Palaeozoic marble, readily weathered Tertiary calcareous rocks, schist, and granite, the last providing the least fertile soils. Table 9.1 portrays six *Chionochloa* associations that reflect the different substrates.

Table 9.1. *Percentage frequencies of main species in six grassland associations in the Matiri and Owen Ranges, classified on the basis of species presence–absence and named according to Chionochloa spp. with > 50% frequency*

Association:	robu.	robu. (pall)	rubr. (robu.)	rubr. (aust)	pall. (aust)	aust.
No. of samples:	15	45	46	7	12	20
Altitudinal range (m):	1150–1450	1000–1700	1000–1500	?–?	1300–1700	1300–1700
Mean no. species/plot:	?	34	37	25	?	?
Chionochloas						
Chionochloa 'robust'	87	80	61	14	0	10
Chionochloa pallens	47	62	7	14	92	40
Chionochloa rubra	7	16	91	100	0	15
Chionochloa australis	0	9	2	57	75	85
Other grasses						
Poa pratensis	73	20	7	0	0	5
Sedges etc.						
Carex sinclairii	20	24	57	29	0	0
Schoenus pauciflorus	20	56	78	57	17	40
Carpha alpina	0	11	37	71	0	10
Empodisma minus	0	0	43	86	0	25
Shrubs						
Hebe topiaria	53	40	61	14	0	0
Aristotelia fruticosa	47	11	4	0	0	0
Hebe coarctata	0	11	0	0	75	35
Dracophyllum uniflorum	0	20	9	43	42	80
Dwarf shrubs						
Coprosma cheesemanii	20	58	89	57	58	10
Gaultheria depressa	13	53	17	14	83	10
Cyathodes colensoi	7	4	2	0	58	10
Pentachondra pumila	7	20	9	86	50	60
Large forbs						
Phormium cookianum	53	49	67	0	25	30
Celmisia semicordata	0	44	89	43	0	10
Celmisia spectabilis	7	9	0	14	67	90
Small native forbs						
Oreomyrrhis colensoi	60	36	43	0	0	5
Ranunculus spp.	54	33	81	0	8	15
Pratia angulata	20	56	87	14	25	0
Brachyglottis bellid.	22	44	63	14	0	25
Anisotome aromatica	13	47	41	86	83	70
Microseris scapigera	0	42	80	14	0	5
Astelia linearis	0	13	20	86	0	0
Drosera arcturi	0	2	2	86	0	5
Donatia novae-zelandiae	0	2	2	71	0	15
Celmisia laricifolia	0	2	0	0	8	65
Adventive forbs						
Trifolium repens	80	18	28	0	0	0
Mycelis muralis	67	20	20	0	0	0
Hypochoeris radicata	20	31	63	0	25	0

Source: Rose 1985

(1) The mainly subalpine *C.* 'robust' association occurs over marble or on landslide debris, debris cones and other recent soils, persisting onto older soils enriched by flushing or downslope movement. In the least disturbed stands, the deep-rooting (> 90 cm) *C.* 'robust' co-dominates with *Phormium cookianum, Dracophyllum longifolium, Aciphylla ferox, A. colensoi, Celmisia semicordata* and *C. verbascifolia*. Such stands grade into scrub dominated by *Dracophyllum longifolium* and *Phyllocladus alpinus*.

In modified communities, which can result from past burning and grazing, *C.* 'robust' is at densities around only 5%, and can be accompanied by *C. pallens* or red tussock, according to drainage. Most cover is provided by swards of species indicating high fertility, notably *Trifolium repens* and *Poa pratensis*. Rhizomatous carices are abundant on wet ground. *C. conspicua* appears below 1200 m, and below 1000 m is likely to be the only *Chionochloa*.

(2) The *C.* 'robust' (*C. pallens*) association occurs on moderately steep slopes on marble and Tertiary calcareous rocks over a wide altitudinal range. Good drainage and recent, non-leached soils are indicated by plants such as *Poa cockayneana, Aciphylla ferox, Helichrysum bellidioides* and *Ranunculus insignis*.

(3) The *C. rubra* (*C.* 'robust') association mostly occupies slopes < 20°, on tablelands underlain by calcareous rocks. Soils are relatively fertile. Red tussock prevails generally, whereas *C.* 'robust' grows mainly on steeper slopes, around sink-holes, and on flushed ground. Hybrids between the two are frequent. Among associated species, *Phormium cookianum, Hebe topiaria, Dracophyllum longifolium* and *Coprosma propinqua* reflect the relatively low altitudes, and *Carex sinclairii, Schoenus pauciflorus, Carpha alpina, Empodisma minus, Hebe pauciramosa* and *Microseris scapigera* reflect poor drainage.

(4) *C. rubra* (*C. australis*) is a minor association of calcareous tablelands with slopes < 10°, but is rooted in peat that insulates it from the influence of the underlying rocks. Wet, infertile conditions are indicated by *Dracophyllum uniflorum, Pentachondra pumila, Astelia linearis, Drosera arcturi, Donatia novae-zelandiae* and *Oreobolus pectinatus*.

(5) The *C. pallens* (*C. australis*) association occurs on well drained soils over marble and schist, over a higher altitudinal range than the preceding associations. Both the characteristic species, especially carpet grass, are calcifuges that, according to Bell (1973), dominate on marble only above 1500 m, on stable soils where, presumably, calcium is leached as soon as it is released from slowly weathering rock fragments. Carpet grass is generally absent from steep lower-penalpine slopes, where *Poa colensoi* forms most of the turf between *C. pallens* tussocks.

(6) The *C. australis* association occurs as thick swards on granite, less frequently on schist and rarely on marble (Fig. 9.6). At the lowest altitudes carpet grass grows only in frosty depressions, but with rising altitude it increasingly replaces *C. pallens* on sloping ground and ridge crests. Soils are well drained, but species such as *Celmisia spectabilis, C. laricifolia* and *C. discolor* indicate low fertility.

Similar penalpine *Chionochloa* grasslands occur throughout Western Nelson, although communities depending on moderately fertile soils are much less extensive than on the limestone mountains of Arthur ED. Grassland is also extensive at

Fig. 9.6. Carpet grass sward, with red tussock on wetter ground; 1450 m near Cobb Lake, North-west Nelson (photo G.Y. Walls).

600–900 m on the Gouland and McKay Downs, which are rolling upland of Palaeozoic granite, schist and sandstone. Red tussocks prevail in the depressions and in places their dominance may have been enhanced through burning of shrubs. They provide tall, dense cover on deep soils that are not continuously waterlogged, but become smaller and sparser, and *Empodisma* and *Gleichenia dicarpa* increasingly dominate, as grassland grades into wet heath (p. 336). At higher altitudes red tussock also displaces woody vegetation from broad, exposed spurs and ridges where soils are deep enough.

On the coal measures of the Denniston Plateau in North Westland, red tussock is replaced by the very narrow-leaved *Chionochloa juncea*, which occurs between 500 and 1000 m and likewise grows in association with *Empodisma* and *Gleichenia*, together with *C. australis* on shallow soils. *C. rubra* and *C. australis* grow together on the Mt Davey coal measures near Greymouth, at 880 m (Fig. 9.7).

Westland

The main snow tussocks of Western Nelson extend also to the north-eastern Alps. Fig. 9.8 illustrates the relationship between vegetation, snow cover, altitude and exposure at Lewis Pass.

Carpet grass is absent south of Arthurs Pass, and red tussock is generally absent south of the Hokitika catchment, except on low-altitude heathlands between the Waiho and Cook Rivers, terraces in the Landsborough Valley, and ultramafic terrain in the Red Hills and northern Olivine Range. Their niches are occupied by *C. pallens* and *Poa colensoi* on the more recent and better drained soils, and by *C.* 'westland' and *C. crassiuscula* on more mature, leached, or boggy soils (Figs 9.9 and 9.10).

In Westland National Park (Wardle 1977), *Chionochloa pallens* forms dense

Fig. 9.7. Red tussock on deeper soil among stone pavement; 880 m on Mt Davey, Paparoa Range, North Westland.

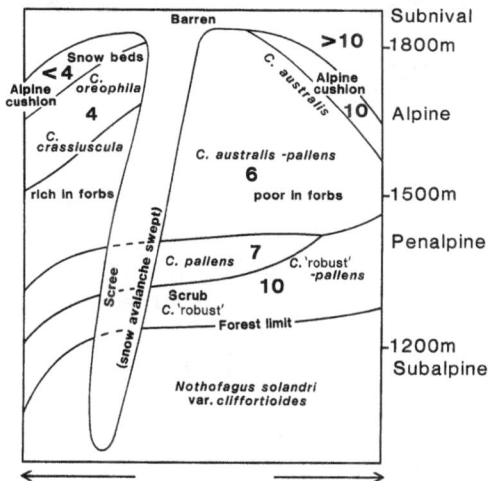

Fig. 9.8. Distribution of major communities in relation to snow cover, exposure and altitude on relatively mature soils of normal drainage. Lewis Pass, north-eastern Alps (from Burrows 1977). *C, Chionochloa*; Numbers are of snow-free months. ⟵, Increasing snow cover (mainly above 1500 m) and shelter (mainly lower altitudes); ⟶, decreasing snow cover and therefore more exposed (mainly above 1500 m) or warmer (mainly lower altitudes).

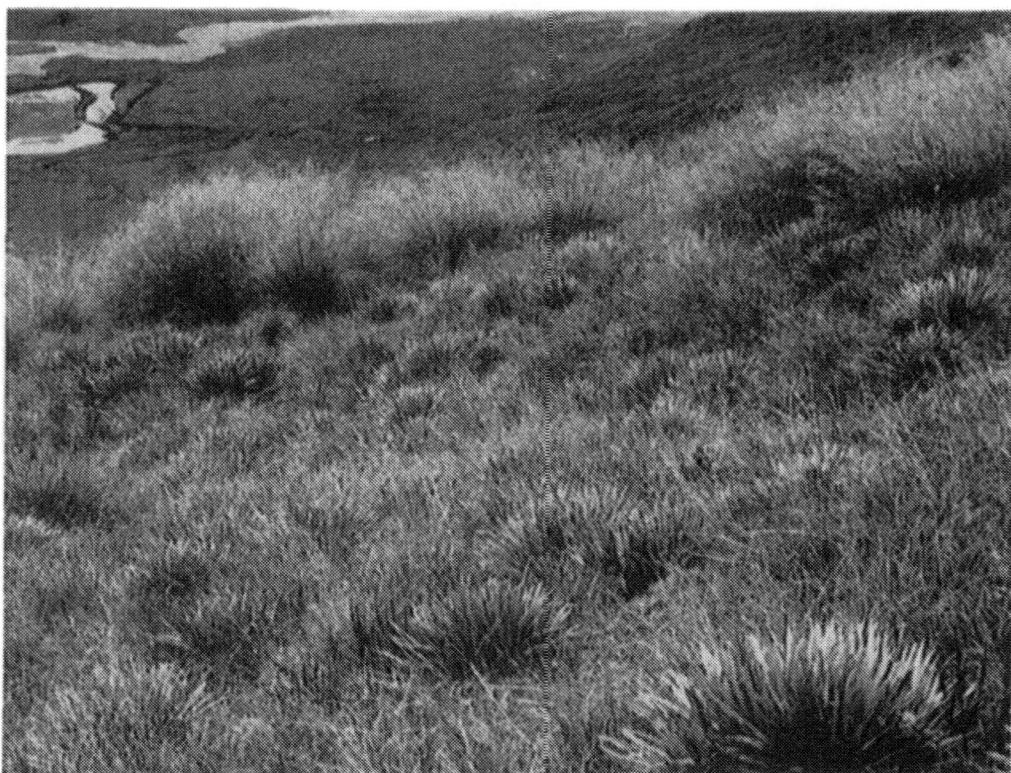

Fig. 9.9. *Celmisia armstrongii, Chionochloa pallens* and *C. crassiuscula* in foreground, taller flowering tussocks of *C.* 'westland' behind; 1450 m, Solution Range, Southwestern Alps.

grassland at 950–1550 m, on weakly weathered, well-drained soils, generally on debris slopes of southerly aspect. *Coprosma cheesemanii, Ranunculus lyallii, Aciphylla crenulata* and *Pratia angulata* can be abundant in openings among the snow tussocks. On sites that are generally over 1150 m a.s.l. and either more exposed, or on shallower, stonier soils, or subject to longer snow-lie, *Chionochloa pallens* reduces to 10–15% cover over a turf mainly of *Poa colensoi*. Other abundant species on these sites include *Microlaena colensoi, Pentachondra pumila, Coprosma perpusilla, Celmisia verbascifolia, C. vespertina, C. sessiliflora* and, where moister or less exposed, *C. armstrongii, Astelia nivicola, Schoenus pauciflorus* and *Lycopodium fastigiatum*.

Chionochloa pallens also occupies fine-textured, recent soils on moraine, alluvium and debris to as low as 750 m but is often eliminated by grazing so that *Poa cockayneana* or *Hypolepis millefolium* dominate instead. Associates include *Festuca matthewsii, Rytidosperma setifolium, Acaena anserinifolia, Celmisia semicordata, C. walkeri, Raoulia glabra* and *Helichrysum bellidioides*. On coarse-textured recent soils, *Chionochloa* 'robust' is more important than *C. pallens*, and species characteristic of seral shrubland appear, including *Olearia moschata, Coprosma rugosa, C. depressa, Coriaria plumosa* and *Uncinia divaricata*.

Fig. 9.10. *Chionochloa oreophila* (left), *C. crassiuscula* (centre foreground) and *C. pallens* (right and background); 1380 m, Arthurs Pass, North-eastern Alps.

Chionochloa 'robust' grassland also grows in very steep troughs, alternating with *Dracophyllum uniflorum* shrub-heath occupying the spurs. *C. pallens*, with *C.* 'robust' in small gullies, forms the highest continuous vegetation in the Park, on a precipitous northerly aspect at 1983 m. Sites such as this are habitats for *Astelia petriei* and the small shrubs *Hebe treadwellii* and *H. ciliolata*.

C. 'westland' grows at 750–1200 m, mainly in mosaics with *Dracophyllum longifolium, D. uniflorum* and other shrubs, on flat or gentle surfaces with gley podzol soils that also favour *Schoenus pauciflorus, Pentachondra* and *Carpha alpina*. Further species include *Coprosma cheesemanii, C. serrulata, Myrsine nummularia, Anisotome haastii, Celmisia semicordata, C. armstrongii, Astelia nervosa, Phormium cookianum, Hebe macrantha, Abrotanella linearis* and *Forstera sedifolia*. In Westland National Park, *Chionochloa* 'westland' grassland without shrubs results from fire, but further south it may be primitive on broad, rolling ridges.

Above 950 m, *C. crassiuscula* tends to occur instead of *C.* 'westland' and its shrubby associates in deep, wet hollows and gullies and where water is forced to the surface by underlying bedrock at 5–15 cm depth. From 1200 m to 1700 m, it occupies similar sites, and also gley podzols on moderate and gentle slopes. *C. crassiuscula* is accompanied by a suite of herbs similar to that listed for *C.* 'westland', to which *Cyathodes pumila, Coprosma perpusilla, Celmisia glandulosa, Gentiana bellidifolia, Phyllachne colensoi, Astelia linearis* and *Oreobolus impar* can be added.

Chionochloa oreophila–Poa colensoi grassland ranges from 1300 to 1914 m, generally on slopes <15°, descending lowest on south aspects where snow lies late, and ascending highest on north and west aspects. *C. oreophila* tolerates neither as much nor as little snow as *P. colensoi*, and is mainly on deeper, more weathered soils. *Carex pyrenaica* co-dominates with *Poa colensoi* in the deepest hollows, and *Marsippospermum* co-dominates on banks and the risers of solifluction steps. *Rytidosperma setifolium*, *Celmisia haastii* and *Anisotome flexuosa* are also frequent.

Celmisia walkeri and other forbs, especially *Helichrysum bellidioides*, dominate steep slopes that are too rocky for *Chionochloa pallens* grassland, although *C. pallens* may form 10–20% of the cover. Up to 1968, large areas of *C. pallens* grassland had been replaced with *Poa colensoi* and other grazing-tolerant or resistant plants, especially *Microlaena colensoi* and *Celmisia walkeri*, but by 1975, with effective control of deer, chamois and thar, *C. pallens* was re-establishing rapidly, initially from persisting tillers and more recently from seed.

Fiordland

South of the Arawata River, where the schists of Westland give way first to ultramafic rocks and then to the weathering-resistant diorites, granites and gneisses of Fiordland, chionochloas tolerating shallow, infertile soils prevail. Above the western fiords, forest and dense scrub ascend highest on steep, rocky slopes of northern aspect, whereas grassland descends into the subalpine belt on gentler slopes with deeper soils. *Chionochloa acicularis*, which has bluish-green, very narrow, sharp-tipped leaves, tends to cover wet slopes, with the tussocks often standing on peaty pedestals. *C.* 'fiord' is most abundant on steeper, better-drained slopes, and *C. crassiuscula* occupies shallow soils in hollows. Where slopes are rocky and uneven there are many shrubs and large forbs (Table 9.2).

At West Cape, *C. acicularis* grassland descends a sloping peneplain to within 100 m of sea level (Fig. 9.11). Where the O–A horizon of peaty silt loam is < 12–20 cm deep, *C. acicularis* grows with shrubs, including manuka and small podocarps; this merges into shrub- and tree-heath where drainage is better. Where the O–A horizon is deeper but waterlogged, the tussock grows with *Empodisma*, *Gleichenia dicarpa* and the spindly shrub *Sprengelia incarnata*.

For a locality close to the high central divide of Fiordland, Table 9.3 defines 11 community types in three groups, distributed along gradients of altitude and fertility–drainage (Fig. 9.12). The *Chionochloa pallens–C. oreophila* group (1) is characteristic of recent soils on debris cones, fans and terraces. The indicator species for the group are *C. pallens*, *Uncinia divaricata*, *Luzula rufa*, *Bulbinella gibbsii* var. *balanifera*, *Cardamine* spp., *Epilobium alsinoides* var. *atriplicifolium*, *Schizeilema haastii* var. *cyanopetalum*, *Oxalis magellanica*, *Geranium microphyllum*, *Viola cunninghamii*, *Oreomyrrhis colensoi* and *Leptinella squalida*; that for the most sampled community (1d) is *Aciphylla takahea*. *C. oreophila* occurs widely except at the lower altitudes and attains full dominance in the high-lying community 1f, which has *Ranunculus sericophyllus*, *Aciphylla congesta*, *Celmisia hectorii*, *Pratia macrodon* and *Carex pyrenaica* as indicator species. *C. pallens* is more important than is usual in Fiordland, probably because land forms favouring it are unusually extensive.

The *Chionochloa crassiuscula* group (2) occupies leached soils that tend to be shallow, steep and wet. Indicator species are *Lycopodium fastigiatum*, *Myrsine*

Table 9.2. *Percentage cover in Chionochloa grasslands, measured by point intercepts at c. 900 m on Secretary Island, western Fiordland*

	31°	0°
Slope:	31°	0°
Aspect:	SE	—
No. of vascular species:	56	44
% cover:	85	89
Snow tussocks		
Chionochloa acicularis	21	42
Chionochloa 'fiord'	7	1
Chionochloa crassiuscula	2	
Other grasses, sedges and rushes		
Rytidosperma setifolium	4	1
Poa colensoi	1	
Luzula sp.	1	
Carex sp.	1	
Shrubs and stunted trees		
Dracophyllum menziesii	5	17
Dracophyllum uniflorum	12	+
Nothofagus menziesii	11	+
Dracophyllum longifolium	1	4
Olearia colensoi	3	1
Hebe odora	1	2
Pseudopanax colensoi v. *fiord.*	f	f
Dwarf shrubs		
Gaultheria rupestris	4	
Coprosma cheesemanii	+	3
Coprosma crenulata	f	1
Myrsine nummularia	f	+
Large forbs		
Celmisia petriei	1	8
Celmisia verbascifolia	2	1
Dolichoglottis scorzoneroides	2	
Aciphylla lyallii		2
Anisotome haastii	1	1
Phormium cookianum		1
Ourisia macrocarpa	f	
Small forbs		
Gentiana montana	f	1
Lycopodium fastigiatum	+	1
Drapetes dieffenbachii	1	
Forstera sedifolia	f	+
Celmisia graminifolia		f
Mosses		
Dicranoloma spp.	2	2
other mosses	1	

Notes:
The table lists species with cover > 0.5% or frequency > 50% (f) in at least one transect; +, present.
Source: Mark & Baylis 1963; Wardle *et al.* 1970

Fig. 9.11. Bands of *Chionochloa acicularis* among tall heath; 50 m a.s.l., West Cape. Fiordland.

nummularia, Aciphylla lyallii, Dracophyllum uniflorum, Corprosma cheesemanii, Forstera sedifolia, Celmisia du-rietzii, Dolichoglottis scorzoneroides, Astelia linearis and *Oreobolus impar*; that for the most-sampled community (2c) is *O. impar*.

Chionochloa acicularis and *C.* 'fiord' belong to the lower altitudes. The former associates with *Schoenus pauciflorus* (community 2a) on steep seepages. *C.* 'fiord' is the main snow tussock in community 1b, growing with species typical of recent, well-drained talus, such as *Helichrysum bellidioides, Polystichum vestitum* and *Hypolepis millefolium*. However, it also co-dominates with *Chionochloa crassiuscula* in community 2b, often on precipitous bedrock slopes. *C. ovata* can be present on broad ridges. Communities 3 and 4 are fellfield and shallow wetland respectively.

East of the Divide, *Chionochloa teretifolia* dominates widely above the tree-limit, ascending to the ridge crests on north aspects and meeting *C. crassiuscula* grassland on south aspects. *C. acicularis* occupies poorly drained, rounded summits in the westernmost headwaters, red tussock clothes valley floors below tree-limit, and *C.* 'fiord' occurs on steep, rocky slopes.

East of the Waiau River, red tussock, *C. teretifolia, C. crassiuscula* and probably *C.* 'fiord' grow together on the boggy summit of the Longwood Range at only 760 m. Penalpine slopes of the Takitimu Range support *C. teretifolia* grassland, passing into *C. crassiuscula* on shaded slopes and at higher altitudes. *C. rigida* grassland commences on the northern spurs.

Stewart Island

The mountains of Stewart Island scarcely exceed the subalpine belt, and mostly comprise broad, exposed, poorly drained ridges supporting a mosaic of stunted scrub,

Table 9.3. *Mean cover to the nearest 5%, in grassland communities at Wapiti Lake, Fiordland*

Community (see text):	1a	1b	1c	1d	1e	1f	2a	2b	2c	3	4
Number of plots:	9	7	6	34	5	5	4	17	30	5	5
Poa colensoi	25	5	+	5	+	+		+	+	10	+
Anisotome haastii	10	10	+	5	+	+	+	5	+		
Chionochloa pallens	15	5	15	20	+	+		+	+	+	+
Rytidosperma setifolium	+	5	15	5	20	+	10	+	+		5
Marsippospermum gracile	+		20	5	+	5			+	+	
Chionochloa oreophila			15	20	5	35			+	5	
Chionochloa acicularis		5					25	5	+		
Schoenus pauciflorus	+	+		+		+	15	5	+		20
Chionochloa 'fiord'		20					5	15			
Chionochloa crassiuscula	10	+		10			+	10	35		10
Centrolepis pallida											25
Celmisia glandulosa	+			+			+	+	+		10
% bare ground (inc. rock)	5	5	10	10	50	30	5	10	15	55	10

Notes:
+, <5% cover
Source: Rose *et al.* 1988

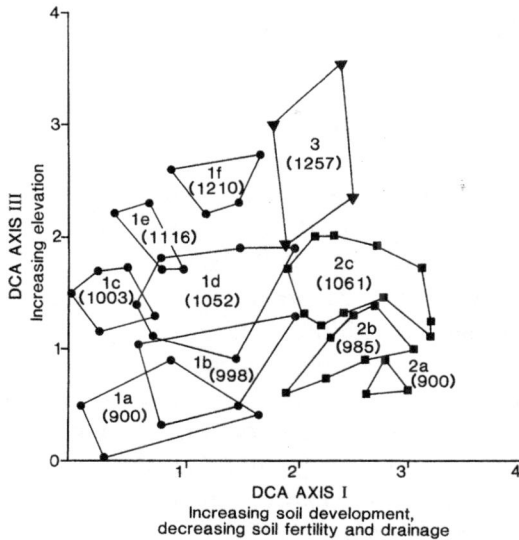

Fig. 9.12. Ordination of nine grassland communities at Wapiti Lake, Fiordland, along two axes in a detrended correspondence analysis (Rose *et al.* 1988). Communities are numbered as in Table 9.3; numbers in brackets are mean altitudes in metres.

open shrubland, dwarf-heath and cushion bogs. *Chionochloa* 'fiord' and *C. lanea* on the northern and southern mountains respectively, and *C. pungens* throughout, form small areas of grassland, but usually share dominance with stunted *Olearia colensoi*, manuka, *Dracophyllum longifolium, Halocarpus biformis* and other shrubs, and low-growing plants of wet, infertile soils (p. 340).

Red tussock forms grassland on damp, leached soils in the lowlands, e.g. among old dunes, but yields to *Empodisma* and *Gleichenia dicarpa* on the least fertile sites. This grassland, like the very limited areas of hard tussock (*Festuca novae-zelandiae*) grassland on the island, is partly maintained by fires killing shrubs that would suppress it. Red tussock also ascends above the limits of continuous forest and scrub, to overlap with the other chionochloas.

Eastern South Island south of the Rakaia River

Through most of this country, *Chionochloa rigida* grassland is the prevailing cover of the mid-mountain slopes. Columns 1–7 in Table 9.4, and Figs 9.13 and 9.14, show examples. Differences among the communities are subtle, and mainly relate to a gradient from high altitudes and cool aspects to lower, warmer sites. *Celmisia spectabilis* is important on the Hunter Hills, which lie just within its southern limit; in the Mackenzie samples, *C. lyallii* is present instead.

On the rolling plateaus of eastern Otago, which receive much easterly fog and drizzle, *C. rigida* grasslands descend to 500 m, and Shag Point has remnants near sea level. The usual species are accompanied by others more typical of red-tussock grassland, such as *Pernettya macrostigma, Herpolirion novae-zeelandiae* and *Pentachondra*, but red tussock itself is confined to wet valley floors (Fig. 9.15). *Celmisia semicordata* is prominent in the subalpine belt. Invasion by shrubs, especially *Dracophyllum longifolium, Cassinia 'vauvilliersii'* and *Hebe odora*, is more prevalent than on less humid inland mountains (p. 552).

Before the intensive farm development of recent decades, red-tussock grassland covered much of the Southland lowlands, which are swept by cold, wet southerlies (Fig. 9.16). It still occupies the higher, wetter parts of inland plains, especially in Mackenzie ER, ascending as high as 1200 m on gentle summits and plateaus.

Column 8 in Table 9.4 represents the dry facies of red tussock grassland. This contains much the same species as other Mackenzie grasslands at comparable altitudes, and becomes transformed into hard-tussock grassland as red tussocks are eliminated. *Cladia retipora, Racomitrium languinosum, Polytrichum commune, Lycopodium fastigiatum, Ranunculus glabrifolius, Gonocarpus aggregatus, Pernettya macrostigma, P. nana, Gnaphalium* aff. *traversii, Celmisia graminifolia* and *Herpolirion novae-zelandiae* appear where soils are wetter and more leached. Local bog pines (*Halocarpus bidwillii*) are presumably relicts from earlier woodland (p. 195). *Schoenus pauciflorus* and *Bulbinella angustifolia* (column 9) indicate flushed rather than leached soils. *Aciphylla subflabellata* is the usual aciphylla, but *A. glaucescens* is prominent on sheltered banks in Southland.

Red tussock can be very dense on deep, fine-textured soils in hollows and on narrow stream-side terraces. The presence of silver tussock (*Poa cita*), *Coprosma propinqua*, *Carmichaelia* spp. or, in maritime districts, cabbage trees (*Cordyline australis*)

Table 9.4. *Composition of tall-tussock grasslands of eastern South Island, based on cover estimates in 10 m × 10 m releves*

Community:	1	2	3	4	5	6	7	8	9	10	11	12
No. of samples:	15	4	4	6	5	4	5	6	4	4	4	3
Mean altitude (m):	857	1336	1063	1290	844	844	1183	859	652	1105	979	1356
Prevailing aspect:	nil	S	SW	NE	nil	S	E	nil	—	NW	S	S
Mean slope (deg.):	22	26	30	27	23	36	28	13	5	29	23	30
Mean percentage cover:	80	90	99	95	90	86	86	99	100	70	96	88
Mean no. of vascular spp.:	?	33	38	29	40	40	46	45	21	34	47	24
Tall tussocks												
Chionochloa rigida	2	3	4	3	4	4	3	+	+			
Chionochloa rubra								3	5			
Chionochloa 'robust'										3	+	
Chionochloa macra										+	3	3
Short tussocks												
Rytidosperma setifolium	f											
Festuca novae-zelandiae	f	1	2	1	2			2		1	1	
Festuca matthewsii						3	2					
Poa colensoi	f	3	2	1	1	2	2	1		f	2	2
Other native grasses												
Rytidosperma gracile	f											
Deyeuxia avenoides	f	f	1	f	+	+	f	+		f	f	+
Agrostis subulata		f										
Festuca multinodis		+		+						+		
Poa lindsayi		+	+	+	+							+
Lachnagrostis filiformis		+	+	+	+	+	+			+	+	
Elymus rectisetus		+	f	f	+	+	f	+		+	+	
Koeleria novo-zelandica		+	+	f	f	f	f			+	f	
Rytidosperma pumilum		1	f	+	f	+	f	f		+	f	f
Dichelachne crinita				+	f			+				
Hierochloe novae-zelandiae											+	+
Introduced grasses												
Holcus lanatus	f			+	+	+			f			
Anthoxanthum odoratum	1		f		1	+	+	+	f	f	f	
Agrostis capillaris	f		+		f			+	1	+	f	
Aira caryophyllea					f	+		+				
Sedges and rushes												
Luzula rufa	f	f	f	f	f	f	f	f		f	f	+
Carex breviculmis	f			+	+	f		f		+	+	+
Carex wakatipu		f	f	f	f	+	f					
Carex colensoi					+		+	+				
Schoenus pauciflorus									1			
Shrubs												
Discaria toumatou			+		1	+		+	+	1	+	
Dracophyllum uniflorum			+				+			+	1	f
Carmichaelia petriei				+	f	f		+				
Melicytus alpinus				+	+	+				+		
Cassinia 'fulvida'							+			+	f	

Table 9.4. (*cont.*)

Community:	1	2	3	4	5	6	7	8	9	10	11	12
No. of samples:	15	4	4	6	5	4	5	6	4	4	4	3
Mean altitude (m):	857	1336	1063	1290	844	844	1183	859	652	1105	979	1356
Prevailing aspect:	nil	S	SW	NE	nil	S	E	nil	—	NW	S	S
Mean slope (deg.):	22	26	30	27	23	36	28	13	5	29	23	30
Mean percentage cover:	80	90	99	95	90	86	86	99	100	70	96	88
Mean no. of vascular spp.:	?	33	38	29	40	40	46	45	21	34	47	24
Dracophyllum pronum								+			+	1
Dracophyllum acerosum										+	1	
Coprosma pseudocuneata											f	+
Dwarf shrubs												
Pimelea oreophila	f	f	f	+	f	f	f	f		+	f	
Coprosma depressa	+										+	
Cyathodes fraseri	f	+	f	f	1	f	1	f		f	+	+
Gaultheria depressa	f	1	+		+	+	1					f
Pentachondra pumila	f	+						+		+	+	+
Drapetes dieffenbachii	f	+	+	+		+	f				f	+
Pimelea traversii		+		+	+					+		
Coprosma petriei		+		+	f	+	+	+			+	
Coprosma cheesemanii			+		+	+						
Muehlenbeckia axillaris			+	+	+		+	+		+		
Cyathodes colensoi			f	+				1		1	1	+
Carmichaelia monroi			+	f	+	+	f	+		f	+	+
Gaultheria crassa									+	1	1	
Large native forbs												
Celmisia spectabilis	3									1	1	f
Gentiana corymbifera	+	+	+	+		+	+	+		+	f	f
Aciphylla aurea		f	f	1	f	+	f	+*		f	f	+
Celmisia lyallii		1	f					f		+	+	2
Celmisia angustifolia											2	+
Mat, cushion and patch-forming herbs												
Nertera setulosa	+											
Oxalis exilis	f*			+	+	+						
Helichrysum filicaule	f				+	+	+	f				
Gnaphalium 'collinum'	+		+	f	+	+	f	+		+		
Scleranthus uniflorus	f	f	f*	f*	f*	+	f	f		f	+	
Hydrocotyle novae-zeelandiae	+		+		f	f	f	f		+		
Lagenifera cuneata	+	+	f	+	f	f	f	+		+	+	
Brachyscome sinclairii	f	1	+	+		f	f	f		+	+	
Raoulia subsericea	f	1	f	f	1	1	2	1		f	1	+
Colobanthus brevisepalus		+		+	+	+	f	f		+	+	
Leptinella pusilla		1		+		+	f	f				+*
Raoulia apicinigra			+		+	+	+					
Galium perpusillum			+		+	+		+	f			
Raoulia hookeri								+			+*	
Other native dicot forbs												
Brachyglottis haastii	f*				+	+	+	f				
Oreomyrrhis rigida	+	+		+	+	+	+	+				

	1	2	3	4	5	6	7	8	9	10	11	12
Community:	1	2	3	4	5	6	7	8	9	10	11	12
No. of samples:	15	4	4	6	5	4	5	6	4	4	4	3
Mean altitude (m):	857	1336	1063	1290	844	844	1183	859	652	1105	979	1356
Prevailing aspect:	nil	S	SW	NE	nil	S	E	nil	—	NW	S	S
Mean slope (deg.):	22	26	30	27	23	36	28	13	5	29	23	30
Mean percentage cover:	80	90	99	95	90	86	86	99	100	70	96	88
Mean no. of vascular spp.:	?	33	38	29	40	40	46	45	21	34	47	24
Gonocarpus montanus	f*		+*		+		+		+			
Geranium sessiliflorum	f	+	f	+	f	+	f	+			f	+
Epilobium glabellum	+	+	f	f	f	f	f	+			f	+
Wahlenbergia albomarginata	f	f	f	f	f	f	f	f	+		f	f
Helichrysum bellidioides	f	f	+	+		f	f				+	
Ranunculus multiscapus	f	+	+		f	f	+	f	+		+	
Plantago lanigera	f				+*			1*			+*	
Euphrasia zelandica	+							+			+	
Acaena caesiiglauca	f	+	+	+	f	+	f	+	+	+		+
Celmisia gracilenta	+	f	1	+	+	f	f	f	+	+	f	+
Viola cunninghamii	f	f	f	f	f	f	f	f	1	+	f	+
Geum leiospermum	f	+*									+*	+*
Epilobium alsinoides	f	f	+	+	+	+	f	+		+	+	f
Anisotome flexuosa	f*	1	1	+	+	2*	+	f*	+*	+	1	f
Stellaria gracilenta		+	+		f	+						
Kirkianella novae-zelandiae		+			+	+		+				
Brachyglottis bellidioides		f	1	f			+	+		f	1	+
Aciphylla monroi		f								+		f
Craspedia spp.*			f					f		+		
Craspedia lanata			+	+	+	+	f	+		+	+	
Microseris scapigera			+			+	+	f		+		
Gingidia decipiens			+			f	+				+	+
Anisotome filifolia			+	+	+					+		
Ranunculus gracilipes			+			f	+				+*	+*
Senecio glaucophyllus				f	+	+						
Vittadinia australis				f	1	+		+		+		
Lobelia linnaeoides						+	+					
Myosotis australis						f	+				f	
Celmisia densiflora							f					
Geranium microphyllum							+		+			
Forstera tenella											+	+
Epilobium chlorifolium											+	+
Other native monocots												
Prasophyllum colensoi	+		+		+			f		+		
Thelymitra longifolia	+						+			+		
Bulbinella angustifolia									1			
Adventive forbs												
Trifolium repens	+									f		
Trifolium dubium	+									+		
Hieracium pilosella	+	f	f	+	f			+		+		
Rumex acetosella	f	f	+	1	1	f	f	f	+	f	f	
Cerastium fontanum	+				+	f	+	+	f		+	
Hypericum perforatum		1			+							
Hypochoeris radicata			f	f	+	+	f	1	+	f	f	

Table 9.4. (*cont.*)

Community:	1	2	3	4	5	6	7	8	9	10	11	12
No. of samples:	15	4	4	6	5	4	5	6	4	4	4	3
Mean altitude (m):	857	1336	1063	1290	844	844	1183	859	652	1105	979	1356
Prevailing aspect:	nil	S	SW	NE	nil	S	E	nil	—	NW	S	S
Mean slope (deg.):	22	26	30	27	23	36	28	13	5	29	23	30
Mean percentage cover:	80	90	99	95	90	86	86	99	100	70	96	88
Mean no. of vascular spp.:	?	33	38	29	40	40	46	45	21	34	47	24
Crepis capillaris				+	f			+				
Linum catharticum					f	f		+				
Hieracium praealtum					1	f		+			+	
Ferns and lycopods												
Blechnum penna-marina	f	+	1	+	+	+	1		f			
Lycopodium fastigiatum		+						+			f	+
Ophioglossum coriaceum				+	+			+			f	+

Notes:

Data in this and following tables are averaged cover values for releves, plots or transects.
Numbers in body of table indicate: 5, ≥75%; 4, 50–<75; 3, 25–<50%; 2, 5–<25%; 1, plentiful, but <5% cover; f, present in >50% of samples; +, present in ≤50% of samples.
Localities and communities:
1. *Chionochloa rigida* and *Celmisia spectabilis* co-dominant, on seaward aspect of Hunters Hills, Pareora ER (Barker 1953).
2–8. Mackenzie ER (Connor 1964): 2–5, *C. rigida* associated with *Festuca novae-zelandiae*; 6–7, *C. rigida* associated with *F. matthewsii*, mainly at higher altitudes or shadier aspects than 2–5; 8, *C. rubra* dominant.
9–12. Rakaia Valley, Puketeraki ER (Connor 1965): 9, dense *C. rubra* at low altitudes; 10, *C.* 'robust' dominant on sunny aspects; 11–12, *C. macra* dominant on shaded aspects and at high altitudes.
* Species listed are replaced respectively by *Aciphylla subflabellata, Oxalis magellanica, Leptinella pectinata, Raoulia monroi, Brachyglottis lagopus, Gonocarpus aggregatus, Plantago spathulata, Geum uniflorum, Anisotome aromatica, Ranunculus enysii. Craspedia* spp. include *minor* and *viscosa. Scleranthus brockiei* also present in communities 3–5.

indicates fertile conditions. *Carex coriacea* or *C. virgata* occur in communities transitional to fertile swamp, but more often red-tussock grassland grades into infertile swamp with scattered red tussocks (pp. 320, 334).

In these southern regions, *Chionochloa macra* occupies the transition from *C. rigida* to low, open, alpine vegetation. It descends lowest on exposed ridges and snowy south aspects, sometimes forming a sharply differentiated belt and sometimes merging into *C. rigida* grassland *via* a zone of hybrids. Towards its upper limit *C. macra* is patchy, the tussocks appear unthrifty, and seedlings are rare. This reflects the extreme environment, as well as sensitivity to fire and grazing which, on many mountains, may have reduced once-continuous belts of *C. macra* to scattered remnants.

At 1350 m on the Ohau and Ben Ohau Ranges *C. macra* forms 5% cover among *Festuca matthewsii* and *Poa colensoi*, as well as *Hieracium praealtum, H. pilosella* and sorrel (*Rumex acetosella*), which are typical of modified tussock grasslands. On a

Fig. 9.13. *Chionochloa rigida* grassland, 1130 m, Bald Hill, Mavora ER. *Aciphylla aurea* in left foreground.

Fig. 9.14. *Festuca matthewsii* and *Poa colensoi* (foreground) and *Chionochloa rigida*; 1200 m, Ben More, Mackenzie ER.

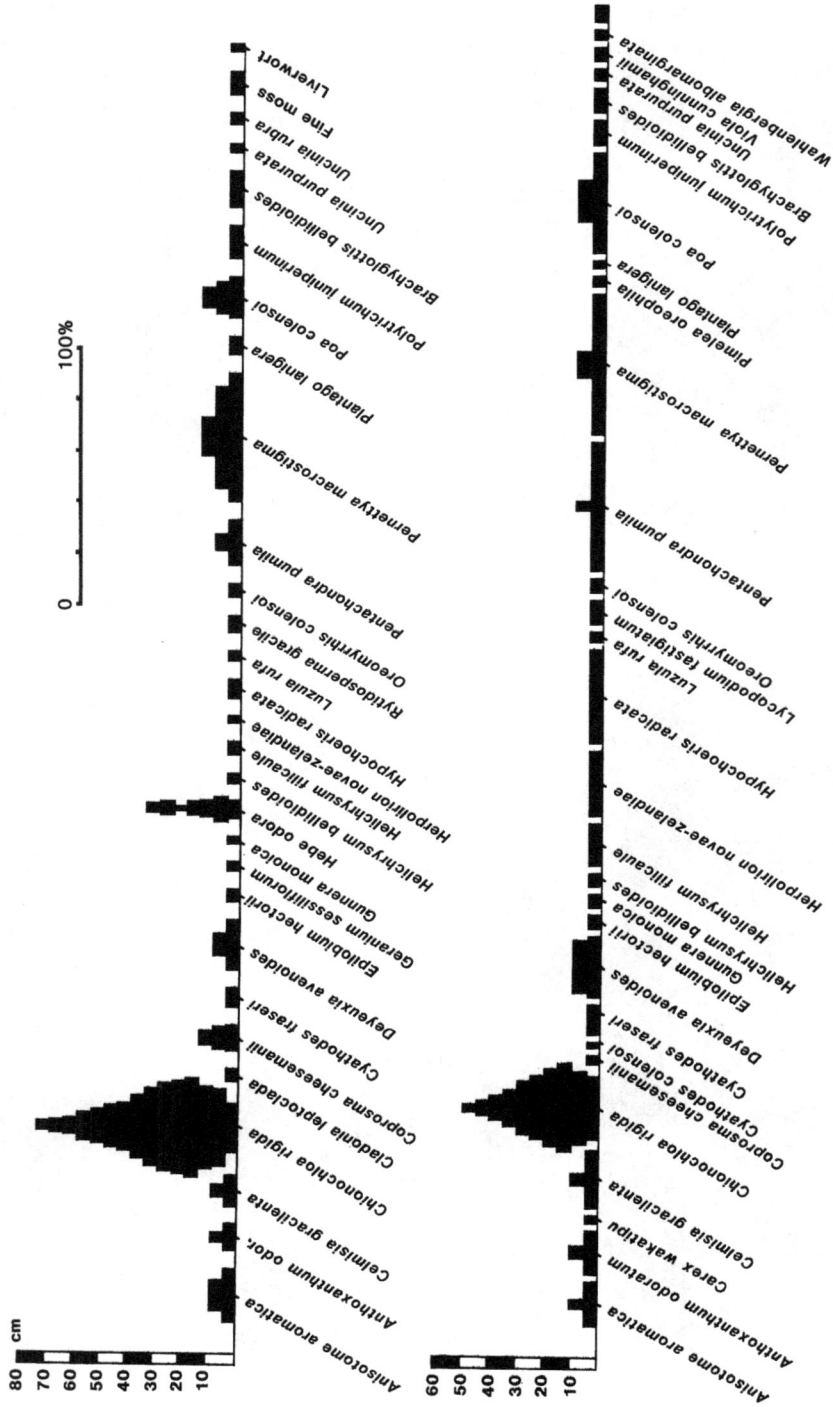

Fig. 9.15. Structure of two low-altitude *Chionochloa rigida* stands at *c*. 700 m, Lammerlaw ER. The histograms show percentage frequency, determined as presence of plant parts in 5 cm diameter cylindrical quadrats at 5 cm increments in height above the ground (250 sampling positions) (Bulloch 1973).

Fig. 9.16. Red tussock persisting on railway reserve near Mossburn, Gore ER.

sheltered slope at 1500 m, it shares dominance with *Celmisia lyallii*, over a lower tier of *Poa colensoi* accompanied by species such as *Rytidosperma setifolium*, *R. pumilum* and *Lycopodium fastigiatum*. There are also scattered tussocks of *Chionochloa rigida*, which dominates a similar assemblage further down slope. In a late-snow hollow at 1680 m, *C. macra* forms 5% cover over turf that includes *Poa colensoi*, *Psychrophila novae-zelandiae*, *Coprosma perpusilla* and *Celmisia sessiliflora*. Finally, at 1850 m, there are small colonies of *Chionochloa macra* and *Aciphylla dobsonii* where large rocks anchor the prevailing scree.

At 1350 m on the wet, southern flank of the Old Man Range, *C. macra* with some *C. rigida* dominates dense penalpine grassland containing *Aciphylla* aff. *horrida*, *A. scott-thomsonii*, *Celmisia semicordata*, *C. prorepens*, *Brachyglottis revoluta*, *Hebe pauciramosa* and *Schoenus pauciflorus*. Where rainfall increases sharply close to the Main Divide, the partition of habitats among *C. rigida*, *C. rubra* and *C. macra* gives way to the western pattern with *C. pallens*, *C. crassiuscula*, *C. oreophila*, and broad- and narrow-leaved tussocks in the *flavescens–rigida* complex.

Northern Canterbury and Marlborough

Chionochloa grasslands on the Canterbury mountains are interrupted by extensive screes and outcrops of crumbling greywacke. North of the Rakaia catchment, *C. rigida* is absent, so that other snow-tussocks prevail. *C. macra* is the main species to as far north as the head of the Awatere River. It grows in the alpine belt and on the more mature, leached soils at lower altitudes, descending to 800 m on deforested foot-hills in northern Canterbury (Fig. 9.17). *C.* 'robust' and *C. pallens* occur on weakly weathered

Fig. 9.17. *Podocarpus nivalis* on a patch of scree in *Chionochloa macra* grassland; 1280 m, Craigieburn Range, Puketeraki ER.

soils throughout, the former mainly on bluffs and warm upper-subalpine slopes that often have a veneer of loose stones, and the latter on cool, moist penalpine slopes. The most prominent indicators of altitude, in view of the wide range of *C. macra*, are large celmisias: *C. spectabilis* in the montane and subalpine belts, *C. lyallii* in the penalpine belt, and *C. viscosa* in the penalpine–alpine transition. In Table 9.4, columns 9–12 represent communities from the Rakaia Valley, where *Chionochloa macra* and *C.* 'robust' already dominate most of the mid-slopes. Red tussock and *C. rigida* alternate on lower slopes according to drainage, with both species descending to the Canterbury Plains.

In the Waimakariri and Rakaia catchments *C. crassiuscula* and *C. oreophila* extend sparingly to the eastern mountains in stable penalpine gullies and alpine cirques respectively. Carpet grass communities similar to those of Western Nelson are extensive on rounded and flat penalpine surfaces with mature soils on the inner ranges of Marlborough, the main associated species being *Celmisia spectabilis, C. incana, C. allanii, Anisotome aromatica, Gentiana* spp., *Ranunculus verticillatus*, the cushion plant *Phyllachne colensoi*, and the small shrubs *Hebe lycopodioides, Coprosma* spp., *Dracophyllum pronum* and *Pentachondra pumila* (P.A. Williams, unpublished).

On the Kaikoura Ranges, where the greywackes are younger than those to the west and south-west and weather more finely, screes and eroding surfaces occupy more of the slopes than intact grassland. The limited areas with *Chionochloa* contain only *C. pallens* and *C.* 'robust', each on its appropriate substrate.

High country in Sounds–Nelson province

In the ultramafic country of the Richmond Range, continuous subalpine vegetation consists of *Chionochloa defracta* growing with *Empodisma* in valley-floor swamps, or with *Poa colensoi*, other herbs, and shrubs such as manuka, *Aristotelia fruticosa*, *Coprosma propinqua*, *Melicytus alpinus* and *Cassinia 'vauvilliersii'* on bouldery slopes. Patches of *Chionochloa defracta* ascend over 1600 m in otherwise desolate fellfield. Elsewhere on the range, *C. pallens* and carpet grass form high-altitude grasslands, and red tussock grows in wet openings in beech forest.

Mt Stokes in Sounds ED is capped at 1200 m by snow tussock and herbfield that botanically link the Richmond Range and the Tararua Range across Cook Strait. The main communities are: (1) rocky, well-drained areas with *Chionochloa pallens* dominant, merging into scrub of *Olearia colensoi*; (2) poorly drained areas with *Carpha alpina*, *Poa colensoi*, and hybrids between *C. pallens* and the North Island species *C. flavescens*; and (3) *Donatia–Phyllachne–Oreobolus pectinatus* bog (Park 1968).

Tararua and Rimutaka Ranges

The grasslands on the crest of the Tararua Range are dominated almost entirely by *Chionochloa pallens* and *C. flavescens*. *C. pallens* grows mainly at the higher altitudes, on soils that are recent or moving on steep slopes. It tends to co-dominate with *Astelia nervosa* and *Aciphylla colensoi*, with *Coprosma depressa*, *Pratia angulata*, *Bulbinella gibbsii*, *Uncinia caespitosa*, *Polystichum vestitum* and *Blechnum* 'mountain' as characteristic species. *C. flavescens* stands share *Astelia nervosa* and *Uncinia caespitosa*, but the other associates, especially *Carpha*, *Astelia linearis* and *Pentachondra*, reflect wetter, less fertile, more humic soils, and there are gradations to wet heath (Williams 1975). Intermediate sites support mixed stands, but on the whole *C. pallens* represents an earlier successional stage that can lead either to *C. flavescens* grassland or to *Olearia colensoi* and other shrubs.

In the southern part of the Tararua Range, where subalpine beech forest is nearly continuous, snow-tussock grassland is generally penalpine, but in the northern part where subalpine beech is lacking, *C. flavescens* and associated wet-heath species displace scrub from broad ridges to as low as 760 m; similar vegetation occupies the gentle summit of Mt Clinie in the Rimutaka Range to the south, at 820 m. In these ranges, *C. flavescens* fills the niche that usually belongs to red tussock, but on outlying Mt Kaiparoro hybrid chionochloas occupy the broad summit at 730–800 m, approaching red tussock on wet ground and *C. flavescens* on better-drained sites (Druce 1957a).

Northern axial ranges

Excepting *C. conspicua* communities that have replaced subalpine forest (p. 581), only two *Chionochloa* species form the tall-tussock grasslands of the axial ranges between the Manawatu Gorge and East Cape. On well-drained slopes *C. pallens* grasslands extend from the forest limit to the summits except where destroyed by fire, grazing or erosion, or on broken terrain where they merge into herbfield or low scrub. Associated plants are essentially as in the Tararua Range, except that *Dracophyllum recurvum* replaces *D. uniflorum*. From 1500 m downwards, red tussock occupies poorly drained ground, especially the lower terraces of broad, tephra-blanketed valley floors in the ranges bordering the Volcanic Plateau. It gives way to silver tussock grassland on dry upper terraces, and to *Phyllocladus alpinus*, *Hebe parviflora* and bog pine along stream margins (Elder 1962).

The high volcanoes

The flanks of the central volcanoes support red-tussock grasslands, which until recently were continuous with those on the axial ranges. The communities have been classified and mapped by Atkinson (1981); Scott (1977) has also classified communities above tree-limit, and defined their ecological parameters (Fig. 9.18). The raw andesitic soils of the steep upper slopes, and recent tephra that mantles extensive surfaces at lower altitudes, support short, patchy vegetation. Continuous vegetation is confined to gentle slopes derived from lava flows and lahars, and has upper limits between 1550 m on western slopes and as low as 1050 m in the south-east where tephra is deepest.

From about 1200 m down to 800 m (and even lower before conversion to pasture) red-tussock communities are seral towards forest or tree-heath; the proportions of red tussock are greatest after fire, and subsequently manuka, *Dracophyllum longifolium* or *Olearia nummulariifolia* increase. This pattern has changed where heather (*Calluna vulgaris*) has invaded, as this adventive captures burnt surfaces more rapidly than either red tussock or native shrubs. From 1200 m to the limit of continuous vegetation, especially above the limit of beech forest and tree-heath at 1500 m, red tussock appears to represent the furthest stage of development in vegetation. Dense stands are confined to deep, better-drained soils, whereas on wet ground the grassland grades into wet-heath or flush communities (p. 349). The grassland, shrub-heaths and open areas are invaded by *Pinus contorta*, which will continually threaten the native vegetation unless the seed-sources are removed.

On Mt Taranaki red tussock dominates a 500 m wide band between 1400 and 1600 m, in association with silver tussock. *Hebe odora* and other shrubs enter at lower altitudes where grassland merges into scrub. Forbs are important, especially where grassland merges into herbfield at higher altitudes or with poorer drainage, and include *Ranunculus nivicola*, *Coriaria plumosa*, *Geranium microphyllum*, *Anisotome aromatica*, *Coprosma depressa*, *Celmisia major* var. *brevis*, *Helichrysum* 'alpinum' and *Ourisia macrophylla*.

On the summit of neighbouring Mt Pouakai, a much older, subdued volcanic remnant, poorly-drained red-tussock grassland around 1300 m includes *Schoenus pauciflorus*, *Bulbinella hookeri* and *Coprosma* '*parviflora* var. *dumosa*', which do not occur on Mt Taranaki. Red tussock also dominates in the Ahukawakawa Swamp, at 920 m between these mountains, and is accompanied by *Carex coriacea*, *C. echinata* and *Bulbinella hookeri*, with small areas being dominated by *Schoenus pauciflorus*, *Oreobolus pectinatus* or sphagnum (Clarkson 1986).

Short-tussock and related grasslands of inland districts

History

Short-tussock grasslands grow under annual rainfalls ranging from 350 mm to >10000 mm, but are mostly in subhumid districts formerly supporting forest that was burnt by early Maori (p. 84). Probably, the only significant areas of temperate grasslands existing a thousand years ago were in intermontane basins in Central Otago, Mackenzie and Marlborough under annual rainfalls less than 500 mm, extending on to north-west aspects receiving as much as 800 mm, and on frosty valley floors, especially on recent soils. On the Volcanic Plateau, high, frosty plains blanketed by recent tephra may have supported primitive grassland, but Maori fires at least helped to maintain and extend this.

Early European pastoralists and explorers, therefore, encountered grasslands that had been greatly extended by Maori activity. Their descriptions are brief and tantalising. We can assume that the vegetation was usually dense, but perhaps not conspicuously tussocky because of the abundance of shrubs, forbs and other grasses. There would have been deep litter, except for a few years after fire. Large *Chionochloa* tussocks were widely dominant on the older, more acid soils, to lower altitudes than today. Silver tussock prevailed on lower-altitude terrain near the coast, extending inland and upwards on recent, coarse-textured soils. Curiously, hard tussock, the most important tussock of inland short-tussock grasslands, is scarcely mentioned in early accounts, although it must have been common on dry, fine-textured soils. The now-scarce, tussock-forming blue wheatgrass *Elymus apricus* dominated in the driest valleys of Central Otago.

Records are vivid about the abundance of large 'spaniards' or 'speargrasses' (aciphyllas) that impeded the progress of people and horses. They benefit from disturbance, the breaks in the grass cover being provided by fires that burned hot enough on accumulated litter to kill tussocks, by larvae which still destroy large patches of tussock, and through scratching by weka (*Gallirallus*), flightless rails that were then abundant in the grasslands. Other herbs, such as 'aniseed' (*Gingidia montana*), gentians and *Celmisia gracilenta*, also would have colonised bared areas. The poisonous tutu (*Coriaria sarmentosa*), a rhizomatous, summer-green, semi-woody herb forming metre-tall colonies on moist, well-drained areas, was a hazard to stock.

Sedge and *Phormium* swamps occupied the lower parts of valleys, and lines of toetoe (*Cortaderia richardii*) followed sandy stream banks. On recent stony flats and fans, matagouri grew as thickets up to 4 m tall, with scattered bushes persisting into mature grassland, and was usually accompanied by native brooms (*Carmichaelia* spp.). Broad, low clumps of *Melicytus alpinus* were frequent on dry, stony ground. Steep, rocky slopes and stream banks supported grey scrub, and on less frosty lower slopes tussock grassland merged into bracken (*Pteridium esculentum*) fernland. Cabbage trees grew in milder districts, and except in the driest districts there were remnants of bush or forest in gullies and other sheltered places.

On taking up runs, sheep farmers burnt the grasslands and further extended them at the expense of scrub and forest. This killed *Aciphylla* and much of the matagouri and exposed fresh herbage to stock, but eventually it weakened or killed the tussocks. By 1878, sheep had increased to a number – 9.5 million – that native grasslands have never carried since (O'Connor 1982). Palatable herbs and grasses were eaten out, the first casualty being *Gingidia montana*. Snow tussocks gave way to hard and blue (*Poa colensoi*) tussocks, and the latter were depleted in turn. Conditions were perfect for rabbits, which were foolishly introduced in the 1870s and quickly became a plague. They even crossed alpine passes to colonise river-flats in South Westland, which have never been reached by sheep.

It was soon realised that the 'English' grasses and clovers were more productive and

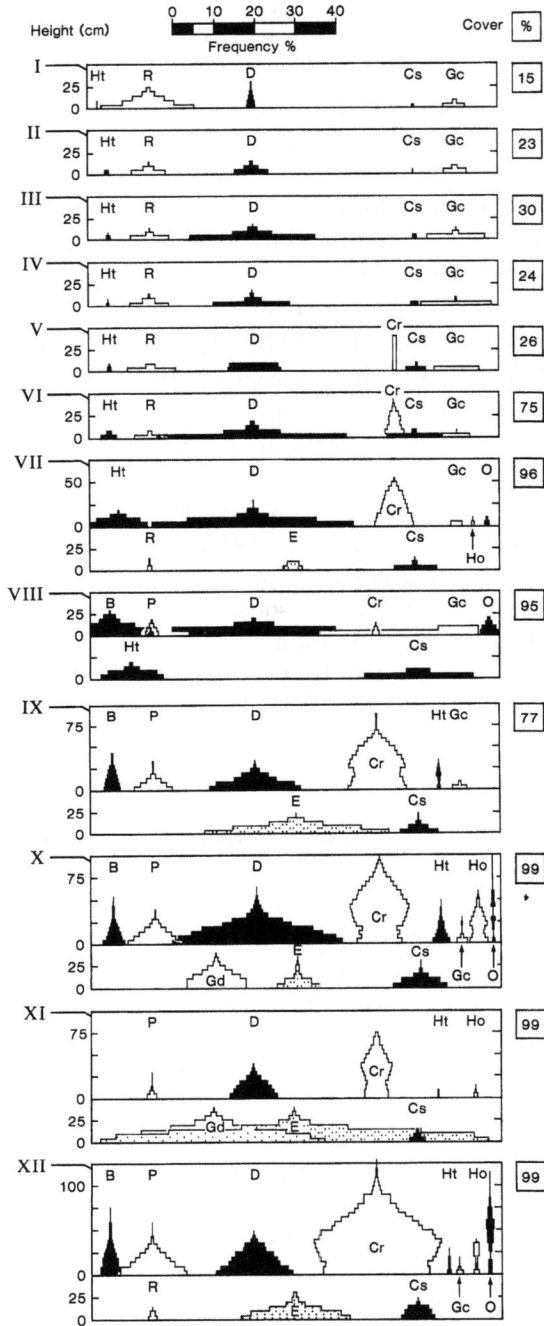

generally more tolerant of grazing than native grasses, and they have been over-sown ever since. Establishment and spread of adventives, both wanted and unwanted, has been facilitated by depletion of the native cover.

By the 1950s, something of an equilibrium had been reached. In the driest districts, tussock grassland on alluvial terraces and sunny slopes had been replaced by open communities of low-growing perennial dicots (mainly native), and vernal annuals and tall biennials (mainly adventive). Terraces, flats and colluvial slopes with deeper, moister, finer-textured soils tended to be occupied by adventive grasses, especially in valleys with moderate to high rainfall. Shrubs, both native and adventive, were often abundant on steeper slopes, especially of shady aspect, though the more palatable were kept in check through browsing of seedlings. Short tussocks continued to prevail on mid-slopes and older, more leached terraces, ascending highest on sunny aspects before they yielded dominance to snow tussocks, and descending lowest on shady aspects before they yielded to vegetation induced through overgrazing.

Recently, the pace of change has quickened. Grazing has diminished over large areas because a measure of control over rabbits and feral deer has been achieved, and some remote and difficult country has been de-stocked. At the same time, there is more intensive use of accessible lower terrain, involving aerial top-dressing with fertilisers, sowing of high-producing pasture species on arable land, closer fencing, and greater use of cattle, domesticated deer and goats. Fertility transfer by livestock (p. 278) has widened the impacts, which have led to increase of adventive species, especially grasses and clovers, at the expense of native species. Where grazing has been reduced, the most vigorous response has generally been by adventive plants hitherto kept in check; notably briar, grasses such as browntop (*Agrostis capillaris*) and sweet vernal (*Anthoxanthum odoratum*) and, often to a spectacular degree, hawkweeds (*Hieracium*). Pines and other planted conifers are also spreading. Over wide areas native species, including tussock grasses, have become less prominent than when the grasslands were overgrazed.

The main grasses, sedges and rushes

Hard tussock is a light brown, scabrid-leaved grass that dominates extensively on older terraces in inland valleys and basins, but where drainage is impeded red tussock

Fig. 9.18. Structure of fell-field (I–V), shrub-heath (VI–VIII), wet heath (XI), red-tussock grassland (XII) and transitional communities on Ruapehu at 1290–1610 m; major species only. Method as in Fig. 9.13 (100 sampling positions) (Scott 1977). Abbreviations: B, *Brachyglottis bidwillii*; Cr, *Chionochloa rubra*; Cs, *Celmisia spectabilis*; D, *Dracrophyllum recurvum*; E, *Empodisma minus*; Gc, *Gaultheria colensoi*; Gd, *Gleichenia dicarpa*; Ho, *Hebe odora*; Ht, *Hebe tetragona*; O, *Olearia nummulariifolia*; P, *Podocarpus nivalis*; R, *Rytidosperma setifolium*.
Other frequent species (no. shows community where maximum cumulative frequency was recorded) are as follows.
VI: *Pentachondra pumila*, *Oreobolus pectinatus*, *Racomitrium lanuginosum*; VII: *Epacris alpina*, *Wahlenbergia pygmaea*, *Poa colensoi*; VIII: *Leptinella pusilla*, *Euphrasia cuneata*, *Ourisia vulcanica*, *Anisotome aromatica*; IX: *Epacris alpina*, *Celmisia incana*; X: *Lepidothamnus laxifolius*, *Leptospermum scoparium*; XI: *Phyllocladus alpinus*, *Cyathodes empetrifolia*, *Schoenus pauciflorus*; XII: *Coprosma cheesemanii*.

often occurs instead. Although hard tussock may owe its prominence to elimination of snow tussocks it is itself vulnerable to grazing and infestations of grass grubs (p. 278); with modest enhancement of fertility it is out-competed by silver tussock and introduced grasses; and recent years have seen hawkweeds oust it over large areas.

Silver tussock ranges from the sea coast in central and southern New Zealand to the penalpine belt. It has finer roots than hard tussock (p. 50) which it replaces on coarser-textured, friable, moderately fertile soils, such as those of recent terraces. In both semi-arid and wet climates, silver tussock tends to dominate in hollows, probably because nutrients are flushed into them. It continues to thrive where competition from adventive grasses has eliminated most other native plants.

The *Poa colensoi* complex comprises plants with filiform leaves coated with bluish wax, the thickness of which correlates with dryness of the habitat (Daly 1964). It is almost universal in inland and mountain grasslands, with forms ranging from short, diffuse tufts at high altitudes to 20 cm tall plants of intermontane basins in the Mackenzie and Central Otago ERs.

Tall, laxly tufted native grasses fill the spaces between tussocks only in localities that escape regular grazing; elsewhere they are largely confined to inaccessible places, such as within tussocks and beneath sprawling shrubs of *Melicytus alpinus*. *Elymus rectisetus* (mainly erect-culmed, outcrossing, native forms) and *Dichelachne crinita* are the most characteristic. Moist sites such as occur on stream banks, slips, forest margins and in scrub patches may contain other *Elymus* spp., *Trisetum antarcticum* or *T. youngii*. Shorter grasses include species of *Deschampsia* (on damp ground), *Agrostis*, *Deyeuxia*, *Koeleria*, *Lachnagrostis* and danthonia (*Rytidosperma*). The last have increased in response to depletion of taller grasses, with the fine-leaved *R. thomsonii* and the low-growing *R. pumilum* being especially characteristic of inland grasslands.

Most of the cool-temperate introduced grasses have become established in inland regions, sweet vernal being almost universal. Browntop, existing as several ecotypes, has displaced native grasses from both moderately deep, moist, fertile soils and dry terraces. *Poa pratensis* swards are mainly on sites with enhanced fertility, such as heavily grazed flats and within matagouri thickets. Cocksfoot (*Dactylis glomerata*), *Festuca* spp. and *Holcus lanatus* are locally abundant. On dry, depleted sites, annual species of *Bromus*, *Aira* and *Vulpia* can dominate in spring, leaving the ground bare by late summer. *Hordeum murinum* is frequent where stock concentrate.

Small carices are universal and being either unpalatable or tolerant of grazing, can persist where perennial grasses have been eliminated. *Luzula rufa* is present generally, whereas true rushes are mainly on stream banks and flushes, common species being *Juncus gregiflorus*, *J. distegus* and *J. effusus*.

Dicot and monocot forbs

Aciphyllas are still impressive in some areas. Below 600 m, the prevailing species is *A. subflabellata*, which has slender, glaucous pinnae. *Celmisia* is widely represented only by *C. gracilenta*, which grows as tufts of linear leaves up to 5–10 cm long. *Aciphylla aurea* and larger celmisias, especially *C. spectabilis*, enter at higher altitudes where snow tussocks may once have dominated. The commonest gentian is the biennial *G. corymbifera*, which has scapes up to 40 cm tall bearing masses of white flowers. In some areas *Bulbinella* is abundant, and in Central Otago there can be a band of *B.*

angustifolia between depleted semi-desert on the lowest slopes and relatively intact tussock grasslands higher up.

Raoulia subsericea is important in fairly open tussock grasslands. Other mat and cushion plants grow mainly on bare ground, such as stony rises and recent gravel flats (p. 367). One of the commonest small forbs is the rhizomatous harebell *Wahlenbergia albomarginata*, which has flowers in shades of pale blue. The laxly branching *W. gracilis* replaces it at low altitudes. The white-flowered *Viola cunninghamii*, yellow buttercups of the *Ranunculus multiscapus* complex, rosette species of *Brachyglottis*, and species of *Geranium, Epilobium, Craspedia* and native daisy (*Brachyscome* and *Lagenifera*) are almost universal.

Two characteristic, although seldom abundant, crucifers of inland grasslands are the monotypic *Ischnocarpus novae-zelandiae*, which grows mainly among rocks and shrubs, and the deeply tap-rooted *Lepidium sisymbrioides*, which grows on very dry, often stony ground from Puketeraki ER to Central Otago. There is a wide range of dwarf shrubs; species additional to those listed in Table 9.5 include *Pimelea prostrata, Carmichaelia enysii, Pernettya nana, Coprosma atropurpurea* and *Hebe pimeleoides*. Small bulbous orchids include *Prasophyllum colensoi, Microtis unifolia, Thelymitra longifolia* and several species of *Pterostylis*.

Piripiri or bidibids, notorious for their spherical burrs, are the main native economic weeds, the widespread *Acaena anserinifolia* being usually less common than the upland, more pilose *A. caesiiglauca*. In depleted areas the Australian *A. agnipila*, which has achenes on elongated spikes, is common and hybridises with native species.

Two native flatweeds, *Taraxacum magellanicum* and the monotypic *Kirkianella novae-zelandiae*, have become vastly outnumbered by adventive relatives. Catsear (*Hypochoeris radicata*) became ubiquitous during the first decades of pastoral occupation, and the annual *Crepis capillaris* is almost as common. Hawkweeds (*Hieracium*) may have arrived as late as the 1940s, but since the reduction of rabbits have become dominant over wide areas. *H. pilosella*, which forms dense patches of small, flat, stoloniferous rosettes, is a serious weed that can provide up to 70% of the plant cover, or cover more than half of the total surface (Fig. 9.19). In depleted grassland it aggressively invades and supplants established species, including fescue tussocks. Within patches, reproductive and vegetative phases compete, and the best way to reduce the vigour of the plants may be to allow them to flower freely (Scott 1984). *H. praealtum* and *H. caespitosum* develop taller patches of smaller diameter. Both increase dramatically in density and vigour when stock are removed, and in Mackenzie ER the latter forms heavy infestations to at least 760 m. The single-stemmed *H. lepidulum* also establishes abundantly on moister sites, but being slender and short-lived does not suppress resident vegetation.

Sorrel spread early and rapidly, over bare, loose ground, providing useful forage. Alsike (*Trifolium hybridum*) and white clover (*T. repens*) are well established on sheltered sites through over-sowing.

Cryptogams

The commonest grassland moss is *Hypnum cupressiforme*, but others can be abundant, especially *Breutelia pendula* and *B. affinis* on shady slopes, *Polytrichum juniperinum* on bare ground in the open, and *Thuidium furfurosum* at margins of forest and scrub.

Table 9.5. *Composition of short-tussock grasslands of inland districts in the South Island*

Community:	1	2	3	4	5	6	7	8	9	10	11	12	13	14
No. of samples:	6	9	7	10	7	4	3	4	4	3	9	5	12	10
Mean altitude (m):	300	470	550	590	680	900	590	550	480	680	720	1048	1054	1038
Prevailing aspect:	NE	NE,S	SE-S	SW-NW	N	W-N	W-NW	W-NW	0	0	0	SW	NW	SE
Mean slope (deg.):	24	20	5	9	31	30	3	13	0	4	4	26	17	16
Mean percentage cover:	100	100	97	98	91	93	87	86	94	92	98	86	82	81
Mean no. of vascular spp.:	?	?	46	50	50	40	23	29	29	35	43	?	31	29
Tall tussocks														
Chionochloa rubra				f										
Chionochloa 'robust'						f							+*	+*
Chionochloa rigida												f		
Short tussocks														
Poa cita	1	2	+	+	f				3	3	3	f	f	+
Festuca novae-zelandiae	3	3	3	3	3	3	3	3	3	3	3	4	3	3
Poa colensoi	2	1	1	1	f	3	2	1	2	1	1	2	1	2
Other native grasses														
Elymus rectisetus	1	1	f	f	f	f	f	1	f	1	1	f	f	f
Dichelachne crinita	f	f	+	f	f	f	+	+	+	f	f	+	f	f
Rytidosperma spp.	f	f	+	+	1	+		+					f	+
Deyeuxia avenoides		+										+	+	+
Pyrrhanthera exigua			f	+	f	f				+	f	f		
Rytidosperma pumilum			f	f	f			+	1	1	f	f	f	+
Lachnagrostis spp.			f	f	f	f					f	+	f	1
Koeleria spp.				+	+		+			+	1	f		+
Agrostis muscosa							f		+					
Poa maniototo							1	+		+				
Poa lindsayi									+			+	+	+
Adventive grasses														
Vulpia bromoides	+	f			+		f	1	1	f	+			
Dactylis glomerata	1	f						+					+	
Anthoxanthum odoratum	3	2	1	1	2	1		+	+			+	+	+

Species													
Agrostis capillaris	2	+	1	+	f	+	1	f	1	f	f	+	+
Holcus lanatus	f	+	2	+	f	f	f	+	f	+	+	1	1
Aira caryophyllea		f	f	+	f	f	f	3	2	1	f	+	
Bromus hordeaceus			+				+	1	1	2			
Bromus tectorum						f		2	1				1
Poa pratensis								2	f			+	
Sedges and rushes													
Carex breviculmis	f	f	f	f	f	+	f	f	f	f	f	f	f
Luzula*		+	f	f	f		+	f	f	f	+	f	f
Schoenus pauciflorus			+	+	f				f	f			
Carex colensoi			+	+	+	f	+	+	+	+	+	+	+
Carex wakatipu			+		+	+	+	+	+	f	+	+	
Carex resectans						+		+	+				
Tall shrubs													
Discaria toumatou	3		1	1	1		+		+	+	+	+	+
Carmichaelia*			+	+	+	+		+	+	+	+	+	+
Melicytus alpinus			+	f	f	f		+	f	+	+	+	+
Muehlenbeckia complexa			+	+	+	+		+	+	+	+	+	+
Rosa rubiginosa					+		+		+	+	+	+	+
Dwarf shrubs													
Cyathodes fraseri	f	f	1	f	f	f		f	+	f	f	f	f
Pimelea (1) oreophila	+*	+*	+	f	f	f		f	+	+	f	f*	+*
Pimelea (2) pulvinaris			+*	+	+	+		+	+	+	+	+	+
Gaultheria depressa v. *nov-z.*			+	+	+		+		+	+	+		
Coprosma petriei			+	1	1	3		1	f	+	1		
Muehlenbeckia axillaris			+	f	+	+		+	+	f	+		+
Cyathodes colensoi			2	+	+	+		+	+	+	+	+	+
Carmichaelia monroi			+	+	+	+		+	+	f	+	+	+
Large native forbs													
Celmisia spectabilis			+	+	+	1	+		+	+	+	+*	+*
Aciphylla aurea			+	f	f	f	f		+	+	+	+*	+

Table 9.5. (cont.)

	1	2	3	4	5	6	7	8	9	10	11	12	13	14
Community:														
No. of samples:	6	9	7	10	7	4	3	4	4	3	9	5	12	10
Mean altitude (m):	300	470	550	590	680	900	590	550	480	680	720	1048	1054	1038
Prevailing aspect:	NE	NE,S	SE-S	SW-NW	N	W-N	W-NW	W-NW	0	0	0	SW	NW	SE
Mean slope (deg.):	24	20	5	9	31	30	3	13	0	4	4	26	17	16
Mean percentage cover:	100	100	97	98	91	93	87	86	94	92	98	86	82	81
Mean no. of vascular spp.:	?	?	46	50	50	40	23	29	29	35	43	?	31	29
Cushion, mat and patch-forming herbs														
Nertera setulosa	+	f	+	f										
Dichondra repens	f	+	f								+	+		
Hydrocotyle novae-zeelandiae	+	f	f	f	f	f					f	f		
Helichrysum filicaule	+	f	f	f	f	f					f	f		
Oxalis exilis	f	f			+	+				+	+	+	+	+
Scleranthus uniflorus	f*	+*	f	f	f	f				f	f	f	f	f
Gnaphalium 'collinum'	f	+	f	f	1	f		f		1	f	f	f	f
Raoulia subsericea		+	1	1	f	1	+	f	f	+	1	f	+	f
Raoulia monroi			+		f									
Mentha cunninghamii		+	+	f		f								
Stackhousia minima			+		+									
Leptinella pusilla			f	+					f*	+	f	+	+*	
Raoulia hookeri			f	+	+		1	1	1	f	f	+	f	1
Brachyscome sinclairii			1	+	f						f	+		
Colobanthus spp.			+	+	+	+	+			+	+	f	+	+
Lagenifera spp.				+		+					+			
Convolvulus verecundus							1	+						
Acaena microphylla							f*	+		f	+	+	+	
Raoulia australis										f	+	+	+	+
Raoulia apicinigra										+	+	+	+	
Other native dicot herbs														
Hypericum gramineum	+	+												
Acaena anserinifolia	+	f								f	+			
Wahlenbergia gracilis	+	+								+	+			

Species													
Plantago spp.	+	f	f	+	f	f	1	1	+	f	f	f	f
Geranium sessiliflorum	f	f	f	f	f	f	f	f	+	f	f	f	f
Epilobium glabellum	f	f	f	f	+	+	f	+	+	f	f	f	f
Geranium microphyllum	+	f	+		+	+	+	+	f	+	+	+	+
Vittadinia australis	f	f	f	f	f	f	+	+	f	+	f	f	f
Acaena caesiiglauca	+	+	+	f	f	f	+	+	1	f	f	1	1
Ranunculus multiscapus	+		f	f	f	+	+	+	+	+	+	+	+
Viola cunninghamii			f	f	+	f	f	f	+	+	+	f	f
Epilobium alsinoides			+	f	+	f	+	1	1	+	+	f	f
Wahlenbergia albomarginata			f	f	f	f	f	f	f	f	f	f	f
Celmisia gracilenta	+	+	f	f	f	+	+	+	f	+	+	f	f
Oreomyrrhis spp.	+	+	f	f	+		+	f	+				
Euphrasia zelandica				+	+	+							
Gentiana spp.	f	f	f	f	+		+	+	+	+	+	+	+
Brachyglottis haastii	f	+	f	f	f		+	+	f	+	+	+	+
Craspedia spp.	+		+	f	f		+	+		+	+	+	+
Stellaria gracilenta			+	+		+	+		f	+	+	f	f
Anisotome filifolia					+								
Epilobium rostratum					f				+	+		f	
Microseris scapigera			+	+	f	f	f	+	+		f	f	f
Epilobium hectorii			+	+	+		+	+	+		+	f	f
Native monocot forbs													
Prasophyllum colensoi	+		f	f	+	+	+	+	+				
Thelymitra longifolia			+	+	f	f			f		+	f	f
Bulbinella angustifolia			+	f					+		+	f	
Adventive flat-weeds													
Hieracium pilosella	+	+	f	f	+	f	f	f	f	+	f	f	f
Crepis capillaris	f	f	+	+	+	1	+	1	f	f	f	f	f
Hypochoeris radicata	1	f	f	f	f	1	1	1	1	f	1	f	f
Hieracium lepidulum		f	+	+	+	+	+	1		1	f	f	
Hieracium praealtum			+	+	f	f	f	f	f	+	f	f	f
Hypochoeris glabra			+	+	+	+	+						+

Table 9.5. (cont.)

	1	2	3	4	5	6	7	8	9	10	11	12	13	14
Community:	1	2	3	4	5	6	7	8	9	10	11	12	13	14
No. of samples:	6	9	7	10	7	4	3	4	4	3	9	5	12	10
Mean altitude (m):	300	470	550	590	680	900	590	550	480	680	720	1048	1054	1038
Prevailing aspect:	NE	NE,S	SE-S	SW-NW	N	W-N	W-NW	W-NW	0	0	0	SW	NW	SE
Mean slope (deg.):	24	20	5	9	31	30	3	13	0	0	4	26	17	16
Mean percentage cover:	100	100	97	98	91	93	87	86	94	92	98	86	82	81
Mean no. of vascular spp.:	?	?	46	50	50	40	23	29	29	35	43	?	31	29
Tall adventive forbs														
Cirsium vulgare	f	+					1	f		+	+		+	+
Verbascum thapsus	+	+	+							+		+	+	+
Other adventive dicots														
Prunella vulgaris	f	f	f		f						f	f		
Linum catharticum	f	f	f	f	f	f					f			
Rumex acetosella	f	f	f	+	f	f	2	1	2	2	1	f	f	f
Cerastium spp.	+			f	f		+	+	f	f	+	+	+	+
Arenaria serpyllifolia			+	+				+		+	+	+	+	+
Dianthus armeria						+								
Aphanes arvensis							+	+		+				
Gypsophila australis							1	f			+			
Clovers														
Trifolium dubium	f	f	+	f	f	+		f			f	+	+	
Trifolium repens	f	f	+	f	f	+		+			f	+	1	+
Trifolium arvense					f			f						
Trifolium hybridum							+	+	+					
Ferns														
Pteridium esculentum	f	+	+	+	f	+								
Ophioglossum coriaceum		+	+	+							+	+		

Notes:
Table derived as Table 9.4.

Localities and communities:

1–2 Hunters Hills, Pareora ER; 1 occupies sunnier, drier slopes than 2 (Barker 1953).

3–6 Rakaia Valley, Puketeraki ER (Connor 1965).

3 Moraine downs and outwash terraces; dry facies.

4 Moraine downs and terraces; derived from *Chionochloa rubra* grassland.

5 Mountain slopes of sunny aspect.

6 Mountain slopes higher than 5; derived from *Chionochloa* 'robust' grassland.

7–12 Mackenzie ER (Connor 1964).

7 Flats or gently sloping fans exposed to north-west winds, and somewhat depleted.

8 Flat to moderate slopes facing north-west with introduced annual grasses prominent.

9 Flats with deeper soil than 7.

10 Exposed sites or thin stony soils.

11 Moraine downs and outwash terraces; the best developed phase of hard-tussock grassland.

12 Higher-altitude community derived from *Chionochloa rigida* grasslands.

13–14 Molesworth ER, sunny and shaded aspects respectively (Moore 1976).

* *Chionochloa* sp. not stated in columns 13–14; *Luzula* probably *L. rufa*; *Carmichaelia petriei* or *C. robusta*; *Pimelea* (1) is *P. prostrata* in columns 1–2, *P. concinna* in columns 13–14; *Pimelea* (2) is *P. sericeovillosa* in column 3; *Aciphylla* sp. not stated in columns 13–14; *Scleranthus* sp. uncertain in columns 1–2; *Leptinella pectinata* in column 9, *L. squalida* in column 13; *Acaena buchananii* in column 7.

Fig. 9.19. *Hieracium pilosella* mat in short-tussock grassland at 630 m, near Mavora Lakes, Mavora ER.

Cladonia spp. are always present on firm, bare soil, and *Cladia retipora* and *Cladonia* spp. can cover stony soils, which are probably leached at the surface (Fig. 9.20).

Short-tussock and related grasslands of southern Canterbury

The short-tussock grasslands of southern Canterbury, as they were in the 1950s and 1960s, and as they still are over diminishing areas, have been portrayed by sociological analysis (Table 9.5). The localities are the seaward flanks of the Hunters Hills, which are subject to maritime influences, especially a steep altitudinal precipitation gradient and frequent cloud on the crests; the middle reaches of the Rakaia valley, where nor'westers alternately bring föhn conditions and heavy rain, and which is also open to southerly and easterly weather crossing the Plains; and the Mackenzie intermontane basin, which experiences low humidity and rainfall and extremes of temperature.

Hard tussock dominates all the vegetation types illustrated. Matagouri is present at all localities, whereas tall native brooms were not listed from the Hunters Hills although they occur there. The presence of *Chionochloa* tussocks in two of the Rakaia communities reflects the intergrading of short- and tall-tussock grasslands where rainfall and altitude are higher. Silver tussock is important on the Hunters Hills, whereas on the Mackenzie plains it is largely confined to recent soils and was rarely listed. The distribution of bracken is similar.

The Hunters Hills provide a link with highly modified grasslands nearer the coast,

Fig. 9.20. Lichens in grassland on a stony terrace near Lake Rotoiti (Spenser ER): (a) *Cladia retipora*; (b) *Cladina leptoclada*.

with perennial adventive grasses and the coarser danthonias being prominent. In contrast, lower-tier grasses of the Mackenzie stands reflect the dry climate and more depleted condition; fine-leaved danthonias, *Pyrrhanthera exigua*, *Agrostis muscosa*, *Poa lindsayi*, *Poa maniototo* and adventive annual grasses are characteristic.

Likewise, some of the native dicots are practically ubiquitous, such as *Cyathodes fraseri*, *Geranium sessiliflorum* and *Gnaphalium 'collinum'*, whereas others were recorded from particular localities or habitats, e.g. *Geranium microphyllum* and *Acaena anserinifolia* in the Hunters Hills, *Cyathodes colensoi* and *Mentha cunninghamii* in the Rakaia, *Celmisia spectabilis* and *Brachyscome sinclairii* at higher altitudes, and *Raoulia australis* and *Convolvulus verecundus* in depleted habitats in Mackenzie ER. Among adventive forbs *Crepis capillaris*, catsear, *Hieracium pilosella*, sorrel and

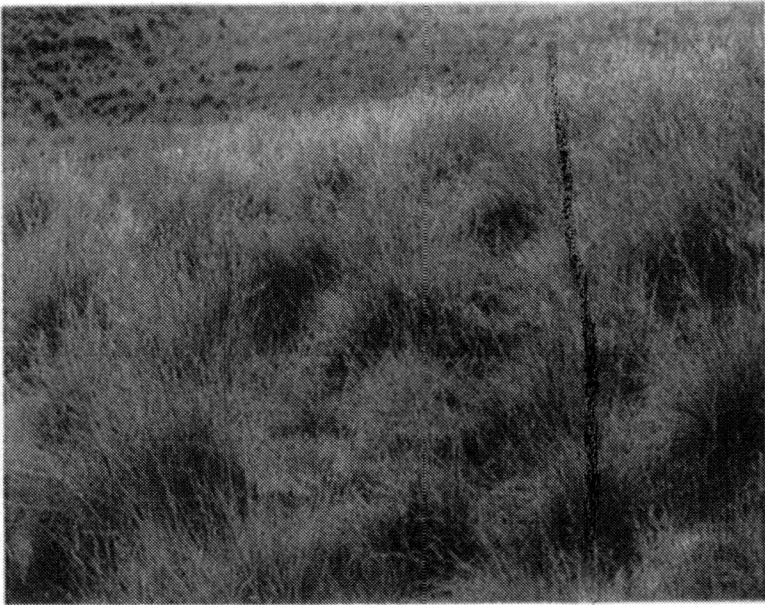

Fig. 9.21. *Festuca novae-zelandiae/Poa colensoi* grassland; 1000 m, Mt Difficulty, Central Otago.

Cerastium spp. are practically ubiquitous, the vernal *Gypsophila australis* and *Aphanes arvensis* are most characteristic of the Mackenzie grasslands, and clovers are evident in the two moister localities.

Short-tussock and related grasslands in other regions

Hard-tussock grasslands show considerable uniformity throughout the eastern South Island (Fig. 9.21); for inland Marlborough this is illustrated by columns 12 and 13 in Table 9.5.

Silver-tussock grassland is best developed at low altitudes on steep, stony colluvial slopes of northerly aspect, and tends to merge into grey scrub, matagouri or bracken. Scattered silver tussocks often grow in vegetation depleted of native plants but strongly invaded by adventives.

The *Elymus apricus* tussock grassland of the Central Otago valleys is now virtually extinct. Grassland dominated by a large form of blue tussock on terraces in Central Otago and Waitaki ERs seems to have developed on stony or deflated soils where droughts, possibly exacerbated by grass grubs, have eliminated hard tussock, which, however, may persist in the lee of rises.

On steep slopes in Marlborough, where comparatively low rainfall, eroding greywacke, burning and grazing preclude continuous grass cover, snow and fescue tussocks and even beech forest have been replaced by the short, yellow-green bristle tussock (*Rytidosperma setifolium*), forming a cover that is often so open as to pass as fellfield (Williams 1989). This extends from 850 m to 1800 m, especially on warm aspects. Depending on altitude and degree of modification, bristle tussock can share dominance with browntop, *Elymus rectisetus, Celmisia monroi, C. spectabilis, Festuca matthewsii* or *Chionochloa pallens*. Less extensive bristle-tussock communities occur on other eastern mountains of the South Island.

At high altitudes or even as low as 550 m on steep, shaded aspects and in cool mountain valleys, hard tussock yields to the smooth-leaved *Festuca matthewsii* co-dominating with *Poa colensoi*; most of this community results from depletion of *Chionochloa* (especially *C. macra*) grassland. Near the upper limits of continuous vegetation on eastern mountains, there are *Celmisia viscosa/Poa colensoi* herbfields. In Central Otago, where *Chionochloa oreophila* is absent, *P. colensoi* is the sole dominant grass in late-snow hollows. Similar *P. colensoi* swards occupy alpine sink-holes in Arthur ED (Western Nelson).

Towards the Main Divide, as altitude rises and annual rainfall increases beyond 2500 mm, slopes of the major valleys become clad in forest or, where this has been destroyed, in bracken, grass and shrubs. The combination of föhn winds and coarse-textured soils allows short-tussock grassland and communities derived from it to penetrate even where annual rainfall exceeds 4000 mm, especially on recent alluvial and colluvial surfaces with sunny aspect. On cooler sites, these grade into thick swards of grasses such as *Poa cockayneana* (p. 30), *Festuca matthewsii, Elymus narduroides, E. tenuis* and *Trisetum antarcticum*. These are seral to snow-tussock grassland or woody vegetation; that some swards have arisen from depletion of taller vegetation is shown, for instance, by relict *Chionochloa* tussocks. An abundance of adventives such as white clover and *Poa pratensis* indicates fertile soil. Where avalanches preclude closed grassland or herbfield in Mt Cook National Park, Wilson (1976) describes open communities with *Rytidosperma setifolium, Poa colensoi, Epilobium pycnostachyum, E. glabellum, Wahlenbergia albomarginata, Helichrysum bellidioides, Blechnum penna-marina* and, at higher altitudes, *Agrostis subulata, Poa novae-zelandiae* and *Epilobium porphyrium*. The shrub *Olearia moschata* occurs locally.

West of the Main Divide, bristle tussock succeeds raoulias on alluvium and fine moraine above 900 m, and pioneers on slips and fractured rock to as high as 1550 m. Associates include *Poa colensoi, Microlaena colensoi* and *Helichrysum bellidioides*, which are also important in grassland modified by grazing (p. 230), and species characteristic of fellfield. On descending the western valleys, seral high-altitude grasslands grade into grassland on recent alluvial flats (p. 264).

Surviving remnants of short-tussock grassland on the tephra plains of the Volcanic Plateau form mosaics with *Dracophyllum subulatum*. Grasses prevail on finer textured topsoils, and for a time after fire (p. 193). Silver tussock generally dominates but hard tussock occurs widely. The main associated plants are as in South Island short-tussock grasslands and, as in the south, the tussocks have often given way to danthonias, browntop, sweet vernal or *Festuca rubra*. On high volcanic slopes, open bristle tussock communities are extensive (p. 430).

Adventive-dominated grasslands and herbfields of inland districts

This section discusses communities where adventives have largely replaced short tussocks, generally without passing through stages of extreme depletion (Fig. 9.22). The first example (columns 1–5 in Table 9.6) is from Puketeraki ER. Earlier in the

Fig. 9.22. Bushes of *Olearia odorata* (left) and *Rosa rubiginosa*, and *Poa cita* tussocks in a largely adventive sward; 640 m, Brands Creek, Mackenzie ER.

pastoral phase there may have been a fire-controlled alternation between matagouri scrub and hard-tussock grassland. Today, floristic content is similar to the hard-tussock communities described for southern Canterbury and matagouri remains prominent, but hard tussock is a minor element, which is consistent with its general decline in this region (Scott *et al.* 1988). The main grasses are now browntop and sweet vernal generally, with *Poa pratensis* prominent in the vicinity of matagouri bushes and *Festuca rubra* on the deeper soil of the terraces.

Old fluvioglacial terraces near Mavora Lakes (columns 6–9) are more exposed to southerly rain. There are areas of dense red tussock, but in the samples this and other native grasses are now unimportant. Instead, *Festuca tenuifolia* dominates on eroded, stony patches, and *F. rubra* dominates on deeper, wetter soils. Browntop, sweet vernal and *Poa pratensis* co-dominate where burnt beech stumps remain, presumably because decaying forest residues provide nutrients. Although most accompanying species are shared with the Puketeraki sites, others indicate wetter, more acid conditions. Mosses are prominent at both localities, with *Breutelia* sp. and *Racomitrium lanuginosum* each dominating one sample. *Hypnum cupressiforme* or *Polytrichum* spp. also form a cover in places.

Table 9.7 documents increases of browntop, sweet vernal and hawkweeds at the expense of native species in a tributary catchment of the Rakaia River over a period of reduced grazing. Individual transects, including some that were set up in 1955, show that browntop and *Hieracium lepidulum* increased most rapidly during early years, levelling off as bare ground became occupied. *H. pilosella* continues to increase,

Table 9.6. *Cover estimates averaged from five or ten 0.5 × 0.5 m plots in tussock grasslands that have changed to dominance by adventive grasses and herbs*

Site:	1	2	3	4	5	6	7	8	9
Aspect:	S	S	S	S	—	—	W	E	—
Slope (deg.):	3	3	30	<3	0	—	5	—	—
Percentage cover:	84	58	89	80	68	84	91	84	56
No. of vascular species:	31	16	39	14	19	19	32	19	16
Tussock grasses									
Festuca novae-zelandiae	3	+	+		+	+	+		
Poa colensoi	2	+	f			+	+	+	+
Poa cita					+				
Chionochloa rubra						+	+	+	
Other native grasses									
Elymus rectisetus			+			+	+		
Rytidosperma sp.						+	+		
Adventive grasses									
Agrostis capillaris	2	2	3	3	2	2	f	f	3
Anthoxanthum odoratum	2	f	2	3	2	+	+		2
Dactylis glomerata		+							
Poa pratensis		2							2
Holcus lanatus			+				+		+
Festuca tenuifolia						4	3	2	
Festuca rubra					+		2	3	1
Sedges and rushes									
Carex breviculmis	+	+							
Luzula rufa	+		+	+				+	
Shrubs									
Discaria toumatou	f	3*	+	s	+				
Hebe rakaiensis	+	+	+						
Melicytus alpinus	+		+						
Coprosma propinqua		+	+						+
Rosa rubiginosa		+	+						
Pimelea traversii			+					+	
Dwarf shrubs									
Cyathodes fraseri	1		+			+	f	+	
*Coprosma petriei**	1		+				+		
Gaultheria depressa						+	f		
Large native forbs									
Aciphylla subflabellata	+				+				
Small native forbs									
Prasophyllum colensoi	+			+					
Geranium sessiliflorum	+		+	f	+	+			
Oreomyrrhis rigida	+		+				+		
Celmisia gracilenta	+			+		+	f	+	
Brachyglottis bellidioides	+		+					+	
Wahlenbergia albomarginata	f		f	+	+	+	f	+	
Acaena caesiiglauca	+	+	+				+*	+*	
Epilobium alsinoides		+					+		

Table 9.6. (*cont.*)

Site:	1	2	3	4	5	6	7	8	9
Aspect:	S	S	S	S	—	—	W	E	—
Slope (deg.):	3	3	30	<3	0	—	5	—	—
Percentage cover:	84	58	89	80	68	84	91	84	56
No. of vascular species:	31	16	39	14	19	19	32	19	16
Ranunculus multiscapus						f	+		
Gonocarpus aggregatus							+	+	
Mat- and patch-forming herbs									
Raoulia subsericea	2		+			+	+		
Helichrysum filicaule						+	+		
Adventive rosette plants									
Crepis capillaris	+	+	+						
Hieracium praealtum	2		+		+				
Hypochoeris radicata	+		+	2	+	+	f	+	
Hieracium pilosella			+		+	+	+		
Clovers									
Trifolium repens	+	+		f	+				+
Trifolium dubium				+	+				
Other adventive forbs									
Cerastium sp.	+	+	+	+	+				
Linum catharticum	+	+					+		
Rumex acetosella			+						+
Senecio jacobaea							+		+
Ferns									
Blechnum penna-marina						+	+		
Mosses									
Racomitrium lanuginosum	+		+	+		+	3		
Hypnum cupressiforme	f		2	2		2	f	f	
Breutelia spp.	+		3	+		+	+	f	
Polytrichum spp.			+	+		+		f	

Notes:

Sites 1–5: *c.* 600 m, Winding Stream, Puketeraki ER, on stony debris slope, fan and terrace; 4 and 5 probably disced and drilled.

Sites 6–9: 760 m, rolling glacial terraces, Mavora Lakes, inland Southland; 6, rise; 7, vegetated parts of eroding slope; 8, broad depression; 9, gently rolling.

* *Discaria* noted as up to 1 m tall; and/or *Coprosma atropurpurea*; *Acaena* listed as *A. anserinifolia.*

s, Seedlings; other symbols as in Table 9.4.

Further species, recorded from only one site:

Site 1: *Hydrocotyle novae-zeelandiae* var. *montana*, *Cassinia* 'fulvida', *Carex colensoi*.

    3: *Ophioglossum coriaceum*, *Aristotelia fruticosa*, *Muehlenbeckia axillaris*, *Anisotome filifolia*, *Pimelea oreophila*, *Celmisia spectabilis*, *Hieracium lepidulum*, *Deyeuxia avenoides*.

    5: *Trifolium pratense*, *Achillea millefolium*, *Taraxacum officinale*.

    6: *Campylopus clavatus*.

    7: *Scleranthus* sp., *Viola cunninghamii*, *Dracophyllum uniflorum*, *Gnaphalium audax*, *Prunella vulgaris*.

    8: *Plantago triandra*, *Olearia virgata*, *Carex echinata*, *Juncus effusus*, *J. antarcticus*, *Sphagnum cristatum*.

    9: *Nothofagus solandri* var. *cliffortioides*(s), *Cirsium arvense*, *Carex coriacea*, *C. ?testacea*, *Juncus gregiflorus*.

Table 9.7. *Mean percentage frequency of species over a 15 year period in grassland of the Harper catchment, Puketeraki ER*

species	1965	1975	1980
Agrostis capillaris	1	13	14
Hieracium pilosella	8	18	31
Undetermined hawkweeds†	18	31	41
Hieracium lepidulum	12	22	23
Anthoxanthum odoratum	39	41	54
Elymus rectisetus	10	17	12
Poa colensoi	14	11	16
Poa cita	16	13	14
Festuca novae-zelandiae	35	26	27
Hypochoeris radicata	31	23	21
Holcus lanatus	20	9	12
Rytidosperma setifolium	18	7	8
Crepis capillaris	17	6	7
Deyeuxia avenoides	9	3	2
Dichelachne crinita	8	2	1

Notes:
Frequencies recorded as presence in 15 cm diameter circular plots, lying at 40 cm intervals along 26 transects; most transects are 40 m long. Altitudes 610–1370 m, aspects various, slopes 5–35°.
†, Mostly *H. lepidulum* and *H. pilosella*.
Source: Rose 1983

although it is not yet present in all transects. Cover ranged up to 21% for *H. lepidulum* and 24% for *H. pilosella*.

Maritime grasslands

History

In northern lowlands, grassland must have been virtually absent before European settlement, except for steep, unstable, coastal slopes with trailing swards of *Poa anceps*, cliffs with *Chionochloa bromoides*, and dunes and sand banks with *Cortaderia*. Nevertheless, *Microlaena stipoides* and species of *Dichelachne, Rytidosperma, Lachnagrostis* and *Elymus* would have been widespread in open fern- and shrubland, and on slips and other temporary clearings. When clearing of forest and secondary vegetation for grazing began, these grasses increased and some are still significant in thin, unimproved pastures.

Further south, there were large areas of lowland grassland, which included scattered trees and shrubs, and graded into swamps, fernland, and remnant and regenerating patches of bush and scrub. Practically all of this grassland was greatly modified or destroyed before its composition could be recorded; historical reconstruction depends on surveyors' notes, early botanical collections, and such native species as remain after one and a half centuries of agriculture. On eastern hill country from southern Hawkes Bay to Otago, wherever annual rainfall is below 750 mm, forest and bush destroyed by Maori fires was likely to have been replaced by grassland instead of fernland. This included both short- and tall-tussock grasslands, the latter occupying leached soils. Only in the droughtiest localities could any grassland on coastal hills have been truly primitive.

On recent alluvial terraces and coastal sand plains, seral grasslands resisted invasion by forest through the competitiveness of the sward, reinforced by frost and periodic drought. Early records from the Wairarapa Plain, where grassland is said to have covered 80000 ha, point to the importance of *Poa anceps*, *Agrostis 'perennans'*, *Deyeuxia quadriseta*, *Festuca rubra*, *Chionochloa beddiei*, *Rytidosperma* spp., *Elymus multiflorus* and *E. rectisetus*. Forbs included *Aciphylla squarrosa*, *Gingidia montana*, *Scandia geniculata* and *Coriaria* (Hill 1963). Esler (1978a) suggests that *Microlaena stipoides*, danthonias and *Poa anceps* were prominent on sand plains in Manawatu ER.

In Westland, grassland existed on river-flats under rainfalls as high as 10000 mm annually. Tussocks of *Poa cita* or *P. cockayneana*, as well as *Festuca matthewsii*, would have been scattered through a sward containing *Rytidosperma gracile*, *Lachnagrostis lyallii* (on the most recent soils), *Elymus rectisetus*, *E. nardurioides*, species of *Hierochloe*, *Trisetum* and *Deyeuxia* and, probably, *Agrostis dyeri* and the native form of *Festuca rubra*. Turf-forming plants such as *Muehlenbeckia axillaris*, *Gonocarpus micranthus*, *Galium perpusillum*, *Plantago* spp., *Pratia angulata*, *Leptinella squalida* and *Helichrysum filicaule*, and the dwarf shrubs *Coprosma brunnea*, *Pimelea prostrata* and *Pernettya macrostigma* grew on stony patches without dense grass cover. *Gingidia montana* and *Coriaria* spp. were probably also common.

The present, far more extensive grasslands of maritime districts are overwhelmingly dominated by introduced grasses and other herbs, most of which occur almost throughout New Zealand. Origins are mainly European and temperate Australian, with some residual native species. Adventives from warm-temperate and subtropical climates are important in warmer districts, resulting in greater diversity in northern pastures.

The grass species

Microlaena stipoides is a rhizomatous native grass of mesic, moderately infertile soils that locally dominates in grazed swards or beneath partial shade, such as open manuka heath provides. It becomes rarer towards the south. *Elymus* species are often abundant on drier, more fertile soils in lightly grazed areas, but do not tolerate heavy grazing. The most widespread are cleistogamous forms of *E. rectisetus*, with procumbent culms, which are probably Australian introductions. *Dichelachne* spp. are also most frequent in places where they escape close grazing, such as steep, dry banks and among open bracken and manuka.

Danthonias are tufted grasses that form low turf under grazing, which they tolerate very well. At first, they dominated most induced grassland, but as pastures improve and the number of competing species increase, they retreat to the driest and least fertile sites. The main species are *Rytidosperma gracile*, *R. racemosum*, *R. unarede* and *R. clavatum*; some are considered to be introductions from Australia.

Poa anceps is mainly in the northern and central zones, where it still covers many steep, ungrazed slopes. The closely related silver tussock persists in central and southern hilly pastures long after most other native plants have disappeared. Introduced Australian tussocks in the same complex occupy similar niches. There are also remnant populations of native fescues, including *Festuca novae-zelandiae* on dry, fine-textured soils, *F. multinodis* and, on bluffs on Banks Peninsula, *F. 'petriei'*. Other native grasses include the shade-tolerant *Oplismenus hirtellus* in northern and central districts and *Echinopogon ovatus* in rough pasture on steep slopes. The tall, broad-leaved, palatable *Hierochloe redolens* was once abundant, especially on swamp margins and flushes, but has been mostly replaced by robust adventive grasses.

The main introduced cool-temperate grasses are the rhizomatous browntop, redtop (*Agrostis stolonifera*), *Poa trivalis*, and *P. pratensis*; ryegrass (*Lolium perenne* and cultivars), cocksfoot, *Bromus willldenowii*, *Phleum pratense*, *Holcus lanatus*, tall fescue (*Festuca arundinacea*) and *F. rubra*, which form broad, long-lived tufts; and *Cynosurus cristatus* and sweet vernal, which form smaller, shorter-lived tufts.

Nassella trichotoma is a totally unpalatable tussock from temperate South America. Because it resembles native tussocks, it escaped notice until well established in Marlborough and northern Canterbury; infestations now range from Northland to Central Otago. *Nassella* invades open grasslands, especially those depleted by drought and overgrazing on dry northern aspects and shallow soils. Once established, seedlings can spring up many years after tussocks have been destroyed, even on land that has been ploughed and drilled. *Stipa* spp. are also naturalised on dry eastern hills in the South Island.

Annual or short-lived perennial grasses are abundant where long-lived plants fail to give a close sward. *Poa annua* establishes on bare patches on fertile soils, and species of *Bromus*, *Vulpia*, *Briza* and *Aira* can provide most of the cover on dry slopes in spring and early summer, before soils dry out. *Hordeum murinum* dominates sheep camps.

Warm-temperate grasses are most widespread and abundant in the northern zone, although they occur further south, especially in coastal districts. Kikuyu (*Pennisetum clandestinum*) spreads by 5 mm thick stolons, and can develop metre-deep swards. It extends from fertile flats to all but the driest and most exposed slopes, but is most aggressive near the sea and on mid-slopes. The even more robust buffalo grass (*Stenotaphrum secundatum*), like kikuyu, was introduced to stabilise eroding land, but is confined to the coast. *Axonopus affinis* is also stoloniferous but is smaller and hardier and can dominate infertile hill pastures. *Paspalum dilatatum* and *Cynodon dactylon* have slender stolons or rhizomes, the former being in good pastures and the latter in thin pastures on dry or saline soils. *Botriochloa maxima* forms dense mats on dry, coastal slopes of sunny aspect, but is still local. Ratstail (*Sporobolus africanus*) is an unpalatable tufted perennial that is increasing on dry, infertile soils. *Digitaria sanguinalis*, *Echinochloa crus-gallii*, *Eragrostis brownii* and *Setaria* spp. are annual or short-lived grasses, mainly of wasteland.

Sedges and rushes

Grazing and trampling by cattle have led to gradual replacement of native wetland by damp pastures, beginning from the margins. Draining, ploughing and sowing effect the conversion more rapidly and thoroughly. Few resident native plants survive the change, although sedges such as *Carex coriacea, C. secta* and *C. virgata* can persist for many years on unploughed land. Near the sea, the tussocky *Cyperus ustulatus* and the rush-like *Isolepis nodosa* also persist.

Rushes become weeds in damp pasture, the most abundant being the native *Juncus australis, J. distegus, J. gregiflorus, J. sarophorus* and the 2 m tall *J. pallidus*, and the adventive *Luzula congesta, Juncus articulatus, J. effusus, J. tenuis* and, on bare ground, *J. bufonius*. Among native tussock-forming sedges *Carex comans* is a serious weed even on dry ground, and *C. flagellifera* and *C. testacea* are occasionally troublesome. The 'Australian sedges' *Carex inyx* and *C. longibrachiata* are 60 cm tall, unpalatable tussocks, which are spreading to dominate rough pastures in the northern zone and more locally further south. *Cyperus congestus, C. eragrostis, C. polystachyos* and *C. tenellus* are also common adventives in damp pastures, mainly in northern and central New Zealand.

Other native plants

Native dicot herbs are frequent in unimproved pastures. *Acaena* spp. are of greatest economic importance because their burrs cling to wool, the main species being *A. anserinifolia* and, especially in coastal sites, *A. novae-zelandiae*. Small, creeping herbs that can be important in the 'bottom layer' and form extensive patches in hard-grazed pasture include several species of *Hydrocotyle, Dichondra brevifolia, D. repens, Oxalis exilis, Nertera setulosa* and, on damp ground, *Pratia angulata*. Thin pastures are at risk from invading native seral shrubs and ferns (pp. 540–3). Mosses are important in thin, lightly or intermittently grazed pastures of shaded or southerly aspect where fertility has declined, *Bryum* species, *Brachythecium albicans, Stokesiella praelonga, Thuidium furfurosum, Triquetrella papillata* (on dry ground) and *Hypnum cupressiforme* (on rough ground) being the most frequent.

Adventive dicots

Clovers are an integral component of New Zealand pastures, because of their nitrogen-fixing ability (p. 496). On fertile soil white clover is the favoured species; red clover (*Trifolium pratense*) is always included in seed mixes, but decreases after a few years. *Lotus pedunculatus* grows on moist but less fertile ground. On lighter, drier soils, these are largely replaced by spindly annual clovers of low agronomic value, i.e. *Trifolium arvense, T. striatum, T. dubium, Lotus angustissimus, L. suaveolens* and, on sandy soils, medicks (*Medicago arabica, M. lupulina, M. nigra*). The low-growing *Trifolium subterraneum*, on the other hand, provides useful cover on dry land during spring and early summer.

Flatweeds usually make up a large proportion of pastures and add variety to diet. Dandelion (*Taraxacum officinalis* aggr.) and *Plantago major* are most common on moist, fertile soils, whereas on drier or less fertile soils *P. lanceolata, Leontodon taraxacoides, Crepis capillaris, Hypochoeris radicata* and *H. glabra* prevail. Thistles can be abundant, especially where pastures have been opened up by overgrazing or drought. The more prolific single-stemmed, monocarpic species are Scotch thistle (*Cirsium vulgare*), swamp thistle (*C. palustre*), variegated thistle (*Silybum marianum*), winged thistle (*Carduus pycnocephalus* and *C. tenuiflorus*), and nodding thistle (*C.*

nutans), which is considered especially noxious. The rhizomatous, so-called 'Californian thistle' (*Cirsium arvense*) can be abundant on alluvial soils. Ragwort (*Senecio jacobaea*) is regarded as the worst weed of cattle pastures because of its toxicity and rapid increase, but sheep control it fairly effectively. Tall composites of similar habit are *Conyza* spp. and *Senecio bipinnatisectus*.

Bare spots, such as tracks and eroded patches, support a host of small forbs, some of which are also common on arable and waste land. They include *Ranunculus parviflorus, Anagallis arvensis, Cerastium* spp., *Silene gallica, Sagina procumbens, Polycarpon tetraphyllum*, sorrel, *Linum trigynum, Capsella bursa-pastoris, Montia fontana* subsp. *chondrosperma, Geranium molle, Erodium* spp., *Modiola caroliniana* (in northern and central districts), *Vicia sativa* and other vetches, *Torilis arvensis, Sherardia arvensis*, Bellis daisy (*Bellis perennis*), *Gnaphalium coarctatum, Mentha pulegium, Veronica arvensis* and *Prunella vulgaris*.

Widespread medium-sized herbs include *Galium aparine, Parentucellia viscosa* and, on moderately fertile soil, docks (mainly *Rumex crispus* and *R. obtusifolius*) and buttercups (*Ranunculus repens* and other species). On rough, wet pasture *Galium palustre*, water pepper (*Polygonum hydropiper*) and related species including the native *P. salicifolium* are common. Sheep camps and other places influenced by trampling and dung support *Sisymbrium officinale, Chenopodium pumilio, Malva* spp., *Coronopus didymus, Arctotheca calendula* and *Marrubium vulgare*. Other plants appear in seldom-grazed areas such as roadsides, e.g. *Pelargonium inodorum*, carrot (*Daucus carota*), fennel (*Foeniculum vulgare*), salsify (*Tragopogon porrifolius*) and yarrow (*Achillea millefolium*).

Several vigorous weeds are mainly northern. *Ageratina adenophora* and *A. riparia* form metre-tall colonies in rough gullies and on steep, southerly aspects. Inkweed (*Phytolacca octandra*), *Solanum linnaeanum* and *Oenanthe pimpinelloides* are weeds of deteriorating pastures, the first two being toxic. Alligator weed (*Alternanthera philoxeroides*) forms rampant colonies, both in water and on adjoining dry land. However, no herbaceous plants pose as great an economic threat as gorse (*Ulex europaeus*), blackberry (*Rubus fruticosus* aggr.) and, potentially, other shrub weeds that are as yet localised.

Maritime grassland communities

Productive lowland grasslands cover some 30% of New Zealand, as intergrading communities that vary according to environment, origins and management; but few details have been published. The next few pages glimpse this variation through five regional windows.

Canterbury

By the end of the nineteenth century, most of the Canterbury Plains had been ploughed and sown to crops or pasture. After 1945, the lower, gentler slopes also fell to the plough. Other native vegetation, whether forest, bush, shrubland, grassland or swamp was steadily modified through grazing practices.

Table 9.8 illustrates grassland dominated by *Rytidosperma clavatum*, in which species of the preceding short-tussock grassland are still prominent. Table 9.9

Table 9.8. *Composition of* Rytidosperma clavatum
grassland on hills in Lowry ER and Canterbury Plains

Community (no. of relevés):	1(3)	2(4)	3(3)
Mean altitude:	249	247	376
Prevailing aspect:	NW	N	NE
Mean slope (deg.):	27	33	21
Mean percentage cover:	98	89	98
Mean no. of species:	29	30	32

Tussock grasses			
Nassella trichotoma	+	f	
Poa cita	3	f	1
Festuca novae-zelandiae	1	+	2
Other native grasses			
Rytidosperma clavatum	3	4	4
*Elymus rectisetus**	f	1	f
Dichelachne crinita	+	+	+
Other adventive grasses			
Lolium perenne	f	f	
Dactylis glomerata	1	1	+
Anthoxanthum odoratum	1	+	1
Festuca rubra	1		f
Holcus lanatus	f	+	f
Agrostis capillaris	f		+
Vulpia bromoides		+	f
Sedge			
Carex breviculmis	+	+	f
Shrubs and lianes			
Carmichaelia ovata	1	+	
Discaria toumatou	+	1	f
Muehlenbeckia complexa		f	
Dwarf shrubs			
Pimelea prostrata	+		+
Cyathodes fraseri	+		f
Native forbs			
Helichrysum filicaule	f		
Oxalis exilis	f	f	
Dichondra repens	1	+	f
Gnaphalium 'collinum'	f	f	f
Crassula sieberiana	+	1	+
Vittadinia australis	+	f	f
Acaena novae-zelandiae	+	+	+
Cotula australis	r		r
Hydrocotyle novae-zeelandiae	+		+
Geranium sessiliflorum	+		+
Convolvulus verecundus		1	
Scleranthus uniflorus		+	+
Daucus glochidiatus		+	r

Community (no. of relevés):	1(3)	2(4)	3(3)
Mean altitude:	249	247	376
Prevailing aspect:	NW	N	NE
Mean slope (deg.):	27	33	21
Mean percentage cover:	98	89	98
Mean no. of species:	29	30	32
Wahlenbergia gracilis		+	f
Ranunculus multiscapus			f
Adventive legumes			
Vicia sativa	f	+	
Trifolium subterraneum	+	f	
Trifolium dubium	f		1
Trifolium repens	f	+	+
Adventive annual forbs			
Crepis capillaris	f	+	+
Silene gallica		f	
Polycarpon tetraphyllum		f	+
Other adventive forbs			
Hypochoeris radicata	f	f	1
Rumex acetosella	f	f	f
Cerastium fontanum	f	+	+
Cirsium vulgare		f	+
Verbascum thapsus		f	+
Hieracium pilosella			f
Ferns			
Pteridium esculentum	1		+
Cheilanthes sieberi		+	+

Notes:
Table derived as Table 9.4.
* Includes introduced forms.
Recorded as + in one community only: 1: *Linum bienne*; 2:
Microlaena stipoides, Stipa nodosa, Bromus hordeaceus,
Sophora prostrata, Trifolium arvense, Arenaria serpyllifolia,
Erodium cicutarium.
Source: Connor & MacRae 1969

illustrates variation in sheep-grazed grassland on the Port Hills near Christchurch, in which native plants have yielded to adventives although the land has never been ploughed. Nearly 80 species were identified. Ryegrass is more important in the moister sites, and densest on the well-manured Site 1. Other north-temperate grasses, as well as the native *Microlaena* and silver tussock, are also on the moister sites. Australian or doubtfully native danthonias, *Elymus rectisetus* and, especially, *Stipa* spp. are most prominent in open grassland on the drier sites. Annual grasses are important throughout.

White clover and vetches are confined to moister ground, whereas the three vernal clovers are wide-ranging, although contributing little cover on dry sites. Other forbs,

Table 9.9. *Cover estimates averaged from five or ten 0.5 × 0.5 m plots on modified coastal slopes of Port Hills, Banks Peninsula ER*

Site:	1	2	3	4	5	6	7	8	9	10	11
Aspect:	SE	N	SSE	S	NW	NE	NE	NNE	NW	NE	
Mean slope (deg.):	13	30	35	15	24	15	34	32	20	35	
Percentage cover:	76	92	93	86	96	100	60	50	36	20	
No. of vascular species:	21	17	33	29	29	17	23	19	17	13	21
Native grasses											
Rytidosperma spp. (i)	*	3	1		3	1	3	3	3	2	f
Poa cita		+	+	1							
Elymus rectisetus (ii)				+	1	1	f	+	f		f
Microlaena stipoides					1						
Perennial adventive grasses											
Stipa nodosa	*								+	+	
Dactylis glomerata	1	2	1	1		+	+				
Lolium perenne	3	2	2	2	2	2	2	+	+	+	+
Holcus lanatus		+	1	2	1		+				
Cynosurus cristatus			2	2							
Poa ?trivialis			+	+	+						
Anthoxanthum odoratum			2	1	1	f	1	f	+		f
Stipa bigeniculata								2			
Annual adventive grasses											
Hordeum ?murinum	2										
Aira caryophyllea	+		+		+		+				
Bromus hordeaceus	1	1	f	+	1	1	1	+	f	+	f
Bromus diandrus	+	1	+	+	f	1	+			+	+
Vulpia sp.	+		+			+	1		+		+
Briza minor						+	+		+		
Sedges and rushes											
Eleocharis acuta				2							
Carex virgata				2							
Juncus gregiflorus				1							
Juncus filicaulis				+							
Shrubs, lianes and Phormium											
Sophora prostrata	R										
Muehlenbeckia complexa	R										+
Phormium tenax			+	+							
Native forbs											
Crassula sieberiana	*					+				f	+
Acaena novae-zelandiae		+	1	+	+						
Oxalis exilis		+	+		+	1	f	f	+	+	+
Microtis unifolia			+	+							
Wahlenbergia gracilis			+		+				+		
Dichondra repens					+	1		+			
Convolvulus verecundus						+	+	f	+	+	
Geranium ?sessiliflorum											+

Site:	1	2	3	4	5	6	7	8	9	10	11
Aspect:	SE	N	SSE	S	NW	NE	NE	NNE	NW	NE	
Mean slope (deg.):	13	30	35	15	24	15	34	32	20	35	
Percentage cover:	76	92	93	86	96	100	60	50	36	20	
No. of vascular species:	21	17	33	29	29	17	23	19	17	13	21
Adventive legumes											
Trifolium glomeratum	4	2			2	1	2	2	1	1	+
Vicia sativa		+	f	f	f			+			
Trifolium dubium		+	2	1	2	2	+	1	+		+
Trifolium repens			2	2							
Vicia hirsuta			+	f							
Trifolium subterraneum			2	1	1	3		+		f	
Small annual forbs											
Geranium molle ɛ		1		+							
Cerastium glomeratum	f		+	f	f	f				+	+
Silene gallica					+		1	+		+	+
Cotula australis (iii)							+	+		+	+
Polycarpon tetraphyllum											
Other adventive forbs											
Marrubium vulgare	*										
Carduus pycnocephalus	f	+			+						
Hypochoeris radicata	*	+	1	+		+	+	+	+		
Hypochoeris glabra	*				+				+		
Silybum marianum		+			+						
Cirsium vulgare		+	+		f	+	+		+		
Sonchus oleraceus		+							+		
Taraxacum officinale				f							
Centaurium erythraea							+	+	f		
Mosses											
Hypnum cupressiforme	R		1								
Acrocladium chlamydop.			1				1				
Stokesiella praelonga			1	1							

Notes:

R, among rocks; *, on thin soil.

(i) *R. clavatum* and other, probably adventive species; (ii) includes adventive forms; (iii) probably native.

Other symbols as in Table 9.4.

Sites:

1, sheep camp on rocky ridge crest; 2, steep debris above the sea; 3, bouldery debris; 4, damp debris; 5, near bottom of broad gully; 6, broad hollow on slope; 7, 8, slope; 9, dry rise; 10, dry brow; 11, crevices in rock outcrop.

The following species were recorded as + in only one site.

1: *Stuartina muelleri, Erodium cicutarium, E. moschatum.* 2: *Bromus willdenowii, Crepis capillaris, Rumex brownii.* 3: *Festuca 'petriei', Echinopogon ovatus, Cynosurus echinatus, Carmichaelia robusta, Ptychomnion aciculare, Triquetrella papillata.* 4: *Agrostis capillaris, Leptospermum scoparium.*

although many in species, contribute little cover. Tall native rushes or *Eleocharis* persist on the two dampest sites. Rocky ground favours ferns, and mosses are significant on southerly aspects. Woody native plants and *Phormium tenax* are present, although only *Muehlenbeckia complexa* is thriving.

On arable land on the South Canterbury downlands (Table 9.10), new pastures are dominated by the sown species (ryegrass, white and red clovers), together with weeds typical of recently ploughed fields. Composition changes as pastures age, most profoundly on the south aspect where *Cynosurus cristatus* and *Brachythecium albicans* provide most of the cover. In the unploughed remnant, the dominance of ryegrass, cocksfoot and white clover reflects enhanced fertility, with less demanding species, including several natives, persisting on dry rises.

Western South Island

Except where high-fertility grasses have been recently sown, dominance in western South Island lowland grasslands is largely partitioned among four species: tall fescue in partly-drained swamps and wet hollows; *Holcus lanatus* on moist, well-drained sandy soils; sweet vernal on over-drained dune crests and gravel ridges; and browntop where leaching is under way. Clovers are similarly distributed, with *Lotus pedunculatus* abundant on damp recent soils and *Trifolium dubium* on over-drained sites, with *T. repens* being intermediate. All these, as well as *Festuca rubra*, grow on the river flats that provide most of the pasturage. Where ungrazed, the grass cover becomes rank and exclusive, but heavier grazing allows a rich florula to persist among the grass. Over 90 species of herbs and dwarf shrubs were listed from river flats in Central Westland, many of which would have been present in the original native grasslands (p. 264).

The grasslands of recent valley flats represent a stage in succession from open flood-plain, and are invaded by shrubs such as gorse and *Coprosma propinqua*. In grassland established through clearing forest, these are joined by blackberry and rhizomatous ferns.

Table 9.11 illustrates some of these trends. Recently sown pastures show high proportions of ryegrass and white clover, but ryegrass is not prominent in older pastures. Site 3, being on a steep slope, is less uniform and supports more species, including persisting natives, than the flatter sites where pasture was established through cultivation. Site 5, in old pasture presumably established through over-sowing after forest clearance, is the 'weediest', despite growing on colluvium derived from limestone. On the road verge ungrazed growth of a few species excludes most others.

Manawatu (Esler 1978*a*)

Sand country supports summer-dry pastures. As marram (*Ammophila arenaria*) declines on stabilised dunes, dominance passes to the vernal grasses *Bromus diandrus, Vulpia bromoides, V. myuros* and *Lagurus ovatus*, the annual clovers *Medicago lupulina, Melilotus indicus* and *Trifolium arvense*, and forbs such as catsear, *Leontodon taraxacoides* and *Geranium molle*. On older dunes, the major grasses are browntop, cocksfoot, sweet vernal, *Vulpia bromoides*, danthonias and *Microlaena stipoides*.

Table 9.10. *Cover estimates averaged from five 0.5 × 0.5 m plots in sown pasture, on downlands at Otaio, Pareora ER*

Site:	1	2	3	4
Aspect:	N	N	S	—
Slope (deg.):	3	2	5	0
Percentage cover:	66	89	94	99
No. of vascular species:	13	10	12	13
Perennial grasses				
Lolium perenne	3	4	1	2
Anthoxanthum odoratum	+		f	f
Cynosurus cristatus		2	3	+
Poa pratensis		+		f
Agrostis capillaris			+	+
Dactylis glomerata				2
Annual grasses				
Poa annua	1			
Bromus hordeaceus	f	f		1
Hordeum murinum		1		
Rush				
†*Juncus distegus*				1
Clovers				
Trifolium pratense	2			
Trifolium dubium	f		2	
Trifolium repens	2	2	2	3
Forbs				
Sherardia arvensis	f			
Hypochoeris radicata	+		f	
Cirsium arvensis	+			f
Taraxacum officinale	+	+	+	+
Achillea millefolium		+	1	2
Crepis capillaris			f	
Moss				
†*Brachythecium albicans*			3	

Notes:
† Native species. Other symbols as in Table 9.4.
Sites:
1, sown 12 months ago; 2, 3, sown > 3 years ago; 4, unploughed remnant between gully and fence.
Recorded as + in only one site: 1: *Rumex acetosella, R. obtusifolius*; 2: *Carduus nutans, Capsella bursa-pastoris*; 3: *Plantago lanceolata*; 4: †*Poa cita*.

Table 9.11. *Abundance estimates averaged from three, four or five 0.5 × 0.5 m plots in pastures near Cape Farewell, North-west Nelson*

Site:	1	2	3	4	5
Altitude (m):	5	60	50	60	90
Aspect:	—	NW	SW	—	NE
Slope (deg.):	0	2	30	±0	10
Percentage cover:	95	95	95	100	95
No. of vascular species:	15	16	20	13	30
Shrubs					
Ulex europaeus					a
Grasses					
Lolium perenne	a	d	d	+	+
Holcus lanatus	a	a	a	a	+
Anthoxanthum odoratum	+	+	a	a	+
Dactylis glomerata		+	+	a	+
Poa annua		+	+		
Festuca rubra			a	d	
Agrostis capillaris			a	a	d
Rushes					
†Juncus gregiflorus	a				
Juncus tenuis	a				
Clovers					
Trifolium repens	a	d	d	a	a
Lotus pedunculatus	+		a	a	+
Flatweeds					
Plantago major	+				+
Taraxacum officinale	+	+			+
Hypochoeris radicata	+	+	+	+	a
Plantago lanceolata	+	+	+	+	+
Other forbs					
Veronica arvensis	+	+	+		+
Gnaphalium coarctatum	+				+
Bellis perennis		+	+		+
Rumex obtusifolius		+			
Achillea millefolium		+		+	+
Cerastium sp.		+	+		+
†Acaena anserinifolia			+	+*	
Prunella vulgaris			+		+
Mosses					
†Stokesiella praelonga			a		+
†Brachythecium albicans			+		+

Notes:

Sites

1. Aorere River terraces; pasture recently developed from manuka, gorse and *Gleichenia* on pakihi.
2. Kaihoka lakes; pasture on wind-blown sand overlying cemented gravels; developed by ploughing.

Hordeum murinum, Aira caryophyllea, Eragrostis brownii, Cirsium vulgare, C. arvense, Carduus tenuiflorus, gorse, bracken and *Isolepis nodosa* are also common.

High-producing pastures on fertile soils are nominally composed of ryegrass cultivars, cocksfoot, *Phleum pratense*, and white and red clovers, but include over 40 weedy species. Few are without *Cerastium* spp., *Rumex conglomeratus, Trifolium dubium, Sherardia arvensis, Plantago lanceolata, P. major*, dandelion, catsear, *Leontodon taraxacoides, Bellis perennis, Veronica serpyllifolia, Poa annua, P. trivialis, Holcus lanatus*, sweet vernal, *Cynosurus cristatus* and the moss *Thuidium furfurosum*. On wet ground *Ranunculus repens, Rumex conglomeratus, Mentha pulegium, Lotus pedunculatus, Agrostis stolonifera, Juncus articulatus*, and the tall rushes *J. effusus, J. gregiflorus, J. sarophorus* and *J. australis* are abundant. On sheep-treaded slopes, composition is related to microtopography, with ryegrass and white clover on the treads, browntop or *Rytidosperma clavatum* on the banks below treads, and sweet vernal and *Leontodon* on the slopes (Fig. 9.23).

Pastures on leached terrace and hill soils consist largely of browntop, *Festuca rubra* var. *commutata* and *Rytidosperma clavatum*, with numerous minor species including the native *Dichondra repens, Hydrocotyle* spp., *Nertera setulosa, Helichrysum filicaule, Acaena novae-zelandiae, A. anserinifolia, Rytidosperma gracile*, and, on wet ground, *Eleocharis gracilis, Centella uniflora* and *Gonocarpus micranthus*. Adventive species, additional to some listed in preceding paragraphs, include *Linum catharticum, Myosotis discolor, Potentilla anglica* and *Alopecurus geniculatus*, the last on wet ground.

Hawkes Bay

Levy's (1970) description of hill pastures in this region, which experiences occasional severe summer droughts, illustrates the profound changes that follow changes in management. At first drought-tolerant pastures were composed of species such as danthonias, *Microlaena stipoides*, sweet vernal, *Poa pratensis*, ratstail, *Vulpia* and annual clovers. Top-dressing and introduction of subterranean clover has led to high-yielding pasture on the better sites; but as a result of generally increased carrying capacity, crowding of sheep camps has broken up the former danthonia-dominated turf on the ridges and knolls. Much of the ground there is bare during summer, and with the onset of autumn rains becomes colonised by thistles, *Hordeum, Erodium cicutarium, Trifolium subterraneum, T. glomeratum* and *T. striatum*, these being mainly annual plants which persist only into the next spring.

3. Near 2; pasture on side of gully, traversed by stock treads; developed by oversowing.
4. Near 2; ungrazed road verge.
5. Rough fan with limestone boulders at foot of Mt Burnett; unimproved pasture reverting to scrub and bush.

d, Dominant or abundant in most plots; a, abundant on at least one plot; +, present; †, native; *, probably *Acaena novae-zelandiae* on site 4.

Recorded as + in only one site:
1: *Juncus effusus, Sagina procumbens*; 2: *Poa pratensis, Ranunculus repens*; 3: †*Coprosma rhamnoides*, †*Nertera depressa*, †*Gunnera monoica*; 4: *Rumex acetosella*; 5: †*Kunzea ericoides*, †*Carex comans, Trifolium dubium*, †*Hydrocotyle ?novae-zeelandiae*, †*Galium ?perpusillum*, †*Oxalis exilis, ?Linum, ?Torilis*, †*Thuidium furfurosum*, †*Breutelia pendula*.

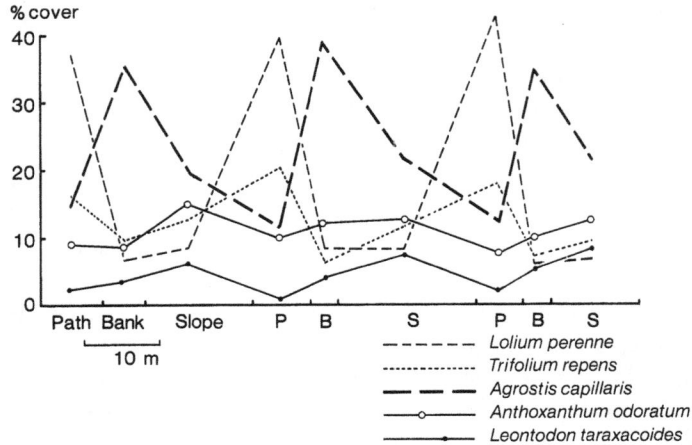

Fig. 9.23. Segregation of major species on different portions of a sheep-treaded slope (Rumball & Esler 1968).

Northland

Table 9.12 compares four pastures at different altitudes at latitude 35° 20′. Other than the increasing importance of warm-temperate grasses with decreasing altitude, the range of species scarcely differs from that in South Island samples. Omissions from the table are danthonia grasslands on dry, upper slopes, the highly productive meadows on flats (which match communities widespread in New Zealand), and the dense kikuyu swards of gentler coastal slopes, which grade into buffalo grass by the sea.

Patterns and processes affecting pasture composition

Topographic gradients (Fig. 9.24)
Grasses grade from those requiring moist, fertile soils to those that grow on dry, infertile, north-facing knolls. This is also a gradient of decreasing productivity and palatability: *Lolium* spp. = *Bromus willdenowii* = *Phleum pratense* > cocksfoot = *Holcus lanatus* > *Cynosurus cristatus* = *Arrhenatherum elatius* = *Poa pratensis* > sweet vernal = browntop > *Festuca rubra* > danthonias = *Stipa* spp. (Levy 1970). On wet ground, additional species are *Alopecurus geniculatus*, tall fescue, *Poa trivalis*, *Glyceria* spp., *Juncus articulatus* and, where salty, *Agrostis stolonifera*.

Grazing

Grasses range from those with tillers initiated at or below the ground surface, which persist in the face of close grazing (e.g. sweet vernal, ryegrass, danthonias), to those with vulnerable growing points that form dense, tall cover where lightly grazed (e.g. *Bromus willdenowii*, tall fescue, cocksfoot, and the robust, stoloniferous warm-temperate grasses). Sheep graze patches closely, avoiding rank grass; goats graze closely and evenly, and browse more woody plants; cattle graze less closely, but trample heavily.

Table 9.12. *Abundance estimates in grassland communities at lat. 35°20′S, Northland*

Site:	1	2	3	4
Altitude (m):	10	250	200	630
Aspect:	W	NE	SW	S
Slope (deg.):	35	5	2	2
Percentage cover:	50	70	95	90

Shrubs

Ulex europaeus		+	s	

Grasses

Pennisetum clandestinum	d			
Sporobolus africanus	a			
Axonopus affinis	a	d	+	
Briza minor	+	+	+	
Dactylis glomerata	+		+	
Anthoxanthum odoratum	+	d	d	a
Lolium perenne	+		a	d
Poa annua	+			+
Holcus lanatus		a	a	a
Bromus hordeaceus			a	
Agrostis capillaris			a	a

Sedges and rushes

Juncus tenuis	+	a	+	
†*Juncus sarophorus*		a		
†*Juncus gregiflorus*		a		
†*Schoenus apogon*		a		
Luzula congesta		+	+	
Juncus bufonius		+	+	
†*Juncus ?australis*		+	+	

Clovers

Trifolium dubium	+	a		
Lotus pedunculatus		a	a	
Trifolium repens	+		a	a

Other forbs

Oenanthe pimpinelloides	a			
†*Gnaphalium delicatum*	+	+		
Galium uliginosum	+		+	
Cirsium vulgare	+		+	
Ranunculus parviflorus	+		+	
Prunella vulgaris	+	+		+
Plantago lanceolata	+		a	a
Hypochoeris radicata	d	a	a	a
Bellis perennis	+	+	+	+
Sagina procumbens		+		+
Cerastium sp.		+	+	+
Sherardia arvensis			+	+
Ranunculus repens			+	a

Table 9.12. (*cont.*)

Site:	1	2	3	4
Altitude (m):	10	250	200	630
Aspect:	W	NE	SW	S
Slope (deg.):	35	5	2	2
Percentage cover:	50	70	95	90
Mosses				
†*Thuidium furfurosum*	+	+		
†*Brachythecium albicans*		a		
†*Stokesiella praelonga*		a	+	+

Notes:
s, Seedlings; other symbols as in Table 9.11.
Sites: 1, on steep clay coastal slope with prominent animal
treads, near Kawakawa; 2, pasture developed from silica-pan
gumland, Kaikohe; 3, lightly grazed pasture on basaltic hills,
Kaikohe; 4, developed pasture on clay ridge, Mataraua.
Recorded as + from only one site:
1: *Rytidosperma caespitosum, Poa trivialis, Vulpia* sp., *Lotus
suaveolens, Modiola caroliniana,* †*Apium prostratum,* †*Oxalis
exilis, Veronica arvensis*; 2: †*Microtis unifolia,* †*Gonocarpus
aggregatus, Plantago major, Parentucellia viscosa*; 3:
†*Microlaena stipoides, Rubus fruticosus, Trifolium pratense,
Geranium molle, Sisymbrium officinale, Cirsium arvense,*
†*Pteridium esculentum*; 4: *Juncus effusus, Montia fontana*
subsp. *chondrosperma, Mentha pulegium.*

Insects are also major consumers of grassland biomass, and some native to
indigenous grasslands are major pests in the modified systems. Of most concern are
larvae of the grass-grub beetle *Costelytra* and porina moth *Wiseana* which respectively
feed on roots and tiller bases during autumn, winter and spring, so that large patches of
grass die, especially when dry conditions follow. Bared areas are colonised by weeds,
although pasture grasses usually return eventually.

Transfer of nutrients
Well-stocked pastures in which fertility levels are maintained remain relatively
uniform, excepting that deposition and decay of excreta lead to small-scale patterns
through high-fertility species, notably ryegrass and cocksfoot, responding to
increased nitrogen while white clover is temporarily suppressed (Harris 1970*a*).

On hill country, much of the soil nutrient is concentrated in the top few centimetres,
and tends to be washed downhill by rain, both in solution and with soil particles.
Upper slopes therefore become leached, and lower slopes and depressions enriched,
with marked effects on composition of swards. Transfer upslope is effected by sheep
camping on high ground. Sheep treads are also enriched compared with the rest of the
slope (Radcliffe 1968). At a certain level of enrichment ryegrass dominates, but
beyond this dominance passes to *Hordeum murinum* and, at still higher levels, to herbs

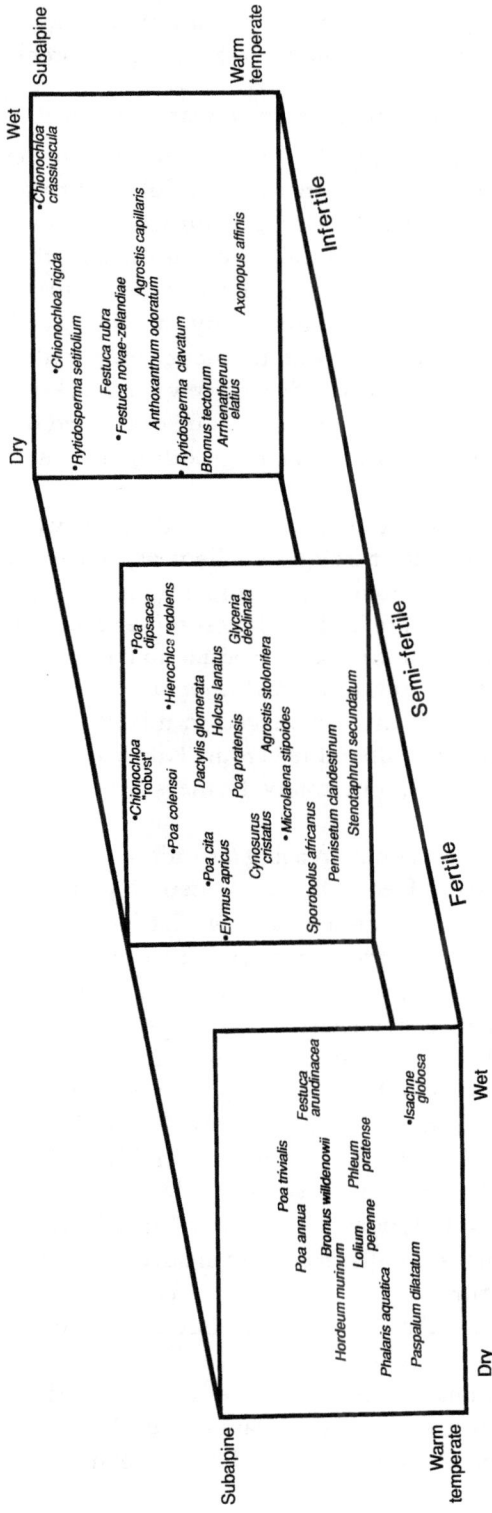

Fig. 9.24. Relative modal distributions of 38 common grasses along gradients of fertility, temperature and moisture. Dots indicate native species.

such as *Marrubium vulgare*. Nettle (*Urtica urens*) can be prominent on heavily manured areas that are only occasionally grazed, such as stock yards.

Ecology of pasture management

Although high-producing pastures of 'English' grasses have long been the agronomic ideal in New Zealand, at first it was neither feasible to bring the quantities of seed required directly from the British Isles, nor possible to obtain pure seed. Instead, mixed seed was stripped from Australian pastures, and then from New Zealand pastures as these became established. Early 'bush-burn' seed mixtures were exceedingly impure, and account for many of the weeds that thrive today. Subsequently, there has been increasing insistence on certified seed, and reliance on selection and breeding among strains of a very few species, especially ryegrass and red and white clovers. Some farmers, however, value the varied diet provided by pasture weeds. Natural selection also occurs, among both grasses and weedy herbs (Harris 1970*b*).

Management of maritime grassland is described by Levy (1970). The primary aim is to achieve high productivity throughout the year, but the seasons set constraints (Fig. 14.3). In the northern zone, cool-temperate grasses grow from late autumn through to early summer but are inhibited by summer drought, especially on shallow soils, whereas warm-temperate grasses produce maximum growth in summer and early autumn. Among the latter, only *Paspalum dilatatum* is of universally accepted value, as it is palatable and can be managed in mixed stands with the better cool-temperate species, being of comparable vigour. Kikuyu also provides useful summer herbage if kept well grazed, but many farmers are reluctant to encourage this aggressive grass.

Winter growth decreases with increasing latitude and altitude; accordingly, there is an increasing need for green feed to be supplemented with silage, hay and root crops. Summer drought can also be intense, and on flat land in the east of both islands, irrigation has been used extensively in recent decades.

The main management tools are superphosphate to enhance nitrogen fixation by clovers, and appropriate concentration and rotation of livestock. High herbage production supports dense, even stocking resulting in high return of nitrogen-rich animal manures. Build-up and maintenance of nitrogen also depends on clovers, which in turn depend on adequate phosphorus. For agricultural needs, phosphorus varies from suboptimal to acutely deficient in New Zealand natural soils, so superphosphate is added, thereby also correcting a prevalent sulphur deficiency. Trace elements, notably molybdenum, selenium or cobalt, are also applied for plant and animal health. Lime is usually needed to ensure optimal pH, legume nodulation, and a good soil structure. To some extent, weeds are kept in check by grazing management and maintenance of fertility, but direct control is also undertaken, usually with selective herbicides.

Topsoils under good pastures have become far more fertile than the original natural soils, which largely explains why native plants other than rushes are virtually absent. However, repeated applications of phosphorus are needed to balance the losses through leaching, removal of animal products and, especially, occlusion (p. 86). Nitrogen fertilisers played little part until about 1970, but since then they have been used to boost production in early spring before clovers are fully active.

In the past, large areas of difficult hill country reverted to secondary vegetation once the initial flush of fertility from bush burns declined. In the post-1945 period, most of this land and much more was reclaimed, because of buoyant economic conditions and incentives. Since around 1975, poorer returns have made it difficult to prevent pastures on 'marginal' lands from reverting to less productive species and being invaded by weeds.

Lawn and turf communities

Lawns, parks and playing fields cover considerable areas, both in urban New Zealand and around rural homesteads. In contrast to productive pastures, the usual aim is to maintain short turf of perennial grasses, preferably the low-fertility *Festuca rubra* var. *commutata* and browntop. This ideal can be achieved with adequate but not excessive moisture throughout the year, moderately fertile, slightly acid soil, control of weeds, and maintenance of a level surface without areas of local compaction. Where conditions favour vigorous growth, robust species such as perennial ryegrass take over; where growth is too contrained grasses cannot suppress mosses and mat-forming plants, mostly natives such as *Oxalis exilis* and species of *Leptinella, Hydrocotyle* and *Dichondra* that are prominent in other close-cropped systems (p. 266). Bare areas resulting from summer drought or infestation by grass-grub are colonised by ephemeral weeds, such as *Aphanes arvensis, Soliva sessilis* (mainly in warm districts), *Poa annua* or *Vulpia* spp. Even dense turf is prone to invasion by flatweeds, bellis daisy and clovers.

Natural turf communities maintained by close grazing share many species with managed lawns. They are most characteristic of moist, relatively fertile soils, especially on recently disturbed ground where succession towards taller vegetation is prevented, but can also develop where trees and understoreys are being destroyed by browsing and bark-stripping (Figs 9.25 and 9.26). The considerable number of native plants that thrive under close grazing is possibly a legacy from the largely extinct herbivorous avifauna (p. 6).

Coastal tussock grassland of Rakiura province
(based on Wilson 1987)

Poa foliosa, a wide-leaved tussock 1 m or more tall, occurs on the Solander Islands and some of the islands north-east of Stewart Island, as well as being widespread in Campbell province (Chapter 13). The similar *P. tennantiana* is almost confined to the Snares and south-western Muttonbird Islands. Both grow on deep coastal peats, covering slips and areas disturbed by seals, before they are excluded by dense scrub (p. 191). The main associates are *Blechnum durum, Asplenium obtusatum, Anisotome lyallii* and *Carex trifida*. There are also mosaics of tussocks and shrubs

Fig. 9.25. *Hoheria glabrata* grove severely damaged by deer; 700 m, Gorge River, Olivine ER. The fence is part of a live-trap for deer.

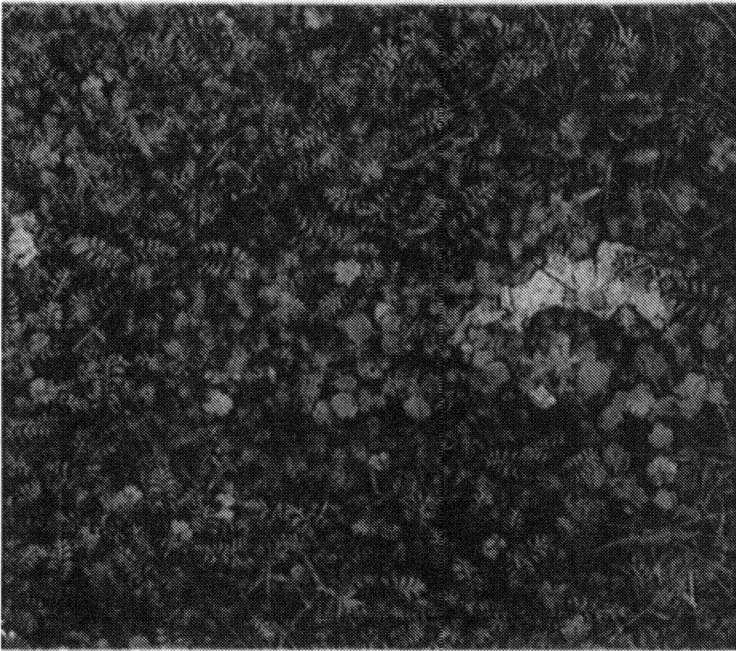

Fig. 9.26. Ground cover of the *Hoheria* stand in Fig. 9.25, showing turf of *Leptinella squalida, Hydrocotyle novae-zeelandiae* var. *montana* and *Ranunculus reflexus*.

Poa astonii, a much smaller, narrow-leaved, bluish-green tussock, is abundant on coastal cliffs throughout Rakiura province, extending up the Fiordland coast and sparingly along eastern coasts to Banks Peninsula. It forms grassland, sometimes with *Rytidosperma setifolium*, on steep, exposed coastal slopes with active soil movement, and also on peat on southern mainland coasts and on islands where *P. foliosa* and *P. tennantiana* are absent naturally or because of grazing.

WETLAND VEGETATION

Wetland vegetation is extremely varied, yet conforms to well-defined gradients. The most apparent of these reflect altitude and nutrient supply. Accordingly, the contents of this chapter follow a sequence from saline habitats that are almost wholly coastal, through lakes and rivers, low-lying swamps that are flushed with nutrients, infertile wetlands that tend to be higher or further from streams, and wet heaths that are mainly on higher terraces and plateaus, to mountain wetlands. No hard and fast line separates wetland from dryland vegetation; kahikatea (*Dacrycarpus dacrydioides*) forest, tree-heath dominated by small podocarps, and red tussock (*Chionochloa rubra*) grassland are treated in Chapters 7–9 although they often grow on wet ground, whereas vegetation of rush-like sedges, restiads or *Gleichenia* fern, although often on ground that is dry much of the time, is included in the present chapter. First, however, physical, botanical and historical aspects relevant to all wetlands are considered.

Physical characteristics of wetlands

The main physical factors determining the composition of wetland vegetation are as follows.

(1) Standing or moving water that varies in depth, duration, aeration, fertility, salinity, opacity, temperature and turbulence. Both standing water and soil water may represent surfacing of the water table, or water overlying a shallow, impervious horizon.

(2) Substrates ranging from permanently submersed to intermittently droughty, and varying in texture, stability, fertility, acidity and aeration. Anaerobic conditions in soils and deep in lakes can lead to toxic gases, especially H_2S.

(3) Temperature, rainfall, evaporation and seasonality, which affect plants directly, and through their influence on water and soil; in lakes, wind is also important.

(4) The nature of the contributing catchments, which determines the amount and regularity of inflow, and its load of solutes and sediments.

(5) Geological processes, whether local or regional, or continuous or catastrophic, that influence wetlands and water bodies.

Landform and drainage allow several kinds of wetland to be defined:

(1) Aquatic systems, i.e. lakes, large ponds and streams, in which the substrate is immersed most or all of the time.
(2) Swamps: surfaces largely covered in shallow water, at least seasonally. Eutrophic or 'fertile' swamps gain nutrients from through-flowing water or other sources. Oligotrophic or 'infertile' swamps are sustained mainly by rainfall; drainage is outwards and water is more acidic. Swamps may also be divided into fens and marshes according to presence or absence of peat.
(3) Flushes: formed where ground-water emerges on hillsides.
(4) Bogs: oligotrophic wetlands with little surface water, that have accumulated peat.
(5) Mires: any wetland, whether swamp or bog, with substantial peat.
(6) Wet-heaths, characterised by ultra-infertile soils with an impervious horizon and little or no peat. Major examples are *pakihi* in the west of the South Island and *gumlands* in the northern zone.

These wetland types can be qualified in terms of vegetation, using prefixes that are structural, such as shrub-, reed-, or cushion-, or floristic, e.g. *Phormium*, sedge, restiad.

Wetness combined with infertile, anaerobic or cold conditions leads to slow, incomplete decay of organic matter, which may be dispersed among inorganic particles as in lake muds and humus-rich silt loams, or accumulate as peat on the surface, sometimes to depths of many metres. Subsequently, peat may become buried and preserved under mineral sediments. Peatlands develop distinctive drainage and vegetation patterns, although most plant communities on deep peat also occur on surfaces with little or no peat. Four main categories of peat are recognised.

(1) Swamp peats are soligenous, in that they are nourished by water flowing over or through soil. They are widely distributed through New Zealand, and when drained provide some of our most productive farmland. Soligenous peats also form on hillside flushes and over springs.

(2) Domed peats (or raised bogs) are ombrogenous, i.e. nourished only by precipitation. The centre is higher than surrounding land and the margins slope down to sluggish streams containing more fertile water. Domed peats accumulate on poorly drained surfaces, which may be interfluves within swamps, terraces with podzolised soils, or plateaus. They are most common in southern New Zealand, but there are extensive domes in the Waikato basin.

(3) Blanket peats extend over varied topography, even quite steep slopes. By and large they are ombrogenous, but include soligenous peat in gullies and depressions. Largely confined to the Chatham and far-southern islands, southern Fiordland and Southland, they form under cloudy skies, persistent cold winds, frequent mist, drizzle, or drifting sea spray but only moderate total rainfall, and cool-temperate to subalpine temperatures.

(4) A category of very wet, highly organic, acid soils develops in wet climates under slow-growing vegetation, including dense rimu (*Dacrydium cupressinum*) forest and primary heaths. Beneath the humus layer, the organic matter is structureless and its silt content increases with depth until, at depths of a few centimetres to a metre or more, it grades into leached silt loam, which in turn overlies an iron-humus pan. It seems best to regard this mantle as an A horizon that has been expanded and diluted by incorporation of colloidal organic matter.

Peat formations can be complex. Over distances of 100 m or more, they may include all the elements described. At much smaller scales, there can be intricate, slowly changing patterns of peat growth and drainage, such as growing hummocks and degrading water-filled hollows. 'String bogs', in which dams of growing peat separate narrow tarns, are a feature of some mountain and southern wetlands.

Adaptations to aquatic habitats

Submerged leaves of water plants experience reduced light, take up oxygen and carbon dioxide from aqueous solution, and are usually thin or finely divided. Floating leaves are usually broad, entire, spongy in texture, and have numerous stomata on their upper surface. Most floating plants have only floating leaves, but bottom-rooted plants may have either kind or both. Although nutrients are absorbed over the whole plant surface, nearly all aquatic vascular plants have roots; bottom-rooted plants anchor in soft substrates and are sparse on coarse gravel or rocky beds. Most submerged angiosperms bring their flowers to the surface for pollination, although some have never been seen to reproduce sexually in New Zealand.

Algal macrophytes lack special conducting tissues and the whole plant surface is photosynthetic. Except in very still water or below the depth of wave action in lakes, most are anchored through intimate contact with objects such as rocks and other plants. Submerged bryophytes thrive only in clear, well aerated water. Being attached by fine rhizoids, they require rocky beds when growing in flowing water.

Facultative aquatics can grow either fully submerged, or with their tops fully exposed. Submerged plants have the etiolated habit, finely divided leaves and well-developed aerenchyma of true aquatics, but as conditions grade towards dry land, they commonly form low, compact turf. Such species are especially characteristic of gently sloping lake shores with fluctuating water levels, and shallow or temporary pools.

Emergent aquatics hold their leaves above the water and photosynthesise and transpire in the manner of dryland plants, but aerenchyma is always present since their roots and rhizomes lie in poorly aerated mud or peat. Several are fast-growing reeds that die down in winter, even in the mildest localities.

Many rushes, several genera of sedges and restiads have stiff, cylindrical, hollow or pithy perennial green stems; leaves are similar, vestigial or absent. They mostly grow where water is shallow or intermittent.

The history of wetlands

Wetlands form where barriers to lateral or vertical drainage arise or while sedimentation or uplift are transforming aquatic habitats to dry land, and their existence in such a tectonically active land as New Zealand is transient. Charcoals in Waikato, eastern Otago and Chatham Island peats show that wetlands have always been subject to fire, as their vegetation and even the peat itself are flammable during dry weather. The greatly increased incidence of fire after the arrival of Polynesians 1000 years ago transformed wooded swamps into herbaceous swamps, and increased sedimentation through deforestation of catchments, thereby creating new wetlands and converting others into dry land. The Maori also manipulated wetlands for fishing, horticulture and defence.

Drainage of lowland swamps for agriculture began in the earliest days of European settlement, and has continued ever since. In recent years the ecological value of wetlands has become recognised, but at the time of writing, attitudes still weigh towards continuing drainage. The remaining 89 000 ha of freshwater wetlands (Newsome 1987) are a fraction of their original extent.

Anthropogenic changes in wetlands proceed along four main paths. Firstly, there is the introduction of adventive plants, whether deliberate or spontaneous, the latter being inevitable where there are upstream sources of propagules or when grazing animals bring in seed *via* hooves or dung. Secondly, dewatering and, in estuarine wetlands, desalination affect hydrology, soil and biota. Thirdly, trampling, especially by cattle, compacts the soil and favours plants that thrive on soil pedestals. Finally there is nutrient enrichment, whether on-site through fertilisers, excreta and decomposing animals, or through wash from surrounding pastures. Sphagnum harvesting (p. 322) and peat mining also affect wetlands.

Today, fertile wetlands that have not been substantially modified are rare. Infertile wetlands have fared better, especially in less developed parts of the country. Coastal salt marshes, except where reclaimed or where *Spartina* has been introduced, usually retain native vegetation, even in districts with few other vestiges of their original cover. This is probably because there has been less disturbance, and because most of the species are either cosmopolitan or vicariant taxa within widespread complexes.

The obverse of drainage has been storage of water for domestic, industrial and agricultural use, electricity generation, and amenity. Hundreds of water bodies have been created or enlarged, ranging from small dams to large lakes. These seldom resemble natural wetlands in their ecology. Small ponds are usually highly eutrophic. Man-made or controlled lakes are often subject to draw-downs so large, erratic and prolonged that shore vegetation does not develop; draw-down is also a means of controlling aquatic weeds. The damming of lakes Manapouri and Te Anau has provided a precedent, in that they are held within levels that cause minimal disturbance to shore vegetation.

Saline wetlands

All large saline wetlands in New Zealand lie within estuaries and lagoons that are partly or completely separated from the open sea by barriers of sand or gravel, but small areas occupy dune slacks and exposed headlands where spray maintains salt-marsh species, often on steep slopes. In estuaries where tides ebb and flow freely, there is a consistent sequence of vegetation zones, i.e. (1) below mid-tide, (2) above mid-tide, (3) reached only by spring tides, and (4) reached only by storm tides; 2–4 are referred to as lower-, middle- and upper-marsh respectively. Estuarine vegetation is also influenced by ponding, salinity, and texture of sediments. In lagoons, bars forming across the mouths during settled weather impound water that decreases or increases in salinity according to whether inflow exceeds evaporation. Heavy floods cut the bars, so that a lagoon may experience tidal fluctuation for a time. Even slightly brackish lagoons develop hypersaline areas where stranded pools evaporate.

The nature of salt-tolerance

Eel-grasses (*Zostera*) are the only vascular plants in New Zealand that grow in undiluted sea water. They metabolise while immersed and cannot tolerate long exposure; as with seaweeds, salt tolerance is a matter of controlling ion exchange between plant tissues and the surrounding water. Plants of salt marshes, however, have their roots in saline soil, and conduct gas-exchange while their photosynthetic organs are exposed to the air. To maintain turgor, they must develop an osmotic tension exceeding that of the soil solution. To prevent internal salt rising to lethal concentrations, halophytes either dilute salts in succulent tissues or excrete them through glands.

Fig. 10.1 shows that most salt-marsh plants grow best at low salinities and, presumably, survive incursions of salt water mainly by suspending metabolism. However, *Suaeda novae-zelandiae* grows best at a salt concentration half that of sea water.

Plants of saline habitats

(1) Plants that are usually submerged
Zostera muelleri throughout New Zealand and *Z. capricornii* in the north form beds on stable mud and sand flats, descending to subtidal depths. They are usually associated with the finely-divided red alga *Gracilaria*. Less stable sand and mud below mid-tide generally lack macrophytes other than the green algae *Enteromorpha*, and, in nutrient-rich estuaries, *Ulva*.

Brackish lagoons support a range of submerged angiosperms, most of which also grow in fresh water. *Potamogeton pectinatus, Ruppia polycarpa* and *R. megacarpa* are the most important; the last, together with *Lepilaena bilocularis*, is apparently confined to brackish water. Characean algae (p. 300) are also abundant.

Many facultative aquatics, including species of *Glossostigma, Elatine gratioloides, Lilaeopsis novae-zelandiae* and *Myriophyllum* (*votschii, pedunculatum* and *propinquum*), tolerate both fresh and slightly brackish water. The more salt-tolerant *Triglochin striatum* is commonest in brackish habitats. Moss-like *Crassula* spp. grow in coastal pools and wet turf, *C. helmsii* extending into tidal water. Sea musk (*Mimulus repens*), which has conspicuous light-purple corollas with yellow throats, is confined to brackish pools and lagoon margins.

(2) Emergent or regularly exposed monocots
Schoenoplectus pungens, with leaves that are sharply triangular in cross-section, is the most salt-tolerant summer-green reed. It is usually abundant on estuarine mud, extending to the upper tidal limit. *S. validus* grows both coastally and inland, rooted in

Salinity (% NaCl)

Triticum aestivum +
Lolium perenne +
Plagianthus divaricatus
Mimulus repens
Carex flagellifera
Poa cita
Lilaeopsis novae-zelandiae
Agrostis stolonifera
Festuca arundinacea
Puccinellia fasciculata
Apium prostratum
Isolepis cernua
Atriplex prostrata
Puccinellia stricta
Schoenus nitens
Lachnagrostis filiformis
Elymus pycnanthus
Plantago coronopus
Polypogon monspeliensis
Leptinella dioica
Cotula coronopifolia
Selliera radicans
Juncus maritimus
Spartina anglica
Schoenoplectus pungens
Samolus repens
Triglochin striatum
Puccinellia novae-zelandiae
Spergularia media
Sarcocornia quinqueflora
Suaeda novae-zelandiae

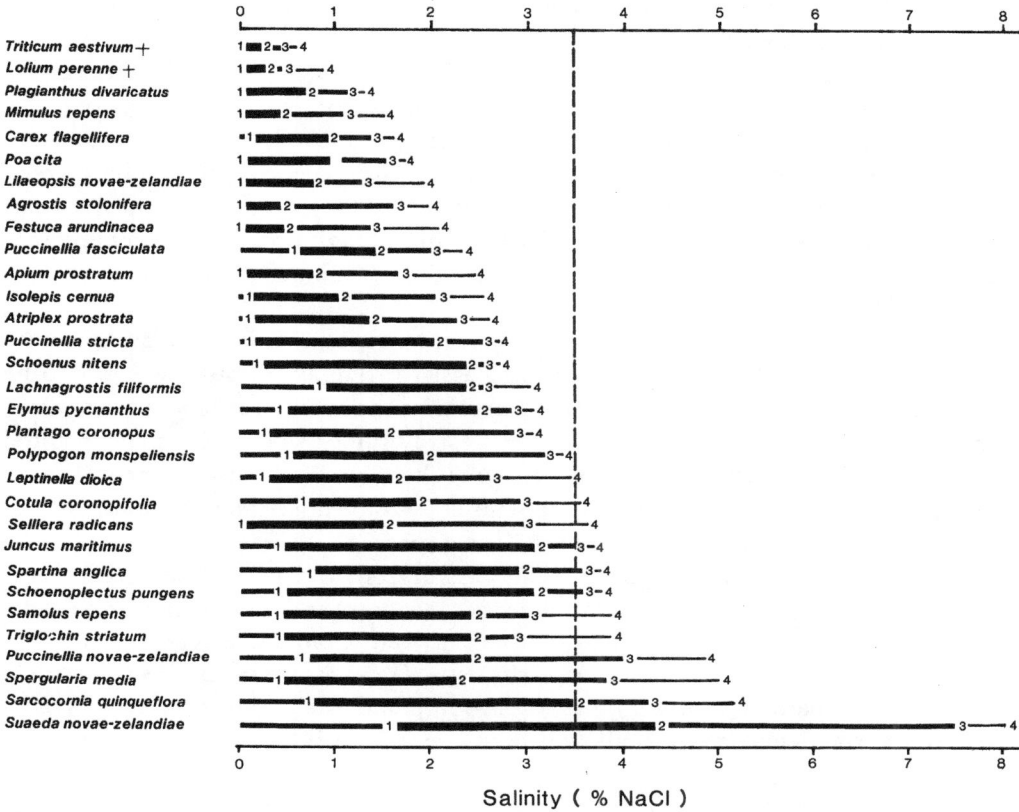

Fig. 10.1. Growth and survival of species over a range of salinity (Partridge & Wilson 1987). Numbers indicate: 1, maximum growth; 2, growth half of maximum; 3, most plants die; 4, no growth. The heavier bars between 1 and 2 indicate the zone of main occurrence; +, not salt-marsh plants. The vertical line represents salinity of sea water.

mud that is seldom exposed. *Bolboschoenus caldwellii*, *B. fluviatilis* and *B. medianus* form dense beds where brackish lagoons merge into fertile freshwater swamps, in northern and central New Zealand, with *B. caldwellii* extending down the east coast to Otago.

The small, tufted sedge *Isolepis cernua* grows on upper-marsh disturbed by trampling, erosion, deposition, etc. *Schoenus nitens*, of similar stature, forms mats on sandy ground. The tussocky *Carex litorosa* also grows on upper tidal flats. Other sedges appear where tidal marshes merge into freshwater wetlands, especially *C. flagellifera* and *Cyperus ustulatus*.

The rush-like sedge *Baumea juncea* is abundant where northern lagoons meet infertile swamps, and extends sparingly to Golden Bay. A true rush, *Juncus maritimus*, can dominate on middle- to upper-marshes to as far south as Okarito and Lake Ellesmere, with an outlier at Purakanui Inlet in Otago. It thrives where continuous wetness keeps salinities lower than on free-draining flats, and can colonise recently eroded or aggraded surfaces. A rhizomatous adventive, *J. gerardii*, forms dense masses in some tidal depressions.

The most abundant tall plant on New Zealand salt marshes is the restiad *Leptocarpus similis*, which grows to 1.8 m as clumps or continuous cover. It requires stable, permanently damp ground with largely organic topsoil, but does not descend quite as low as *Juncus maritimus* on tidal flats. The densely-packed tillers resist erosion, so that beds must be deeply undercut before they break away. *Leptocarpus* also grows on banks of tidal creeks regularly flooded at high tide, coastal marshes that are reached only by storm tides, inland lake shores subject to wave erosion and fluctuating levels, and in infertile swamps.

Spartina (cord grass) is the most important introduced estuarine plant. Until the 1970s, it was planted to trap sediment and assist reclamation, but has come to be regarded as a nuisance as the importance of estuaries and their native biota has become appreciated. Herbicides have been used to destroy minor spontaneous and planted colonies, but the extensive beds of *S. anglica* in the New River (Southland), Pelorus (Sounds ER) and Manawatu estuaries are unlikely to be eradicated. In warm North Island harbours *S. alterniflora* is the usual cord grass.

In the New River estuary, *Spartina* is successful from low tide up to where it meets dense *Leptocarpus*, thereby displacing short salt meadows and the upper *Zostera* beds, as well as occupying the tidal muds that lie between. Colonies extend up to 5.3 m annually and accrete 3–12 mm of sediments (Lee & Partridge 1983). A C4 photosynthetic pathway (p. 485), salt-excretion glands, tall stature, and air spaces continuous from leaves to roots allow it to utilise shorter intertidal periods than any emergent plant other than mangrove.

Densely tufted, fine-leaved *Puccinellia* grasses grow on dryish parts of salt marshes, the native *P. stricta* being the most common and widespread. Still drier areas with salt levels enhanced by evaporation support annual adventives, i.e. sea-barley (*Hordeum marinum*) which completes its vegetative phase while salts are still winter-diluted, and the 'sea hard-grasses' *Parapholis incurva* and *Hainardia cylindrica*.

The number of grasses increases at upper tidal fringes rarely reached by sea water. *Agrostis stolonifera* and tall fescue (*Festuca arundinacea*) are usually abundant on damp ground and descend into the upper tidal zone to colonise recent alluvial fans and reclaimed land, probably owing their success to vigorous regrowth after salt damage. The former dominates grazed areas, but is suppressed by tall fescue elsewhere. The tussock *Stipa stipoides* is common at the tidal fringe in the northern zone, and occurs locally as far south as Tasman Bay. Further south, silver tussock (*Poa cita*) descends to the same level. *Lachnagrostis lyallii* is a common pioneer on upper-marsh, and *Elytrigia pungens* is becoming well established. Other adventive grasses include *Polypogon monspeliensis* and *Lagurus ovatus*.

(3) Regularly exposed dicot herbs

The chenopod *Sarcocornia quinqueflora* (glasswort) has stems that appear jointed through being invested by opposite, fused, succulent leaves. It is the most characteristic plant of saline flats on substrates of intermediate texture, being absent from both poorly aerated mud and unconsolidated sandy gravel. *Suaeda novae-zelandiae* is even more salt-tolerant. *Apium prostratum* and the probably adventive

Atriplex prostrata have an open, sprawling habit and, although scarcely succulent, commonly grow with *Suaeda* and *Sarcocornia*, as well as frequenting stony beaches within reach of spray. *Chenopodium glaucum* also grows in the upper parts of some estuaries.

Samolus repens, a wiry, loosely tufted or creeping plant with white, regular flowers, is widespread on tidal flats, often as pure communities on coarse or accreting substrates. It is also one of the main plants in the spray zone on coastal cliffs. *Selliera radicans*, a mat-forming plant with a white corolla split to the base on one side, grows on moist, stable substrates in mid-marshes and, with *Samolus*, can form an understorey to open stands of *Leptocarpus* or *Juncus maritimus*. Inland forms grow on the Volcanic Plateau and in the east of the South Island. *Leptinella dioica* forms turf in southern upper-marshes.

The rosette-forming *Plantago coronopus*, the sprawling *Spergularia media*, the more or less erect *Cotula coronopifolia* and, on northern marshes, the tall bushy *Aster subulatus* are locally abundant on ground disturbed by erosion, deposition, grazing, trampling, or passage of vehicles.

(4) Woody plants

Avicennia resinifera, the only New Zealand mangrove, extends southwards to at least Raglan and Ohiwa at latitude 38°, occupying estuarine muds to a lower tidal level than any angiosperm other than *Zostera* and *Spartina*. These muds are poorly aerated, especially where distant from channels, but the roots bear peg-like pneumatophores, which are exposed at low tide (Fig. 10.2). These carry out respiration and some photosynthesis (Dromgoole 1988). The leaves develop high osmotic tensions to maintain transpirational flow, and secrete excess salt. Trees are up to 9 m tall, but decrease to as little as 2 m towards the southern limit and with increasing distance from creek channels; the height gradients are probably responses to temperature and root aeration respectively.

Mangrove flowers mainly in autumn, and the large seeds germinate as they fall early in the following summer. With radicles protruding 2 cm and fleshy cotyledons still folded together, embryos can take root where they fall or be washed to and fro for some time. Seedlings grow up within canopy gaps and colonise new surfaces. Shoots grow 5–6 cm annually, but there is no information on longevity except that trees die prematurely when tidal flats are either drained or subjected to prolonged flooding (Burns & Ogden 1985).

Plagianthus divaricatus (shore ribbonwood) is the only other woody plant confined to coastal wetlands. It grows along the larger estuaries throughout the main islands and also on Chatham Island, as filiramulate bushes up to 2 m tall that are concentrated on rises and the banks of channels within the uppermost levels of salt marshes. *Coprosma propinqua* is common under less saline conditions.

The salt-marsh communities

Regional differences among saline wetlands are somewhat obscured by peculiarities of individual salt marshes, related to disturbance or vagaries of species-distribution (Partridge & Wilson 1988). However, as with lowland terrestrial vegetation, the

Fig. 10.2. Mangrove and pneumatophores; Bay of Islands, Northland.

northern zone has species that do not reach the southern zone; among these, mangrove imparts a distinctive character to northern estuaries. The present section begins with northern marshes, and concludes with tidal areas in the west of the South Island where prodigous rainfall diminishes the saline influence, and the inland salt pans of Central Otago.

Northern zone

Much of the northern coastline is indented by large estuaries. For the Waitemata and Manukau harbours, Fig. 10.3 portrays the following sequence of increasing elevation and decreasing salinity.

1 *Zostera muelleri* or *Z. capricornii.*
2 Scattered colonising mangroves, associated with *Z. muelleri.*
3 Mangrove associated with *Zostera*, glasswort, *Juncus maritimus* and *Samolus.*
4 *Samolus*; usually pure but there may be isolated plants of glasswort, *Selliera, Apium prostratum, Juncus maritimus* and *Triglochin striatum.*
5 Glasswort, as scattered colonising plants.

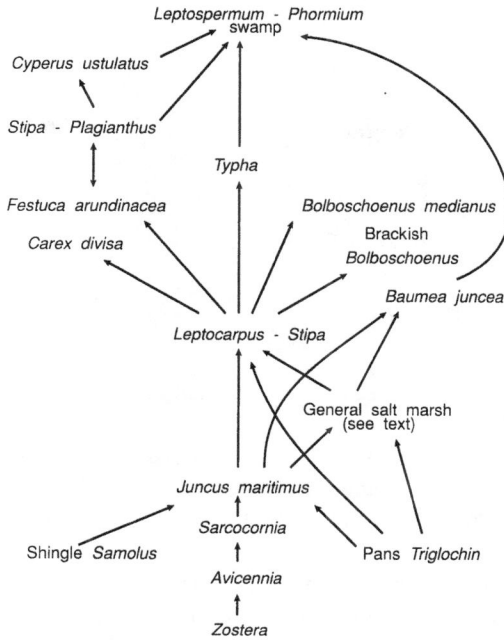

Fig. 10.3. Relationships among estuarine communities on the Auckland isthmus (from Chapman & Ronaldson 1958).

6 Glasswort with mangrove, *Samolus, Cotula coronopifolia, Stipa stipoides, Juncus maritimus* and *Triglochin.*

7 *Juncus maritimus* as scattered colonising plants.

8 *Juncus* with *Plagianthus divaricatus, Samolus, Apium, Selliera, Aster subulatus, Cotula, Leptocarpus, Stipa, Baumea juncea* and *Triglochin.*

9 *Leptocarpus* associated with *Plagianthus, Apium, Selliera, Samolus, Aster, Stipa, Juncus* and *Festuca arundinacea.*

10 *Juncus–Leptocarpus*: Combines the two preceding communities.

11 General salt-marsh community, containing *Plantago coronopus, Selliera, Samolus, Cotula, Aster, Isolepis cernua* and *Triglochin.*

12 *Triglochin striatum* colonies; usually pure, but with *Paspalum dilatatum* in pans.

13 *Bolboschoenus medianus* or *Schoenoplectus pungens*, either alone or co-dominant, in association with *Baumea juncea, Leptocarpus* and *Juncus maritimus.*

14 *Carex divisa*: usually pure.

15 *Cyperus ustulatus*: usually pure.

16 *Stipa stipoides* associated with *Plagianthus divaricatus, Muehlenbeckia complexa, Ranunculus* sp., *Sarcocornia, Plantago coronopus, Selliera, Samolus, Apium, Aster, Leptocarpus, Juncus, Baumea* and *Festuca arundinacea.*

17 *Baumea juncea* associated with *Leptocarpus* and *Juncus.*

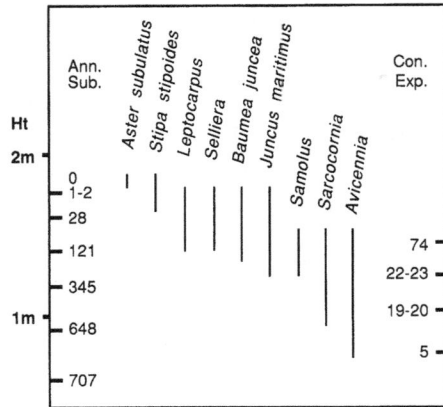

Fig. 10.4. Tidal range of species in northern estuaries (from Chapman & Ronaldson 1958). Abbreviations: Ht, height above mean sea level; Ann. Sub., total number of annual submergences; Con. Exp, longest continuous exposure (days).

Fig. 10.4 shows submergence and emergence periods for important species. Fig. 10.5 maps a typical pattern with mangrove on mud flats bordering tidal creeks, being joined by glasswort at higher levels or replaced by the latter away from the creeks. *Juncus* and *Leptocarpus* grow on the upper-marsh, largely as isolated patches that may have survived erosion of more continuous beds. Wave action has straightened a beach facing the harbour entrance, and in places built it high enough to support terrestrial vegetation.

Central zone

South of Auckland province, tidal flats are of limited extent around the North Island, although in Hawkes Bay 2 m uplift in the 1931 earthquake transformed much of the 3000 ha Ahuriri Lagoon into salt marsh (Aston 1933). This was reclaimed for farmland as salinity decreased. In the north of the South Island, extensive tidal flats lie between Nelson and West Wanganui Inlet, whereas the drowned valleys of Sounds ER have only small tidal flats except where the Pelorus River enters Pelorus Sound. Because the sounds are sheltered, freshwater tends to linger and succulent halophytes are generally unimportant, in contrast to Lake Grassmere (Kaikoura ER) where salt is concentrated in polders for commercial recovery.

Lake Ellesmere, on the edge of the Canterbury Plains, is the largest lagoon in New Zealand. Although impounded behind a spit of sand and shingle and only occasionally open to the sea, its expanse (180 km²) and shallowness (< 2 m) combine with high evaporation, low precipitation and low freshwater inflow to create hypersaline areas around its shores. The maximum lake level has been artificially controlled since pre-European times, with progressive reduction from 3 m above mean sea level to 1 m at present, and drainage has reduced surrounding freshwater and saline wetlands by at least 80% to a marginal zone up to 1.5 km wide. Fig. 10.6 interrelates the saline communities and Fig. 10.7 summarises zonation of their species and environmental factors:

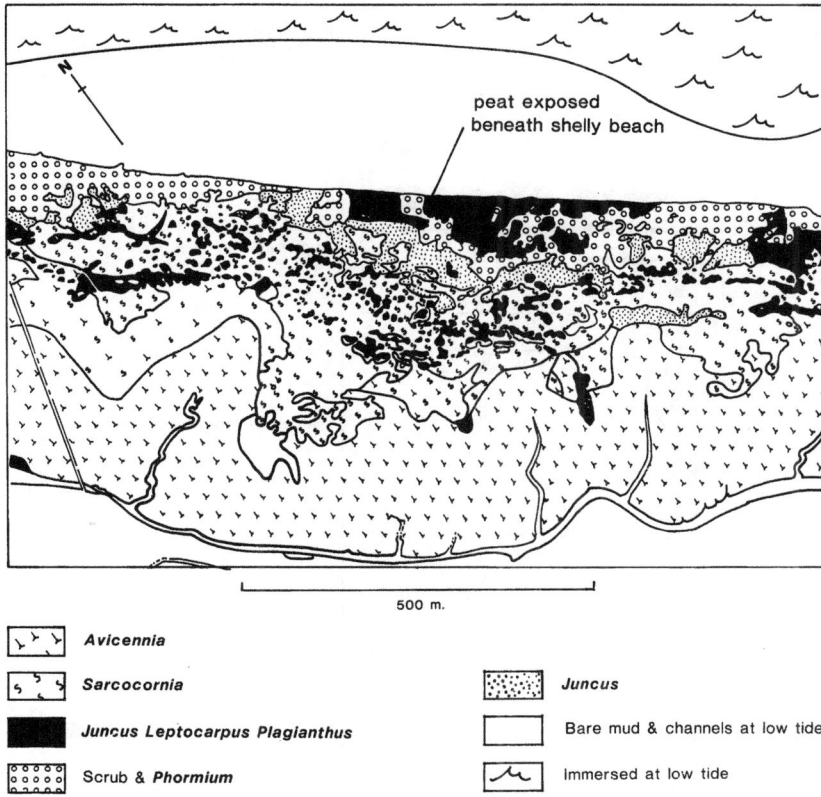

peat exposed
beneath shelly beach

500 m.

Avicennia		Juncus	
Sarcocornia		Bare mud & channels at low tide	
Juncus Leptocarpus Plagianthus		Immersed at low tide	
Scrub & Phormium			

Fig. 10.5. Zonation of communities on Pollen Island, Waitemata Harbour, Auckland ER. (Chapman & Ronaldson 1958).

(1) *Ruppia* beds, associated with green and blue-green algae, are seldom emersed. The great numbers of waterfowl on the lake largely depend on these beds and when the storm of April 1968 temporarily destroyed them, numbers plummetted.

(2) *Mimulus repens*, accompanied by *Lilaeopsis* and *Triglochin*, grows in a zone between the preceding community and salt-pan vegetation that tolerates greater exposure; it also occupies wave-hollowed pools within the pans. *Cotula coronopifolia* is common in stagnant water.

(3) *Schoenoplectus pungens* grows on mud, as pure stands or with *Lilaeopsis, Mimulus, Cotula* and *Triglochin*. At higher levels it passes into a *S. pungens–Agrostis stolonifera* community.

(4) Glasswort prevails on the salt flats. At the lowest levels, its main associate is *Triglochin*. Other species are *Spergularia, Atriplex prostrata, Mimulus, Cotula, Juncus maritimus* and *Puccinellia stricta*. Glasswort and *Puccinellia* can form open cover on sandy ground. Higher sandy areas support glasswort and *Hordeum marinum*, accompanied by *Spergularia, Atriplex,*

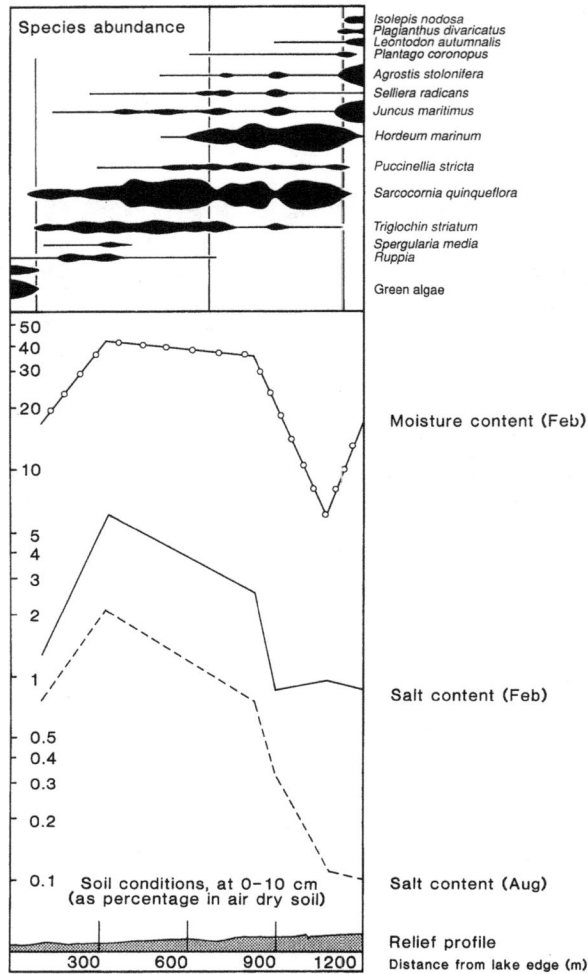

Fig. 10.6. Zonation of species in relation to salt and water content of the soil; Lake Ellesmere, Canterbury (Evans 1953).

Cotula, Puccinellia, Parapholis incurva, Hainardia cylindrica and Agrostis.

(5) Selliera mats occur where freshwater seeps from springs or bores. Other species include Plantago coronopus, Leptinella dioica, Triglochin, Isolepis cernua and Agrostis.

(6) Juncus maritimus often dominates between saline flats and more inland vegetation, forming dense communities in the absence of grazing. It may associate with Plagianthus divaricatus (on higher ground), Isolepis nodosa (on drier, sandier areas), Leptocarpus (on moister, peatier areas), and Agrostis stolonifera. Leptocarpus beds are of limited extent, and some are being invaded by Salix cinerea.

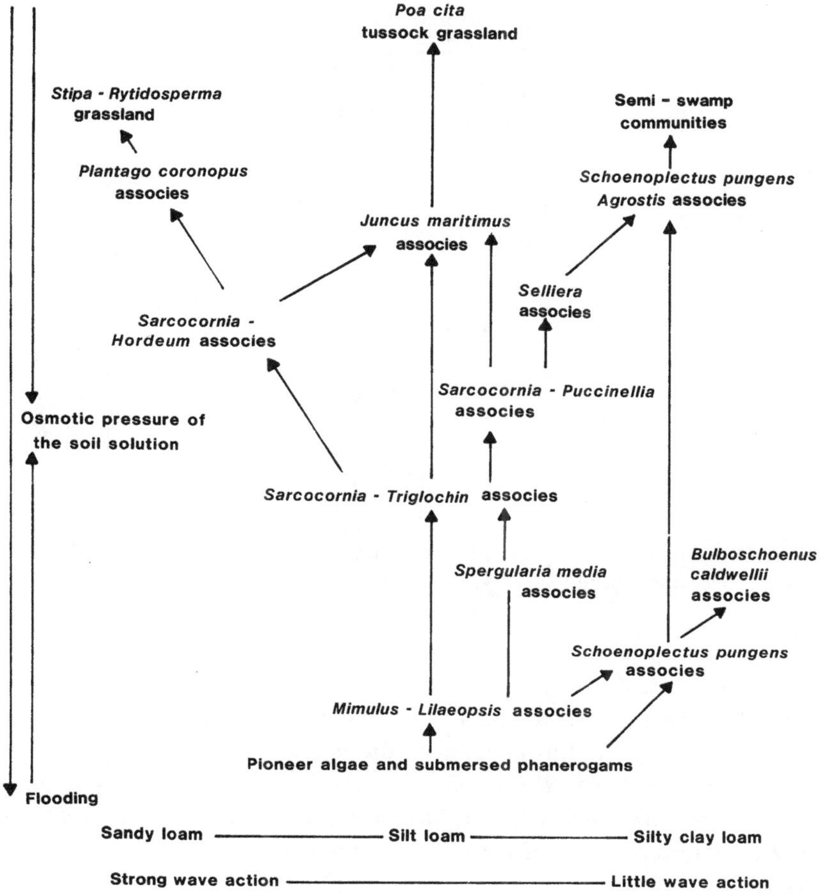

Fig. 10.7. Distribution of plant communities at Lake Ellesmere in relation to soil texture, immersion and salinity (Evans 1953).

(7) Beds of *Schoenoplectus validus*, and on less saline ground, *Bolboschoenus caldwellii* and raupo (*Typha orientalis*) grow on mud near stream entrances.

Southern coasts

The largest tidal flats in the south are near Dunedin, along the northern shore of Foveaux Strait and in Freshwater Inlet in Stewart Island. Fig. 10.8 shows the tidal range of species in eastern Otago. The ordination axis in Fig. 10.9 represents a gradient from lower- (left) to upper-marsh. The species in group A1 extend down to middle-marsh or below. A2 and A3 comprise upper-marsh species, and those in A4 belong to the marsh fringe. B has lower-marsh species most characteristic of lagoons. Group C consists mainly of less common species with varied ranges.

Salt marsh can merge into shoreline turf (p. 303) where there is strong coastal

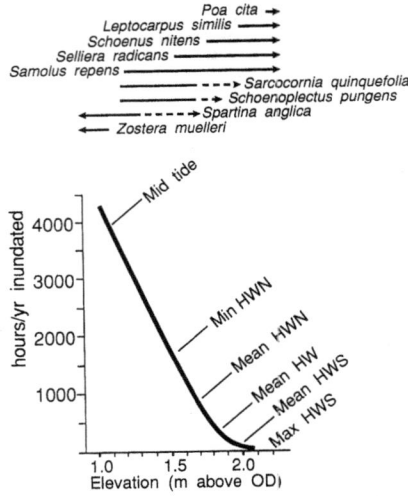

Fig. 10.8. Range of species in relation to tidal inundation in Otago (Partridge & Wilson 1989). OD, Ordnance datum; HWN, high water of neap tides; HWS, high water of spring tides.

exposure or close grazing. From the edge of Waituna Lagoon in Southland, Kelly (1968) lists 24 turf species including *Gonocarpus aggregatus, Myriophyllum pedunculatum, Lilaeopsis, Schizeilema cockaynei, Chenopodium glaucum, Plantago triandra, Leptinella dioica, Selliera, Glossostigma elatinoides, Limosella lineata, Euphrasia repens, Triglochin, Lachnagrostis striata, Schoenus nitens* var. *concinnus, Eleocharis gracilis, Isolepis cernua* and *I. basilaris.*

West coast of the South Island

Heavy surf along western coasts creates long gravel and sand spits, that impound lagoons in which low salinities prevail because of heavy rainfall and large volumes of river water. Glasswort and other succulent halophytes are rare or absent south of Karamea. In the Fiords surface water is fresh, especially in the upper reaches.

Okarito is the largest western lagoon (Fig. 10.10). Its aquatic vegetation consists of Characeae, *Ruppia megacarpa*, colonies of *Schoenoplectus validus* and, in shallow water, *Lilaeopsis*. There are *Zostera* beds near the mouth, and *Schoenoplectus pungens* and *Mimulus repens* grow on tidal mud. On marshes scarcely reached by tidal water *Leptocarpus* and *Juncus maritimus* stand over turf of *Myriophyllum pedunculatum, Centella uniflora* and *Selliera*. Scattered shrubs are mostly *Coprosma propinqua*, but *Plagianthus divaricatus* is also present. Other plants include species of both coastal and freshwater wetlands: *Ranunculus amphitrichus, Eryngium vesiculosum, Lilaeopsis, Hydrocotyle novae-zeelandiae, Plantago triandra, Lobelia anceps, Leptinella squalida, Triglochin, Juncus articulatus, Carex flaviformis, Deschampsia caespitosa, Lachnagrostis lyallii* and *Poa* sp., with *Utricularia monanthos* common in boggy hollows.

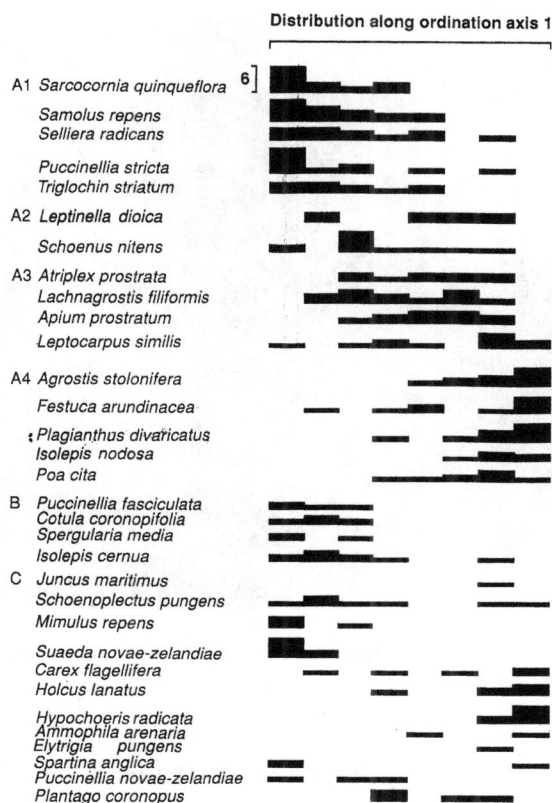

Fig. 10.9. Distribution of species in relation to spatial gradients in eastern Otago salt marshes. Species are tabulated on the left according to results of cluster analysis, and their frequency of distribution along the first ordination axis of individual marshes is charted on the right. Analysis is based on presence or absence in plots (Partridge & Wilson 1988).

Inland saline habitats

Saline pans in the driest Central Otago valleys develop mainly around the bases of hills where salt is being leached from Tertiary sediments. Never extensive, they have been modified by farming and their survival is precarious. Instead of distinctive communities, there are haphazard assemblages of salt-tolerant plants which, with the exception of *Lepidium kirkii*, a rare, slender, annual, also grow on Otago coasts. Typical species, in order of decreasing salt-tolerance, are *Atriplex buchananii*, glasswort, *Puccinellia stricta*, *Plantago coronopus*, *Selliera*, *Hordeum hystrix*, *H. jubatum*, *Schoenoplectus pungens* and *Agrostis stolonifera*.

On the banks of the Blind River, 17 km inland in Wairau ER, salt springs support marsh of *Triglochin striatum*, *Schoenoplectus pungens* and *Cotula coronopifolia*.

Fig. 10.10. Inland edge of Okarito Lagoon, Central Westland, with *Leptocarpus* in front of *Phormium tenax*, cabbage trees and shrubs (mainly manuka).

Lakes, ponds and streams

Freshwater macrophytes are usually most abundant in clear, well-aerated streams flowing gently over fine sediments. Native plants tend to be small and in oligotrophic environments. Today, in many places, aquatic vegetation is dominated by vigorous adventives. Some of these grow rampantly when first introduced to a waterway and later stabilise at a lesser level, possibly because they exploit and eventually exhaust nutrient pools inaccessible to native species.

Following an ecological classification of freshwater plants, communities are illustrated through specific examples. The first is a deep, oligotrophic lake with a catchment clothed in native vegetation. The next example is also of large, deep lakes, but their surrounds are now given to farming and urban development, and they have been invaded by adventive water weeds. In a shallow, eutrophic lake surrounded by farmland, adventive plants are even more prominent. The section concludes with accounts of upland and thermal waters.

Aquatic plants

(1) Characeae (*Nitella* and *Chara*) have complex, tubular thalli that form swards in still water, descending to greater depths than vascular plants. Filamentous green algae are most evident as epiphytes that can choke small ponds in summer, but *Ulothrix* colonises stones in rapid streams during intervals between freshes. Planktonic algae are numerous, diverse and ecologically important, but outside the scope of this book.

(2) The moss *Drepanocladus fluitans* forms long, dark tassels in still, clear water up to 3 m deep, whereas *Sphagnum falcatulum* grows submersed in small wetland pools. Stable rocky beds of fast-flowing streams are mainly exploited by mosses, including *Andreaea* spp., *Eurhynchium austrinum* and *Fissidens rigidulus*, the last thriving in powerful currents; these aquatic communities grade into moss-rich vegetation on rocky stream banks. On peatlands, the thallose liverwort *Aneura* forms mats in shallow streams. *Riccia fluitans* and *Ricciocarpus natans* have unattached, floating forms.

(3) Bottom-dwelling vascular plants include quillwort (*Isoetes*) and the small fern *Pilularia novae-zelandiae*, which grow in shallow, still water. Among angiosperms, *Zannichellia palustris* grows mainly in running water, *Lepilaena bilocularis* is mainly in brackish lagoons, and *Hydatella inconspicua* grows in sandy shallows in Northland.

(4) Among tall submerged angiosperms, *Ruppia polycarpa* extends down to 3 m in still water. Species of shallower or flowing water include *Myriophyllum robustum, M. aquaticum, Ranunculus trichophyllus, Potamogeton pectinatus, P. ochreatus*, and *P. crispus* which extends to brackish water. The adventive 'oxygen weeds', all in the family Hydrocharitaceae, are notorious invaders of waterways. *Elodea canadensis* is widespread; *Lagarosiphon major* is even more aggressive but as yet local; *Egeria densa* grows as large colonies in some northern rivers, especially the Waikato.

(5) The only bottom-rooted native aquatics with entire floating leaves are *Myriophyllum triphyllum*, which also has finely-divided submerged leaves, and *Potamogeton cheesemanii* and *P. suboblongus*, which have thin, linear submerged leaves. Two vigorous but as yet localised adventives, *Ottelia ovalifolia* and *Aponogeton distachyus*, show similar differentiation. Robust adventives with thick, entire floating leaves and few or no submerged leaves include *Hydrocleys nymphoides*, which massively infests a few northern lakes, and the water lily *Nymphaea alba*, which so far remains close to points of introduction.

(6) Unattached floating plants thrive only in eutrophic waters from which they are unlikely to be removed by winds and currents. The minute *Lemna, Wolffia* and *Spirodela* can cover small ponds or become trapped among reeds; *Lemna* tolerates the least fertile conditions. The floating fern *Azolla filiculoides* thrives in eutrophic ponds that are more enriched. Three warm-temperate adventives that pose a threat in northern lakes are the fern *Salvinia molesta* and water-hyacinth (*Eichhornia crassipes*) which have thick floating leaves, and *Ceratophyllum demersum* which has finely divided leaves that spread just beneath the surface. Although often attached to muddy bottoms, *Ceratophyllum* lacks roots, a condition that it shares with *Wolffia*.

(7) Some plants that grow in submerged, etiolated form in shallow water and more compactly on damp ground have already been mentioned for brackish habitats. Further examples are the native *Ranunculus amphitrichus, R. macropus, R. limosella, Limosella lineata, Hypsela rivalis, Montia fontana* subsp. *fontana, Callitriche petriei, Gratiola* species and the very local *Tetrachondra hamiltonii*, and the adventive *Callitriche stagnalis, Ludwigia palustris, Veronica anagallis-aquatica, Myosotis laxa* and *Juncus bulbosus*.

(8) Emergent aquatic plants are mostly native monocots. Raupo, *Bolboschoenus* and *Schoenoplectus* (p. 288) are tall, leafy and summer-green. The large, soft reed

Fig. 10.11. *Eleocharis sphacelata* in a shallow lake; Ohinemaka, South Westland.

Eleocharis sphacelata (Fig. 10.11) and the more slender *E. acuta* form beds of leafless stems that persist into winter. Water plantain (*Alisma plantago-aquatica*) is the most widespread adventive in this group, but the robust grass *Zizania latifolia* is naturalised in some northern waters.

(9) As well as fully aquatic species there are wetland plants that can spread from their terrestrial base into adjacent water. Through trapping sediment and organic matter, these can advance the margins of the land and eventually obliterate shallow lakes; treacherous 'quaking bog' can be a transitional phase. Some of the more conspicuous of these marginal plants are *Sphagnum* mosses, rhizomatous carices including *Carex coriacea* and, in mountain tarns, *C. gaudichaudiana*, *Isolepis distigmatosa*, *I. aucklandica* (mainly subalpine), and the introduced *Juncus articulatus*, *Agrostis stolonifera* and *Glyceria* spp. Common dicots include water-cress (*Rorippa nasturtium-aquaticum*) and *Mimulus* spp. *Carex secta* tussocks also thrive in open water (p. 310).

(10) Crack willow (*Salix fragilis*) is by far the most important woody coloniser of edges of lakes and streams. Spontaneous spread is almost entirely from broken branches, and therefore normally downstream from the many points where it has been planted. Crack willow now lines banks of rivers and lakes for many kilometres, colonises islands in flood plains, and has negative impacts on river control through impeding the discharge of floods and forcing changes of course. Through tolerance of deep, prolonged immersion and aggradation, it exploits a niche not occupied by native trees, and by casting dense summer shade and heavy leaf litter, and extending mats of roots several metres into standing water, it suppresses smaller native plants on both sides of the land–water boundary.

Manapouri: a near-pristine lake with a mountainous catchment

Lake Manapouri in Fiordland is a large (143 km²), deep (444 m), cold lake of glacial origin, which is oligotrophic but faintly peat-stained. On broad, gentle shores species have characteristic depth ranges (Fig. 10.12), to form the following zones.

(1) Aquatic vegetation to 30 cm tall extends upwards from 6 m below mean lake level. *Isoetes kirkii, Myriophyllum triphyllum* and, locally, *Elodea* are dominant, and *Chara corallina, C. fibrosa* and *Nitella* are present. In shallower water *Potamogeton cheesemanii* and *Myriophyllum propinquum* become important.

(2) From a metre below mean lake level, aquatic vegetation grades into turf with some 45 species of matted, creeping herbs. At the lowest level *Isoetes* gives 18% cover and *Myriophyllum propinquum* 8%. At mean level *M. propinquum* (12%) shares dominance with *Selliera* (8%) and *Centrolepis pallida* (7%). Above this dominance shifts to *Eleocharis acuta* (11%) and *Carex gaudichaudiana* (8%). Species additional to those in Fig. 10.12 include *Pilularia, Crassula sinclairii, Epilobium komarovianum, Schizeilema cockaynei, Limosella, Leptinella pusilla, Juncus pusillus*, and the bryophytes *Tridontium tasmanicum, Fissidens asplenioides* and species of *Chiloscyphus*.

(3) *Carex gaudichaudiana* forms a sward up to 50 m wide above the turf zone. In addition to the species in Fig. 10.12, *Ranunculus amphitrichus, R. flammula* and *Myosotis laxa* border runnels, and *Carex berggrenii* and *Juncus bufonius* join the sward around a sheltered lagoon. *J. gregiflorus, Gunnera dentata* and *Potentilla anserinoides* are also present.

(4) Towards high-water level, the sward meets either *Leptocarpus* or manuka (*Leptospermum scoparium*), with the latter grading into forest.

Rotorua Lakes

The lakes near Rotorua were originally clear and oligotrophic, but have been enriched from surrounding farmland and housing and invaded, first by *Elodea*, and more recently by *Lagarosiphon*. Table 10.1 compares depth ranges in two lakes with different light penetration, and Fig. 10.13 shows the three main zones of submersed macrophytes.

(1) Short mixed communities at depths of 0.1–1.8 m on sheltered, sandy, gently sloping shores. Samples from Lake Rotoiti contained *Glossostigma elatinoides* and *G. submersum* (49% of total dry mass), *Lilaeopsis novae-zelandiae* (20%), *Myriophyllum propinquum* (20%), *Limosella* (4%), *Elatine gratioloides* and *E.* sp. (2%), *Lagarosiphon* (1%) and *Nitella hookeri* and *N. pseudoflabellata* (5%). Other species are *Chara corallina, Enteromorpha nana, Ulothrix subtilis, Spirogyra* sp., *Anacystis cyanea, Drepanocladus fontinaliopsis, Isoetes kirkii, Elodea canadensis, Ludwigia palustris, Myriophyllum triphyllum, Potamogeton cheesemanii, P. crispus, P. ochreatus, Ranunculus trichophyllus* and *Utricularia australis*. Fig. 10.14 shows how colonisers trap sediment, to build up mounds which enlarge 6–32 cm per month in radius.

(2) Tall mid-depth communities of *Myriophyllum* and *Potamogeton* that have been invaded and displaced in varying degrees by *Elodea* (before 1900) or *Lagarosiphon* (since 1950). *Elodea* outcompetes *Lagarosiphon* under relatively fertile conditions and, if eutrophication of the lakes continues, should regain dominance. Table 14.1 (p. 468)

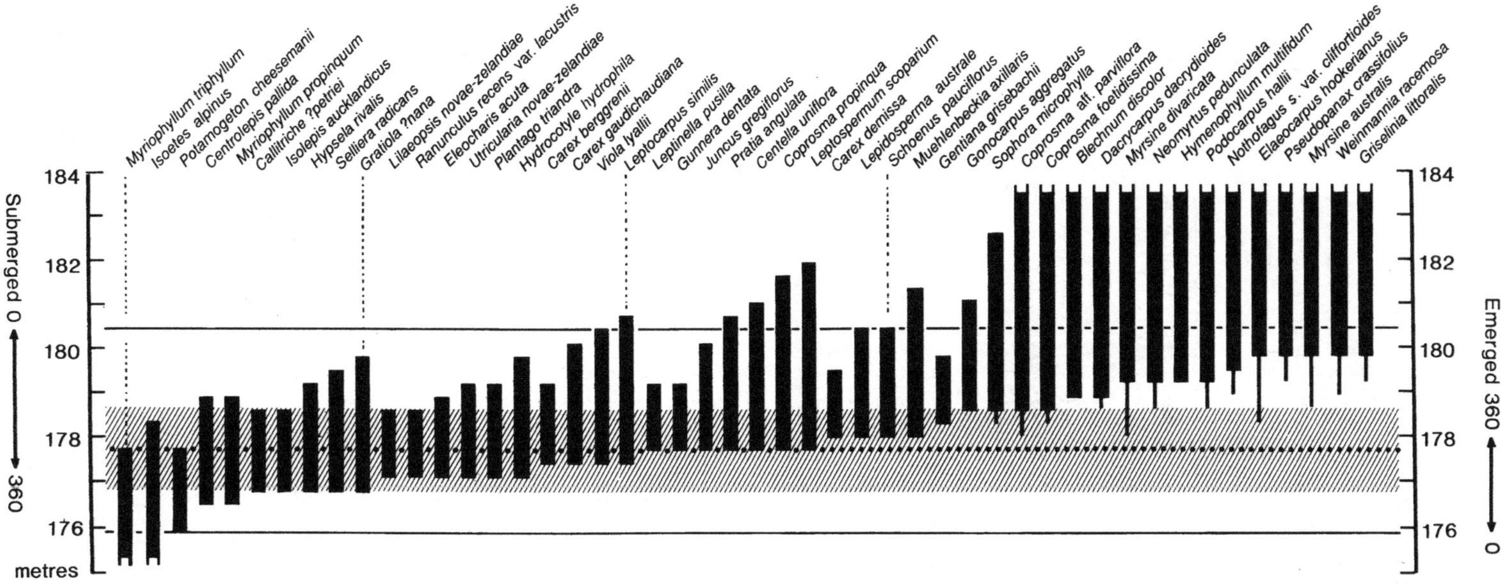

Fig. 10.12. Depth ranges of species on non-rocky shores at Lake Manapouri, Fiordland. Vertical axes show altitude (m a.s.l.), and longest recorded periods (days) of submergence (left) and emergence (right). Horizontal lines show mean and extreme water levels. Thick bars show maximum extent of species as mature plants, and lines show any further extent as seedlings only. Shaded area represents depth range of the turf community (from Johnson 1972).

Table 10.1. *Depth ranges (m) of species in two Rotorua lakes*

	Okataina	Rotoiti
Lagarosiphon major	0.8–6.5	1.0–6.5
Elodea canadensis	1.0–15.0	6.5–7.5
Potamogeton crispus	0.2–6.0	0.1–5.4
Potamogeton ochreatus	2.0–5.0	0.8–4.6
Potamogeton cheesemanii	1.0–4.0	1.8–3.9
Myriophyllum propinquum	0.8–6.0	0.3–4.0
Myriophyllum triphyllum	1.0–3.4	1.2–3.6
Characeae	0.1–19.0	0.1–7.5
No. of spp. of Characeae	5	2
Depth of 5% surface light	16.5	7.0
Inorganic phosphorus (μg/litre)	8.5	10

Notes:
Values for *Lagarosiphon* and *Elodea* exclude occasional plants
in shallow water.
Source: Brown 1975

shows standing dry matter; some New Zealand values for *Elodea* are said to be the highest measured for freshwater macrophytes.

(3) Deep charophyte beds completely covering soft substrates.

Submerged rock and shingle support seasonal filamentous algae, including *Spirogyra, Ulothrix* and *Cladophora. Drepanocladus* is recorded on rock.

A eutrophic lake

Pukapuka Lagoon (Manawatu ER) drains coastal farmland and has gradually shrunk to its present area of 15 ha. It is eutrophic, warm and has a maximum depth of 1.2 m, sometimes becoming almost dry in summer.

(1) Aquatic plants include bottom-anchored *Potamogeton cheesemanii, P. crispus, P. pectinatus, Ruppia polycarpa, Zannichellia, Myriophyllum triphyllum, Ranunculus trichophyllus* and *Chara* spp., and floating *Lemna, Wolffia, Spirodela* and *Azolla filiculoides*. Mud and sand exposed in summer is colonised by *Callitriche stagnalis, Rorippa nasturtium-aquaticum, Ranunculus sceleratus, Polygonum hydropiper, P. salicifolium* and *Veronica anagallis-aquatica*, which persist as a semi-floating mat when the water rises in winter. Fig. 10.15 shows the changes in proportions of species following almost complete drying of the lake in summer 1970. The disappearance of macrophytes in late summer 1972 was caused through drastic reduction of light penetration by a phytoplankton bloom (indicated by the chlorophyll *a* values) that may have been initiated by enriched flood-water.

(2) *Typha* and *Carex secta* surround the open water, the former invading from drier margins, whereas the latter apparently establishes as seedlings on surfaces exposed at times of low water.

(3) Further from open water, where the bottom is emersed more frequently, dying

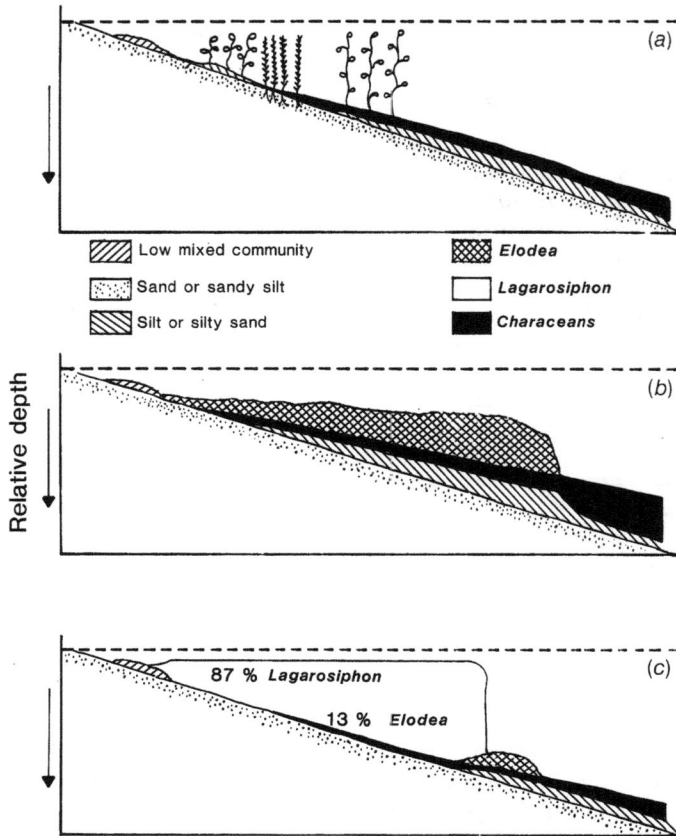

Legend	
▨ Low mixed community	▩ *Elodea*
▦ Sand or sandy silt	☐ *Lagarosiphon*
▧ Silt or silty sand	■ *Characeans*

Fig. 10.13. Displacement of native tall mid-depth communities (*a*) by *Elodea* (*b*) and then *Lagarosiphon* (*c*) in Rotorua lakes (Brown 1975).

Carex secta tussocks are overtopped by vigorous *Typha* and colonised by *Phormium tenax* seedlings. Other plants in this phase are *Cyperus ustulatus*, *Galium palustre* and vines such as *Calystegia sepium* and *Solanum dulcamara*. The next phase is a *Phormium*–cabbage tree (*Cordyline australis*) association that includes *Muehlenbeckia complexa* and shrubs of *Coprosma robusta*, *C. propinqua* and *Olearia solandri* (Ogden & Caithness 1982).

Upland tarns and pools

Glacial kettleholes in the Mackenzie region lack outlet channels, and most have widely fluctuating water levels. Turf, interrupted by stones and soil bared by wave or frost action, extends from the lowest water-level to where hard-tussock (*Festuca novae-zelandiae*) grassland begins. In the tarn represented by Table 10.2 there are 42 species, ranging from permanently submerged through facultatively aquatic to dry-land species with short life cycles or some tolerance of submergence. Assemblages vary.

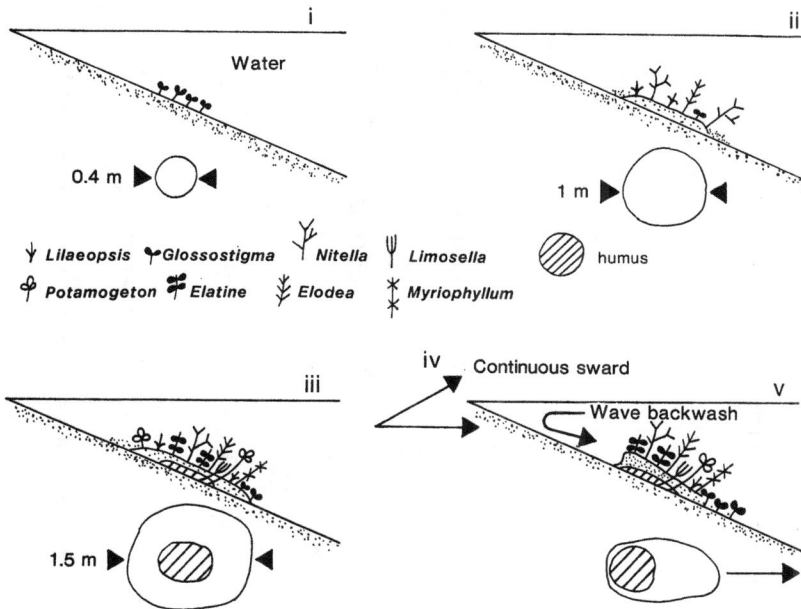

Fig. 10.14. Development of short mixed communities in the Rotorua lakes, Volcanic Plateau. Initial colonisation (i) leads to mounds (ii–iii) and then to either continuous swards in sheltered areas (iv), or erosion of the shoreward end of the mound and 'migration' into deeper water (v) (Brown 1975).

One kettlehole has a zone of the erect weeds *Cirsium vulgare* and *Centaurium erythraea*; in others, incipient bogs are indicated by *Oreobolus pectinatus* and *Herpolirion novae-zelandiae*.

The number of aquatic species decreases with increasing altitude. On sub- and penalpine terrain of hard Cambrian sedimentary rocks in North-west Nelson, hollows inundated for long periods support *Crassula sinclairii*, *Limosella* and *Myriophyllum propinquum*, whereas less frequent submersion favours richer turf communities with *Deschampsia chapmanii*, *D. tenella*, *Agrostis muscosa*, creeping willow-herbs, *Plantago* spp., *Neopaxia australasica*, *Galium propinquum*, *Viola cunninghamii*, *Gnaphalium mackayi* and *Raoulia grandiflora* (P.A. Williams, unpublished). The most abundant macrophyte between 1000 and 1200 m in Westland National Park is *Drepanocladus fluitans*, followed by *Potamogeton cheesemanii* and *Myriophyllum propinquum*. Floating mats of *Isolepis aucklandica* extend from margins. The only other species noted were *Myriophyllum triphyllum* and *Isoetes alpinus* (Wardle 1977).

Thermal water

Thermal waters occur in numerous places in northern New Zealand, and hot springs also follow the Alpine and Hope Faults in the north of the South Island. No macrophytes have been reported from these, except the moss *Campylium polygamum* growing at 20–28 °C in the outflow from springs in the Hurunui Valley (Spenser ER)

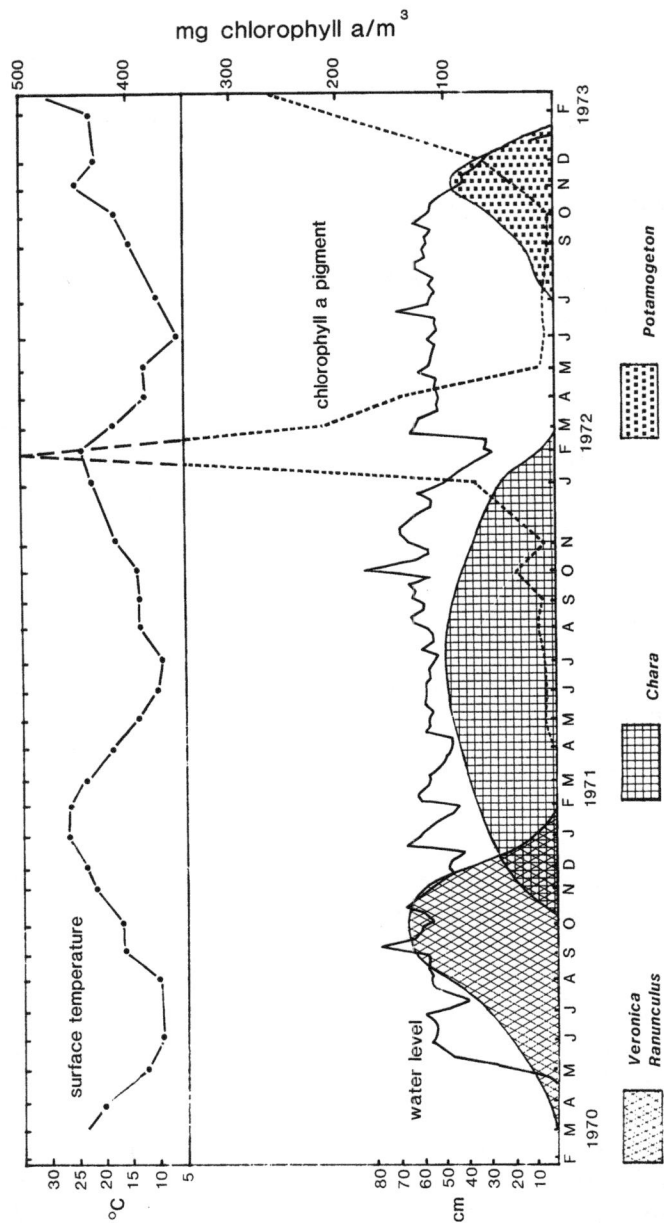

Fig. 10.15. Changes in macrophytes and phytoplankton during three years after near-drying of Pukepuke Lagoon, Manawatu ER (Gibbs 1973).

Table 10.2. *Percentage cover of species in shoreline turf at a kettlehole tarn near Lake Tekapo, Mackenzie ER*

Area of tarn 250 × 200 m; depth of permanent water *c*. 1 m; vertical and horizontal distance from permanent water level to tussock grassland 3 m and 16 m respectively.

Distance from edge of water to far edge of zone (m):	−6	−3	3	8	13	16
Myriophyllum triphyllum	50	30				
Potamogeton ochreatus	10					
Glossostigma sp.	10	+				
Lilaeopsis sp.		10	+			
Myriophyllum pedunculatum		10				
Agrostis 'canina'		10	20	+		
Isolepis aucklandica		10	15	20	20	
Epilobium angustum		10	15	20	10	
Parahebe canescens			20			
Hydrocotyle hydrophila			15	+		
Neopaxia australasica			5			
Galium perpusillum			+	15	20	
Poa lindsayi			+	10	15	
Rumex acetosella			+	+		10
Selliera radicans				35		
Hieracium pilosella				+		40
Wahlenbergia albomarginata				+		10
Luzula sp.				+		10
Carex breviculmis				+		10
Stellaria gracilenta				+		5
Scleranthus uniflorus						5

Source: P.N. Johnson, unpublished data.

(Stark *et al.* 1976). However, the microflora is distinctive; filamentous and mat-forming blue-green algae tolerate 60–65 °C (Winterbourn 1973), and anaerobic, sulphur-metabolising bacteria have optima at 85–90 °C (Patel *et al.* 1986).

Eutrophic lowland swamps

Most lowland swamps in New Zealand occupy former beds of rivers forced to aggrade and change course because of the vast yield of sediments from the ranges. Because of their agricultural potential, all but a fraction have been drained, and most of that has been greatly modified through entry of livestock and weeds. Adventive plants, therefore, figure as prominently as the native. The autecological outline is followed by examples of swamps in Auckland and Wellington provinces, and western, eastern and inland parts of the South Island. Although human activity has been especially

destructive of lowland swamps, drainage and irrigation channels constitute a new
wetland habitat, which is described last.

Monocot herbs

'New Zealand flax,' *Phormium tenax*, forms giant clumps, with leaves up to 3 m tall
that are overtopped by scapes. It grows best on river banks and hillside seepages.
Although abundant in swamps, it thrives only close to channels, being sparse, stunted
and yellowish on poorly drained interfluves. Established plants can persist with
submerged bases but seedlings establish only on emersed ground. *P. tenax* also
tolerates fires and responds to drainage; most 'flax swamps' have been induced by
clearing forest and scrub from wet soils, or partly draining deep swamps. It is
vulnerable to grazing, especially by cattle after fire, and establishment is inhibited by
rank growth, especially of tall fescue.

P. tenax provides one of the world's strongest natural fibres and was cultivated by
the Maori, who recognised special-purpose varieties. Trade in 'flax' began with
European contact, and remained important until the 1920s. The industry then
declined because of competition from countries with lower labour costs, an insect-
transmitted virus (Boyce & Newhook 1953) and drainage of swamps to provide
dairying land. Industrial harvesting virtually ceased by 1975, but use in Maori crafts
has revived.

Sedges are the most abundant swamp plants, and *Carex* is the largest genus. Most
swamp carices are rhizomatous or form loose clumps. They grade from robust species
such as *C. maorica* and *C. lessoniana* that grow in fertile lowland swamps in the warmer
districts, through the very widespread *C. coriacea*, to *C. sinclairii* and *C. gaudichaudiana* in less fertile and mountain swamps. *C. trifida* occurs in southern coastal
regions.

Among species forming dense tussocks, *C. dissita* is widespread, *C. flagellifera* often
dominates transitions to brackish swamp, and *C. diandra* grows mainly in mountain
swamps. *C. virgata* is up to 1 m tall and thrives in deep swamps. The pedestalled
tussocks of *C. secta* begin life on emersed ground or on stumps and logs, but once
established tolerate permanent flooding of their bases. Especially when moribund they
are colonised by tree seedlings and other plants, thereby facilitating succession to
dryland vegetation. *C. ovalis*, which grows in shallow swamps and wet pastures, is the
most widespread adventive sedge.

Cyperus ustulatus is a robust tussock, with conspicuous umbels of dark-brown
spikes, that grows on estuary margins, dune hollows and coastal flushes to as far south
as Fiordland and mid-Canterbury. Other *Cyperus* spp. are naturalised, mostly in wet
pasture, ditches and wasteland in northern and central lowlands. The largest native
sedge, *Gahnia xanthocarpa*, usually grows in wet, warm-temperate forest, but survives
fires that transform forest to secondary swamps.

The rush-like sedges *Baumea juncea* and the 2 m tall *B. articulata* are abundant, and
Schoenus carsei less so, in deep swamps in the northern zone, whereas *S. pauciflorus* is
mainly a mountain sedge that dominates shallow swamps and flushes as well as
growing in wet grassland and on dripping cliffs. *Eleocharis gracilis* also grows in

shallow swamps. Small sedges sprawling on the mud of temporary pools range from the delicate *Schoenus maschalinus* and *Isolepis reticularis* to the rather stout *I. prolifer*.

Some 40 species of true rush (*Juncus* spp.) are prominent in damp pasture and fertile wetlands, especially those that have been grazed and compacted. The native *J. gregiflorus* ranges from virgin swamps to quite well-drained grassland, the introduced *J. effusus* is almost as wide-ranging although usually in more modified vegetation, and the 2 m tall *J. pallidus* grows mainly in deep swamps and drains. *J. articulatus*, named for the externally visible septa of its leaves, is a soft, procumbent adventive rush, abundant in wet pastures, modified swamps and lake margins. Its stiffly erect relative *J. canadensis* widely dominates marshy forest clearings in the west of the South Island. Raupo enters both fertile and infertile swamps, *Sparganium subglobosum* is in lowland swamps of the North Island and sparingly on western lowlands of the South Island, and the orchid *Spiranthes sinensis* is locally common.

Toetoe (*Cortaderia* spp.) are giant tussock grasses that extend plume-like panicles to a height of 3–4 m. They flower in spring, unlike the autumn-flowering introduced cortaderias or pampas grasses. All toetoe except *C. splendens* (p. 354) grow in swamps, but are more characteristic of stream banks.

Hierochloe redolens is locally abundant in fertile, shallow swamps, but in most places has succumbed to grazing and competition, especially from tall fescue. *Isachne globosa* is still common in northern swamps, whereas *Amphibromus fluitans* is rare. *Deschampsia caespitosa*, an otherwise north-temperate species, is indigenous in New Zealand swamps, mainly in the mountains. Adventive grasses become dominant as swamps are converted into pasture.

Trees, shrubs, dicot herbs, ferns and mosses

Filiramulate shrubs, chiefly coprosmas, are often abundant in shallow swamps on alluvial flats. Manuka frequents drier ground, especially bordering estuarine swamps. Cabbage trees are invariably prominent where wooded lowland swamps have been burnt. Floods carry crack-willow debris into swamps, some of which are massively infested; open kahikatea forest has also been invaded in this manner. Grey willow (*Salix cinerea*), despite its smaller stature, can also suppress native swamp vegetation, and is potentially an even greater threat as both sexes are established and there is spread by seed.

Dicot herbs are generally insignificant in unmodified fertile swamps. *Epilobium pallidiflorum, E. insulare* and the scarce *Urtica linearifolia* are obligate swamp dwellers. *Callitriche petriei* grows mainly on wet mud. *Gunnera prorepens* and *Mazus radicans* appear where swamp merges into grassland. Most other native species are incidental on dry enclaves. Introduced dicots, however, are prominent in modified swamps. *Veronica anagallis-aquatica, Mimulus guttatus, M. moschatus, Ranunculus flammula* and *R. sceleratus* grow in pools and on wet mud. *R. repens, R. acris, Polygonum* spp. (including the native *P. salicifolium*), *Rumex* spp., *Galium palustre, Stellaria alsine* and *S. graminea* are abundant in transitions to wet pasture.

The most widespread ferns of fertile wetlands are in the *Blechnum* 'capense' complex: *B.* 'black spot' forms dense, tall colonies along channels, and the smaller *B.*

minus grows in more stagnant parts. Some northern swamps contain the tropical ferns *Thelypteris confluens* and *Cyclosorus interruptus*. *Sphagnum* is important in transitions to infertile swamps and mires. Especially in mountain districts, *Polytrichum commune* and *Breutelia pendula* form dense mats in seasonally-wet depressions and flushes respectively. Other common swamp mosses are *Climacium dendroides* and species of *Bryum*, *Brachythecium* and *Drepanocladus*. *Camptochaete angustata*, *Hypnodendron marginatum* and *Hypopterygium* spp. grow in wet, shaded places.

Northern zone

Two off-shore islands in Auckland Province provide contrasting examples, one of an almost pristine swamp, and the other greatly modified by farming. The third example represents the swamplands of the Waikato basin, in which both native and adventive plants have prominent roles.

Mayor Island (Coromandel ER)

Swamps surrounding two lakes in the volcanic crater of Mayor Island are in near-primitive condition, except for local invasion by *Salix alba* (Bayly et al. 1956). *Typha* prevails in depths of 75–90 cm. Where water averages 70 cm deep, *Baumea rubiginosa*, *B. articulata* and *B. juncea* are dominant, with *Polygonum salicifolium* and *Eleocharis acuta* locally abundant. *Eleocharis sphacelata* has colonised a lake edge in water 1 m deep. Towards the swamp margin, *Carex virgata* and *C. secta* appear, followed by *Cyperus ustulatus* in less than 25 cm of water. *Phormium tenax* is common on one shallow margin. *Isolepis inundata* occupies wet mud and shallow water at the forest edge.

In an area with water 50 cm deep there are cabbage trees, *Blechnum minus* and *Aster subulatus*. Death of some *Carex secta* tussocks and manuka bushes that colonised them suggests that water levels had been higher than usual for some time.

Thick, semi-floating sphagnum covering an inlet supports plants of bogs and successions to forest: abundant manuka, *Astelia* sp., *Phormium* and *Drosera binata*, together with *Coprosma* spp., *Myrsine australis*, five-finger (*Pseudopanax arboreus*), kamahi (*Weinmannia racemosa*), *Dracophyllum strictum*, *Baumea juncea*, *Carex secta*, *Hypolepis* sp., *Schizaea fistulosa*, *Blechnum 'capense'*, *Gleichenia dicarpa*, *Cyclosorus interruptus*, and the mosses *Dicranoloma billardierei*, *Ptychomnion aciculare* and *Thuidium furfurosum*.

Motutapu Island (Hauraki Gulf)

In modified wetlands on Motutapu Island, species fall along a moisture gradient (Fig. 10.16). In an ungrazed, permanently flooded area, the pink-flowered form of *Calystegia sepium* twines among raupo, and quaking swamp is formed by *Schoenoplectus validus*, *Isolepis prolifer*, *Eleocharis acuta*, *Isachne globosa*, *Agrostis stolonifera*, *Holcus lanatus*, *Poa trivalis*, *Epilobium pallidiflorum* and *Polygonum* sp. Trampling and grazing favour species such as *Isolepis prolifer*, *I. sepulcralis*, *Juncus articulatus*, *Glyceria declinata*, *Rorippa nasturtium-aquaticum* and *Callitriche stagnalis*, that vigorously colonise bare mud by seed or vegetative spread. Margins with pasture are the prime habitat for *Juncus articulatus*, *Mentha pulegium* and *Ranunculus repens*.

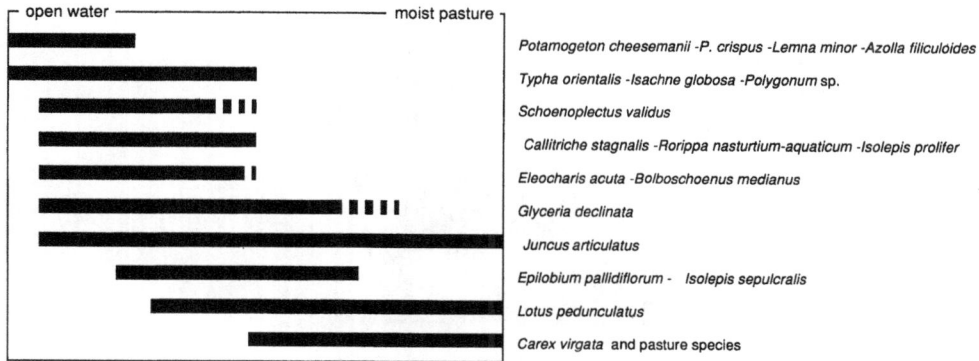

Fig. 10.16. Position of plants along a moisture gradient in a wetland on Motutapu Island, Auckland ER (Esler 1980).

Waikato Basin

The lower Waikato wetlands are still the most extensive in the North Island, although reduced from their original extent of nearly 200000 ha to less than 34000 ha in 1978. The largest single system is the 7000 ha Whangamarino Swamp. Ogle & Bartlett (1981) map 22 vegetation types that fall into three groups. The first is mostly swamp near streams, on recent, gleyed, alluvial soils with pH averaging 4.2 and underlain by peat and logs. *Salix cinerea* or *S. fragilis* form either pure stands, or overstoreys to remnants of native vegetation (*Baumea, Phormium* or *Carex*) or introduced herb and grasses. For similar vegetation on the margins of the Kopuatai peat dome to the east, species recorded in at least two sample plots are the dominant willows, cabbage tree, the shrubs *Ligustrum sinense* and *Solanum pseudocapsicum*, blackberry (*Rubus fruticosus* aggr.), the terrestrial herbs *Bidens frondosa, Hydrocotyle novae-zeelandiae, Ludwigia peploides, Myriophyllum propinquum, Polygonum hydropiper, Ranunculus flammula, R. amphitrichus, Solanum nigrum, Baumea tenax, Carex fascicularis, C. scoparia, Isolepis prolifer, Blechnum 'capense'* and *Histiopteris incisa*, the epiphytes *Asplenium flaccidum, Phymatosorus diversifolius* and *Pyrrosia serpens*, and the floating *Lemna minor* and *Azolla pinnata* (Irving *et al.* 1984).

The second group of communities in the Whangamarino Swamp is native-dominated, and ranges from deep swamps with reeds such as *Baumea articulata*, through shallower swamps with manuka, cabbage tree and *Phormium tenax*, to rain-fed, actively-growing peat bogs with pH around 3.4 supporting manuka, *Baumea teretifolia, B. huttonii, Schoenus brevifolius, Empodisma* and *Gleichenia* spp.

The third group occupies swamp margins and includes pasture, cabbage tree groves and a few remnants of kahikatea forest. Stumps indicate that the last occupied a third of the swamp, until overcome by flooding and peat growth. At Kopuatai, rough pasture on alluvial soils at the margin of the peat dome consists mainly of adventive species reflecting varied drainage; for example, dryland plants such as white clover (*Trifolium repens*), gorse (*Ulex europaeus*), *Holcus lanatus* and *Vicia sativa*, wet-ground plants such as *Rumex conglomeratus* and *Ranunculus repens*, swamp plants such as

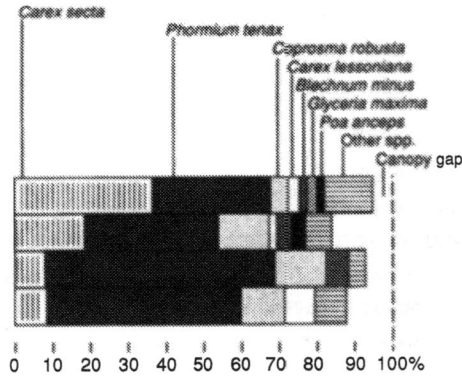

Fig. 10.17. Percentage canopy cover in four *Phormium tenax* stands at Plimmerton swamp, near Wellington, derived from intercepts at each of 50–60 random points (from Bagnall & Ogle 1981).

Glyceria spp., *Juncus articulatus* and the native *Baumea articulata*, and water-plants such as *Ceratophyllum*, *Myosotis laxa* and *Myriophyllum aquaticum*.

Southern North Island

Between the Volcanic Plateau and Cook Strait, lowland swamps are few and mostly highly modified. Those bordering Lake Wairarapa are the most extensive, but a 30 ha semi-coastal swamp on peaty silt near Plimmerton is the best known. Bagnall & Ogle (1981) recognise 24 vegetation types, the most important being dominated by *Phormium tenax* and carices. *Phormium* mostly grows as dense stands with scattered *Corprosma robusta*, *Hebe stricta* and *Carex secta*, but in wetter variants co-dominates with *Cortaderia toetoe*, *Carex secta*, *C. lessoniana*, raupo or *Baumea rubiginosa* (Fig. 10.17). The sedgeland largely consists of dense *Carex lessoniana*, but variants have abundant rushes (*Juncus sarophorus, J. gregiflorus*), or lower tiers of *Ranunculus repens* or *Agrostis stolonifera*. Vascular species total 165, 82 being adventive.

Thirty years earlier, the swamp had been grazed, and seemed likely to be taken over by introduced plants. Since then, a rising water table necessitated removal of livestock, dense *Phormium* has tended to replace communities which it co-dominated with sedges, toetoe or *Typha*, and grazed mats of *Centella uniflora*, *Schoenus nitens* var. *concinnus*, *Eleocharis gracilis*, *Gnaphalium 'collinum'*, *Lagenifera cuneata* and *Hydrocotyle moschata* have disappeared. The adventives *Lonicera japonica*, *Glyceria maxima* and, on very wet ground, *Bidens frondosa* are still locally aggressive, but dense herbaceous vegetation is preventing increase of broom (*Cytisus scoparius*), gorse, blackberry, *Salix cinerea* and most native shrubs.

Western South Island

Much of lowland western Nelson and Westland consists of wetlands occupying infilled estuaries and former flood plains. North of latitude 43° 20′ most have been modified by drainage and grazing, but to the south there have been few attempts at

Fig. 10.18. *Phormium tenax* and carices in a shallow swamp, with young kahikatea forest behind. Near Paringa, South Westland.

drainage, grazing although widespread is only locally intensive, and adventive plants are generally unimportant.

Westland

In central Westland fertile swamps with pH in the range 5.7–6.0 occur in areas with moving water or influx of silt, or that have been recently impounded (Wardle 1977). The most abundant plant is *Carex coriacea*, but on less fertile sites *C. gaudichaudiana* joins or replaces it. Two other rhizomatous carices, *C. sinclairii* and *C. geminata*, also occur. *Phormium tenax* is present throughout, and although generally stunted, is tall and dense where drainage is good. *Astelia grandis* can be important, especially in very wet wooded swamps (Fig. 10.18). *Carex secta* co-dominates with *C. coriacea* in deeper water, and in depths over 0.6 m can form pure colonies. *Coprosma propinqua* or *C.* aff. *parviflora* give cover up to 20%, the former mostly in fertile *Phormium* swamps, the latter in less fertile sedge swamps.

Sphagnum, Blechnum 'black spot', *B. minus, Myrsine divaricata, Lotus pedunculatus, Viola lyallii*, cabbage tree and *Juncus gregiflorus* can be abundant. *Juncus articulatus* fills pools, *Carex gaudichaudiana* extends into ponds from the margins, and *Eleocharis acuta* forms colonies in deeper water. Other common species include *Ranunculus flammula, Hydrocotyle novae-zeelandiae* (pilose form), *Hebe salicifolia* var. *paludosa, Elatine gratioloides, Glossostigma elatinoides, Gratiola sexdentata* and *Carex virgata.*

Fertile swamp often borders kahikatea forest, and fire has extended sedge-dominated areas at the expense of forest-swamp transitions. Young plants of kamahi, kahikatea, lancewood (*Pseudopanax crassifolius*) and, in frost hollows, *Libocedrus bidwillii* indicate primary or secondary successions to forest.

Rhizomatous carices, especially *C. coriacea*, are often a major component of rough pastures on open flats. Although they occupy wet hollows and flushes in natural seral grassland, their wide extent today results from tolerance of burning, grazing and silting, which tend to eliminate *Phormium* and potential woody successors.

Karamea (North-west Nelson)

A partly modified swamp near Karamea illustrates the floristic richness possible in a small area: 56 vascular species were listed from 100 m², as follows.

Shrubs and trees (5% cover): *Coprosma* aff. *parviflora, C. tenuicaulis, C. propinqua × robusta, Carpodetus serratus, Hebe salicifolia, H. gracillima, Myrsine divaricata*, cabbage tree; seedlings of kahikatea, *Elaeocarpus hookerianus* and *Griselinia littoralis*.

Rushes and rush-like sedges: *Baumea rubiginosa* (10%), *Eleocharis acuta, E. gracilis, Juncus bulbosus, J. articulatus, J. canadensis, J. effusus, Schoenus maschalinus*, a small tufted *Isolepis*, and *I. prolifer*.

Other monocots: raupo (10%), *Carex sinclairii* (60%), *C. maorica, C. coriacea* aggr., *C. secta, Gahnia xanthocarpa, Holcus lanatus, Glyceria fluitans, Phormium tenax, Astelia grandis, Wolffia, Potamogeton suboblongus* and *Microtis unifolia*.

Dicot herbs: *Ranunculus amphitrichus, R. repens, R. acris, Stellaria alsine, Sagina procumbens, Epilobium pallidiflorum, E. chionanthum, E. insulare, E. pedunculare, Lotus pedunculatus, Hydrocotyle novae-zeelandiae* var. *montana, Plantago australis, Nertera depressa, Myosotis laxa, Gnaphalium limosum, Senecio minimus, Prunella vulgaris, Mimulus moschatus* and *Parentucellia viscosa*.

Ferns: *Blechnum* 'black spot' and *B. minus* (20%), *Pteridium esculentum* (5%) and *Paesia scaberula*.

The main bryophytes are *Philonotis pyriformis, Sphagnum cristatum* and *Marchantia* ?*berteroana*.

Eastern South Island

Before European colonisation, swamps were extensive on eastern South Island lowlands, especially along the seaward margin of the Canterbury Plains (Fig. 10.19) and over the Taieri Plain near Dunedin. Surveyors' maps refer to raupo, flax (*Phormium*), tussock (presumably *Carex secta*) and toetoe (*Cortaderia richardii*). Buried logs reveal that much of this had replaced podocarp forest, as a result of either flooding and aggradation, or Maori fires. These wetlands have been reduced to modified remnants, those around Lake Ellesmere being the most extensive and varied. Communities are described by D.J. Clark & T.R. Partridge (unpublished) as follows.

(1) *Juncus articulatus* dominant, often on quaking swamp and associated with
Mimulus guttatus, Holcus lanatus, Cynosurus cristatus and *Alopecurus*

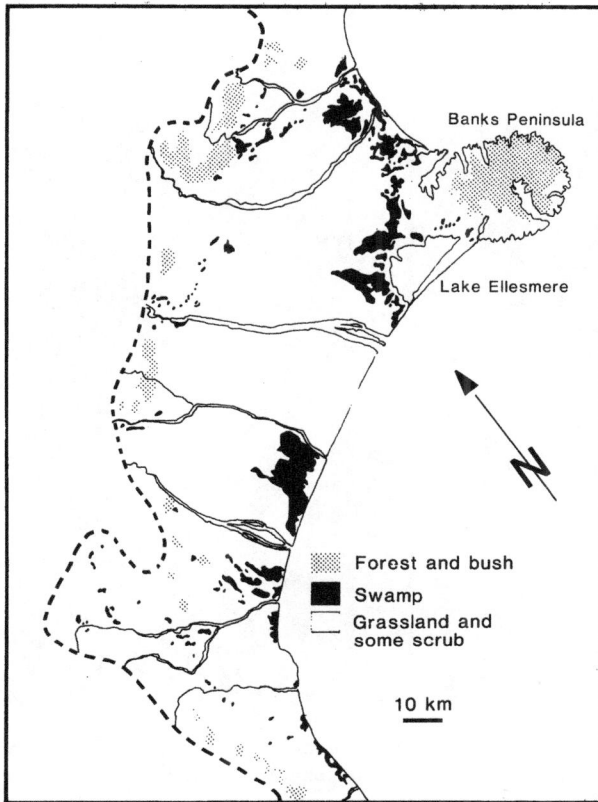

Fig. 10.19. Distribution of swamp, bush and forest on the Canterbury Plains and downland and Banks Peninsula at the beginning of European settlement (Johnston 1961).

geniculatus. Less common are *Eleocharis acuta, Microtis unifolia, Carex maorica, C. sinclairii, Juncus caespiticius, J. effusus, J. bufonius, Mentha × piperita* and *Trifolium repens.*

(2) *Schoenus pauciflorus* with *Holcus, Cynosurus, Agrostis stolonifera, Alopecurus geniculatus, Anthoxanthum odoratum* and *Juncus articulatus.*

(3) *Carex sinclairii* over a lower tier of *Agrostis* and *Potentilla anserinoides.*

(4) *Carex coriacea* over *Agrostis* with *Trifolium fragiferum, Potentilla, Rumex crispus,* and on boggy ground *Juncus articulatus, J. caespiticius, Polygonum hydropiper, Myosotis laxa, Lotus pedunculatus* and *Mimulus guttatus.*

(5) *Carex secta* up to 2 m tall on very wet ground.

(6) Raupo with either *Agrostis* or *Mimulus* and *Mentha.*

(7) *Phormium tenax* with *Juncus gregiflorus, J. distegus, Agrostis, Holcus* and *Anthoxanthum.*

(8) *Phormium* with *Carex secta, Juncus articulatus, Mimulus* and grasses.

(9) *Juncus gregiflorus* with *Agrostis; J. distegus* is locally dominant. This grades into wet pasture with scattered rushes.

Fig. 10.20. Shallow lake with mainly *Carex diandra* in foreground, *C. secta* in deeper water, and invading willows; 500 m, Mackenzie ER. Ben Ohau Range is in the background.

(10) *Cortaderia richardii* with *Mimulus* or pasture grasses; other plants include *Polygonum, Lotus* and *Ranunculus repens.*

Superimposed on or more-or-less supressing these communities are thickets of *Salix fragilis* to 10 m tall and *S. cinerea* to 4 m tall; these may be open or dense, and pure or mixed.

Inland South Island

Alluvial and fluvioglacial plains and broad terraces among the eastern South Island mountains support numerous wetlands (Figs 10.20 and 10.21). Until recent decades most were in near-primitive condition, but latterly draining, top-dressing and, especially, increased use of cattle have greatly affected their quality and extent. Many lowland plants are absent (e.g. *Baumea* spp.) or altitudinally limited (e.g. *Phormium tenax*). Eutrophic or mesotrophic systems are favoured in the lower and drier valley reaches; infertile swamps and bogs become more prominent as rainfall or altitude increase. For the Cass basin, lying at 600 m a.s.l., Dobson (1977) relates the distribution of major species to pH and conductivity (a measure of ionic concentration) (Fig. 10.22) and describes the following communities.

Fig. 10.21. A kettlehole tarn at 980 m, Mackenzie ER. The elongated strips of vegetation are mainly *Schoenus pauciflorus*, *Sphagnum* and *Carex* spp. and probably represent interaction between strong north-west winds and a slowly rising water table.

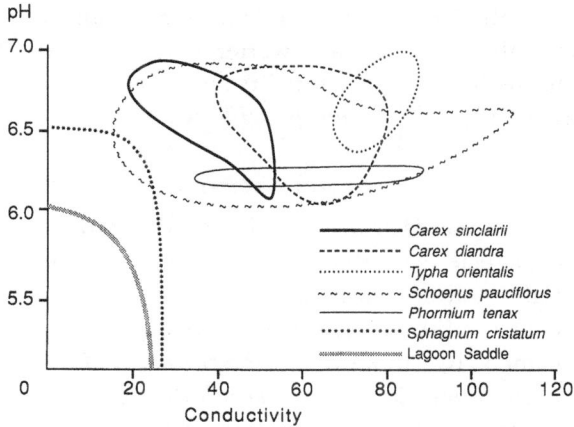

Fig. 10.22. Distribution of principal wetland species at Cass (Puketeraki ER) in relation to pH and conductivity (micromhos/cm). The Lagoon Saddle curve refers to all main communities there (p. 341) (Dobson 1977). Copyright Plant Science Department, University of Canterbury, Christchurch, N.Z.

(1) Raupo is most abundant as pure stands in slowly moving, relatively fertile water, extending to depths of 2 m.

(2) *Phormium* is confined to very wet, less acid areas, in association with *Carex secta, C. diandra, C. sinclairii* and *Eleocharis acuta.* Other common plants are *Juncus articulatus, J. gregiflorus, Holcus lanatus* and *Myosotis laxa.*

(3) *Carex secta* can form pure stands where water levels fluctuate. The pedestalled tussocks are healthy in open water, but elsewhere are colonised by *Coprosma propinqua, Blechnum* spp., and several grasses including *Hierochloe redolens.*

(4) *Carex coriacea* dominates on wet, flood-deposited gravel and silt.

(5) *Schoenus pauciflorus* dominates extensively in shallow valley-floor swamps and flushes on the slopes, which it ascends to the penalpine belt. *Breutelia pendula* may form a near-continuous mat beneath. Associated small herbs include *Blechnum penna-marina, Cerastium fontanum, Linum catharticum, Viola cunninghamii, Hydrocotyle sulcata, Celmisia gracilenta, Myosotis laxa, Mentha cunninghamii, Mimulus moschatus* and pasture grasses. *Carex diandra* increases to dominance on wetter sites transitional to raupo swamp, whereas *C. sinclairii* and *Sphagnum cristatum* increase on less fertile sites. *Juncus articulatus* seems associated with disturbance.

(6) Shallow, infertile swamps are invaded by *Sphagnum cristatum* (forming mounds) and *S. falcatulum* (in pools), which rise to dominance by raising acidity (p. 322). *Polytrichum commune* is also important, and seems to increase during droughty summers.

Dobson's account applies to most wetlands in wide inland valleys, except that in many localities, especially those colder, wetter or higher than Cass, red-tussock grassland merges into medium- to low-fertility swamps, in which red tussocks stand over a tier consisting mainly of *Schoenus pauciflorus.*

Ditches and water-races

Ditches de-watering swamplands or bringing irrigation water to dry lowlands contain variable florulas. Most species are adventive, but these channels also support native wetland plants in totally modified landscapes.

In Manawatu ditches *Callitriche stagnalis* grows on wet mud or in water. Taller communities are formed by *Rumex conglomeratus, Polygonum salicifolium, P. hydropiper, Bolboschoenus fluviatilis, Cyperus eragrostis, Glyceria declinata,* or the very robust grasses *G. maxima* and *Phalaris arundinacea,* which can exclude other plants. Aquatic assemblages contain *Potamogeton crispus, Elodea, Egeria* and *Myriophyllum aquaticum; Alisma plantago-aquatica* grows on the banks (Esler 1978a). Along irrigation channels on the Canterbury Plains, the native *Blechnum 'capense'* forms linear colonies among *Agrostis* and *Bromus* spp., cocksfoot (*Dactylis glomerata*), *Festuca rubra, Trifolium hybridum* and *Mentha ×piperita* (Healy 1961); bracken (*Pteridium esculentum*), *Polystichum vestitum, Phormium tenax, Carex secta* and *Cortaderia richardii* also occur.

Oligotrophic lowland mires and wet heaths

The species

The characteristic sedges of infertile wetlands are rush-like. *Baumea rubiginosa* and *B. teretifolia* are abundant in lowland mires, the former in pools throughout and the latter on firm ground almost as far south as Fiordland. *B. huttonii* is widespread in Waikato wetlands, but in the South Island is reported only from coastal mires. *Schoenus brevifolius* is confined to the northern zone, whereas the very slender *Tetraria capillaris* extends to north Westland. On gumlands these are joined by *Schoenus tendo* and *Lepidosperma filiforme*, which are rush-like, *L. laterale*, which has flattened, leaf-like stems, and *Morelotia affinis*, which is a short, leafy tussock related to *Gahnia*.

Other leafy sedges mostly belong to better-aerated or more fertile parts of the habitat complex. The most widespread carices are *Carex gaudichaudiana* (in reduced form) and the slender *C. echinata*. The tufted *Carpha alpina* is abundant in mountain and southern mires where there is some water movement. The large, scabrid-leaved tussock *Gahnia rigida* abounds in many infertile swamps and secondary wet heaths in Western Nelson and Westland, with rare North Island occurrences.

Restiads can be prominent. *Empodisma minus* and *Leptocarpus similis* occur throughout the main islands, the former extending to the subalpine belt. Where most vigorous, *Empodisma* has dense 50–100 cm tall, scrambling, wiry stems that suppress most competitors, but as sparse, 2–4 cm tall shoots it extends on to cushion-bogs. The slender rhizomes produce exceedingly fine roots with copious hairs, that cover the ground with a dense felt holding up to 15 times its own mass of water (Campbell 1964). *Empodisma* can be killed by fire; circular, spreading colonies are evident in the re-invasion stage. *Sporadanthus traversii* has thicker, stiffer stems up to 2 m tall. Thick roots with large air-spaces descend from an interweaving system of 10 mm thick rhizomes, which form a rigid raft 5–60 cm below the bog surface but just above the water-table. It is confined to Chatham Island and a few peat domes in the Waikato basin.

Red tussock or vicarious relatives (p. 217) are important in many upland and southern mires and wet heaths. There are usually also *Rytidosperma* spp., and other grasses may be incidentally present. Although cushion plants are most characteristic of far-southern and mountain wetlands (p. 338), *Oreobolus pectinatus* and *Centrolepis ciliata* are often common in South Island lowland mires and at higher altitudes in the North Island. Usually mat or turf forms are represented only by *Nertera scapanioides* and the small sedge *Oreobolus strictus*, but in Southland lowland mires also support mat plants characteristic of higher altitudes. The least fertile wetlands contain the moss-like umbellifer *Actinotus novae-zelandiae* and *Liparophyllum gunnii*, in which stout rhizomes, submersed in peaty pools, bear rosettes of tiny leaves.

Infertile wetlands are the main habitat for insectivorous plants. *Drosera binata*, with forked leaves up to 30 cm tall, grows in lowland swamps. *D. stenopetala*, *D. arcturi*, and the small, rosette-forming *D. spathulata* and *D. pygmaea* grow on bare peaty ground in wet shrub-heath, subalpine bogs, and herbaceous mires respectively. *Utricularia monanthos* or *U. novae-zelandiae* are common in shallow peaty or muddy pools and lake edges. *U. lateriflora* is a northern plant of similar habit, whereas *U. australis*, also northern except for an occurrence near Westport, is a tall, slender aquatic growing in deep holes in peat.

Bogs and wet heaths are prime habitat for orchids, especially thelymitras. Celmisias in the *C. gracilenta–graminifolia* complex are abundant in mires and wet heaths except on North Island lowlands. *Coprosma* aff. *intertexta* has linear leaves and slender branches that weave through sedge and fern in lowland infertile swamps in the South Island. *Cyathodes empetrifolia* and *Lepidothamnus laxifolius* adopt a similar habit on dryish southern and upland mires.

Gleichenia, a rhizomatous fern with wiry, branching fronds that can straggle to 1 m in tall vegetation or reduce to a few centimetres in cushion-bog, is one of the most ubiquitous plants of infertile wetlands. *G. dicarpa* usually predominates in open places, but in partial shade is replaced by *G. microphylla*, which has pinnules that are flattened instead of being minute and hooded. In openings among taller vegetation, firm ground supports tufts of comb-fern (the small *Schizaea fistulosa* generally, the larger *S. bifida* mainly in the northern zone), and mats of *Lycopodium laterale* or *L. ramulosum*. *L. serpentinum* is found on a few northern bogs and wet heaths.

Sphagnum mosses store water in inflated leaf cells, and because of the large surface area presented for ion exchange, take up nutrients at very low concentrations. Through satisfying their water and nutrient requirements from rainfall, they can outcompete vascular plants in habitats where the latter must draw water and nutrients from cold, infertile, poorly aerated soils; and the mosses themselves induce an acid, saturated environment that promotes rapid accumulation of sphagnum peat, thereby transforming swamp into bog.

Although sphagnum is abundant throughout the wetter and cooler parts of New Zealand, it dominates only over small areas. The main habitats are wet, sheltered depressions, within vegetation ranging from lowland forest and swamp to subalpine grassland and bog. *S. falcatulum* growing in shallow water and *S. cristatum* forming hummocks on emersed ground (Fig. 10.23) are by far the most abundant of the eight species. *S. cristatum* is gathered for export, the most productive sites being in logged forest where numerous wet hollows provide ideal conditions. Growth of sphagnum in harvesting areas is estimated as 3–9 cm or 1–5 wet tonnes/ha annually (J.A. De Goldi, unpublished).

On wet heaths and bogs *Campylopus* spp. form dense, dark brown mats, and *Dicranoloma robustum* forms light brownish-green cushions on drier ground. *Pulchrinodus inflatus* is also common. The minute liverwort *Cephaloziella* often forms a weft over peaty ground. *Marchantia berteroana* colonises surfaces where mosses have been killed by drought, fire or trampling.

Low fertility and waterlogging inhibit erect woody plants. Where there has been no burning, trees and shrubs grade from low and sparse near the centres of wetlands, to increase in height, density and number of species towards margins where there is better aeration and flux of nutrients. Manuka is the most tolerant and adaptable, often growing as rhizomatous mats among cushion plants. On less extreme sites, erect manuka bushes grow among the prostrate ones, the former having thickened, aerenchymatous cortex just below ground level (Cook *et al.* 1980). Genetically

Fig. 10.23. *Sphagnum cristatum* invading an *Oreobolus pectinatus* cushion; 630 m, Mavora Lakes, inland Southland.

prostrate manuka on wet heaths in Western Nelson attests to the antiquity of the habitat. Several slender dracophyllum species and *Epacris pauciflora* are confined to infertile wetlands and equivalent heathlands, the latter being common in Western Nelson and parts of the North Island.

Oligotrophic mire communities

The largest oligotrophic mires, whether infertile swamp or raised bog, lie on alluvial plains. Some occupy older surfaces than their eutrophic equivalents, whereas other wetlands grade from eutrophic margins to oligotrophic central areas that are not flushed by nutrient-bearing water. Communities are described in a sequence from Northland through the Waikato Basin and the Volcanic Plateau, to Westland and Southland. No significant examples occur in the dry eastern lowlands of the two main islands.

Northland

One of the northernmost mires is Kaimaumau Swamp in Aupori ER. The deepest part, where water is impounded by high coastal dunes, is mostly covered by dense *Baumea juncea*, with sparse raupo and occasional sphagnum cushions supporting *Empodisma*, stunted manuka, *Lycopodium serpentinum* and *Drosera binata*. An area of tall *B. articulata* and raupo contains *Thelypteris confluens* and *Cyclosorus interruptus*. The swamp margin is dominated by *B. juncea* and *B. rubiginosa*, with smaller amounts of *B. huttonii*, *Leptocarpus* and, on drier ground, *B. teretifolia*.

On the much drier peat surface further inland, manuka dominates over a mosaic of *Gleichenia dicarpa, Schoenus brevifolius* and *Baumea teretifolia*. Other species include *Lycopodium deuterodensum, Cyathodes fasciculata, Hakea sericea, Dianella nigra, Microtis parviflora* and the only New Zealand population of the large Australian orchid *Cryptostylis subulata*. Kanuka (*Kunzea ericoides*) grows among bracken on mounds. Pollen analysis indicates that forest was extensive until removed by early Maori fires (M.S. McGlone, personal communication); wood occurring from the surface down includes large stumps and fallen trunks of kauri. Leached sand ridges interrupting the peat carry similar vegetation, with the addition of *Dracophyllum lessonianum, Pomaderris* spp., *Cyathodes fraseri, Pimelea prostrata, Lepidosperma laterale* and *Morelotia affinis*, plants typical of dryish gumlands.

Waikato Basin

On the 8765 ha Kopuatai peat dome, *Sporadanthus* covers 2000 ha where the water table is at or above the surface during summer (Table 10.3, column 1). Where dense, it excludes other plants. In more open stands described from the Moanatuatua dome (Campbell 1964), clumps of *Sporadanthus* and sparse *Epacris pauciflora* form an upper tier. *Empodisma* grows as a dense, 0.6 m tall sward between, and straggles to 1.2 m among *Sporadanthus*. A ground cover of *Lycopodium laterale, Campylopus acuminatus* var. *kirkii* and liverworts is densest where the taller plants thin out.

A low-growing *Schoenus/Empodisma* community (column 2), with areas dominated by *Campylopus acuminatus*, adjoins the *Sporandanthus* stands and presumably indicates a low nutrient status. A *Baumea/Empodisma* community (column 3) is on wetter ground; *Sphagnum falcatulum* grows in pools, and *Marchantia berteroana* rosettes colonise bare, dried peat. A *Baumea/Gleichenia* zone (column 4) is generally further from the dome centre and drier than the preceding types, and the dominants are often dense enough to exclude other plants. It grades into dense manuka up to 3 m tall occupying the mesotrophic fringe of the dome (column 5).

Volcanic Plateau

Near Pureora mires have formed in depressions in tephra erupted in AD 177 from Lake Taupo to the east (p. 525). One at 550 m a.s.l. is dominated by stunted manuka, *Baumea rubiginosa, Lepidosperma australe* and *Gleichenia dicarpa*. Shallow pools support *Juncus planifolius* and *J. articulatus* (which may reflect wallowing by deer), *Oreobolus pectinatus, Centrolepis ciliata, Drosera spathulata, D. binata* and *Utricularia* aff. *monanthos*. Shrubby margins include *Coprosma tenuicaulis, Hebe stricta, Carex secta* and *Gahnia rigida*. The presence of *Carex secta, Baumea teretifolia, Spiranthes* and the aquatic *Sparganium, Eleocharis sphacelata, Potamogeton suboblongus* and *Myriophyllum aquaticum* indicates lowland affinities, whereas *Oreobolus, Centrolepis, Cyathodes empetrifolia* and *Celmisia graminifolia* have mainly higher-altitude distributions. Further species include *Gleichenia microphylla, Nertera scapanioides, Gunnera prorepens* and *Thelymitra venosa* (Wallace 1984).

A mire in a shallow basin at 820 m has peat with a pH range of 4.4–5.7 and depth generally < 50 cm overlying the tephra. Four vegetation zones are recognisable (Fig. 10.24).

Table 10.3. *Constancy in 10 m² quadrats of species on the Kopuatai peat dome*

	Community no.				
	1	2	3	4	5
Trees and shrubs					
Leptospermum scoparium	1	2	3	3	*5*
Epacris pauciflora	1	1		2	1
Salix cinerea (stunted)				1	
Dracophyllum lessonianum				1	1
Insectivorous plants					
Utricularia lateriflora	3	3			
Drosera spathulata	1	2			
Drosera binata	3	3	3		1
Utricularia monanthos		1	2		
Other herbs					
Sporadanthus traversii	*5*				
Empodisma minus	4	*5*	*5*		
Schoenus brevifolius	2	*5*	3		
Baumea teretifolia		3	*5*	*5*	1
Dianella nigra					2
Nertera scapanioides					2
Ferns and lycopods					
Lycopodium serpentinum	1	2			
Lycopodium laterale	2	3		2	
Gleichenia dicarpa	2	1	1	4	2
Schizaea fistulosa		1		1	
Gleichenia microphyllum					2
Hypolepis distans					2
Bryophytes					
Campylopus acuminatus var *kirkii*	3	5	1	1	
Sphagnum cristatum	2	3	2	1	
Sphagnum falcatulum	1	1	3	1	
Bryum ?pseudotriquetrum		1	1		
Campylopus introflexus				1	1
Polytrichum commune					2
Aneura sp.	2				
Goebelobryum unguiculatum	4	3	2		
Aneura ?palmata	2	2	2	1	
Telaranea tetradactyla	2	2	1	1	
Telaranea gottscheana	1	1	1	1	
Telaranea herzogii	1	2	1	1	
Lethocolea squamata			1		1
Lophocolea semiteres					2

Notes:
1, Present in 1–20% of quadrats; 2, 21–40%; 3, 41–60%; 4, 61–80%; 5, 81–100%. Italicised values show community dominants.
Species recorded in only one community, at presence level 1, as follows.
Community 2: *Thelymitra venosa, Dicranoloma billardierei, Thuidium furfurosum.* 3: *Tetraria capillaris, Baumea huttonii, Polytrichum* sp., *Sphagnum subsecundum.* 4: *Sematophyllum amoenum.* 5:
Muehlenbeckia australis, Rubus fruticosus aggr., *Coprosma rigida, C. areolata, C. propinqua, C. tenuicaulis, Hebe stricta, Baumea tenax, Carex lessoniana, C. virgata, Isolepis inundata, Phormium tenax, Cordyline australis, Chiloglottis cornuta, Corybas rivularis, Blechnum 'capense', Histiopteris incisa, Hypolepis ambigua, Pteridium esculentum, Achrophyllum quadrifarium, Breutelia pendula, Ptychomnion aciculare, Sematophyllum contiguum, Chiloscyphus compactus, Riccardia* sp.
Source: Irving *et al.* 1984

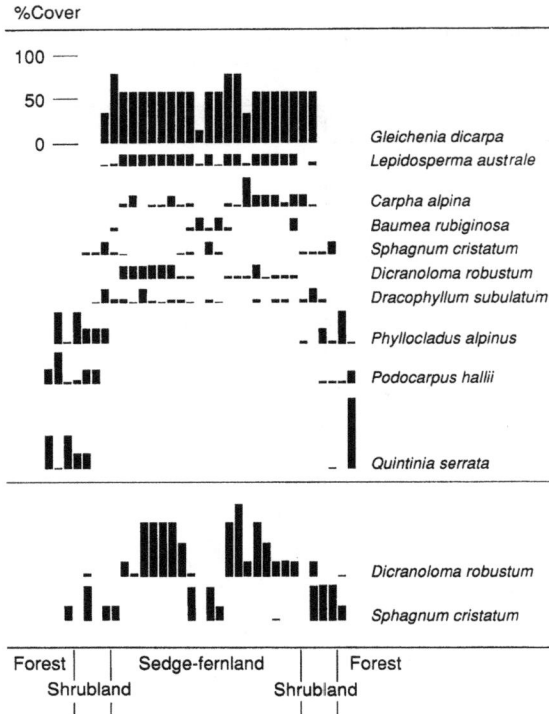

Fig. 10.24. Distribution of common species on a 320 m transect across an upland mire near Mt Pureora, western Volcanic Plateau (Clarkson 1984). Upper panel represents the canopy, lower panel the understorey.

(1) Hummocks of *Gleichenia dicarpa*, associated with *Lepidosperma* and *Dicranoloma robustum*, covering most of the central area where no water movement is apparent.

(2) *Carpha alpina* occupying local flushes between the hummocks.

(3) Abundant *Sphagnum* at the shrubland margin and beside the central drainage stream, but *Baumea rubiginosa* dominant in faster-moving water.

(4) Marginal shrubland of *Phyllocladus alpinus*, bog pine (*Halocarpus bidwillii*) and *Coprosma* spp., accompanied by manuka and *Dracophyllum subulatum*, which also grow on hummocks in the mire.

Western South Island

The infertile swamps and bogs of western lowlands have pH down to 3.8, and substrates varying from peaty silt to wet, structureless peat over 8 m deep. Sequences in Central Westland are described by Wardle (1977). Dominants include *Carex gaudichaudiana*, *Baumea rubiginosa*, *B. teretifolia* on slightly drier ground, and *Gleichenia dicarpa* and *Empodisma* on central areas furthest from drainage channels. *Leptocarpus* and *Lepidosperma australe* are usually present, and can dominate by coastal lagoons. *Sphagnum* varies from isolated hummocks to continuous cover. Other

Fig. 10.25. Gradation from infertile swamp with *Empodisma*, *Baumea* spp. and *Gleichenia dicarpa* growing on peaty silt filling a former lagoon, through *Leptospermum*-dominated tall heath, to forest mainly of *Dacrydium cupressinum* on a former dune ridge. Ohinemaka, South Westland.

frequent or locally important species are *Blechnum 'capense'*, *Drosera binata*, *Centella uniflora*, *Dracophyllum palustre*, *Coprosma* aff. *intertexta*, *C.* aff. *parviflora*, *Utricularia monanthos*, *Euphrasia disperma*, *Centrolepis ciliata*, *Baumea tenax*, *Gahnia rigida* and raupo. Stunted *Phormium tenax* marks transitions to fertile swamp.

Shrubs can occupy much of the surface, manuka up to 3 m tall being the commonest. Silver pine (*Lagarostrobos colensoi*) is also common, whereas bog pine and pink pine (*Halocarpus biformis*) are local. Such woody communities grade into tree- and shrub-heath, and are fire-prone. Manuka, coprosmas and *Olearia virgata* var. *laxiflora* soon re-establish after fire, but the podocarps can be completely eliminated, and their return probably requires a nurse stand of manuka.

At some margins the sequence is from *Baumea–Empodisma–Gleichenia* swamp, through stunted tree-heath, to co-dominance of rimu and silver pine, before forest of rimu and kahikatea on alluvium is reached (Fig. 10.25). The transition may exhibit stages of succession after fire, from herbaceous swamp through manuka to regenerating tree-heath and forest (Fig. 10.26). In a mire in the Paringa Valley, the transition from *Empodisma* with scattered manuka, bog pine and small silver pines on 2 m deep peat to dense, mature kahikatea forest on wet silt loam occupies only 6 m.

New Zealand's most pristine lowland wetland is at Big Bay, around a shallow lagoon separated from the sea by dunes. Fig. 10.27 illustrates relationships between the aquatic habitat, swamp, bog, heath and forest. Profile A extends 200 m from the centre

STAGE

	1	2	3	4	5	6
Canopy height (m)	0.3	1	4	12	20	30
Tree basal area (m²/ha)				46	66	66

Dracophyllum longifolium
Leptospermum scoparium
Lagarostrobus colensoi
Phyllocladus alpinus
Coprosma foetidissima
Coprosma colensoi
Neomyrtus pedunculata
Psuedopanax colensoi
Myrsine divaricata
Dacrydium cupressinum
Podocarpus hallii
Nothofagus sol. var. cliffortioides
Dacrycarpus dacrydioides
Pseudopanax crassifolius
Weinmannia racemosa
Nothofagus menziesii
Elaeocarpus hookerianus
Dicksonia squarrosa
Prumnopitys ferruginea
Pseudowintera colorata

□ 10 %

Relative density: ▭ shrub <5m tall ▨ tree <5cm dbh ■ tree >5 cm dbh

Empodisma minus
Baumea teretifolia
Baumea tenax
Gleichenia dicarpa
Gahnia rigida
Blechnum procerum
Blechnum "black spot"
Uncinia rupestris
Hymenophyllum scabrum
Astelia fragrans
Microlaena avenacea
Blechnum discolor

□ 10 % cover
■ 10 % relative density

Dicranaloma billardierei
Sphagnum falcatulum
Sphagnum cristatum
Hypnodendron sect. Mniodendron
Ptychomnion aciculare

Riccardia ?striolata
Riccardia alcicornis
Kurzia hippuroides
Zoopsis caledonica
Lepidozia glaucophylla
Telaranea gottscheana
Lepidozia kirkii
Tetracymbiella decipiens
Chiloscyphus billardierei
Riccardia oppositifolia
Lepidozia pendalina
Trichocolea mollissima
Lepidozia microphylla
Schistochila nobilis
Schistochila glaucescens

Not recorded

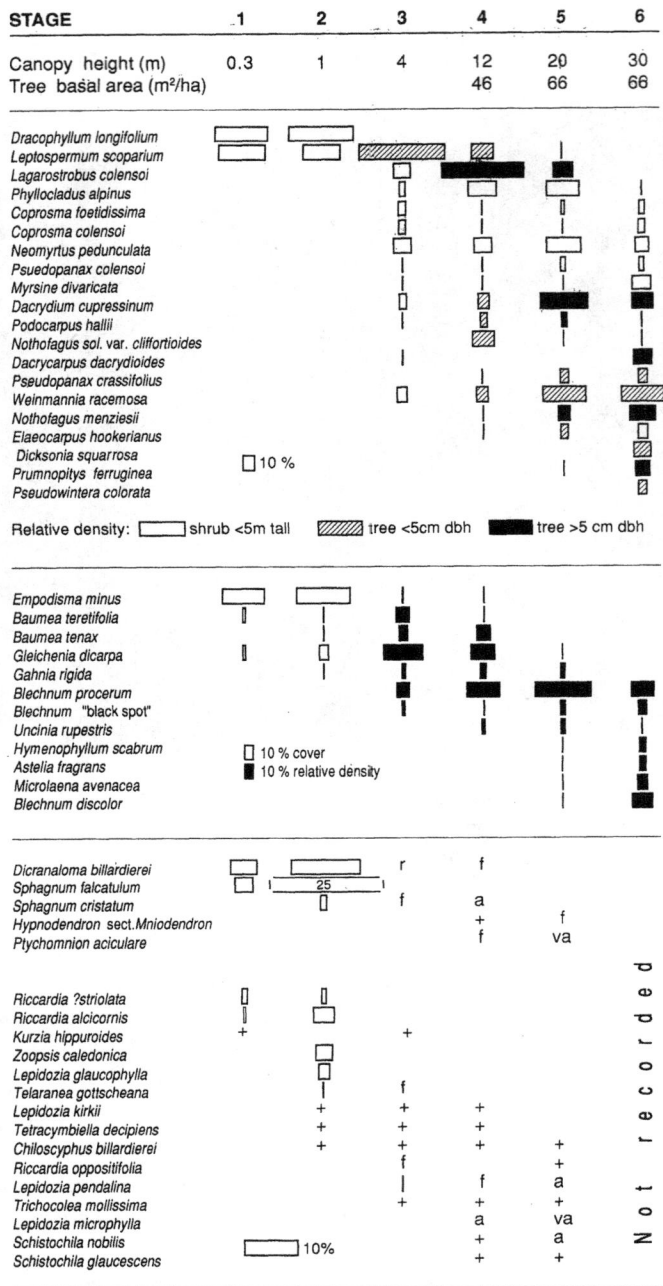

Fig. 10.26. Contributions of the major species to six stands representing a gradation from mire to forest in the Haast district, South Westland. (From Mark & Smith 1975; Scott & Rowley 1975).

Trees and shrubs are shown only in the tier in which they are most important. Terrestrial mosses and hepatics are shown as either cover values based on 5000 points or estimates of abundance: +, rare or occasional; f, frequent; a, abundant; va, very abundant or dominant. Further bryophytes, recorded as + or f in one stage only, are: 4: *Bazzania novae-zelandiae, Lepicolea scolopendra*; 5: *Achrophyllum quadrifarium, Hypnodendron menziesii, Hypopterygium rotulatum, Plagiochila gigantea, Tylimanthus saccatus.*

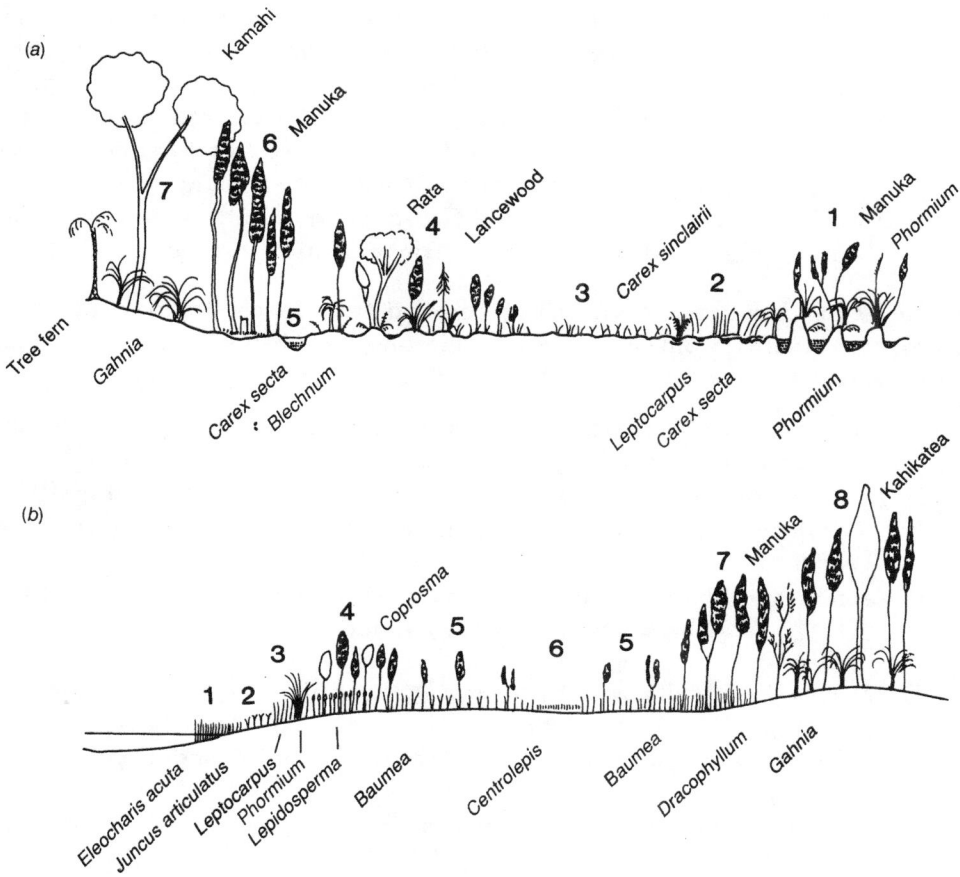

Fig. 10.27. Profiles along wetland transects at Big Bay (Olivine ER). (From drawing by P.N. Johnson.) See text for explanation.

of a peat-filled inlet to the edge of a dune. The following succession from herbaceous swamp to forest is complicated by fertility and drainage gradients, and moving water that delays infilling of pools.

(1) Steep, vegetated pedestals above deep pools in quaking peat; *Carex secta* 2.5 m tall and 30% cover, *Phormium* (20%), manuka 4 m tall and 40% cover.

(2) Hummocks lower and pools shallower; small *Carex secta* (20%), *Blechnum minus* (20%), *Phormium* (10%), *Leptocarpus*, *Baumea teretifolia* and *Carex sinclairii* (10%), *Coprosma propinqua* and manuka (20%).

(3) *Carex sinclairii* providing 80% cover on a flat surface of peat and sediments.

(4) Similar to (1), but further species on pedestals are *Cortaderia richardii*, the shrubs *Carmichaelia arborea*, *Myrsine divaricata*, *Coriaria arborea* and *Olearia avicenniifolia*, and young trees of *Pseudopanax crassifolius* and *Metrosideros umbellata*.

(5) A stream with *Juncus articulatus* and *Potamogeton cheesemanii*, and muddy channels with *P. suboblongus, Isolepis inundata, I. reticularis, Triglochin striatum, Hydrocotyle sulcata, Callitriche petriei* and *Ranunculus* spp.

(6) Manuka to 8 m tall, above remnant pedestals of *Carex secta* and *Phormium*, and a ground cover of stunted *Blechnum minus, Nertera depressa, Schoenus maschalinus, Cardamine debilis* and *Epilobium nerteroides*.

(7) A zone of 10–12 m tall kamahi and *Griselinia littoralis* trees over tree ferns, *Gahnia rigida* and *G. xanthocarpa* fringes mature forest on the dune.

Profile B extends 400 m across firm, peaty ground, from open water to young forest. It represents a gradient from lake-edge to cushion-bog, and beyond that, a succession towards kahikatea forest that is probably occasioned by improving drainage.

(1) *Eleocharis acuta* 0.4 m tall and 70% cover, emergent at the lagoon edge; also *Juncus articulatus* (20%), *Hydrocotyle sulcata* (10%).

(2) *Juncus articulatus* 70% cover in peaty puddles; also *Callitriche petriei* (10%), *Myriophyllum triphyllum* (10%), *Montia fontana* and *Potamogeton cheesemanii*.

(3) *Leptocarpus* 40% cover among *Phormium* (15%) and *Coprosma* aff. *parviflora* (15%); *Carex gaudichaudiana* (10%), and many small herbs on hummocks.

(4) Manuka 2–4 m tall and 30% cover, *Coprosma* aff. *parviflora* (10%) and *C. propinqua* (10%), over *Lepidosperma australe* (20%) and *Blechnum minus* (10%). Other shrubs include *Myrsine divaricata* and *Coprosma* aff. *intertexta*.

(5) Manuka 1–3 m tall and 20% cover, over 1 m tall *Leptocarpus* (40%), *Baumea rubiginosa* (20%), *Lepidosperma* (10%) and *Blechnum minus* (10%). *Sparganium subglobosum* finds its southern limit in this community.

(6) *Centrolepis ciliata* reaches 60% cover in cushion bogs 2–3 m across, which support *Drosera binata* and other small herbs.

(7) Manuka rising to 5–6 m tall and 60% cover with *Dracophyllum longifolium*, over 1.5 m tall *Baumea rubiginosa* (50%) and *B. teretifolia* (30%); occasional shrubs of *Neomyrtus pedunculata* and *Myrsine divaricata*.

(8) Manuka 6–7 m tall and 70% cover, with young kahikatea trees growing through it. *Phyllocladus alpinus, Metrosideros umbellata*, kamahi and *Pseudopanax colensoi* var. *ternatus* are 3–5 m tall and total 20% cover. Pink pine, *Pseudopanax anomalus, Cyathodes juniperina* and *Coprosma foetidissima* are 1–2 m tall and total 20%. There are tussocks of *Gahnia* spp. (2 m tall, 40%) and *Carex secta* (15%) above *Baumea teretifolia* (20%), *Carex dissita* (15%), *C. maorica* (5%), *C. geminata* (10%), *C. sinclairii* (20%), *Juncus gregiflorus* (10%) and *Blechnum minus* (20%).

Other vegetation around the lagoon includes kahikatea forest developing where an alluvial fan is aggrading, and a sequence from *Empodisma* mire through tree-heath with *Lepidothamnus intermedius*, silver pine, *Phyllocladus* and *Dracophyllum longifo-*

lium to dense rimu with mountain beech (*Nothofagus solandri* var. *cliffortioides*), that is correlated with decreasing water over a flat surface underlain by deep peat.

Eastern Fiordland

Mires on low, gentle terrain near lakes Manapouri and Te Anau are floristically similar to the infertile mires of Westland, but structurally approach cushion bog. Communities (Table 10.4) grade from fertile swamp on low-lying areas such as abandoned river channels (sample 3) and margins of domes (samples 2, 4), through low-lying parts of bogs (samples 5, 7), to higher bog surfaces (samples 8–15). Sample 6 is a flat *Leptocarpus* mire in a former shallow lake basin. Sample 1 represents a peat surface that may be becoming drier, thereby allowing dense, tall manuka to develop. Although this may be seral to forest, manuka can also form cyclical mosaics on peat, as mature trees fall over or die and saturated conditions are re-established in the gaps.

Analysis of cover data along a transect from the edge to the centre of one domed bog yielded eight community types, characterised by (1) bog pine with abundant *Empodisma*; (2) *Empodisma* and *Campylopus acuminatus*; (3) *Campylopus* without *Empodisma*; (4) a variable mixture including *Baumea tenax*; (5) manuka and *Empodisma*; (6) samples with sphagnum or *Pulchrinodus*; (7) similar to (6), but with less manuka, more moss and, locally, *Oreobolus pectinatus*; and (8) *Cladina leptoclada* (Mark *et al.* 1979). Peat depth increased from 2.7 m at the edge to 5.3 m in the centre, and depth to the water table (during a dry spell) decreased from 70 to 17 cm along the same gradient.

A complex of hummocks and pools occupies all but the driest surfaces of the domes (Burrows & Dobson 1972). Typically, *Baumea rubiginosa*, *Utricularia monanthos* and algal mud occupy the floor of each pool, and *Sphagnum falcatulum* extends into the water from the sides. At the base of the hummock, there are *S. cristatum*, *Drosera binata* and *D. spathulata*, and above this, *Empodisma*, *Baumea tenax* and *S. cristatum; Dicranoloma* ?*robustum* caps the hummock. Alternatively, hummocks may consist mostly of *Empodisma* and *B. tenax*. Shallow pools that frequently dry up may be colonised by cushions of *Oreobolus pectinatus* and *Centrolepis ciliata*. However, effective peat growth is practically confined to hummocks, so that pools become deep and steep-sided. String bogs arise in which numerous pools are elongated along the contours of the bog, and separated by strips and islands of vegetated peat that may harbour the fire-sensitive *Lepidothamnus intermedius*, *L. laxifolius* and bog pine.

Peatland tarns are also usually steep-sided, and most have a border of *Baumea rubiginosa* and *Eleocharis sphacelata* with some *Phormium* and *Carex secta*. Where mires meet tall forest on drier ground there is, in the absence of fire, a shrub zone with manuka, *Phyllocladus alpinus*, bog pine and, on one mire, silver pine growing far south of its main range, which extends to Martins Bay in northern Fiordland.

Southland

Most of the peatlands of lowland Southland are now drained and farmed except southeast of Invercargill, where drainage of the low-lying Awarua Plain is impeded by bars and beach ridges. Over 5000 ha retain native vegetation, on 2–2.5 m of peat overlying

Table 10.4. *Percentage cover in vegetation samples on Manapouri–Te Anau mires, measured by point intercepts*

Sample no.	1	2	3	4	5	6	7a	7b	7c	7d	8a	8b	9	10	11	12	13	14	15
Peat depth (cm)	350		0	(494*)		>220	(	390	)	)	(340)		940		>500	>490	500	500	494
Trees and shrubs																			
Nothofagus solandri v. *cliff.*	s																		
Dacrycarpus dacrydioides	s																		
Leptospermum scoparium	d																	+	
Coprosma aff. *intertexta*							6						+						
Dracophyllum longifolium		12												8			5	32	23
Halocarpus bidwillii																	8		6
Leafy monocots																			
Carex dissita	10																		
Carex secta	10	11	13	23															
Carex maorica			9	9															
Carex diandra				8	46		5												
Carex sinclairii						5	12	6											
Phormium tenax							7												
Carex echinata															5				
Rush-like monocots																			
Eleocharis acuta		5	5																
Juncus articulatus		11			39		9												
Juncus gregiflorus			13																
Leptocarpus similis						28													
Lepidosperma australe								20	6		8								
Baumea rubiginosa													5	5					
Baumea tenax									20	31	10	7	10		5				
Empodisma minus									15		56	17	19	7	16	6	13	16	11
Dwarf shrubs and cushion plants																			
Oreobolus strictus																5	9		
Cyathodes empetrifolia													+					5	
Oreobolus pectinatus																5			

	1	2	3	4	5	6	7	8	9	10	11	12	13	14	15	16	17	18
Lepidothamnus laxifolius																	5	
Pentachondra pumila																		7
Ferns and lycopods																		
Blechnum 'capense'						12												
Lycopodium ramulosum								8										
Gleichenia dicarpa										+								
Aquatic plants																		
Potamogeton suboblongus	17	5																
Azolla filiculoides			16															
Lemna minor				5														
Callitriche stagnalis					8													
Bryophytes																		
Sphagnum australe	18	18											61					
Drepanocladus sp.			5															
Marchantia berteroana				32														
Campylopus spp.								(	)							5		
Pulchrinodus inflatus								(+)	+									
Sphagnum falcatulum								8	16				18			42		
Sphagnum cristatum								19	+				19			37	13	
Dicranoloma ?robustum								9	22							13	27	
Lichen																		
Cladina aff. *leptoclada*								(+)	+									
Cladia sullivanii															5			
Other species	22	13	12	4	10	8	9	4	8	7	3	11	1	12	5	11	7	25
Water	29	5								10			43				7	
Mud	23	29	11			8	9	6										15

Notes: Peat depths refer to centre of mire, not the sample location; only cover values ⩾5% shown; s, sapling; d, dominant (not included in cover percentage); +, present.

Source: Burrows & Dobson 1972.

bleached quartz gravel. Very wet ground and stream margins support *Phormium* and *Carex* spp., but mainly there is cushion bog, notable for species that usually grow at much higher altitudes. Kelly (1968) describes the following assemblages.

(1) Aquatic plants including *Utricularia monanthos* in standing water.
(2) *Baumea huttonii* in shallow tarns.
(3) Concave surfaces about the level of the water table with sphagnum, *Centrolepis ciliata* and *Oreobolus pectinatus*, together with *Gleichenia dicarpa*, *Schoenus pauciflorus*, *Empodisma* and, on bare peat, plants such as *Lycopodium ramulosum*, *Schizaea fistulosa*, *Drosera spathulata*, *D. binata*, *Actinotus novae-zelandiae*, *Pernettya nana*, *Nertera scapanioides*, *N. balfouriana*, *Herpolirion novae-zelandiae*, *Centrolepis* and *Carpha*.
(4) Large green *Donatia novae-zelandiae* hummocks, commonly with *Penta-chondra pumila* around their edges, where the peat is unlikely to be submerged.

Higher, drier parts of the bog support scattered shrubs; further inland, red-tussock grassland merges into open *Dracophyllum longifolium* – manuka scrub; alternatively there may be dense *Empodisma* and *Gleichenia*, which is possibly fire-induced. Repeated firing destroys the cushion bog and encourages dense manuka.

Wet heath communities

Wet heaths are best exemplified by the western South Island pakihi, which are expanses of manuka, rush-like sedges and *Gleichenia* fern on severely leached soils with perched water tables. Like the northern gumlands (p. 198), they usually develop towards dominance by manuka, but under the cooler climate the herbaceous phase is more enduring. The organic horizon, whether shallow or deep, grades into bleached E horizons, but true peat is local.

Most pakihi have been derived from forest and tree-heath during the millennium of human occupation, and the plant communities closely resemble those on infertile lowland swamps and bogs. However, many have expanded from cores of primitive vegetation growing on soils too intractable to support closed scrub or forest. Especially in South Westland, 'natural pakihi' still exist, usually embedded in virgin woody heathland.

Typical pakihi are underlain by fluvial or glacial gravel, but hard, siliceous bed-rock can carry similar vegetation, especially in Stewart Island, Fiordland and Western Nelson. Other than gumlands, the only significant temperate wet heaths in the North Island are on the frosty tephra plains of the Volcanic Plateau.

Moraines and fluvioglacial terraces in Westland and Western Nelson

The poorest soils on rolling moraines between sea level and 450 m between the Waiho and Cook Rivers support mosaics of herbaceous vegetation and scrub. On unburnt herb-dominated areas the main species is *Empodisma*, and *Gleichenia dicarpa* is also abundant. The most common woody plant, manuka, is semi-rhizomatous and often less than 30 cm tall. *Dracophyllum palustre* is always present and *Lepidothamnus laxifolius* and bog pine are locally common. The shallow gley podzols usually have little surface peat, and roots and a perched water table are confined within a 20–80 cm thick E horizon of leached silt, overlying an impervious iron-humus pan. However, there can be a structureless, slightly silty organic layer up to 1.5 m thick (p. 286).

Very poorly drained central areas can pass into *Donatia* bog. Where there is much surface water, *Empodisma* forms hummocks and *Sphagnum* is common; the pools have *Centrolepis ciliata* and sometimes *Utricularia monanthos* and *Liparophyllum*. Where water moves in channels, *Carex gaudichaudiana* rises to dominance and *C. echinata* can be present. Woody plants are taller and denser where drainage is better. In transitions to tree-heath, manuka usually dominates but is accompanied and sometimes replaced by silver pine. Dense *Gahnia rigida* also occurs in these transitions. Other wet-heath species include *Lycopodium ramulosum*, pink pine, *Cyathodes empetrifolia*, *Gentiana* aff. *spenceri*, *Celmisia graminifolia*, *Bulbinella modesta*, *Carpha alpina*, red tussock, *Thelymitra venosa* and the rather uncommon *Actinotis novae-zelandiae* and *Astelia linearis* var. *linearis*.

Similar natural wet heaths occur among shrub- and tree-heath on the largely ultramafic Cascade and Gorge River moraine plateaus further south in Westland, and on fluvioglacial terraces in scattered localities that have escaped fire throughout Westland and Western Nelson. Fire-induced pakihi on upper Pleistocene terraces near Westport have been described by Rigg (1962).

(1) The main community is a mosaic of *Gleichenia dicarpa* and *Baumea teretifolia*, the former being absent from drainage depressions. Sphagnum, *Campylopus introflexus* and *Lycopodium ramulosum* form a mosaic beneath.

(2) Low-growing communities occupy short slopes of 1–10° and the centres of large, flat pakihi. Three variants have, as their main species, *Euphrasia disperma* and *Herpolirion* with much bare ground; *Campylopus introflexus*, *Liparophyllum* and *Gonocarpus micranthus*; and *Rytidosperma* (probably *R. gracile*). Other plants include *Schizaea fistulosa*, *Gleichenia dicarpa*, *Drosera spathulata*, *D. binata*, *Celmisia alpina*, prostrate manuka (Fig. 10.28), *Centrolepis ciliata*, *Empodisma*, *Baumea teretifolia*, *Tetraria*, *Oreobolus pectinatus*, thelymitras and sphagnum.

(3) Thick sphagnum and some peat may develop in wet basins.

(4) On areas that are better-drained or supported forest within the last hundred years, *Gleichenia* and *Baumea teretifolia* are joined by bracken, *Blechnum 'capense'*, *Lepidosperma australe*, *Baumea rubiginosa* (on very wet ground) and *Gahnia* tussocks. There is vigorous invasion by erect manuka.

Other notable species of the Westport pakihi include the moss *Pleurophascum grandiglobum*, named for its very large capsules, *Pulchrinodus*, *Donatia*, *Oreostylidium subulatum*, *Gentiana townsonii*, *Mitrasacme novae-zelandiae*, *Carpha alpina*, another *Carpha* that seems identical to the South American *C. schoenoides*, *Dianella nigra*, *Gaimardia setacea*, *Bulbinella modesta* and orchids such as *Spiranthes sinensis* and

Fig. 10.28. Prostrate form of *Leptospermum scoparium*, among *Empodisma* on a pakihi near Westport, western Nelson.

Calochilus paludosus. Similar fire-induced pakihi occur on fluvioglacial gravels, coal-measure sandstones, and the harder granites from the Cascade Plateau to Collingwood in North-west Nelson. Southern examples are less rich floristically, especially in respect of orchids, whereas those near Collingwood are distinguished by abundant *Lepidosperma filiforme*, which is otherwise confined to the northern zone.

Fiordland and Stewart Island

In these regions infertile granite country covered in tree-heath includes openings with wet heath. *Gleichenia dicarpa* and *Empodisma* are the usual dominants, but *Sprengelia incarnata* and *Chionochloa acicularis* (p. 230) are important near West Cape and *C. rubra* is important on Stewart Island.

Plateaus of Western Nelson

Heath covers much of the coal-measure plateaus near Greymouth and north of Westport and the granite peneplains that extend northward from the Kohaihai River. Herb-dominated areas grade into infertile swamps in depressions, and into woody heaths on steeper slopes and more fertile soils, but the latter have been reduced by fire, especially on the coal plateaus. As these heaths mainly lie between 500 and 1000 m a.s.l., they are intermediate between lowland and upland in character, this being reflected in the absence of *Baumea* spp. except at the lowest altitudes, and the presence of high-altitude species. There are several endemics including *Chionochloa juncea*, the moss-like *Coprosma talbrockiei, Celmisia similis, C. parva*, and on the Paparoa Range, *Mitrasacme montana*.

In herb-dominated wet heath occupying a wide terrace at 600 m a.s.l. on the Gouland Downs, red tussock and some *Lepidosperma australe* provide 10% cover over a turf of *Donatia*, small *Empodisma, Carpha*, and *Campylopus ?clavatus*. There are scattered, small bog pines; these and red tussock are taller and denser in depressions.

Other species are *Cassinia 'vauvilliersii'*, *Dracophyllum palustre*, *Actinotus*, *Gonocarpus micranthus*, *Celmisia dallii*, *Pentachondra*, *Herpolirion*, *Oreobolus strictus*, *Thelymitra venosa*, *Rytidosperma nigricans*, *Lycopodium ramulosum*, *Dicranoloma robustum*, *Cladia retipora* and *C. aggregata*.

This community is on 15–30 cm of greyish silt loam with occasional boulders, overlying either weathering shale, or up to 30 cm of subangular granite or gravel that probably was derived from higher terrain under periglacial conditions. The severe modern climate can be gauged from knolls near the western edge of the basin where exposed quartz-metamorphic rocks at only 700 m are frost-riven, and there are solifluction lobes in bare soil. Vegetated areas have mainly prostrate manuka and carpet grass (*Chionochloa australis*); further species include *Lepidothamnus laxifolius*, stunted pink pine, *Celmisia similis* and *Rhacocarpus purpurascens*.

Volcanic Plateau

The largest surviving temperate wet heath in the central North Island is on thin peat overlying Taupo pumice, at 730 m on the Rangitaiki Plain. The vegetation is a mosaic of *Lepidosperma australe* and *Gleichenia dicarpa*, with scattered *Dracophyllum subulatum* and manuka, patches of *Empodisma*, and a ground tier of *Dicranoloma robustum*, *Campylopus* sp. and localised sphagnum. This merges into *Dracophyllum subulatum* and silver tussock where soils are dry and porous (p. 193), or into shallow swamps in which *Gleichenia* and *Lepidosperma* are accompanied either by *Baumea* spp. or, in flushed areas, by *Carex dipsacea*, *C. coriacea*, *Cortaderia toetoe* and *Hierochloe redolens*. There are also turfy patches, some dominated by *Oreobolus pectinatus*, and others occupying temporary pools in which *Carex rubicunda* is accompanied by plants such as *Lachnagrostis* sp., *Gnaphalium delicatum*, *Dichondra repens* and *Stackhousia minima*.

Mountain wetlands

Except on the schist plateaus of Central Otago, mountainous terrain supports few extensive wetlands because of steepness and rapid geomorphic evolution, but numerous minor areas develop where landslides block valleys, where ground water emerges on slopes, and on formerly glaciated passes. In high-rainfall regions, there are wet heaths on shallow soils with impeded drainage, but blanket peats are infrequent except in southern regions.

Small-scale variations in depth and drainage of mountain wetlands make consistent community types even harder to recognise than in the lowlands. Further, distinctions between wetland and wet grassland or herbfield are blurred; many communities with red tussock, *Chionochloa crassiuscula* and other snow tussocks could be described under either heading. At very high altitudes, wetlands merge into grassland or fellfield kept wet by melting snow (Chapter 12).

In this section, the autecological outline emphasises the importance of cushion and mat plants in mountain wetlands, and communities are described in a south-to-north sequence, in keeping with the greater extent and diversity in the south.

Cushion and mat plants

Wetland with cushion-forming angiosperms appears unique to the southern hemisphere and wet tropical mountains, largely replacing the sphagnum bogs of the northern hemisphere. The plants are slow-growing and their success probably relates to being able to draw on soil water and to control transpiration, thereby avoiding the desiccation that afflicts bryophytes during dry, sunny weather.

Oreobolus pectinatus (comb sedge), the most widespread wetland cushion plant, colonises wet gravel, grassy terraces and interfluves with deteriorating fertility and drainage, and firm, peaty surfaces on ageing mountain swamps. In established cushion bogs, it tends to colonise depressions where seasonal flooding has killed other plants.

Centrolepis ciliata forms globose cushions up to 30 cm across, and is common in peaty, periodically flooded depressions in bogs and wet heaths. *C. pallida* is a pioneer of muddy tarn-shores and eroded peats that occurs very locally on South Island mountains. The related *Gaimardia setacea* is locally common in wet cushion bogs.

Phyllachne colensoi forms hard, light green, semi-woody cushions up to 30 cm in diameter, and occurs in high-mountain habitats, including bogs, throughout mainland New Zealand excepting Mt Taranaki. *Donatia*, which forms cushions up to 1 m across and 50 cm high, extends from the Tararua Range to Stewart Island, being mainly subalpine but descending to low altitudes in Westland and beside Foveaux Strait. With *Phyllachne*, it could be regarded as a 'climax' dominant of upland bogs (Lough *et al.* 1987).

The mat-plants *Coprosma perpusilla*, *Pentachondra pumila* and *Astelia linearis* are present in most mountain bogs, whereas *Cyathodes pumila* and the minute podocarp *Lepidothamnus laxifolius* are more local. The mat-forming grass *Microlaena thomsonii* and the cushion plant *Astelia subulata* are southern except for occurrences in Western Nelson. *Dracophyllum prostratum* and *D. politum* grow as open mats or dense, woody cushions, the former in eastern and central Otago and the latter further west and south. *Celmisia argentea* is also a cushion plant of southern bogs.

Plants with other growth forms

Carex coriacea ascends to the penalpine belt, mainly on damp recent alluvium and fertile flushes, but the most common leafy sedges at high altitudes are *C. gaudichaudiana*, *C. sinclairii* and *C. echinata*. Several carices have disjunct (e.g. *C. lachenalii*, *C. libera*) or restricted distributions (e.g. *C. trachycarpa* in Western Nelson, extending to the north-eastern Alps, and *C. edgariae* in Otago). *Schoenus pauciflorus* is the only rush-like sedge prominent at high altitudes. The main grasses, other than chionochloas, are *Rytidosperma nigricans* and sometimes *Deschampsia caespitosa*.

Juncus antarcticus is a small cushion plant of shallow mires and wet heath. *J. novae-zelandiae* is usually semi-floating in muddy flushes, whereas its much smaller relative *J. pusillus* grows mainly in less fertile sites. *J. articulatus* commonly ascends to the subalpine belt in modified wetlands, as do a few tall rushes, such as *J. gregiflorus* and *J. effusus*. *Rostkovia magellanica* forms open tufts about 15 cm tall in mountain mires in Otago and Southland.

Bulbinella spp. occur in shallow fertile swamps, whereas *Empodisma* is often abundant in infertile mires, although scarcely reaching the penalpine belt. Orchids are usually present in less fertile communities, and include *Aporostylis bifolia*, *Caladenia lyallii*, *Lyperanthus antarcticus*, *Prasophyllum colensoi* and *Thelymitra venosa*.

Springs and flushes often support *Montia fontana* and, on stony ground, the robust, creeping *Epilobium macropus*. *Myriophyllum propinquum* and occasionally *Crassula multicaulis* grow as semi-floating forms in shallow water and as mats in seasonally dry tarns.

Mires also support small, apparently annual gentians in the *G. grisebachii–G. matthewsii* complex, and mat-forming gnaphaliums such as *G. paludosum* and *G. traversii*. Tufted celmisias in the *C. gracilenta–graminifolia* complex are always present, and the rhizomatous, glabrous *C. glandulosa* is abundant except in Stewart Island, often covering small areas. *Nertera balfouriana* is widespread, creeping through sphagnum cushions. Many herbfield species grow on the drier habitats within high-altitude wetlands.

With increasing altitude, shrubs become less prominent although manuka can be abundant, especially on wet heaths. *Hebe odora* is practically throughout, forming thickets on the margins of flushes and swamps; the related *H. pauciramosa* is more local, and tends to grow on peatier, less fertile sites. *Dracophyllum longifolium* s.l., *D. uniflorum* and regional species such as *D. recurvum* (Volcanic Plateau) and *D. pearsonii* (Fiordland and Stewart Island) grow on bog margins and wet heaths. Some plants in the *Olearia virgata* complex, such as var. *rugosa* in the east of the South Island, are mainly on high-altitude flushes. *Cassinia 'vauvilliersii'* can also be a wetland shrub.

Blechnum 'capense' is widespread, whereas *Gleichenia dicarpa* scarcely reaches the penalpine belt. Bogs on wet mountains contain *Schizaea fistulosa* and *Lycopodium ramulosum*, and *Polystichum vestitum* can dominate sheltered flushes in treeless eastern districts. *Sphagnum*, *Dicranoloma*, *Campylopus* and *Marchantia* are important in mires, *Polytrichum commune* covers seasonally flooded depressions, and *Drepanocladus fluitans* is the commonest aquatic moss. *Rhacocarpus purpurascens* is more abundant than at lower altitudes, forming mats where water seeps over bare peat or rocks.

Mountain wetlands of Stewart Island (based on Wilson 1987)

Herbaceous vegetation on wet uplands in Stewart Island covers about 10 km², mainly in the south, of peaty soils that grade into shallow blanket peat. The nodal community is dense turf in which prostrate manuka shares dominance with the mat- or cushion-forming *Lycopodium ramulosum*, *Dracophyllum politum*, *Pentachondra*, *Donatia*, *Celmisia alpina*, *Astelia linearis*, *Oreobolus pectinatus*, *O. impar*, *Carpha* and *Microlaena thomsonii*. *Drosera stenopetala* and *Thelymitra venosa* are always present. Such turf often forms lanes through taller manuka-dominated vegetation, which Wilson (1987) attributes to the interplay of prevailing wind and waterlogging.

This community grades towards tree-heath *via* stands in which manuka 1–2 m tall grows among the turf; additional species here are *Gleichenia dicarpa*, *Schizaea fistulosa*, *Cyathodes empetrifolia*, *Centrolepis ciliata* and *Gahnia procera*. On more benign sites other shrubs appear, especially pink pine, *Dracophyllum longifolium*, *Cyathodes juniperina* and *Olearia colensoi*.

Donatia and *Oreobolus pectinatus* co-dominate on flat, very wet upland peat, where further species are *Drosera spathulata, Actinotus, Mitrasacme novae-zelandiae, Plantago uniflora, Liparophyllum, Euphrasia* aff. *dyeri, Astelia subulata, Isolepis aucklandica, Gaimardia setacea, Centrolepis* and *Carpha*. *Chionochloa pungens* increases towards flushed, better-drained or higher sites and may dominate at upper subalpine and penalpine levels. At the wetter end of the gradient it is scattered among *Dracophyllum politum, Donatia, Oreobolus* spp. and other cushion plants, and prostrate *Hebe odora*. The drier end has abundant *C. pungens* among large hummocks of *Dracophyllum politum*. As *C. pungens* increases with altitude, it is joined by plants of rock and herbfield, i.e. *Raoulia goyeni, Dolichoglottis lyallii, Celmisia clavata, Phyllachne colensoi, Gentiana lineata, Bulbinella gibbsii* and narrowly endemic species of *Aciphylla* and *Celmisia* (p. 110). The most prominent cryptogams in the mountain mires of Stewart Island are *Dicranoloma* and *Campylopus* species. *Breutelia elongata, B. pendula, Rhacocarpus, Jamesoniella colorata, Isotachis montana, Riccardia cochleata, Siphula* spp. and, locally, sphagnum are also important.

Cushion bogs on southern borders of the Otago plateaus

As southerly winds sweep from the Southern Ocean towards Central Otago, cloud and precipitation increase with rising altitude on the plateau mountains, the latter to as much as 2500 mm annually, before decreasing abruptly in the rain shadow. The only extensive blanket peats in the South Island are on gentle, exposed summits, especially around 1000 m on the Blue Mountains, 900 m on Maungatua and, as eroding remnants, at only 700 m a.s.l. on Swampy Hill near Dunedin. On the Blue Mountains, *Dracophyllum longifolium* and *Chionochloa rigida* (including hybrids with *C. rubra*) each provide *c.* 5% cover over cushion bog, which is a mosaic of five elements.

(1) *Sphagnum cristatum* hummocks invading open *Dracophyllum longifolium* scrub at the bog margins. Elsewhere, moribund sphagnum is being invaded by *Dicranoloma robustum, Dracophyllum prostratum, Pernettya macrostigma, Cladia retipora* and *Cladonia* spp.

(2) Dryish flat areas with *Dicranoloma* (80% cover), *Dracophyllum prostratum* (10%) and *Pentachondra* (10%).

(3) Wetter flat areas with *Oreobolus pectinatus* (90%), *Dicranoloma, Dracophyllum prostratum* and sphagnum.

(4) Flat areas, probably where previous cover has died, with *Racomitrium lanuginosum* (70%) and *Dracophyllum prostratum* (30%).

(5) A community of *Donatia* (70%), *Dracophyllum prostratum* (15%), *Celmisia sessiliflora* (5%), *Racomitrium* (5%), *Drosera arcturi, Pernettya, Pentachondra, Phyllachne rubra, Astelia linearis, Dicranoloma* and *Thamnolia*.

Further plants on the bog are *Cyathodes pumila, Myrsine nummularia, Forstera* sp., *Oreostylidium, Cassinia* 'vauvilliersii', *Hebe odora, Astelia nervosa, Gaimardia setacea, Centrolepis ciliata, Isolepis aucklandica*, and *Dracophyllum* hybrids that appear to involve *D. uniflorum* and *D. politum* as well as the species already mentioned. Bog pine grows on eroding peat, perhaps having been eliminated from vegetated bog by fire.

Soligenous wetlands of the Central Otago mountains

The wetlands of Central Otago lie mainly in broad valleys above 900 m, but ascend above 1500 m on flushes. For the Lammerlaw mountains, where valley mires are more than 2 km across, Table 10.5 arranges communities along a gradient of decreasing wetness. The *Marchantia* community (1) occupies stream courses and places where water under pressure erupts through the mire. Communities 2 and 4–7 form part of a cycle controlled by the growth and degeneration of *Sphagnum cristatum*, which builds fans of peat that advance as waves down the mire (Fig. 10.29). *Carex gaudichaudiana* (3) and *Carpha* (11) grow on sloping surfaces, the former in wet swamp, the latter on firm bog; communities 9 and 12 are transitions from dry cushion bog dominated by *Celmisia argentea* to the surrounding *Chionochloa* grassland. Adventives prevail in 10, which has been grazed and compacted.

For the Old Man and adjoining ranges Brumley (1986) recognises four wetland groups corresponding respectively to the wetter end of the Lammerlaw sequence (1–4), the cushion-dominated communities (5–11), modified eutrophic areas (10), and flushes dominated by *Schoenus pauciflorus* and *Carpha*. These wetlands mostly lie at higher altitudes than the Lammerlaw mires, and include penalpine species such as *Abrotanella caespitosa*, the whipcord *Hebe propinqua*, *Ranunculus gracilipes*, *Rostkovia* and *Carex lachenalii*.

Canterbury and Marlborough

Except for *Schoenus*-dominated flushes, high-altitude wetlands are a minor feature of the relatively dry, steep, crumbling greywacke mountains. Mires as large and complex as that on Lagoon Saddle, at 1150 m between the Waimakariri and Rakaia watersheds, are rare except on Main Divide passes. The Lagoon Saddle mire is a string bog with peat up to 5.4 m deep, pH values between 4.7 and 6.0, and very low conductivity (Fig. 10.22); fertility differences relate to rates of water movement rather than nutrient contents. Dobson (1975) describes three main communities.

(1) Flushes and overflow channels support *Schoenus pauciflorus* and *Carpha*, accompanied by *Hebe pauciramosa*, *Gentiana corymbifera*, *Anisotome aromatica*, *Viola cunninghamii* and the mosses *Campylium stellatum* and *Bryum laevigatum*.

(2) A *Sphagnum cristatum–Empodisma* community grows close to deep pools, on the steep (10–30°) rises separating them, and around the lagoon edge. It forms water-retentive peat that dams the pools, causing them to deepen.

(3) On flat ground *Donatia* forms dense hummocks to 1 m across that may eventually coalesce. These are penetrated only by the rhizomatous epacrids *Pentachondra* and *Cyathodes pumila*. *Gaimardia* is on wetter sites, together with *Sphagnum falcatulum* and *Utricularia monanthos*. *Oreobolus pectinatus* and infrequent *Centrolepis ciliata* are scattered throughout. As peat formed by the cushion plants does not retain water, pools are shallow, dry in summer and often occupied by *Rhacocarpus*.

Table 10.5. *Mean percentage cover of major species in 12 vegetation types at Teviot and Red Swamps, 975 m, Lammerlaw Range, Otago*

Community (see text):	1	2	3	4	5	6	7	8	9	10	11	12
No. of plots:	3	7	4	18	7	9	9	1	3	4	4	9
Total spp.*:	20	20	14	31	35	18	18	14	18	25	20	57
Shrub												
Hebe odora												3
Cushion plants												
Colobanthus sp.	7										3	
Juncus antarcticus	3	+	3	+	1		+		+			
Centrolepis ciliata		12		+								
Luzula leptophylla		1		1								
Gaimardia setacea		1			1							
Oreobolus pectinatus		18	9	2	22	71	9		7		14	3
Dracophyllum prostratum			5	+	8	3	44		10			1
Phyllachne colensoi				+	1			40				
Donatia novae-zelandiae					4	+						2
Celmisia argentea					2					32		+
Coprosma perpusilla												1
Sedges and rushes												
Isolepis aucklandica	8	31	+	+	2	2	7					
Carex echinata	10	8		7	1	2		+		3	8	+
Carex gaudichaudiana	15	3	58	10	14	3	3		17	6	+	3
Eleocharis acuta		1								7		
Carex sinclairii				1	2							
Carpha alpina					2	+					41	2
Carex ovalis										14		
Juncus effusus										18		1
Carex coriacea										8		5
Grasses												
Agrostis stolonifera	2									13		1
Poa pratensis	3									6		1
Agrostis pallescens		+	3	+	1							
Rytidosperma australe		+		+	1			5				1
Anthoxanthum odoratum			+		+			25				3
Poa colensoi			+		1		2	10	2			
Chionochloa rigida					2				23		5	24
Chionochloa rubra					1	+				8	4	21
Agrostis capillaris								10				7
Insectivorous plants												
Utricularia monanthos	5	2				+					+	+
Drosera arcturi	1	5	1	2	+	3	1		4		4	+
Other dicot herbs												
Stellaria alsine	2									+		
Montia font. ss. *fontana*	+			+	1					+		
Craspedia uniflora	3											1
Gentiana matthewsii		2	+	+	+	2					+	
Euphrasia dyeri		+		+	1						3	

Community (see text):	1	2	3	4	5	6	7	8	9	10	11	12
No. of plots:	3	7	4	18	7	9	9	1	3	4	4	9
Total spp.*:	20	20	14	31	35	18	18	14	18	25	20	57
Gentiana bellidifolia				2	12	+	1	5	2			+
Epilobium alsin. v. *atriplex*					+				+			+
Ranunculus foliosus										+		+
Acaena ?profundeincisa												2
Mosses												
Bryum sp.	10				4	1				3		
Sphagnum cristatum			4	15	66	12	11	12	2	10	12	2
Breutelia pendula			3	2								
Polytrichum juniperinum				1								
Polytrichum longiset.				3						4		
Dicranoloma ?billardierei					1		2		+			+
Racomitrium lanuginosum							6					
Lichens												
Cladonia sp.					+	1	3					
Cladina mitis					3		6					
Litter												3
Bare		5	8									

Notes:
* Includes bryophytes
Source: P.N. Johnson, unpublished data.

Red tussock is scattered through the three communities and dominates adjacent grassland. Other species include *Lycopodium fastigiatum, Lepidothamnus laxifolius, Drosera arcturi, Nertera balfouriana, Coprosma perpusilla, Celmisia glandulosa, C. gracilenta, Carex echinata, C. gaudichaudiana, C. sinclairii, Juncus novae-zelandiae, Rostkovia* (the only record north of Otago), *Rytidosperma nigricans, Poa colensoi,* the mosses *Polytrichum formosum* and *Racomitrium lanuginosum,* and the lichens *Cladia aggregata* and *Cladina leptoclada.*

Wetlands on western ranges of the South Island

Significant peat deposits on the western ranges are confined to patches of blanket peat on gentle ridges, bogs in broad saddles and passes, and swamps impounded by moraines. Elsewhere peats are thin or lacking, so that the wetlands are best described as flushes and wet heaths. The latter are extensive on the granite mountains of the Paparoa Range and Fiordland, where rounded glacial topography has remained intact. On the schist mountains of Westland the occurrence of wet heath is limited by steeper, rapidly wasting slopes with more fertile regolith, the largest areas being in the upper reaches of the Karangarua valley (p. 178).

Communities are mostly dominated by *Donatia, Oreobolus* spp., *Carpha, Schoenus pauciflorus* (on sloping swamps) and, rather locally, *Gleichenia dicarpa* and *Empodisma.* Carpet grass is important in wet heaths in Western Nelson. Carices enter

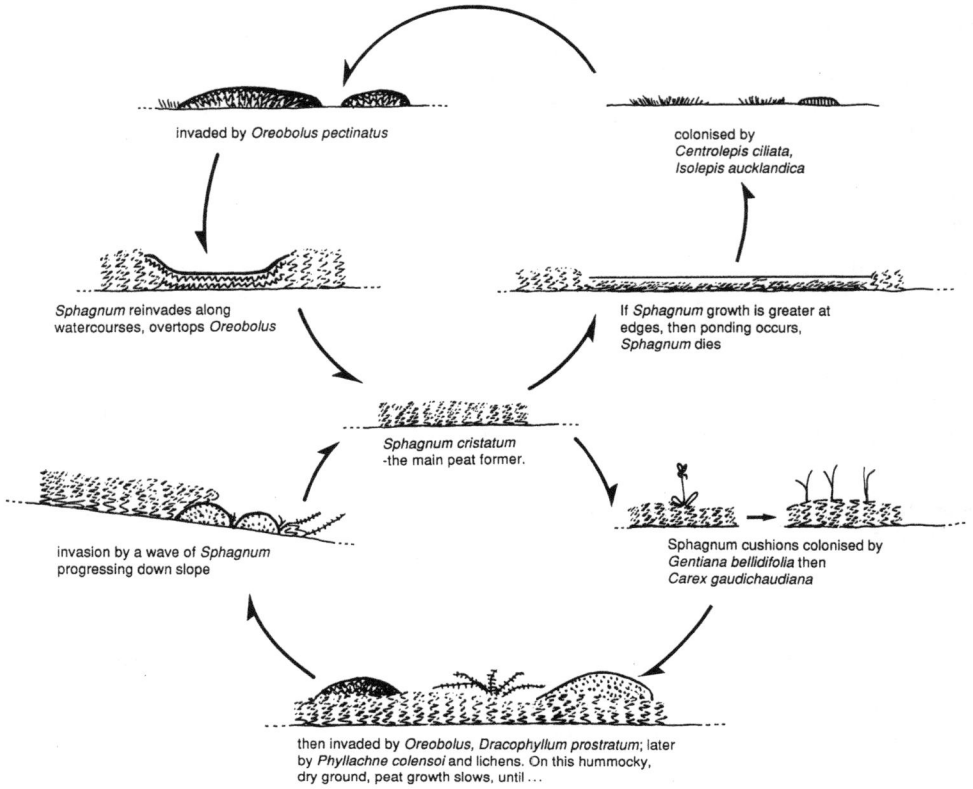

Fig. 10.29. Vegetation cycles in sphagnum bogs on Lammerlaw Range, Otago (from drawing by P.N. Johnson).

where substrates are deepest and wettest. At the higher altitudes there are scattered *Chionochloa* tussocks, and below tree-limit stunted shrubs, especially manuka, *Hebe odora*, *Dracophyllum longifolium* and pink pine (Fig. 10.30). Such vegetation grades, on the one hand, into wet *Chionochloa* grassland or tree-heath, and on the other into partly vegetated rock pavement.

Nearly all these elements are present in a single, especially diverse wetland on a once-glaciated saddle near the head of the Gorge River in South Westland. Although it lies at only 700 m, high-altitude plants are prominent. The following communities are in order of decreasing surface water (Figs 10.31–10.34):

(1) Beds of *Myriophyllum triphyllum* and *M. propinquum* grow in a tarn, and streams support *Myosotis laxa*, *Lilaeopsis* and *Potamogeton cheesemanii*.

(2) *Eleocharis acuta* is emergent at the tarn margin, accompanied by *Deschampsia caespitosa* and *Epilobium chionanthum*.

(3) *Carex secta* and *Phormium tenax* fringe the tarn, both above their usual altitudinal limit.

(4) Emergent rocks in a small stream support *Montia fontana*, *Epilobium brunnescens*, *Hydrocotyle sulcata* and *Leptinella squalida*.

Fig. 10.30. Tussock-shrubland with *Lepidothamnus laxifolius* mat (foreground), rounded shrubs of *Halocarpus biformis* and *Chionochloa rubra* tussocks at 1150 m, Arthurs Pass, Hawdon ER.

(5) *Carex gaudichaudiana* gives 30–80% cover in wet runnels and on sloping terraces beside streams, often growing with *Breutelia pendula, Drepanocladus* sp. or *Sphagnum cristatum*. Associated species include *Psychrophila novae-zelandiae, Ranunculus gracilipes, R. maculatus* and *Gnaphalium paludosum*.

(6) *Schoenus pauciflorus* grows along the tops of stream banks, and reaches 30–60% cover on extensive areas with moving water. Other plants are *Chionochloa* 'westland', *Carpha alpina, Gunnera prorepens, Coprosma* aff. *intertexta* and sphagnum.

(7) *Carpha* is widespread on bogs and achieves 50–80% cover on thin peat at the foot of concave slopes. Frequent companions are *Donatia, Celmisia glandulosa, C. gracilenta, Brachyglottis bellidioides* and *Oreobolus pectinatus*.

(8) *Empodisma* grows on peat, mainly against scrub fringes, as patches 10–20 m across and 40 cm tall, among sphagnum, *Hebe odora, Carpha*, etc.

(9) *Donatia* hummocks 10–15 cm tall provide 40–80% cover on level or gently sloping peat without much water movement. Other common species are *Pimelea prostrata, Pentachondra, Mitrasacme novae-zelandiae, Coprosma perpusilla, Celmisia gracilenta, C. glandulosa, Oreobolus pectinatus* and *O. strictus*.

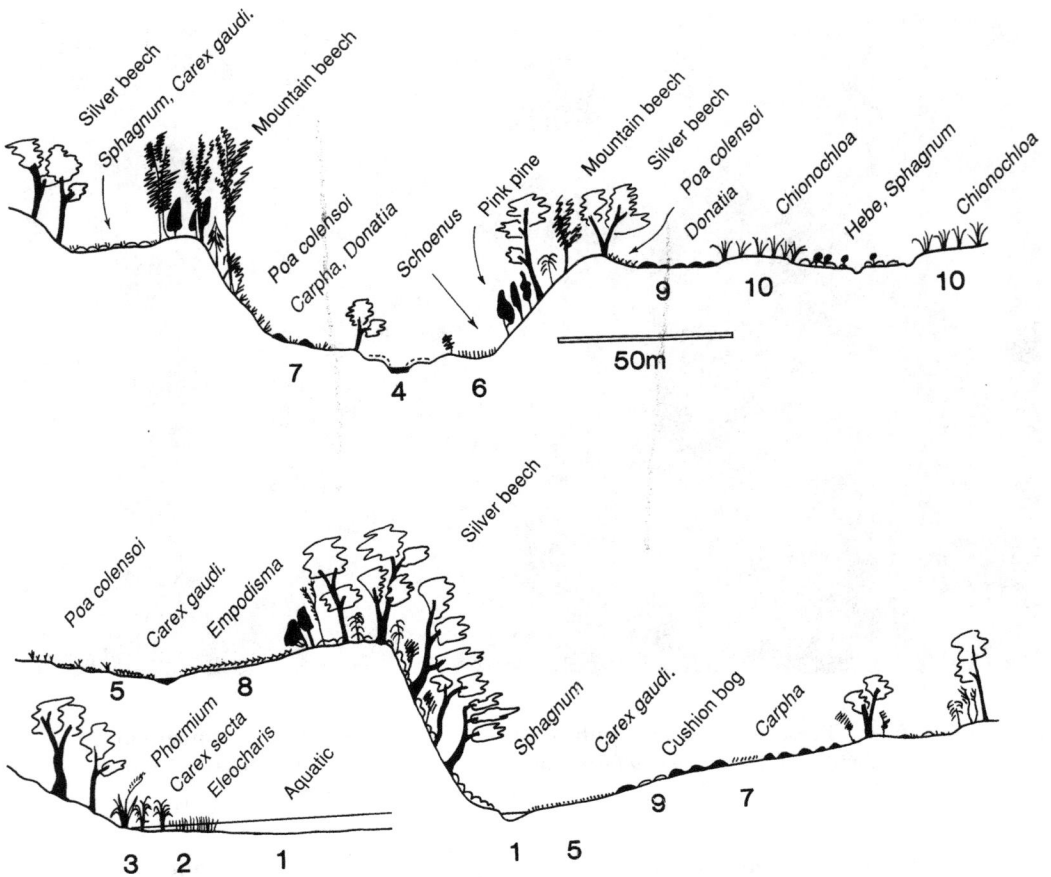

Fig. 10.31. Profiles of transects across shallow bogs at 700 m, Gorge River, Olivine ER (from drawing by P.N. Johnson). See text for numbered segments.

(10) *Chionochloa* tussocks are scattered across the bogs, and increase to 30–60% cover on drier sites to form a grassland of *C*. 'westland', *C. acicularis* and *C. rubra*, this probably being the only locality where the three meet. Other species include *Carpha*, cushion plants, and shrubs of *Hebe odora* and *Dracophyllum longifolium*.

Despite the boggy ground, peat nowhere exceeds a depth of 50 cm. Where the surface rises to moraine knolls and ridges, herbaceous vegetation gives way to woody communities variously dominated by mountain beech (on concrete-like till), silver beech (*Nothofagus menziesii*), pink pine, *Myrsine divaricata*, *Dracophyllum longifolium*, *Coprosma propinqua* or *Hebe odora*, the last being accompanied by *Poa colensoi*. Within the surrounding silver beech forest, there are also small swamps with sparse *Carex coriacea* among mats of sphagnum and *Breutelia pendula*.

Fig. 10.32. Cushion bog mainly of *Donatia* (left), *Empodisma* (lower right), and *Chionochloa* tussocks, mainly *C. rubra* (centre right); locality as in Fig. 10.31.

Fig. 10.33. Cushions of *Donatia* (above centre), *Phyllachne colensoi* (lower left) and *Oreobolus pectinatus* (top left); locality as in Fig. 10.31. Other species are *Mitrasacme novae-zelandiae* (small white flowers; foliage inconspicuous), *Carpha alpina*, *Brachyglottis bellidioides* and *Pentachondra pumila* (centre).

Fig. 10.34. *Ranunculus gracilipes* on a cushion bog; locality as in Fig. 10.31. Other species include *Brachyglottis bellidioides* (lower right), *Celmisia gracilenta* (top) and *Dicranoloma robustum* (lower left).

North Island mountain wetlands

Wet dwarf heaths occur on broad ridges and summits of the Tararua and Rimutaka ranges, mainly in the subalpine belt where exposure to prevailing westerlies precludes continuous scrub or forest. Soils are shallow and leached, with drainage impeded by rock or iron–humus pans. *Donatia* is rare and some species common in the west of the South Island and further north in the North Island are absent, notably *Lepidothamnus laxifolius* and *Empodisma*. At the higher altitudes, these heaths have scattered, small tussocks of *Chionochloa flavescens*, and grade into wet grassland (p. 243).

The Ruahine Range stands apart in the North Island in supporting extensive blanket bogs on gentle slopes and flat summits, at altitudes between 1140 m and 1430 m. Peat is up to 5 m deep, and in one bog the uppermost 75 cm encompasses 4000 years of growth (Elder 1965). In places this is eroding to reveal much wood, mainly of pink pine, which is infrequent on the bogs as a living tree. Important species include bog pine, *Carpha*, *Gleichenia dicarpa*, *Empodisma*, *Centrolepis ciliata*, red tussock, *Oreobolus pectinatus*, *Schoenus pauciflorus*, *Carex gaudichaudiana*, *C. echinata*, *C. coriacea* and sphagnum.

High-altitude variations of wet heath, grading into infertile swamp, cover 1535 ha of the ring-plains of the central volcanoes up to an altitude of 1000 m. The communities are (*Phormium tenax*)–red tussock/*Lepidosperma australe; Lepidosperma*; and *Schoenus pauciflorus*. In wetlands ascending to 1550 m on Ruapehu, red-tussock grassland grades into *Gleichenia dicarpa* or *Empodisma* communities on peaty loam, or flushes dominated by *Schoenus pauciflorus*, as *Phormium* and *Lepidosperma* are lacking at these higher altitudes (Atkinson 1981). Transitions to dwarf heath of *Dracophyllum recurvum, Lepidothamnus laxifolius* and *Gleichenia dicarpa* grow on less saturated soils.

Communities may be close mosaics, an example at 1220 m on Ruapehu having *Gleichenia, Dracophyllum recurvum, Empodisma, Schoenus* and *Racomitrium lanuginosum* each dominating discrete patches. Further species include *Pentachondra, Cyathodes empetrifolia, Celmisia gracilenta, C. spectabilis, C. incana, Craspedia* sp. and *Gahnia procera*. As this site is below the tree-limit, there are also stunted manuka, mountain beech, silver pine and *Phyllocladus alpinus*.

Similar wet heath occurs on the Kaimanawa Range, which received much Taupo pumice. At Ngamatea Swamp, it surrounds semi-aquatic *Eleocharis sphacelata* beds at the unusually high altitude of 880 m (Elder 1962). Red tussock dominates wetlands near Mt Taranaki (p. 244).

Outliers of high-altitude wetland occur on mountains further north that are otherwise forested to their summits. The most isolated are one at 760 m on the Herangi Range in northern Taranaki, containing *Schizaea fistulosa, Lepidothamnus laxifolius, Pentachondra, Herpolirion, Empodisma* and *Oreobolus pectinatus* (B.D. Clarkson, unpublished), and another at 890 m on Te Moehau on the Coromandel Range, consisting of *Oreobolus pectinatus* and *Carpha*.

11

OPEN OR PATCHY VEGETATION ON PRIMARY SURFACES AND DEPLETED LANDS

Sparse vegetation on raw mineral substrates often represents early stages of primary succession. The florulas include plants specialised for growing in open habitats, young plants of species that may later form continuous vegetation, and an assortment favoured by random dispersal and lack of competition. Other sparse communities are permanent in severe environments, or arise from depletion of other vegetation, especially dry grassland. This chapter covers varied habitats from the seashore to the penalpine belt, but discussion of open alpine vegetation is held over to Chapter 12.

Coastal sand and gravel

The environment

Dunes occur where on-shore winds sweep sand from beaches and deposit it further inland, being absent where beaches are too stony, and best developed where the supply of sand to the beach is continually replenished. Dunes can be confined by steep hinterlands, or drift many kilometres inland on coasts of low relief. Large quantities of tephra washed into the North and South Taranaki Bights augment the extensive dunes of the Manawatu and western Northland coasts. Near Cape Farewell, dunes have built to heights of 60 m, and extend northwards as the 27 km long Farewell Spit. In the rest of the South Island, the largest dune areas are on drier, more subdued eastern coasts, especially adjoining the Otago and Banks peninsulas. The trough of recent sediments that bisects Stewart Island meets the west coast as a belt of dunes 5 km long.

The *foredune* is first to receive sand blown from the beach, and builds up until sand supply, wind velocity, and vegetation set an equilibrium. Further sand or stronger winds result in sand being redistributed to *rear dunes*. On rapidly prograding coasts, new foredunes may appear in front of the original one. Dunes migrate inland when sand blown from their seaward slopes is redeposited on their inner slopes, and even vegetated dunes are subject to breaches or *blow-outs*. Dunes may contain remnants of preceding vegetation, such as partly buried trees, and be separated by hollows deep enough to intersect the water table, forming wet and sometimes saline *slacks* which can expand into wide *sand plains*.

Gravel and cobble beaches occur where rivers deliver large quantities of gravel to the coast, or where gravel is being eroded from coastal cliffs, and are especially characteristic of Canterbury and Westland. They often rise to a ridge that is disturbed so rarely by waves that plants begin to establish; a lagoon may be impounded behind. On prograding coasts, stranded beach ridges parallel the outermost, active one.

Adverse environmental factors on coastal dune sands are too-rapid accretion or deflation, abrasion, unweathered substrates with few nutrients, high surface temperatures, summer droughting of shallow-rooted plants, and influx of salt; but for deep-rooted plants sand provides an easily penetrated, well-aerated medium in which low water-holding capacity is offset by large exploitable soil volumes and low moisture tensions. Coarse deposits are more stable, but as particle size increases water-holding capacity decreases, and colonisation of cobble and boulder beach ridges generally has to await accumulation of fines and humus.

The plants of unconsolidated coastal deposits

Woody plants

Three small, sprawling shrubs are confined to coastal sand that is not too mobile. *Coprosma acerosa* is filiramulate but of very open habit, with prostrate main stems that root adventiously when buried. It colonises sand all around the main islands, and is probably conspecific with the inland *C. brunnea*. *Pimelea arenaria* and *P. lyallii*, in keeping pace with moderate accretion, often form mounds with only the leafy tips showing above the sand. The former grows around the North Island and at Farewell Spit, and the latter around Stewart Island and along Foveaux Strait. On other South Island beaches the widespread *P. prostrata* may fill the role of these obligate sand-plants.

Woody plants establishing on stable surfaces include several filiramulate species. The liane *Muehlenbeckia complexa* is almost universal in both sandy and stony habitats, the shrub *M. astonii* occurs sparingly on the shores of Cook Strait and as far south as Lake Ellesmere, and the prostrate, almost leafless *M. ephedroides* grows on eastern South Island beach ridges. *Coprosma propinqua* is widespread, but tends to be replaced in drier districts by species such as *C. crassifolia* and *C. rigida*. The low-growing, thick-stemmed *Melicytus crassifolius* grows on the shores of Cook Strait; similar plants on stony Canterbury beaches appear to be *M. alpinus*. The near-leafless lawyer *Rubus squarrosus* is also locally prominent.

Several larger-leaved lianes also trail on beaches, especially on stony ground, taking root where they encounter soil and climbing over bushes that they meet. These include *Muehlenbeckia australis*, blackberry (*Rubus fruticosus* aggr.), the strictly coastal *Tetragonia* spp. and *Calystegia soldanella* × *C. tuguriorum*, and garden escapes such as *Vinca major* and *Lonicera japonica*.

Cassinia leptophylla, *Haloragis erecta* and poroporo (*Solanum aviculare* or *S. laciniatum*) grow on most beaches, whereas manuka (*Leptospermum scoparium*) is local. Pohutukawa (*Metrosideros excelsa*) occurs on northern beaches, *Dodonaea viscosa* and *Olearia solandri* extend to the central zone, *O. avicenniifolia* is common on the west coast of the South Island, and southern rata (*Metrosideros umbellata*), as a sand plant, is confined to rainy southern localities. Ngaio (*Myoporum laetum*) also occurs among dunes. In sheltered hollows or later in successions, these plants are joined by mesic species such as mahoe (*Melicytus ramiflorus*) and wineberry (*Aristotelia serrata*).

Tree lupin (*Lupinus arboreus*) has been widely sown to promote succession through its nitrogen-fixing ability, and is now the main shrubby pioneer on most coastal sands. It is soft-wooded, fast-growing and short-lived. Seedlings germinate when the sand is moist, and the radicles descend vertically to about 17 cm. A rapidly spreading fungal disease, first noted in 1987, is killing mature lupins over large areas. Gorse (*Ulex europaeus*) is often abundant, especially on stony beaches and older sands. *Pinus radiata* seedlings establish on stable sand wherever there is a seed-source, under annual rainfalls from 500 to 6000 mm; planted and spontaneous forests of this tree will increasingly become part of the coastal scene. Other common adventives include elder (*Sambucus nigra*) and common broom (*Cytisus scoparius*). The Tasmanian *Myoporum insulare*, widely planted instead of the native ngaio, is scarcely naturalised.

Sand binders

Foredunes support one or more of three robust plants with far-extending rhizomes, i.e. pingao (*Desmoschoenus spiralis*), *Spinifex sericeus* and the introduced marram (*Ammophila arenaria*). Each builds dunes that differ in slope and height as follows: pingao, 8–14°, < 3 m; spinifex, 14–16°, 6 m; marram, 24–28°, ≥8 m (Esler 1978a).

Pingao is a golden-green sedge with stout rhizomes bearing densely leafy but well-spaced tillers that allow sand to spread widely. It once covered active and recently stabilised dunes throughout New Zealand, but on most coasts has been almost completely ousted by marram. There is now awareness of the plight of this endemic, monotypic genus, partly because of the revival of Maori weaving that uses the leaves, and efforts are being made to propagate and protect it.

Spinifex extends further from foredunes on to the beach than pingao and marram. Although fast-growing, its foliage is open, so that it builds dunes of moderate height. It was greatly reduced by grazing, but has recovered in many areas through protection, sometimes assisted by artificial seeding. *Spinifex* once grew as far south as Christchurch, but its present limits are Farewell Spit and Cape Campbell.

Marram has been planted on a large scale and now dominates most New Zealand dunes, spreading vegetatively and through seed. Its competitiveness rests in part on rapid growth; on recently deposited sands near Westport colonies of marram and pingao had attained diameters of 5 m and 2 m respectively. It is also assisted by densely spaced tillers that allow it to trap sand rapidly, roots that descend 1–2 m, and upward growth that can match an accretion rate of 1 m per year (Huiskes 1979). A disadvantage of high, steep marram dunes is that the grass is liable to be undercut by blow-outs, leaving it stranded on tall pedestals. Although most vigorous on recently colonised sand, marram can form very dense cover on fixed dunes before it is replaced by other species. Pingao seems able to coexist with it where sand is coarse and in too-limited supply to form fast-growing dunes.

Carex pumila has tough, slender leaves that spread as an open lattice over bare sand flats and hollows, which are often damp and brackish. It grows on most New Zealand coasts, and seldom has to share its niche with other sand-binders. *Austrofestuca littoralis* is distributed around the New Zealand coast, but is rare and local in most districts. It is practically confined to the lower part of the foredune and top of the beach, growing as discrete tussocks which hold their position through rapid upward growth of buried culms.

The cosmopolitan *Calystegia soldanella* is found on nearly all New Zealand sandy coasts, usually from the top of the beach to the crest of the foredune. Its well-spaced leaves lie flat on bare sand between colonies of the taller sand-binders, but where it persists into denser vegetation, stems twine up supporting plants. *Euphorbia glauca* is a leafy spurge, with 50 cm tall stems that rise from a network of rhizomes. Once abundant along New Zealand coasts, in the east of the South Island there are now only five known localities. In the North Island and north-west of the South Island, it grows mostly on cliff ledges. However, it is still widely distributed on sand and shingle beaches in South Westland and Fiordland, sometimes as large colonies. The reasons for its decline are unknown, as it can survive among dense marram and seems undeterred by the presence of deer.

Several rhizomatous grasses colonise unstable, unweathered coastal sand or fine shingle through invading from stable areas, or keep pace with sand burying established vegetation. *Pennisetum clandestinum*, *Stenotaphrum secundatum* and *Cynodon dactylon* are especially effective sand-binders of this kind. There are also the native *Zoysia* spp. and *Poa pusilla*, and the adventive *Poa pratensis*, *Agrostis stolonifera* spreading from damp ground, and *Elytrigia repens*, the familiar couch-grass of waste and cultivated land. Other plants with rhizomes are bracken (*Pteridium esculentum*), which is locally dominant on fixed dunes, and *Libertia peregrinans*. The tufted *L. ixioides* can also be a sand plant.

Succulents

Sea rocket (*Cakile edentula*, accompanied in the northern zone by *C. maritima*) is often abundant on damp sand beaches, especially near estuaries. The rare *Theleophyton billardierei* grows on drier beaches just beyond the reach of normal tides, as loose clumps that gather sand into low mounds. *Carpobrotus edulis* has spread widely from plantings on dunes and stony beach ridges; the native ice-plant *Disphyma australe*, with which it hybridises, is uncommon in these habitats. *Atriplex prostrata*, *Plantago coronopus* and *Chenopodium glaucum* grow mainly low on the back slopes of beach ridges where there is saline seepage. *C. ambrosioides* is common on some northern beaches.

Other herbs

Several native tussocks are important colonisers of stabilised sand. The rush-like sedge *Isolepis nodosa* is the most widespread, and extends furthest on to foredunes. It competes well with marram and, being unpalatable, can be the only tall plant where the rest of the vegetation has been reduced to open turf. The leafy sedge *Cyperus ustulatus* can be abundant on weathered sand. Carices include *C. testacea* and, on boulder beaches in Fiordland, *C. pleiostachys*. *Poa cita* (silver tussock) is an early colonist on stable sand and gravel in the South Island, and can dominate on fixed dunes in eastern Otago and along Foveaux Strait. Wilson (1987) records *Festuca novae-zelandiae* on Stewart Island dunes.

The largest toetoe grass, *Cortaderia splendens*, grows on sand country in the northern zone, establishing on less consolidated dunes than the adventive *C. selloana*. In central and southern districts respectively, *C. fulvida* and *C. richardii* also extend to coastal sand. As a sand plant, *Phormium tenax* grows mainly in dune hollows, but in the west of the South Island, it is widely distributed over dunes and forms dense stands on beach ridges. Where the shore is retreating before marine erosion – as it is along much of this coast – *P. tenax* clumps are usually the last bastion of terrestrial vegetation, withstanding breaking storm waves and dumping of shingle, and succumbing only as they are destroyed by the sea.

Non-succulent, tap-rooted forbs include adventive annuals or biennials, and native species with prostrate shoots that spread out from a multi-headed, perennial crown. Among medium-sized herbs, *Senecio elegans* is mainly on dunes, whereas horned poppy (*Glaucium flavum*), *Salsola kali* and the native *Eryngium vesiculosum*, *Apium prostratum* and, in southern localities, *Rumex neglectus* grow mainly on stony beaches. Smaller native plants common on Foveaux Strait coasts are *Geranium sessiliflorum* var. *arenarium*, *Gentiana saxosa*, *Plantago raoulii* and a form of *Myosotis pygmaea*.

Coastal sand and shingle support many short-lived or fast-maturing plants, mainly adventives of dry grassland that can complete their life cycles before the substrate moves or summer drought sets in. Perennial weeds of arable and waste land also find niches, especially on stony beaches; examples are *Phytolacca octandra*, *Rumex* spp., fennel (*Foeniculum vulgare*), and thistles such as *Cirsium vulgare*, *C. arvense*, *Silybum marianum* and *Sonchus* spp.

A number of low-growing native herbs are especially adapted to deflation hollows and sand flats, and range from drought-tolerant species that thrive well above the water table, to species of wetter ground that include halophytes; those confined to the coast are marked†. The most xeric are the cushion plants *Colobanthus muelleri†*, *Scleranthus* spp. including *S. biflorus†*, and *Raoulia australis*, and the mat-plants *R. hookeri* s.l. (Fig. 11.6) and *R. glabra*; all these are reported only from Cook Strait southwards. *Ranunculus acaulis†*, *Oxalis exilis*, *Hydrocotyle heteromeria*, *Cyathodes fraseri*, *Helichrysum filicaule*, *Leptinella dioica*, *Wahlenbergia congesta†*, *Mentha cunninghamii* and the adventive *Rumex acetosella* have extensive fine rhizomes and act as minor sand binders, as can stoloniferous acaenas and gunneras (including *Gunnera hamiltonii*, now reduced to four known colonies in Stewart Island and one near Invercargill).

Non-vascular plants

Neither bryophytes nor lichens appear before substrates are stable. Stones then develop a cover of lichens and to a lesser extent mosses such as *Schistidium apocarpum*, whereas open sand is colonised by mosses, which are conspicuous only during the cooler months. The most prominent are *Tortula princeps* and *Triquetrella papillata*, and in wetter districts, *Campylopus* spp. Soil-encrusting lichens also become common, especially *Cladonia* and *Xanthoparmelia* spp.

Communities on coastal sand and gravel

A selection of North Island dune systems

The dunes of the Manawatu district, the most extensive in New Zealand, are described by Esler (1978*a*) as follows.

(1) The seaward slope of the foredune is occupied by spinifex, marram being mainly on the steeper crest. Pingao is on lower dunes but is decreasing. All three are on the inland slope. *Cassinia leptophylla, Coprosma acerosa* and *Pimelea arenaria* are present near the crest, but often are more abundant at the foot of the inner slope. Locally there are *Hypochoeris radicata, Conyza albida, Lagurus ovatus* and *Bromus diandrus*.

(2) Behind the foredune, there are mobile and fixed low dunes, and flat or deflated surfaces lying above or below the water-table. Marram and pingao occupy the highest of these inner dunes, whereas sand within 30 cm of the summer water table supports *Isolepis nodosa, Lagurus ovatus, Oenothera stricta, Hypochoeris radicata, Leontodon taraxacoides, Melilotus indicus, Medicago lupulina* and tree lupin, the last mainly where it has been sown. Seedlings of marram and pingao also germinate here, but must collect sand if they are to develop further. *Isolepis nodosa* seedlings are less dependent on proximity to the water-table.

Depressions intersecting the water-table are girdled by *Carex pumila*. This tends to build low dunes, on which it is eventually supplanted by marram and pingao. Wetland turf of *Myriophyllum, Limosella, Eleocharis, Ranunculus* and *Selliera* develops in the depressions (Fig. 11.1), but *Leptocarpus similis* usually dominates eventually, in company with *Schoenus nitens* and other plants. After some 50 years, the vegetation may be characterised by *Cortaderia toetoe, C. selloana*, cabbage tree (*Cordyline australis*) and, sometimes, manuka and *Olearia solandri*.

(3) Moving sand eventually gathers on rear dunes up to 12 m high, which can advance over vegetated land behind. Marram and pingao are the main plants.

Esler (1974, 1975) also describes two beaches west of Auckland.

(1) At Whatipu Beach spinifex and marram are the main colonisers, but *Cynodon dactylon* is the primary sand-binder on one dune area. Sand-plain depressions are similar to those of the Manawatu. *Muehlenbeckia complexa* generally dominates older dunes, but is joined by *Solanum linnaeanum* and blackberry where topsoil is deeper. Lupin establishes both on young sand, and on older sand where topsoil is thin or obliterated by trampling or burial. Dry flats support grassland of *Microlaena stipoides*, cocksfoot (*Dactylis glomerata*), *Sporobolus africanus, Axonopus affinis* and *Eragrostis brownii*, with annual forbs such as *Ranunculus parviflorus, Lotus suaveolens* and *Cotula australis*.

(2) At Piha, in 1948, spinifex was succeeded by *Cassinia 'retorta'*, and eventually by

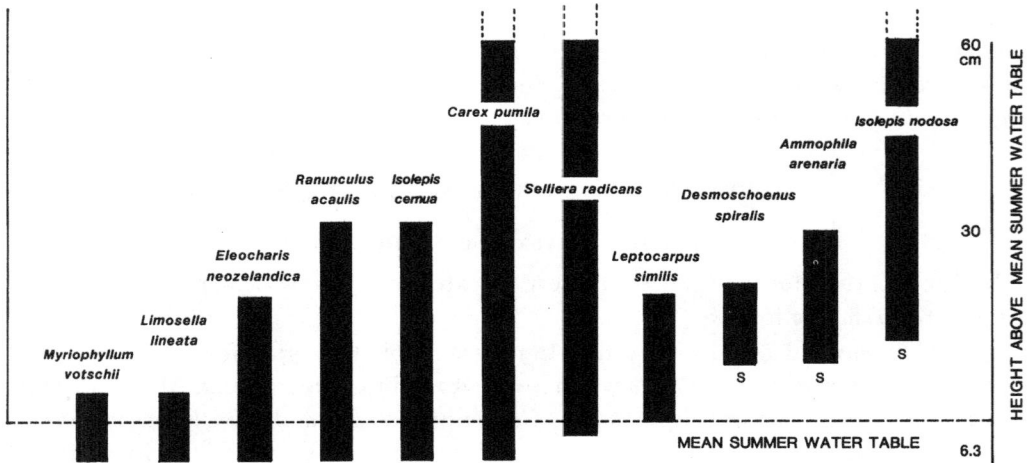

Fig. 11.1. Distribution of species in relation to summer water tables across low rear dunes and sand flats in Manawatu ER (Esler 1978a); s, seedlings.

Muehlenbeckia complexa and *Tetragonia trigyna*. By 1975, the *Cassinia* zone had been eliminated, and *Muehlenbeckia* and *Pennisetum clandestinum* were vying for dominance behind the spinifex zone.

Most of the large western dunes in Northland, including those of Ninety Mile Beach, have been thoroughly planted in marram, but on some eastern dunes of Aupori ER the native assemblage of *Spinifex*, pingao, *Austrofestuca, Pimelea arenaria* and *Coprosma acerosa* still prevails (Fig. 11.2).

South Island dunes

The Kaitorete dunes, which impound Lake Ellesmere in Canterbury, support the largest remaining area of pingao in New Zealand. On the foredune, it typically provides about 30% cover. Other species, which together contribute only about 5%, are *Lagurus, Hypochoeris radicata, Calystegia soldanella* and *Rumex acetosella*. Unfortunately, scattered marram colonies close to the beach are outgrowing pingao. Further from the sea, greater stability allows *Muehlenbeckia complexa* colonies and the locally endemic, prostrate broom *Carmichaelia appressa* to establish. Blow-outs with stony floors support *Raoulia australis* and some *Zoysia minima, Pimelea prostrata* and *Pseudognaphalium luteoalbum*. Still further inland, bracken or silver tussock dominate fixed sand hills, whereas sand and shingle flats support dry grassland with *Stipa nodosa, Rytidosperma unarede*, annual grasses and the inconspicuous *Hypoxis hookeri* (p. 30). Rare ngaio and *Dodonaea* trees may be the last vestiges of coastal bush that had almost disappeared by the time of European settlement.

Pingao still dominated on Christchurch dunes into the late 1920s, but now only a few colonies survive among the prevailing marram. Localised adventive sand-binders are *Carpobrotus edulis, Thinopyrum junceiforme* and *Leymus racemosus*. Non-rhizomatous plants include *Hypochoeris radicata, Senecio elegans* and the native *S. glomeratus*.

Fig. 11.2. Pingao (dark plants, mainly on right), *Spinifex* (mainly on left) and *Austrofestuca* (right foreground), Rarawa Beach, Aupori ER.

Occasional *Coprosma acerosa*, *Cassinia 'fulvida'*, ngaio and *Dodonaea* persist in the lupin–marram phase. Most other eastern and northern South Island dunes, including those of Farewell Spit, are also marram- and lupin-dominated; native sand-binders, other than *Calystegia soldanella* and *Carex pumila*, are sporadic relicts.

Marram and tree lupin clothe most western South Island dunes to as far south as latitude 43°30′, but pingao is usually present and dominates locally, perhaps because beaches are generally stony and do not yield enough sand for marram to prevail universally (Figs 11.3 and 11.4). From South Westland to Stewart Island, marram occurs mainly as isolated, recently planted or self-established pockets in otherwise native communities. The beach at Fortrose in Southland spans the transition from western indigenous to eastern marram–lupin dune vegetation (P.N. Johnson, unpublished). The western portion is coarse sand, which has formed low dunes in the following sequence.

(1) The crest of the foredune supports a 40% cover of pingao with some *Austrofestuca*, and colonies of marram building their own mounds.

(2) A low crest behind (1) has mostly pingao and silver tussock, other species being *Pimelea lyallii*, *Geranium sessiliflorum* var. *arenarium*, *Myosotis pygmaea*, *Hypochoeris radicata* and *Leontodon autumnalis*.

(3) The grazed gentle back slope has an 8% cover, in which the species of (2) are joined by pasture grasses and *Raoulia hookeri*. This meets *Leptocarpus* marsh.

Fig. 11.3. *Lupinus arboreus* on fixed dunes, Five-Mile Beach, Central Westland. The moraine ridge in the distance is truncated by marine erosion.

Fig. 11.4. Marram (foreground), clumps of pingao in middle ground, *Phormium tenax* and *Cortaderia richardii* behind; Five-Mile Beach, Central Westland.

The prevailing westerly wind blows fine sand to the eastern end of the beach, to form higher dunes. The sequence on these is:

(1) Foredune with 70% marram cover and only 1% pingao.

(2) Dune crest with almost total cover of 30% lupin/75% marram/40% *Hypochoeris, Leontodon, Cerastium fontanum,* cocksfoot, *Holcus lanatus, Trifolium repens* and *Cardamine debilis.*

(3) An inner slope with lupin and marram among pasture grasses.

(4) Bare sand hollow.

(5) Rear dune, with 40% lupin/40% marram – 20% pingao, as well as gorse and *Phormium tenax.* The persistence of pingao probably indicates that it was already dominant before marram arrived.

Shingle beaches

At Cape Turakirae east of Wellington, sudden earth movements over the past 6900 years have raised a flight of five stony beaches and shore platforms, the oldest being 27 m a.s.l. The sequence of soils and plant communities (the latter modified by fire and grazing) has been studied by Bagnall (1975).

(1) On the younger ridges and platforms the halophytes *Apium prostratum, Atriplex prostrata, Sarcocornia quinqueflora, Samolus repens, Juncus maritimus, Isolepis nodosa, Selliera radicans* and *Triglochin striatum,* rooted in pockets of fines, cover less than 2% of the surface. Patches of *Plagianthus divaricatus* grow in the lee of large boulders.

(2) On older ridges and platforms cover is more or less complete where peat and alluvium have accumulated. Open stony areas, which persist even on the oldest surface, have a 70–80% shrub cover up to 1.2 m tall, mainly of *Coprosma propinqua, Muehlenbeckia complexa* and *Melicytus crassifolius* rooting deeply among the stones (Fig. 11.5). *Cassinia leptophylla* and *Olearia solandri* can also be abundant. Other woody plants include *Coprosma crassifolia, Plagianthus divaricatus, Pennantia corymbosa, Urtica ferox,* and the lianes *Calystegia tuguriorum, Clematis forsteri* and *Rubus squarrosus.* A grassy understorey consists of *Anthoxanthum odoratum,* cocksfoot, *Lolium perenne, Rytidosperma unarede,* adventive poas, *Silybum marianum, Acaena agnipila* and clovers. Lichens and bryophytes cover the stones.

In Canterbury, crests of shingle beach ridges support *Muehlenbeckia ephedroides,* sometimes accompanied by *Plantago lanceolata, Bromus* spp., *Calystegia soldanella* and *Poa pratensis* (Mason 1969). On the inland slopes these can be joined by *Muehlenbeckia complexa,* tree lupin, gorse, *Galium aparine, Cirsium arvense, C. vulgare, Achillea millefolium, Hypochoeris radicata, Sonchus* spp. and *Holcus lanatus.* At the inner margin, *Plagianthus divaricatus* and *Plantago coronopus* appear where there is saline ground-water.

The prograding coast at Birdling's Flat south of Banks Peninsula has parallel low shingle ridges that support hummocks of *Coprosma propinqua, C. rigida, Melicytus alpinus, Muehlenbeckia complexa, Rubus squarrosus* and, where protected from grazing, *Carmichaelia violacea.* Herbaceous plants, including *Asplenium flabellifolium, Halora-*

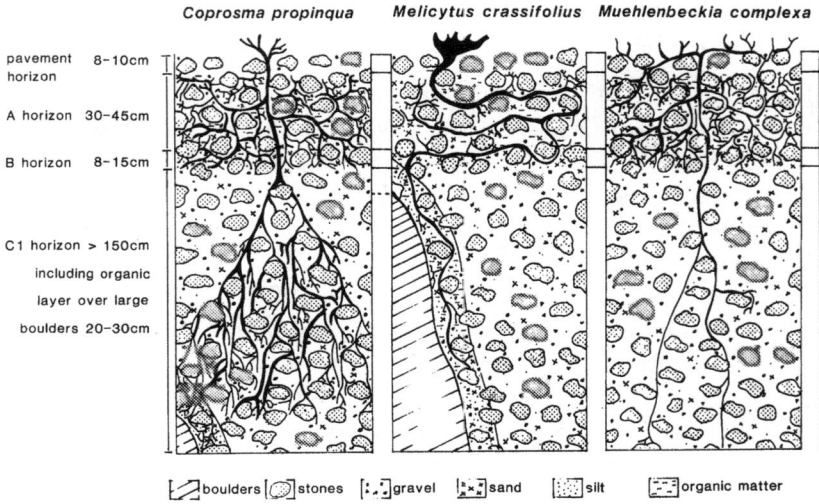

Fig. 11.5. Rooting patterns of shrubs on a stony beach ridge at Cape Turakirae near Wellington (Bagnall 1975).

gis erecta, Einadia allanii, Oxalis exilis, Geranium microphyllum, Vicia sp., *Senecio quadridentatus, Leptinella pusilla, Luzula* sp., and the grasses *Stipa nodosa, Rytidosperma* sp., *Holcus lanatus* and cocksfoot grow where there are fines. Beaches with finer gravel support vegetation more like that of dunes, with pingao, marram, *Carpobrotus edulis*, etc.

In the west of the South Island coastal erosion is throwing fresh gravel against cliffs or into older vegetation which, more often than not, is dense *Phormium tenax*. More stable beach ridges have sparse, varied cover, with *Calystegia soldanella* always present and *Muehlenbeckia complexa* (or occasionally *M. australis* or *M. axillaris*) often appearing further back. Cappings of sand support marram and less frequent pingao. There is invasion by wide-ranging pioneers, especially *Phormium tenax*, tall fescue (*Festuca arundinacea*), gorse, *Lotus pedunculatus, L. suaveolens*, blackberry, *Olearia avicenniifolia, Hypochoeris radicata, Sonchus* spp., and coprosmas, especially *C. propinqua* and *C. robusta*, as well as the more typically coastal *Cyperus ustulatus, Isolepis nodosa, Agrostis stolonifera, Polycarpon tetraphyllum, Fuchsia perscandens*, lupin and, on Western Nelson coasts, *Pennisetum clandestinum*. Coarse, well-drained sand may support *Rumex acetosella, Raoulia glabra, R. hookeri* (Fig. 11.6), *Festuca rubra, Poa pusilla* and *P. pratensis*. A shingle beach at Saltwater Lagoon in central Westland has *Navarretia squarrosa* which otherwise is a weed of dry cultivated land, and *Polygonum prostratum* which may have arrived spontaneously from Australia.

Inland flood-plains, dunes, moraines, tephra and depleted lands

The land-forms

Most large New Zealand rivers and many smaller ones rise in steep, eroding catchments, and carry massive loads of gravel and silt which they spread across wide

Fig. 11.6. *Raoulia hookeri* on a shingle beach, Saltwater Lagoon, Central Westland.

beds during floods. During normal flow they cut braided channels, leaving islands and flats available for colonisation, and years can pass before a further flood terminates the succession. Where rivers are degrading, at rates determined by tectonic uplift and downcutting in rock-bound gorges, flats become flights of terraces that may support progressively older soil and vegetation.

During dry, windy weather, dust blown from river-beds is trapped by vegetation to form loess. In the east of the South Island, where beds are widest and the prevailing westerlies strongest and driest, loess accumulates at rates approaching 0.2 mm annually (Ives 1973), and locally there are low dunes. During glacial phases, these features developed on a far larger scale; on the western flanks of Banks Peninsula, loess is up to 15 m deep, and ancient fixed dunes extend over many hectares on the Canterbury and Mackenzie plains. There are also active dunes on the Rangipo Desert, a tephra plain lying at 1050–1400 m on the eastern flank of Mt Ruapehu. Except for some planted marram, inland dunes support no specialised sand-binders, being colonised by the same species that colonise river-bed sand.

On moraines, substrates vary from expanses of large boulders to pockets of debris and water-sorted fines, but the colonising plants are mostly the same as on nearby river-beds. Although recent moraines are a minute portion of the present landscape, they are valuable markers for studies of succession (Chapter 15).

Quaternary eruptions deeply blanketed the central North Island in ignimbrite and tephra that form plains or gentle slopes. Most present surfaces derive from the massive Taupo eruption (p. 525), but significant quantities of tephra were erupted from

Tarawera in 1886, and there has been accretion from lesser eruptions. Some materials lie where they fell, whereas others have been eroded or re-deposited by water and wind. Pioneering species are much the same as on other unconsolidated surfaces, but succession tends to follow somewhat different courses because the parent material is mostly of low inherent fertility, and because much of it is lightly welded.

Native pioneers

Dicot herbs

Circular *Raoulia* mats, or scabweeds, are a distinctive feature on sand and gravel flood plains. *R. tenuicaulis*, the species most characteristic of river-beds, grows radially at rates up to 20 cm/yr. Its outer, extending shoots are less closely packed and have broader leaves than those in the centre and there is a corresponding colour gradient from silvery to green. *R. tenuicaulis* belongs mainly in mountain valleys with moderate to high rainfall, on moist recent alluvium that its roots exploit to depths of more than 30 cm. Being damaged by frosts of about −15 °C, it is infrequent above the subalpine belt.

The greyish mats of *R. hookeri* colonise older, drier alluvium. *R. haastii* forms brownish, dense cushions up to a metre across and 30 cm high, and mostly grows on river-beds just east of the South Island Divide. *R. monroi* and *R. parkii* (in northern and southern districts of the South Island respectively) and the more wide-ranging, very compact *R. australis* also grow on stony alluvium in dry districts, but today depleted grasslands are their usual habitat. The cushion-forming *Myosotis uniflora* grows in similar habitats from inland Canterbury to the upper Clutha valley in Central Otago.

Willow-herbs are also early pioneers. Among creeping species, *Epilobium brunnescens* is widespread on damp sand and fine gravel, whereas *E. nerteroides* is usually among rocks close to streams. *E. microphyllum*, the most widespread tap-rooted willow-herb on river-beds, is about 15 cm tall. The more robust *E. melanocaulon* has close-set, reddish leaves and wiry black stems; although mainly on stony river-beds in the eastern South Island mountains, it extends west of the Divide and to the North Island. Forms of *E. glabellum* are also common on river-beds, mainly in wetter mountain valleys.

Species of *Scleranthus* and *Colobanthus* forming broad, soft cushions and small, hard cushions respectively, and the erect, wiry *Stellaria gracilenta* grow on dry river gravel in the east of the South Island. Under higher rainfall, *Gunnera* forms mats of leafy rosettes on damp river sand, *G. dentata* being almost restricted to this habitat, and *G. monoica* being a wide-ranging pioneer. Other small native forbs that colonise river-beds include species of *Acaena*, *Gnaphalium* and *Craspedia*, *Nertera* in wetter districts, *Hydrocotyle* and *Pratia* on damp ground, and *Geranium sessiliflorum* which has red and green morphs in leaf colour.

Monocot herbs

The most widespread pioneering native grasses are *Lachnagrostis lyallii* and *L. filiformis*, which form broad, low, fine-leaved tufts. The fine-leaved, tussocky

Fig. 11.7. *Carmichaelia monroi* in eroded *Chionochloa* grassland. Other species are *Brachyglottis lagopus* (left) and *Wahlenbergia albomarginata* (right). Porters Pass, Puketeraki ER.

Rytidosperma setifolium and, in high-rainfall districts, small, densely tufted forms of *Poa novae-zelandiae* are common on subalpine and penalpine stony river-beds. The rhizomatous *Poa pusilla* and *Pyrrhanthera exigua* spread through sandy alluvium, the latter mainly on eastern South Island river terraces.

Toetoe (*Cortaderia* spp.) tussocks are a striking feature of many riverine sand banks in the lowlands and large mountain valleys. *Chionochloa conspicua* fills a similar role in higher, narrower, cooler valleys. Other pioneering grasses persist as dominants of closed grassland, the most consistent native species in this role being silver tussock and its mountain relative *Poa cockayneana*.

Woody Plants

Many shrubs important in closed vegetation, such as coprosmas, carmichaelias, matagouri (*Discaria toumatou*) and manuka, and even seedling trees (e.g. beeches, southern rata) can colonise alluvium within a few years of deposition. A few low shrubs are practically confined to the early open phase, the most widespread being *Coprosma brunnea*; female plants, normally undistinguished, acquire beauty in autumn when spangled with translucent blue drupes. *Helichrysum depressum*, which grows on eastern riverbeds of the South Island and Hawkes Bay, has prostrate, cupressoid, dead-looking grey stems. The sprawling *Carmichaelia nigrans* grows in South Westland and at Makarora (Lakes ER) and hybridises with the erect *C. grandiflora*. Several clump-forming and rhizomatous dwarf carmichaelias also colonise river gravels, some persisting into stony grassland (Figs 11.7 and 11.8). In the smallest plant in the genus, *C. uniflora* of inland Canterbury, only slender, grass-like cladodes are visible above ground, except when the disproportionately large, purplish flowers appear.

Fig. 11.8. *Carmichaelia corrugata* among grasses on a stony terrace of the Rakaia River, Heron ER.

Cryptogams

Common mosses on fine alluvial gravel are *Ceratodon purpureus, Campylopus* spp., especially *C. clavatus, Racomitrium lanuginosum, Bryum* spp., *Triquetrella papillata, Hypnum cupressiforme, Chrysoblastella chilensis* and *Polytrichum juniperinum*. Silty areas also support *Weissia controversa* and *Hypnum cupressiforme* if dry, and *Brachythecium rutabulum, Stokesiella praelonga,* species of *Breutelia, Tortula, Philonotis,* and the liverwort *Marchantia* if moist (A.J. Fife, personal communication). Larger stones attract the same cryptogams as rock surfaces (p. 519).

Adventive pioneers

The adventive colonists of recent alluvium far exceed the native in respect of numbers of species, rates of increase, and final height and density, and have practically ousted native plants from lowland river-beds through most of the country. Excepting urban wasteland, river-beds may be the richest habitat for naturalised plants, but only some of the more prominent are listed below.

Herbaceous species

On dry, porous sand and gravel the commonest adventive grasses are annual *Aira, Vulpia* and *Bromus* species, which are joined or replaced by *Festuca rubra* and other perennial species as closed grassland develops. On moister ground, especially deep silts, tall fescue (Fig. 11.9) or *Holcus lanatus* rapidly establish dense cover. The most widespread adventive forb is catsear (*Hypochoeris radicata*), but other 'dandelions' such as *H. glabra, Crepis capillaris*, and on damp ground, *Taraxacum officinale* are also

Fig. 11.9. *Festuca arundinacea* (left) pre-empts damp raw soil, whereas *Carex coriacea* occupies the saturated slope on the right; 600 m, Fyffe River, North-west Nelson.

important. Thistles, especially the rhizomatous *Cirsium arvense* on silt, are also abundant. Coltsfoot (*Tussilago farfara*), first detected about 1966 in the upper Waimakariri catchment, now occurs on both sides of the Main Divide near Arthurs Pass and is probably spreading. Its rapid growth, thick rhizomes, dense leafy cover, ability to survive transport by flood and burial by gravel, and pappus-bearing achenes may give it the potential to exclusively colonise large areas of river bed. Other introduced composites include ragwort (*Senecio vulgaris*), *Conyza* spp. and *Gnaphalium coarctatum*.

Various clovers are also early pioneers, notably *Trifolium arvense* and *T. dubium* on dry soils and *Lotus pedunculatus* on damp sites. Viper's bugloss (*Echium vulgare*), mulleins (*Verbascum thapsus* and *V. virgatum*) and Californian poppy (*Eschscholzia californica*) provide colourful displays in the east of the South Island. Docks (*Rumex* spp.) and rushes such as *Juncus articulatus* and *J. effusus* colonise wet hollows.

The horsetail *Equisetum arvense* is present in a few localities, resembling coltsfoot in its robust growth, thick rhizomes and ability to survive burial and fragmentation. The spread from its area of greatest abundance, i.e. the Mokihinui catchment in North-west Nelson, to neighbouring catchments has presumably been through spores. Colonies occur on river-beds and in swamps that receive flood water.

Woody plants

Common broom, gorse and, especially, *Lupinus arboreus* colonise very young surfaces, even those still disturbed by floods. In some places, especially in the Mackenzie region, Russell lupin (*L. polyphyllus*) forms dense stands of startling floral brilliance. Most plants are deep blue, but there are light-flowered hybrids where *L. arboreus* is present. Whereas lupins are short-lived, broom and gorse persist to form dense scrub. These legumes spread rapidly downstream from points of introduction, but upstream spread is slow; Allan's (1927) observation, that the Rangitata riverbed supported dense gorse whereas side streams were clear, is still widely valid. *Buddleja davidii* has reached many localities, including narrow, remote mountain valleys, and once established increases rapidly, forming thickets up to 3 m tall. Willows (*Salix* spp.), which are also vigorous pioneers, are discussed on p. 302.

Pioneering vegetation

The assemblages of species on these recent surfaces vary according to environment and the source of propagules. In three of the regional examples, the upstream catchments are almost wholly covered in native vegetation, and pioneering species are mostly native. These are discussed further in Chapter 15, as starting points for succession.

Since grazing can prolong open, seral stages, the flood-plain assemblages of the inland South Island lead on to discussion of semi-arid grasslands that have been degraded to a sparse cover, largely of pioneering species. The final example involves mainly adventive plants, and is now more typical of lowland flood-plains.

Lowland flood-plains of Central Westland

Except where gorse is well established, Westland flood-plains are still dominated by native plants (Wardle 1977). A *Raoulia* community on alluvial gravels below 200 m in Westland National Park forms 10–80% cover, but is subject to flooding and destruction. *R. tenuicaulis* is the main species, and *Epilobium brunnescens* is always present. *R. hookeri, E. glabellum, Uncinia divaricata, Lachnagrostis lyallii*, and *Poa cita* or *P. cockayneana* are present at most sites. *Racomitrium crispulum* and lichens are also important.

There are always seedlings of shrubs, especially *Carmichaelia grandiflora* (the commonest and most vigorous), *Olearia avicenniifolia, O. arborescens, Coriaria arborea, Hebe salicifolia* and *Aristotelia serrata*. In the valleys of the Fox and Franz Josef Glaciers these can be abundant enough to lead towards seral shrubland, although on many sites *Carmichaelia* dies out after a few years (presumably because of drought on porous gravel) and succession deflects towards grassland.

Most of the 82 species recorded also occur in river-flat grassland (p. 264) but others such as kamahi (*Weinmannia racemosa*) seedlings and sporelings of *Dicksonia squarrosa* anticipate development of forest. The following species were *not* recorded from neighbouring more advanced communities: *Oxalis magellanica, Gentiana* sp., *Ourisia caespitosa* and *Poa novae-zelandiae*, which are normally found at higher altitudes; *Carmichaelia nigrans, Epilobium microphyllum, Nertera ciliata, Pseudognaphalium luteoalbum, Aira caryophyllea* and *Vulpia myuros* (all pioneers); *Anaphalis trinervis* (a plant of cliffs); and the parasites *Cuscuta epithymum* and *Parentucellia viscosa*.

Similar communities extend to higher altitudes, although the number of species decreases and proportions change in response to cooler, moister conditions. At 900–1250 m, raoulias (*R. tenuicaulis* or *R. glabra*) dominate only on fine-textured alluvium,

Fig. 11.10. *Raoulia australis* dominating after depletion of grassland; 640 m near Omarama, Mackenzie ER. Other plants include *Poa cita* tussocks and, in the background, *Verbascum thapsus* and *Rosa rubiginosa*.

with *Rytidosperma setifolium* prevailing elsewhere. *Racomitrium lanuginosum* and *R. ptychophyllum* can also provide extensive cover. Most other species, especially *Festuca matthewsii, Poa colensoi, Hydrocotyle novae-zeelandiae* var. *montana* and *Wahlenbergia pygmaea*, represent the succeeding grassland, although *Carmichaelia grandiflora, Coprosma rugosa* and other shrubs are present at the lower altitudes.

Dry river flats and severely depleted lands in inland regions of the South Island

In the wide eastern valleys, alluvial flats formed by braided rivers are droughty even near the Main Divide where annual precipitation reaches 4000 mm, because of porous gravels and strong, dry winds. Early 'epilobium' and 'raoulia' stages (p. 517) are followed by open scrub of matagouri and briar (*Rosa rubiginosa*). Between shrub patches, *Raoulia australis* usually dominates on fine gravel, *Helichrysum depressum* may be the only common vascular plant on coarse gravel, and silver tussocks occupy silty areas. There are many shallow-rooted, short-lived adventives. Surfaces without vascular plants support sheets of *Polytrichum juniperinum* on sand and *Racomitrium* spp. on fine gravel.

This herb- and moss-field phase is prolonged through heavy grazing, especially by rabbits. In semi-arid districts, similar vegetation has developed through degradation of grassland in the valleys and on lower slopes, especially of northern to western aspect (Fig. 11.10). Grazing-tolerant grassland species persist and are joined by scabweeds, short-lived adventive forbs and grasses, whereas susceptible species, notably *Elymus rectisetus* and *Dichelachne crinita*, survive only within remnant tussocks and *Melicytus alpinus* bushes. These degraded communities vary considerably according to site factors, management, region, and continuing accretion of adventive plants.

Table 11.1 illustrates some community types. Scabweeds are especially important in the earlier samples, chiefly *Raoulia hookeri* on average sites, *R. australis* on very dry convex slopes and terrace brows, and *R. subsericea* persisting from the grassland on better sites. Sorrel (*Rumex acetosella*) is abundant in all types except 8. *Polytrichum juniperinum* forms large patches, and the pale yellow, unattached lichen *Chondropsis semiviridis* is abundant on the barest ground. Two southern localities (columns 2 and 3) illustrate dominance of the short, erect, semi-woody *Vittadinia gracilis* and *Thymus vulgaris*. Since the reduction of rabbit numbers from the 1950s, adventive grasses and briar have increased, hawkweeds have irrupted and scabweeds have been suppressed by taller plants that find them effective seed beds (Moore 1976).

Small, vernal annuals are the main adventive component, further species with this habit being the native *Myosurus minimus* and *Ceratocephalus pungens* and the possibly native *Gypsophila australis*. The tall adventive forbs occupy moister, less stable ground. They are mostly tap-rooted rosette plants that die after producing tall flower scapes, but St John's wort (*Hypericum perforatum*) and 'Californian thistle' (*Cirsium arvense*) are rhizomatous perennials.

Further native species of dry river-flat and depleted communities include *Pyrrhanthera exigua* (on sand), *Carex muelleri, C. albula, C. resectans, Lepidium sisymbrioides, Brachyglottis haastii* (persisting from grassland), the mat-forming herbs and dwarf shrubs *Leptinella pusilla, L. maniototo, Stackhousia minima, Cyathodes fraseri* var. *muscosa, Coprosma atropurpurea, C. petriei, Raoulia parkii* (Otago), *R. monroi* and *R. apicinigra* (Marlborough and Canterbury), cushions of *Pimelea pulvinaris, Myosotis uniflora, Agrostis muscosa, Luzula celata* and *L. ulophylla*, and the small rhizomatous broom *Carmichaelia enysii*. The silvery, 50 cm tall shrub *Pimelea aridula* is common in severely depleted areas in Central Otago and Waitaki ER.

Through cluster analysis of quadrats in the upper Clutha valley (Central Otago) Hubbard & Wilson (1988) define eight communities, named for the most frequent species and differentiated along a moisture gradient. At one end are *Salix fragilis* and miscellaneous wetland assemblages, and dry pastures with cocksfoot and *Cirsium vulgare*. The next five communities are characterised, respectively, by *Verbascum thapsus, Thymus vulgaris* and sorrel; briar and *Cyathodes fraseri; Echium vulgare* and *Verbascum virgatum; Bromus tectorum*; and *Trifolium arvense* and *Echium vulgare*. The dry end of the gradient is taken by a *Raoulia australis* community. Only 37 of the 85 important species are native, and several of these, including shrubs such as *Melicytus alpinus* and kanuka (*Kunzea ericoides*), are most prominent on rocky ground.

Unconsolidated tephra

At low altitudes, fresh tephra is rapidly colonised by plants such as *Coriaria arborea* and *Cortaderia fulvida* (p. 524), but with increasing altitude successions are retarded, and communities are more like those of upland river-beds. Above 900 m on the summit dome of Mt Tarawera, bared by the eruption of 1886, Clarkson & Clarkson (1983) describe an 8% shrub cover, mainly of *Dracophyllum subulatum*, over a 44% low cover of *Racomitrium lanuginosum* (14%), *Raoulia glabra* (5%), *R. albosericea*

Table 11.1. *Composition of vegetation replacing depleted tussock grassland*

Community:	1	2	3	4	5	6	7	8
No. of samples:	25	15	15	4	9	3	7	3
Mean altitude (m):	420	320	240	510	980	870	920	930
Prevailing aspect:	—	N	N	—	NE–SE	—	W–NE	NNW–E
Mean slope (deg.):	0	<30	<30	0	8	0	21	23
Mean percentage cover:	46	53	53	54	82	65	41	40
Mean no. of vascular spp.:	(61)	(80)	(56)	16	28	20	13	14

Short tussocks

	1	2	3	4	5	6	7	8
Poa cita	+	+	1	+				
Festuca novae-zelandiae	+				1			

Other native grasses

	1	2	3	4	5	6	7	8
Rytidosperma spp.	1	2	1					
Poa maniototo	1	1	1	f				
Elymus rectisetus	1	+		+			+	
Rytidosperma thomsonii	1		+	+				
Poa colensoi	1			+	1		+	
Poa lindsayi				f	f			

Adventive perennial grasses

	1	2	3	4	5	6	7	8
Agrostis capillaris	1							
Anthoxanthum odoratum	1							
Poa pratensis	+	1	1		+	f	+	
Holcus lanatus				f		+		

Adventive annual grasses

	1	2	3	4	5	6	7	8
Bromus tectorum	2	1	1	1				
Vulpia spp.	1	2	2	f				
Aira caryophyllea	1	1	1	1	+			
Bromus hordeaceus	+	1	1	+				
Bromus sterilis		1						
Hordeum murinum		1	+					

Sedges and rushes

	1	2	3	4	5	6	7	8
Carex breviculmis	1	1	1	+	1	f	+	+
Luzula rufa					f			

Tall woody plants

	1	2	3	4	5	6	7	8
Rosa rubiginosa	1	+	1		+		+	
Melicytus alpinus		1		+			+	
Muehlenbeckia complexa		1						
Discaria toumatou					+	+	f	+

Small semi-woody shrubs

	1	2	3	4	5	6	7	8
Vittadinia gracilis		1	2					
Vittadinia australis		1	+	+	+			
Thymus vulgaris			1					

Dwarf shrubs

	1	2	3	4	5	6	7	8
Muehlenbeckia axillaris	1	+		+	+		+	
Carmichaelia monroi		1						
Cyathodes fraseri			+		f			

Table 11.1. (*cont.*)

Community:	1	2	3	4	5	6	7	8
No. of samples:	25	15	15	4	9	3	7	3
Mean altitude (m):	420	320	240	510	980	870	920	930
Prevailing aspect:	—	N	N	—	NE–SE	—	W–NE	NNW–E
Mean slope (deg.):	0	<30	<30	0	8	0	21	23
Mean percentage cover:	46	53	53	54	82	65	41	40
Mean no. of vascular spp.:	(61)	(80)	(56)	16	28	20	13	14

Native mat, cushion and patch-forming herbs

	1	2	3	4	5	6	7	8
Convolvulus verecundus	1	1	+	+				
Leptinella pectinata	1	2			f*			
Raoulia subsericea	1		+		+			
Scleranthus uniflorus	+	1			f			
Colobanthus brevisepalus	+	+		+	+*			
Oxalis exilis	+	1	1		+	f	f	
Raoulia hookeri	+	+	1	2	4	f	+	
Raoulia australis	1	1	1	+	+	3	f	f
Dichondra repens		1			+			

Other native herbs

	1	2	3	4	5	6	7	8
Stellaria gracilenta	+	+	1	+	+			
Epilobium spp.	1			1	f		f	+
Geranium sessiliflorum	1	1	1	+	f	+	+	
Crassula sieberiana		1	+					
Wahlenbergia albomarginata		+	1			f		
Acaena caesiiglauca			+		+	f		
Viola cunninghamii					f			
Galium perpusillum						f		

Adventive rosette herbs

	1	2	3	4	5	6	7	8
Hieracium pilosella	1							
Hieracium praealtum	1							
Hypochoeris glabra	1	1						
Taraxacum officinale	+	1						
Hypochoeris radicata	1	1	1	f	f	f		
Crepis capillaris	1	1	1	f	f	f	+	

Tall adventive forbs

	1	2	3	4	5	6	7	8
Echium vulgare	1	1	1		+	f	+	+
Verbascum thapsus	+	1	1		+		+	3
Reseda luteola		1	1					
Cirsium vulgare		+	+		+	f	f	2
Verbascum virgatum			1		+	+	f	f
Cirsium arvense			+			+	f	
Carduus tenuiflorus								f

Small annual adventive forbs

	1	2	3	4	5	6	7	8
Logfia minima	1	+						
Silene gallica	+	1						
Trifolium dubium	+	1						
Trifolium arvense	1	1	1					
Erophila verna	1	+	+					

Community:	1	2	3	4	5	6	7	8
No. of samples:	25	15	15	4	9	3	7	3
Mean altitude (m):	420	320	240	510	980	870	920	930
Prevailing aspect:	—	N	N	—	NE–SE	—	W–NE	NNW–E
Mean slope (deg.):	0	<30	<30	0	8	0	21	23
Mean percentage cover:	46	53	53	54	82	65	41	40
Mean no. of vascular spp.:	(61)	(80)	(56)	16	28	20	13	14
Gypsophila australis	+	1	1	+	+			
Aphanes arvensis	+	+	1		+			
Veronica verna	1	1	1	+			+	
Myosotis discolor	1	1	1		+		+	f
Capsella bursa-pastoris		1						
Dianthus armeria		1						
Spergula arvensis		1						
Linum catharticum		1	1					
Navarretia squarrosa		1	+					
Euphorbia peplus		+	1					
Arenaria serpyllifolia		1	1		f	f	f	+
Anagallis arvensis			+			+		1
Cerastium glomeratum					f			
Other adventive forbs								
Salvia verbenaca	+	1						
Rumex acetosella	2	1	1	3	1	1	3	+
Cerastium fontanum	1	+	+	+			+	
Erodium cicutarium	1	1	1	+			+	
Sedum acre		+	1					
Marrubium vulgare		1	1					+
Fern								
Cheilanthes sieberi		+	1					

Notes:

Symbols as in Table 9.4.

Localities and communities:

1. Sorrel-dominated communities on low river terraces, Omarama, Mackenzie ER.
2. *Vittadinia gracilis* communities at Otematata, Waitaki ER.
3. *Vittadinia gracilis* communities in Central Otago.
    1–3 From point analyses on 2 × 10 m plots, recorded in 1977 and 1978 (Williams 1980).
4. Sorrel-dominated communities on flat terraces, Mackenzie ER, from cover estimates recorded in 1963 (Connor 1964).
5–8. Molesworth ER, from cover estimates recorded in 1951–53 (Moore 1954, 1976): 5–6, scabweed dominant; 7, sorrel dominant; 8, tall biennial forbs prominent.

Parentheses indicate total number of species in community type.

* *Leptinella squalida*, *Colobanthus strictus* in community 5.

(3%), *Rytidosperma viride* (3%) and other species. Open communities extending to the upper limits of plant growth on the central volcanoes are described on p. 430.

Moist river flats dominated by adventive herbs

The Horokiwi Stream near Wellington is typical of small streams with shingle beds and silt banks, that drain pasture on hill country formerly clad in podocarp/broad-leaved forest (Croker 1955). Within two months of a destructive flood the annual forbs *Polygonum hydropiper* and *Spergula arvensis* established complete cover, the former mainly on silt and the latter mainly on shingle. Within 3–4 years there was a dense sward of foxglove (*Digitalis purpurea*), *Rumex obtusifolius*, *Leucanthemum vulgare*, *Tanacetum parthenium*, *Juncus bufonius*, *J. articulatus*, cocksfoot and sweet vernal (*Anthoxanthum odoratum*). The grasses, which also included *Holcus lanatus* and *Poa annua*, were more prominent in silty areas. Stony ground admitted *Geranium molle*, *G. robertianum*, *Acaena anserinifolia* and *Epilobium ?nummulariifolium*, the last two being among the few indigenous plants. There were *Agrostis stolonifera*, *Juncus effusus*, *Rumex obtusifolius* and *Cotula coronopifolia* in wet depressions and *Glyceria fluitans*, *Mimulus guttatus* and *Callitriche stagnalis* in shallow water. Other prominent species were *Bromus hordeaceus*, *Ranunculus repens*, *Stellaria media*, sorrel, *Trifolium repens*, *T. dubium* and catsear.

Debris and boulder fields

Material eroded from the upper parts of slopes can accumulate as debris further down, and take the form of scree if it is being delivered faster than it can be fixed by vegetation and reduced through weathering. Unlike alpine screes continuously fed by frost-riving of rock outcrops, most temperate screes are temporary erosional features. Slope debris and scree grade into boulder fields, especially where gravity carries the large rocks to the base of the slope. Boulders also accumulate below cliffs of jointed rock, on banks of torrents, and as landslides, moraine and volcanic ejecta; they may also remain where finer material has been washed away.

Slope debris

Mobile low-altitude debris supports few, if any, specialised plants, but is readily colonised when it comes to rest. On fine debris in sheltered, mild, moist environments, seedling trees and shrubs, including hebes, coprosmas, carmichaelias, fuchsia (*Fuchsia excorticata*), wineberry and *Coriaria arborea*, and the large tussock grasses *Cortaderia* (Fig. 11.11) and *Chionochloa conspicua* appear as soon as, and quickly shade out, the pioneer mosses, small grasses and forbs that prevail on other stony sites such as river beds; broad-leaved bush develops rapidly. Rhizomatous, summer-green *Coriaria* species form large colonies on moist debris, these being *C. pottsiana* in the north of Gisborne province, *C. pteridoides* in Taranaki and Volcanic Plateau provinces, *C. kingiana* mainly east of the North Island axial ranges, *C. sarmentosa* mostly in drier or warmer South Island districts, *C. angustissima* on South Island mountains, and *C. plumosa* in cool, wet districts throughout. Beech seedlings often colonise raw debris in cool districts, especially on concave surfaces where there is less competition.

Fig. 11.11. *Cortaderia fulvida* on debris in Motu gorge, Raukumara ER.

Poa anceps covers slips on dry coastal and inland cliffs in the North Island, and in the Bay of Plenty has been joined by the garden escape *Miscanthus nepalensis*. Debris in drier, colder districts, especially inland parts of the eastern South Island, supports plants of the grey scrub community. Matagouri, the xeric broom *Carmichaelia petriei* and the lianes *Parsonsia capsularis* var. *rosea*, *Rubus schmidelioides* and *Muehlenbeckia complexa* are abundant at temperate altitudes, together with filiramulate coprosmas, *Melicytus alpinus*, *Muehlenbeckia axillaris*, *Hypolepis millefolium* and *Rytidosperma setifolium* which ascend to the penalpine belt.

Low, broad thickets of *Podocarpus nivalis* on subalpine and penalpine screes seem to be remnants of more extensive cover, an impression often supported by yellow-brown subsoil showing through the debris. *P. nivalis* seedlings are not found on mobile screes, existing plants having either persisted from a stable phase, or grown on to the scree. Growth of marginal shoots of up to 5 cm annually allows plants to spread downhill and laterally, compensating for burial of uphill shoots. Excavation of a mature plant showed that the aerial portion was 16 m downslope from an earlier position indicated by buried stems.

Transects across subalpine scree margins showed a mean gain by the bordering vegetation of grasses, herbs and low shrubs of only 2.5 cm in 10 years (Wardle 1972). Some 22 species were recorded, including *Muehlenbeckia axillaris* (the most frequent), *Wahlenbergia albomarginata*, *Galium perpusillum*, *Hydrocotyle novae-zeelandiae*, *Ourisia caespitosa* (on sand), *Poa colensoi*, *Podocarpus nivalis*, *Racomitrium lanuginosum* and three species characteristic of sparsely vegetated, stable debris, i.e. *Acaena glabra*, *Hebe epacridea* and *Epilobium crassum*.

None of the specialised plants of high-altitude scree (Chapter 12) grow in these margins, although they can descend to mid-altitude screes to mingle with a few species that scarcely reach the alpine belt, such as the rhizomatous *Convolvulus fractosaxosa* and *Anisotome filifolia*, the sprawling *Acaena glabra*, and the low shrub *Swainsona novae-zelandiae*, a legume belonging to a genus that is otherwise Australian.

Because debris slopes support palatable species, they attract domestic and feral herbivores, so that succession is deflected towards open vegetation of filiramulate and other browse-resistant shrubs, unpalatable tussocks such as *Poa cita*, *Uncinia* spp. and *Carex* spp., and turf of browse-tolerant grasses (e.g. *Poa breviglumis*, *Festuca rubra*) and creeping forbs (e.g. *Blechnum penna-marina* and species of *Hydrocotyle* and *Acaena*). Tall weeds such as thistles, *Verbascum* spp. and foxglove are prominent, and on bush margins the giant nettle *Urtica ferox* can form thickets. These communities resemble some depleted grasslands.

Although influences of vegetation or land use on the amount of erosion in New Zealand's steeplands are ultimately insignificant compared with the rate of geological erosion, grazing must have greatly increased the extent of superficially mobile screes. In particular, bare screes beneath bush canopies are predominantly induced, as is shown by the regrowth of continuous understoreys once animals are excluded.

Boulder fields

Boulder fields and very coarse debris are unpromising habitats, especially where composed of weathering-resistant rocks that lack crevices holding soil and water. For many years lichens and mosses may be the only colonists (p. 519). Debris that is not too coarse can be invaded from the margins by rhizomatous plants, including the ferns *Histiopteris incisa* and *Paesia scaberula*, and sprawling adventitiously rooting lianes such as species of *Muehlenbeckia* and native and introduced *Rubus*. The creeping ferns *Pyrrosia serpens* and *Phymatosorus diversifolius* can spread over rocks in mild localities or in partial shade. Coastal boulder fields support the semi-woody lianes *Tetragonia trigyna*, *Calystegia tuguriorum* and, in the northern zone, *C. sepium* and *Sicyos angulata*. Direct establishment of seedlings is mostly restricted to pockets of fines, although in wetter districts, plants with epiphytic traits, notably arborescent species of *Metrosideros*, can establish in small crevices or among moss.

In the montane and subalpine belts, especially in drier inland areas, vascular plant cover on boulder fields that have been stable since the end of the last glaciation may still be very sparse, whereas similar boulder fields in humid districts may support forest, with roots, moss and humus blanketing a cavernous substrate. Boulders appear to become covered most rapidly on south-western coasts, where high humidity and continual drenching by rain and sea-spray allow caps of peat to build up under bryophytes. Peat-capped boulders on raised beaches at Awarua Point (Olivine ER) support patches of salt marsh, in which *Leptocarpus similis*, *Samolus*, *Apium prostratum* and other wide-ranging species are accompanied by the tall, southern forb *Anisotome lyallii*.

Coarse debris and boulder fields provide fire-breaks that allow remnants of earlier vegetation to survive on enclaves of fine material. These include patches of forest and bush in the east of the South Island, in country otherwise converted to grassland and shrubs a thousand years ago. Boulder fields are also refuges for plants intolerant of grazing. Pockets of fines sheltered by boulders provide shaded, mesic habitats where delicate plants thrive, such as *Cystopteris tasmanica, Asplenium trichomanes* and *Stellaria parviflora.*

Along the Southern Alps, from Arthurs Pass southwards, *Olearia moschata* forms patches of low scrub on penalpine bouldery debris. The procumbent *Brachyglottis revoluta* occurs south of latitude 44°, and its relative *B. adamsii* grows in similar habitats in Marlborough and the Tararua Range, but mainly close to the upper forest limit.

Cliffs, bluffs and rock outcrops

The rupestral habitat

Most cliffs are steepened by streams or waves cutting at the base, but bold outcrops also arise through faulting or more rapid erosion of surrounding or underlying softer strata. Soil and plants tend to be removed by wind, frost, rain or avalanching snow and debris, and any mantle formed is apt to slide off periodically to re-expose the rock. In terrain scraped by glaciers, even gently rounded outcrops of hard rock may remain almost bare for many thousands of years.

Cliffs of soft or non-cohesive rocks, such as mudstone or gravel, may crumble continuously, offering neither firm surfaces for bryophytes and lichens, nor crevices and ledges for vascular plants. Stable cliffs, however, can rapidly develop continuous vegetation, at least in wet districts. Hard-rock cliffs and outcrops provide ideal surfaces for many lichens, and where moist for bryophytes as well, and crevices provide secure root holds. Spread of rooted vegetation across rock surfaces is slow, except where debris, soil, or bryophytic and algal humus accumulate.

Most New Zealand rocks are siliceous, and differences in plant cover relate more to rates of weathering than to chemical differences. Although limestones offer distinctive chemistry and distinctive surfaces formed by solution, fewer than 30 named taxa are strongly calciphile; moreover, soils over limestone are often acidic, especially where the rock is hard and precipitation high. On basic igneous and metamorphic rocks, including basalt and certain types of schist, rupestral florulas seem subtly different from those on siliceous rocks, but this has yet to be documented. The special features of ultramafic rock are discussed on pp. 396–400.

Cliffs and rock outcrops provide the most varied habitats of any landform, from uncolonised or uncolonisable surfaces to soil supporting seral or climax vegetation; from fully exposed to sheltered or shaded; and from waterless to permanently wet. On sheltered coasts, salt-intolerant species may descend to high-tide mark, whereas on exposed coasts waves may surge tens of metres up sloping rocks and vegetation 100 m above the sea may be drenched with spray. Many coastal cliffs are forbiddingly barren, whereas others harbour an unequalled diversity of habitats and species. Inland cliffs and rock outcrops tend to support more continuous but less varied vegetation because they lack the salt-influenced and frost-free habitats.

Among New Zealand mainland species that have ranges no larger than a single province, but excluding those restricted by climate to northern districts, at least two thirds grow only on steep, rocky terrain. This is partly because precipitous terrain provides isolation that may encourage speciation, and diverse habitats that buffer against major oscillations in climate and regional vegetation. However, bluffs flanking the valleys and fiords of Westland and Fiordland support no regional endemics other than stragglers from higher altitudes, presumably because they were beneath glacial ice only 14 000 years ago. Cliffs also give refuge to plants that browsing mammals or fire have exterminated from surrounding areas, although their value as refuges for indigenous species is compromised where the surrounds provide an overwhelming seed source of weeds that pre-empt new surfaces.

The rupestral flora

Many species that grow on cliffs and rock outcrops are more common in closed vegetation, especially grassland, shrubland and bush. Some forest epiphytes are also at home on ledges, and many penalpine plants descend to low altitudes, even to the coast, on sparsely-vegetated rock. This section concentrates on species that are mainly or entirely rupestral.

Trees, shrubs and lianes

Pohutukawa is superbly adapted to northern coastal cliffs. It also ascends maritime gorges, often as hybrids with northern rata (*Metrosideros robusta*), and grows on lakeside cliffs on the Volcanic Plateau. The light seed can blow into any crevice, and the roots spread widely over rock faces, seeking fissures and pockets of soil. The canopy moulds to the wind and tolerates salt spray, and aerial roots descend from the trunks to provide further anchorage. Northern rata and southern rata are almost as well-adapted to rupestral environments, and the latter can flower when a few centimetres tall. *M. parkinsonii*, which is confined to Western Nelson, the adjacent part of the Grey Valley, and the summits of Great and Little Barrier Islands, is a straggling shrub that mainly grows on the sides of forested ravines.

Brachyglottis includes many rupestral shrubs. *B. monroi* grows on bluffs and ledges in Marlborough and Lowry ER. The well-known garden shrub *B. greyi* has its natural habitat on greywacke rocks on the northern side of Cook Strait. Similar plants in northern Marlborough link with the probably conspecific *B. laxifolia*, which grows on limestone and dolerite in North-west Nelson ER. *B. bifistulosa*, a rare subalpine shrub with linear, revolute leaves, is confined to south-western Fiordland.

B. hectorii, which has large, soft, semi-deciduous leaves and white ray florets, grows around sink holes and other moist calcareous habitats in Western Nelson. *B. turneri*, a semi-woody plant with glabrous, cordate leaves and a habit intermediate between the shrubs and the rosette-forming herbs of the genus, is confined to wet mudstone cliffs in the east of Taranaki province. *Traversia baccharoides* grows on ledges and in rocky gullies in the northern third of the South Island.

Many olearias are abundant in rocky places, and the three species of the related *Pachystegia* (Marlborough rock daisy), a sprawling shrub with thick, oblong leaves up to 12 cm long, and white-rayed capitula to 8 cm in diameter, are exclusively rupestral (Fig. 11.12).

Fig. 11.12. Young *Pachystegia insignis* in the Waima gorge, Kaikoura ER.

The cupressoid species of *Helichrysum*, small, rupestral shrubs with terminal discoid capitula, are confined to the South Island. *H. intermedium* occurs throughout eastern and northern parts, but there are distinct local forms, notably some with bright yellow instead of the usual pale flowers, and var. *tumidum* on coastal cliffs on Otago Peninsula. The slender, yellow-flowered *H. parvifolium* is common at sub- and penalpine levels from eastern Nelson and Marlborough to North Canterbury, whereas the very stout *H. coralloides* is confined to eastern Marlborough. The lianoid *H. dimorphum* scrambles through scrub on the walls of the Waimakariri and tributary gorges, but has recently been decimated through spraying herbicide on common broom that invades its habitat.

Pink brooms (*Notospartium* spp.) mainly grow in gorges in eastern Marlborough, but *N. torulosum* extends sparingly to South Canterbury. Several other native brooms are also mainly rupestral, although this may reflect palatability as much as preference. *Clianthus puniceus* is a tall leafy legume, known as kaka-beak because of its large red (or in one variety, white) flowers shaped like a parrot's beak. Although familiar in cultivation, wild shrubs were confined to scattered localities around northern coasts and cliffs by Lake Waikaremoana. It has been further depleted by browsing, and the Waikaremoana population is now the only thriving one known.

Hebes, by virtue of effective dispersal and establishment, are important in most rupestral habitats. They include tall shrubs such as *Hebe stricta* (mainly North Island),

H. salicifolia and *H. subalpina* (South Island) and *H. elliptica* (coastal) that are common in non-rupestral habitats, as well as exclusively rupestral species. Most of the latter are narrowly endemic, small, spreading shrubs with short, broad, glaucous or ciliate leaves. Examples include *H. colensoi* (Hawkes Bay), *H. albicans* (Nelson and adjacent districts), *H. gibbsii* (Richmond Range), *H. rupicola* (Marlborough and northern Canterbury), *H. amplexicaulis* and *H. pareora* (South Canterbury), and *H. pimeleoides var. rupestris* (Central Otago). *H. rigidula* is mainly in gorges in Richmond ER, but a variety is confined to limestone in King Country ER. *H. townsonii*, which has rows of domatia on its leaves, is a calcicole disjunct between Western Nelson and the King Country. *H. cockayneana* is sub- to penalpine on bluffs and in low scrub in South Westland and Fiordland, as well as North-west Nelson.

The very distinct section *Paniculatae*, comprising the 'New Zealand lilac' *H. hulkeana* and its smaller relatives, are serrate-leaved, sprawling shrubs with lax panicles. All are rupestral and contained within an area extending from eastern Marlborough to mid-Canterbury. *H. lavaudiana* grows only on Banks Peninsula, and *H. raoulii* var. *maccaskillii* is known only from limestone in Lowry ER.

Many epacrids and other heath-like shrubs such as manuka and small podocarps extend to pockets of dry, leached soil on rock outcrops. *Archeria* spp. are characteristic of rocky terrain within closed forest and scrub. Three small shrubs confined to the central North Island, *Dracophyllum strictum, Gaultheria oppositifolia* and *G. paniculata*, grow mostly in open, rocky places. *G. subcorymbosa* (North Island) and *G. rupestris* (South Island) grow mainly on rocky river banks; *G. crassa*, which may be a penalpine form of the latter, extends from ledges and crevices to rocky debris.

The glaucous *Dracophyllum kirkii* forms espaliers on penalpine ledges on both flanks of the Southern Alps. In western Nelson, it is replaced by *D. pubescens*, which seems conspecific with *D. trimorphum*, a lowland plant of steep banks near West Wanganui Inlet. South of Paringa ED, *D. kirkii* grades into *D. politum*; typical plants of the latter grow on bogs, and intermediates grow on rock outcrops. The erect, unbranched *D. fiordense* is confined to ledges and clefts at the upper tree limit in South Westland and Fiordland.

Species of *Melicytus* characteristic of stony ground, i.e. *M. alpinus, M. crassifolius* and *M. obovatus*, are also commonly rooted in crevices. *Pittosporum dallii*, a handsome small tree with elliptic, serrate leaves 5–10 cm long and umbels of white flowers, grows only on rocky ledges and bouldery ground in subalpine beech forest in the Takaka catchment (North-west Nelson). Another small tree, *Pseudopanax macintyrei*, grows on calcareous outcrops from Nelson to North Westland. The umbellifer *Scandia rosifolia* is a small, leafy shrub that grows on cliffs from the Rangitikei region northwards.

Root-climbing rata vines can cling to sheer rock walls, especially *Metrosideros perforata* on the brows of coastal cliffs and *M. colensoi* as curtains pendant from limestone overhangs. Many other lianes, both native and adventive, spread over cliffs and other rocky habitats.

Succulent dicots

Several salt-marsh halophytes thrive in soil pockets on spray-drenched cliffs and headlands. The ice-plant *Disphyma australe*, which has thick, trailing stems and terete, succulent leaves 2–4 cm long, carpets sloping ground and hangs from ledges; large pink flowers make it one of the showiest native plants. *Crassula sieberiana* is a small plant common in dry crevices near the sea and inland, whereas the mainly southern, patch-forming *C. moschata* is confined to coastal rocks. *Peperomia urvilleana* creeps on sheltered coastal ledges to as far south as Sounds ER and the Kohaihai River (Northwest Nelson). *Sedum acre* (stone crop) is becoming universal where rubble thinly covers rock, concrete or bitumen, and through the summer provides splashes of bright yellow in otherwise barren habitats. Less succulent plants include the scrambling *Tetragonia trigyna* and *Einadia* species and, mainly around Cook Strait and Otago coasts, the sprawling *Atriplex buchananii*. In Westland and Fiordland, the near-absence of succulent plants on salt-marshes applies also to coastal rock.

Other dicot herbs

The yellow-rayed *Senecio lautus* and its close relatives are somewhat succulent groundsel-like plants that grow on most coastal cliffs except on western coasts south of Greymouth. In the same group, *S. sterquilinus* is mainly known from bird colonies on islands in Cook Strait, *S. glaucophyllus* subsp. *glaucophyllus* is a subalpine calcicole in Western Nelson, and the tall, semi-woody *S. rufiglandulosus* is widespread on damp cliffs and stream banks to as far south as North Westland. Its glaucous relative *S. banksii* grows mainly on coastal cliffs from Mayor Island (Coromandel ER) to Castlepoint (Eastern Wairarapa ER).

The herbaceous, rosette-forming species of *Brachyglottis* are often on rock, the larger forms of *B. lagopus* being confined thereto. In some districts palatable leafy herbs, such as the large mountain ranunculi, *Gingidia montana* and the two species of *Dolichoglottis*, are almost restricted to cliffs inaccessible to grazing animals (Fig. 11.13).

Some penalpine *Celmisia* species descend on coastal cliffs to sea-level, notably *C. monroi* in Marlborough, *C. major* west of Auckland city, *C. verbascifolia* in Fiordland and *C. semicordata* north of Greymouth. There are also low-altitude, locally endemic rupestral species, i.e. *C. adamsii* with separate varieties on the coast near Whangarei and on the summits of the Coromandel Peninsula, *C. morganii* in the Ngakawau Gorge north of Westport, *C. mackaui* on Banks Peninsula, *C. hookeri* in the Palmerston district of Otago, and *C. lindsayi* on the Catlins coast.

In *Craspedia*, taxa with large rosettes and capitula and short scapes are generally rupestral. Plants on coastal and lowland cliffs are mostly referred to *C. uniflora* and *C. robusta*, but nomenclature is confused. *Sonchus kirkii* grows on coastal cliffs, often in association with the adventive *S. oleraceus* and *S. asper* which, however, are more abundant on unstable ground.

Leucogenes is an endemic genus of wiry, procumbent herbs in the *Helichrysum* complex, known as New Zealand edelweiss by virtue of the white, woolly, petalloid bracts surrounding the compact inflorescence. *L. grandiceps* occurs throughout the higher mountains of the South Island and Stewart Island, ascending to the alpine belt and descending to low altitudes in rocky gorges. *L. leontopodium* is more compact, larger-leaved, and grows mainly on exposed ridges, from the Richmond Range to the Raukumara Range; similar plants from Inland Marlborough and Mt Peel (Pareora ER) are undescribed polyploids (Beuzenberg & Hair 1984). *Ewartia sinclairii* in the

Fig. 11.13. *Dolichoglottis scorzoneroides, D. lyallii* (top left) and *Ranunculus lyallii* among damp rocks; 1380 m, Arthurs Pass, North-eastern Alps.

same complex, and the only New Zealand member of an Australian mountain genus, is a dwarf shrub that grows in dry gorges in Marlborough.

Anaphalis trinervis trails down river banks, damp coastal cliffs and road cuttings, being especially abundant in Westland. The smaller-leaved *A. rupestris* of the south-west of the South Island and Stewart Island is mainly coastal, and scarcely distinguishable from hybrids between *A. trinervis* and *Helichrysum bellidioides*. *A. subrigida*, which has narrow, involute leaves, occurs through the North Island, mainly on limestone. The *A. keriensis* complex comprises smaller plants that take root in crevices of flood-swept rocks in shady gorges in the North Island and west of the South Island; this is also the main habitat for *Parahebe catarractae*, which grows in the North Island, northern South Island and, as a distinct subspecies, in Fiordland. Other parahebes, including *P. hookeriana* (North Island), *P. lyallii* (South Island) and the penalpine *P. linifolia*, tend to grow among drier, more exposed rocks.

From the Manawatu Gorge northwards, shaded ravines contain masses of the soft, leafy herb *Elatostema rugosum*. Stony ground beyond the reach of flood water in gorges support *Jovellana* or native calceolaria, *J. sinclairii* being confined to the Gisborne district, and the smaller *J. repens* extending to Central Westland. Similar habitats in lowland gorges in and near Nelson ER support *Scutellaria novae-zelandiae*, one of the only two native labiates, and *Poranthera microphylla*, which grows also in Australia.

Many *Epilobium* species extend to rock and some are mainly rupestral, including the wide-ranging *E. glabellum*, *E. matthewsii* confined to western Fiordland and Stewart Island, and *E. wilsonii* restricted to limestone in eastern Marlborough. *Linum monogynum*, a larger plant than the adventive flaxes, with white flowers 2 cm across, is common on coastal cliffs and extends several kilometres inland in gorges.

Lepidium oleraceum, a robust, rather fleshy crucifer, gained historical notice when it was gathered from coastal cliffs by Cook's sailors as an antiscorbutic. It has since become almost extinct on mainland coasts. The reasons seem that it is highly palatable to herbivorous mammals; it is attacked by the cabbage white butterfly, which became established in New Zealand in 1930; and it benefits from manuring by sea birds, which have been mostly driven by predators from accessible coasts (Ogle 1987). 'Cook's scurvy grass' still thrives on scattered off-shore islands where fertilisation by birds may compensate for attack by insects. Other rare lepidiums include the coastal *L. flexicaule* and *L. sisymbrioides* subsp. *kawarau* of schist bluffs in Central Otago. The Australian *L. desvauxii* has become abundant where rubble gathers on coastal cliffs.

Most of the 32 described species of *Myosotis* native to the main islands are scarce with highly local or disjunct distributions. Although ranging from wet to well-drained soils, and from coastal sand and lowland forest to exposed alpine slopes, they all grow on sites with relaxed root competition, especially pockets of alluvium, debris below cliffs, ledges, crevices, and edges of lakes, streams and roads. *M. tenericaulis*, however, is sometimes on bogs. About 13 species occur mainly below 1200 m, and 4–5 species are confined to limestone.

Monocot herbs

Three large chionochloas with restricted distributions on North Island cliffs are *C. bromoides* on eastern coasts and islands of Northland, *C. flavicans* between sea-level and 1000 m between the Coromandel Peninsula and Hawkes Bay, and *C. beddiei* on either side of Palliser Bay. In the South Island, penalpine snow tussocks descend to low altitudes on steep, sheltered, rocky sites, notably *C.* 'robust', which reaches sea-level in gorges in eastern Marlborough. *Rytidosperma setifolium* descends to the coast on exposed bluffs in the west and south of the South Island. Cliffs can be refuges for other native grasses (p. 265) in districts where grasslands are now adventive-dominated.

Simplicia buchananii and *S. laxa*, constituting an endemic genus, are slender, tufted grasses that grow beneath rock overhangs. The only recent collections of the former are from Central Otago, but the latter is possibly not uncommon in Western Nelson; both may have been overlooked through their resemblance to *Poa* and *Deschampsia* species that occupy similar habitats.

The broad, strap-like leaves of the sedge *Machaerina sinclairii* hang down wet cliffs, mainly of soft calcareous rock, throughout the North Island north of the Manawatu Gorge. *Schoenus pauciflorus* hangs from crevices, and can be the main plant in the spray of waterfalls; in these niches, it is widely distributed in the South Island.

The slender, yellow-sepalled *Phormium cookianum* subsp. *hookeri* is abundant on inland cliffs throughout the North Island. On coastal cliffs it is more local, but extends south of Kaikoura to North-west Nelson. The stouter, red-sepalled subsp. *cookianum* is usually subalpine but descends on rocky cliffs to the South Island coast. The two subspecies mingle on both shores of Cook Strait. Only *P. tenax* is present on long stretches of coast in both main islands, but where the two species meet on coastal cliffs, *P. cookianum* occupies crevices and ledges, whereas *P. tenax* occupies deeper soils.

Xeronema callistemon has the habit of *Phormium*, but bears shorter scapes of crowded red flowers and is confined to the Poor Knights and Hen Islands. *Arthropodium cirratum*, a lily with broad, soft leaves and graceful panicles of small white flowers, is abundant on many coastal cliffs to as far south as Golden Bay and Kaikoura. It maintains a lush appearance on apparently waterless ledges and crevices, presumably because the roots effectively locate moisture. The much smaller *A. candidum* is locally distributed among dry, shaded rocks, and extends to inland districts including Central Otago.

Ferns and fern-allies

Ferns in the *Blechnum 'capense'* complex are the most characteristic plants of vertical banks in moist districts. *B.* 'black spot', which has hanging fronds up to 2 m long, prevails in temperate habitats, but in the North Island and north-east of the South Island is often replaced, especially on limestone, by the equally robust *B.* 'green bay', which differs in that the pinnae do not diminish in length towards the base of the frond. At sub- to penalpine levels, the shorter *B.* 'mountain' occupies similar niches. *B. vulcanicum*, which has smaller triangular fronds, is less common. *B. colensoi* is confined to deeply shaded banks overhung by trees.

   Asplenium oblongifolium grows throughout maritime districts except in the west south of Hokitika, mainly on moist limestone. *A. lyallii* has smaller, thicker fronds, and grows sparingly in drier limestone crevices. *Pellaea* spp. and small spleenworts such as *A. terrestre*, *A. hookerianum*, *A. trichomanes* and *A. flabellifolium* are widespread in crevices, the last two often in dry habitats. *Cystopteris tasmanica* grows in deep, dry crevices and under boulders, whereas *Cheilanthes* spp. and the rare *Pleurosorus rutifolius* grow in dry, exposed crevices.

   The epiphytic polypods *Pyrrosia serpens*, *Phymatosorus diversifolius*, *P. scandens* and *Anarthropteris lanceolata* are all abundant on rock, the first often being on coastal cliffs. Filmy ferns, including *Trichomanes reniforme*, extend to shaded rocks; *T. strictum* and *T. elongatum* are confined to deeply shaded banks, and *Hymenophyllum minimum* is most common in bryophyte cushions at the top of the littoral splash zone. *Lycopodium varium* is abundant on rock ledges, and *Psilotum nudum* is frequent on semi-coastal cliffs in Northland.

Non-vascular plants

Racomitrium crispulum often forms a brownish-green carpet over dry rocks in the open, the greyish *R. lanuginosum* being more characteristic of flatter rocks with a veneer of rubble. Exposed rock supports small cushions of *Schistidium*, *Hedwigia* and *Grimmia* species or, if wet, mats of *Rhacocarpus purpurascens*. Species of *Zygodon*, *Breutelia*, *Bartramia*, *Amphidium*, *Bryum* and *Distichium* grow on dry rocks in partial shade. Mosses on wet shaded rocks include *Breutelia pendula*, *B. elongata*, *Cratoneuropsis relaxa* and *Philonotis* spp. *Bryum blandum*, *B. laevigatum*, *Fissidens rigidulus*, *Tridontium tasmanicum* and *Cratoneuropsis* grow in the splash of waterfalls. Gravel cliffs may be dominated by the 20 cm tall *Polytrichadelphus magellanicus* (A.J. Fife, personal communication).

   Thick carpets of leafy liverworts, especially species of *Lophocolea*, *Lepidozia*, *Plagiochila*, *Jungermannia* and *Schistochila*, can drape shaded, moist rocks. *Isotachis lyallii* forms conspicuous, reddish masses on vertical banks in high-rainfall areas. Thallose species tend to be on flatter, wetter surfaces; *Monoclea* forms a thick crust in deep shade, as does *Anthoceros* on more open sites. *Marchantia* spp. are more wide-ranging.

Lichens prevail over bryophytes on dry, exposed rock, some of the more common genera being indicated on p. 395. The 'red' alga *Trentepohlia* is usually the first plant to cover boulders in sheltered gorges, and the blue-green alga *Nostoc* can form a gelatinous crust on dripping ledges.

The vegetation of cliffs and similar land forms

This class of vegetation is illustrated as a north-to-south sequence. Although many of the examples are coastal, inland examples from lowland river gorges to penalpine bluffs are also described.

Eastern coasts and islands in the northern zone (based on Esler 1978*b*, and reports in *Tane*, including Wright 1977)

The cliff vegetation on off-shore islands from the Bay of Plenty to the Bay of Islands is rich and varied. On coastal rock, a 'black zone' of the marine lichens *Arthopyrenia sublitoralis*, *Verrucaria maura* and *Lichina confinis* extends up to high tide mark. This is followed by a bare zone and then by one that is often conspicuously yellow from the presence of *Xanthoria*; other lichens here are *Buellia*, *Pertusaria* and *Caloplaca holocarpa*. Finally, from 2 to 10 m above high tide, according to degree of exposure to spray, there begins a 'green zone' characterised by *Parmelia* spp. (Hayward & Hayward 1974).

Halophytes descend to high tide mark on seepages, the main species being *Leptocarpus*, *Apium prostratum*, *Selliera*, *Lobelia anceps*, *Isolepis cernua* and *I. nodosa*. Drier, more exposed surfaces support *Disphyma australe*, whereas *Senecio lautus* and sometimes *Spergularia media* are common on unstable ground. Other plants common in this zone are *Lachnagrostis filiformis*, *Deyeuxia billardierei*, *Dichondra repens*, *Sarcocornia*, *Einadia allanii*, *E. triandra* and *Sagina procumbens*. *Lepidium oleraceum* survives on some islands.

These communities merge upwards into rupestral and pioneering vegetation on ledges, crevices and patches of debris, that is potentially seral to coastal bush. Typical species are *Psilotum nudum*, *Asplenium flaccidum* var. *haurakiense*, *A. obtusatum*, *Poa anceps*, *Cortaderia splendens*, *Arthropodium cirratum*, *Phormium tenax*, *Astelia solandri*, *A. banksii*, *Peperomia urvilleana*, *Linum monogynum*, *Scandia rosifolia*, *Haloragis erecta*, *Parietaria debilis*, *Carmichaelia aligera*, *Coriaria arborea*, taupata (*Coprosma repens*), *Hebe stricta*, *H. speciosa*, kanuka and pohutukawa. Local endemics include *Carmichaelia williamsii*, which is largely confined to islets from the Coromandel Peninsula to East Cape, several hebes including *H. bollonsii*, *Xeronema callistemon*, and, representing the deepest incursion of mountain genera into the warm-temperate belt, *Celmisia major*, *C. adamsii* var. *rugosula* and *Chionochloa bromoides*.

Drier ledges can support manuka, *Cyathodes fraseri*, *C. fasciculata* and *Cassinia* '*retorta*'. The lianoid *Einadia triandra, Tetragonia trigyna, Muehlenbeckia complexa* and locally, *Sicyos* can be abundant on bush margins. Further from the sea, rock outcrops support *Cheilanthes sieberi, C. distans, Crassula sieberiana* and, if surrounded by pasture, adventives such as *Hypochoeris glabra, Linum trigynum* and the vernal *Aira* spp., *Vulpia* spp., *Anagallis arvensis, Polycarpon tetraphyllum, Avena barbata* and *Bromus diandrus*.

A North Island gorge

Along the Motu River (Gisborne province), reaches where forest descends to gravel flats alternate with gorges where floods periodically sweep the vegetation from the rock walls, to a height that can be several metres above the river bed. Dripping vertical banks above half-way to the forest edge have sheets of *Anaphalis keriensis* and *Blechnum* 'black spot', together with *Pratia angulata, Hydrocotyle moschata, Oxalis magellanica* and *Adiantum cunninghamii. Jovellana sinclairii* grows in shaded clefts just below the forest. Lower or drier crevices support *Phormium cookianum, Parahebe catarractae, Dichelachne crinita* and *Carmichaelia flagelliformis*, which has slender cladodes that offer little resistance to flood water.

Pockets of silt are colonised by *Rytidosperma unarede, Carex coriacea, Nertera depressa, Gunnera monoica, Haloragis erecta* and the adventive cocksfoot, tall fescue, *Juncus articulatus, Ranunculus repens* and *Prunella vulgaris*. Seedlings of the seral shrubs *Coprosma robusta, Coriaria arborea*, kanuka, *Hebe stricta* and an unnamed, locally endemic hebe are frequent. Ledges high above the river support *Scandia rosifolia* and *Senecio banksii*.

Hawkes Bay

At Waipatiki north of Napier, there are near vertical coastal cliffs of soft limestone. Although supporting two eastern North Island endemics, marked†, their vegetation is typical of limestone and calcareous mudstone cliffs between here and Tainui ER. *Machaerina* (60%), *Phormium cookianum* (25%), *Blechnum* 'green bay' (10%) and *Chionochloa flavicans*† (5%) form a dense cover. Where this has slipped away, other species occur, including *Adiantum cunninghamii, Samolus, Lobelia anceps, Coriaria arborea, Linum monogynum, Pimelea prostrata, Centaurium erythraea, Senecio banksii*†, *Anaphalis subrigida, Conyza* sp., *Poa anceps* and *Lachnagrostis richardii*.

Eastern Wairarapa (Park 1967)

On coastal limestone cliffs at Castlepoint, *Samolus, Disphyma* and *Apium prostratum*, accompanied by *Senecio lautus, Puccinellia stricta* and *Spergularia media*, occupy benches close to the sea, and also form a 3 m wide strip back from the brows of the cliffs. Ledges on the cliff faces support *Poa cita, P. anceps, Disphyma*, taupata, and *Matthiola incana*, whereas crevices hold *Senecio banksii* and the local endemic *Brachyglottis compacta*.

Communities on limestone debris slopes are variously dominated by *Poa cita, Isolepis nodosa, Phormium cookianum* and *Brachyglottis compacta*. These potentially succeed to coastal bush, as represented by remnant patches of karaka (*Corynocarpus laevigatus*), ngaio, *Macropiper excelsum* and *Solanum aviculare*. Sand blown on to limestone debris supports *Poa cita, Isolepis nodosa, Zoysia minima, Phormium cookianum, Coprosma acerosa, Pimelea arenaria* and *Cassinia leptophylla*.

Vicinity of Cook Strait

Near Cape Turakirae steep spurs of solid greywacke support a 10% cover of *Chionochloa beddiei*. Where the rock is more fissured, there is open scrub of the prostrate Cook Strait form of *Sophora microphylla, Melicytus crassifolius, Brachyglottis greyi, Olearia paniculata*, kanuka, manuka, *Coprosma propinqua, Hebe parviflora* and *Phormium cookianum*, with herbaceous species such as *Asplenium terrestre, Pyrrosia serpens, Colobanthus strictus, Haloragis erecta, Linum monogynum, Sonchus* sp., *Senecio lautus, Vittadinia australis, Wahlenbergia gracilis, Rytidosperma* sp., *Dichelachne crinita, Briza major, Agrostis capillaris*, a fine-leaved fescue and *Thelymitra longifolia*. The main moss is *Campylopus clavatus*. The bluffs are separated by coarse debris being colonised mainly by *Muehlenbeckia complexa*, with further species including *M. australis, Clematis forsteri, Calystegia tuguriorum, Convolvulus verecundus, Aciphylla squarrosa, Craspedia uniflora* var. *grandis, Poa anceps, Polystichum richardii* and bracken.

Corresponding inland vegetation grows on sandstone cliffs in the Aorangi Range, between 350 m and 850 m a.s.l. (Druce 1971). Dominant species are *Chionochloa beddiei, Phormium cookianum* and *Brachyglottis greyi* below 600 m, and *Celmisia spectabilis, Poa colensoi, Hebe venustula* and *Blechnum 'capense'* above 600 m. Others are *Pyrrosia serpens, Polystichum richardii, Carmichaelia arborea, Linum monogynum, Melicytus crassifolius, Aciphylla squarrosa, Coprosma crassifolia, Helichrysum aggregatum, Brachyglottis lagopus, Craspedia uniflora* var. *grandis* and *Libertia grandiflora* below 600 m, and *Ctenopteris heterophylla, Hymenophyllum multifidum, Pimelea gnidia, P. longifolia, Pseudopanax colensoi* and *Griselinia littoralis* above 600 m.

The small islands of Cook Strait are extremely exposed to westerly and southerly gales; Little Brother Island, for example, can be drenched with spray to its summit at 74 m. Fig. 11.14 illustrates vegetation along a gradient related to salinity and pH (Table 11.2) and damage by salt storms (Table 11.3).

Kaikoura Ecological Region

The spectacular limestone gorge of the Waima River is rich in species endemic to Marlborough or extending only to northern Canterbury or Arthur ED in North-west Nelson (marked[1]) or confined to limestone (marked[2]). On ledges and in crevices, the main plants are *Carmichaelia ovata, Notospartium carmichaeliae*[1], *Linum monogynum, Clematis foetida, Pachystegia insignis*[1], *Celmisia monroi*[1] and *Phormium cookianum*. Others include *Pleurosorus rutifolius, Adiantum cunninghamii, Blechnum* 'green bay', *Clematis afoliata, Ranunculus insignis, Haloragis erecta, Cardamine debilis, Carmichaelia astonii*[1,2], manuka, *Dodonaea, Gingidia montana, Corokia cotoneaster, Olearia paniculata, O. coriacea*[1], *Senecio glaucophyllus* aff. subsp. *toa*[1], *Brachyglottis monroi*[1], *Wahlenbergia matthewsii*[2], *Myosotis arnoldii*[1,2], *Gentiana astonii*[2], *Hebe hulkeana*[1], *H. traversii*[1], *Parahebe catarractae* subsp. *martinii*[1], *Poa acicularifolia* subsp. *acicularifo-*

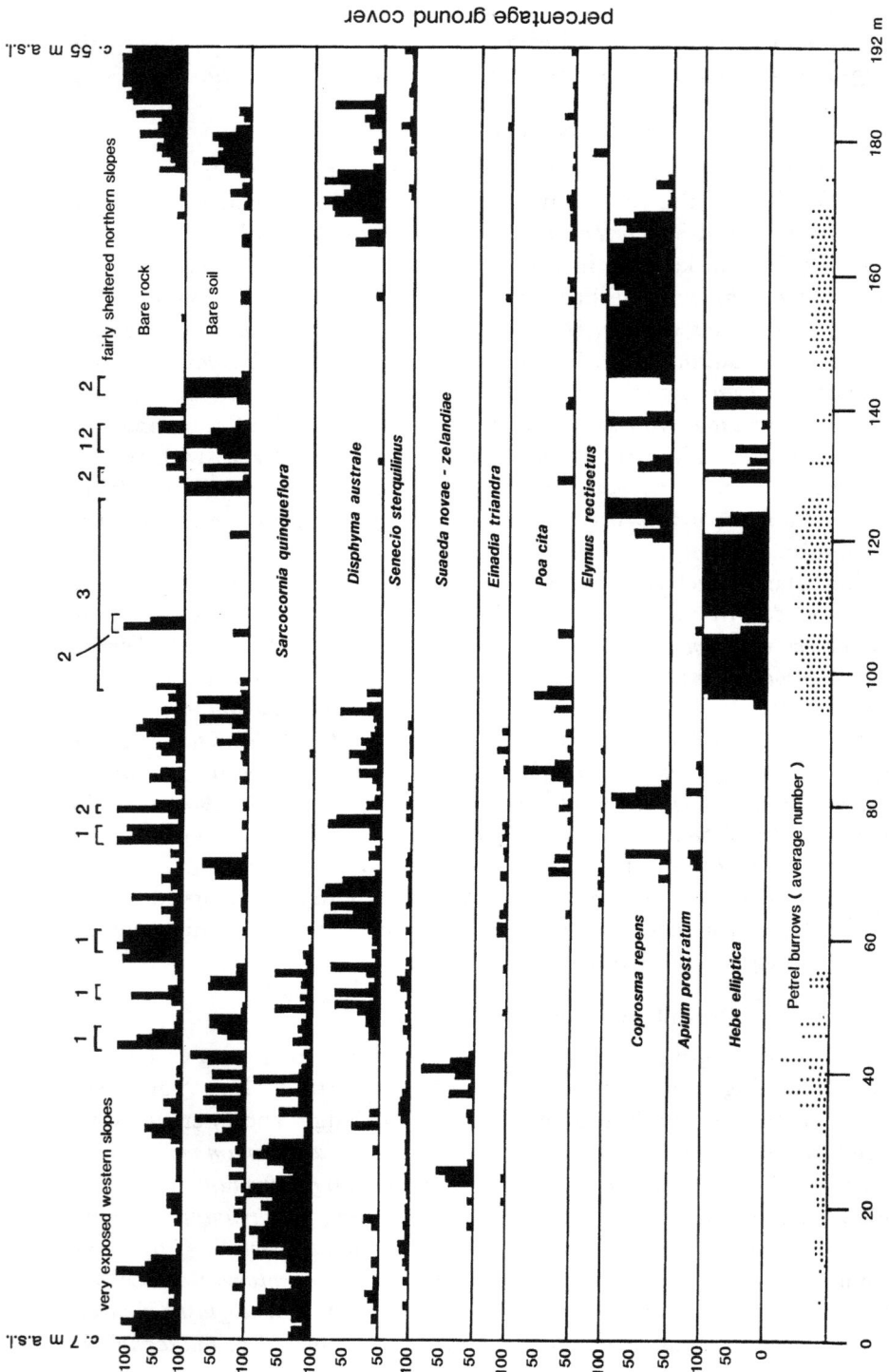

Fig. 11.14. Transect across Little Brother Island, Cook Strait, from the very exposed western slope to the more sheltered northern slope (Gillham 1960b). Numbers at top indicate: 1, vertical; 2, paths or tracks; 3, sheltered by lighthouse buildings.

Table 11.2. *Salinity and pH on sheltered eastern slopes of Stephens Island, Cook Strait*

Chlorides as percentage of mass of air-dry soil; pH measured on moist soil. All communities are burrowed by prions or petrels, except the first, which supports a colony of gulls.

soil type	plant community	chlorides	pH
organic	Halophytes and crevice plants	1.06	7.8
organic	*Poa cita* tussock grassland	0.52	4.6
organic	*Lolium perenne* pasture	0.44	4.8
mineral	*Poa cita* tussock grassland	0.25	7.0
mineral	*Lolium perenne* pasture	0.13	6.8
organic	*Myoporum laetum* on cliff	0.10	3.6
organic	Bush further inland with mixture of dominants	0.07	3.6
		0.01	3.6

Source: Gillham 1960*b*

Table 11.3. *Damage by salt gales in Cook Strait*

The upper portion of each column shows the effect of a strong gale on exposed Little Brother Island. The lower portions show damage at the forest margin on the lee shore of Kapiti Island.

severe	medium	slight to nil
Sonchus oleraceus	*Kirkianella novae-zelandiae*	*Poa cita*
Lepidium oleraceum	*Elymus rectisetus*	*Melicytus obovatus*
Einadia triandra	*Wahlenbergia* sp.	*Senecio lautus* s.l.
	Coprosma repens	*Disphyma australe*
	Hebe elliptica	*Sarcocornia quinqueflora*
	Apium prostratum	
	Hypochoeris radicata	
Anagallis arvensis	*Griselinia lucida*	*Arthropodium cirratum*
Beilschmiedia tawa	*Hebe speciosa*	*Asplenium oblongifolium*
Coprosma propinqua	*Olearia paniculata*	*Cassinia leptophylla*
Macropiper excelsum		*Coprosma repens*
Muehlenbeckia complexa		*Hebe elliptica*
Pseudopanax arboreus		*Peperomia urvilleana*
Pittosporum sp.		*Phormium cookianum*
Brachyglottis repanda		*Einadia triandra*
Stellaria media		*Senecio lautus* s.l.
Myrsine australis		*Disphyma australe*
Trifolium dubium		*Hypochoeris radicata*

Source: Gillham 1960*b*

lia[2], *Rytidosperma setifolium* and *Libertia ixioides*. On wet ledges *Schoenus pauciflorus* is abundant, accompanied by *Cortaderia richardii*, *Chionochloa* 'robust', *Craspedia* spp. and *Epilobium wilsonii*[1,2].

Dry, open limestone debris supports many of the same plants, but weedy adventives are also common, including *Catapodium rigidum*, *Cirsium vulgare*, *Cerastium* sp., *Linum bienne*, *Arenaria serpyllifolia*, *Echium vulgare*, *Conyza* sp. *and Euphorbia peplus*, together with the native *Asplenium trichomanes*, *A. lyallii*, *Oxalis exilis*, *Galium trilobum*, *Leptinella pyrethrifolia*, *Vittadinia australis*, *Pseudognaphalium luteoalbum*, *Senecio quadridentatus*, *Elymus rectisetus*, *Poa breviglumis* and *P. cita*. Further species listed by Druce & Williams (1989) from cliffs in this district include *Coriaria kingiana*, *C. sarmentosa*, *Epilobium brevipes*[2], *Aciphylla aurea*, *Anisotome filifolia*, *Schizeilema roughii*, *Helichrysum intermedium* var.[1,2], *Hebe decumbens*[1] and *Festuca multinodis*.

The debris shows ragged successions to closed woody vegetation. Manuka, *Dodonaea*, *Coprosma propinqua*, *Brachyglottis monroi* and *Cassinia leptophylla* are abundant in an early stage. On moister or more mature areas, totara (*Podocarpus totara*) saplings and *Pseudopanax arboreus* become prominent, together with small broad-leaved trees such as *Griselinia littoralis*, *Coriaria arborea*, wineberry, ngaio, *Alectryon excelsus* and cabbage tree; two uncommon plants are *Coprosma obconica* and *Pseudopanax ferox*. Patches of forest are dominated by totara and matai (*Prumnopitys taxifolia*) on calcareous debris, and by black beech (*Nothofagus solandri* var. *solandri*) on argillite (Fig. 7.29).

The more exposure-tolerant species of the gorges, such as *Pachystegia insignis* and *Phormium cookianum*, also grow above the halophyte zone on stable coastal cliffs. At Cape Campbell, gullied shelly mudstone supports sparse turf of *Plantago spathulata*, accompanied by *Senecio glaucophyllus*, the local endemic *S. hauwai*, *Spergularia media*, *Pimelea prostrata*, *Samolus*, *Glaucium flavum* and *Microseris scapigera*, usually an inland species of swampy grassland.

Spenser Ecological Region

A north-facing bluff at 1050 m in Hopeless Creek in the Travers Range interrupts the prevailing mountain beech (*Nothofagus solandri* var. *cliffortioides*) forest, and supports a mosaic of *Griselinia littoralis* and southern rata trees, shrubby areas with *Brachyglottis adamsii*, *Helichrysum intermedium* or *Hebe subalpina*, and herbaceous vegetation consisting mainly of *Rytidosperma setifolium*. There are a few mountain beech trees, and manuka grows where water seeps over the rock. Other species listed are *Asplenium flabellifolium*, *A. terrestre*, *A. richardii*, *Blechnum penna-marina*, *B.* 'mountain', *Podocarpus nivalis*, *Clematis paniculata*, *Ranunculus insignis*, *Stellaria parviflora*, *Rubus cissoides*, *Gingidia montana*, *Schizeilema roughii*, *Muehlenbeckia axillaris*, *Aristotelia fruticosa*, *Aciphylla aurea*, *Oreomyrrhis rigida*, *Dracophyllum longifolium*, *D. uniflorum*, *Gaultheria crassa*, *Cassinia* 'vauvilliersii', *Olearia avicennii-folia*, *Coprosma* aff. *parviflora*, *C. rugosa*, *Gentiana* ?*patula*, *Hypochoeris radicata*, *Celmisia monroi*, *Leptinella pyrethrifolia*, *Luzula* ?*banksiana*, *Carex* ?*cockayneana* and *Phormium cookianum*.

Many similar bluffs enclose gorges that descend from hanging valleys east of the zone of maximum rainfall, from Spenser ER to inland Southland. As well as rupestral species, they support trees and shrubs that cannot compete in continuous beech forest, or which have been removed from gentler terrain by fire. Where shaded and moist, these gorges also support mesophytic plants such as *Gingidia montana*, *Craspedia* aff. *major*, *Dolichoglottis lyallii*, *Celmisia bellidioides*, *Ourisia macrophylla* s.l. and *Schoenus pauciflorus*.

Western Nelson

Little Wanganui Head

This high bluff overlooking the Tasman Sea consists of bands of limestone, volcanic rocks, and siliceous sedimentary rocks. Shallow debris trapped by ledges supports patches of *Rytidosperma setifolium*, *Phormium cookianum* and *Olearia arborescens*. Deeper debris supports seral communities, ranging from *Cortaderia richardii* and *Coriaria arborea* to dense bush of *Dodonaea*, *Macropiper*, *Coprosma arborea* and nikau (*Rhopalostylis sapida*) palms. Other species include *Pimelea prostrata*, *Juncus caespiticius*, *Haloragis erecta*, *Lagenifera pumila*, adventives such as *Plantago australis* and cocksfoot, and in the bush, *Pteris macilenta* and *Uncinia uncinata*. There are many goats.

Mt Burnett (886 m)

Dolomite bluffs on this isolated summit near Collingwood support diverse vegetation. The most open part of the summit slopes east at 45°. *Rytidosperma setifolium* covers 50% of this area, and litter and bare rock each 20%. The other main species are *Phormium cookianum*, *Hebe albicans* and *Gingidia montana*. Fluted rock sloping north at 45° has a 20% cover of shrubs 0.5–2 m tall, the main species being *Olearia avicenniifolia*, southern rata and *Pimelea longifolia*, with less *Brachyglottis laxifolia*, *Coprosma propinqua*, *Hebe townsonii*, *H. albicans* and *Phormium cookianum*.

Less precipitous surfaces have open, stunted forest of silver beech (*Nothofagus menziesii*), southern rata and kamahi, with a remarkable mixture of associated plants. Typical karst species include most of the plants listed above, together with *Asplenium oblongifolium*, *Pseudopanax macintyrei*, *Melicytus obovatus* and *Brachyglottis hectorii*. Species normally found on recent soils are *Blechnum fluviatile*, *Pittosporum tenuifolium*, mahoe, *Hoheria sexstylosa*, *Olearia arborescens*, *Brachyglottis repanda*, *Coprosma grandifolia* and *Libertia* sp. *Phyllocladus alpinus*, *Dracophyllum filifolium* and *Gahnia pauciflora* usually indicate leached, acidic soils. *Coprosma linariifolia* and *Dianella nigra* are typical of dry lowland forest. *Astelia trinervia*, *Cordyline banksii* and *Ascarina lucida* represent a warm-temperate element, whereas *Cyathea colensoi*, *Podocarpus hallii* × *nivalis*, *Dracophyllum traversii* and *Astelia nervosa* are usually at higher altitudes. The usually epiphytic *Lycopodium varium*, *Astelia solandri* and *Earina mucronata* are growing on rock. A shrub related to *Myrsine divaricata*, but with branches that are not flexuous and interlacing, may be endemic to the mountain. Remaining species are normally associated with beech forest at this altitude, i.e. *Podocarpus hallii*, *Prumnopitys ferruginea*, *Coprosma rhamnoides*, *C. foetidissima*, *Cyathodes fasciculata*, *Pseudopanax crassifolius*, *Griselinia littoralis* and *Libertia pulchella*.

Fig. 11.15. Penalpine limestone bluff on Turks Head Range, North-west Nelson.
Species listed in the text.

Subalpine and penalpine bluffs

On north-facing limestone bluffs at 1450 m on the Turks Head Range, the main
species on ledges are *Helichrysum intermedium, Hebe topiaria, Celmisia monroi,
Phormium cookianum* and *Chionochloa* 'robust' (Fig. 11.15). Others include *Ranunculus insignis, Muehlenbeckia axillaris, Aciphylla ferox, Olearia avicenniifolia, Senecio
glaucophyllus, Rytidosperma setifolium, Festuca matthewsii* and *Elymus rectisetus*. Steep
debris cones between the outcrops support a herb-rich grassland with cover estimated
as 20% *Phormium* – 10% *Aciphylla glaucescens*/10% *Chionochloa pallens*/40% *Elymus
rectisetus* – *E. narduroides, Uncinia* sp. – *Festuca matthewsii*.

North-facing outcrops of Cambrian sedimentary rocks at 1200–1500 m in North-
west Nelson support low scrub, in which *Dracophyllum pubescens* provides 20% cover
in association with *Pseudopanax colensoi* var. *ternatus, D. uniflorum* and *Brachyglottis
bidwillii*. There are also *Podocarpus nivalis, Coprosma* 'penalpine', *Exocarpus bidwillii*,

Myrsine nummularia, Aciphylla ferox, Forstera mackayi, Celmisia spp., *Dolichoglottis scorzoneroides, Hebe albicans, Phormium cookianum, Chionochloa* 'robust' and *Astelia skottsbergii* (P.A. Williams, unpublished). Granite outcrops on the Paparoa Range, at 1050 m, have further species derived from woody heath and infertile penalpine grassland, especially *Halocarpus biformis, Olearia colensoi, Chionochloa australis* and *Carpha alpina.*

Westland and the western Alps

Schist bluffs in the Alps (Wardle 1977)
Bluffs below 1000 m present a great variety of habitats. Lower (< *c.* 600 m) and higher altitude variants are recognisable, with the latter grading into penalpine communities. The main species at low altitudes are *Blechnum* 'black spot', *Coriaria arborea* (damp bluffs below 300 m), southern rata, *Griselinia littoralis, Olearia avicenniifolia, O. arborescens, Chionochloa conspicua* (to 1050 m) and manuka (on a few bluffs below 600 m in the Karangarua catchment). Similar vegetation develops on rock recently scoured by glaciers (p. 511). *Blechnum* 'mountain', *Olearia colensoi* (dominant on many bluffs above 700 m), *Chionochloa* 'robust' and *Rytidosperma setifolium* (on dry rocks) are characteristic of higher altitudes. *Dracophyllum longifolium, Phormium cookianum* and *Schoenus pauciflorus* are common at all altitudes, the last often being the main plant behind temporary waterfalls and beside steep cataracts. Trailing *Poa subvestita* can dominate where water drips continuously. *Gunnera monoica* and *Chionochloa conspicua* are also common on wet rock.

Also frequent are *Hymenophyllum multifidum, Carmichaelia grandiflora, Gingidia montana, Pseudopanax colensoi* var. *ternatus, Gaultheria rupestris* grading to *G. crassa, Coprosma rugosa, Forstera tenella, Parahebe lyallii* and *Astelia nervosa* (all widespread); *Myrsine divaricata* and *Anaphalis trinervis* (lower altitudes); *Ranunculus lyallii, Coriaria plumosa, Geum parviflorum, Dracophyllum traversii, Coprosma serrulata, Celmisia semicordata, C. verbascifolia, C. bellidioides, Ourisia macrocarpa, Lachnagrostis richardii, Poa novae-zelandiae* and *P. colensoi* (mainly higher altitudes); and *Dracophyllum fiordense* above 900 m.

Vertical cliffs and cuttings in lowland alluvium and moraine first develop a cover of *Polytrichadelphus magellanicus* or the leafy liverwort *Isotachis lyallii*, followed by vascular plants with fast-growing stolons or rhizomes, especially *Lycopodium volubile, Blechnum* 'black spot', *Gunnera monoica* and *Anaphalis trinervis. Blechnum vulcanicum* dominates locally. Seedlings of kamahi, *Quintinia acutifolia, Coprosma foetidissima,* etc. usually become established but are pulled out by their own weight before they become large. Filmy ferns such as *Hymenophyllum flabellatum* or *Trichomanes strictum* can cover banks in shaded ravines.

On penalpine bluffs, the main sites are ledges and slopes with pockets of soil, crevices, and rock surfaces, the last being occupied almost exclusively by cryptogams. Sites may be damp and sheltered, or dry and exposed. The species are mostly characteristic of other communities, but bluffs are the main habitat of some. The following grow between 1250 and 1600 m, the most abundant being marked†.
Soil pockets: *Geum parviflorum*†, *Dracophyllum kirkii*†, *D. uniflorum, Celmisia*

verbascifolia†, *C. semicordata*, *Schoenus pauciflorus*†, *Poa colensoi*†, *Microlaena colensoi*†. *Festuca matthewsii* and *Chionochloa pallens*†.

Sheltered crevices: *Polystichum vestitum, P. cystostegia* and *Ourisia caespitosa*.

Dry or exposed crevices: *Gaultheria crassa, Coprosma serrulata, Leucogenes grandiceps, Parahebe linifolia* and *Rytidosperma setifolium. Poa novae-zelandiae*† was listed from both exposed and sheltered crevices but in the latter is probably *P. subvestita*.

Further species are *Grammitis poeppigiana, Ranunculus lyallii, Epilobium glabellum, Schizeilema haastii, Anisotome pilifera, Gaultheria depressa, Gentiana* aff. *divisa, Wahlenbergia pygmaea, Celmisia vespertina, Dolichoglottis scorzoneroides, Craspedia* aff. *major, Ourisia macrocarpa* and *Hebe ciliolata*. As the amount of soil increases, open vegetation merges into continuous scrub and grassland.

Zonation in a narrow gorge

The Gorge River provides the following zonation, within a flood range of 5 m.

0–1 m: above the stony river bed: bryophytes growing on trapped silt.

1–2 m: *Gunnera monoica* 40% cover, *Lophocolea* 20%; others are *Gunnera dentata, Gonocarpus aggregatus, Oxalis magellanica, Rumex flexuosus, Neopaxia australasica, Hydrocotyle novae-zeelandiae* var. *montana, Leptinella squalida, Anaphalis* aff. *keriensis, Lagenifera petiolata, Nertera depressa, Mazus radicans, Juncus novae-zelandiae, J. antarcticus, Poa pusilla,* and seedlings of *Hoheria glabrata* and beech.

2–3 m: *Gunnera monoica* dominant, species additional to those already listed being *Blechnum fluviatile, B.* 'black spot', *Ranunculus reflexus, Plantago raoulii, Acaena anserinifolia, Helichrysum bellidioides, Mentha cunninghamii, Pratia angulata, Luzula picta, Microlaena avenacea, Corybas macranthus, Coprosma rugosa* and seedlings of *C. propinqua, Coriaria plumosa* and *Aristotelia fruticosa*.

3–5 m: *Chionochloa conspicua* 40%, *Coprosma ciliata* 30%, *Microlaena avenacea* 20%, *Blechnum fluviatile* 10%; further species are *B. penna-marina, Epilobium pedunculare, Poa colensoi* and *Uncinia uncinata*.

At other points in the gorge *Epilobium nerterioides, Anaphalis trinervis, Parahebe lyallii, Schoenus pauciflorus* and seedlings of *Coriaria arborea* are also present in the flood zone. Forest of silver beech and kamahi commences above high flood level.

Coastal cliffs

In Westland, coastal cliffs are mostly cut in moraine and support stages of succession to bush dominated by mahoe and other small broad-leaved trees. Differences from similar inland cliffs lie mainly in the abundance of *Phormium tenax* and, especially low on the cliffs, presence of coastal plants such as *Hebe elliptica, Sonchus kirkii, Isolepis nodosa, Samolus* in the splash zone, and *Blechnum banksii* in dark clefts. *Lobelia anceps* and the adventive gorse, *Sonchus* spp., *Plantago australis* and *Sagina procumbens* are also common. Where solid rock is exposed, as on the coast near Paringa, more typical rupestral species appear, especially above the splash zone. Crevices support *Phormium cookianum* or, where more exposed, *Rytidosperma setifolium,* and sheets of *Adiantum*

Fig. 11.16. *Cordyline australis* and *Phormium tenax* on bouldery coastal debris, near Paringa, South Westland.

cunninghamii grow beneath overhangs. Successions where these cliffs crumble into debris are much the same as on moraine cliffs (Fig. 11.16), but on spurs of solid rock a zone of *Dracophyllum longifolium*, together with southern rata and silver pine (*Lagarostrobos colensoi*), intervenes between the splash zone and tall forest above.

Fiordland

Cliffs facing the open sea generally support dense *Olearia oporina* scrub, but on more exposed rock this is replaced by *Phormium cookianum*, accompanied by *Dracophyllum longifolium*, *Hebe elliptica*, *Olearia oporina* and stunted plants of southern rata and *Griselinia littoralis*. *Anisotome lyallii*, *Poa astonii* and *Rytidosperma setifolium* occupy the most exposed rocks. When slips expose rock, bryophyte encrustations spread from crevices and build up peaty humus. The crevices, moss and peat are then colonised by seedling shrubs and other vascular plants including *Nertera depressa* and *Anisotome lyallii*.

On the precipitous granite and gneiss walls that enclose the fiords, successions towards dense forest are frequently initiated where vegetation slides off scarcely weathered rock. Glacially rounded slopes mostly support tree- and shrub-heath, but knolls may have only sparse patches of dwarf heath.

Rupestral vegetation is better developed on upland cliffs. Plants listed near Lake Shirley mainly grow in crevices (Given 1971):

(1) 800 m a.s.l. on a near-vertical cliff below the forest limit: *Asplenium richardii*, *Blechnum chambersii*, *Grammitis billardierei*, *G. givenii*, *G. patagonica*, *Epilobium*

glabellum, E. brunnescens, Griselinia littoralis, Cyathodes juniperina, Nertera depressa, Parahebe catarractae, Luzuriaga parviflora, Phormium cookianum, Agrostis dyeri var. *delicatior, Poa* sp., and a few plants of *Dracophyllum fiordense*, mountain beech and *Pseudopanax colensoi* var. *fiordense*.

(2) About 1020 m (just above the forest limit) on damp overhanging gneiss/schist: *Asplenium richardii, Blechnum penna-marina, Grammitis poeppigiana, Ranunculus* cf. *lyallii, Geum parviflorum, Anisotome haastii, Celmisia bonplandii, C. holosericea, C. verbascifolia, Craspedia robusta, Forstera tenella, Ourisia caespitosa* and *O. macrophylla* var. *lactea*.

(3) South-facing 70° slope of large, angular marble blocks: mainly scattered *Grammitis poeppigiana, Gingidia decipiens, Myosotis lyallii, Myosotis* sp., *Celmisia bonplandii, C. inaccessa* (a local endemic), *Chionochloa acicularis* and *C. ovata*.

Inland Canterbury

Waimakariri Gorge

The large Canterbury rivers emerge from the mountains through deep gorges cut into fluvioglacial gravels and greywacke bed-rock. At the lower end of the Waimakariri Gorge, the vegetation on the walls is an amalgam of species seral to black beech forest and mainly adventive plants that have dispersed from pastorally modified tussock grassland on terraces above the gorge. Except for *Rytidosperma setifolium*, distinctively rupestral species are absent.

(1) On a greywacke face sloping 65° to the north-west, plants are largely confined to rubble in fissures. Gorse up to 80 cm tall provides *c.* 2% cover; other vascular plants, i.e. manuka, *Corokia cotoneaster, Coprosma propinqua, Olearia avicenniifolia, Sedum acre, Rytidosperma setifolium*, browntop (*Agrostis capillaris*) and cocksfoot are only scattered. About 1% cover is provided by mosses, including the usually rare *Ptychomitrium australe*.

(2) Gravel cliffs sloping at 70–90° are bare, except where patches of *Coriaria arborea* and occasional plants of gorse and *Olearia avicenniifolia* occupy debris resting on the least steep pitches.

(3) As soon as gravel debris comes to rest, it is colonised by gorse, *Coriaria arborea*, and herbs among which browntop, *Plantago lanceolata* and *Verbascum thapsus* are the most abundant. Others are *Reseda luteola, Sagina procumbens, Arenaria serpyllifolia, Fragaria vesca, Plantago major, Trifolium repens, T. dubium, Galium aparine*, catsear, *Crepis capillaris, Cirsium vulgare, Sonchus asper, Tanacetum parthenium, Pseudognaphalium luteoalbum, Prunella vulgaris, Holcus lanatus, Festuca rubra, Elymus rectisetus, Dichelachne crinita, Rytidosperma buchananii* and, on damp gravel, *Epilobium brunnescens, Leucanthemum vulgare, Centaureum erythraea, Poa cita, Juncus articulatus* and *J. bulbosus*. The next phase is thickets of gorse with some *Coriaria*. At present, older surfaces support scrub of manuka, *Coprosma propinqua, Helichrysum aggregatum* and *Corokia cotoneaster* that contains saplings of black beech, *Griselinia littoralis*, kanuka, *Pittosporum tenuifolium* and kowhai (*Sophora microphylla*), but it is likely that, in future successions, gorse and other adventive shrubs and trees will become increasingly important.

Mt Cook National Park (Wilson 1976)

Flushed penalpine and alpine rocks in the Park support mats of *Poa novae-zelandiae* s.l. with *Epilobium glabellum* (or less frequently *E. macropus*), *Philonotis* and other mosses. Other plants are *Colobanthus apetalus, Geum parviflorum, Gingidia montana, Gentiana divisa, Helichrysum bellidioides, Leucogenes grandiceps, Leptinella pectinata* subsp. *willcoxii, Ourisia caespitosa, Parahebe linifolia, Uncinia divaricata*, and sometimes *Celmisia bellidioides* or *Schoenus pauciflorus*. Similar communities occur down to the montane belt.

Dry penalpine and alpine rocks are often almost completely covered by vegetation in which the more frequent species are *Podocarpus nivalis, Geum parviflorum, Gaultheria crassa, Dracophyllum kirkii, Coprosma* aff. *pseudocuneata, Raoulia grandiflora, Leucogenes grandiceps, Celmisia angustifolia, C. lyallii, Rytidosperma setifolium, Chionochloa* tussocks and the cryptogams *Racomitrium crispulum, Andreaea, Grimmia, Rhizocarpon, Placopsis, Pertusaria, Umbilicaria, Parmelia, Alectoria, Thamnolia, Stereocaulon, Buellia* and *Physcia*. Others include *Colobanthus buchananii* and *Raoulia eximia* (north aspects and rubbly ridges); *Grammitis armstrongii, Anisotome flexuosa, A. pilifera* and *Hebe ciliolata* (north aspects); and *Epilobium pycnostachyum, E. crassum, Hebe buchananii* and *H. epacridea* (rubbly ridges). Even near-barren rock supports the mosses *Pohlia* (in crevices), *Andreaea, Grimmia* and *Dicranoweisia antarctica*, and scattered *Epilobium glabellum, E. porphyrium, Agrostis subulata* and, in damper, shadier places, *Ourisia caespitosa* and *Poa novae-zelandiae* s.l.

Volcanic cliffs of the eastern South Island coast

On coastal cliffs of late Tertiary volcanic rocks below urban settlements, garden escapes and weeds form communities that are 'living relics of horticultural subjects fashionable with earlier generations of gardeners' (Healy 1959). Species mainly hail from the Canary Islands, Europe, South Africa and Australia, with some persisting natives.

On north-facing sea-cliffs of the Port Hills near Christchurch, gentler slopes show a shrub tier of *Chamaecytisus palmensis* and *Cytisus scoparius* with occasional *Paraserianthes lophantha, Teline monspessulana* and *T. stenopetala*, the soft-wooded shrubs *Pelargonium* ×*hortorum* and *Argyranthemum frutescens* at margins, and a lower stratum including *Cheiranthus cheiri, Tetragonia trigyna* and *Aptenia cordifolia*. More exposed and steeper slopes have scattered plants and colonies of *Pelargonium* ×*hortorum, Echium candicans, Chrysanthemoides monilifera* and *Argyranthemum frutescens*, with the lianes *Lathyrus* aff. *tingitanus, Muehlenbeckia complexa, Pelargonium peltatum, Senecio angulatus, S. mikanioides*, and *Tropaeolum majus*. Succulents are represented by *Aptenia, Cotyledon orbiculata, Drosanthemum floribundum, Othonna capensis, Sedum album, S. praealtum, S. reflexum* and *Carpobrotus* sp.

Steep to perpendicular, often overhanging cliffs, with vegetation confined to pockets, shelves, fissures and the prominent basal debris, carry occasional shrubs of *Teline stenopetala, Cytisus scoparius, Lycium ferocissimum, Echium candicans* and *Chrysanthemoides*, with scrambling *Muehlenbeckia complexa, Pelargonium* and *Senecio cineraria*, and occasional colonies of *Gazania* sp. There are extensive colonies of succulents, which include *Aeonium arboreum, A.* ×*velutinum, Aptenia, Cotyledon, Crassula tetragona, Sedum album, S. praealtum, S. reflexum, Carpobrotus edulis, Disphyma australe* and *C. edulis* × *D. australe.* A wet fissure on one steep face bears *Apium prostratum, Sonchus* sp., *Juncus maritimus* and *Isolepis nodosa.*

Cliffs at Lyttelton, on the south aspect of the Port Hills, lack succulents, but have some further species, namely *Lavatera assurgentiflora,* gorse, *Foeniculum vulgare, Centranthus ruber, Petroselinum crispum* and *Agapanthus orientalis.* Further species listed from Timaru, Oamaru and Port Chalmers (near Dunedin) include *Hedera helix, Clematis vitalba, Einadia nutans, Sedum acre* and the native *Geranium retrorsum, Poa cita* and *Scandia geniculata.*

Rakiura province (based on Fineran 1966; Kennedy 1978; Wilson 1987)

The islands of Foveaux Strait and headlands of Stewart Island share many species with more northern localities, e.g. *Disphyma australe, Linum monogynum, Lepidium oleraceum, Sarcocornia, Pimelea prostrata, Apium prostratum, Samolus, Selliera, Pseudognaphalium luteoalbum, Sonchus* spp., *Senecio lautus, Isolepis nodosa, I. cernua* and, on the cliff tops, *Tetragonia trigyna. Crassula moschata* close to the shore, *Carex trifida* on peaty ledges, *Asplenium obtusatum, Colobanthus muelleri, Hebe elliptica, Luzula banksiana* var. *acra* and *Isolepis praetextata* have mainly southern distributions. The remainder, i.e. *Blechnum durum, Gentiana saxosa, Anisotome lyallii, Myosotis rakiura, Poa astonii* and seedlings of *Olearia oporina, Brachyglottis rotundifolia* and *Stilbocarpa lyallii* are strictly southern. Taupata is naturalised on some islands. *Pertusaria, Verrucaria* and *Xanthoria* encrust littoral rocks where scattered tufts of *Crassula moschata* are the only vascular plants.

Patches of vegetation on exposed granite, diorite or schist in the cool-temperate and subalpine belts of Stewart Island mostly consist of species drawn from neighbouring continuous vegetation. *Dracophyllum politum,* manuka, *Drapetes lyallii, Pentachondra pumila, Astelia linearis, Oreobolus impar, Carpha alpina, Chionochloa lanea* and *Microlaena thomsonii* represent grassland and dwarf heath; *Gentiana lineata, Plantago uniflora, Gingidia flabellata, Anisotome haastii, Schizeilema haastii* var. *cyanopetalum, Brachyglottis bellidioides* var. *crassus, Raoulia goyeni, Abrotanella muscosa, Leucogenes grandiceps, Phyllachne colensoi, Deyeuxia aucklandica, Rytidosperma setifolium* and, on Mt Anglem, *Chionochloa* 'fiord' represent rocky herbfield. The most frequent mosses and lichens are *Rhacocarpus purpurescens* and species of *Andreaea, Campylopus, Siphula* and *Usnea.* In the extreme west, the usually coastal *Celmisia rigida, Luzula banksiana* var. *acra, Poa astonii, Asplenium obtusatum* and, rarely, *Grammitis rigida* grow also on upland outcrops.

Ultramafic surfaces

Ultramafic rocks crop out at North Cape, in Sounds–Nelson Province, and in a belt extending from South Westland to inland Southland. They present bluffs, boulder fields, debris slopes and alluvial surfaces, but colonisation is slow and even early seral stages tend to be dominated by ericoid heathland species (p. 527). Plants characteristic of fertile soils, especially filiramulate shrubs such as *Coprosma propinqua,* can also be prominent, perhaps reflecting the neutral or alkaline reaction of the soils. Ultramafic local endemics have diverged from non-ultramafic close relatives, and there are ultramafic ecotypes of widespread species (Lee *et al.* 1983*b*).

Surville Cliffs near North Cape

The Surville Cliffs are 200 m high, and slope at 30–90°. Druce *et al.* (1979) describe three communities. Fifteen taxa are considered endemic; of these, the woody plants marked[1] have a trailing habit that seems genetically fixed, the remainder being marked[2].

(1) Patches of pohutukawa forest.

(2) Variable open communities dominated by *Phyllocladus trichomanoides* var.[1], manuka, kanuka, *Pseudopanax lessonii* var.[1], *Pittosporum crassifolium* var.[1], *Myrsine australis*, *Melicope simplex*, *Melicytus micranthus*, *Hebe ligustrifolia*, *Coprosma rhamnoides* subsp.[1], *Geniostoma rupestre* var. *crassum*[1], *Parsonsia capsularis*, *Cassytha paniculata*, *Cortaderia splendens*, *Phormium tenax* and *Astelia banksii*. Other taxa are *Corokia cotoneaster* var.[2], *Pomaderris oraria* var. *novae-zelandiae*[2], *P. prunifolia* var. *edgerleyii*, *Pimelea prostrata* var. *erecta*, *Cyathodes juniperina*, *C. parviflora* var.[2] (the species is also in Australia and the Chatham Islands), *Coprosma* aff. *obconica*[1] and *Hebe macrocarpa* var. *brevifolia*[2]. *Silene gallica*, *Helichrysum lanceolatum*, *Olearia albida*, *Cassinia 'amoena'*[2], catsear and *Morelotia affinis* are less common.

(3) Crevices contain *Adiantum hispidulum*, *Doodia media*, *Haloragis erecta* subsp. *cartilaginea*[2], *Wahlenbergia gracilis* aggr., *Carex spinirostris*, *Elymus multiflorus*, *Deyeuxia billardieri*, *Rytidosperma* sp., *Stipa stipoides* and *Arthropodium cirratum*.

Further endemic taxa are *Pittosporum pimeleoides* var. *major*[1], *Parsonsia capsularis* var.[1], and *Coprosma spathulata*[1], in which the stems may trail for 5 m.

Sounds–Nelson

The 'mineral belt' is 8 km wide in the Richmond Range, narrows northwards into the Bryant Range and then becomes discontinuous and lower, reaching sea-level on D'Urville Island. Ultramafic influences have been obscured at low altitudes, because most of the primitive vegetation has been burnt and replaced by manuka-dominated shrubland. However, an unusually dense stand of *Phyllocladus trichomanoides* in the Bryant Range, although secondary, suggests that the original vegetation may have had distinctive features.

Above 800 m, the ultramafic influence becomes increasingly apparent in the vegetation, which contains some 20 taxa restricted to the mineral belt. In addition to the species marked† below, these are *Colobanthus* aff. *wallii*, *Clematis 'australis* var. *rutifolia'*, *Cardamine* aff. *debilis*, *Myosotis monroi*, *M. laeta*, *Gentiana* aff. *tenuifolia*, *Coprosma* sp., *Hebe urvilleana*, *H.* sp. and *Pterostylis humilis* (A.P. Druce, unpublished).

At 850–1100 m, the prevailing vegetation is mostly manuka and *Chionochloa defracta*†, with much bare, unstable soil. Other species include *Phyllocladus alpinus*, *Pimelea suteri*†, *Melicytus alpinus*, *Dracophyllum longifolium*, *Cyathodes fraseri*, *Gentiana* sp.†, *Olearia serpentina*, *Cassinia 'vauvilliersii'*, *Celmisia spectabilis*, *Euphrasia monroi*, *Hebe venustula*, *Astelia graminea*, *Phormium cookianum*, *Carex devia*†,

Fig. 11.17. Manuka, *Dracophyllum longifolium* and *Chionochloa defracta* dominant on ultramafic debris in foreground, stunted mountain beech forest with *Libocedrus bidwillii* overstorey on melange in middle distance; 1070´m, Red Hills, Nelson ER.

Rytidosperma setifolium and, on eroding ground, *Colobanthus* sp.†, *Notothlaspi* aff. *australe*† and *Helichrysum bellidioides*. This probably replaced fire-sensitive tree-heath, which still forms transitions between open vegetation and beech forest on siliceous substrates (Fig. 11.17).

In these transitions, which can occupy 'melanges' and 'knockers' where ultramafic influence is diluted, the dominant trees and shrubs are mountain beech, *Libocedrus bidwillii*, *Phyllocladus alpinus* and *Dracophyllum longifolium*; pink pine (*Halocarpus biformis*) and southern rata occur on the Bryant Range. *Dicranoloma robustum* forms most of the ground tier. Further species include *Podocarpus hallii*, manuka, *Cyathodes juniperina*, *C. fasciculata*, *Pittosporum crassicaule*, *Myrsine divaricata*, *Coprosma* aff. *parviflora*, *Phormium cookianum* and *Gahnia procera*, which are typical of infertile soils, and *Coprosma propinqua*, *Griselinia littoralis*, *Corokia cotoneaster* and *Hebe vernicosa* which, in non-ultramafic situations, indicate greater fertility.

Barren upper slopes culminate in Red Hill (1785 m). At 1600 m closed *Chionochloa defracta*† grassland is confined to pockets of loess. Edges of boulder fields and stony steps of solifluction terraces support scattered *Chionochloa* and lattices of *Dracophyllum pronum* (Figs 11.18 and 11.19). There are also *Racomitrium lanuginosum*, *Melicytus alpinus*, *Coprosma cheesemanii*, *Drapetes* sp., *Anisotome flexuosa* var.†, *Gentiana bellidifolia* var., *Celmisia spectabilis*, *Hebe carnosula*, *Carex devia*†, *C.*

Fig. 11.18. Espaliers of *Dracophyllum pronum*, and tussocks of *Chionochloa defracta* on deeper soil. Ultramafic terrain at 1620 m, Red Hills, Nelson ER.

Fig. 11.19. *Chionochloa defracta* with *Celmisia spectabilis* on steps of solifluction terraces in ultramafic gravel at 1620 m, Red Hills, Nelson ER.

traversii†, *Luzula* sp., and the grasses *Rytidosperma setifolium* and *Poa acicularifolia* var. *ophitalis*†. Level, frost-worked finer material has a sparse cover that includes the two grasses, *Colobanthus*† and *Notothlaspi*†. Turf around tarns consists of *Schoenus pauciflorus*, *Carpha alpina*, *Oreobolus pectinatus*, *Isolepis aucklandica* and *Leptinella pyrethrifolia* var. *linearifolia*†.

South Westland

The Red Hills of South Westland were heavily glaciated; unlike other ultramafic areas, they have no endemics. The lower debris slopes carry tree-heath like that on the moraines and fluvioglacial terraces towards the coast (p. 179), with stunted rimu (*Dacrydium cupressinum*) to 400 m, *Lepidothamnus intermedius*, pink pine, mountain beech, manuka, and some stunted silver pine, kamahi and southern rata. By 600 m, this opens out to rocky barrens with scattered patches of heath (p. 527) which, with decreasing slope and increasing fines, merges into *Chionochloa* 'westland' tussock grassland, or turf of *Oreobolus pectinatus*, *Carpha*, *Schoenus pauciflorus* and *Donatia novae-zelandiae* on wet ground. At higher altitudes, as on the slopes of Fiery Peak (1600 m), there are very few plants, *Neopaxia australasica*, *Rytidosperma setifolium*, *Poa colensoi* and *Luzula crinita* var. *petrieana* being the most frequent (Mark 1977).

Inland Southland

Ultramafic rocks crop out in the southern portions of the Eyre and Livingstone mountains. *Celmisia spedenii* is confined to these, extending over an altitudinal range of 450–1400 m. At West Dome mountain beech forest, fire-induced manuka scrub and red tussock (*Chionochloa rubra*) grassland grow on acid soils, where the ultramafic influence has been subdued through weathering and leaching. Grey scrub of matagouri, *Melicytus alpinus*, *Coprosma propinqua* and *Corokia cotoneaster* occupy coarse scree with high values for calcium and base saturation, which is largely derived from basalt and other non-ultramafic rocks. *Celmisia spedenii*, providing 23% ground cover among very open manuka, occupies a broad ridge where wind erosion maintains a stony soil in which ultramafic influence results in high magnesium levels (Table 11.4).

Vegetation of thermal areas

The geothermal district in the northern part of the Volcanic Plateau is the largest in New Zealand. At Karapiti, there are five zones of vegetation on heated ground (Fig. 11.20; Table 11.5). Zone 1 is bare, with temperatures reaching 97 °C within 5 cm of the surface and steam issuing through small vents. In zones 2 and 3 mats of moss lie on the hot surface. Kanuka dominates zone 4, but is < 30 cm tall, whereas in zone 5 it is > 50 cm tall, giving almost complete cover even though soil temperatures are 35–40 °C below 5 cm. Kanuka root penetration is limited by the 50 °C isotherm, i.e. to about 4 cm in zone 3, and 5–8 cm in zone 4. Soil temperatures around fumaroles are similar to those in zones 4 and 5, but constant steam allows frost-sensitive species to thrive, including the otherwise tropical ferns *Dicranopteris linearis*, *Christella dentata* and *Nephrolepis* aff. *cordifolia* (Fig. 11.21).

Table 11.4. *Mean physical and chemical properties of soils (0–7.5 cm depth) at West Dome, Southland*

community (no. of samples)	stones %	clay %	Ca me %	Mg me %	TEB me %	BS %	pH (H₂O)
beech forest (5)	50	37	5.4	3.9	10.3	26	4.5
tussock grassland (2)	64	34	6.5	2.6	10.5	31	5.4
tall manuka (5)	24	32	8.5	10.0	19.4	59	5.5
open manuka (5)	45	36	5.6	7.3	13.6	50	5.6
Celmisia spedenii (5)	31	25	3.0	14.9	18.3	82	6.1
grey scrub (2)	73	16	15.3	6.0	22.5	80	6.1

Notes:
TEB, total exchangeable bases; BS, base saturation; me, milliequivalents; % clay refers to <2 mm soil fraction.
Source: McIntosh & Lee 1986

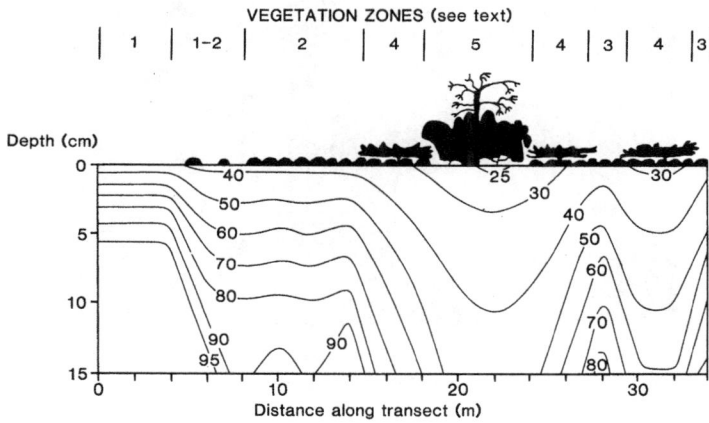

Fig. 11.20. Transect across stable heated ground at Karapiti (Central Volcanic Plateau) showing profiles of temperature (°C) and vegetation. Vertical scale 40× horizontal (Given 1980).

Campylopus holomitrium is endemic to the district, and the kanuka is a true-breeding prostrate form (Fig. 11.22). *Psilotum nudum* and *Lycopodium cernuum* also grow on heated ground. While heat is the main factor restricting most plants and favouring a few, hydrothermally altered soils also have toxic qualities. Given (1980) found that kanuka seedlings, which grew normally on zone 4 soils, died within two weeks on zone 2 soils and after several days on zone 1 soils. This correlates with a gradient of decreasing Fe, Mn, Ca, K, P, Al, Mg, Na, pH (6.0 to 1.4) and loss on ignition, and increasing Ti and Si.

Fig. 11.21. *Nephrolepis* aff. *cordifolia* near a hot pool at Waiomangu, Northern Volcanic Plateau ER.

Volcanic hot springs exist in many other places in northern New Zealand, and in the South Island hot springs are associated with the Alpine and Hope faults, but none of these emit water at pressure, or at temperatures as high as around Rotorua and Taupo. The only extant plants of *Baumea complanata* are in gumland on the shore of the hot lake at Ngawha in Northland, and *Phormium tenax* and *Carex secta* grow beside the hot pools in the Copland Valley at 450 m, at a higher altitude than is usual in Westland.

Table 11.5. *Zonation of species on heated ground at Karapiti*

Shrubs and trees	zones
Kunzea ericoides	3–5
Pinus spp.	5
Coprosma robusta	F

Dicot herbs	
Gnaphalium sphaericum	3–5,F
Portulaca oleracea	3–5
Conyza bonariensis	3,F
Gnaphalium subfalcatum	F
Hypochoeris radicata	F

Monocot herbs	
Digitaria sanguinalis	2,5
Dichelachne crinita	3
Cortaderia sp.	F
Eragrostis sp.	F
Dianella nigra	F
Paspalum dilatatum	F

Ferns and lycopods	
Lycopodium cernuum	2–4,F
Cheilanthes sieberi	3–4
Dicranopteris linearis	3–5,F
Nephrolepis aff. *cordifolia*	4
Histiopteris incisa	5,F
Paesia scaberula	5,F
Hypolepis dicksonioides	F
Pteridium esculentum	F

Bryophytes	
Campylopus clavatus	2
Campylopus pyriformis	2
Campylopus introflexus	2–3
Lepidozia glaucophylla	2–3
Campylopus holomitrium	2–4
Sphagnum cristatum	5

Lichens	
Cladonia capitellata	2
Cladonia solida	2–3
Cladina leptoclada	3
Cladia aggregata	3
Cladia retipora	3
Parmotrema perlatum	3
Parmotrema reticulata	3
Usnea sp.	3

Notes:
2–5, Zones of decreasing soil temperature (Fig. 11.20); F, fumarole margins.
Source: Given 1980

Fig. 11.22. Hot lake at Waiomangu surrounded by a depressed form of *Kunzea ericoides*.

12

THE ALPINE AND NIVAL BELTS

Terrain above the limits of continuous penalpine grassland is largely barren because of rapid rock-wasting, limited soil development, and a short, harsh growing season. Frequent, intense freeze–thaw cycles result in unstable substrates and special microtopography, including solifluction lobes and patterned ground. Presence and composition of vegetation depend on substrate texture, stability, aspect, exposure to wind and, in contrast to lower altitudes, duration of the snow-free period. Expanses of broken rock occur as slope debris, moraine and rock-glaciers; texture and stability range from boulder fields to fine, mobile scree. Such terrain has, at most, a scattering of hardy opportunists, and even these are practically absent in the subnival belt of the drier mountains. Partial cover develops on stable alpine substrates with adequate proportions of fines, contrasts being provided by sheltered hollows where snow persists, and exposed rocky areas that are often blown free of snow, even in winter; the term fellfield will be restricted to the latter. Alpine grassland (discussed in Chapter 9), short herbfield and cushion-field signify more or less continuous vegetation.

Rock outcrops that are not continually exfoliating support mosses and lichens. Vascular plants grows in crevices, and with increasing altitude become confined to northerly aspects where some, along with lichens and mosses, ascend into the nival belt.

Species of high altitudes

The alpine flora consists in part of species of wide altitudinal and geographical distribution, but mostly of species that are geographically limited or practically confined to high altitudes. Although some of the latter descend on sparsely vegetated outcrops, debris, moraine, or stream gravels to intermingle with the pioneers of lower altitudes (Fig. 12.1), few enter closed vegetation. The South Island mountains hold all but a small proportion of New Zealand's alpine terrain and nearly all of the true alpine species. The central portion of the Alps has the highest mountains, but few locally endemic plants; ranges to the north and south are much richer in this respect (p. 109).

Fig. 12.1. *Leucogenes grandiceps* and *Brachyglottis bellidioides* (lower right) on greywacke debris; 1130 m, Crow Valley, Eastern Alps.

In the North Island, there are probably only two alpine endemics, and only the three central high volcanoes and Mt Taranaki have true alpine belts. In contrast, the small alpine areas of southern Fiordland support distinctive endemic plants.

Scree plants

Stable patches on most alpine screes support scattered individuals of widespread species such as *Epilobium glabellum, Poa novae-zelandiae* and *Neopaxia australasica*, but only about 16 species grow on mobile material and 12 of these grow almost nowhere else. This specialised group is found mainly on dry greywacke mountains from Marlborough to Waitaki ER, with outlying occurrences in North-west Nelson, the Richmond Range and inland Southland. Their scarcity in wet climates suggests that a threshold of warmth or radiation must be crossed for them to maintain growth and set seed in face of destructive forces. Also, as plant densities are remarkably low, it may be that only the eastern greywacke screes are large enough to maintain viable populations.

Screes move only in their uppermost layers, except when deeply undercut at the base. Studies of rock weathering (Whitehouse *et al.* 1980) show that most are long-standing features, many having existed as long as the Holocene. Greywacke screes have an average slope of 32°. A 7–15 cm layer of mobile stones overlies more stable debris with a high proportion of sand and silt, the separation being effected through

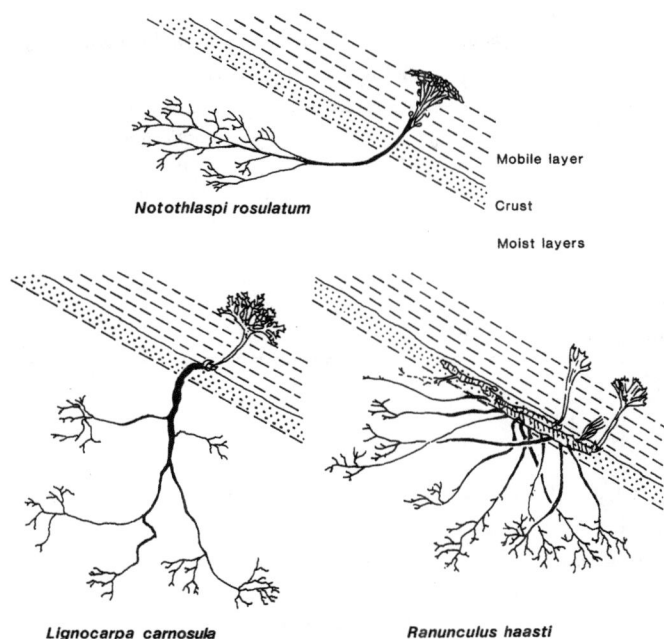

Fig. 12.2. Root systems of scree plants (Fisher 1952).

gravity in conjunction with freeze–thaw cycles in spring and autumn; significantly, scree plants also grow on nearly flat surfaces where frost-sorting into stone stripes and nets is active. During dry summer weather, a relatively stable crust forms immediately beneath the mobile layer, but below this scree is continually moist.

The morphology of scree plants is strongly convergent. Short rootstocks or rhizomes elongated in a downhill direction lie in the subsurface crust (Fig. 12.2). From these, roots descend some 45 cm into the moist layers (deeper in *Lignocarpa carnosula*) and bear root hairs for most of their length. Most species have greyish wax or tomentum, and several are summer-green. High leaf water contents comparable to those of halophytes (Table 12.1), low osmotic concentrations in the sap, and often finely-divided leaves with concentric mesophyll and numerous, shallow stomatal pits suggest a strategy of maintaining photosynthesis under conditions of high evapotranspiration, when surface temperatures can exceed 50 °C.

Some species have fine, tufted, branching stems, that produce adventitious roots in response to burial. One of these, *Epilobium pycnostachyum*, has yellow-green or reddish leaves crowded on stems 4–18 cm tall. It is the most widespread obligate scree plant, extending from eastern greywacke mountains to the Main Divide, and occurring also on Ruapehu and the nearby Kaweka, Kaimanawa and Ruahine Ranges, in Western Nelson and in Southland. *E. forbesii* is a rare, viscid-glandular endemic of Marlborough. *Senecio glaucophyllus* subsp. *discoideus*, which differs from the typical subspecies in lacking ray florets, is less branched than the willow-herbs. It occurs sparingly through the eastern South Island ranges, but on North Island greywacke

Table 12.1. *Water in leaves of scree plants (left) and halophytes (right), as percentage of dry mass*

Stellaria roughii	875	Disphyma australe	1202
Ranunculus haastii	655	Suaeda novae-zelandiae	811
Leptinella atrata	594	Sarcocornia quinqueflora	642
Senecio glaucophyllus	514	Selliera radicans	610
Lobelia roughii	488	Samolus repens	358
Notothlaspi rosulatum	436		
Lignocarpa carnosula	382		
Epilobium pycnostachyum	318		

Source: Fisher 1952

ranges is replaced by subsp. *toa*. It also occupies other loose, bare substrates, and is sub- and penalpine rather than truly alpine.

Parahebe cheesemanii occupies a triangle defined by the Arthur and Richmond Ranges and the north-eastern Alps, growing on fine scree under higher rainfall than most scree plants. Its very fine stems form mats up to 10 cm across. The more robust *P. spathulata* grows on stable and loose scoria on the central North Island volcanoes and greywacke screes to the east.

Ranunculus haastii, which has a thick, elongated root-stock that produces palmately lobed, summer-green leaves, is one of the most characteristic scree plants of eastern mountains. The smaller *R. crithmifolius* grows on patches of loose shingle rather than extensive screes, with one population occupying limestone rubble at only 800 m at Castle Hill (Puketeraki ER); *R. scrithalis* represents it in Southland (p. 108).

The umbellifer genus *Lignocarpa* comprises *L. diversifolia* confined to Marlborough and mountains overlooking Lake Rotoiti (Fig. 12.3), and *L. carnosula*, which extends southwards to Puketeraki ER. The leaves, divided into linear segments, are borne on a short root-stock. *Notothlaspi rosulatum*, a crucifer with a similar range to *L. carnosula*, also has a short root-stock, that bears a flat rosette of grey, elliptical, serrate leaves and produces a dense, cylindrical raceme up to 25 cm tall. On limestone rubble it descends as low as 800 m.

Stellaria roughii, which grows from Marlborough to inland Canterbury and in inland Southland, produces loose clumps of delicate aerial shoots each summer. *Lobelia roughii* and *Wahlenbergia cartilaginea* have thicker, possibly evergreen leaves; the former extends from North-west Nelson to the Hawkdun Range (Waitaki ER) whereas the latter is confined to Marlborough and eastern Nelson. *Leptinella atrata* and *L. dendyi* have thicker rhizomes, and grey, pubescent, pinnately lobed leaves. The former, notable for its black, discoid capitula, grows on eastern greywacke screes from the seaward Kaikoura Range to the Kakanui Range; the latter has a more northerly distribution, from the Richmond Range to Mt Torlesse (Puketeraki ER). *Myosotis traversii* is a deeply tap-rooted, often multi-crowned rosette plant with leafy scapes up to 20 cm long, that occurs sparingly on screes from North-west Nelson to Mt Cook, mainly east of the Divide.

Fig. 12.3. *Lignocarpa diversifolia* on greywacke scree, Wairau catchment, Molesworth ED (photo G.Y. Walls).

Fig. 12.4. *Haastia pulvinaris* at 1690 m, Travers Saddle, North-eastern Alps.

Cushion and mat plants

These are the most characteristic growth forms of the alpine belt; the woody 'vegetable sheep' are the most spectacular. The largest, *Haastia pulvinaris* (Fig. 12.4) and *Raoulia eximia*, form hard mounds up to 1 m across. *Haastia*, with tightly packed, 2 cm diameter shoots obscured by pale-fawn hairs, is the 'woollier' of the two. It grows to as high as 2700 m, mainly on fractured rock and stable debris on relatively snow-free sites

Fig. 12.5. *Raoulia buchananii* on a rock ridge at 1490 m, Douglas Range, South-western Alps. Other species are *Ranunculus lyallii*, *Aciphylla multisecta*, *Dracophyllum* aff. *kirkii* and *Astelia linearis*.

on the ranges of Marlborough and eastern Nelson. A form with thinner shoots may be a distinct species. *Raoulia eximia*, which also has thinner shoots and leaves with short, greyish hairs, tends to anchor in greywacke crevices from Marlborough to Mt Ida in Waitaki ER. *R. rubra* of the Tararua Range and western Nelson may be conspecific. *R. bryoides* and *R. mammillaris* are smaller vegetable sheep with overlapping distributions based on Marlborough and Canterbury respectively; in Central Otago they are represented by an unnamed species. *R. buchananii* grows from South Westland to Fiordland (Fig. 12.5); *R. goyenii* is confined to Stewart Island. Woody raoulias occasionally hybridise with *Leucogenes*.

Other high-altitude raoulias are herbaceous mat plants. *R. grandiflora*, which forms open mats in stony penalpine and alpine grassland and short herbfield, ranges from Mt Hikurangi to Stewart Island, being absent only from Taranaki. *R. cinerea* is restricted to fine, frost-patterned gravel in central Marlborough, whereas *R. hectorii*, *R. petriensis* and *R. youngii* extend through Otago and South Canterbury, the last reaching the Craigieburn Range. *R. subulata* grows in late-snow hollows in the South Island. Among the true scabweeds, *R. apicinigra* is characteristic of gravel ridges in Western Nelson, Marlborough and Canterbury, and *R. albosericea* is an important pioneer on volcanic sand and gravel in the North Island. Other scabweeds sometimes attain the alpine belt by virtue of ready dispersal and establishment.

Celmisia sessiliflora forms firm, low cushions or more open mats in alpine and penalpine fellfield, stony grassland and low herbfield throughout the South Island. In Central Otago it is usually accompanied by the mat-forming *C. argentea*, which descends to subalpine mires.

Dracophyllum muscoides dominates fellfield or cushion-field east of the Divide in Otago and southern Canterbury. *Coprosma perpusilla* has woody rhizomes, is generally mat-forming, and grows throughout the New Zealand mountains in a wide range of mountain sites, although edges of streams and flushes are preferred. The orange fruit, which contain 3–4 pyrenes instead of the two that are usual in coprosmas, probably need two summers to form and ripen. *C. niphophila* is a little-known plant of rocky alpine terrain in the South Island. *Pernettya alpina* has a similar habit, and overlaps in dry fellfield with the lower-altitude, more robust *Gaultheria depressa*.

The small, exclusively alpine genus *Chionohebe* consists, with one or two exceptions (p. 415), of dwarf shrubs that form dense, ciliate cushions 10 cm or more in diameter. *Myosotis pulvinaris*, which is abundant on the mountains of Central Otago, could be taken for a *Chionohebe* except when in flower.

Phyllachne colensoi (p. 338) grows in rock crevices, fellfield, cushion-field, snow hollows and on bogs, from Mt Hikurangi to Stewart Island. *P. rubra* is a smaller plant confined to fellfield and snow-banks on southern mountains of the South Island. *Hectorella caespitosa* grows on gravelly alpine ridges, mainly along and west of the Main Divide from Arthurs Pass to Fiordland, but in Central Otago extends east to the Rock and Pillar Range. The moss-like but tap-rooted *Abrotanella inconspicua* is confined to Central Otago and, unlike the rest of its genus, grows mainly in cushion-field rather than bog.

Several alpine species of *Colobanthus* are tap-rooted cushions less than 5 cm in diameter. They mostly grow in rock crevices and on exposed terrain, but some tolerate late-lying snow. The widely distributed *C. apetalus* s.l. and its close relative *C. affinis* s.l. have shorter stems and longer leaves, and are best described as tufted. Other small cushions are formed by reduced plants in the widely distributed *Agrostis subulata–A. magellanica* complex, *Poa pygmaea* which is endemic to Mt Pisa and a few other summits in Central Otago, and several species of *Luzula*, the most wide-ranging being *L. colensoi* and *L. pumila*.

Other plants forming broad mats include *Leptinella goyenii*, which is confined to exposed sites in Central Otago, *Drapetes* aff. *lyallii*, which grows throughout the South Island in a range of alpine habitats, and *Oreoporanthera alpina*, which grows in North-west Nelson, mainly on limestone. Small, tight mats are formed by *Anisotome imbricata*.

Many other herbs form low, leafy patches rather than mats. Tomentose species include *Ourisia sessilifolia*, and its close relatives *Anisotome flexuosa*, *Plantago lanigera* and *Gnaphalium mackayi*, the last usually being on flushed ground. Glabrous plants, mostly growing in damp fellfield, include small forms of the willow-herb *Epilobium brunnescens*, *E. tasmanicum* on eastern ranges, the sweet-scented *Pratia macrodon*, *Ourisia caespitosa* and, in flushes, *Psychrophila novae-zelandiae* or the more local *P. obtusa*. Dry, gravelly soils support leptinellas such as *L. pyrethrifolia*, whereas wet, loose gravel usually supports patches of *Neopaxia australasica*, an alpine plant that also descends to the lowlands on comparable sites.

Other herbaceous dicots

The major penalpine herbaceous genera have some fully alpine species as well as those that ascend higher than usual in favoured habitats. The former includes *Celmisia haastii* in snow hollows, and *C. viscosa* at the upper limits of grassland east of the South Island Divide.

Ranunculus sericophyllus, a plant with silky, finely divided leaves, grows in late snow hollows and on other damp, stony sites along the flanks of the Alps. *R. pachyrrhizus* represents it in Central Otago. *R. grahamii* is confined to nival crevices in the Mt Cook region. The glaucous, 50 cm tall, white flowered *R. buchananii* grows on moist talus and bluffs from Fiordland to the south-western Alps. In the central Alps, similar habitats are occupied by *R. godleyanus*, a plant with glossy, entire green leaves, that is closely related to the more northern *R. insignis* which also ascends to high altitudes. *R. nivicola*, which appears to be a polyploid hybrid between *R. insignis* and *R. verticillatus*, extends from the Kaweka and Kaimanawa ranges to Mt Taranaki, being the only high-altitude buttercup on that mountain.

The tufted willow-herb *Epilobium glabellum* s.l. is widespread in the alpine zone through most of New Zealand. Plants are mostly procumbent on loose debris, and in the South Island are usually accompanied by *E. porphyrium*, which has dull green or reddish leaves. *E. crassum*, with glossy green leaves and a creeping habit, grows mainly on moist scree margins east of the Divide.

Anisotome pilifera is a robust, glaucous, pinnate-leaved herb, which grows mainly where protected among boulders. On south-western mountains it may be accompanied by the fine-leaved *A. capillifolia*. Alpine aciphyllas belong to the section with broad, compound umbels. The *Aciphylla multisecta–A. divisa* complex grows mainly in rock crevices although also extending on to open herbfield on high penalpine spurs. *A. congesta* develops loose hummocks up to 50 cm across on boulder fields in South Westland and Fiordland (Fig. 12.6). *A. dobsonii*, with leaves reduced to a simple, stout, 10 cm long lamina flanked by sharp stipules, is confined to fellfield in the western mountains of the Mackenzie region. In the western mountains of Central Otago, it is replaced by *A. simplex*, in which the stipules are suppressed. *Schizeilema haastii*, with tufts of rounded, shining leaves, is one of the most characteristic plants of coarse debris, especially on wet mountains from the Ruahine Range southwards. The slenderer var. *cyanopetalum* seems more eastern.

Haastia sinclairii, a sprawling herb with overlapping, densely silky leaves 1 cm long, grows among loose rocks, mainly east of the Main Divide but extending a little to the west; it can hybridise with the vegetable sheep *H. pulvinaris*. *H. recurvum*, which has recurved, woolly leaves, is more common on north-eastern mountains of the South Island, to as far south as the Rangitata catchment. *Craspedia* includes robust forms that seem restricted to.high altitudes. One, with rather sparsely glandular-hispid leaves and glomerules up to 2.5 cm across, grows on ledges and in flushes, mainly west of the Divide. *C. incana*, which has white, flocculose tomentum, grows on debris on eastern ranges of the South Island.

Myosotis macrantha is mainly penalpine and common through most of the South Island, on rough, gravelly surfaces such as moraine. Its scapes are up to 30 cm tall and

Fig. 12.6. *Aciphylla congesta* at 1220 m, Gorge Plateau, Olivine ER.

the flowers range from yellow through brownish-orange and blue to purple. Strictly alpine *Myosotis* species are more compact plants with white flowers, and generally narrow ranges (Fig. 12.7). Perhaps the most beautiful New Zealand gentian, *G. divisa*, grows in crevices and stony herbfield, chiefly along and west of the Main Divide. Its numerous white flowers are densely massed on erect scapes arising from a tap-rooted rosette.

Parahebe, in addition to the scree species, includes *P. birleyi*, which grows on nival ledges from the Mt Cook district to the Olivine Range, *P. planopetiolata* which grows on debris in the south-western Alps and Fiordland, and *P. trifida*, a plant of late-snow areas on several mountains of Central Otago. Small, tufted euphrasias are common in short alpine vegetation; *Euphrasia revoluta* ranges widely in the main islands whereas *E. townsonii* is nearly confined to Western Nelson. *E. petriei* grows on rocky ground from the central Alps southwards. Some penalpine euphrasias also reach the alpine belt.

South Island alpine vegetation includes three genera of robust, tap-rooted crucifers, but all are absent from the western Alps. *Notothlaspi* contains *N. australe*, which is confined to fellfields from North-west Nelson to Marlborough, as well as species on scree and on ultramafic debris. In southern districts, *Pachycladon* occurs instead; *P. novae-zelandiae* grows mainly in Central Otago but extends sparingly to Mt Cook, whereas *P. crenata* is confined to Fiordland. *Cheesemania* has branched inflorescences up to 50 cm tall. *C. fastigiata* and *C. enysii* range from Marlborough to about Lake Wakatipu, and three other species have narrower ranges. Small species of *Cardamine* grow in high-alpine debris. Some seem reduced states of the wide-ranging *C. debilis*, but other taxa are strictly high-altitude.

Fig. 12.7. *Myosotis pygmaea* and *Drapetes* aff. *lyallii* (left); 1408 m, Bald Hill, Mavora ER.

Other grasses, sedges and rushes

Poa colensoi s.l. (including the sward-forming, western *P. hesperia*) is the most widespread and abundant alpine vascular plant. A small, tufted form of *P. novae-zelandiae* grows on loose alpine gravel, and the firm-leaved *P. buchananii* grows on stable portions of greywacke screes. *P. dipsacea* and several small poas with limited ranges occupy alpine flushes and late-snow hollows. *Koeleria cheesemanii* and *Trisetum spicatum* are small, tufted grasses growing mainly in fellfield. *Rytidosperma setifolium* and most of the dominant penalpine *Chionochloa* species ascend patchily into the alpine belt, but *C. oreophila* is largely confined to snow hollows (p. 218).

The rush *Marsippospermum gracile* grows on damp, snowy sites, whereas both cushion-forming and taller luzulas grow mainly on exposed sites. *Carex pyrenaica* is the commonest alpine sedge, growing mainly in late-snow areas throughout the South Island and also occurring on Mt Taranaki and locally on axial ranges of the North Island. Distributions of other carices of wet alpine habitats, such as *C. pterocarpa* of cushion communities in Central Otago and Heron ED, are reported as extremely disjunct, but this may represent inadequate collection. The widespread *Uncinia divaricata* extends to stable alpine debris. The slender, rhizomatous *U. drucei* is widely distributed on moist ground, and is the closest in its genus to being exclusively alpine.

Shrubs

Small shrubs of relatively open habit grow mostly in the shelter of rocks on snow-free ridges and north-facing bluffs. *Hebe pinguifolia* is robust and procumbent, with spreading, glaucous leaves, and occurs widely on rocky penalpine and alpine terrain on eastern mountains, especially in Canterbury. It overlaps with its close relatives *H. decumbens, H. buchananii* and *H. treadwellii* which, respectively, have ranges in the north-east of the South Island, the south-east, and along both flanks of the central Alps. *H. haastii*, which is distributed from Mt Cook northwards, is prostrate with thick, overlapping, almost ericoid leaves. Its relatives are *H. epacridea*, which ranges widely on the drier mountains, *H. ramosissima* (Inland Kaikoura Range) and *H. petriei* (Otago Lakes to Fiordland). A group of cupressoid species grows mainly in crevices: *H. ciliolata* extends from Nelson along both flanks of the Southern Alps, grading eastwards into *H. tetrasticha* on the mountains of Canterbury; *H. cheesemanii* grows on dry eastern ranges of Canterbury and Marlborough; and *H. tumida* is largely confined to the Richmond Range.

The minute shrub *Chionohebe densifolia* is common in the fellfields of Otago east of the Divide. *C. armstrongii* is probably a rare hybrid between this and cushion-forming chionohebes.

Dracophyllum pronum forms lattices appressed to exposed rocky surfaces, mostly east of the Main Divide but extending to North-west Nelson and southern Westland, and descending to lower altitudes to hybridise with *D. uniflorum*, especially on frosty flats. *D. recurvum* on the Volcanic Plateau, *D. kirkii* in the Southern Alps and *D. politum* in Fiordland and Stewart Island, although mainly penalpine, also ascend to alpine fellfield and rock crevices. *Melicytus alpinus* ascends to rocky alpine ridges, including those of western mountains. *Helichrysum intermedium* attains the alpine belt only in the north of the South Island.

Ferns and lycopods

Lycopodium australianum is a small, tufted plant common in penalpine and lower alpine fellfield. The robust, summer-green fern *Polystichum cystostegia* has a similar altitudinal range, but grows protected among large boulders where it overlaps and hybridises with *P. vestitum*. *Grammitis poeppigiana*, which has short, creeping stems and crowded, oval leaves up to 12 mm long, grows on shaded ledges and in crevices.

Bryophytes and lichens

Racomitrium crispulum, R. ptychophyllum and *R. lanuginosum* var. *pruinosum* often cover extensive areas of moderate exposure, the first mainly on rock and the other two on gravelly or sandy surfaces. Flushed rock and spray zones are often dominated by mosses, especially *Rhacocarpus purpurascens, Tortula robusta, Bryum blandum, B. laevigatum, Philonotis pyriformis* and species of *Andreaea* and *Grimmia*. Mosses also dominate where duration of snow cover exceeds the tolerance of vascular plants, and form small cushions on subnival and nival rock. Hepatics are inconspicuous in the alpine zone, but the very reduced *Goebelobryum* occurs as thread-like growth in *Andreaea* cushions on exposed rock, and *Gymnomitrion* is recorded from late-snow areas.

Lichens, in considerable variety, are the most important subnival and nival plants, but there is little information as to the communities they form. Most are crustose or foliose on rock, but fruticose lichens in *Usnea* and related genera are also conspicuous;

the most striking among these are the black tufts of *Neuropogon* that densely cover rocks on ridge-crests fully exposed to moisture-bringing gales. *Cetraria islandica* and soil encrusting *Thamnolia* and *Cladonia* spp. are important in fellfield.

Alpine communities

The highest portion of the Southern Alps has the most extensive, but not necessarily the richest, alpine vegetation, and the national parks on either side of the Divide present a cross-section of the greywacke and schist steeplands. Discussion opens with these parks, and moves on to the plateau mountains of Central Otago, which possess unique terrain and marked endemism. I next consider the ranges of northern Canterbury and Marlborough, which present great expanses of crumbling greywacke with alpine vegetation that is quite varied despite its patchiness. The chapter concludes with North-west Nelson, Sounds–Nelson and the North Island axial ranges, which have only fragmentary alpine vegetation, and the much larger but floristically poor alpine areas on the high volcanoes.

Eastern Alps

In Mt Cook National Park (Wilson 1976) the higher altitudinal belts are: upper nival > 2750 m; lower nival > 2150 m; subnival > 1850 m; alpine > 1550 m; and penalpine > 1300 m.

Alpine grasslands are dominated by *Chionochloa oreophila*, *C. pallens*, *C. crassiuscula*, *Rytidosperma setifolium* or *Poa colensoi*, with *Marsippospermum gracile* being extensive on cool, damp sites. These give way to continuous mat- or cushion-field, in which *Celmisia sessiliflora* is usually dominant, with *Anisotome flexuosa*, *Phyllachne colensoi*, *Raoulia grandiflora* and *Poa colensoi* abundant. Other species are *Drapetes* aff. *lyallii*, *Coprosma perpusilla*, *Gentiana corymbifera*, *Agrostis subulata*, and less consistently, *Celmisia lyallii*, *C. haastii*, *C. angustifolia*, *Chionochloa oreophila* and *C. crassiuscula*.

On exposed, stony ground, such herbfield may grade downwards into penalpine dwarf heath with *Dracophyllum pronum* usually, and *D. kirkii* occasionally, dominant, other important species being *Lycopodium fastigiatum*, *Anisotome flexuosa*, *Gaultheria depressa*, *Celmisia angustifolia*, *C. lyallii*, *Brachyscome sinclairii*, *Poa colensoi*, *Rytidosperma setifolium*, *Racomitrium lanuginosum* and *R. crispulum*.

Stable alpine and subnival gravel free of snow for at least five months supports an open cushion community in which *Hectorella* is usually the main species. Others (those marked† also grow sparingly in the nival belt) are *Ranunculus sericophyllus†*, *Colobanthus buchananii†*, *Anisotome flexuosa*, *Raoulia grandiflora*, *R. youngii†*, *Chionohebe pulvinaris*, *Luzula pumila†*, *Carex pyrenaica* and *Agrostis subulata*. Disturbed alpine debris has very sparse vegetation, consisting of *Andreaea*, *Racomitrium crispulum*, *Grimmia* and crustose lichens on rocks, scattered plants of *Epilobium porphyrium*, *E. glabellum*, *Poa colensoi*, *P. novae-zelandiae*, *Agrostis subulata*, *Rytidosperma setifolium* and, in the shelter of rocks, *Blechnum penna-marina*. Only *Agrostis subulata*, *Poa novae-zelandiae†*, *Epilobium porphyrium†* and the cryptogams ascend to unstable subnival or nival† debris, where they are joined by

Epilobium rubromarginatum†, *E. tasmanicum* and, in stable pockets, *Haastia sinclairii* and *Raoulia youngii*.

Mobile screes, which can extend from the alpine to the montane belts, lack specialised scree plants other than *Epilobium pycnostachyum*; the only other species listed are *Blechnum penna-marina* (where protected by large rocks), *Myosotis traversii* var. *cantabrica*, *Poa novae-zelandiae* and *Agrostis subulata*.

In alpine hollows kept wet by melting snow, *Raoulia subulata* and *Carex pyrenaica* are the main contributors to a cover of 35–50%, others being *Drapetes* aff. *lyallii*, *Epilobium tasmanicum*, *Celmisia haastii*, *Luzula colensoi*, *Agrostis subulata*, *Chionochloa oreophila* and, less commonly, *Ranunculus sericophyllus* and *Plantago novae-zelandiae*. On similar subnival sites *Raoulia subulata* and *Ranunculus sericophyllus* co-dominate.

Penalpine vegetation on rock outcrops (p. 395) becomes attentuated with increasing altitude. Flushed subnival rocks can support lush growth of *Poa novae-zelandiae* s.l., bryophytes and algae, or where exposed, sheets of *Andreaea* (mainly *A. australis*), usually mixed with *Conostomum*, *Polytrichum*, etc. Rare colonies of *Ranunculus godleyanus* grow on alpine rock continuously flushed by melt-water. Dry subnival rock supports fragmentary cover, in which *Poa novae-zelandiae* is the most widely distributed vascular plant, others being *Colobanthus buchananii*, *Epilobium rubromarginatum*, *E. porphyrium*, *Anisotome flexuosa*, *Pratia macrodon*, *Raoulia youngii*, *Leucogenes grandiceps*, *Luzula pumila* and *Agrostis subulata*. There are also the mosses *Andreaea*, *Grimmia*, *Bartramia papillata*, *Ceratodon purpureus*, *Polytrichum juniperinum*, *Pogonatum alpinum* and *Racomitrium crispulum*, and many lichens. *Raoulia eximia* and sometimes *Dracophyllum kirkii* ascend to the subnival belt on north-facing rocks, as does *Ourisia caespitosa* on south-facing rocks.

Nival rock supports the lichens *Neofuscelia* sp., *Rhizocarpon* spp., *Umbilicaria* spp., *Pseudephebe miniscula*, *Neuropogon ciliatus* and *Pertusaria dactylina* and the mosses *Grimmia* spp., *Andreaea* spp. and *Racomitrium crispulum*. *Bartramia papillata*, *Pohlia cruda*, *Dicranoweisia antarctica* and *Lophocolea* spp. are common in damp crannies, and *Philonotis tenuis* and *Andreaea* on flushed rocks. Vascular plants include five confined to very high altitudes; *Hebe haastii* (the most frequent), *Parahebe birleyi* and *Ranunculus grahamii* have been found up to 3000 m. The others are *Cheesemania enysii* and *Myosotis suavis*. Species ascending from lower altitudes are *Poa novae-zelandiae* (especially on flushed rocks), *Agrostis subulata*, *Colobanthus buchananii*, *Epilobium rubromarginatum*, *E. porphyrium*, *Raoulia youngii* and, less commonly, *Grammitis poeppigiana*, *Anisotome flexuosa*, *Leptinella pectinata* subsp. *willcoxii*, *Raoulia eximia*, *Chionohebe* sp. and *Luzula pumila*.

The lichens *Acarospora* sp., *Alectoria nigricans*, *Neofuscelia* sp., *Neuropogon ciliatus*, *Rhizocarpon geographicum*, *Stereocaulon caespitosum*, *Umbilicaria hyperborea* and *U. vellea* have been identified from upper-nival rock, at least some of them being present on the highest north-facing rocks of Mt Cook (3764 m). Névés support *Chlamydomonas* spp. and *Chodatella brevispina*, algae that redden snow during summer.

Solifluction terraces at 1710 m on a rocky lateral moraine of the Tasman glacier, which probably was deposited about 250 years ago, show the following partition of vegetation (Archer *et al.* 1973):

(0) The stony tread supports sparse *Agrostis subulata*, *Raoulia grandiflora*, *Poa colensoi* and *Luzula pumila*.

(1) The brow has a narrow band of *Anisotome flexuosa*, *Cyathodes fraseri*, *Gaultheria depressa*, *Pimelea oreophila*, *Poa colensoi* and *Raoulia grandiflora*.

(2) On the vertical riser, there is a band of *Chionochloa pallens* and *Celmisia lyallii*.

(3) On the lower part of the riser, *Dracophyllum kirkii* is associated with most of the species of (1) together with *Celmisia angustifolia*, *C. gracilenta*, *C. verbascifolia*, *Hebe lycopodioides*, *Euphrasia petriei*, *Gentiana corymbifera* and *Brachyscome sinclairii*.

(4) On the slope between the riser and the tread below it, *Coprosma perpusilla* dominates with some *Gaultheria depressa* and *Wahlenbergia albomarginata*.

On the Ben Ohau Range, Archer (1973) describes a catena in which grasslands dominated by *Chionochloa macra* and *Celmisia lyallii* grade to *Dracophyllum pronum* heath on exposed rises, and through short *Chionochloa oreophila* grassland into fellfield communities with increasing persistence of snow. Within the snow-tolerant vegetation there is a succession depending on time since deglaciation, in the order *Andreaea*, *Racomitrium* and *Gymnomitrion*→*Celmisia hectorii* or *Marsippospermum* →*Marsippospermum* and *Celmisia sessiliflora* (*Poa colensoi*, *Drapetes* aff. *lyallii*, *Phyllachne colensoi*)→*Chionochloa oreophila* (*Carex pyrenaica*, *Celmisia haastii*, *Marsippospermum*) (p. 528).

Western Alps (Wardle 1977)

Westland and Mt Cook National Parks are contiguous along the Main Divide. Although their alpine communities mostly correspond, there are differences related to the wetter and much less sunny climate in the west. The tree-limit lies about 100 m lower, and permanent snow descends as low as 1700 m, some 450 m lower than on ranges east of the Divide. Continuous alpine vegetation on north aspects can overlap altitudinally with permanent snow on south aspects, so that a subnival belt is scarcely developed.

Between 1500 and 1750 m, fragmentary mat and cushion communities replace grassland of *Chionochloa* spp. and *Poa colensoi* where soils are shallow and stony, especially on knolls and ridges. Vascular plant cover varies from 40 to 100%, the remainder consisting of loose or anchored stones, or soil covered with mosses (usually *Racomitrium lanuginosum*) or soil-encrusting lichens. Species that can dominate are *Celmisia sessiliflora* (most frequently), *Marsippospermum*, *Poa colensoi*, *Pernettya alpina*, *Coprosma perpusilla* and *Raoulia grandiflora*. *Anisotome flexuosa*, *A. imbricata*, *Leptinella pectinata* subsp. *willcoxii*, *Carex pyrenaica* and *Chionochloa oreophila* may share dominance. Other frequent species are *Lycopodium fastigiatum*, *Hectorella*, *Psychrophila novae-zelandiae*, *Drapetes* aff. *lyallii*, *Gentiana patula*, *G. bellidifolia*, *Celmisia vespertina*, *C. du-rietzii*, *C. haastii*, *Raoulia grandiflora*, *Phyllachne colensoi*, *Forstera sedifolia*, *Euphrasia* (usually *E. revoluta*), *Luzula pumila* and *L. colensoi*.

Sparser communities grow on stable gravelly or rubbly soil on crests of spurs and ridges between 1600 and 2050 m. At the lower altitudes they occur on south aspects, but above 1750 m are only on north and west aspects and occupy increasingly restricted areas. Plant cover is usually 1–5% but can be up to 50%. The main species form dense cushions; *Colobanthus monticola*, *Hectorella*, *Chionohebe ciliolata* and *Poa colensoi* provide most of the cover. Other species include *Ranunculus sericophyllus*, *Drapetes* aff. *lyallii*, *Anisotome flexuosa*, *Gentiana divisa*, *Raoulia grandiflora*, *R. subulata*, *Luzula crinita* var. *petrieana*, *Marsippospermum*, *Carex pyrenaica*, *Poa novae-zelandiae* and *Agrostis magellanica*.

The most extensive alpine habitat consists of unstable, angular rock fragments subject to solifluction. It mostly lies between 1500 and 2050 m but species typical of the habitat occur as low as 1200 m in a deeply shaded, recently deglaciated valley, and on rocky, usually dry water-courses. Plant cover is usually 1–5% but occasionally up to 40% where the debris is more stable, e.g. through being anchored by large rocks. Slopes are between 10° and 35° and occupy all aspects except at the highest altitudes, where plants are confined to north and west aspects. The most frequent species are *Poa novae-zelandiae* and *Ranunculus sericophyllus*. *Neopaxia australasica*, *Colobanthus* (*monticola* and/or *canaliculatus*), *Epilobium glabellum*, *Schizeilema haastii*, *Anisotome pilifera*, *Gentiana divisa*, *Myosotis* (*suavis*, aff. *lyallii*, *pygmaea*), *Dolichoglottis scorzoneroides*, *Luzula crinita*, *Marsippospermum*, *Poa colensoi*, *Microlaena colensoi* and *Agrostis magellanica* are also frequent. This is also the main habitat of *Ranunculus godleyanus*.

Flat to gently sloping hollows between 1300 and 1800 m, where snow lies till January or later, support a 30–50% cover of vascular plants, mainly *Raoulia subulata*, *Carex pyrenaica*, and *Ranunculus sericophyllus* which tends to dominate on less stable soils. Mosses and lichens often provide considerable cover. *Colobanthus canaliculatus*, *Drapetes* aff. *lyallii*, *Ourisia sessilifolia* and *Agrostis magellanica* are also present.

No macrophytes were found in tarns above 1200 m, but periodically inundated margins are dominated by mosses, together with *Neopaxia australasica*, *Raoulia subulata*, *Gnaphalium paludosum*, *Carex pyrenaica*, *Isolepis aucklandica* and *Chionochloa oreophila*. Flushes also support mosses (including species of *Brachythecium*, *Bryum* and *Bartramia*) together with *Neopaxia* and *Marsippospermum*.

Alpine bluffs support fragmentary communities, *Poa novae-zelandiae* and *Ourisia caespitosa* being present on most sites between 1600 and 2050 m. *Parahebe birleyi* occurs on a Main Divide pass, on north- to west-facing rock crevices between 2050 and 2150 m; 24 further species listed are otherwise characteristic of fellfield. Other species occasionally present in alpine terrain are *Polystichum cystostegia* (among boulders), *Cardamine* sp., *Leucogenes grandiceps*, *Pratia macrodon*, *Hebe ciliolata*, *Euphrasia petriei*, *Rytidosperma setifolium* and *Uncinia divaricata*.

Central Otago

The alpine belt is well expressed in Central Otago but, uniquely within New Zealand, consists predominantly of gently rolling plateaus with occasional steep-sided tors. The high-grade schists weather to potentially fertile regoliths that are deeper than on granite and more stable than on greywacke, and soils can include a substantial amount of loess. These mountains are relatively dry (although far wetter than the semi-arid basins they overlook) and experience more severe frosts than steeper mountains. Intense freeze–thaw cycles have led to subdued surfaces with hummocks, stripes, and solifluction lobes and terraces. Mark & Bliss (1970) describe the vegetation in relation to microtopography.

Sparsely vegetated rock-fields are developed mainly on greywacke and non-foliated schist which fractures into angular blocks, and are therefore most extensive on the Hawkdun and St Bathans Ranges in the adjoining Waitaki ER. The lichen *Umbilicaria cylindrica* is the most important plant. The characteristic vascular species are *Aciphylla dobsonii, Raoulia petriensis, Hebe epacridea, Ranunculus haastii* and *R. crithmifolius*; the last three grow also on greywacke screes. Where the rocks form stone nets 2–3 m across that enclose 'frost boils' of finer material, plants are concentrated towards the periphery of the latter, where frost activity is less intense. On the summit of Ben More (1585 m; Mackenzie ER), the centres of inactive stone nets support continuous cushion-field of *Dracophyllum muscoides* and its associates. *Raoulia eximia* is present on steep but stable rocks near Mt Ida.

Rock-fields are also extensive on moderate to steep slopes in the Remarkables and Hector Mountains bordering Lakes ER. The species include some of wide habitat-range, such as *Celmisia viscosa, C. brevifolia, C. laricifolia, C. haastii* and *Chionohebe thomsonii*, as well as *Haastia sinclairii, Aciphylla simplex, A. similis* and *A. lecomtei*, which are confined to rock-fields. On rocky surfaces at the highest altitudes on these south-western mountains there are also *Parahebe birleyi* and *Ranunculus buchananii*, the latter growing where snow persists.

Herbfields (a in Fig. 12.8) are characteristic of relatively low altitudes, where loamy soils exceed 40 cm in depth. *Celmisia viscosa* and *Poa colensoi* dominate, the proportion of the former increasing with exposure. Probably, this community results from depletion of *Chionochloa macra* grassland, which persists in areas least influenced by fire and grazing; to some extent, the same may apply to categories (b) and (c) below.

On the plateau summits, *Dracophyllum muscoides* dominates an assemblage of mat or cushion plants except on the most exposed sites (b; Fig. 12.9). Species include *Anisotome imbricata, Phyllachne* spp., *Abrotanella inconspicua, Raoulia hectorii, Celmisia laricifolia, C. argentea, Leptinella goyenii, Luzula pumila, Poa colensoi* and *Agrostis subulata*. On highly exposed sites, erosion pavement increases and *Raoulia youngii, R.* aff. *mammillaris* and *Poa pygmaea* reach their maximum importance. Exposed cushions become crescentic, with eroding windward edges and growing leeward edges, resulting in down-wind movement averaging 8 mm per year.

Well-vegetated soil hummocks and stripes (c) occur where underlying rock is strongly foliated. Hummocks are up to 48 cm high and 2.2 m across, and stripes are slightly less; they are much larger than actively forming stone nets and stripes in the same localities, but unlike the latter are formed entirely within loamy horizons that overlie the stony regolith (Fig. 12.10).

Microsites are classified as tops, sides and hollows. *Poa colensoi* and *Luzula pumila* show no obvious preferences; *Celmisia haastii, Psychrophila obtusa, Drapetes* aff. *lyallii, Coprosma perpusilla* and *Rytidosperma pumilum*, otherwise characteristic of late-snow areas (e), grow in the hollows; *Celmisia viscosa* grows mainly on sides; and

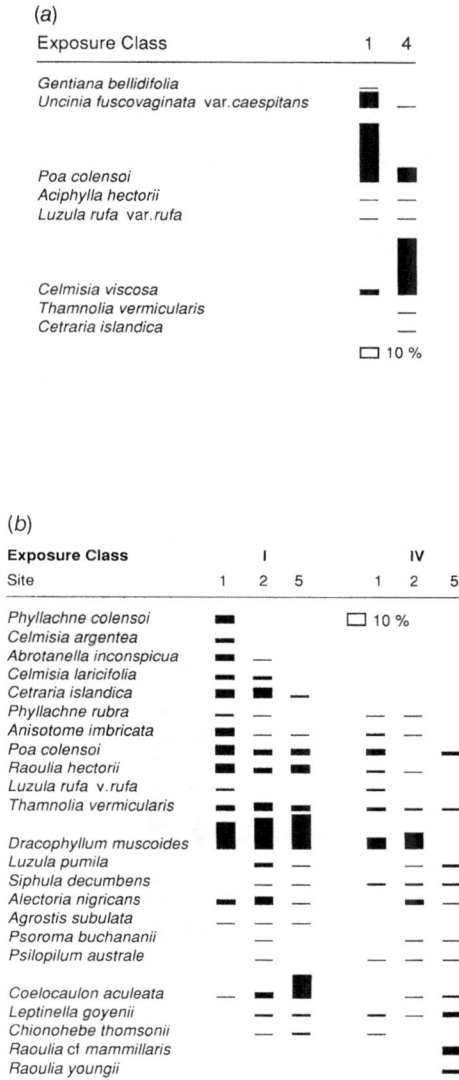

Fig. 12.8. Percent cover of the main species on five Central Otago ranges between 1370 m (Rock and Pillar Range in the east) and 2040 m (Remarkables Range in the west). Lichen cover is expressed as percent total plant cover (Mark & Bliss 1970); †, occurs on only one of the five mountains studied. (a) Low-altitude herbfield; site 1 is less exposed than site 4. (b) Cushion-field; exposure increases from I to IV, and altitude increases from 1 to 5. (c) Vegetation on soil hummocks; protection and micro-relief decrease from 1 to 6. (d) Zonation on solifluction terraces (see text). (e) Vegetation zones associated with snow banks, in order of increasing tolerance of snow cover.

(c)

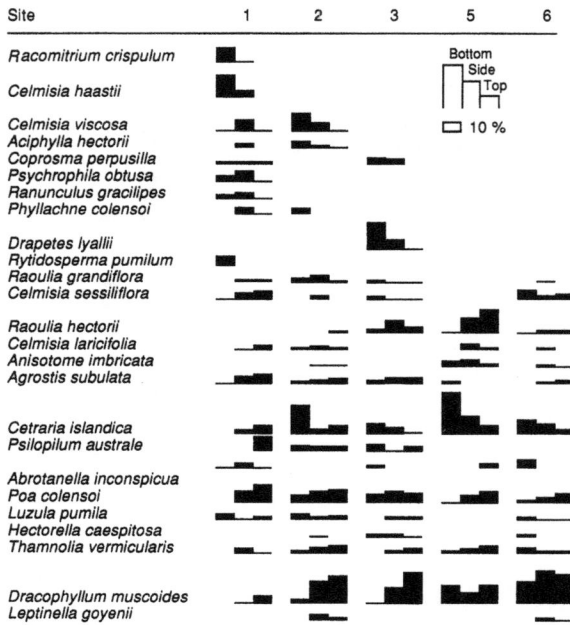

(d)

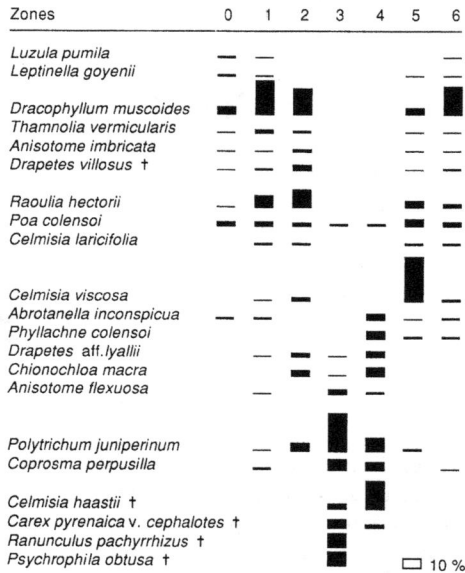

Fig. 12.8 (*cont.*)

(e)

| Zones | 1 | 2 | 3 | 4 | 5 | 6 | 7 | 8 | 9 | 10 |

Celmisia prorepens
Ourisia caespitosa †
Pernettya alpina
Celmisia viscosa
Raoulia grandiflora
Abrotanella caespitosa
Abrotanella inconspicua
Phyllachne rubra
Anisotome flexuosa
Gnaphalium mackayi
Dracophyllum muscoides
Poa incrassata
Phyllachne colensoi
Rytidosperma pumilum
Carex kirkii †
Carex hectorii †
Psychrophila obtusa
Luzula rufa v. *rufa*
Poa colensoi

Coprosma perpusilla
Plantago lanigera
Drapetes aff. *lyallii*

Celmisia haastii
Carex lachenalii †
Polytrichum juniperinum
Colobanthus canaliculatus
Ranunculus pachyrrhizus
Poa pygmaea †
Carex pyrenaica v. *cephalotes*
Luzula pumila
Agrostis subulata
Buellia sp.
Raoulia subulata
Colobanthus buchananii
Neopaxia australasica
Grimmia trichophylla †

Andreaea acutifolia †

□ 10 %

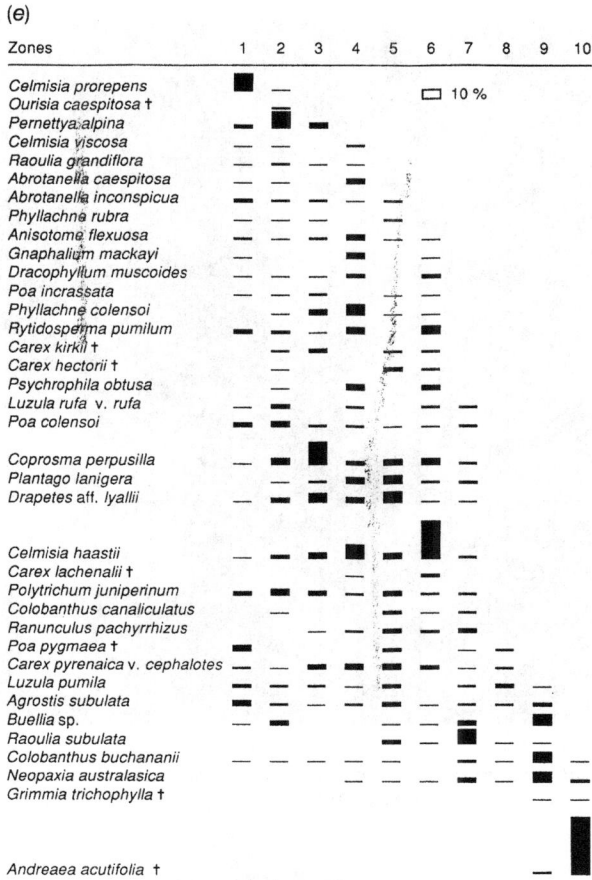

Dracophyllum muscoides is usually the main species of tops and sides, but increases in the hollows as exposure increases. The lichen *Cetraria islandica* occupies upper sides in sheltered localities and hollows where more exposed.

Solifluction terraces (d) develop on leeward slopes below extensive snow banks, with risers up to 1.4 m high and several metres apart. They result from down-slope movement of soils saturated by melt water; unlike solifluction lobes (g), this movement is restrained by vegetation. Fig. 12.8 shows zones of vegetation as follows.

0: Tread with less than 20% cover of *Dracophyllum muscoides, Leptinella goyenii* and *Luzula pumila* among erosion pavement and bare soil.

1: Terrace brow with about 80% cover, mainly of *Dracophyllum* and *Raoulia hectorii*.

2: Steep to vertical riser, in which Zone 1 plants are joined by those benefiting from protection.

3: Immediate lee with species requiring protection, such as *Polytrichum juniperinum, Coprosma perpusilla* and *Chionochloa macra*, and snow-

Fig. 12.9. *Dracophyllum muscoides* and *Raoulia hectorii* (lighter patches) with *Celmisia viscosa* in the hollow behind; 1680 m, Old Man Range, Central Otago.

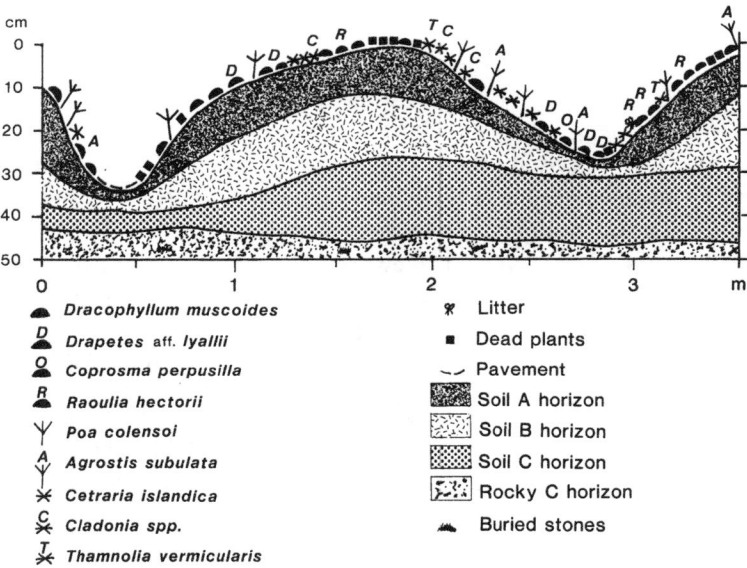

Fig. 12.10. Profile of a soil hummock at 1625 m, Carrick Range, Central Otago (Mark & Bliss 1970).

tolerant plants such as *Ranunculus pachyrrhizus*, *Psychrophila obtusa* and *Celmisia haastii*.

4: Mid-lee; similar to preceding zone, but including species less tolerant of snow, such as *Phyllachne colensoi* and *Abrotanella inconspicua*.

5: Far-lee of terrace, dominated by *Celmisia viscosa*.

6: Beyond lee effects; *Dracophyllum* cushion-field.

Snow-banks (e) persist mainly above 1400 m, in cirques along the leeward edges of the plateaus. The vegetation is graded into ten zones according to tolerance of snow-cover; species in the outer zones require some shelter (e.g. *Celmisia prorepens*) or have wide tolerances (e.g. *Poa colensoi*). *Psychrophila obtusa*, *Ranunculus pachyrrhizus*, *Celmisia haastii* and *Raoulia subulata* reach maximum importance in zones 6 or 7. In Zone 10, where snow is near-permanent, mosses provide nearly all of the cover. On the Garvie and Hector ranges, *Celmisia hectorii* is prominent in areas of medium snow-lie.

Solifluction lobes develop within snow banks on slopes of about 20°, forming narrow crescents 18–51 cm high at their near-vertical fronts. They have less vegetation than terraces (48% cover on their tops, 68% on the fronts). Snow-bank species are distributed throughout although predominantly on the fronts and lees, whereas cushion-field species are restricted to the tops, and do not include *Dracophyllum* or other woody plants. Lichens are well represented by *Cetraria islandica* (the main species), *Thamnolia vermicularis*, *Alectoria nigricans*, *Solorina crocea*, *Psoroma buchananii*, *Placopsis trachyderma*, and species of *Siphula*, *Buellia*, *Cladonia* and *Sticta*.

Waimakariri catchment, northern Canterbury

All high country in the Waimakariri catchment consists of greywacke. On the Craigieburn Range, which lies midway between the Main Divide and the easternmost ranges and rises to 2195 m, screes occupying most of the alpine landscape support very low densities of the special scree plants; for Mt Bailey, Fisher (1952) lists *Ranunculus haastii*, *Stellaria roughii*, *Notothlaspi rosulatum*, *Epilobium pycnostachyum*, *Lignocarpa carnosula*, *Leptinella atrata*, *Senecio glaucophyllus* var. *discoideus* and *Lobelia roughii*. Stable areas with fines, as well as shattered ridge crests, support scattered *Poa buchananii*, *Hebe epacridea* and *Haastia recurva*. In places, there are patches or isolated representatives of continuous vegetation, especially *Chionochloa macra* tussocks.

Moraine-like topography in small cirques supports more vegetation. On loamy soils, open *C. macra* grassland on rises grades into *C. oreophila* grassland in hollows, *via* a zone of hybrids. On coarser, less-weathered material, *Celmisia viscosa* dominates on flat rises, *Dracophyllum pronum* on brows, *Coprosma perpusilla* and *Drapetes* aff. *lyallii* in depressions, and *Carex pyrenaica* in deeper hollows; *Poa colensoi* and *Anisotome flexuosa* are common throughout (Table 12.2). On exposed spurs *Dracophyllum pronum* and its associates can descend to 1500 m. Rock buttresses support a variety of rupestral plants, including the cushion-shrubs *Raoulia mammillaris* and *Chionohebe pulvinaris*, and other plants rooted in pockets of soil.

The Torlesse Range overlooking the Canterbury Plains carries similar alpine vegetation except that the large vegetable sheep *Raoulia eximia* is conspicuous. At

Table 12.2. *Cover estimates from five or ten 0.5 × 0.5 m plots in alpine fell-field; 1770 m, Craigieburn Range, Puketeraki ER*

Site:	1	2	3	4	5	6
Aspect:	NE	—	—	—	NE	SE
Slope (deg.):	5	0	0	0	30	40
Mean percentage cover:	64	40	75	71	40	40
No. of vascular species:	11	14	10	13	22	16
Dwarf shrubs						
Dracophyllum pronum	3	+		+	2	
Pernettya alpina	+					+
Drapetes aff. *lyallii*		+	2	1		
Coprosma perpusilla		+	3	2		
Raoulia mamillaris					1	
Grasses, sedges and rushes						
Carex pyrenaica	+	f	1	2		
Rytidosperma pumilum	+			+	+	
Poa colensoi	2	1	1	2	1	2
Luzula pumila	+	f	f	1		+
Agrostis subulata		+	f		+	+
Marsippospermum gracile				1		
Luzula traversii					f	f
Mat- and cushion-herbs						
Phyllachne colensoi	2	1		1		2
Anisotome flexuosa	1	1	f	1	1	1
Raoulia grandiflora	1	+	+	+	1	f
Celmisia laricifolia	f	+			1	1
Colobanthus buchananii		+			+	
Other forbs						
Celmisia viscosa	1	2	1	2	+	+
Celmisia angustifolia					+	1
Leptinella pyrethrifolia					+	+
Schizeilema haastii						1
Cryptogams						
Neuropogon ?ciliatus					1	
Other lichens	2	1	f	1	1	2
Polytrichaceae	f	1	1	1		+
Racomitrium lanuginosum					1	1

Notes:

Symbols as in Table 9.4.

Communities: 1–4, brow, top, flat depression & hollow in a small moraine; 5–6, sides of a spur of fissured greywacke.

The following species were recorded as present in only one community. Community 2: *Chionochloa oreophila*; 3: *Epilobium tasmanicum*; 4: *E. brunnescens*; 5: *Celmisia lyallii*, *Rytidosperma setifolium, Wahlenbergia albomarginata, Chionohebe pulvinaris, Trisetum spicatum*; 7: *Anisotome imbricata, Ourisia caespitosa, Brachyscome sinclairii*.

Fig. 12.11. *Celmisia sessiliflora, Phyllachne colensoi, Raoulia grandiflora* (top left) and *Carex pyrenaica* (left) in cushion-field, 1750 m, Travers Saddle, North-eastern Alps.

Temple Basin on the Main Divide, several eastern species are replaced by western equivalents, e.g. *Chionohebe pulvinaris* by *C. ciliolata, Hebe tetrasticha* by *H. ciliolata,* and *Luzula rufa* by *L. crinita* var. *petrieana. Chionochloa macra, Celmisia viscosa* and woody raoulias are absent. Zonation of vegetation is the same as at Lewis Pass (Fig. 9.8), except that *Chionochloa pallens* and *C. crassiuscula* are important on less snowy slopes in the absence of *C. australis.*

In hollows at Temple Basin where the snow-free period is less than four months, *Chionochloa oreophila* grassland merges into cushion herbfield almost identical with that of the Western Alps, with *Celmisia sessiliflora, Phyllachne colensoi, Raoulia grandiflora* and *Anisotome imbricata* as dominant species (Fig. 12.11). There are also high-altitude fellfields with *Hectorella.* Widely-scattered plants on the poorly sorted debris that occurs instead of graded screes include *Hebe haastii, Anistome pilifera,* and *Parahebe cheesemanii* at its southern limit (Fig. 12.12). As on more eastern ranges, however, *Dracophyllum pronum* is extensive.

Marlborough (Williams 1989)

The Inland Kaikoura Range includes the highest peaks north of the Southern Alps, culminating in Tapuaenuku which, despite a height of 2885 m, carries no permanent snow. Greywacke and argillite predominate but the spine is an alkaline mafic–ultramafic complex that contains abundant calcium phosphate; both rock types are

Fig. 12.12. *Chionochloa pallens, Hebe haastii* and *Anisotome pilifera* (top right) on debris; 1500 m, Arthurs Pass, North-eastern Alps.

heavily intruded by dolerite dykes. Inherent fertility of the regolith therefore varies considerably, although influences on the vegetation are suspected rather than proven. The alpine landscape is typical of Marlborough, being dominated by disintegrating rock and debris. Active rock glaciers are prominent.

The uppermost patches of continuous vegetation consist of the following assemblages.

(1) *Rytidosperma setifolium* with some *Celmisia spectabilis* and *C. monroi*, ascending to 1800 m, mainly on stony terraces and steep debris cones.

(2) *Chionochloa pallens* – *R. setifolium* grassland reaching 1700 m on patches of well-developed soil on steep fans, with clumps of *C. pallens* ascending to 1850 m on north aspects.

(3) Low swards of *Poa colensoi* on benches and lobes of colluvium up to 2000 m; on exposed sites *P. colensoi* co-dominates with *Agrostis* aff. *subulata* and *Celmisia sessiliflora*. Further species include *Epilobium cockayneanum, Anisotome imbricata, Gentiana bellidifolia, Leptinella pyrethrifolia, Raoulia bryoides, Chionochloa pallens, Carex wakatipu* and *Luzula pumila*.

(4) A variable assemblage occurring above 2200 m on stable, well-watered ground such as late snow hollows. It includes *Epilobium tasmanicum, Drapetes dieffenbachii, Aciphylla monroi, Celmisia allanii, Agrostis* aff. *subulata, Rytidosperma pumilum* and *Luzula rufa*, which grow also in (3), together with *Cardamine* sp., *Cheesemania fastigiata, Colobanthus affinis, Stellaria gracilenta, Neopaxia australasica, Epilobium porphyrium, Oreomyrrhis colensoi, Coprosma perpusilla, Myosotis pygmaea* var. *drucei,*

Chionohebe pulvinaris, Hebe ramosissima, Craspedia lanata var. *elongata, Haastia sinclairii, Raoulia grandiflora, Poa buchananii, P. dipsacea, P. cita, Rytidosperma setifolium, Trisetum* aff. *spicatum, Isolepis aucklandica* and mosses.

At penalpine levels, common species on stony seepages are *Ranunculus foliosus, Cerastium fontanum, Epilobium macropus, Geranium sessiliflorum, Viola cunninghamii, Acaena saccaticupula, Aciphylla glaucescens, Hydrocotyle novae-zeelandiae* var. *montana, Celmisia monroi, Microseris scapigera, Taraxacum magellanicum, Chionochloa pallens, Festuca* sp., *Carex wakatipu* and *Luzula traversii.*

Widespread plants on mobile screes are *Epilobium pycnostachyum, Lobelia roughii, Myosotis traversii, Stellaria roughii, Wahlenbergia cartilaginea, Leptinella dendyi* and *Poa buchananii. Lignocarpa diversifolia* grows mainly where surface rocks are relatively large, *Epilobium forbesii* where screes are fine-grained, and *Notothlaspi rosulatum* where they are relatively stable. *E. forbesii, Leptinella, Lobelia, Notothlaspi* and the local endemic *Raoulia cinerea* also grow on gentle slopes of fine-textured scree showing frost-stripes and polygons. *Acaena glabra* grows on screes with underlying soil. Rock fields support *Blechnum penna-marina, Polystichum cystostegia, Hebe cheesemanii, H. epacridea, Haastia sinclairii, Colobanthus buchananii* and *Poa buchananii,* the last two reaching 2600 m. *Haastia sinclairii, Hebe epacridea* and, where moist, *H. ramosissima* persist where the rocks are loose. *Oxalis exilis* and *Convolvulus fractosaxosa* are on subalpine screes and *Echium vulgare* colonises talus below 1500 m.

Crevices in alpine rocks support *Rytidosperma setifolium, Colobanthus acicularis, C. buchananii, Epilobium glabellum, Hebe epacridea, H. cheesemanii, Myosotis traversii, Haastia recurva, H. sinclairii,* and rare *Leucogenes grandiceps. Haastia pulvinaris* is abundant up to 2200 m, with some as high as 2700 m. Lichens are abundant on the summit of Tapuaenuku.

Western Nelson

Only some 13 summits in this province exceed 1600 m. These are all north of the Buller River, mainly in the eastern part, and have alpine florulas that reflect isolation and varied lithology. The marble massifs of Mt Owen (1875 m) and Mt Arthur (1826 m) consist of exposed ridges and summits, with some deep sink-holes where snow can lie into February. Around 1500 m *Chionochloa pallens* tussocks scattered among turf of *C. australis* or *Poa colensoi* give way to sparsely vegetated outcrops, stone pavements and talus. Bell (1973) lists the following above 1500 m on the peaks and ridges of Mt Owen:

Cystopteris tasmanica, Ranunculus gracilipes × *insignis, Cardamine* sp., *Oreoporanthera alpina, Cheesemania latisiliqua, Notothlaspi australe, Colobanthus* aff. *canaliculatus, C. masoniae, Muehlenbeckia axillaris, Epilobium glabellum, E. rubromarginatum, Anisotome pilifera, Gingidia decipiens, Celmisia allanii, C. semicordata, C. laricifolia, C. sessiliflora, C. spectabilis, Craspedia lanata, Helichrysum intermedium, Raoulia apicinigra, Senecio glaucophyllus, Gentiana filipes, G.* aff. *montana, Phyllachne colensoi, Myosotis angustata, M. concinna, M. pygmaea* var. *drucei, Hebe ciliolata, H. haastii* var., *Parahebe linifolia, Chionohebe pulvinaris, Pterostylis mutica, Juncus antarcticus, Luzula rufa, L. pumila, Carex pyrenaica, Rytidosperma setifolium* and *Poa novaezelandiae.* Below 1500 m, these species are either accompanied or replaced by others typical of penalpine marble bluffs, such as *Aciphylla ferox* and *Chionochloa* 'robust'.

On Mt Arthur further alpine species include *Leucogenes grandiceps* (common) and *Chionohebe ciliolata*. The epacrids *Cyathodes fraseri, C. pumila, C. colensoi, Dracophyllum uniflorum* and *D. pronum*, noted at 1580 m, presumably indicate slow weathering and rapid leaching.

Non-calcareous outcrops (mainly granite, schist or greywacke) do not exceed 1775 m, but support a fragmentary alpine flora. Bell (1973) lists *Aciphylla monroi, Anisotome imbricata, Celmisia lateralis, Schizeilema haastii, Raoulia rubra* and *Dracophyllum pronum* as calcifuges, although the last three are present on Mt Arthur. Further calcifuge species are *Cheesemania gibbsii* and, on screes, *Lobelia roughii, Haastia sinclairii, Epilobium porphyrium, E. pycnostachyum, E. glabellum* and *Neopaxia australasica* (P.A. Williams, personal communication).

Northern mountains

From the Richmond Range and along the North Island axial ranges to Mt Hikurangi no summits are truly alpine, although local factors of exposure and geology lead to occurrences of alpine species or communities of alpine appearance; notable are the ultramafic barrens of the Richmond Range, the presence of *Raoulia rubra* on the Tararua Range, and greywacke screes on the Ruahine and Kaweka Ranges that support *Epilobium pycnostachyum*.

On the central volcanoes, scoria slopes and material redistributed by water or wind carry sparse vegetation, with low heath extensive below 1850 m, and grasses or herbs prevailing on finer material and at higher altitudes. These communities can have a lower limit against closed red-tussock grassland or wet heath around 1500 m, but descend below 1100 m on south-eastern slopes of Ruapehu. Species combine in varying proportions, according to slope, substrate texture and aspect. Atkinson (1981) maps the vegetation as follows.

(1) *Racomitrium lanuginosum* with scattered shrubs on moist gravel (1400–1650 m).

(2) *Dracophyllum recurvum* on coarse gravel dominating alone or, at subalpine levels, with *Gaultheria colensoi* or *Podocarpus nivalis* (1100–1850 m).

(3) Widely spaced *Rytidosperma setifolium* tussocks on sand and fine debris (1050–1950 m, especially above 1500 m).

(4) Scattered *Poa colensoi* with *Gentiana bellidifolia, Anisotome aromatica* and *Ranunculus nivicola*, on gravel summit ridges of Tongariro (1800–1950 m).

(5) *Raoulia albosericea* on gravel fans, especially around the base of Ngauruhoe (1250–1500 m).

(6) Debris slopes above 1700 m, with scattered *Parahebe hookeriana, P. spathulata, Ranunculus insignis, Helichrysum* 'alpinum' and *Anisotome aromatica*, the last reaching 2020 m.

(7) Permanent snow and small glaciers capping the summit of Ruapehu (2797 m).

Cycles of regeneration and decay are prevalent, under the influence of solifluction and showers of fine andesitic tephra. These may commence with *Dracophyllum*

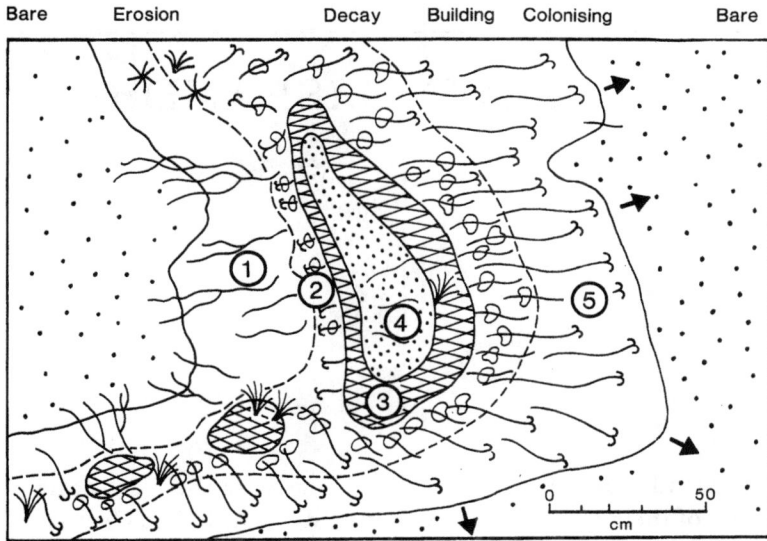

Fig. 12.13. Segment of an old circle of *Dracophyllum recurvum* in process of developing into a daughter circle; 1280 m, saddle between Ruapehu and Ngauruhoe (Chambers 1958). Numbers indicate: 1, dead *Dracophyllum* roots; 2, stems colonised by lichen; 3, live *Dracophyllum* stems; 4, *Racomitrium lanuginosum*, dead *Dracophyllum*; 5, advancing *Dracophyllum*.

colonising bare ground (Fig. 12.13). As its stems age, they are colonised by *Pannoparmelia angustata* and, on death, replaced by a *Racomitrium* mat. The patch of vegetation accumulates tephra, but when the moss mat dies from the middle in its turn, bare ground and dead roots are exposed to erosion and, eventually, recolonisation. Patches of heath at the base of nearby screes undergo similar cycles, except that *Podocarpus* colonises bare surfaces ahead of *Dracophyllum*.

On Mt Taranaki (2518 m) red-tussock grassland merges at its upper limit of about 1600 m into short vegetation variously dominated by *Racomitrium* spp., leafy herbs or cushion plants, especially *Celmisia major* var. *brevis*, *C. glandulosa*, *Helichrysum 'alpinum'*, *Anisotome aromatica* and *Forstera bidwillii*. Other common species include *Coprosma perpusilla*, *Gaultheria depressa* var. *novae-zelandiae*, *Poa colensoi*, *Lycopodium fastigiatum*, and, in pools, *Juncus novae-zelandiae*, *Oreobolus pectinatus* and *Plantago lanigera*. On the western side of the mountain, similar vegetation descends almost to 1000 m on former debris flows, but is seral to taller vegetation.

Above 1650 m, plant cover becomes patchy among volcanic gravel, boulders or lava. From 1700 to 1900 m *Poa colensoi* provides around 5% cover, with other common species being *P. novae-zelandiae*, *Carex pyrenaica* in late-snow areas, and *Colobanthus* aff. *affinis*, which ascends to 2400 m, higher than any other vascular plant on the mountain. Mobile scoria supports *Helichrysum 'alpinum'*, *Epilobium glabellum* and *Neopaxia australasica*. Alpine rock, whether in cliffs or lava flows, supports mainly *Rhizocarpon*, *Stereocaulon*, *Racomitrium*, *Andreaea* and *Grimmia*, with some vascular plants rooted in crevices. There are nival communities of mosses and lichens, and also snow algae (Clarkson 1986).

13

OUTLYING ISLANDS

The outlying islands have close geographical links with the New Zealand mainland, and share most of their plant species with it (Table 13.1). The Chatham Islands and the far-southern Bounty, Antipodes, Auckland and Campbell groups form part of the block of continental rocks that includes the New Zealand mainland, although intervening seas are over 500 m deep. The Kermadec group rises from a volcanically active submarine ridge that extends NNE from the central North Island volcanic zone, and Macquarie Island lies on a ridge that extends SSE from Fiordland.

The Chatham Islands and the islands of the far-southern or Campbell province each support distinctive endemic plants that imply long evolutionary histories. In the Kermadec group and Macquarie Island endemism is weak but there are subtropical or circum-antarctic species otherwise absent from New Zealand. Further afield, Norfolk and Lord Howe Islands also lie on submarine rises extending from New Zealand, and occurrence of genera otherwise restricted to New Zealand (*Phormium* and *Rhopalostylis* on Norfolk, *Carmichaelia* on Lord Howe) may reflect Gondwanan connections. They share many herbaceous species with New Zealand, but endemic, Australian and tropical Pacific elements are larger among woody plants.

Kermadec Islands
(based on Sykes 1977)

History and flora

The Kermadec Islands form two clusters separated by 100 km, Raoul in the northern cluster being the highest and largest. All are volcanic, Raoul and Curtis being active, and no rocks predate the Pleistocene (Fig. 13.1). Among indigenous taxa, 95, 55 and 44 are identical with or closely related to species in New Zealand, Norfolk and Lord Howe Islands, and tropical Polynesia, respectively; these numbers include all 23 endemic taxa, which have a low order of distinctiveness.

Although the Kermadecs were uninhabited when first seen by Europeans, there is evidence of past occupancy of Raoul Island by Polynesians, who may have introduced

Table 13.1. *Statistics for outlying islands*

group	latitude	distance km[1]	area km²	maximum alt.	submarine structure	total indig. vasc. spp.	endemic spp.	adventive spp.
Kermadec	29°16'	976*	33	518*	Kermadec Ridge	113	12	152
Three Kings	34°10'	55	8	296	Reinga Ridge	175	12	22
Chatham	44°00'	863	950	287	Chatham Rise	319	31	153
Bounty	47°42'	700	1.3	70	Bounty Platform	0	0	
Snares	48°02'	110	2.6	130	Campbell Plateau	17	1	2
Antipodes	49°42'	750	21	366	Bounty Platform	69	1†	3
Auckland	50°44'	460	625	668	Campbell Plateau	187	4†	41
Campbell	52°33'	595	113	569	Campbell Plateau	126	2†	86
Macquarie	54°30'	900	118	434	Macquarie Ridge	40	1†	5

Notes:

Latitude is median for group.

* Statistic refers to Raoul Island.

† No. of species endemic to the Campbell province as a whole is 35.

[1] Approximate distances from the nearest point of North, South or Stewart Island.

Sources: Baylis 1958; Fineran 1969; Madden & Healy 1959; Sykes 1977; Godley 1989; A.P. Druce, unpublished.

Fig. 13.1. *Metrosideros kermadecensis* two years after the 1964 eruption from the crater of Green Lake, Raoul Island (photo W.R. Sykes).

candlenut (*Aleurites moluccana*) and *Cordyline terminalis* from the tropics, as well as *Rattus exulans*. Archaeological material, including obsidian that is probably of New Zealand mainland origin, raises a question as to the status of karaka (*Corynocarpus laevigatus*) on the island, as its large drupes are unlikely candidates for transoceanic dispersal. From 1879, the flatter coastal parts of Raoul Island have been intermittently farmed, and there is a permanent meteorological station. The vegetation on most of Macauley Island has been burnt. The event of greatest significance, however, was the introduction of goats to Raoul and Macauley in 1836.

Many of the adventive species are common on the New Zealand mainland, whereas others reflect the subtropical latitude better than most of the native flora. Most occupy open habitats, plantations or pastures, but three invade and significantly modify closed native vegetation. The following paragraphs concern native and adventive species with subtropical or tropical affinities.

Homalanthus polyandrus and *Boehmeria australis* var. *dealbata* are medium-sized endemic trees with their closest relatives in the tropical Pacific and Norfolk Island respectively. *Myrsine kermadecensis* has its closest relative in the Cook Islands, and nikau palm (*Rhopalostylis baueri* var. *cheesemanii*) is represented on Norfolk by var. *baueri*. Kermadec ngaio (*Myoporum kermadecense*) belongs to a species cluster that includes Polynesian species as well as those of Norfolk Island and the New Zealand mainland. *Olea europaea* subsp. *africana* is an adventive tree with a potential for rapid spread.

Scaevola gracilis, Canavalia rosea and *Ipomoea pes-caprae* var. *brasiliensis* are trailing, intermittently rooting plants of open habitats, the first being closely related to and the others identical with abundant strand plants of the tropical Pacific. Mysore thorn (*Caesalpinia decapetala*), a subtropical climbing legume, has spread aggressively in bush and forest on Raoul in the wake of damage by goats and humans. Its thick, thorny stems weigh down the host trees. There are no native lianes, and *Peperomia urvilleana* is the only habitually epiphytic flowering plant.

The large aroid *Alocasia macrorrhiza*, which has massive, edible rhizomes, has spread rapidly throughout Raoul Island forests since the nineteenth century, inhibiting regeneration of native species. It has also vigorously colonised ground bared by recent eruptions. *Furcraea foetida*, an agavaceous plant indigenous to Central America, is naturalised in an old crater. The grass *Imperata cheesemanii*, endemic to Raoul Island, but closely related to *I. exaltata* of the tropical western Pacific, is locally abundant in open areas.

Communities

Metrosideros kermadecensis, a tree >15 m tall closely related to pohutukawa (*M. excelsa*), dominates most of the forest on Raoul Island, forming either low-altitude 'dry' forest or 'wet' forest on ridge crests, with *Myrsine kermadecensis* or *Ascarina lucida* var. *lanceolata* dominating the respective subcanopies. Wet forest is more luxuriant and also includes *Melicytus ramiflorus*, nikau (as a pure stand on one slope) and *Papillaria crocea* festooning twigs. Other trees are *Pseudopanax arboreus* var. *kermadecensis, Homalanthus, Coprosma acutifolia, Boehmeria* and karaka. Most ferns are shared with the New Zealand mainland, but the tree ferns *Cyathea kermadecensis* and *C. milnei* are endemic, although related to *C. cunninghamii* and *C. dealbata* respectively.

Up to 1966, regeneration of many species, including *Metrosideros*, was being prevented by goats and dense *Alocasia*, and Mysore thorn was spreading rapidly. However, by 1983 the goats had been hunted to the point of extinction, resulting in substantial increase of all palatable native species, including some that had become scarce, such as *Pseudopanax arboreus* and *Coprosma acutifolia*. Even the endemic *Hebe breviramosa*, last seen in 1908, reappeared. In places, young nikau thickets are suppressing *Alocasia* (Parkes 1984). All mature plants of Mysore thorn have been killed by herbicides, but the durable seeds will pose a threat for many years.

Coastal bush on Raoul comprises stunted *Metrosideros, Myoporum kermadecense*, rare *Pisonia brunoniana*, and the endemic *Coprosma petiolata*, which is related to *C. repens. Myoporum* appears to have been dominant on Macauley Island before fire and goats induced close grassland of *Microlaena stipoides*, with lesser proportions of *Cyperus brevifolius, Rytidosperma racemosum* and *Vulpia bromoides*. Two decades after removal of goats, the grassland has changed in composition; for example, *Cyperus ustulatus* and *Solanum americanum* are now abundant, but thick turf and litter have restricted woody regeneration to a slow increase of *Myoporum*. On Raoul Island, there are grasslands of *Stenotaphrum secundatum* and *Sporobolus africanus*.

As on the mainland, weedy or salt-tolerant plants of exposed coastal debris include

Asplenium obtusatum, Disphyma australe, Portulaca oleracea, Polycarpon tetraphyllum, Samolus repens, Lobelia anceps, Conyza bonariensis, Pseudognaphalium luteoalbum, Leontodon taraxacoides, Sonchus kirkii, S. oleraceus, Cyperus ustulatus, Isolepis nodosa, Bromus hordeaceus, Lachnagrostis filiformis var. *littoralis* and *Polypogon monspeliensis.* Gullies on inland cliffs have been important refuges from goats on Raoul and Macauley; *Adiantum* and *Asplenium* species, *Pellaea falcata, Pyrrosia serpens, Coprosma petiolata, Lobelia anceps, Carex kermadecensis* and *Poa anceps* are typical. There is little free water on the islands, other than the crater lakes on Raoul. In 1966, the sole patch of swamp was dominated by *Juncus usitatus.*

Chatham Islands

Environment and history

The Chathams group consists of Chatham Island (referred to here by its early name, Rekohu), Pitt, two smaller islands and scattered islets and rocks. Schist crops out in the north of Rekohu, but the islands are formed mainly of Eocene and Pleistocene lava flows and tuffs, interbedded with shallow-water sediments including limestone. Most of the northern part of Rekohu is occupied by Te Whanga lagoon, which is enclosed by rock outcrops linked by aeolian sand.

Rainfall on the Chatham Islands is only about 800 mm annually and summer droughts occur. Nevertheless, frequent showers, spray-laden gales, cool, equable temperatures and subdued topography have led to thick blanket peats covering most of the northern part of Rekohu, almost the whole of its southern tableland which lies around 250 m above sea-level, and smaller areas elsewhere. In places they exceed 10 m, the deeper layers being almost lignitic. Volcanic ash erupted from Lake Taupo about 20 000 years ago is several centimetres thick in places. As this usually lies within 2 m of the surface, basal peats must be several times older. Although growth is currently inactive in many places, peat is still accumulating over most of the southern tableland and part of northern Rekohu. Annual upward growth of the undecomposed surface is estimated as 1.2–8.7 mm, but this may correspond to only 0.1 mm for the compact peat beneath.

The primitive vegetation of the islands is broad-leaved bush grading to tree-heath, with woody and restiad mires on active peat domes. Coastal areas support heaths, herbfields, dune vegetation and wetlands. Habitation began at least 800 years ago by Moriori; on the evidence of archaeological obsidian, these Polynesians came from the New Zealand mainland (Leach *et al.* 1986). The population grew to about 2000, subsisting mainly on sea-food, seals and birds. Several species of flightless and other birds were exterminated, and fernland probably replaced much woody vegetation, especially around Te Whanga Lagoon. Possibly, karaka was introduced from the mainland, although it is now ubiquitous in suitable habitats.

In the 1830s the islands were colonised by Europeans and Maori, and the Moriori all but disappeared as a separate people through ill-treatment and neglect. Grazing of livestock led to the replacement of most of the bush and tree-heath by grass and fern. Pigs, sheep and cattle became feral, and weka (*Gallirallus*) and rats were naturalised on Rekohu.

Fig. 13.2. Nikau palms in swamp-bush remnant, and rank pasture with rushes. Nikau Reserve, Rekohu.

Species of bush and tree-heath

Broad-leaved bush originally occupied most peat-free surfaces, but as these are required for farming, it now exists only as modified fragments. Tree-heath occupies the drier, firmer peats, grading towards bush where peat is thin or patchy, as on ridge crests, in incised upland valleys and on swampy ground. Tree-heath has disappeared from the northern half of Rekohu, but is still extensive in the south-west of the island. Degraded transitional stands occupy the south-west of Pitt Island.

Apart from karaka, ngaio (*Myoporum laetum*) and nikau (*Rhopalostylis sapida*; but said to differ from the mainland plant in leaf shape) (Fig. 13.2), the bush is formed by endemic trees with glossy leaves larger than those of mainland relatives: *Melicytus chathamicus*, which is most similar to *M. novae-zelandiae*; *Pseudopanax chathamicus*, which lacks the linear-leaved juvenile of *P. crassifolius*; *Myrsine chathamica*, also found locally in Rakiura province; and *Coprosma chathamica*, related to *C. repens* but, with heights up to 15 m and stems to 1 m diameter, the largest of all coprosmas. *Olearia traversii* is likewise taller and stouter than any mainland olearia and, excepting a few filiramulate species, the only one with opposite leaves. *Hebe barkeri*, which is most closely related to coastal hebes of northern New Zealand, can also be a canopy tree. Chatham Islands *Plagianthus regius* and kowhai (*Sophora microphylla*) lack the filiramulate juveniles of mainland populations.

Fig. 13.3. Remnant *Dracophyllum arboreum* tree, Pitt Island.

Tarahinau (*Dracophyllum arboreum*) (Fig. 13.3) dominates tree-heath. At 5–12 m, it is probably the tallest member of its genus. Unlike its close relative *D. longifolium*, the leaves are fringed by long hairs which, in the bud, possibly deflect salt spray. Leaves of juvenile plants and reversion shoots are far larger than adult leaves (e.g. 180 × 13 mm *versus* 90 × 1.5 mm). Seedlings establish mainly on tree fern trunks, and on the ground in the open. Trees 40 cm diameter can be 200 years old. *Brachyglottis huntii* is a rare heathland tree with a leaning sinuous trunk and compact crowns covered by yellow flowers in late summer.

Two endemic shrubs grow among tall, woody vegetation on wet ground. *Coprosma propinqua* var. *martinii* is leafier and less divaricating than mainland *C. propinqua*. *Myrsine coxii*, like the mainland *M. divaricata*, has small, obovate leaves, but forms rhizomatous colonies of erect stems up to 4 m tall that are not filiramulate. The coprosma prefers more fertile sites than *M. coxii*.

Tall woody communities

Broad-leaved bush

On optimal sites, such as colluvial slopes of volcanic hills and weakly podsolised sands, karaka is usually the most abundant tree (Fig. 13.4); it possibly was encouraged by the

Fig. 13.4. Interior of karaka-dominated bush, Haupupu Reserve, Rekohu.

Moriori, whose bark carvings still exist (Fig. 13.5). Karaka is accompanied by variable proportions of all the other broad-leaved trees, although ngaio is common only on the east of Pitt Island, and kowhai survives only as a few groves on the shore of Te Whanga Lagoon. Tree ferns abound in reasonably intact bush; *Cyathea cunninghamii*, *C. smithii*, *C. dealbata*, *Dicksonia squarrosa* and *D. fibrosa* can all be present. Nikau is widespread in the south and east of Pitt Island (almost no bush remains in the north and west) but is local on Rekohu.

Olearia traversii dominates what remains of the seaward fringe of the bush because, in the absence of grazing, it colonises stable dunes and coastal debris. With *Coprosma chathamica*, it also dominates on swampy ground. *Melicytus* forms low, wind-swept bush on small islands. *Myrsine chathamica* produces root suckers, enabling it to form thickets where sand builds up and to invade margins of drying swamps. *Plagianthus* grows mainly in moist gullies. In swamps and on sharp ridges, there are usually some tarahinau trees.

Ripogonum scandens was probably the only abundant liane inside dense bush, although *Muehlenbeckia australis* would have proliferated on margins and in gaps. The undergrowth would have consisted mainly of seedlings, young tree ferns, *Macropiper excelsum* and ground ferns, including endemic relatives of *Polystichum vestitum* and *Asplenium terrestre*. The robust, tussocky, silver-leaved *Astelia chathamica* may have also been important, but is now confined to inaccessible places. The ground layer included *Australina pusilla* and *Nertera depressa*.

The present condition of most of the bush is very different. Partial clearing, death of

Fig. 13.5. Moriori carving on karaka bark, Haupupu Reserve, Rekohu.

exposed, old and short-lived trees (notably *Melicytus*) and browsing of seedlings and undergrowth has led to open groves with pasture beneath (Fig. 13.6). In denser stands *Solanum aviculare, S. laciniatum* and *Parietaria debilis* occur, but otherwise the floor is almost bare except for tufts of fern persisting under fallen logs.

Since 1980 some areas have been reserved and fenced, and two off-shore islands were destocked about 1960. Tree regeneration under intact canopies has been spectacular in places, the most vigorous responses being by *Melicytus* in the north of Rekohu, and *Plagianthus* on South-east Island. Very dense patches of karaka seedlings have grown up near parent trees. In open glades and on margins, rank grasses or bracken usually confine tree seedlings to rotting logs and stumps, and in many places *Muehlenbeckia* or blackberry (*Rubus fruticosus*) are smothering seedlings and saplings.

Tarahinau–broad-leaved mixtures
In transitional communities, tarahinau shares dominance with broad-leaved trees, especially *Coprosma chathamica* and lesser numbers of *Corokia macrocarpa, Pseudo-*

Fig. 13.6. Broad-leaved trees, mainly karaka, dying from exposure where isolated in pasture. Haupupu, Rekohu.

panax and *Myrsine chathamica*. There are also *Hebe barkeri, Melicytus, Brachyglottis huntii, Cyathea medullaris* and *C. cunninghamii*. On the broad, southern ridges of Pitt Island, nikau emergent above the main canopy are the world's most poleward wild palms. *Dicksonia squarrosa* is abundant as a lower tier, and is nearly always accompanied by *D. fibrosa*. Locally, *Ripogonum* forms dense thickets. The ground cover is dominated by *Blechnum* 'black spot', *Histiopteris incisa, Rumohra adiantiformis, Phymatosorus diversifolius, Hymenophyllum demissum, H. multifidum* and *Trichomanes reniforme*; all but the first two also form thick carpets over logs and tree trunks. Other species include *Hypolepis rufobarbata, Asplenium polyodon* and the epiphyte *Earina mucronata*.

Most of the canopy trees begin as epiphytes on tree ferns, which is fortunate in view of disturbance of the ground and understorey by the ubiquitous pigs, cattle and sheep. On the southern tableland the canopy is collapsing in places, giving way to glades containing *Marchantia, Histiopteris, Epilobium* spp., *Hydrocotyle* sp., Scotch thistle (*Cirsium vulgare*) and blackberry. On Pitt Island, tree fern stands with grassy openings, scattered old canopy trees, and little regeneration represent further deterioration, in which animals, fire and exposure of crowns to wind have all played a role.

Tarahinau tree-heath

This is still extensive in the south-east of the southern tableland. Tarahinau provides up to 80% canopy cover. Other trees, mainly *Myrsine chathamica* and *Corokia*, tend to be scattered in the subcanopy. *Pseudopanax* is usually present, but sometimes only as seedlings. *Dicksonia squarrosa* up to 3 m tall and 30% cover dominates an uneven lower tier. *Ripogonum* occurs locally. *Blechnum* 'black spot' is usually present in the tall herb tier, and sometimes abundant. A discontinuous ground cover is mainly *Ptychomnion aciculare, Dicranoloma* spp., *Hymenophyllum multifidum* and/or *H. demissum*. Other plants include hepatics, *Trichomanes reniforme, Rumohra adiantiformis, Ctenopteris heterophylla, Histiopteris incisa, Coprosma chathamica, Myrsine coxii, Uncinia rupestris* and orchids.

Tarahinau heath grows on convex surfaces of firm, dryish peat, that is thickening very slowly if at all. Most stands described from the southern tableland are clearly even-aged, with the tarahinau trees having established on the ground, probably after fire. In Pitt Island stands, on the other hand, most tarahinau trees began as epiphytes on tree ferns.

Coastal tall heath

This community, never extensive in the first place, is now intact only on islets. Remnants indicate that it grew mainly on brows of steep spurs and ledges overlooking the sea, on soils too shallow or peaty for broad-leaved bush. The characteristic species are the purple-rayed *Olearia chathamica* and tarahinau. Ferns and robust herbs such as *Myosotidium hortensia* probably filled the openings.

Shrubland and fernland

Vegetation of peat domes

Growing peat domes of Rekohu support distinctive shrub-heath (Fig. 13.7). *Dracophyllum scoparium*, the most important shrub, is confined to Rekohu and Campbell Island. Although it hybridises with *D. arboreum*, it seldom exceeds 3 m in height, has narrower, strict adult leaves, and lacks a wide-leaved juvenile stage. Seedlings come up thickly after fire. The pink-rayed bog aster (*Olearia semidentata*) hybridises with the coastal *O. chathamica*, but is slender, sparingly branched, and usually less than 1 m tall. The large restiad *Sporadanthus traversii*, the most abundant herbaceous plant on peat domes that have not been severely modified, is described on p. 321. *Aciphylla traversii*, which has softer leaves than most of its genus, and the small, white-flowered *Gentiana chathamica* are endemic herbs.

On the most 'natural' peat domes, *Dracophyllum* 2.5–4 m tall forms 10–50% cover. *D. scoparium* dominates, but stunted *D. arboreum* can be present or co-dominant. The canopy merges into a tier of *Olearia semidentata* and young dracophyllum; *Coprosma propinqua* var. *martinii* and *Myrsine coxii* are often also present. *Coprosma chathamica, Corokia, Pseudopanax* and *Dicksonia squarrosa* were recorded only as seedlings and stunted plants. *Sporadanthus* is always present, usually as discrete clumps with stems 1–2 m tall, but can be nearly continuous. *Gleichenia dicarpa* and *Blechnum ?procerum* provide up to 10% cover. Sphagnum is continuous between *Sporadanthus* clumps on wet areas, but on drier ground *Dicranoloma robustum* and *Ptychomnion aciculare* largely replace it. Other species include *Myriophyllum propinquum, Nertera depressa* and *Drosera binata*.

Mature stands appear self-perpetuating, having young and old dracophyllum plants, but are vulnerable to fire. Fire-regenerated stands are typically dense, less than

Fig. 13.7. *Dracophyllum scoparium* shrub-heath with *Sporadanthus* beneath. Southern tableland, Rekohu.

4 m tall and almost exclusively formed of *D. scoparium* with continuous *Sporadanthus* understories. There are large areas of dense *Sporadanthus* with regenerating dracophyllum; with more frequent disturbance even these species can be lost, and the vegetation reduced to fernland.

Fernland

Bracken (*Pteridium esculentum*) fernlands are extensive on thin or dry peats and all but the most fertile mineral soils, having replaced tree-heath and bush. Dense stands over 1 m tall fill gullies avoided by stock, but shorter bracken providing 30–60% cover is more usual. *Isolepis nodosa* and *Cyathodes robusta* are the commonest tall plants among bracken, the latter being an endemic shrub related to *C. juniperina* but with wider leaves, that re-sprouts from burnt bases and forms open thickets, mainly on sandy ridges.

Gleichenia dicarpa* and *Baumea tenax* dominate on the wet peat of depressions and domes, where the vegetation was once shrub-heath. Bracken is nearly always present, and there is probably succession to bracken as peats become compacted and mineralised under pastoral use. *Blechnum 'capense'* can survive the initial burning of woody vegetation, but usually disappears from grazed land except for *B. minus*, which persists in wet areas where its rhizomes are less likely to be trampled or burnt.

The ground tier of fernlands on wet peat includes sphagnum in the wettest areas (but decreasing where trampled), *Campylopus ?introflexus, Lycopodium ramulosum, L.*

varium, Lindsaea linearis, Lepidosperma australe, Schoenus pauciflorus, Luzula sp. and *Libertia peregrinans*. In drier or less peaty, bracken-dominated fernland these are replaced by grassland species. *Cladia retipora* can be common in dry openings.

Gorse scrub

Both main islands have patches of gorse (*Ulex europaeus*) that could spread.

Grassland

Pastures now cover much of the northern parts of Rekohu and Pitt. Most have been induced through stocking and oversowing, and consist of low-producing adventive and native grasses, small, mainly native forbs, and mosses. Characteristic species include danthonias (*Rytidosperma* spp.), *Microlaena stipoides*, Yorkshire fog (*Holcus lanatus*), sweet vernal (*Anthoxanthum odoratum*), *Helichrysum filicaule, Centella uniflora, Dichondra repens, Lobelia anceps, Gnaphalium ?audax, Cerastium* sp., *Hypochoeris radicata* and *Plantago lanceolata*, with ryegrass (*Lolium perenne*), cocksfoot (*Dactylis glomerata*) and white clover (*Trifolium repens*) on more fertile or drilled areas. *Poa pratensis* prevails where stock concentrate, for instance, around bush remnants. On Pitt Island silver tussock (*Poa cita*) has been introduced to shelter sheep.

Wetlands

Sphagnum fens occur in headwater basins on the southern tableland. The largest area is mainly flat, wet, quaking moss without vascular plants other than *Isolepis inundata*, occasional *Drosera binata* and *Carex chathamica* tussocks where the surface rises a few millimetres above the water table. It also includes places where sphagnum is overwhelming mature tree-heath, and other places where sphagnum has formed mounds that are being colonised by *Sporadanthus, Dracophyllum scoparium* and *D. arboreum*.

Peat around the larger lakes is mostly eroded into scarps and gullies, giving better drainage. There are usually dense fringes of *Phormium tenax*, accompanied by *Carex ternaria* and the pedestalled *C. sectoides*, which occurs also on Antipodes Island. *Leptocarpus similis* colonies occur at some lakes, and at a few there are tussocks of the endemic *Cortaderia turbaria*. *Phormium* and carices also grow in flushes and swamps where there is water movement, together with *Eleocharis acuta* and *Blechnum 'capense'*. In part, these communities have replaced swamp bush of *Olearia traversii*, tarahinau and *Coprosma chathamica*. They in turn are degraded by trampling and grazing to *Carex sectoides* among rank grass (mainly *Holcus* and *Agrostis capillaris*), *Eleocharis acuta*, native and introduced rushes, *Potentilla anserinoides, Acaena anserinifolia, Blechnum 'capense'*, thistles, blackberry, and other weedy plants.

In places, especially in the north of Rekohu, trampling has reduced fern- or grassland to isolated pedestals among puddled peat. *Marchantia* seems the first coloniser of the bare surfaces, but eventually turf develops, characteristic species being *Myriophyllum pedunculatum, Isolepis habra, Selliera radicans* and *Centella*. On drier ground these are joined or replaced by plants such as *Pratia arenaria, Holcus* and sweet vernal. *Myriophyllum* also forms mats on shelves exposed during low levels of peatland lakes.

Near the sea, there are close-cropped salt meadows. On sand accumulated over schist outcrops near Ohira Bay, *Selliera* and *Samolus* each give 20% cover, other species being *Dichondra brevifolia*, *Rumex ?neglectus*, *Disphyma*, *Cerastium* sp., *Epilobium komarovianum*, *Apium prostratum*, *Lilaeopsis*, *Oreomyrrhis colensoi*, *Centella*, *Pratia*, *Hypochoeris radicata*, *Pseudognaphalium luteoalbum*, *Leptinella potentillina*, *L. squalida*, *Triglochin striatum* and *?Aira*. Under more saline conditions, *Sarcocornia* dominates in association with *Selliera* and *Puccinellia chathamica*. The shore of the almost fresh Taia Lagoon has turf of *Limosella lineata* and *Chenopodium glaucum* between clumps of *Leptocarpus* and *Juncus pallidus*; other species include *Lilaeopsis*, *Eleocharis acuta*, *Juncus articulatus*, *Rorippa palustris* and *Cotula coronopifolia*.

Sand dunes and cliffs

Distinctive species

Open coastal habitats in the Chatham Islands support four of the most notable plants of the New Zealand flora. *Myosotidium hortensia* and *Embergeria grandifolia* represent monotypic genera; *Aciphylla dieffenbachii* and *Leptinella featherstonii* stand apart in their genera. All are large, palatable herbs. *Myosotidium*, the Chatham Island forget-me-not, has glossy, ribbed, cordate leaves with blades up to 30 cm long that rise from shallowly buried rhizomes 5–8 cm in diameter. These leaves die back in early winter, but by then the next season's leaves are appearing. The bright-blue flowers, each 12–15 mm across, are massed in broad cymes and appear in late spring. The winged nuts are 15 mm in diameter. *Myosotidium* grows best in loose sand with organic mulch such as seaweed, a recipe that also suits its cultivation.

In *Embergeria*, a giant milk-thistle, rosettes of leaves up to 1 m long rise from stout rhizomes, and scapes up to 2 m tall bear capitula 2.5–4 cm across. Its main habitat is at the foot of dunes on sheltered beaches. *Aciphylla dieffenbachii*, unlike other large aciphyllas, has broad inflorescences and its divergent leaf-segments are flaccid and blunt. On Rehoku and Pitt Island it is confined to inaccessible cliffs, but on Mangere Island it has responded to removal of stock; there are several hundred plants on rocky shores. *Leptinella featherstonii* differs from the rest of its genus in being procumbent, woody and up to 1.2 m tall. It grows on exposed headlands where stock do not have access. The fleshy groundsel *Senecio radiolatus* is also vulnerable to grazing. In contrast, the robust, leafy, summer-green nettle *Urtica australis*, which extends to the islands of Foveaux Strait and Campbell province, is little damaged by stock.

The inland small tree *Hebe barkeri* grades into shrubby forms on coastal rock. More-or-less erect plants up to 1 m tall, with leaves 2–3 cm long, are referred to *H. dieffenbachii*, whereas prostrate plants with subdistichous leaves up to 12 mm long that grow in very exposed, spray-swept places are referred to *H. chathamica*. Like *H. barkeri*, they have purplish flowers that fade to white.

On dunes the largest New Zealand geranium, *G. traversii*, forms compact silver-leaved rosettes that spread by stolons as well as seed, and bears white or pink flowers up to 25 mm across. *Disphyma papillatum* instead of the mainland *D. australe* trails down steep slopes. The chenopod *Theleophyton billardieri* is more common on northern beaches of Rekohu than anywhere else in New Zealand. The tussocky *Festuca coxii* and the trailing *Poa chathamica* grow on coastal cliffs, the latter being also common in grassland and fernland away from the coast.

Coastal communities
Active and stabilised dunes extend around the northern half of Rekohu, and there are also dunes on Pitt Island. Marram (*Ammophila arenaria*) and *Hypochoeris radicata* are abundant everywhere, but native plants remain prominent, especially *Isolepis nodosa*, *Cyathodes parviflora*, *Pimelea arenaria*, *Coprosma acerosa*, *Acaena novae-zelandiae* and *Geranium traversii*. Bracken invades older dunes. *Myosotidium*, *Embergeria*, *Desmoschoenus* and *Euphorbia glauca* would have once been abundant on foredunes and hollows. They are now exceedingly rare in these habitats, but are all thriving in one locality that a landholder has fenced in the north-west of Rekohu, despite abundant marram.

The endemic coastal plants still thrive on small islands that lie up to 50 km from the two main islands. G.C. Kelly (unpublished) distinguishes four communities: *Myosotidium*, *Embergeria* and *Urtica australis; Leptinella featherstonii*, *Carex trifida* and *Senecio radiolatus; Hebe* spp., *Poa chathamica* and *Festuca coxii*; and *Selliera*, *Samolus*, *Disphyma* and *Leptinella potentillina*. Remnants on Pitt and Rekohu indicate that these are respectively centred on sandy or stony shores; sheltered clefts and seepages; dry ledges and fissures; and spray-drenched debris and soil pockets.

On precipitous slopes and ledges at Cannister Cove on Pitt Island, there is the altitudinal sequence *Sarcocornia < Disphyma < Poa chathamica* and some *Festuca coxii < Phormium tenax <* degenerate patches of *Olearia chathamica* or, on more sheltered sites, *O. traversii*. *Myosotidium* grows on loose debris and *Carex trifida* on seepages. Other coastal species here include *Blechnum durum*, *Geranium traversii*, *Tetragonia trigyna*, *Chenopodium glaucum*, *Apium prostratum*, *Lobelia anceps*, *Selliera*, *Samolus*, *Senecio lautus*, *Sonchus oleraceus*, *Hebe chathamica*, *H. dieffenbachii* and *Isolepis nodosa*. Other localities have similar assemblages, although on headlands and exposed ridges the two hebes are more prominent, alternating with species of salt turf such as *Disphyma* and *Sarcocornia*. Further species noted on headlands are *Asplenium chathamense*, *A. obtusatum*, *Blechnum banksii*, *Crassula moschata*, *Chenopodium glaucum* and *Puccinellia chathamica*.

Auckland Islands

(based on Cockayne 1909; Moar 1958; Taylor 1971; Rudge & Campbell 1977; Lee et al. 1983a; P.N. Johnson, unpublished; C.D. Meurk, personal communication).

Environment and history

The Auckland Islands, the largest group in Campbell province, are eroded remnants of two Pliocene volcanoes on a granite basement. The southern crater is penetrated by drowned valleys, and the northern is reduced almost to its eastern rim. Quaternary glaciers have left subdued cirques, widened valleys, and fiords on the eastern coasts. Cool, equable temperatures, westerly winds, low sunshine and high humidity prevail, and rain is frequent but seldom torrential. Below the alpine belt, all but the steepest terrain is blanketed in peat.

Like other far-southern groups, the Auckland Islands support large colonies of penguins and other pelagic birds and, until the arrival of people, immense concentrations of seals, including sea lions and sea elephants. These mammals and birds damage or kill vegetation around their colonies, especially that growing on peat, through trampling, sliding, wallowing, burrowing, building nest platforms, and excreting. The few plants adapted to enriched ground can form the first stage of revegetation when animals move on.

Discovery in 1806 was followed by hunting of seals. Shore gangs burnt vegetation, introduced rodents, and hunted birds for food. The sealing industry had practically collapsed by 1840, but with complete protection since 1946, the colonies are gradually recovering. A new threat to seals and sea-birds may be modern fishing fleets competing for food.

The islands were settled by Maori and Europeans in the 1840s, and their cattle, goats, pigs and rabbits became feral when the settlers left in 1856. Today, Auckland Island is extensively influenced by pigs and goats, Rose and Enderby islands have had both cattle and rabbits, but only Enderby still has cattle. Unmodified vegetation is therefore restricted to Adams Island and some smaller islands.

Far-southern plants present on the Auckland Islands

Campbell province has 35 endemics and some more-or-less circum-antarctic species that do not reach mainland New Zealand. The Auckland Islands flora includes most of these far-southern species, including all the provincial endemics excepting a few weakly differentiated taxa confined to other groups.

Dracophyllum longifolium var. *cockayneanum* extends from the shore to the subalpine belt on the Auckland and Campbell groups, differing from mainland forms in its larger, more pubescent leaves. In Campbell Province, *Olearia lyallii* is confined to several points around Ross harbour. Possibly it arrived there from a Rakiura coast as recently as the ninteenth century, either through human agency, or spontaneously as much of the woody flora has done during the Holocene (Godley 1965). *Hebe benthamii* belongs with mainland alpine species such as *H. haastii*, but has blue flowers and is erect where sheltered. It grows on stony ground and shallow peat in the Auckland and Campbell groups.

Chionochloa antarctica is the climax tussock grass on peats that are least influenced by salt spray or guano. *Poa litorosa* is abundant through most of Campbell Province, but has established at only one point on Macquarie Island. On coastal slopes large tussocks grow on peaty pedestals up to 2 m tall, with the ground between often worn by passage of seals and penguins. *P. litorosa* also dominates inland and higher areas where *Chionochloa* is absent (Antipodes Island) or has been depleted (Campbell Island). Tussocks are much smaller than near the sea, and apt to separate into rings of daughter tussocks as peat growth forces tillers upwards and outwards. With $2n = c.\ 266$, the highest number recorded in grasses, *P. litorosa* is the end-point of the polyploid sequence *P. anceps* ($2n = 28$ as in most native poas), *P. cita* (84), *P. cockayneana* and *P. chathamica* (112).

P. foliosa is a robust tussock with broad, flat leaves, which grows throughout Campbell province and on small islands in Rakiura province. On Macquarie Island tussocks reach maturity in ten years, and live a further 40 years. It is vulnerable to

Fig. 13.8. *Poa ramosissima* at the edge of a mollymawk colony, *Stilbocarpa* and *Polystichum* on the bank behind; Campbell Island.

grazing. *P. ramosissima* forms bright green swards of trailing culms on the cliffs of the Auckland and Campbell groups, especially around bird colonies (Fig. 13.8). *Hierochloe redolens* and most of the smaller grasses grow also on the New Zealand mainland, but *Poa aucklandica* subsp. *aucklandica*, *Hierochloe brunonis* and *Lachnagrostis leptostachys* are provincial endemics although closely related to mainland taxa.

The circum-antarctic rush *Rostkovia magellanica* is abundant in subalpine and alpine bogs and flushes on the Auckland and Campbell groups; it also occurs locally in Stewart Island, Central Otago and Fiordland. *Juncus scheuchzerioides* is circum-antarctic and common on all the islands of Campbell province. It grows in shallow water and as turf on eroded peat. Only two uncinias are present, *U. hookeri* being confined to Campbell province and *U. aucklandica* extending to the South Island. Three large carices, the tussocky *Carex trifida* and the rhizomatous *C. appressa* and *C. ternaria*, are centred on the province but extend to southern parts of the New Zealand mainland and the Chatham Islands.

Large-leaved, endemic forbs are the most striking botanical feature of the far-southern islands (Fig. 13.9) (p. 38) and are generally abundant except among dense woody vegetation or where depleted by feral mammals. *Pleurophyllum*, which seems related to the macrocephalous olearias, has large leaf-rosettes arising at ground level from a thick, more-or-less vertical root-stock. The capitula are borne in racemes on tall scapes. All three species occur on the Auckland and Campbell groups; *P. hookeri* extends to Macquarie Island and *P. criniferum* to the Antipodes.

In *P. hookeri*, which grows mainly on upland herbfields, the capitula lack rays. The

Fig. 13.9. Tussock-herbfield with *Poa litorosa*, *Pleurophyllum speciosum*, *Anisotome latifolia* and *Bulbinella*; 200 m, Campbell Island.

leaves form rather lax tufts, and although older leaves die in autumn, the plant retains young leaves that expand during the following summer. Long, thick, elastic adventitious roots extend outwards at steep angles, being most abundant in the top 30 cm. These are markedly aerenchymatous, in keeping with the wet, poorly aerated peat. Macquarie Island plants live some 40 years (Fig. 13.10) but axillary buds at depth produce rhizomes 5–28 cm long, which can lead to circles of daughter plants replacing the parent. Flowering is markedly periodic (p. 57); heavy seeding is followed by mass germination in the next spring.

In the splendid *P. speciosum*, long light-purple or white rays surround dark purple disc florets. In habitat and phenology it is similar to *P. hookeri*, although rhizomes are not reported and flowering seems less strictly periodic. Its habitats are generally lower, less exposed and more fertile. The ribbed and corrugated leaves bear stiff, water-repellent hairs, and lie flat as overlapping tiles, with a 'greenhouse' space between where leaf temperatures rise as much as 15 °C above air temperature. *P. criniferum*, which grows in fens, has rayless capitula and summer-green leaves with nearly glabrous upper surfaces. The monotypic *Damnamenia vernicosa*, a relative of *Celmisia*, forms clumps of hard, glossy, rosulate, 2–5 cm long leaves in herbfield, fellfield and bog on the Auckland and Campbell groups.

Anisotome latifolia is the largest of its genus, with broad leaf segments and stout scapes up to 1 m tall. Male plants bear purplish flowers. It grows in the Auckland and Campbell groups, often forming almost pure stands on coastal ledges and cliff tops. The smaller *A. antipoda* has narrow leaf segments that spread in three dimensions. On the Auckland and Campbell groups it is usually alpine, whereas on Antipodes Island it is widespread except close to the sea.

The herbaceous araliad *Stilbocarpa* bears bright-green, round leaves up to 60 cm

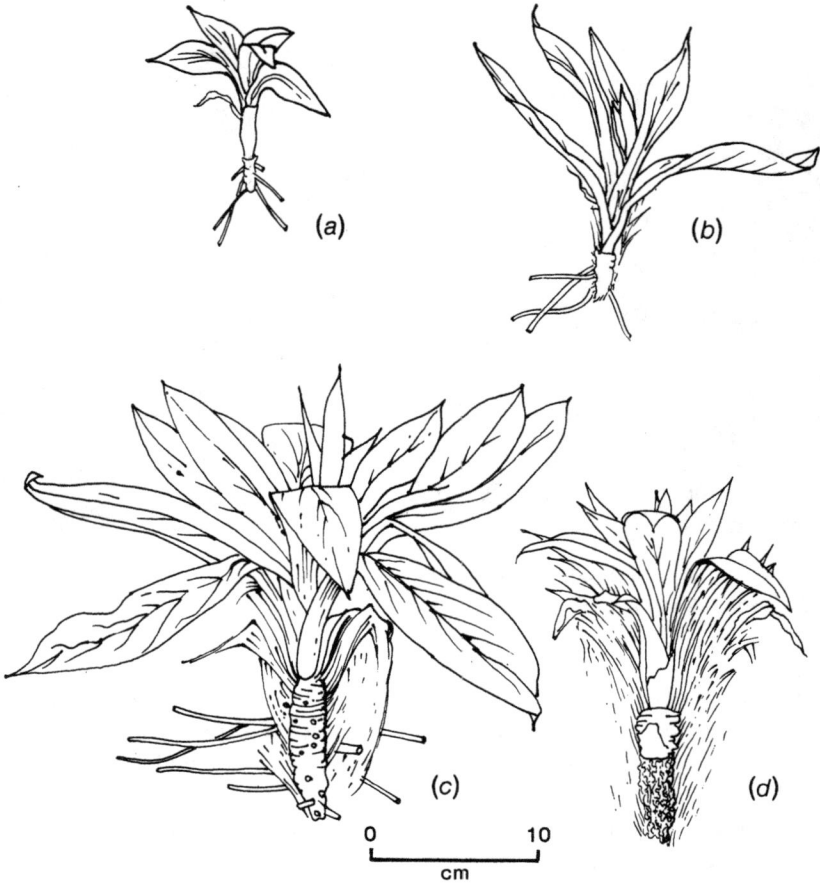

Fig. 13.10. Stages in development of *Pleurophyllum hookeri* at 170 m a.s.l. on Macquarie Island (Jenkin & Ashton 1979): (*a*) pioneer, 7 years; (*b*) building, 10–11 years; (*c*) mature, >23 years; (*d*) degenerate, showing litter sheath and fabric of rhizome.

across on long, stiff petioles, and greenish-yellow flowers in large, globose, compound umbels. Short, densely clumped stems rise from rhizomes over 4 cm thick. *S. polaris* occurs over a wide range of altitude on all the islands of Campbell province; two other species are confined to small islands in Rakiura province. The summer-green, yellow-flowered geophyte *Bulbinella rossii* of the Auckland and Campbell groups is larger than mainland relatives, with 5 cm wide leaves and scapes up to 1 m tall. It grows on uplands and fens, abundantly where grazing has depleted grassland or herbfields, but pigs have depleted it on Auckland Island.

Subantarctic gentians are monocarpic, and widespread on bogs, herbfields and fellfields. On the Auckland Islands they grade from the small, annual *G. concinna* to the larger, perennial *G. cerina*. Gentians endemic to the Antipodes and Campbell islands are probably segregates from the Aucklands complex. Petal colour in these gentians ranges from the white that prevails in mainland congeners to shades of pink and violet (p. 54). This is weakly linked with variations in leaf and stem colour, from green to purplish (Godley 1982).

Epilobium confertifolium is endemic to the Auckland and Campbell groups, but differs little from the Australasian alpine species *E. tasmanicum* except in occupying a wide range of altitude and habitat, and in that many plants have rose-purple instead of white flowers. Several other willow-herbs are shared with the mainland. *Myosotis capitata* is endemic to the same groups, where it grows mainly in fellfield; unlike mainland species its flowers are deep blue. *Ranunculus pinguis*, like related mainland plants in Section *Epirotes*, grows mainly at high altitudes, whereas *R. subscaposus* in Section *Ranunculus* descends to the coast.

Leptinella plumosa is the largest creeping leptinella, forming thick patches near the sea on far-southern islands of New Zealand and the Indian Ocean. *L. lanata* grows in low turf on wind-exposed coastal rocks and cliff brows in the Auckland and Campbell groups. *Plantago triantha* and *P. aucklandica* are endemic to the Auckland Islands, the former growing on coastal rocks and the latter, a robust plant unrelated to other New Zealand plantains, growing in wet fellfield.

The number of adventive plants on far-southern islands is related to size and disturbance (Table 13.1). *Poa annua* and *Stellaria media* have reached all the islands.

Vegetation gradients

The Auckland group, with the largest area and flora and lying at a relatively low latitude, has the most diverse vegetation in Campbell province. Although other island groups lack some ecologically important species, their vegetation shares, with the Auckland Islands, features that reflect the exigencies of a cool, highly oceanic environment.

Fig. 13.11 demonstrates two principal gradients that relate to altitude (temperature), slope, shelter, wind, distance from the sea (soil Na), other soil attributes (K, Mg, depth, pH, moisture) and grazing. A third gradient reflects species diversity, which is greatest in low, open, disturbed vegetation with many adventive species, and decreases towards littoral and high-altitude environments and in tall, closed, undisturbed vegetation.

Forest and shrubland

Forest along sheltered shores and valleys is the southernmost in New Zealand. Southern rata (*Metrosideros umbellata*) dominates, with main trunks leaning or prostrate because the peat is unable to support them (Fig. 13.12). The wind-compacted canopy reaches 6–9 m, and includes *Pseudopanax simplex, Myrsine divaricata, Coprosma foetidissima, Dracophyllum longifolium* and, at one locality, *Fuchsia excorticata*. There is a patchy undergrowth of ferns, especially *Histiopteris incisa, Polystichum vestitum, Blechnum* 'mountain' and, near the sea, *B. durum* and *Asplenium obtusatum. Phymatosorus diversifolius* ascends trunks. Local *Cyathea smithii* are probably the most southern tree ferns. The floor is generally bare, but there are bryophyte cushions and seedlings in places least disturbed by marine animals.

With increased altitude or exposure, forest grades into scrub, much of which is also dominated by rata. On exposed sites, *Cassinia 'vauvilliersii'* forms 1.5 m tall scrub with

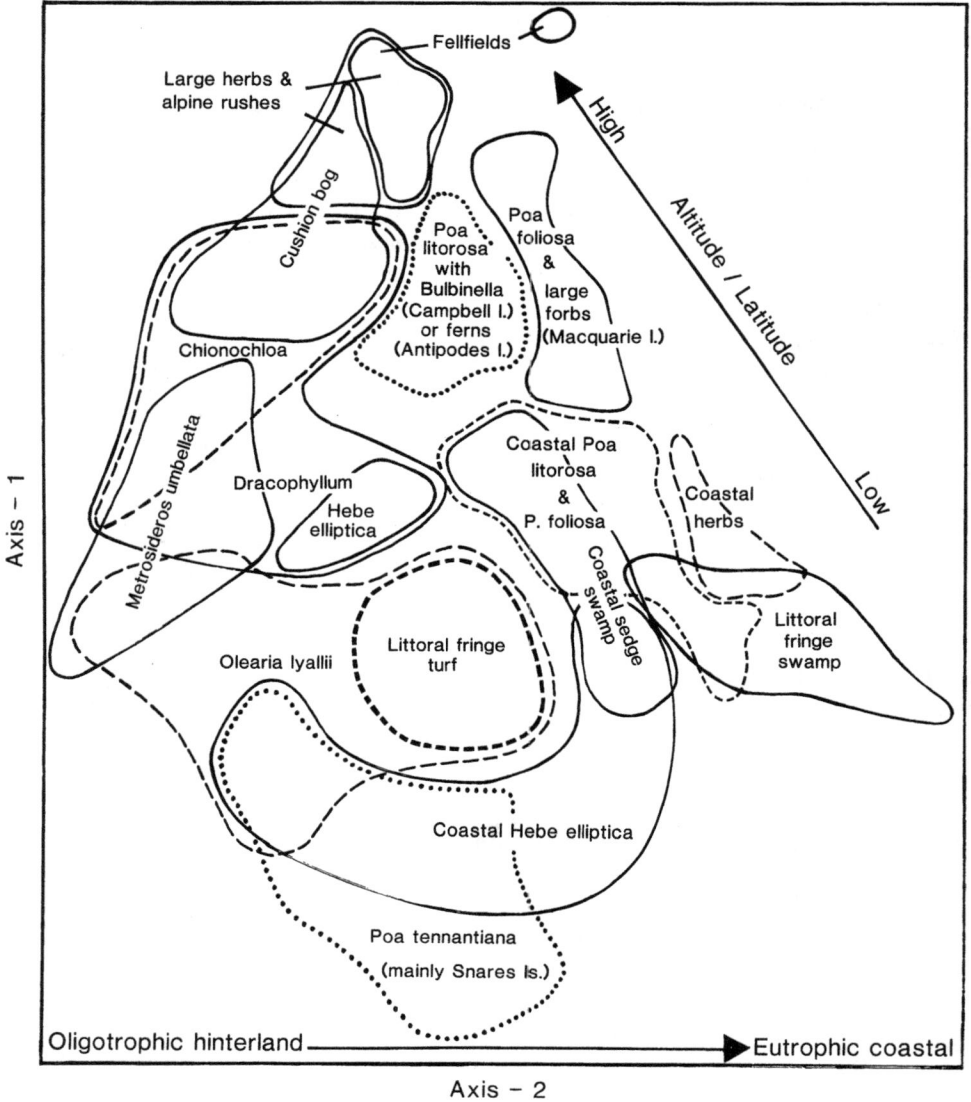

Fig. 13.11. Ordination and clustering of major communities on the far-southern and Snares islands. Overlaps indicate mixtures of dominant elements (from Meurk & Foggo 1988).

Dracophyllum, Myrsine and stunted rata. On valley slopes, *Myrsine* forms taller scrub with *Dracophyllum, Coprosma foetidissima, Pseudopanax* and *Cassinia*, with dense *Blechnum* 'mountain' and *Polystichum* beneath; *Coprosma cuneata* and *Phymatosorus* are also present.

In patches of low, windshorn scrub of coprosmas and *Myrsine* above 300 m, the canopy supports a bryophyte succession: first, *Frullania* and *Metzgeria* creep on twigs, then *Lepidolaena* strands form a network, and finally, cushions of *Hypnum*

Fig. 13.12. Southern rata on Enderby Island, Auckland group; understorey depleted by feral cattle (photo F.J.F. Fisher).

cupressiforme, Acrocladium chlamydophyllum and *Chandonanthus squarrosus* develop, which can be compact enough to support seedlings. 'Caverns' beneath the canopy support delicate plants where not too dark.

Around Ross Harbour *Olearia lyallii*, being relatively light-demanding and favouring soils high in sodium and potassium, is spreading slowly into a zone seaward of the rata forest, displacing scrub of *Hebe elliptica, Myrsine, Dracophyllum* and some stunted rata, or *Poa litorosa – Stilbocarpa* herbfield. It is no threat to rata forest, which occupies soils higher in magnesium and calcium. With rata, *O. lyallii* has invaded the abandoned settlement clearing at Erebus Cove, but although it reaches 8–13 m tall, 1.1 m in diameter and an age of 140 years, rata outlives it.

Grassland, herbfield and fen of the lower altitudes

Where feral mammals are absent, as on Adams Island, lush herbaceous vegetation occupies exposed coastal terrain. *Poa litorosa* prevails on drier ground, with abundant *Polystichum vestitum, Pleurophyllum speciosum, Bulbinella*, and some *Chionochloa* and *Poa foliosa*. This tussock grassland is interspersed with fens supporting *Carex trifida, C. appressa, Hierochloe redolens* and *Pleurophyllum criniferum*, and stream channels supporting *Polystichum vestitum. Poa foliosa* with *Stilbocarpa* and *Anisotome latifolia* forms a seaward fringe, above a band of *Leptinella plumosa*.

Smaller plants, growing mainly in gaps, include *Gentiana concinna, Leptinella plumosa, Epilobium confertifolium, E. pedunculare, Acaena minor, Nertera depressa, Coprosma perpusilla, Helichrysum bellidioides, Isolepis aucklandica* and, on swampy ground, *Ranunculus subscaposus, Montia fontana* and *Juncus scheuchzerioides*.

Near Ross Harbour, pigs and goats have destroyed the large leafy herbs and *Poa foliosa*, so that *P. litorosa* and sedges dominate alone. Also in this area, clearing and grazing have given rise to turfy swards in which *Isolepis aucklandica* provides 20–40% of the cover. Other plants in the swards are *Marchantia berteroana*, leafy liverworts, mosses, *Blechnum penna-marina, Colobanthus apetalus, Epilobium confertifolium, Nertera depressa, Leptinella lanata, Helichrysum bellidioides, Lagenifera petiolata, Juncus scheuchzerioides, Deschampsia chapmanii, Poa antipoda, P. breviglumis* and the adventive *Cerastium fontanum, Stellaria media, Sagina procumbens, Agrostis capillaris* and *Poa annua. Rumex neglectus* dominates on sandy or peaty shores; associates include *Ranunculus subscaposus, Crassula moschata* and *Isolepis cernua*. The swards are most extensive on Enderby Island and seasonally become *Bulbinella* herbfields, but on Rose Island they have been decreasing since removal of cattle because rabbits cannot prevent the return of tall vegetation.

Chionochloa grassland, low scrub and bogs

Chionochloa supersedes the scrub belt above 200 m, via a usually gradual transition with stunted rata and shrubs among the tussocks. On exposed western slopes, dense woody vegetation is absent and *Chionochloa* grassland descends to meet coastal *Poa litorosa* grassland and herbfield. In places, fire may have extended grassland at the expense of woody plants. *Chionochloa* is accompanied by *Pleurophyllum speciosum* and *Bulbinella*, and often there are *Coprosma cuneata, C. ciliata, C. foetidissima, Dracophyllum, Myrsine, Cassinia* and *Hebe odora*. Mossy openings support *Breutelia pendula, B. elongata* and *Acrocladium*, and *Marchantia berteroana* colonises ground bared by pigs and albatrosses.

Bogs are dominated by *Oreobolus pectinatus*, other mat and cushion plants being *Phyllachne clavigera* (doubtfully distinct from *P. colensoi*), *Coprosma perpusilla, Astelia linearis, A. subulata, Centrolepis ciliata* and *Dicranoloma robustum* s.l. Sphagnum is rare. There are also *Isolepis aucklandica*, orchids, *Cyathodes empetrifolia, Drosera stenopetala, Lycopodium ramulosum, Schizaea fistulosa*, scattered tufts of *Chionochloa, Poa litorosa* and *Bulbinella*, and stunted plants of *Dracophyllum, Cassinia, Myrsine* and rata. Rata seedlings may establish on and suppress the *Oreobolus* cushions so that hollows develop. Eventually *Oreobolus* reinvades, to form mounds with projecting rata shoots and later-established shrubs.

Wherever drainage is better, as on steeper slopes, on rises, and along channels, cushion bog grades to dense scrub. On exposed sites, bogs often form 'lanes' between strips of scrub, that tend to align with the prevailing wind, either following contours or crossing over ridges. The lane shown in Fig. 13.13 occupies a shallow trough 30 m wide, between scrub-covered borders 10–15 m across. The pattern seems self-reinforcing, in that wind is channelled along the lanes, and the shrubs provide mutual protection and probably accumulate litter and humus faster, thereby providing more 'freeboard' above the water-table for their roots. Other lanes occupy crests of interfluves, separated by woody vegetation in depressions.

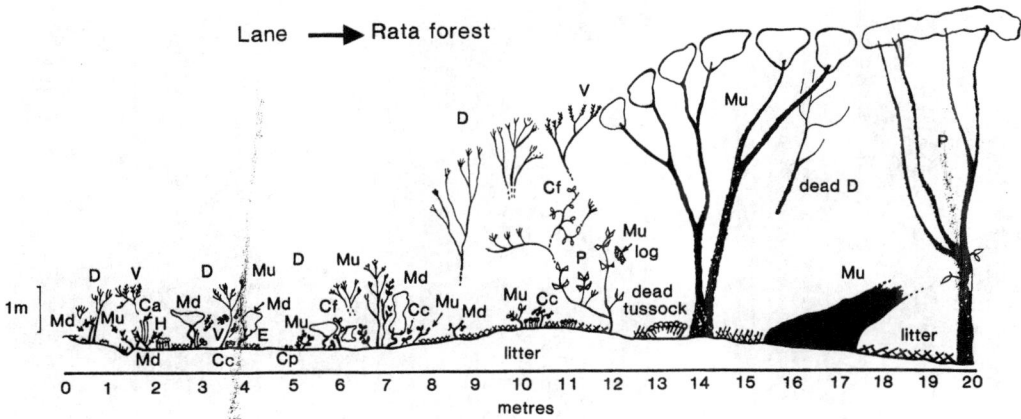

Fig. 13.13. Profile through a transition from an open lane to tall scrub and rata forest near Deas Head, Auckland Island (Rudge & Campbell 1977). Ca, *Chionochloa antarctica*; Cc, *Coprosma cuneata*; Cf, *Coprosma foetidissima*; Cp, *Coprosma perpusilla*; D, *Dracophyllum longifolium*; E, *Cyathodes empetrifolia*; H, *Helichrysum bellidioides*; Md, *Myrsine divaricata*; Mu, *Metrosideros umbellata*; P, *Pseudopanax simplex*; V. *Cassinia 'vauvilliersii'*.

Vegetation of the hill tops

Grassland ceases about 500 m, and peat becomes thin and patchy. A vegetation mosaic is determined by exposure, drainage, and peaty versus stony surfaces. The far-southern endemics (marked†) are concentrated on the hill tops. In wet herbfield, open or dense *Pleurophyllum hookeri*† stands above turf- or cushion-forming plants (Fig. 13.14). Common species include *Hymenophyllum multifidum*, *Hebe benthamii*†, *Bulbinella*†, *Carpha alpina*, *Oreobolus pectinatus*, *Centrolepis ciliata* and, on stony ground, *Ranunculus pinguis*†, *Cardamine subcarnosa*†, *Plantago aucklandica*†, *Damnamenia*† and *Myosotis capitata*†. On flushed ground herbfield merges into pure *Marsippospermum gracile*. There is also *Rostkovia* flush grading to *Centrolepis pallida* bog.

Further species include *Anisotome antipoda*†, *Stilbocarpa*†, *Gentiana cerina*†, *Phyllachne clavigera*, *Abrotanella spathulata*†, *Helichrysum bellidioides*, *Pleurophyllum speciosum*†, *Agrostis subulata*, *Luzula crinita* var. *crinita*†, and, in crevices and on debris, *Polystichum cystostegia*, *Grammitis poeppigiana*, *Colobanthus hookeri*†, *Cardamine depressa*, *Geum parviflorum* var. *albiflorum*† and *Schizeilema reniforme*. *Rhacocarpus purpurascens* is the commonest moss on rocks, ?*Campylopus* and hepatics form cushions in herbfield, and species of *Lepidolaena* and *Metzgeria* thrive 8–10 cm below the surface among broken rock.

Coastal cliffs

Rocks by the sea support green cushions of *Colobanthus muscoides*. *Crassula moschata* and *Isolepis aucklandica* grow on these cushions and in crevices. There are also stunted plants of *Blechnum durum*, *Leptinella lanata*, *L. plumosa* and *Plantago triantha*. All but the most vertical cliffs carry grassland of *Poa litorosa* and *P. foliosa*, with *Urtica australis*, both leptinellas, *Isolepis praetextata*, *I. cernua*, *Asplenium obtusatum*,

Fig. 13.14. Upland herbfield on Adams Island, Auckland group. The main plants are *Pleurophyllum hookeri*, *Damnamenia* and scattered *Bulbinella rossii* (photo P.N. Johnson).

Blechnum durum and, where secure from grazing animals, *Anisotome latifolia* and *Stilbocarpa*. Dripping cliffs have sheets of *Poa ramosissima* and abundant *Montia fontana* and *Callitriche antarctica*.

Campbell Island

Environment, history and flora

Campbell Island is the eastern part of a Pliocene volcano, in which a gabbro basement, Tertiary limestone and other sediments are exposed above sea level. High western cliffs represent the eastern rim of the caldera and glacially-sculptured valleys descend the eastern slopes, some ending in fiords. There is an almost complete blanket of peat to 350 m a.s.l. Depth on steeper slopes is limited to 1–2 m by slipping and gullying during storms, but on gentle rises in the valley floors peat in raised bogs exceeds 5 m; a basal radio-carbon age of 13 350 years has been obtained. The water table during dry weather varies from 5 cm below the surface in bogs and < 10 cm in fen, to 1–2 m under grassland and scrub (Meurk 1980). Table 13.2 indicates fertility levels in the peats.

The island experienced the usual ruthless hunting of seals. Sheep and cattle were farmed from 1895 to 1931, thereafter becoming feral. In 1970, they were fenced into

Table 13.2. *Analysis and bioassays on four peat soils from Campbell Island*

| soil | alt. (m) | pH | cations (mg/l soil vol.) | | | | bioassay | |
			Ca	K	Na	Mg	I	II
A	23	4.1	107	36	106	125	0.16	0.03
B	200	4.3	164	96	208	208	0.77	0.18
C	69	4.6	394	86	308	265	1.09	0.33
D	1	4.7	392	57	669	317	1.36	0.58

Notes:
A   Peat dome under cushion bog with *Chionochloa* and *Dracophyllum*.
B   Blanket peat with induced *Poa litorosa* grassland.
C   Thick peat under 4 m tall *Dracophyllum*.
D   Thin peat under coastal *Poa litorosa* grassland.
I, II Mean mass (g) per plant in bioassay of *Avena sativa* and *Agrostis capillaris* respectively.
Source: Foggo & Meurk 1983

the southern half of the island, and since then depleted vegetation in the northern half has been recovering steadily (Meurk 1982).

Campbell Island holds almost the full complement of far-southern endemics other than *Plantago* spp. and those represented by vicariant taxa. Among the latter, only the annual *Gentiana antarctica* has the rank of species. *Myosotis antarctica* may be endemic to the island (although reported from Patagonia) but scarcely differs from *M. pygmaea* except in the blue or white polymorphy of its flowers. However, many mainland species that reach the Auckland group are lacking, especially in the forest element, so that woody vegetation is less complex.

Tall heath and scrub

The tallest (up to 5 m), densest heath, which tends to be pure *Dracophyllum longifolium*, occupies sheltered coastal slopes and narrow valleys. There is a subcanopy of *Myrsine divaricata* and a sparse shrub tier of *Coprosma ciliata* and *C. cuneata*; these filiramulate species can dominate small openings. The herb tier is usually sparse and consists of *Polystichum vestitum*, with *Blechnum* 'mountain' on wet ground and *Histiopteris incisa* in gaps. Beneath, dracophyllum leaf litter grades into blanket peat.

Younger, shorter and often more open stands of *D. longifolium* and its hybrids with *D. scoparium* cover most of the lower slopes of the island, up to 180 m. They began to establish after sheep farming commenced (Zotov 1965) and are still growing denser and taller. It is uncertain whether they represent regeneration of original *D. longifolium* stands that were burnt (fires occurred at least as early as 1840) or whether they have invaded grassland and herbfield damaged by burning and grazing.

Small patches of *Coprosma cuneata* and *C. ciliata* are dispersed among grassland over the same range of altitude. Being reduced by exposure to as low as 30–50 cm, they merge imperceptibly into surrounding vegetation, occupying slopes too exposed for *Dracophyllum* or too steep for closed grassland, or forming a seral stage on slips. Bryophytes can be dense in the canopy.

Fig. 13.15. *Bulbinella* in sheep-grazed turf near Penguin Bay, Campbell Island.

Herbaceous vegetation

The original grassland, herbfield and bogs were essentially the same as they are today on the Auckland Islands. In 1984, on the part of the islands where sheep remained, *Poa litorosa* coastal grassland and sedge- and fern-fens had lost their palatable, large herbs. Tall herbfields, *P. ramosissima* swards around accessible bird colonies, and slopes formerly carrying *Chionochloa* grasslands but not yet invaded by dracophyllum had been replaced by close-cropped turf, and *Chionochloa* was severely depleted on bogs. *Polystichum vestitum, Blechnum* 'mountain' and low coprosma thickets remained on rough ground in gullies. The predominant tall plants had become *Bulbinella* and *Poa litorosa*, with even the latter being absent from areas most favoured by animals (Fig. 13.15).

The sheep turf contains all the species recorded from Auckland Island turf, but being much more extensive and diverse in origin, greater differentiation is apparent. *Poa annua* is especially abundant on fertile sites such as around bird colonies, whereas wet upland turf contains sphagnum, *Phyllachne clavigera* and *Coprosma perpusilla*. Species further to the Aucklands list include *Hymenophyllum multifidum, Geranium microphyllum, Stellaria decipiens, Epilobium brunnescens, E. pedunculare, Uncinia hookeri, Luzula crinita* and *Agrostis magellanica*. Common bryophytes are *Thuidium furfurosum, Ptychomnion aciculare, Hypnum cupressiforme* and a *Dicranoloma*. In turf colonising peat slides, the main species are *Breutelia pendula, Epilobium pedunculare, Isolepis habra, Agrostis magellanica, Lachnagrostis leptostachys, Poa breviglumis* and, on

Fig. 13.16. Bogs with *Chionochloa* on interfluves, *Dracophyllum* heath in gullies and on steeper slopes. Campbell Island.

wet ground, *Callitriche antarctica, Juncus scheuchzeroides* and *Deschampsia chapmanii* (C.D. Meurk, unpublished).

The floor of Northeast valley is dissected by deep gullies choked with scrub (Fig. 13.16) and seems not to have been penetrated by animals. Broad interfluves support cushion bog with *Cladia retipora, Dicranoloma robustum, Coprosma perpusilla, Damnamenia, Gentiana antarctica, Astelia subulata, Centrolepis ciliata, Oreobolus pectinatus and Isolepis aucklandica. Chionochloa* and *Bulbinella* are abundant, the former rising to dominance on better-drained ground. Nearly all the shrubs are *Dracophyllum scoparium*, which is tallest and densest on the best drained portions; this species is shared with the peat domes of Rekohu. Sphagnum is the main contributor to borders of fast-growing peat between bogs and the fens occupying shallow depressions, which are dominated by *Carex appressa* and *Bulbinella* with some *Poa litorosa* and *Pleurophyllum criniferum*.

The wide, flat Hooker valley, which also seems little modified, runs parallel to the western coast, and shows a graded influence of sea-spray. *Anisotome latifolia – Bulbinella* herbfield above the western cliffs passes inland into dense *Poa litorosa*. Towards the eastern side of the valley, *P. litorosa* and *Polystichum vestitum* occupy hollows, whereas intervening hummocks support *Chionochloa* and low *Dracophyllum scoparium*, presumably being above the influence of flood water bringing nutrients.

At 270–360 m on the broad, gently rolling Faye Ridge cushion bogs form part of a regeneration cycle on areas of peat deflation.

(1) Intact vegetation consists of *Bulbinella, Poa litorosa*, scattered *Pleurophyllum hookeri*, cushions and mats of *Isolepis aucklandica, Coprosma perpusilla* and *Phyllachne clavigera*, and mosses such as *Breutelia elongata*. Sphagnum grows in depressions on the hummocky surface.

(2) The peat surface becomes ruptured and then deflates, presumably through wind, frost and rain, to a level where the water table imposes some stability.

(3) Colonisation then begins. *Centrolepis pallida* can develop continuous close turf, or occupy the higher parts of gentle slopes while the lower parts that receive drainage and water-borne peat support *Juncus scheuchzerioides*, which is replaced by *Isolepis aucklandica* or *Centrolepis* as the surface becomes fixed.

(4) The *Centrolepis* mat becomes colonised by *Phyllachne*, mosses, liverworts, *Bulbinella*, and other plants; uneven upward growth leads to re-establishment of phase (1).

The upper limit of *Chionochloa* and *Poa litorosa* varies from 260 to 370 m according to exposure. The vegetation above this was probably once identical to that on the Auckland Islands, but on the highest point, Mt Honey (569 m), where sheep were still present in 1984, the larger herbs were confined to ledges and fissures. They were only slowly returning to accessible ground on northern tops where stock had been excluded since 1970. Another result of grazing is the presence of browntop (*Agrostis capillaris*) in *Marsippospermum* communities about 350 m a.s.l.

Stony solifluction terraces on the summits of Mt Honey and Fizeau (504 m) indicate vigorous freeze–thaw cycles. Their low, patchy vegetation includes *Stereocaulon*, soil-encrusting lichens, *Bulbinella*, *Luzula crinita*, *Agrostis subulata* and *Epilobium confertifolium*. Further high-altitude plants are *Andreaea*, *Racomitrium crispulum*. *Breutelia elongata*, *Dicronoloma robustum*, *Lycopodium australianum*, *Polystichum cystostegia*, *Hymenophyllum villosum*, *Ranunculus subscaposus*, *Acaena minor*, *Coprosma perpusilla*, *Myosotis antarctica*, *Craspedia uniflora*, *Uncinia hookeri*, *Deschampsia tenella*, *Hierochloe brunonis* and *Poa aucklandica*, the last as a variety endemic to the island.

Offshore islets

Campbell Island is surrounded by islets and stacks with pristine samples of the grassland and herbfield of precipitous coastal slopes, and florulas that vary in size and content according to area, topography and distance from the main island (Table 13.3). On Jacquemart Island, the summit plateau (200 m a.s.l.) and upper ridges support almost continuous cover of *Poa litorosa* tussocks 0.8–1.4 m tall, on deep peat undermined with petrel burrows. The only other tall plants are occasional *Bulbinella* and *Polystichum vestitum*. Minor contributors, chiefly in small canopy openings or on old petrel nest sites, are *Stellaria decipiens*, *Cardamine subcarnosa*, *Epilobium confertifolium*, *Acaena minor*, *Gentiana antarctica*, *Uncinia hookeri*, *Marchantia berteroana*, *Lophocolea* cf. *novae-zelandiae*, *L.* cf. *amplectens*, *Metzgeria furcata*, *Campylopus pallidus* and *Cladonia campbelliana*. On steep flanks of ridges, in gullies and as a continuous skirt around the vegetated part of the island are densely matted stands of *Poa foliosa* to 0.5–0.6 m tall, with *Stilbocarpa* particularly common in the gullies. Around bird nesting sites and on seepage banks *Poa ramosissima* and *Leptinella plumosa* are characteristic.

Table 13.3. *Size of vascular florulas of Campbell Island's offshore islets and stacks*

Area (ha):	7.1	27.4	8.1	18.8	3.6	11.2	1.6	0.5
Elevation (m):	75	220	175	250	150	135	70	50
*Distance (m):	50	1440	450	810	140	810	410	50
No. of spp.:	39	32	23	22	20	15	15	10

Notes:
* Distance from the main island.
Source: C.D. Meurk, unpublished data.

Rocks and ledges stand out from the grassland as buttresses whitened by the lichen *Ochrolechia ?parella, Pertusaria graphica* and *Opegrapha diaphoriza*, accompanied by a pyrenocarpous sp., *?Buellia* sp., black *?Verrucaria* sp. and *Psoroma ?microphyllizans*. Other common cryptogams are *Macromitrium* sp., *Muelleriella angustifolia, Dicrano-weisia antarctica* and *Bryum billardierei* (all interwoven with occasional *Metzgeria furcata, Lophocolea* cf. *amplectens* and a member of the Lejeuneaceae), *Candelariella vitellina* and *Pseudoparmelia pseudosorediosa*. Ledges and crevices support the *Poa* spp. already mentioned, together with *Trisetum spicatum, Luzula crinita, Uncinia hookeri, Ranunculus pinguis, Cardamine depressa*, and rare appressed shrubs of *Coprosma cuneata*. On exposed peaty ledges there are *Crassula moschata, Colobanthus muscoides* and *Cardamine depressa* (Foggo & Meurk 1981).

The summit plateau of Folly Island, although less than 50 m high, supports dense grassland with *Chionochloa* tussocks on low peat pedestals providing 40–60% cover. The understorey is mainly *Bulbinella*, and either *Coprosma cuneata* and *C. ciliata*, or *Pleurophyllum speciosum* and *Anisotome latifolia*. *Polystichum vestitum* increases towards gully bottoms. Dominance passes to *Poa foliosa* and *Anisotome latifolia* in a giant-petrel colony and on relatively sheltered cliff brows which slope up to 60°. On windward brows there is turf of *Leptinella lanata* and *Puccinellia chathamica*. *Poa ramosissima* occupies ledges where birds roost.

Antipodes Islands
(based on Godley 1989, and personal communication)

On the main island, coastal basalt cliffs rise to an undulating plateau wholly blanketed in peat. Most of the small flora is shared with the Auckland and Campbell groups, but *Apium prostratum* grows on mainland New Zealand but on no other far-southern islands, and *Coprosma rugosa* var. *antipoda* is an endemic form of a mainland species. The perennial *Gentiana antipoda* and *Senecio radiolatus* var. *antipodus* are endemic, although the latter is represented on the Chatham Islands by var. *radiolatus*. The two groups also share *Carex sectoides*.

Coastal cliffs and rocks are mostly covered in crustose lichens, but crevices support the vascular plants *Crassula moschata* and *Puccinellia antipoda* (growing nearest the sea), *Blechnum durum* (plentiful in caverns), *Asplenium obtusatum, Colobanthus muscoides, C. apetalus, Lepidium oleraceum, Apium prostratum* and *Leptinella plumosa*.

Poa litorosa tussocks up to 2 m tall clothe gentle coastal slopes. Between, there is

Fig. 13.17. View south to Mt Galloway (366 m), Antipodes Island. *Poa litorosa* and *Polystichum vestitum* in the foreground. On the slopes, dark patches are *Coprosma rugosa* and *Polystichum*, and light patches are lichen-covered slips (photo E.J. Godley, 1969).

luxuriant growth of forbs, especially *Asplenium obtusatum* and *Blechnum durum*; other species include *Colobanthus apetalus, Anisotome antipoda* and the adventive *Sonchus asper*. The peat is burrowed by petrels. *Poa litorosa* also dominates on steep coastal slopes, or shares dominance with *P. foliosa* (on wet ground) or *Polystichum vestitum*, other notable plants being *Blechnum durum, Histiopteris incisa, Urtica australis, Anisotome, Stilbocarpa polaris* and *Coprosma rugosa*. *Poa litorosa* tussocks on inland slopes are smaller and accompanied by abundant *Polystichum, Histiopteris* and *Anisotome*, but *Blechnum durum* and *Urtica* are absent (Fig. 13.17). *Coprosma rugosa*, the only erect woody plant on the island, forms scrub patches in gullies among larger areas of *Polystichum*.

On inland flats, the peat surface is raised into 30–60 cm tall hummocks that are occupied by stunted *Poa litorosa* or *Polystichum vestitum*. Sides of hummocks support abundant *Hymenophyllum multifidum* and some *Lycopodium varium*, whereas turf between consists mainly of *Coprosma perpusilla*, accompanied by *Hymenophyllum, Gentiana, Epilobium alsinoides, Luzula crinita, Uncinia hookeri* and prostrate *Coprosma ciliata*. Petrel burrows and albatross nesting platforms are everywhere, and are the typical habitat of *Senecio radiolatus*; abandoned platforms show stages of revegetation, with *Marchantia, Epilobium, Acaena minor* and *Stellaria decipiens* as the pioneers.

In coastal swamps, dense *Carex appressa* is up to 1.5 m tall, with *C. sectoides* occurring in the deepest parts. In inland swamps the summer-green *C. ternaria* dominates alone or with *Blechnum* 'mountain'; *Anisotome antipoda* and *Pleurophyllum criniferum* can be present. Depressions among the grassland on inland flats contain interconnecting fens with short, open *Carex ternaria*, which can become quite dry on the surface. Some flushes support tall herbfield of *Anisotome, Pleurophyllum* and *Stilbocarpa polaris*.

Exposed summits above 330 m are peat-covered, but tussock grassland gives way to herbfield in which stunted *Pleurophyllum* and *Stilbocarpa* are the main plants. There are also luxuriant patches of *Coprosma perpusilla*, and most other species of the inland grasslands are present. Slips exposing regolith are covered with *Stereocaulon ramulosum*, with *Lycopodium fastigiatum* or *L. scariosum* locally abundant.

Bounty Islands

Small size, drenching by salt spray, and dense habitation by birds and seals limits the terrestrial plants of these granite islands to a few cryptogams; C.D. Meurk (unpublished) lists the green algae *Prasiola* and *Chlamydomonas*, and the tentatively identified lichens *Verrucaria maura, Pertusaria* and *Candelaria*.

Macquarie Island

Environment, history and flora

Macquarie, a high, narrow island formed from oceanic crust uplifted during the Pleistocene, consists of basalt, diorite, gabbro and other mainly igneous rocks. It has been at least partly glaciated, and is almost encircled by a raised beach-terrace. Peat blankets the slopes to an average depth of 1 m, and exceeds 6 m on the coastal terrace.

Macquarie Island was exploited for seal skins from its discovery in 1810, and penguins and sea elephants were slaughtered for their oil until 1919. It has been a nature reserve since 1933. Rabbits, introduced about 1880, reach plague proportions in different places at different times, and can eat even rank vegetation to the ground, exposing the peat to severe erosion. When rabbits decrease, vegetation recovers to varying degrees. Cats have exterminated two bird taxa including a subspecies of red-fronted parakeet (*Cyanoramphus novaezelandiae*), but they probably hold rats and mice in check and mitigate the effects of rabbits. Weka are abundantly naturalised.

Of the island's 40 indigenous vascular species, 30 are shared with other islands in Campbell province, and of these 20 extend to the New Zealand mainland. *Puccinellia macquariensis* is endemic but has close New Zealand relatives. The remaining nine species do not occur anywhere else in Campbell province, four of these being circum-antarctic: *Azorella selago*, the dominant plant of the uplands, forms cushions from 2 cm to 1 m across that can coalesce into continuous mats, which turn brown as the leaves die in autumn; *Ranunculus biternatus* is ecologically wide-ranging, but thrives best on sheltered, wet sites; *Acaena magellanica* is a semi-woody plant of damp, open ground; and *Festuca contracta* is a widespread tussock grass. *Poa cookii*, also a tussock, extends to Kerguelen and Heard islands, and favours areas enriched by excreta.

Vegetation

Taylor (1955) recognises some 37 communities (excluding seral and coastal sequences), which mostly express combinations of a few dominant species in relation

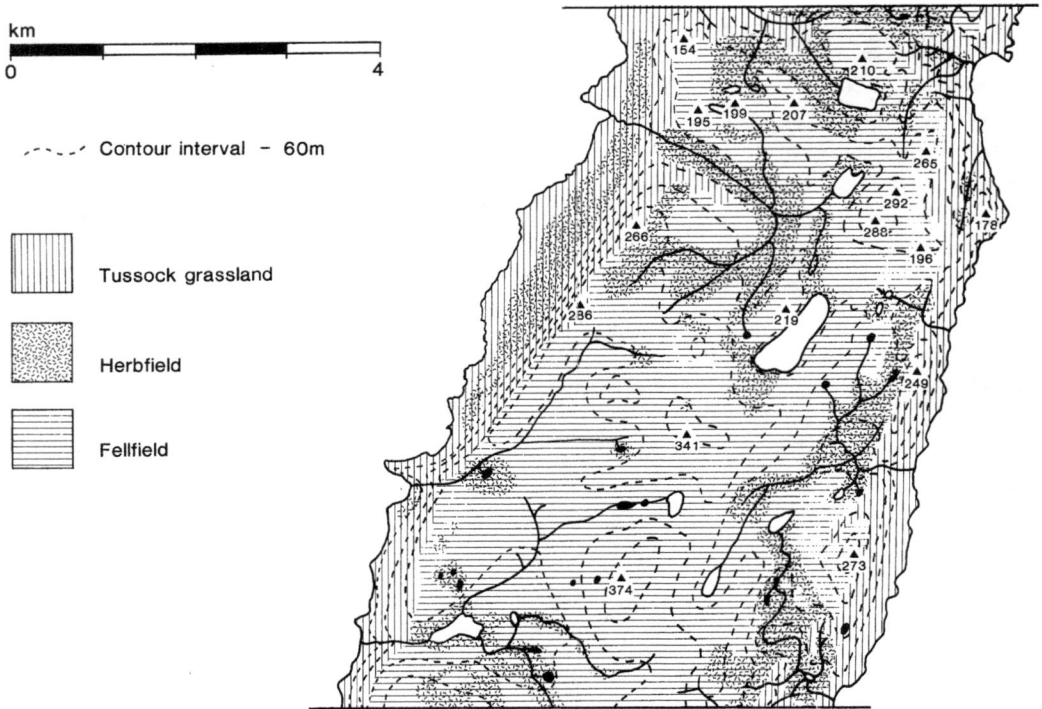

Fig. 13.18. Map of the middle portion of Macquarie Island (from Taylor 1955).

to environmental gradients. They are grouped into five main 'formations', the three most extensive being mapped in Fig. 13.18.

(1) Tussock grassland is the climax on steep, peat-blanketed coastal slopes, and also grows on less waterlogged parts of the coastal terrace, and on some inland slopes and flats. In the most typical community, *Poa foliosa* tussocks form a continuous canopy supported on peaty stools up to 1 m tall. The main associates are *Stellaria decipiens* and *Cardamine corymbosa*. Where wallowing sea elephants have reduced the density and vigour of tussocks, other species become frequent, notably *Leptinella plumosa* and *Poa annua*. *P. cookii* co-dominates with *P. foliosa* wherever penguins congregate. *Stilbocarpa polaris* accounts for up to 20% of the cover on areas subject to peat movement, and is completely dominant locally. *Polystichum vestitum* co-dominates with *P. foliosa* and *Stilbocarpa* or forms dense stands.

(2) Herbfield replaces tussock grassland on peats that are wetter (water table within 45 cm of the surface) and probably slightly more acid and lower in phosphorus, and on more exposed sites. It forms a narrow band between tussock grassland and the fellfield above, and is the main vegetation on the coastal terrace and in inland valleys. Towards its limit at 340 m, herbfield becomes confined to small, sheltered valleys, often where snow persists. Usually *Pleurophyllum hookeri* completely dominates over a lower tier that is mainly *Stilbocarpa*. In better-drained but wind-exposed sites, *Coprosma perpusilla* and *Acaena minor* form the lower tier. Minor communities have *Carex*

trifida or *Festuca contracta* dominant or co-dominant. There are also transitions from *Pleurophyllum* herbfield to tussock grassland, fellfield, bog and fen. Jenkin & Ashton (1979) describe cycles in which *Pleurophyllum* either reoccupies gaps left by death of old plants, or alternates with bryophytes, *Stilbocarpa, Azorella*, small forbs, grasses and grass-like species.

(3) In fairly sheltered fens *Juncus scheuchzerioides* is usually dominant, but *Isolepis aucklandica* co-dominates on ombrogenous fen on the coastal terrace, rising to dominance on the edges of pools. On exposed upland fens, *Juncus* forms combinations with *Agrostis magellanica* and *Deschampsia caespitosa*, with *Ranunculus biternatus* and *Callitriche antarctica* as further species.

(4) Bogs are most extensive on the coastal terrace. The mosses *Breutelia pendula, B. elongata, Drepanocladus aduncus, Dicranoloma robustum, Brachythecium salebrosum* and *Ptychomnion aciculare* form one community in which *Colobanthus apetalus* also dominates small areas. Other vascular plants include *Juncus scheuchzerioides, Isolepis aucklandica, Agrostis magellanica, Uncinia hookeri, Ranunculus biternatus, Hydrocotyle* sp., and *Cardamine corymbosa*. *Sphagnum falcatulum* fills former ponds and channels but is not common. Cushion bogs of *Colobanthus muscoides* occur on the western side of the island. Aquatic communities are formed by *Myriophyllum triphyllum* in deep water, *Ranunculus biternatus* in shallow water, *Callitriche antarctica* in peat pools (including abandoned sea-lion wallows), and green algae.

(5) As most of Macquarie Island lies above 200 m, fellfield is the main vegetation. *Azorella* grows as continuous mats in sheltered lees and on risers of solifluction terraces. On more exposed sites, separate *Azorella* cushions colonise *Racomitrium crispulum* mats, but become eroded on their windward edge, thereby giving rise to migrating double strips with *Azorella* to windward and *R. crispulum* to leeward (Ashton & Gill 1965). *R. crispulum* alone dominates on the most exposed summits and saddles, covering up to 10% of the stony surface. Other common fellfield species include *Dicranoloma robustum, Thuidium furfurosum, Drepanocladus aduncus, Pogonatum alpinum, Grammitis poeppigiana, Luzula crinita, Agrostis magellanica, Festuca contracta, Colobanthus apetalus, Coprosma perpusilla*, stunted *Pleurophyllum* and *Stilbocarpa*, and lichens epiphytic on *Azorella*. Moss buttons grow on rocks and the treads of solifluction terraces; *Ditrichum strictum* is the main species, but *Andreaea acutifolia* is common.

At lower altitudes, steep clay-loam slopes derived from deeply weathered rock are colonised by mosses and several vascular plants, including *Colobanthus apetalus* and *Epilobium* spp. Stony screes are colonised by *Montia fontana* (on seepages), *Ranunculus biternatus, Acaena minor, Poa annua, Agrostis magellanica*, and near the sea, *Colobanthus muscoides*. Both kinds of surface, and also mineral soils exposed by slips, eventually succeed to herbfield and then tussock grassland.

Rocks close to the sea support tufts of *Puccinellia macquariensis* among mosses, lichens and blue-green algae. Above the reach of storm waves *Puccinellia* is joined by *Colobanthus muscoides* cushions up to 45 cm across and, in places, *Crassula moschata*. Still further from the sea, *Colobanthus* cushions support tufts of *Leptinella plumosa* and other species, and grade into continuous herbfield and tussock.

14

BIOMASS, GROWTH, NUTRITION
AND TOLERANCES

Total biomass – plant and animal, living, inert and dead, above and below ground – is the most direct expression of how effectively a community occupies its site. As no New Zealand community has been completely analysed for biomass, Chapter 14 commences with partial comparisons, and proceeds to simpler volumetric, areal and linear parameters.

The next section discusses production, or rate at which living matter accumulates, although information is incomplete, especially for below-ground components. At least in mature, stable communities, annual accumulation must be balanced by loss, and litter-fall provides a partial measure of this. Again, however, comparisons of production between communities rest largely on measuring increment in volume, diameter or height.

Production is usually expressed in terms of dry matter, and by far the largest component of this is carbohydrate and other organic compounds. Measurements of carbon dioxide exchange (pp. 480–5) therefore lead to a physiological understanding of production. Most New Zealand work concerns only leaves, although there are some estimates of carbon balance for whole plants and single-species stands.

Mineral elements and nitrogen form a much smaller proportion of plant material than carbon, but are no less important. Foliar concentrations are considered first, because they provide the clearest relations between species and soil fertility. Nutrients are then discussed on an areal basis, in respect of living plants and litter. Reflecting the prevailing low fertility of New Zealand soils, symbiotic uptake of nutrients is important, especially in relation to phosphorus and nitrogen (pp. 492–7). Although accumulation of potentially toxic (Lyon *et al.* 1971) and economically important (Yates *et al.* 1974) trace elements in native plants has been investigated, their nutritional role in natural vegetation has scarcely been considered, although recognition and correction of deficiencies has been a major success of New Zealand agriculture and commercial forestry.

The final sections of Chapter 14 consider water and temperature stress, which influence C uptake, production and, ultimately, biomass.

Biomass and related parameters

Biomass

A few New Zealand studies of total biomass have been published and several more relate to above-ground biomass, but comparisons are restricted by different methods of partitioning material. In Table 14.1 the only forest value for total biomass is six times higher than the highest grassland value, and values for *Chionochloa* are higher than for short tussocks or alpine herbfield; similar relations hold for living roots and dead above-ground material. Large amounts of dead material are held within *Chionochloa* and *Poa foliosa* tussocks.

Values for live above-ground material in the species-poor subalpine mountain beech (*Nothofagus solandri* var. *cliffortioides*) stand at the upper forest limit exceed those reported for tall, mixed lowland forest; and those in young, even-aged beech stands are almost as high as in mature stands. Values cited for live above-ground herbaceous and aquatic vegetation generally lie between 10 and 1 % of values for forest biomass, which, however, consists mostly of inert wood. Measurements in vegetation that has not attained its potential mass, including pole stands of native or exotic trees, depleted native grassland and grazed pasture, are not directly comparable.

Leaf biomass depends on plant architecture, leaf longevity, and productivity of the site. In young stands of *Pinus contorta*, the rapid increment of biomass seems related to the large quantity of needles, which provide a total projected leaf area per unit of ground surface that is greater than in mountain beech (14 versus 9) (Benecke & Nordmeyer 1982) and intercept more light (Turton 1985). Schoenenberger (1984) points out that a unit quantity of leaves supports more above-ground biomass in mountain beech trees with stunted, gnarled trunks at the upper forest edge than in neighbouring more vigorous stands. In *Chionochloa* grasslands, leaf biomass similar to that in forest supports much less total above-ground biomass.

For a warm-temperate mixed forest, Brockie & Mooed (1986) estimate a wet-mass animal biomass, excluding arboreal invertebrates, of 504 kg/ha. Litter arthropods contribute 145 kg, earthworms 333 kg, and mammals 25 kg, almost entirely as brushtail possum (*Trichosurus vulpecula*). Birds total only 0.6 kg/ha, which is more than that estimated for beech forest (0.12 kg/ha) but much less than the estimate of 1.4–3.9 kg/ha from Kapiti Island, a sanctuary where introduced mammals are held at low numbers. Productive pastures can support 20 stock units per hectare, which approximates to 1000 kg/ha, as well as 1163–3048 kg/ha of earthworms.

Standing volume

Table 14.2 gives representative standing timber volumes for forest, which are related to the number of merchantable trees; stands with low merchantable values might nevertheless contain large volumes of non-merchantable wood. The value shown for pure beech forest (82 m³/ha) is also on the low side, as merchantable volumes in virgin beech forest typically range from 70 to 200 m³/ha, and can be as high as 400 m³/ha (Wardle 1984). Inclusion of industrial wood can raise values to 450 m³/ha in virgin silver beech (*Nothofagus menziesii*) stands and 570 m³/ha in mixtures with podocarps.

Table 14.1. *Biomass measurements* (t/ha)

community	reference	altitude (m)	live above ground foliage	live above ground total	litter	roots	total
Agathis australis	1	WT	11	134			
Podocarp/broad-leaved	2	250	6	254	82†		
Podocarp/beech/broad-leaved	3	250	6	302	48†	147	526°
Nothofagus truncata	4	600	9	327		83	410
N. solandri var. *cliffortioides*	5	1000	13	273		37	
'' '' '' ''	5	1320	7	285		66	
'' '' '' ''	6a	1320	15	323			
'' '' '' ''	6b	1320	10	272			
'' '' '' ''	6c	1320	7	177			
'' '' '' ''	6d	1320	4	135			
Festuca novae-zelandiae	7	1050		3	3*	16	22
Chionochloa macra	7	1200		3	37*	10	51
C. pallens	7	1440		7	47*	10	64
C. rigida	8	884		8	31*	31	70
C. macra	8	1257		4	12*	22	38
C. rigida	9	900	10	21	45*	21	87
C. rigida	9	1000	11	21	26*	15	62
C. macra	9	1300	10	17	22*	25	63
C. macra	9	1900	13	19	47*	21	88
C. macra	9	1400	9	14	29*	23	66
Poa colensoi	9	1300		6	16*	14	37
P. colensoi	9	2000		6	21*	12	39
Chionochloa australis	10	1370					60‡
Herbfield	11a	1220		19		15	
Herbfield	11a	1390		18		14	
Cushion-field	11b	1220		14		7	
Cushion-field	11b	1390		19		11	
Poa foliosa grassland	12	45		9	27*	17	53
Pleurophyllum herbfield	12	235		5	3	7	15
Pinus radiata 22 yr	13		9	316			
P. contorta 20 yr	14	1050	29	224		77	
P. contorta 23 yr	14	1320	22	188		63	
Elodea canadensis	15			1–15			
Myriophyllum	15			<1			
Isoetes	16			10–12			

Notes:

† Includes fallen stems and branches; ° includes 28 t/ha standing dead stems; * includes dead material within tussocks; ‡ roots not included.

WT, IT: Warm- or interior-temperate.

Localities, other details, and references for Tables 14.1 and 14.3–4.

  1 Auckland ER; secondary stand 130 years old (Madgwick *et al.* 1982).

  2 North Westland; (*Dacrydium cupressinum*)/*Quintinia–W. racemosa* (Levett *et al.* 1985*b*).

  3 North Westland; (podocarp)/*Nothofagus truncata/Weinmannia racemosa* (Beets 1980).

  4 Nelson; values for *N. truncata* only, which are estimated as 90% of total biomass (Benecke & Evans 1987).

Even-aged pole stands of red beech (*N. fusca*) vary between 350 and 740 m³/ha. The only published figures for total main stem volume are for mountain beech forest, and range from 406 m³/ha at 900 m a.s.l. to 206 m³/ha at 1200 m.

Over all, there is an increasing trend from high-altitude beech forests to low-altitude beech and mixed forests. The highest figure cited for native forest is 770–840 m³/ha in a 'typical' kauri (*Agathis australis*) stand (Masters *et al.* 1957) but clustering of kauri trees makes this inapplicable to large areas. Tended, 40-year-old *Pinus radiata* stands on good sites range from 560 to 700 m³/ha, with a recorded maximum of 945 m³/ha.

Basal area

Basal area in beech forests increases from cool, dry climates and infertile soils to moister, milder, more fertile sites (reference 20 in Table 14.3). The high value cited for silver beech (reference 21) applies to a 160-year-old stand that had densely colonised a landslide. The data show no clear trends among other types of forest; although large basal areas are recorded from broad-leaved and *Libocedrus* stands in North Westland (reference 22), the sampling method is different and tree ferns are included.

According to Wardle (1984), basal area in pure mountain beech forest increases from 41 m²/ha at altitudes < 800 m to 52 m²/ha above 1200 m, which is consistent with

5 Puketeraki ER; stands 52 years old (Benecke & Nordmeyer 1982).
6 Locality as 5; a, mature stand; b, young stems arising from older bases; c, pole stand; d, stunted trees at upper forest limit (Schoenenberger 1984).
7 Puketeraki ER (Evans 1980).
8 Puketeraki ER (Williams 1977).
9 Central Otago (Meurk 1978).
10 Inland Marlborough (Wraight 1965).
11 Central Otago; a, *Celmisia viscosa* dominant; b, *Dracophyllum muscoides* dominant (Bliss & Mark 1974).
12 Macquarie Island (Jenkin 1975).
13 Volcanic Plateau; planted trees (Madgwick *et al.* 1977).
14 Puketeraki ER; planted trees (Benecke & Nordmeyer 1982).
15 Rotorua lakes (Brown 1975).
16 Lakes in Spenser ER (Brown 1975).
17 Northland (Ahmed & Ogden 1987).
18 Volcanic Plateau; *Dacrydium cupressinum* dominant, only merchantable trees > 30 cm d.b.h. included (Smale *et al.* 1985).
19 Bay of Plenty; *Dacrycarpus*/*Beilschmiedia tawa* (*Laurelia*) (Smale 1984).
20 Composite data from several localities, all tree species > 1 cm d.b.h. (Wardle 1984).
21 Valley floor in northern Fiordland; all woody species (Stewart 1986).
22 An altitudinal sequence from valley floor to upper forest limit, North Westland; tree ferns included (Coleman *et al.* 1980).
23 Wellington ER (Miller 1963).
24 Puketeraki ER (Wardle 1970).
25 Manawatu ER; pasture mown four times per year, with equivalent quantity of dung and urine returned (a) or withheld (b) (Melville & Sears 1953).
26 Manawatu ER; adventive ground layer in podocarp/broad-leaved forest (Kelly & Skipworth 1984).

Table 14.2. *Mean merchantable timber volumes (m³/ha) and tree densities (no./ha) for 10 selected forest types*

	1	2	3	4	5	6	7	8	9	10
Dacrydium cupressinum	396	37	176		9		29	290	49	48
Dacrycarpus dacrydioides	62		53				21		47	378
Prumnopitys ferruginea				37			10		16	
Prumnopitys taxifolia	126		141	32					16	
Podocarpus spp.†			70	30	12				5	
Beilschmiedia tawa		29								
Libocedrus bidwillii					11					
Nothofagus menziesii						42				
Nothofagus fusca						39	14			
Total merchantable volume	653	79	457	106	36	82	83	314	137	442
Merchantable no./ha	158	34	101	69	20	52	57	185	52	208
Total no./ha	175	104	124	94	?	?	106	224	117	282

Notes:
The six columns on the left refer to Volcanic Plateau, the four on the right to Westland. Only species contributing >10% of stand volume are shown.
† *P. totara* and/or *P. hallii.*
Source: New Zealand Forest Service, unpublished data.

Table 14.3. *Basal area (m²/ha) in forests*

forest type	reference (Table 14.1)	total	dominant species
kauri	17	35–177	23–127
dense podocarp	18	63–69	37–42
Dacrycarpus/broad-leaved	19	50	17
podocarp/broad-leaved	2	55	28
podocarp/beech/broad-leaved	3	48	22
Nothofagus truncata	4	50	45
N. fusca–N. solandri var. *cliff.*	20	37	
N. solandri var. *cliffortioides*	20	48	
N. menziesii–N. fusca	20	54	
Nothofagus menziesii	20	60	
Nothofagus spp./*Weinmannia racemosa*	20	73	
N. menziesii (even-aged)	21	165	148
N. menziesii/(*Weinmannia racemosa*)	21	80	51
Weinmannia racemosa†	22	142	93
tree ferns†	22	107	66
Quintinia–Weinnmania	22	105	
Metrosideros umbellata–Weinm. racem.	22	148	
Griselinia littoralis	22	88	49
Libocedrus bidwillii	22	119	49
subalpine tree-heath	22	92	

Notes:
† Residual canopy after podocarps logged.

the high biomass recorded in subalpine stands. Fully stocked young and mature beech stands need not differ greatly in basal area, as the following figures show.

Mean diameter (cm)	Mean basal area (m²/ha)	
	mountain	silver
10–15	47	47
> 30	51	73

Similarly, fully stocked mountain beech stands with 13 000 small and 450 large stems respectively both had basal areas of 61 m²/ha. In contrast, a seral stand dominated by *Pseudowintera colorata* with 24 885 stems/ha produced a basal area of only 32.4 m²/ha (Coleman *et al.* 1980). Relatively young exotic pines can achieve values comparable with mature native forests.

Stand height and complexity

Height is obviously related to quality of habitat, as it usually decreases with increasing altitude and exposure and decreasing fertility. Nevertheless, communities of markedly different height may be contiguous in apparently identical environments. In many cases, the differences can be related to age or seral status, with taller communities usually being older. Stature of herbaceous communities may be affected by grazing or seasonal dieback of the dominant plants.

Genetic differences may outweigh environmental control of height, as where subalpine heaths abut on much taller beech forests. Most native conifers are taller than the broad-leaved trees they grow among; the two elements may form different tiers in the same community, or form a mosaic with two canopy levels, or be segregated as separate communities. Similarly, snow tussocks (*Chionochloa*) and short tussocks can form different types of grassland, or different tiers in the same grassland.

The upper limit of mountain or silver beech forest against penalpine grassland or scrub is related to the altitudinal climatic gradient, but its abruptness results because the limiting factors act at the seedling stage (p. 504); height diminishes gradually only where beech trees are stunted by strong winds, winter die-back above persistent snow drifts, or shallow soils. Smaller subalpine native trees reach about the same upper limit as the beeches, but several exotic trees can grow at higher altitudes; *Pinus contorta* can become a tree at least 150 m above the native forest limit, and persist in stunted form for another 350 m.

Structural complexity and floristic richness follow the same environmental trends as height (Fig. 14.1), partly because taller vegetation is likely to contain more tiers, and partly because lower tiers, lianes and epiphytes are usually better developed in more benign habitats (Fig. 14.2). However, dominant species forming dense canopies can suppress lower tiers, even in benign habitats; examples are exotic conifer plantations, most beech forests, and dense *Chionochloa* grassland.

Production and growth

Dry matter production

Net annual dry matter production is, in effect, the balance between carbon gained by photosynthesis and that lost through respiration. Estimates are usually indirect, and

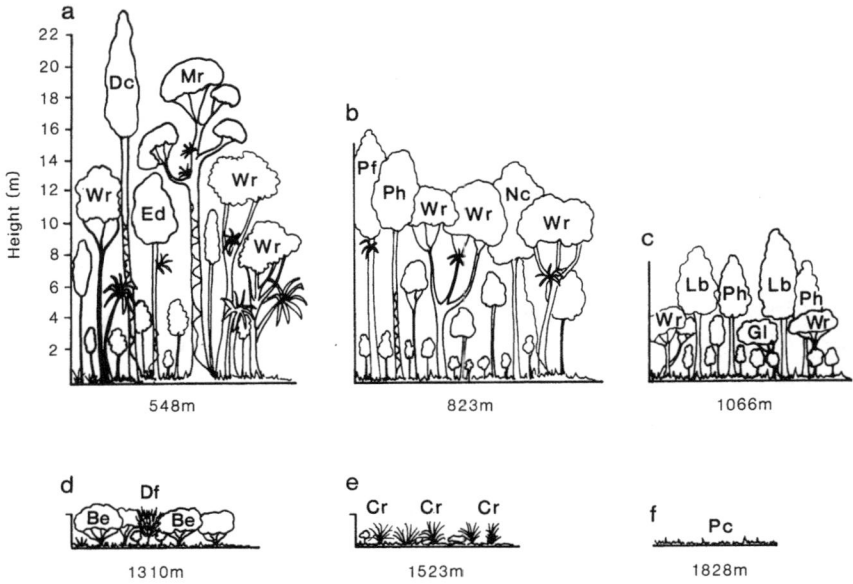

Fig. 14.1. Vegetation profiles from Mt Taranaki, showing diminishing stature and reducing number of tiers with increasing altitude and exposure (Clarkson 1986). Abbreviations: Be, *Brachyglottis elaeagnifolia*; Cr, *Chionochloa rubra*; Dc, *Dacrydium cupressinum*; Df, *Dracophyllum filifolium*; Ed, *Elaeocarpus dentatus*; Gl, *Griselinia littoralis*; Lb, *Libocedrus bidwillii*; Mr, *Metrosideros robusta*; Nc, *Nestegis cunninghamii*; Pc, *Poa colensoi*; Pf, *Prumnopitys ferruginea*; Ph, *Podocarpus hallii*; Wr, *Weinmannia racemosa*.

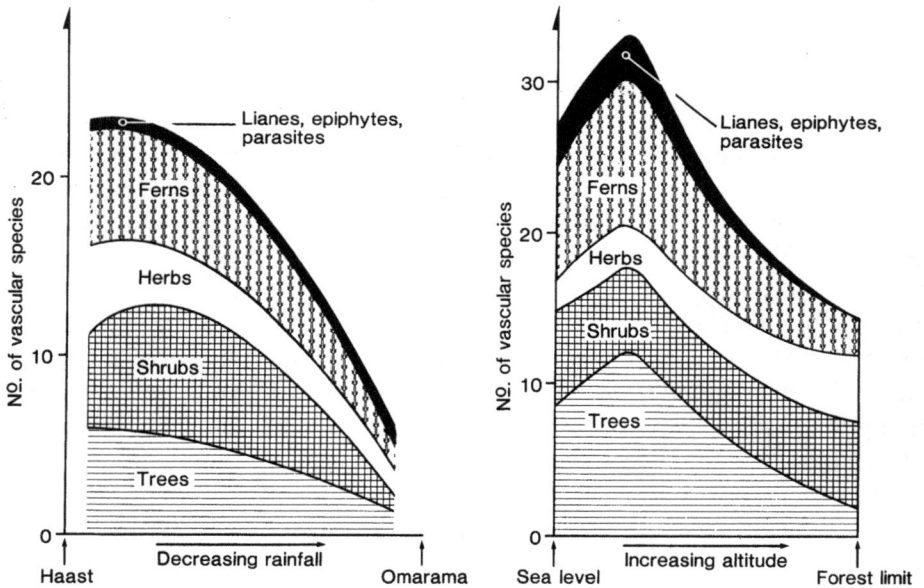

Fig. 14.2. Trends in floristic and structural diversity of beech forest along precipitation and altitudinal gradients (Wardle 1984).

Table 14.4. *Annual dry matter production*

community	reference (Table 14.1)	altitude (m)	main stem	total above ground	roots	total
				annual dry matter production (t/ha)		
Nothofagus truncata	4	600		13.7	6.3	20.0
Nothofagus truncata	23	WT		10.0		
N. solandri var. *cliff.*	24	1040	4.1	7.8		
N. solandri var. *cliff.*	24	1340	2.8	5.8		
N. solandri var. *cliff.*	5	1000	8.8	28.0	5.6	33.6
N. solandri var. *cliff.*	5	1320	4.4	13.0	5.0	18.0
Chionochloa rigida	9	900		5.5	4.1	9.6
C. macra–C. rigida	9	1300		8.4	3.7	12.1
Chionochloa macra	9	1300		7.6	5.0	12.6
C. macra	9	1400		8.3	4.1	12.5
C. macra	9	1900		6.3	4.2	10.5
Poa colensoi	9	1300		7.1	4.2	11.3
P. colensoi	9	2000		2.7	2.0	4.7
herbfield	11*a*	1220		3.3	3.3	6.6
herbfield	11*a*	1390		2.7	1.6	4.3
cushion-field	11*b*	1220		3.0	2.0	5.0
cushion-field	11*b*	1390		2.1	2.4	4.5
*Poa foliosa***	12	45		19.1	36.7	55.8
Pleurophyllum herbfld*	12	235		4.6	5.5	10.1
Pinus radiata 4–8 yr	13	WT	10.7	21.7		
P. radiata 10–22 yr	13	WT	18.0	24.6		
Pinus contorta 20 yr	14	1050	17.4	31.8	9.6	41.4
P. contorta 23 yr	14	1320	7.6	18.0	7.2	25.2
Lolium–Trifolium rep.	25*a*	WT		17.0		
Lolium	25*b*	WT		2.5		
Tradescantia flumin.	26	WT		0.2–12.5		

Notes:

* Difference between seasonal minimum and maximum biomass values.

the varied methods influence results. Direct measurements are either in artificial communities such as even-aged forests, or involve repeated harvesting which itself influences productivity.

Values (Table 14.4) tend to decrease with increasing altitude and are mostly higher in forest than in grassland, but there is much overlap. Those for lowland pastures are high, except where nutrient return is prevented; Fig. 14.3 shows the seasonal course of production under a range of climates. The highest values, other than the estimate for *Poa foliosa* roots, are for dense, young stands of mountain beech and *Pinus contorta*.

The adventive monocot forb *Tradescantia* growing under native forest, at a light intensity only 7% of that in the open, produces almost as much dry matter as pasture (Fig. 14.4). The only data available for indigenous understoreys refer to nikau palm

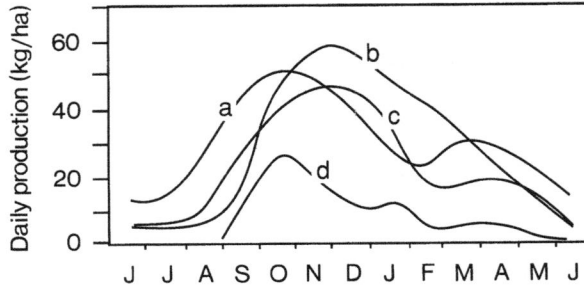

Fig. 14.3. Average above-ground dry matter production (kg/ha/day) from lowland pastures (Radcliffe & Baars 1987): a, North Island (7 sites); b, Southern Zone (13); c, Volcanic Plateau, 380 m (3); d, Dry Central Otago (1).

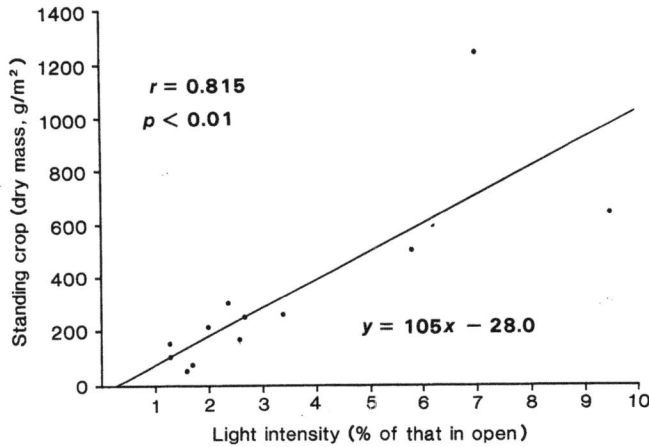

Fig. 14.4. Regrowth of *Tradescantia fluminensis* over 12 months in relation to light intensity in August (Kelly & Skipworth 1984).

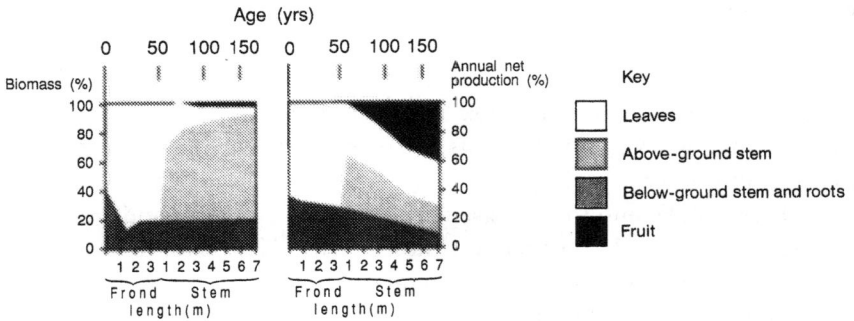

Fig. 14.5. Distribution of (left) total biomass (59.4 kg*), and (right) annual net production (1.7 kg/yr*) among the parts of a nikau palm, in relation to age and stage of growth (Enright 1985). *Derived from a 9.6 m tall palm.

(*Rhopalostylis sapida*) (Fig. 14.5); the net annual production for individual palms should translate to *c*. 1.7 t/ha, assuming a density of 1000 palms/ha.

Litter fall

Litter fall is a useful indicator of annual above-ground production, with two reservations: processes in the canopy, including herbivory, decay and leaching, reduce the mass of litter reaching the ground; and in forests, the contribution of logs and large branches can be difficult to measure, especially when it occurs erratically. Nevertheless, the values selected for Table 14.5 broadly support conclusions from other parameters. Production in beech forests decreases with increasing altitude, but differences between beech, mixed and exotic forests are not of convincing magnitude. Values for *Chionochloa* grassland overlap with those for forest, and are greater than those recorded in alpine *Poa colensoi* grassland. The highest value is from young *Coriaria arborea–Aristotelia serrata* bush, and relates to rapid shoot growth and leaf production.

Volume increment

This statistic is most often applied to tree stands, and may include merchantable wood only, or all main-stem wood. Table 14.6 indicates relatively small differences among even-aged beech and kauri stands, although mountain beech stands at two altitudes (reference 5) differ remarkably. Increment is very low in dense rimu (*Dacrydium cupressinum*) stands; after thinning, residual trees show a decrease instead of the release that might have been expected. Exotic pines show far higher increments than the native trees.

Diameter increment

Annual diameter increment can be measured directly or from growth rings, which are annual in most New Zealand woody plants. Although it varies widely from tree to tree within a stand, depending on age and degree of dominance or suppression, measurements support and extend patterns shown by other parameters of productivity (Table 14.7). Kauri and beeches tend to grow faster than podocarps under similar conditions, but not as fast as short-lived, small, seral trees, or exotics such as *Pinus radiata*.

Growth rates decrease with increasing altitude, both within species, and when species of different altitudinal range are compared; especially slow growth is shown by high-altitude podocarps. In southern rata (*Metrosideros umbellata*) diameter growth may be negligible at the upper limit (Payton 1989a), whereas mountain beech (Fig. 14.6) and silver beech trees at the upper forest limit often grow faster than those just below, presumably because they are less subject to competition from neighbours. Species such as kauri and rimu that tolerate a wide range of soil fertility grow most slowly on poor soils, even though they may achieve their greatest dominance; and species adapted to relatively fertile soils tend to grow faster than species of poor soil, although matai (*Prumnopitys taxifolia*) is an exception.

Year-to-year variations in widths of growth rings that are repeated in different trees enable multiple or missing rings to be identified, ontogenic variations in ring width to

Table 14.5. *Litter fall*

community	reference	alt. (m)	litter fall (t/ha) micro	macro	total
podocarp/broad-leaved	1	100	4.9	1.6	6.5
podocarp/broad-leaved	2	WT	3.6	0.8	4.5
Nothofagus truncata	2	WT	4.1	3.0	7.3
Nothofagus truncata	3	WT	4.3	1.7	6.0
N. solandri var. *solan.*	4a	60	5.3	0.4	5.7
N. solandri var. *solan.*	4b	60	4.5	0.5	5.0
N. solandri var. *cliff.*	5a	1190	4.6		
N. solandri var. *cliff.*	5b	1340	3.1		
Nothofagus menziesii	5c	576	4.7		
Nothofagus menziesii	5c	886	3.5		
Leptospermum scoparium	6	WT	7.8		
Coriaria–Aristotelia	6	WT	10.1		
Chionochloa rigida	7	884	5.2		
Chionochloa macra	7	1257	2.9		
Chionochloa rigida	8	900	2.6		
C. macra–C. rigida	8	1300	3.3		
Chionochloa macra	8	1300	2.1		
Chionochloa macra	8	1400	2.6		
Chionochloa macra	8	1900	1.9		
Poa colensoi	8	1300	1.1		
Poa colensoi	8	2000	1.9		
Pinus radiata	9	WT	3.5	3.9	7.4
Pinus radiata	9	WT	4.2	2.0	6.2

Notes:

WT, IT: warm- or interior-temperate.

'Micro' litter is mainly leaves. Small twigs and bark fragments are mostly included as 'micro' litter except in reference 9, where they are assigned to the 'macro' component.

Altitude as belt or m a.s.l.

Locality and reference:

1 Wellington ER (Daniel & Adams 1984).
2 North Westland (Levett *et al.* 1985b).
3 Wellington ER (Miller 1963).
4 Tararua ER: *a*, lower slope; *b*, upper slope (Bagnall 1972).
5 *a*, Kaweka Range; *b*, inland Canterbury; *c*, inland Southland (Wardle 1984).
6 Sounds ER (M.C. Macdonald, in Daniel & Adams 1984).
7 Heron ER (Williams 1977).
8 Central Otago (Meurk 1978).
9 Volcanic Plateau, means for two years, air-dry only (Will 1959).

Table 14.6. *Annual wood volume increment in tree stands*

dominant species	reference	altitude (m)	annual increment (m³/ha)
Agathis australis	1	WT	4.5–7.5
Agathis australis	1	WT	10.5*
Dacrydium cupressinum	2	660	1.8
Dacrydium cupressinum	2	660	1.2–1.7†
Dacrydium cupressinum	3	WT	1.5
Nothofagus fusca 70–80 yr	4		6–10
N. menziesii 50–140 yr	4		5–8
N. soland. var. *cliff.* 45 yr	4		6–11
N. soland. var. *cliff.* 52 yr	5	1000	17
N. soland. var. *cliff.* 55 yr	5	1200	10
Pinus contorta 13 yr	5	1000	26
Pinus contorta 14 yr	5	1250	35
Pinus radiata: range	6		23–36
Pinus radiata: maximum	6		50

Notes:

References 3, 4, and 6 are merchantable wood only; 5 is total main stem under bark; 1 and 2 are not specified.

* Fertiliser applied; † thinned.

Locality and reference:

1 Northern zone; pole stands (Barton 1983).
2 Volcanic Plateau; dense, all-aged podocarp stand (Herbert 1980).
3 Central Westland; dense podocarp forest on gley podzol (Franklin 1973).
4 Composite data from several localities, all tree species > 1 cm d.b.h. (Wardle 1984).
5 Puketeraki ER (Nordmeyer 1980).
6 Various sources.

be recognised, and tree-ring chronologies to be built up. These in turn allow dating of events such as frosts, storm damage to canopies, fires, insect epidemics or mast years that temporarily reduce or suppress cambial growth, and also 'release events' such as death of competitors that result in increased diameter growth. However, most interest has centred on recognising the climatic signal in ring-width variation. Ring width is positively correlated with growing season temperatures in subalpine *Nothofagus* trees, and with growing-season rainfall in *N. solandri* trees at lower altitudes. In kauri, narrow growth rings are associated with cool, damp spring weather (Norton & Ogden 1987).

Shoot growth and height increment

Mature trees and stable stands have negligible net annual height increment, even though shoots continue to grow throughout the canopy. As shoot growth is correlated with cambial growth but even more variable, the few measurements published for

Table 14.7. *Diameter increments (mm/yr) as means or typical ranges*

species	reference	belt	increment	community
Agathis australis	1	WT	8.2	secondary stands
" "	2	WT	5–8	secondary stands or planted
" "	3	WT	3.2	secondary stands
" "	4	WT	4.1	(large kauri)/broad-leaved
" "	4	WT	2.3	dense kauri on ridges
Dacrydium cupressinum	4	WT	2.5	kauri forest
" "	5	T	0.3–3.1	dense rimu on glacial outwash
" "	6	T	8.1	secondary, thinned
" "	6	T	2.5	secondary, not thinned
" "	7	T	0.9–4.4 ⎫	dense mixed-aged
" "	8	T	0.9 ⎭	podocarps
Podocarpus totara	6	T	9.9	pole stand, thinned
" "	6	T	4.0	pole stand, not thinned
Podocarpus totara, ⎫ *Prumnopitys taxifolia* ⎭	9	T	2.6	young forest on 　alluvial plain
Prumnopitys taxifolia	8	T	0.3	dense mixed-aged podocarps
Phyllocladus trichom.	3	WT	2.6	secondary stands
Phyllocladus alpinus	10	S	0.2–0.4	shrub-heath
Halocarpus biformis	11a	S	0.4–0.8	shrub-heath
Libocedrus bidwillii	11a	CT–S	0.5–1.4	conifer/broad-leaved
" "	12	CT	2.6	young trees on alluvial terrace
" "	12	S	1.2	conifer/broad-leaved
" "	12	S	0.8	severe sites
Beilschmiedia tawa	8	WT	1.0 ⎫	dense mixed-aged
" "	13a	WT	1–3 ⎬	podocarp forest
" "	13b	WT	7–8 ⎭	
Weinmannia racemosa	14	CT	2.4	even-aged on landslide
" "	15	CT	1.3–3.8	seral
" "	15	CT	0.5–1.0	dense podocarp forest
Metrosideros umbell.	16	CT	0.2–1.0	conifer/broad-leaved
Nothofagus all spp.	17	CT	10–>20	young or released trees
Nothofagus fusca	18	CT	3.6 ⎫	dominant trees in
Nothofagus menziesii	18	CT	3.6 ⎬	sample
N. soland. cliffort.	18	CT	3.6 ⎭	
Nothofagus fusca	19	CT	1.2 ⎫	
Nothofagus menziesii	19	CT	1.3 ⎬	all trees in sample
N. soland. cliffort.	19	CT	1.4 ⎭	
Nothofagus menziesii	14	CT	5.5	even-aged on landslide
Nothofagus menziesii	20	CT	1.4	scattered in tree-heath
Entelea arborescens	21	WT	4–40	seral
Aristotelia serrata	22	CT	2–10	seral
Hoheria glabrata	11a	S	2–10	bush
Olearia ilicifolia	23	S	2.4–4.0	bush
Olearia colensoi	11a	S	1–2	shrub-heath
Brachyglottis buchan.	23	S	2.8	shrub-heath
Dracophyllum travers.	23	S	0.6–1.2	tree-heath
Dracophyllum uniflor.	11a	S	0.3–0.6	shrub-heath
Hebe odora	11a	S	0.8–1.0	hebe scrub

species	reference	belt	increment	community
Discaria toumatou	24	IT	1.4	invading grassland
Crataegus monogyna	24	IT	3.0–3.3	invading grassland
Pinus radiata	25	T	21	plantation 26 years old

Notes:

WT, CT, IT = warm, cool-, or interior-temperate (T): S = subalpine

Localities and references for Tables 14.7 and 14.8:

1 Auckland ER (Ogden 1983).
2 Northland ER (Ecroyd 1982).
3 Auckland ER (Pook 1978).
4 Northland ER (Cameron 1960).
5 Westland (Franklin 1968).
6 Volcanic Plateau (Cameron 1960).
7 Volcanic Plateau (Herbert 1980).
8 Volcanic Plateau (Smale *et al.* 1985).
9 Central Westland (McSweeney 1982).
10 Inland Canterbury (Wardle 1969).
11 *a*, Central Westland; *b*, Puketeraki ER (Wardle 1963).
12 Central Westland (Wardle 1978*a*).
13 Volcanic Plateau: *a*, 'average' trees; *b*, wide-crowned trees (Knowles & Beveridge 1982).
14 Fiordland (Stewart 1986).
15 Various localities (Wardle & MacRae 1966).
16 North Westland (Payton 1989*a*).
17 Various localities (Wardle 1984).
18 Volcanic Plateau (Wardle 1984).
19 Fiordland (Wardle 1984).
20 South Westland (Wardle 1980*b*).
21 Northland (Millener 1947).
22 Central Westland (P. Wardle, unpublished).
23 Hawdon ER (Haase 1986*c,d,e*).
24 Puketeraki ER (Williams & Buxton 1986).
25 Volcanic Plateau (Manley & Knowles 1980).
26 Northern South Island (J.R. Bray, W.D. Burke & G.J. Struik, personal communication).

mature plants add little further information; 4–8 cm per year for mountain beech and 1–18 mm for *Halocarpus biformis*, both growing near the subalpine forest limit, indicate how much species in similar environments can differ (Wardle 1963). The most consistent measurements are for young plants at the fastest-growing stage of their life cycle (Table 14.8). As with most other parameters, shoot growth of podocarps is slower, and that of seral species and pines is faster than in most other trees, irrespective of site quality. The values for *Coriaria arborea* refer to cane-like shoots produced near the base of young plants.

In mountain beech saplings, mean annual terminal growth decreased from 40.5 to 14.3 cm between 820 and 1190 m; that of silver beech decreased from 17.3 to 10.9 cm over a similar altitudinal range (Wardle 1984). At the upper forest limit (1340 m), the mean value for mountain beech had fallen to 4.9 cm, probably reflecting winter dieback, as summer growth can attain 60 cm and 19 cm at 1100 m and 1300 m respectively. Response of seedlings to different environments is discussed on p. 561, in relation to regeneration.

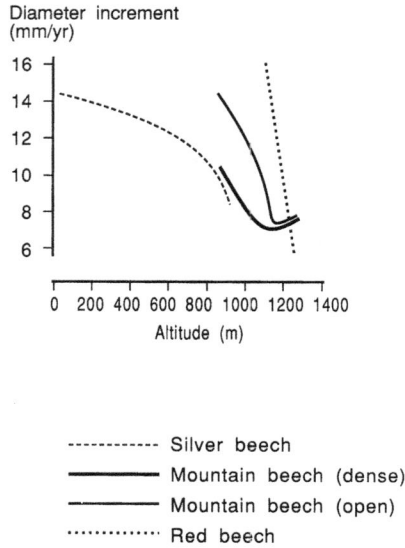

Fig. 14.6. Relation between altitude and stem diameter increment (mm/yr) in beeches. Sources: mountain beech (Wardle 1984); silver beech (Herbert 1973); red beech (Ogden 1978).

Carbon assimilation

The greatest reported difference in maximum photosynthesis (Table 14.9) is between South Island snow tussocks and the far-southern grass *Poa cookii*. Even illumination equivalent to 10% of bright sunshine, approximating to the prevailing cloudy conditions, reduced assimilation in *P. cookii* by only 50%. This accords with the high production measured in *P. foliosa* on Macquarie Island (Table 14.4).

External constraints on maximum assimilation lead to differences in growth and production; prominent among these are water deficits in soil and atmosphere, shading (including self-shading), and suboptimal temperatures, which all reduce stomatal opening. Fig. 14.7 illustrates their influence on hard beech (*Nothofagus truncata*), in which winter assimilation remained over half of that measured during summer. Photosynthetic dormancy occurs for short periods in mountain beech and *Pinus contorta* at the upper forest limit but not at lower altitudes (Benecke & Havranek 1980).

Although summer photosynthetic capacity among snow tussocks varied little over a range of ecotypes and altitudes, winter reductions ranged from 90% in *Chionochloa macra* at 1590 m to only 34% in *C. rigida* growing near sea-level (Fig. 14.8). This correlates with variations in mean annual leaf elongation, from 23 cm at 1590 m to 78 cm at sea-level; both genetic and environmental factors contribute to these differences.

The proportion of assimilated CO_2 that is converted into dry matter is affected by losses through respiration, which in turn are influenced by structural features such as the ratio between photosynthetic surface and respiring mass. Similar considerations

Table 14.8. *Annual height increment of trees at the fastest-growing stage of life, as means or typical ranges*

species	reference	belt	increment (cm)
Agathis australis	2	WT	30–40
Dacrydium cupressinum	5	T	15–30
Dacrydium cupressinum	6	T	24
Podocarpus totara	6	T	29
Beilschmiedia tarairi	26	WT	45*
Beilschmiedia tawa	26	WT	27
Elaeocarpus dentatus	26	WT	64
Laurelia novae-zelandiae	26	WT	26
Knightia excelsa	26	WT	65
Nothofagus fusca	17a	CT	70–100
Nothofagus solandri	17a	CT	90
N. fusca, N. solandri	17b	CT	30–60
Nothofagus menziesii	17b	CT	30–40
Hoheria populnea s.s.	26	WT	79*
Entelea arborescens	21	WT	100–300
Kunzea ericoides	26	WT	56
Leptospermum scoparium	26	WT	47
Melicytus ramiflorus	26	WT	48
Aristotelia serrata	22	T	100–150
Coprosma robusta	26	WT	48
Hebe salicifolia	26	WT	40
Coriaria arborea	22	T	200–450
Solanum aviculare	26	WT	60
Hoheria glabrata	11	S	30
Dracophyllum traversii	23	S	3
Brachyglottis buchananii	23	S	6–8
Olearia ilicifolia	23	S	9–12
Cordyline australis	26	WT	59
Rhopalostylis sapida	26	WT	26
Cyathea medullaris	26	WT	63
Dicksonia squarrosa	26	WT	24
Pinus radiata	25	T	140

Notes:
References numbered as in Table 14.7:
17a, maximum growth; 17b, 'usual' growth; 25, mean growth over 26 years; 26, mean increment measured over ≤ 10 years in closed stands at ≤ 50 m a.s.l.; *, grown south of natural range.

Table 14.9. *Maximum summer net photosynthetic rates under non-limiting conditions*

species	source	rate* $(mgCO_2 \, dm^{-2}h^{-1})$
Nothofagus truncata	1	16
Nothofagus solandri	2	16
Chionochloa rubra	3a	11
C. macra, C. rigida	4	7
Lolium perenne	3b	13
Poa cookii	5	24
Pinus contorta	2	21†

Notes:

* Calculated for a one-sided leaf area.

† Original value × 2.6 to convert to a planar projection.

Locality and reference:

1 Nelson ER, 600 m (Benecke & Evans 1987).

2 Craigieburn Range, 930 m (Benecke & Nordmeyer 1982).

3 *a*, Ruapehu; *b*, cultivar (Scott *et al.* 1970).

4 Otago, 100–1590 m (Greer 1984).

5 Marion Id, South Indian Ocean (Bate & Smith 1983).

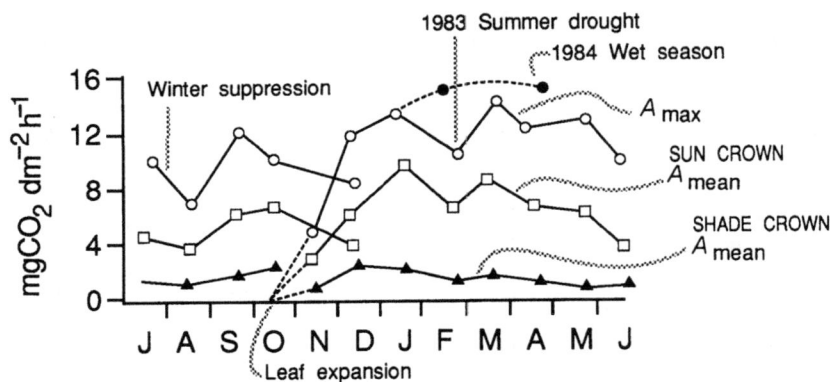

Fig. 14.7. Seasonal trend of net photosynthesis in *Nothofagus truncata* (from Benecke & Evans 1987).

A_{max} represents maximum photosynthesis of sun leaves with light, temperature and humidity non-limiting. A_{mean} represents mean monthly photosynthesis of sun and shade leaves under natural conditions. Values calculated for one-sided leaf surface.

$$\text{mg } CO_2 \text{ dm}^{-2} \text{ h}^{-1}$$

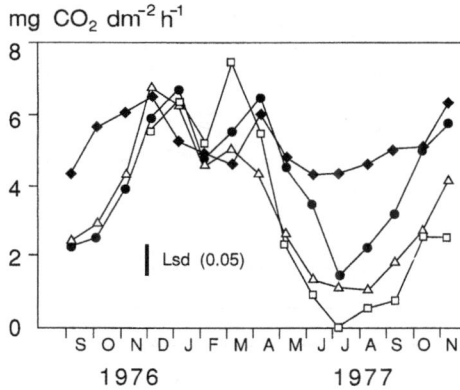

Fig. 14.8. Seasonal trend of the net photosynthetic capacity (one-sided leaf surface) of four populations of *Chionochloa*. The plants were transported to the laboratory each month and measured under constant conditions (Greer 1984). Symbols indicate: ◆, ●, △, *C. rigida* at sea level, 900 m and 1200 m respectively; □, *C. macra* at 1590 m.

may explain why red beech (*Nothofagus fusca*) seedlings grow faster than those of silver beech (*N. menziesii*) even though the latter have higher stomatal conductivity and photosynthetic rates (P. Van Gardingen, unpublished). Benecke and co-workers (references 4, 5, 14 in Table 14.1) calculate that about half of the carbon assimilated is lost to respiration in mountain and hard beech, and about 60% in young *Pinus contorta*. In kauri and rimu, large, apparently fixed resistances to CO_2 flux exist within the leaves, and appear to limit assimilation and growth (J.M. Barton, unpublished; W.M. McEwen, unpublished). Large, old southern rata trees succumb more readily to possum browsing or artificial defoliation than young trees, probably because their low ratio of leaf surface to respiring tissues leads to a precarious carbon balance (Payton 1983).

Expressed on a shoot mass basis, assimilation in pasture species is much greater than in *Chionochloa rubra* and *Celmisia spectabilis* (Fig. 14.9), which probably contain more structural tissue. Accordingly, *Chionochloa* species show lower relative growth rates, and little response to increasing temperature as compared with introduced grasses, which, at 18 °C, grow about five times as fast (Fig. 14.10). *Festuca* and *Poa* tussocks, although growing faster than the snow tussocks, also had subdued temperature responses, whereas the broad-leaved *Hierochloe redolens* is more like the introduced grasses.

In bracken (*Pteridium esculentum*) growing in a *Pinus radiata* plantation and *Blechnum discolor* growing beneath hard beech in the same locality, maximum hourly photosynthetic rates are 15.5 and 5.9 mg/dm² respectively, in conformity with the different light environments that the species are adapted to. Both ferns have high stomatal conductances relative to photosynthetic capacity, and the stomata respond less than those of the canopy trees to changes in light and humidity. While this implies inefficient water-use, it also mean that the ferns can take advantage of sunflecks, which

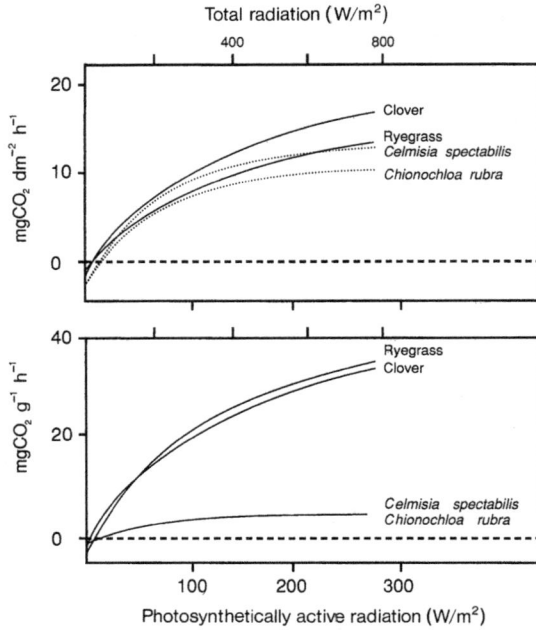

Fig. 14.9. Net assimilation rate at 21 °C and different light intensities (watts/m²) for species of native grassland and exotic pasture, in relation to (above) area of vertical projection of canopy on to a horizontal surface and (below) shoot mass. Upper and lower horizontal scales show total and photosynthetically active radiation respectively (Scott *et al.* 1970).

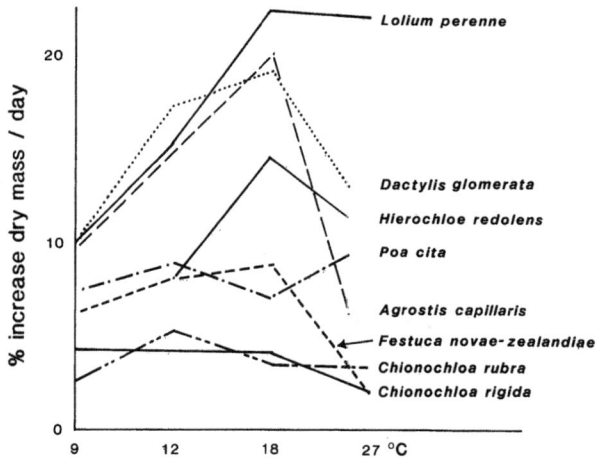

Fig. 14.10. Relative growth rates of grasses under controlled conditions (from Scott 1970).

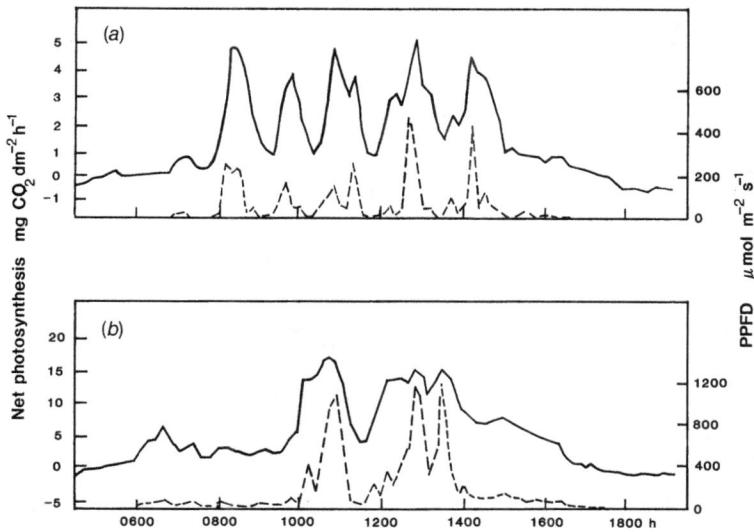

Fig. 14.11. Daily course of photosynthetically active radiation (solid line; measured as photon flux density) and net photosynthesis (broken line; one-sided leaf surface) for (a) *Blechnum discolor* in beech forest and (b) *Pteridium esculentum* in a *Pinus* plantation, at Big Bush, Nelson ER (Hollinger 1987).

contribute a large proportion of the photosynthetically active light reaching the understorey (Fig. 14.11).

The coastal *Atriplex buchananii* and *Theleophyton billardierei* are the only native plants that have been shown to have the C4 photosynthetic pathway that produces more dry matter for a given uptake of water than the usual C3 pathway (Troughton & Card 1974). However, the C4 pathway has been demonstrated for Australian material of the sand grasses *Spinifex sericeus* and *Zoysia pungens* (Prendergast & Hattersley 1987) and is also followed by *Spartina* and all introduced warm-temperate grasses.

'Crassulacean acid metabolism' (CAM) is another water-conserving mode, whereby plants can photosynthesise while stomata are closed by using CO_2 taken up at night. It occurs in many non-halophytic succulents and some aquatic plants. Among native plants, CAM has been demonstrated only in the aquatic *Isoetes kirkii* and *Lilaeopsis novae-zelandiae* (D.R. Webb *et al.* 1988) but could be sought in aquatic and terrestrial *Crassula* spp. and the epiphytes *Peperomia* and *Bulbophyllum*.

Nutrient element contents of plant material

Foliar nutrient concentrations

Nutrient concentrations tend to be consistent within species and even across broader taxonomic groups, being highest in those adapted to fertile soils. However, they also vary seasonally among organs and tissues within a plant, and in response to environmental factors, especially soil fertility. Foliar nutrients provide the most useful comparisons (Table 14.10).

Table 14.10. *Nutrient content of green leaves*

species	reference	miscellaneous	alt. (m)	Na	K	Ca	Mg	P	N	S
Asplenium bulbiferum	1	understorey	⎫	0.5	37.0	11.0	5.3	3.0	27.0	2.2
Coprosma grandifolia	1	understorey	560–620	0.7	22.0	18.0	5.5	1.8	16.0	1.7
Melicytus ramiflorus	1	understorey		1.7	27.0	15.0	5.0	2.6	23.0	3.4
Weinmannia racemosa	1	understorey	⎭	1.3	7.0	7.0	2.8	0.8	9.0	1.4
Coprosma 6 spp.	2	filiramulate	CT	2.9	13.1	16.3	4.8	1.7	20.8	2.8
Coprosma 3 spp.	2	large-leaved	CT	0.5	18.1	13.3	4.1	1.3	13.9	3.2
podocarp/broad-leaved forest ⎰	3	canopy	IT	1.1	6.7	5.2	3.4	0.6	7.0	2.0
⎱	3	understorey	IT	1.6	3.8	3.1	2.4	0.4	4.8	1.8
Agathis australis	4	current lvs	WT		10.4	7.8	0.1	0.9	8.1	
Agathis australis	4	older lvs	WT		0.5	1.5	0.1	0.5	6.6	
Nothofagus fusca	5	recent soil	IT	0.5	5.6	11.0	2.6	1.6	13.0	
Nothofagus fusca	5		IT	0.3	5.2	7.0	1.8	1.0	17.0	
Nothofagus menziesii	5	yellow-brown earth	IT	0.5	4.2	3.0	1.0	1.1	15.0	
Nothofagus truncata	5		IT	0.7	5.6	8.0	1.5	0.6	11.0	
N. soland. var. cliff.	5	boggy site	IT	0.4	4.4	7.0	1.0	0.6	9.0	
N. soland. var. cliff.	6	mountain slope	1250		4.6	8.0	1.3	1.6	13.0	
Pinus contorta	6	1 yr needles	1200		6.0	1.6	0.9	1.3	15.0	
Pinus contorta	6	5 yr needles	1200		5.0	6.5	0.5	1.0	15.0	
Pinus radiata	7	(a) 24 m, 32 cm	WT		13.0	1.0	2.0	2.7	17.0	
(height & d.b.h.)	7	(b) 16 m, 16 cm			13.0	1.9	1.5	1.3	12.0	
Pinus radiata	8	1 yr needles	WT	<0.1	8.3	2.7	1.3	1.7	15.0	
	8	3 yr needles		<0.1	7.3	3.2	1.4	1.5	10.0	
Chionochloa pallens	9	Tararua Range ⎰	1230–	2.2	11.5	0.9	0.9	0.9	10.4	1.4
C. flavescens	9	⎱	1460	0.4	10.5	0.8	0.6	0.5	8.3	1.2
C. pallens	10	Victoria Range, ⎰	1220	2.7	9.2	2.0	1.6	0.8	9.1	1.1
C. rubra	10	N. Westland ⎱	1220	1.8	7.5	1.0	0.9	0.7	7.4	1.0

nutrient content (mg/g dry mass)

C. 'robust'	Seaward	10	1204	0.1	11.4	1.9	0.8	1.0	6.4	1.0
C. pallens	Kaikoura Range	10	1204	0.6	12.7	2.0	1.3	1.4	9.5	1.0
C. 'robust'	Arthurs	10	1067	0.4	11.5	1.5	0.9	0.5	7.5	1.1
C. pallens	Pass,	10	1067	0.6	10.4	1.6	0.9	0.6	9.5	1.5
C. crassiuscula	North-east	10	1067	0.7	10.0	0.6	0.9	0.5	7.0	1.5
C. rubra	Alps	10	1067	0.6	8.2	0.6	0.5	0.5	6.4	1.1
C. rigida	Old Man Range,	10	860	0.2	9.1	1.1	0.8	0.5	2.6	
C. rubra	Central Otago	10	860	0.4	7.5	1.0	0.7	0.4	2.1	
C. rigida	Mid Dome,	10	1420	0.2	9.7	0.9	0.9	1.2	8.5	1.0
C. macra	Waikaia ER	10	1420	0.2	8.8	1.0	1.0	1.0	6.3	0.6
C. pallens	Puketeraki	11	1440		16.0	1.2	1.2	1.0	10.0	
Celmisia spp.	ER	11	1440		14.0	6.3	1.0	0.8	7.0	

Notes:

Altitude as m or belt: WT, CT, IT, warm–, cool– or interior–temperate.

Localities, references and other details for Tables 14.10–14.13:

1 Mt Taranaki, rimu/northern rata–kamahi forest (Mitchell *et al.* 1987).

2 Dunedin (Lee & Johnson 1984). Filiramulate species: mean for *C. rubra, rotundifolia, areolata, rhamnoides, propinqua, crassifolia*. Large–leaved species: mean for *C. grandifolia, robusta, lucida*.

3 North Westland; understorey data includes twigs (Levett *et al.* 1985a).

4 Auckland; 130 year old stand (Madgwick *et al.* 1982).

5 North Westland; canopy foliage (Adams 1976).

6 Puketeraki ER (Nordmeyer 1980).

7 Nelson ER; mean heights and diameters of trees on (*a*) gully floor and (*b*) ridge crest, contents for all live needles (Adams & Walker 1975).

8 Volcanic Plateau (Madgwick *et al.* 1977).

9 Williams *et al.* 1978a.

10 Williams *et al.* 1978b.

11 Main *Celmisia* sp. is *C. lyallii* (Evans 1980).

12 Heron ER (Williams *et al.* 1977).

13 Wellington; yellow–brown earth on fan (Daniel & Adams 1984).

14 North Westland (Levett *et al.* 1985a,b).

15 Sounds ER (M.C. Macdonald, in Daniel & Adams 1984).

16 Heron ER; data refer to annual dead leaf production of *Chionochloa* tussocks (Williams 1977; Williams *et al.* 1977).

17 Manawatu ER; values summed from four harvests at 3–monthly intervals (Melville & Sears 1953); (*a*), equivalent in dung and urine returned to pasture; (*b*), no return of harvested material.

18 Source as 16; values refer to total above–ground dead material.

Concentrations are high in leaves of species preferentially browsed by goats in the understorey of a mixed forest on Mt Taranaki, i.e. *Coprosma lucida, Schefflera* and *Ripogonum* as well as those listed in the table; the lowest values are for kamahi (*Weinmannia racemosa*), which also dominates the canopy. In *Coprosma* spp. from Dunedin, filiramulate species have higher nutrient concentrations than large-leaved species, and Lee & Johnson (1984) suggest that their habit helps to offset a consequent attractiveness to herbivores. For calcium, high values were also found in lowland samples of *Melicytus ramiflorus* (19.3%), *Podocarpus totara* (17.1%) and *Dacrydium cupressinum* (14.3%) in the same district by Thomson & Simpson (1936). In the podocarp/broad-leaved stand from a less fertile site in North Westland, all values are rather low, but are not strictly comparable with other values in the table as species are not separated.

Calcium concentrations are relatively high in beeches and young kauri leaves. In beeches, potassium concentrations are low, and there is also a relation between site preference and nutrient concentration, with red beech and lowland stands of mountain beech occupying the ends of the range. Within red beech, trees growing on a less fertile soil have lower nutrient concentrations in their leaves. Subalpine mountain beech forests are associated with dryness rather than infertility, and this too seems reflected in nutrient concentrations.

In *Pinus radiata* plantations on gravel hills in Nelson (Adams & Walker 1975), nitrogen, phosphorus and manganese concentrations in needles are strongly correlated with those in the soil A horizon. Moreover, foliar concentrations of nitrogen and phosphorus strongly correlate with vigour; positive correlations of growth with iron, and negative correlations with calcium and manganese probably result from parallel correlations between these elements and phosphorus concentration. Except for calcium, nutrient concentrations of pine needles decrease as they age; older kauri leaves show even more marked reductions.

More values are available for *Chionochloa* spp. than any other native plants, and only a selection is cited in Table 14.10. Concentrations of elements other than sodium are consistently higher in *C.* 'robust', *C. rigida, C. pallens* and *C. macra* than in *C. rubra, C. crassiuscula* and *C. flavescens*, which grow on wet, leached soils. Because the main source of sodium is the sea, via precipitation, sodium concentrations are highest in western, high-rainfall localities. Sodium concentrations are also relatively high in *C. crassiuscula*, which, of the species studied, is most centred on such localities. Values for sodium are nearly always negatively correlated with those for potassium, which is derived from the parent rocks and leached more rapidly under high rainfall. Element concentrations in leaves of *Celmisia spectabilis* are similar to those of *C. pallens* in the same community, except for calcium.

In *C.* 'robust', the phosphorus concentration of shoots correlates with calcium-bound soil phosphorus, i.e. the fraction that predominates in the young, rubbly soils favoured by this species. In *C. rigida* and *C. macra*, which occupy more mature soils, it correlates with the aluminium- and iron-bound fractions; and in the edaphically wide-ranging *C. pallens* shoot phosphorus correlates with all the soil phosphorus fractions. Calcium concentration is correlated with soil pH and, in *C. rigida* and *C. pallens*, with soil calcium (Williams *et al.* 1978*b*).

Table 14.11. *Nutrient content of live biomass*

community	reference (Table 14.10)	biomass (t/ha)	nutrient content (kg/ha)						
			Na	K	Ca	Mg	P	N	S
Podocarp/broad-lv.	5	254	118	425	592	206	27	306	120
Pinus radiata	8	316	6	300	180	80	35	280	
Chionochloa rigida	12a	8	2	86	14	9	10	74	6
Chionochloa macra	12a	4	1	29	4	3	4	24	4
Chionochloa rigida	12b	31	10	48	88	14	19	154	24
Chionochloa macra	12b	22	7	30	51	8	16	100	14
Festuca novae-zel.	11a	0.5		4	1	0.4	0.5	4	
Agrostis capillaris	11a	1.6		14	6	2.2	2	14	
Hieracium pilosella	11a	1.2		24	15	2.5	3	12	
total above ground	11a	3.4		42	22	5.0	5.5	30	
total roots	11b	17		65	50	9	16	110	

Notes:

11b and 12b refer to below-ground biomass. All other data refer to above-ground biomass.

Nutrient elements on a land area basis

The quantity of nutrients in a community depends on total biomass, and on the concentrations of elements in the various species and among the plant organs and tissues. Table 14.11 is selected from the limited information concerning living biomass. Total quantities and concentrations in the pine plantation are similar to those in the native forest for potassium and nitrogen, much lower for calcium and magnesium, and higher for phosphorus, the last probably reflecting the rapid growth of the pines. The high sodium content of the native forest may reflect the high rainfall of the locality. Total nutrients above ground are far lower in *Chionochloa* grasslands than in the two forest samples, but concentrations of potassium, phosphorus and nitrogen are higher because forest biomass consists mostly of wood, which is low in these elements.

Total nutrients above and below ground in *C. rigida* are 1.5–3.5 times greater than in *C. macra*, because of biomass and concentration differences. Stands of *C.* 'robust' probably contain even more nutrients, since dry weights of individual shoots are 3.6–7.2 g compared with 2.8–3.7 g in *C. rigida* and 0.7–1.0 g in *C. macra* (Williams *et al.* 1977). Where these species grow in proximity, they occupy a fertility gradient in the order of their shoot weights. The modified fescue–tussock grassland has more nutrients than *C. macra* grassland of considerably greater biomass; this difference is attributable to the greater absolute and proportional nutrient contents of the adventive *Agrostis* and *Hieracium*.

Quantities of nutrient elements returned in litter-fall are broadly proportional to its mass. Thus in Table 14.12 the largest return is by the *Coriaria–Aristotelia* community; the values for phosphorus and, especially, nitrogen doubtless reflect high uptake by *Coriaria arborea*, a vigorous N-fixer. Values are lowest in the *Chionochloa* grasslands, except in respect of potassium in *C. rigida* grassland. The litter of the

Table 14.12 *Nutrient content of annual litter-fall and harvested pasture*

community	reference (Table 14.10)	dry matter (t/ha)	nutrient content (kg/ha)						
			Na	K	Ca	Mg	P	N	S
Podocarp/broad-	{ 13	4.6		20	51	12	2.8	44	
leaved forest	{ 14	4.5	3.3	11	30	9	2.1	31	6.2
Nothofagus truncata	14	7.3	4.2	10	62	10	2.8	40	5.0
Pinus radiata	14	3.9	2.1	7	17	5	2.0	31	4.0
Leptospermum–Kunzea	15	7.8		25	48	12	4.2	49	
Coriaria–Aristotelia	15	10.1		31	117	29	9.3	153	
Chionochloa rigida	16	5.2	1.7	33	12	5	4.6	13	2.4
Chionochloa macra	16	2.9	0.4	9	6	3	3.8	10	2.5
Ryegrass–clover	17	17		648	134		83	738	
Ryegrass	17	2.5		50	15		11	56	

South Island podocarp/broad-leaved forest, which is on an old, leached soil, has a lower nutrient content than the North Island stand, which is on younger alluvium. Per unit of dry mass, hard beech, manuka–kanuka (*Leptospermum–Kunzea*) and *Pinus radiata* litter are poorer in most elements than mixed-forest litter. Although relative nutrient concentrations in litter tend to follow those in green foliage, comparison of Tables 14.10 and 14.12 indicates that absolute concentrations are mostly lower, presumably because of leaching and resorption from senescent foliage; calcium concentrations, however, are higher.

The data for harvested pastures, although not strictly comparable with either standing biomass or litter-fall, show that fertilised, grazed, high-producing grasses and legumes have extremely vigorous nutrient-cycling; but if nutrient-cycling is prevented experimentally, biomass and its element content are no greater than in native plant communities. Litter is rapidly broken down in vigorously cycling systems such as lowland bush and pasture, whereas for *Chionochloa rigida* litter Williams *et al.* (1977) estimate a turnover time of 6.7 years. Where decay of litter is incomplete, peat can accumulate (p. 285). Table 14.13 shows that even modest quantities of litter can hold quantities of nutrient elements as large as those in the living biomass.

Seasonal variations in nutrient content

Element concentrations in shoots of *Chionochloa rigida* and *C. macra* increase through the growing season, reflecting increasing availability and accumulation as soils become warmer and mineralisation more active (Fig. 14.12).

Under mixed forest, hard beech and *Pinus radiata*, quantities of elements being returned in litter vary in proportion to seasonally varying quantities of litter-fall, but there are also seasonal variations in concentration that partly depend on the plant organs being shed (Daniel & Adams 1984; Levett *et al.* 1985a). In leaf litter, concentrations of nitrogen, phosphorus, magnesium and calcium tend to be highest in

Table 14.13. *Nutrient content of accumulated litter*

community	reference (Table 14.10)	alt. (m)	dry matter (t/ha)	nutrient content (kg/ha)						
				Na	K	Ca	Mg	P	N	S
Podocarp/broad-leav.	3	WT	82	88	245	219	119	9	206	40
Chionochloa rigida	18	884	31	3	47	58	22	13	97	10
Chionochloa macra	18	1257	12	2	8	12	3	5	40	7
Chionochloa pallens	11	1440	42		150	95	35	15	230	
Chionochloa macra	11	1200	33		70	180	41	23	160	
Festuca novae-zeland.	11	1050	2		10	10	1	1	20	

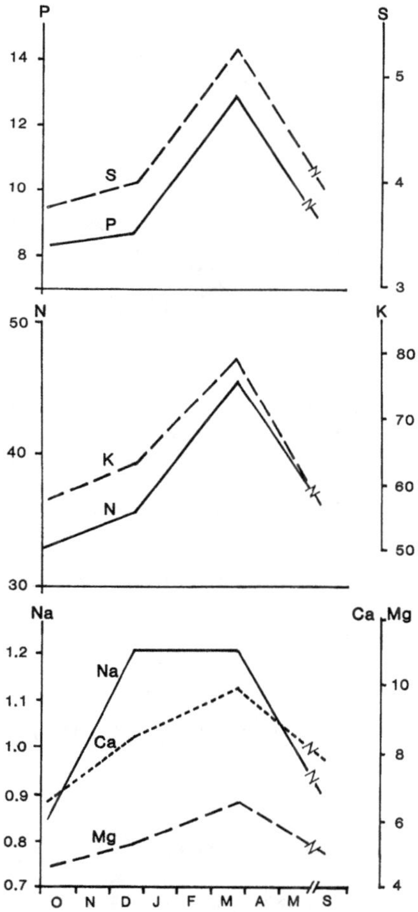

Fig. 14.12. Mass of elements (kg/ha) in live shoots and leaves of *Chionochloa rigida* in different months; 884 m a.s.l., Paddle Creek, Heron ER (Williams *et al.* 1977).

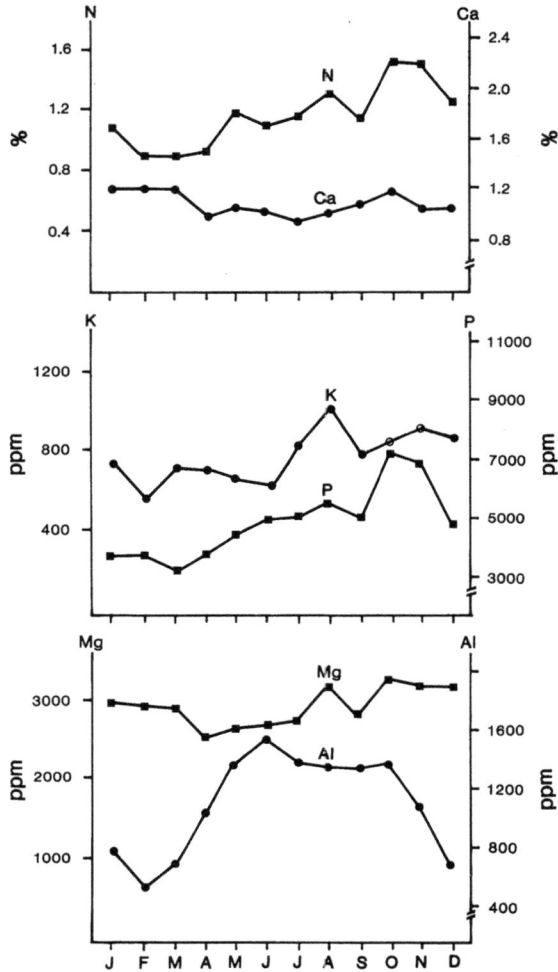

Fig. 14.13. Seasonal variations in element concentrations as proportions of total dry mass in leaf litter of podocarp/broad-leaved forest, Orongorongo valley, near Wellington (Daniel & Adams 1984).

early summer and lowest in late autumn (Fig. 14.13), although details vary among species and years.

Mycorrhizas and phosphorus uptake

Mycorrhizas (symbiotic associations of plant roots with fungi) are probably universal in natural vegetation in New Zealand. Generally, these are endomycorrhizas, formed with vesicular–arbuscular (VA) fungi. Among native plants, only *Nothofagus* is purely ectomycorrhizal; its mycorrhizas take various forms (Mejstrik 1972) according to which of some 170 species of mainly Agaricalean fungi is the partner (E. Horak in

Table 14.14. *Species in a descending series from those mycotrophic in soils relatively rich in available phosphorus to those that are never mycotrophic; no other factors limiting growth*

P (μg/ml)	species	habit	growth rate	rootlet diam. (mm)	root hairs	
					lgth (mm)	frequency
> 32	*Podocarpus totara*	tree	slow	> 1.0	0.1	rare
	Griselinia littor.	tree	slow	> 1.0	0.1	rare
15–32	*Weinmannia racem.*	tree	slow	0.2–0.3	0.07	rare
	Coprosma robusta	shrub	medium	0.2–0.3	0.07	rare
	Solanum laciniatum	shrub	fast	0.2–0.3	0.8	constant
	Leptospermum scopa.	shrub	medium	0.1–0.2	0.3	inconstant
8–15	*Fuchsia excorticata*	shrub	medium	0.1–0.2	0.2	constant
	Histiopteris incisa	fern	medium	0.3–0.4	1.5	constant
	Solanum nigrum	herb	medium	0.2–0.3	0.7	constant
	Metrosideros umbel.	tree	slow	0.1–0.2	0.3	inconstant
	Chionochloa rigida	grass	v. slow	0.1–0.2	0.5	inconstant
	Pteridium esculent.	fern	medium	0.2–0.3	1.6	constant
4–8	*Lolium perenne*	grass	fast	0.1–0.2	1.3	constant
	Poa colensoi	grass	slow	0.1–0.2	0.7	constant
	Anthoxanthum odor.	grass	fast	0.1–0.2	1.8	constant
NM	*Chionochloa macra*	grass	v. slow	0.1–0.2	0.8	constant
	Carex coriacea	sedge	medium	0.1–0.2	1.0	constant
	Juncus planifolius	rush	fast	0.2–0.3	1.1	constant
	Asplenium bulbif.	fern	v. slow	0.2–0.3	2.2	constant

Notes:
P, available phosphorus according to the Truog method of estimation.
NM, never mycotrophic; some mycorrhizal infection may occur.
All other plants mycorrhizal.
Source: Baylis 1975

Wardle 1984). Many introduced trees, especially among conifers, Fagales and Salicales, have ectomycorrhizal associations with introduced fungi such as *Agaricus subperonatus*, *Amanita muscaria* and *Suillus luteus*. Manuka can be either ecto- or endomycorrhizal. Forest ferns range from fully ectomycorrhizal under beech or pine to endomycorrhizal under mixed forest (Cooper 1976).

Mycorrhizas commonly increase nutrient uptake, especially of phosphorus. Most plants in most soils depend on them for survival, i.e. they are mycotrophic. However, mycotrophy ceases when plant-available P exceeds a certain level, which varies primarily according to the ability of a species to produce root hairs (Table 14.14). Where root hairs are highly developed, as in some dicot families, grasses, and especially, sedges and rushes, VA infection is sometimes present but probably has

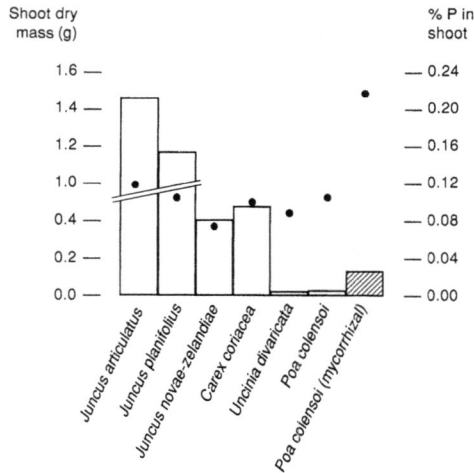

Fig. 14.14. Shoot dry mass and percentage P(dots) of pot-grown rushes, sedges and a grass (Powell 1975). All non-mycorrhizal except one *Poa colensoi* treatment.

little effect (Fig. 14.14). Among tussock grasses investigated by Crush (1973), i.e. *Festuca novae-zelandiae, Poa cita, Chionochloa rigida, C. macra* and *Poa colensoi*, only the last two benefited from mycotrophy, and then only in high-altitude soils with less than 8 ppm of readily available P. Conversely, plants with thick rootlets and few or no root hairs, such as *Griselinia* and all genera in the Magnoliales, are mycorrhizal even in fertile soil (Baylis 1975). The nodular rootlets of podocarps (Baylis 1969) and kauri (Morrison & English 1967) are strongly mycotrophic.

Fast-growing plants tend to be mycotrophic at higher P concentrations than slow-growing plants; compare manuka with its relative southern rata. The latter derives less benefit from its mycorrhizas than any other native woody plant studied (Table 14.15), which probably reflects a low demand for phosphorus, as indicated by the low P concentration of its leaves (0.4 mg/g; cf. values in Table 14.10) and seeds (10% of that in the equally minute seeds of kamahi).

The strongly mycotrophic *Griselinia littoralis* responds only weakly to added phosphorus, which suggests that its mycorrhizal system and growth rate are adapted to the generally low concentrations of available P in its native soils. Mycorrhizal plants of *Coprosma robusta*, on the other hand, respond vigorously to added P, in keeping with ability to colonise fertile soils (Fig. 14.15). Seedlings of relatively slow-growing, shade-tolerant species such as *Griselinia littoralis* have lower percentages of infected roots and lower rates of P-uptake in the shade than those of fast-growing, light-responsive species such as *Leptospermum scoparium*; the high P concentrations in the latter may enable them to respond quickly to increased light (Fig. 14.16).

Beeches resemble most other ectomycorrhizal trees in forming pure stands at high altitudes and on poor soils, and also in producing large amounts of mor-forming litter. Although a factor in the generally slow spread of beech may be difficulty in finding fungal partners beyond the rhizospheres of established beech trees (Baylis 1980),

Table 14.15. *Response to differing phosphorus concentrations in mycorrhizal and non-mycorrhizal seedlings of four woody species, as expressed in shoot dry mass (g).*

	290		315				380			
Duration (days):	4		4				10			
P level in soil[a]:										
Added P:	−		−		$+^b$		−		$+^c$	
Mycorrhizas:	−	+	−	+	−	+	−	+	−	+
Metrosideros umb.	0.07	0.21*	0.12	1.82*	0.14	1.81*	0.28	0.24	0.59†	0.53†
Weinmannia racem.			0.02	0.01		0.07†	0.01	0.26*	0.03†	0.61*†
Leptospermum scop.	0.01	1.56*								
Coprosma robusta	0.03	1.05*	0.03	1.32*	0.02	4.53*†				

Notes:
*† significant effects of mycorrhizas and added phosphorus respectively ($P \leq 0.05$)
[a] µg/ml, determined by the Truog method
[b] as 0.6 g apatite/kg
[c] as 15 ml aliquots $Na_2H_2PO_4$ added at 6 intervals
Source: Hall 1975

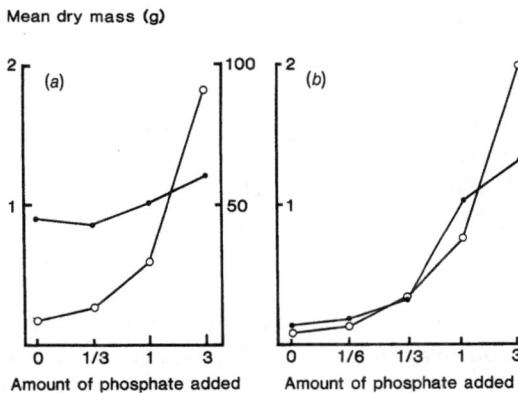

Mean dry mass (g)

Fig. 14.15. Mean dry mass of seedlings of (*a*) *Griselinia littoralis* and (*b*) *Coprosma robusta* at four levels of added phosphate (Baylis 1967). Filled circles, mycorrhizal seedlings in non-heated soil; open circles, uninfected seedlings in steam-sterilised soil.

Allen (1988) found mycorrhizal silver beech (*Nothofagus menziesii*) seedlings established in kamahi–southern rata forest, up to 50 m from the nearest mature silver beech tree.

Nitrogen in ecosystems

The amount of nitrogen in the living part of the ecosystem is broadly proportional to the total biomass, with the greatest concentrations being in green foliage. However, far

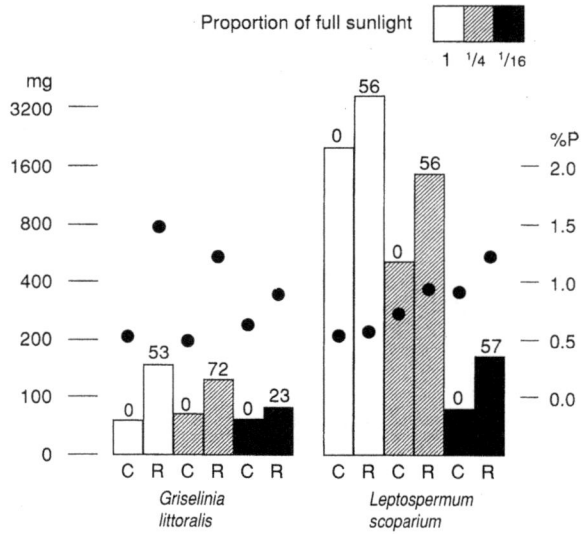

Fig. 14.16. Mean shoot dry mass (mg) and phosphorus content (black circles) of two species grown at three levels of shading, with two inoculum treatments (from Johnson 1976*b*). Numbers above the bars are mean percentage infection; C, control, R, *Rhizophagus tenuis*.

more is stored in decaying organic matter in the soil, so that the 16 t/ha contained in the kauri forest ecosystem (Silvester 1978) is only of the same order of magnitude as in managed pastures (Table 14.16). The high N content and productivity of the latter depends on symbiotic N-fixation by clovers stimulated by S and P in superphosphate (Table 14.17), vigorous N-cycling involving livestock, and in recent years, application of N fertilisers.

Native legumes are woody except for the subtropical strand plant *Canavalia rosea* and the semi-herbaceous *Swainsona* (p. 374), practically confined to recent soils, and associated with acid-producing strains of *Rhizobium*. This contrasts with the many Australian woody legumes that grow on acid heathland and have alkali-producing rhizobia (Greenwood 1978). Four native N-fixing genera are non-leguminous, *Discaria* and *Coriaria* having root nodules that harbour the actinomycete *Frankia* (Torrey 1978), *Gunnera* having the blue-green alga *Nostoc* in glands at its leaf bases, and the aquatic fern *Azolla* being associated with the blue-green alga *Anabaena*.

Introduced legumes are important N-fixers in many successional and disturbed communities, most being herbaceous, but woody forms such as gorse, broom and *Racosperma* spp. tolerate older, more acid soils. In a secondary succession to native forest in the Hutt Valley (p. 539), N content increases rapidly above and below ground during a gorse phase, but levels off or even decreases as the biomass of plants without an N-fixing capacity increases (Fig. 14.17).

Most indigenous ecosystems, including alpine vegetation, tall forests, heathlands and *Chionochloa* grasslands, practically lack vascular species known to fix nitrogen.

Table 14.16. *Amounts of nitrogen (kg/ha) in various ecosystems*

	reference	live above ground	litter	roots	soil excl. roots
Managed pasture	1	50–220		35–70	7000–20000
Marram and lupin on sand	2	59	30	12	1260
Pinus radiata on sand, 6 yr	2	38	47	16	1395
P. contorta 1250 m, 17 yr	3	690	406	269	5900
Chionochloa macra 1250 m	3	32	153	90	8100
Festuca novae-zeland. 1050 m	3	32	10	140	3950*

Notes:
* top 20 cm
References: 1, Ball 1982; 2, Gadgil 1982; 3, Nordmeyer & Kelland 1982.

Table 14.17. *Inputs through biological nitrogen fixation*

community	reference	annual N input (kg/ha)
managed pastures	1	100–680
Lupinus arboreus on fixed dunes	2	160
Ulex europaeus, regrowth after fire	3	100–200
Coriaria arborea on river terrace	2	150
Discaria toumatou on fan	4	70
Gunnera spp.	5	7
native forests { Lichens	6	1–5
Leaf surfaces	6	0.2
Litter	6	5–20
Chionochloa and *Festuca* tussock grasslands	7	0–3

Notes:
    References: 1, Hoglund & Brock 1982; 2, Gadgil 1982; 3, Egunjobi 1969; 4, Daly 1969;
5, Silvester & Smith 1969; 6, Green 1982; 7, Line & Loutit 1973.

The few available figures show that the total amount of nitrogen cycled is far less than where vigorous N-fixing plants are present (Table 14.18), indicating that losses from the system must also be small. Replenishment is from rainfall (4–6 kg/ha annually), free-living N-fixing bacteria and blue-green algae and, in forest, the abundant, large lichens of the Stictaceae, which contain blue-green algal symbionts (Green *et al.* 1980).

Water relations among woody plants

Transpiration, being controlled by stomatal conductance, closely parallels carbon dioxide uptake. In a hard beech forest, transpirational losses showed both daily and seasonal peaks (Fig. 14.18) and amounted to 28% of the annual rainfall of 1510 mm.

Near the end of a severe summer drought in a normally humid lowland district,

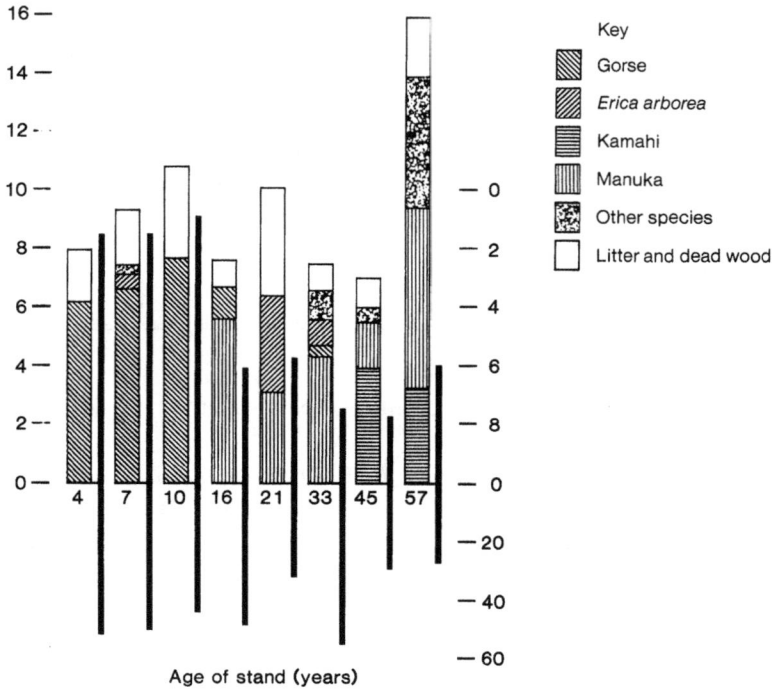

Fig. 14.17. Dry matter and nitrogen accumulation during secondary succession at Taita, near Wellington (from Egunjobi 1969).
Wide bar: total dry matter above ground (unit: 10^4 kg/ha).
Narrow bar: total nitrogen above ground (including litter) and below (unit: 10^2 kg/ha).

woody plants showed three kinds of response. The species governed by the curve in Fig. 14.19 all wilted, those at the top being the most susceptible in that wilting occurred at high water potentials (i.e. small negative values indicating a small degree of dehydration; low water potentials, i.e. large negative values, indicate a high degree of dehydration). Even *Melicytus, Myoporum* and *Griselinia*, which were damaged most severely, produced new growth by the following spring. The species at the top of the column showed drought avoidance: they maintained high water potentials through devices such as deep rooting and closing of stomata. Those low on the column showed drought tolerance; i.e. they tolerated low water potentials without wilting.

During a study in summit cool-temperate forest within and below the cloud belt, soil moisture was consistently above wilting point, and shoot water potential in all species remained above -1.5 MPa. Despite this, stomata closed early on fine days, probably in response to atmospheric vapour pressure deficits (Fig. 14.20). The susceptibility of this forest to die-back is discussed on p. 579.

Excised twigs of seven shrub species from semi-arid Central Otago had varied responses to dehydration (Fig. 14.21). *Rosa* and *Thymus* lost water rapidly while developing low osmotic potentials (i.e. strong affinities for water) whereas *Coprosma*

Table 14.18. *Amounts of nitrogen cycled between living plants and their environment*

community	annual N cycled (kg/ha)	reference
Beech forest via foliage drip	24 ⎫	Green 1982
via litter	26 ⎭	
Chionochloa pallens, 1440 m	24 ⎫	
Chionochloa macra, 1200 m	23 ⎬	Evans & Kelland 1982
Festuca novae-zeland., 1050 m	18 ⎭	

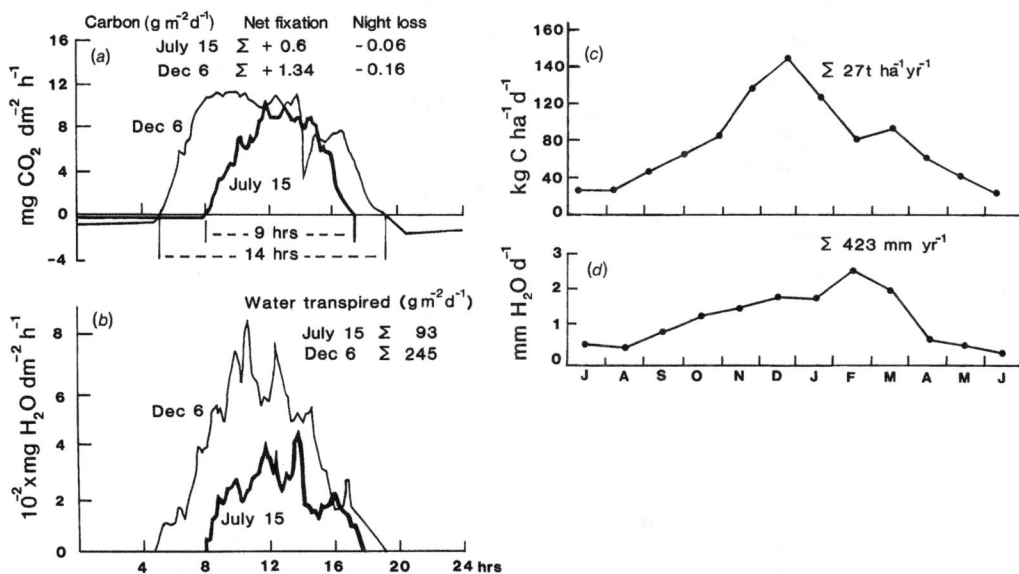

Fig. 14.18. Gas exchange in the canopy of hard beech forest at Big Bush, Nelson (Benecke & Evans 1987). (*a*) Net CO_2 fixation and (*b*) transpiration by sun leaves during summer and winter days, calculated for one-sided leaf surface. (*c*) Monthly and annual photosynthesis and (*d*) transpiration of the total crown. Sigmas (Σ) represent totals.

propinqua had a low osmotic potential at full turgor. The locally endemic native broom, *Carmichaelia compacta*, lost water only slowly.

Temperature tolerances

New Zealand trees and shrubs, and probably herbaceous plants as well, have only a small degree of frost tolerance, in keeping with generally oceanic climates. On the whole, winter frost resistance correlates well with geographic limits (Table 14.19), but provenances can differ considerably (Table 14.20). Some species of low resistance that reach relatively high latitudes (notably *Ascarina lucida*) do so only in mild, sheltered

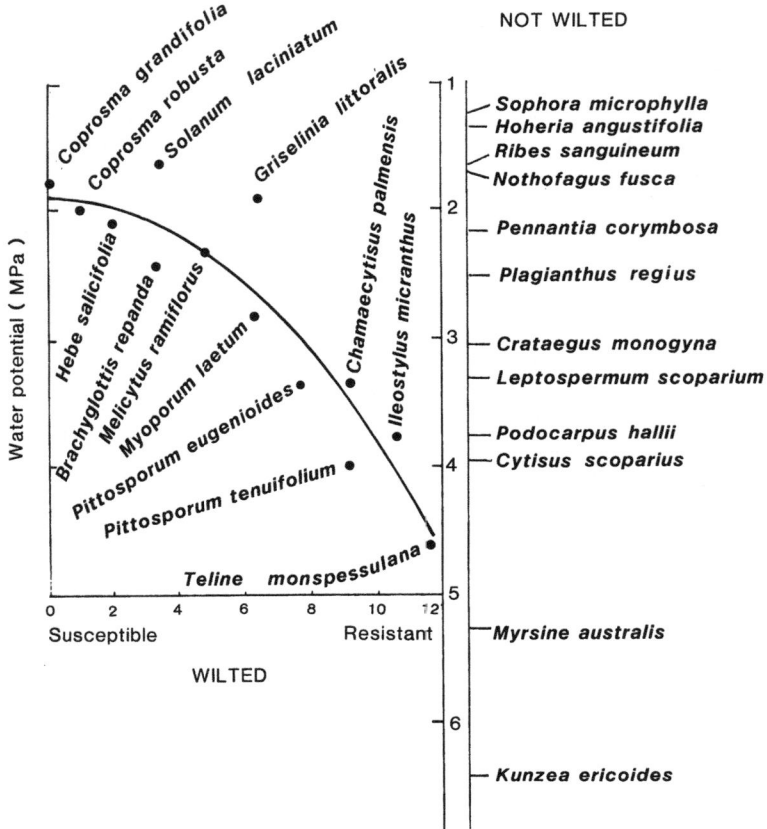

Fig. 14.19. Water potentials (megapascals) in shoots, measured in the field in April 1985, during a severe drought at Dunedin (Bannister 1986a). Left: relationship of maximum water potential in wilted shoots to a drought resistance ranking. Right: water potential in species that did not wilt. MPa values are negative.

habitats. Seven species extend to much colder localities than measured resistances would suggest, where at least two of them (manuka and *Fuchsia excorticata*) are frequently damaged by frost. In *Aristotelia serrata*, the resistance value refers only to the semi-deciduous leaves.

Low freezing resistances leave little room for wide seasonal variations; but the latter are comparatively large in the hardiest subalpine species, and are greater in the adventive *Dryopteris filix-mas* than in native ferns (Figs 14.22 and 14.23).

Killing of expanding foliage by late frosts is common in beeches and occurs also in other species, but recovery is usually rapid. Winter frosts occasionally cause severe damage at low altitudes, as in northern New Zealand in July 1982, when *Cyathea medullaris*, mangrove (*Avicennia resinifera*), pohutukawa (*Metrosideros excelsa*), bracken and other wild and garden plants were severely frosted (Beever & Beever 1983). When air temperatures fell below freezing on 12 consecutive nights and to −6.5 °C on one night near Taumarunui (King Country ER), tawa (*Beilschmiedia tawa*) trees

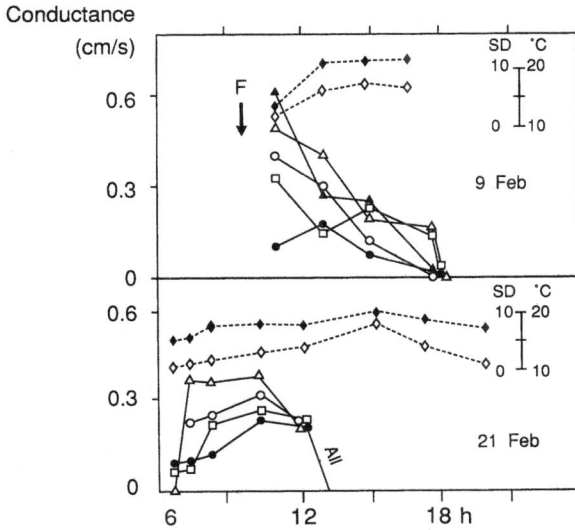

Fig. 14.20. Diurnal patterns of stomatal conductance (continuous lines) for five tree species, in relation to temperature and saturation deficit of air (g/m³) on days with morning fog (9 February) and days without (21 February). Kaimai Range, Coromandel ER. (Jane & Green 1985; copyright 1985 by the University of Chicago. All rights reserved.) F, time of fog clearance. Symbols: ◆, temperature; ◇, Saturation deficit (SD); △, *Quintinia serrata*; ▲, *Weinmannia racemosa*; □, *Myrsine salicina*; ●, *Ixerba brexioides*; ○, *Pseudopanax arboreus*.

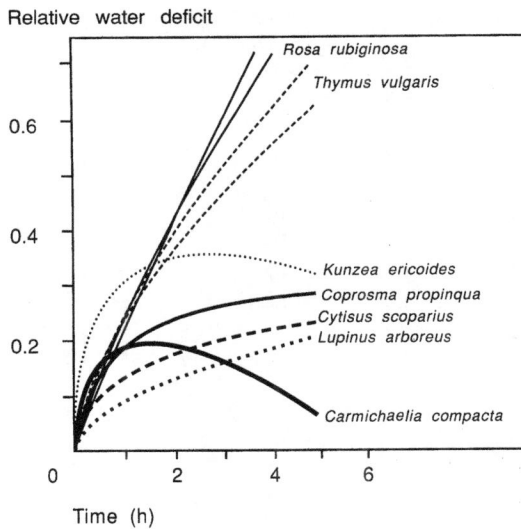

Fig. 14.21. Rates of water loss of twigs of seven shrub species from Central Otago, under constant laboratory conditions (Kissel *et al.* 1987). Relative water deficit expressed on a 0–1 scale.

Table 14.19. *Winter freezing resistance of woody plants and* Phormium *in relation to southernmost latitude and upper- or innermost altitudinal belt attained*

species	lf	b/c		latitude	belt	reference
Elingamita johnsonii	−2			34° 10′	CsWT	3g
Tecomanthe speciosa	−2			34° 10′	CsWT	3g
Meryta sinclairii	−2			36° 00′	CsWT	3g
Avicennia resinifera	> −3	> −3		36° 00′	CsWT	1
Metrosideros excelsa	−3	−3		39° 00′	CsWT	1
Ascarina lucida	−3	−3		46° 00′	MWT	1
Pittosporum eugenioides	−3	−7	↓↓	46° 37′	MCT	1
Planchonella costata	−3	−8	↑	38° 14′	CsWT	1
Pennantia baylisiana	−4		↑	34° 10′	CsWT	3g
Streblus smithii	−4		↑	34° 10′	CsWT	3g
Beilschmiedia tarairi	−4	−5c		38° 00′	MWT	1
Passiflora tetrandra	−4			43° 50′	MWT	4g
Aristotelia serrata	−4		↓↓	47° 17′	IT	4g
Coprosma macrocarpa	−5		↑	34° 10′	CsWT	3g
Elaeocarpus dentatus	−5	−5		43° 30′	MWT	1
Plagianthus regius	D	−5b	↓	46° 55′	IT	1
Fuchsia excorticata	D	−5b	↓↓	50° 44′	IT	1
Knightia excelsa	−5	−8		41° 20′	MWT	1
Cordyline kaspar	−6		↑	34° 10′	CsWT	3g
Kunzea ericoides	−6		↓	46° 00′	IT	5g
Agathis australis	−7	−7b		38° 05′	MWT	1
Libocedrus plumosa	−7	−7b		40° 40′	MWT	1
Metrosideros robusta	−7			42° 47′	MWT	3g
Griselinia lucida	−7			44° 30′	MWT	3g
Dacrycarpus dacrydioides	−7	−7b	↓	46° 55′	IT	1
Leptospermum scoparium	−7	−7b	↓↓	47° 17′	*SA	1
Nothofagus truncata	−7	−10		44° 00′	MCT	2
Quintinia acutifolia	−8	−8		43° 37′	MCT	1
Weinmannia racemosa	−8	−8		47° 17′	MCT	1
Coprosma lucida	−8	−8		47° 17′	MCT	1
Coprosma rhamnoides	−8			47° 17′	IT	6
Nothofagus fusca	−8	−10		45° 50′	IT	1
Dacrydium cupressinum	−8	−10c		47° 17′	MCT	1
Griselinia littoralis	−8	−10	↓	47° 17′	SA	1,3
Pittosporum tenuifolium	−8	−12c		46° 37′	IT	1
Phormium tenax	−8			47° 15′	IT	3g
Phormium cookianum	−8		↓↓	47° 15′	PA	3g
Metrosideros umbellata	−8		↓	50° 44′	IT	1
Hebe albicans	−9		↓	41° 10′	SA	4g
Phyllocladus trichomanoides	−10	−10	↑	41° 20′	MWT	1
Prumnopitys ferruginea	−10	−8	↑	47° 17′	IT	1
Cassinia 'fulvida'	−10	−10	↓↓	46° 00′	IT	1
Dracophyllum longifolium	−10	−10	↓↓	52° 33′	*SA	1
Brachyglottis buchananii	−10	−10		47° 10′	*SA	1
Olearia avicenniifolia	−10	−13c	↓	47° 00′	IT	1
Coprosma 'parvifl. dumosa'	−12			47° 17′	*SA	6
Lagarostrobos colensoi	−13	−13b	↑	45° 35′	MCT	1

species	lf	b/c		latitude	belt	reference
Podocarpus hallii	−13	−13b		47° 17′	SA	1
Libocedrus bidwillii	−13	−13b		46° 30′	SA	1
Halocarpus biformis	−13	−13b		47° 17′	*SA	1
Hebe brachysiphon	−13	−15c		43° 30′	IT	1
Nothofagus solandri	−13	−15		46° 15′	*SA	1
Lepidothamnus laxifolius	−13	−15c		47° 10′	PA	1
Dracophyllum acerosum	−15	−15		43° 50′	*SA	1
Hoheria glabrata	D	−15		46° 10′	*SA	1
Phyllocladus alpinus	−20	−20c		46° 00′	*SA	1
Podocarpus nivalis	−22	−20b		46° 00′	PA	1
Halocarpus bidwillii	−25	−23c		46° 00′	*VCT	1

Notes:

Freezing resistance defined as lowest temperature (°C) at which plant shows little or no damage.

lf: Freezing resistance of leaf

b/c: Freezing resistance of bud (b) and/or cortex (c), whichever is lower

D: Deciduous

↑↓   Measured freezing resistance greater (↑) or less (↓) than the distribution would suggest
      (especially where a double arrow is shown)

*   Species reaches penalpine tree limit

Cs, coastal; M, maritime; I, interior; V, valley; W, warm; C, cool; T, temperate; SA, PA, sub- or penalpine.

References in Tables 14.19 and 14.20: 1, Sakai & Wardle (1978); 2, Sakai *et al.* (1981); 3, Bannister (1984*a*); 4, Bannister (1986*b*); 5, Bannister & Fagan (1989); 6, Bannister & Lee (1989).

g: Cultivated material used.

Table 14.20. *Winter freezing resistance of leaves in relation to origin of material*

species	resistance (°C)	latitude	altitude (m)	reference
Leptospermum scoparium	−6	35° 07′	100	4g
	−6	44° 51′	884†	4g
Weinmannia racemosa	−5	42° 50′	90	1
	−8	42° 50′	760	1
Nothofagus solandri	−8	43° 20′	550	1
var. *cliffortioides*	−10	43° 10′	800†	1
	−13	43° 10′	1200†	1
Blechnum penna-marina	−6	45° 50′	30	5
	−11	45° 50′	530	5
	−19	45° 20′	1200	5
Phyllocladus alpinus	−12	42° 50′	30	1
	−20	43° 10′	910	1

Notes:

† Interior zone. References and other symbols as in Table 14.19.

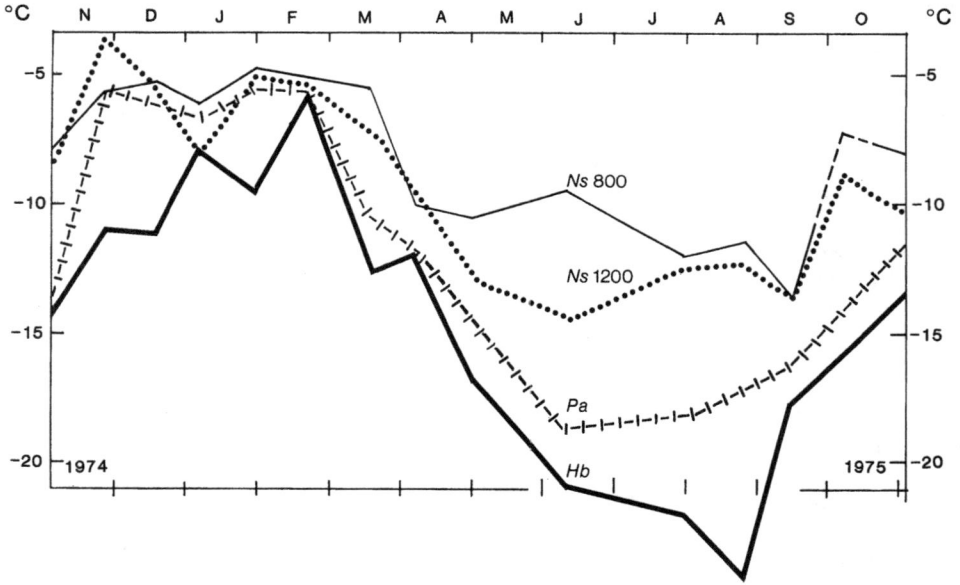

Fig. 14.22. Seasonal changes in the highest temperature at which freezing damage occurred in three subalpine species in the Craigieburn Range, Canterbury (Wardle & Campbell 1976). *Ns* 1200, *Nothofagus solandri* var. *cliffortioides*, 1200 m; *Ns* 800, *Nothofagus solandri* var. *cliffortioides*, 800 m; *Pa*, *Phyllocladus alpinus*; *Hb*, *Halocarpus bidwillii*.

in a forest remnant were badly damaged, especially those exposed to cold air drainage. Some died, and others were resprouting from lower in the crown four years later (Kelly 1987).

Winter frost damages young beech plants on frosty flats and, through setting the length of the growing season, prevents seedlings from establishing above the subalpine forest limit (Wardle 1985). Winter-drying of beech foliage is frequent along the forest limit, as a result of interaction between low temperatures and wind (Baylis 1959; McCracken *et al.* 1985) and is most intense above snow-drifts, where repeated dieback leads to stunted growth forms. In the penalpine belt, freezing winds kill foliage of shrubs and *Chionochloa* tussocks exposed above the winter snow pack. Wind also reduces growth through cooling foliage.

Bannister & Smith (1983) tested heat resistance in more than 40 species of ferns and woody plants. Correlations with latitude and altitude were weak, and some species were more resistant in winter than in summer. All species resisted temperatures over 42 °C and in summer some as much as 52 °C. Since air temperatures have never exceeded 42 °C in New Zealand and leaves are cooled by evaporation and convection, heat damage is unlikely to be important except in short, open vegetation fully exposed to the sun. Nevertheless, evaporative demands occasioned by high temperatures lead to stomatal restriction of CO_2 uptake, especially during föhn weather and in large

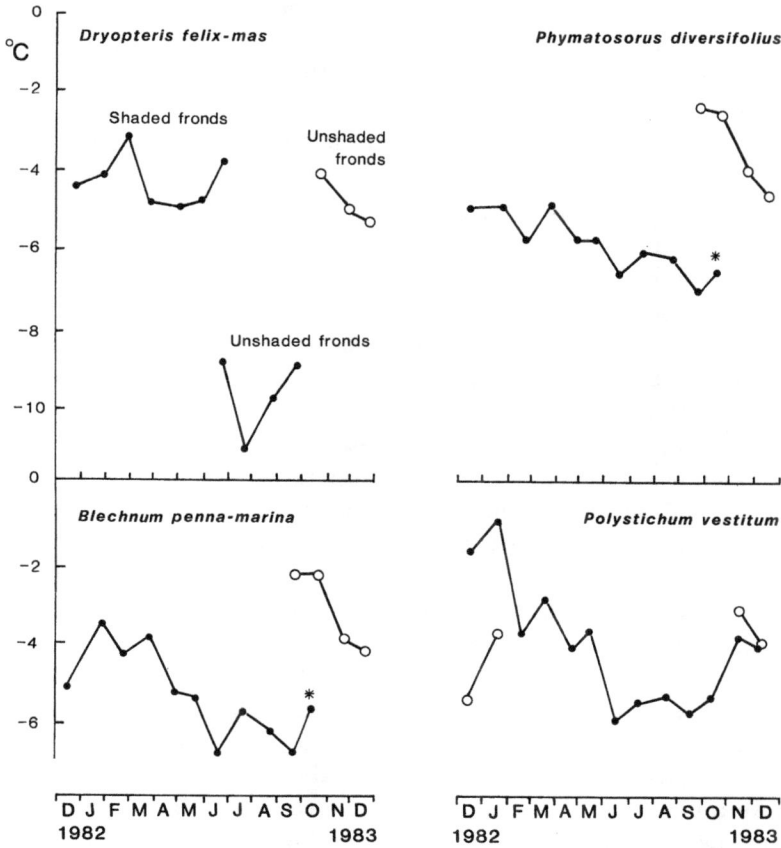

Fig. 14.23. Seasonal changes in the temperature at which 50% of the leaf area is damaged after eight hours exposure, for four ferns (Bannister 1984b). Values for freezing resistance (Table 14.19) are 0.1 to 1.0 °C higher. Symbols: filled circles, growth of the 1982–83 growing season; open circles, the previous or subsequent season's growth; asterisks, fronds senescent.

leaves exposed to full sunlight, which can exceed air temperature by more than 5 °C (U. Benecke, personal communication).

Conclusions

Biomass and production generally decrease along gradients from a warm, moist, fertile optimum, because growth rates and stature decrease within species, and dominance passes to smaller, slower-growing species. However, these trends are irregular, mainly because (1) young or seral stands usually have less biomass than mature or climax stands, but can have more production; (2) factors such as grazing, substrate disturbance and fire can reduce production or increase it where they lead to young, vigorous vegetation; and (3) there can be abrupt transitions between vegetation types of different stature, structure and productivity along apparently gentle gradients, or even within apparently uniform environments. Where tall vegetation abuts on shorter vegetation, as at the subalpine forest limit, reduced intra-specific competition may allow an increase in vigour that runs against the environmental gradient.

Biomass and production values in native terrestrial vegetation are similar to those reported for physiognomically equivalent vegetation in other temperate regions. Some values for aquatic vegetation are unusually high by world standards, and probably arise where hitherto unexploited resources are being tapped. On the other hand, rates of carbon assimilation and production measured in native plants are generally low compared with introduced tree and pasture species that have been selected and managed for optimal productivity. Neither do the growth rates of our taller trees match those that can be achieved by *Pinus radiata*, let alone some willows, poplars and eucalypts with pioneering attributes, although they are probably in the same order as 'climax' forest trees in other temperate regions. However, fast-growing smaller trees and tall shrubs, notably *Aristotelia serrata, Entelea, Coriaria arborea* and some hebes, colonise moist, recent soils at low altitudes.

Concentrations of nutrient elements among species are correlated with their edaphic range; within species, concentration also depends on site fertility. In vegetation on fertile soils, vigorous nutrient cycling involves high nutrient concentration, fast growth, copious litter-fall and quick decay. High-producing pastures are the best example, with slow-growing heath communities providing the opposite extreme. Total amounts of nutrient elements, however, are primarily a function of biomass; forests clearly hold more nutrients above ground than grasslands, even though these are mostly locked up at low concentration in inert wood. Nutrient quantities below ground more closely reflect soil fertility and community vigour.

From an agricultural viewpoint, New Zealand soils are naturally deficient, especially in respect of phosphorus and nitrogen. Among native plants, widespread mycorrhizal infection may be a consequence of low phosphorus levels, whereas remarkably few fix nitrogen symbiotically, and of those only the non-leguminous *Coriaria* and *Discaria* are very important ecologically.

Native plants possess generally low levels of frost and drought resistance; intolerance of extremes, rather than low average temperatures, may explain the extreme reduction of forest cover during glacial episodes (p. 14). New Zealand forest vegetation is clearly well-adapted and competitive in moist, equable, comparatively stable environments, whereas adventive plants are increasingly prominent where there is a premium on pioneering characteristics such as rapid growth, ability to fix nitrogen, and tolerance of extremes. This may be a legacy of geological history in so far as our lowland forests have persisted from the benign era of the mid-Tertiary; the high temperature optimum (27 °C) for assimilation and growth in kauri and several podocarps may reflect this (Hawkins & Sweet 1989). Adaptation to harsher, less stable Quaternary environments has depended on transoceanic dispersal and local evolution; the adventives that succeed in these environments result from an influx hitherto prevented by isolation.

15

SUCCESSION, RETROGRESSION AND INVASION

What is succession?

In a broad sense, succession refers to any sequence in which the structure and floristic content of vegetation change through time, but the concept can be narrowed to mean sequences that have begun on bare surfaces. A succession should be predictable if the climate is unchanging, there is no further significant disturbance, and local sources provide propagules at the appropriate time, in adequate quantity and evenly; although these conditions are never met fully, there is a good measure of predictability in fact.

In primary successions, which begin on raw mineral soils, increase of biomass is accompanied by differentiation of soil horizons, increase of carbon and nitrogen, uptake of nutrient ions and, usually, increase in floristic diversity and structural complexity. Secondary successions begin on soil developed under preceding vegetation, and are mostly initiated by fire. Most of the carbon, nitrogen and phosphorus above ground is lost to the atmosphere during combustion. Although the soil profile may remain intact, there are usually nutrient losses through enhanced leaching and erosion. Nevertheless, early stages of secondary succession can be favoured by a temporary flush of nutrients through mineralisation and decay of roots; for instance, when pastures were established after 'bush burns' (p. 280).

Ideally, successions lead to a stable climax, in which diversity, complexity, and quantities of C, N and other nutrient elements are at a maximum, with resident species maintaining themselves in constant proportions. This stage may be followed by prolonged retrogression, during which biomass, diversity and nutrient content decrease as rates of growth and organic cycling decline.

No community has all the optimal attributes of the 'ideal' climax, which is an asymptote never reached in nature. For instance, floristically simple beech forests often replace shorter but more diverse vegetation. A dense rimu stand on a gley podzol may have retrogressed from 'climax' conifer/broad-leaved forest on a more fertile yellow-brown earth but has greater biomass, and the dead biomass associated with stunted vegetation on peatland can contain more carbon and nutrient elements than forest on younger surfaces without peat.

507

Succession and climax can be considered in terms of intensity of disturbance. Primary succession is initiated by total destruction, whereas the ideal climax can be subject only to disturbances on a scale and intensity compatible with resident species maintaining themselves in constant proportions, without entry of species characteristic of successional stages. For the concepts to be useful, more latitude must be allowed. For instance, a forest might still be considered climax, although wineberry (*Aristotelia serrata*) and *Histiopteris incisa* are establishing on mineral soil exposed in minor slips or the root-plates of fallen trees. Extensive wind-throws that have led to lower-tier small trees, shrubs, tree ferns or ground ferns achieving temporary dominance could be considered part of the regeneration cycle of a climax forest. Even in an outwardly stable community, resident organisms may undergo fluctuations, often in response to weather patterns. Numbers can increase to levels that threaten the stability of the community, especially if the organism is a pathogen or parasite. Outbreaks of phytophagous insects that affect the structure of beech forests, pastures and tussock grasslands are cases in point.

Climax vegetation is highly ordered, a state that can be sustained only while the nutrient supply is adequate. It therefore mostly occurs on moderate slopes where nutrients are replenished through mixing and renewal in the soil profile. In forests, wind-throw also rejuvenates soils by overturning horizons. Fire reduces order and initiates secondary successions. However, few secondary successions begin with a *tabula rasa* as even the hottest fires usually leave some survivors, and areas burnt repeatedly develop fire-tolerant vegetation that can resist displacement by the original vegetation. On infertile soils, repeated fire also hastens retrogression.

Invasion by plants and animals new to a locality can totally alter succession. Humans are the most potent invaders, through direct actions and through the adventive flora and fauna that have accompanied us. There is also spontaneous transoceanic invasion of plants and animals; that of birds during the past 150 years is well documented. The arrival of manuka blight from Australia had marked effects on secondary succession (p. 197). Invasion patterns also develop because organisms spread at different rates to areas from which they were excluded by past environments or events. Species displaced by invaders suffer local or widespread reduction or extinction; the massive changes of the Quaternary, including the arrival of humans, have resulted in many episodes of regional invasion and extinction.

Because of secular climatic change, very long successions began in different environments from those beginning on comparable surfaces today. Also, there is little doubt that the ranges of many species have not kept pace with climatic change, so that similar environments in different localities may support successions involving different species.

Arrangement of material

Discussion of primary succession begins with the very complete sequence leading to lowland conifer/broad-leaved forest and subsequent retrogression that exists in Central Westland. Succession to woody vegetation at higher altitudes in the same region is discussed next, followed by examples of succession to beech forest in the west of the South Island. Rather fragmented successions to beech forest, mixed forest, subalpine heath and tussock grassland are described for the east of the South Island, followed by brief mention of early stages composed of adventive plants. The remaining sections on primary vegetation deal with special substrates (dune sands, volcanic, ultramafic and hard rock surfaces) and, finally, with the penalpine belt.

Secondary successions are taken in order from northern examples that lead to lowland mixed forest, through sequences near Wellington that lead either to mixed or beech forest, to eastern South Island examples involving adventive shrubs where the final outcome is less apparent. Montane and subalpine examples are taken from burns of known age, and conclude with discussion of shrub–grass mosaics where grazing influences the course of recovery. Invasion is discussed mainly in relation to the distribution of *Nothofagus*.

Primary succession

Low-altitude successions in central Westland

Sequences on schist and greywacke till and alluvium deposited by the Franz Josef Glacier, covering at least 100 000 years, have been dated through historical records, tree-rings, radio-carbon measurements and geology. The main markers are successive moraines, that increase in age towards the coast and with distance from the present terminal face of the glacier. Sites for the past 5000 years are contained in a deep mountain valley, and seem reasonably uniform through time and space in respect of climate, which is very wet (5000–8000 mm rainfall annually) and mild, apart from valley-floor frosts. Towards the coast, rainfall decreases to c. 2500 mm, sunshine hours increase, frosts are less severe, and secular changes in climate become more significant with increasing age of the surfaces. Successions on surfaces that differ in topography, texture and drainage are summarised in Fig. 15.1.

(1) On gentle ridges of fine moraine gravels there is a steady progression from open pioneering vegetation (p. 366) through to forest. During the shrub phase *Coriaria arborea* and *Carmichaelia grandiflora* fix nitrogen vigorously. Although southern rata (*Metrosideros umbellata*) and kamahi (*Weinmannia racemosa*) seedlings, destined to be the first forest dominants, establish within five years of the surface becoming available, they are at first overtopped by the faster-growing shrubs. The entry of tree ferns into seral bush at 130 years is important, as the seedlings of several tree species, including kamahi, establish epiphytically on their trunks. Miro (*Prumnopitys ferruginea*), mountain totara (*Podocarpus hallii*) and a very few rimu (*Dacrydium cupressinum*) also appear at 130 years, and are canopy trees by 400–450 years when the original rata and kamahi trees are collapsing. Veteran first-generation rimu trees survive on a 1000-year terrace where the understorey species are still those of semi-fertile soils, such as *Schefflera digitata* and *Asplenium bulbiferum*.

The next ridge in this sequence, dated at 5000 years, supports climax (rimu)/kamahi/*Blechnum discolor* forest. This is supported on yellow-brown earth, which has developed a leached, gleyed horizon over an iron pan on flatter areas. On moraines formed late in the last major glaciation (i.e. over 14 000 years ago), otherwise similar forests contain the lower-fertility species *Quintinia acutifolia*, *Phyllocladus alpinus*, *Neomyrtus pedunculata* and *Sticherus cunninghamii*. Soils are strongly podzolised and

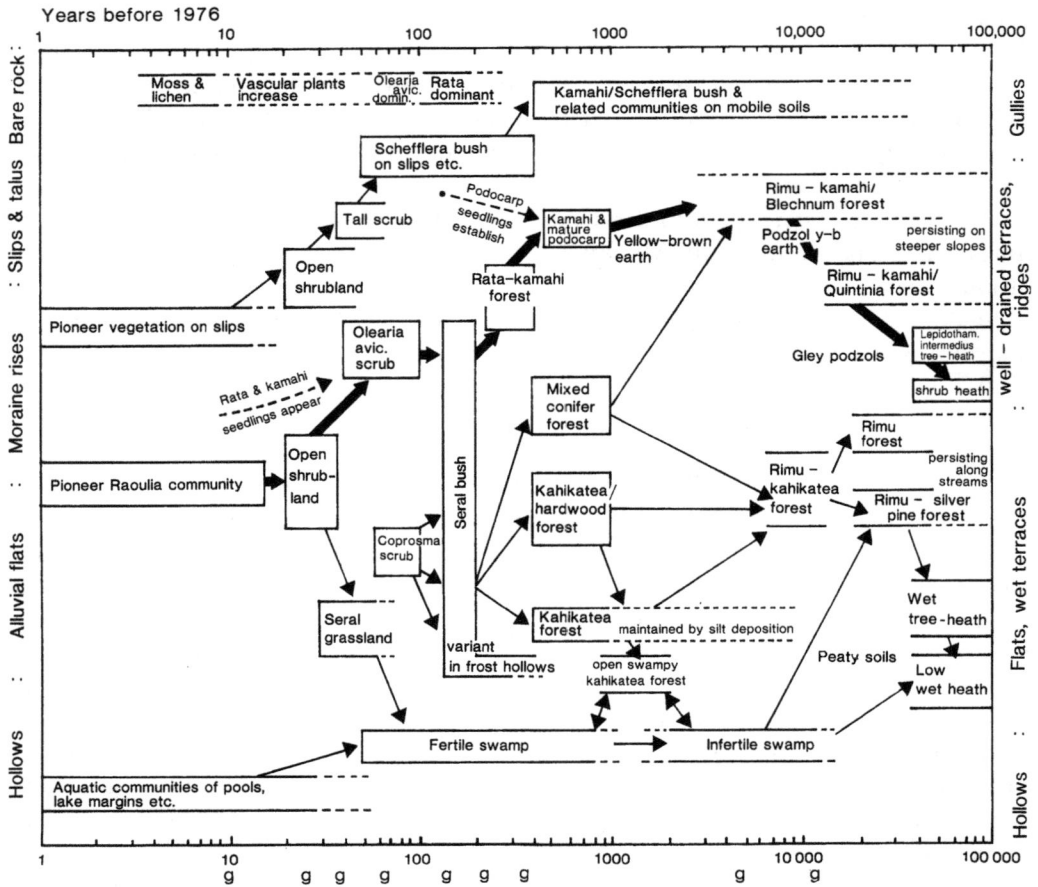

Fig. 15.1. The main lowland communities in Westland National Park arranged according to age of surface and topography. Arrows indicate successional trends, with heavy arrows referring to optimal sites (Wardle 1980a); g, dated advances of the Franz Josef and Fox glaciers since the Last Glaciation.

calcium-bound phosphorus has been exhausted (Fig. 5.8). This marks the beginning of retrogression which, on ridges deposited earlier in the last glaciation, has resulted in tree-heath, or in shrub-heath where there is almost no organic horizon. Still older moraine ridges, some dating from the penultimate glaciation and therefore at least 200 000 years old, have surfaces rejuvenated by erosion and support tall forest.

(2) On alluvial flats, early colonisation by shrubs and trees tends to be restricted to edges of terraces, gullies, bouldery areas and driftwood logs. Elsewhere grassland develops that can resist invasion by forest for centuries. Sooner or later, however, forest develops, often through a phase dominated by *Plagianthus regius*, other trees with filiramulate juvenile forms, and *Podocarpus totara* var. *waihoensis*. By a thousand years or so, there are mature podocarp stands with a mixture of species on well-drained terraces and kahikatea (*Dacrycarpus dacrydioides*) on poorly drained terraces.

However, on surfaces aggrading through floods, such as fans and between the meanders of sluggish rivers, broad-leaved bush persists with only a scattered overstorey of very large, wide-crowned podocarps. Fertile swamp develops on wet ground.

Older Holocene terraces beyond the reach of floods become mantled in wet, humic loam or peat and, depending on subtleties of drainage, support dense rimu (often with silver pine (*Lagarostrobos colensoi*)), rimu–kahikatea mixtures, and gradations to infertile swamps that occupy central, outward-draining areas. Broad depressions on surfaces deposited during the last glaciation support dense rimu along streams, tree-heaths on wet humic loams, and *Gleichenia dicarpa, Empodisma minus*, red tussock (*Chionochloa rubra*), stunted manuka (*Leptospermum scoparium*) and cushion plants on shallow gley-podzols (p. 334).

(3) Moist, stable debris is quickly occupied by small fast-growing broad-leaved trees, especially wineberry (*Aristotelia serrata*). More exposed sites pass through a shrub phase with species such as *Carmichaelia grandiflora, Coprosma rugosa, Hebe salicifolia* and, where water percolates through coarse debris, *Coriaria arborea*. On shattered bed-rock exposed by slips, slow-growing seedlings of rata and kamahi establish directly in the absence of competition and grow into even-aged stands. Large areas of such stands along the Alpine Fault may date a major earthquake around AD 1730–40.

On steep slopes, repeated sliding can maintain seral vegetation, and down-slope drift of regolith may prolong dominance by small broad-leaved trees and understorey species that indicate semi-fertile conditions, whereas ridges and spurs, even where narrow and steep, may support tall forest on podzolised yellow-brown earths.

(4) Retreat of the Franz Josef and other glaciers during the past 400 years has exposed extensive areas of bed-rock. Mosses appear in crevices after 3–4 years. By nine years, *Racomitrium crispulum* and *Stereocaulon* cover the whole of south-facing rock surfaces, and *Epilobium brunnescens, E. glabellum, Ourisia* sp., *Poa novae-zelandiae* and *Olearia arborescens* seedlings are rooted in the larger crevices. On north aspects, crustose lichens up to 15 cm in diameter cover about 70% of the rock and *R. crispulum* 10%.

Vascular plant cover increases very slowly as soil builds up in crevices and hollows. Seedlings of rata were first noted on rock bare of ice for 32 years and *Olearia avicenniifolia* can dominate at 60–80 years. Succession is faster on pockets of moist gravel stranded on the rock, *Poa novae-zelandiae* and *Epilobium glabellum* being common in the early stages; later, dense clumps of *Coriaria arborea* usually develop. At 100–120 years there is patchy scrub dominated by rata, with intervening areas covered in *Blechnum* 'black spot', *Hymenophyllum multifidum, Lycopodium scariosum* and *Racomitrium crispulum*.

Montane and subalpine successions in Central Westland

Kaimata Range

Stewart & Harrison (1987) describe forest stages at 600–1200 m in the north of the region. Slopes are mostly steep and soils frequently rejuvenated by erosion and deposition. *Griselinia littoralis/Pseudowintera colorata (Hebe salicifolia)* and *Hoheria glabrata/Polystichum vestitum* bush grow below and above 900 m respectively. Rata, kamahi and *Quintinia*, with some kaikawaka (*Libocedrus bidwillii*), form a more mature community that mainly occupies yellow-brown earths less than 1000 years old; rata and *Griselinia* co-dominate on similar sites, although only above 770 m. Pink pine (*Halocarpus biformis*) co-dominates with rata and kamahi, or *Olearia colensoi*, or *Dracophyllum traversii* and *D. longifolium* respectively in an altitudinal series of communities on remnant surfaces, mainly with gley podzol soils > 2000 years old.

Westland National Park

After only nine years, hollows in deep, unconsolidated landslide debris in the Copland Valley at 750 m were occupied by dense, 1 m tall scrub of *Hebe salicifolia*, *Olearia ilicifolia* and *Coprosma rugosa*, with *Acaena anserinifolia* as the main plant beneath. Even the driest patches of gravel were largely covered by herbs such as *Gnaphalium ruahinicum*, *Hypochoeris radicata* and *Carex cockayneana*. Probably this site will eventually support *Hoheria/Polystichum* stands.

Succession is slower on upland moraines in the Park, even though seedlings may establish while underlying ice is still melting. At subalpine levels, shrubs that appear early include *Carmichaelia grandiflora*, *Coprosma rugosa*, *Hebe subalpina*, *Olearia moschata*, *O. avicenniifolia*, *O. nummularifolia* and *Dracophyllum longifolium*. The last, by virtue of its longevity, dominates on moraines formed during the seventeenth and eighteenth centuries, although areas of large boulders are still virtually bare apart from lichens. At the Balfour Glacier, at 800 m, rata trees are beginning to overtop *Dracophyllum* on moraines of this age.

At the La Perouse Glacier, tall heath of *Dracophyllum* and *Olearia* species still occupies weakly podzolised soils on moraine crests at 850 m, that probably date between 1450 and 2500 years. On moraines some 5000 years old with strongly podzolised soils, *Phyllocladus alpinus* and kaikawaka are present in the heath, which has become confined to the steeper and more sheltered sides; elsewhere there is *Chionochloa* 'westland' grassland and patches of *Oreobolus pectinatus* bog. Alluvial flats and troughs between moraines usually proceed through the sequence *Raoulia tenuicaulis*→*Poa cockayneana*, *Lachnagrostis lyallii* and *Rytidosperma setifolium* →*Chionochloa pallens* (seventeenth to eighteenth century surfaces)→*C.* 'westland' or infertile swamps.

Hokitika catchment

Successions on slopes at 800–950 m in the Cropp basin, where precipitation averages about 10 m and rates of uplift and compensating lowering of surfaces by erosion about 11 mm annually are summarised in Fig. 15.2. On fans and terrace gravels sloping < 10°, *Chionochloa pallens* and *C. rubra* dominate at < 70 years, with intervening cover consisting largely of early seral herbs such as *Blechnum penna-marina*, *Leptinella squalida* and *Festuca matthewsii*; *Carmichaelia grandiflora* s.l., *Hebe subalpina*, *Coprosma rugosa* and *Muehlenbeckia axillaris* also enter early, especially on coarser substrates.

With time and decline of fertility, *Chionochloa pallens* decreases in abundance and vigour relative to *C. rubra*, and late-successional woody plants increase. *Dracophyllum uniflorum* co-dominates with chionochloas on a surface dated at under 1275 years. At

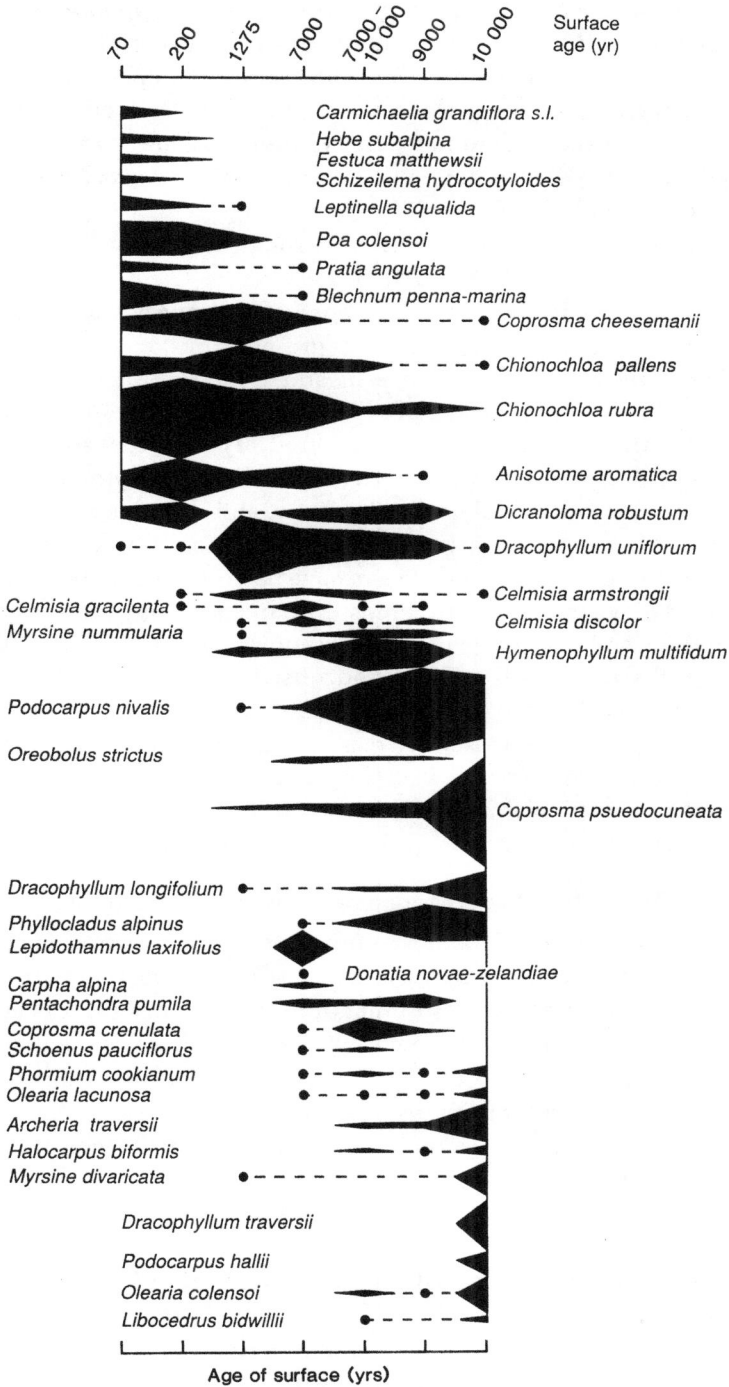

Fig. 15.2. Percentage cover of species on seven sites of increasing age on slopes < 10° at 850–905 m; Cropp River, Western Alps (from L.R. Basher, unpublished).

7000–10 000 years *Phyllocladus alpinus, Dracophyllum longifolium, D. uniflorum* and *Coprosma pseudocuneata*, with an overstorey of kaikawaka, pink pine and *Dracophyllum traversii*, form a mosaic with *Podocarpus nivalis* and openings with *Chionochloa* spp. and herbs of gley-podzol soils, such as *Abrotanella linearis* and *Oreobolus* spp. Wet areas, which at 200 years are dominated by red tussock, *Carex coriacea* and sphagnum, may develop by 2500 years to bog with *Donatia, Celmisia glandulosa, Oreobolus* and *Carpha*, but peat is only local.

Slopes of 30–45° are often rejuvenated through debris avalanches. At 40 years, patchy, 1–2 m tall scrub (*Dracophyllum longifolium–Olearia colensoi–O. ilicifolia/ Blechnum* 'mountain') has already developed. Seral species such as *Carmichaelia grandiflora* and *Chionochloa pallens* are present, but so also are seedlings of *Dracophyllum traversii, Archeria traversii* and *Coprosma pseudocuneata*; this early entry of 'climax' species may reflect proximity of seed source, and a shallow, compact rooting medium that is already partly weathered. By 140 years shrub-heath (*Olearia lacunosa–Dracophyllum traversii)/D. longifolium/Coprosma pseudocuneata/Blechnum* 'mountain') has reached its fullest development, and contains saplings of kaikawaka and pink pine. Mature conifers are present on surfaces 1000 years old. Despite the steep slopes, some spurs and terrace risers have been stable for 7000–10 000 years and have developed gley-podzols, but there is little further vegetation change.

Corresponding soil development is in the sequence recent soils (< 200 years)→yellow-brown earths (500–1000 years)→podzolised and gleyed yellow-brown earths (1000–1500 years)→gley podzols (2500–10 000 years). In the surface horizon, pH, calcium and phosphorus decrease, whereas iron and aluminium sesquioxides initially increase; later, sesquioxides decrease in the A and increase in the B horizon. Reflecting the rapid leaching, available nutrient ions are in low concentrations from the outset, with hydrogen and aluminium being the predominant exchangeable cations.

Beech forest successions in the western South Island

Beech seedlings can establish almost immediately after disturbance, especially at higher altitudes, under drier climates, and on less favourable parent materials. If seed is not limiting, pure stands may result that differ little from climax beech forest. This success is due to growth rates consistently faster than those of the dominant trees of mixed forests and, perhaps, to more efficient cycling of phosphorus. More benign environments or sites further from beech seed-trees, such as the centres of large areas bared by landslides or floods, are pre-empted by seral small trees, shrubs and herbs, and only slowly invaded by beech.

North Westland

In a terrace sequence in Reefton ED (Fig. 5.9) red beech (*Nothofagus fusca*), silver beech (*N. menziesii*), lowland totara (*Podocarpus totara*) and matai (*Prumnopitys taxifolia*) occupy the recent terrace. Earlier Holocene and Last Glaciation terraces support hard beech (*Nothofagus truncata*) with some red and silver beech, and the Penultimate Glaciation terrace supports rimu, silver pine, kamahi and mountain beech (*Nothofagus solandri* var. *cliffortioides*). The fertility sequence for beeches parallels that of conifers, and conforms to the sequence indicated by foliar nutrients (p. 486).

South Westland

In the Arawata valley, terraces situated 1.5 m, 4.5 m and 15 m above the flood-plain at 47 m a.s.l. are tentatively dated at 150, 450 and 1500 years (Smith & Lee 1984). The lowest carries grazed adventive grassland, with patches of silver and red beech, mountain totara and small broad-leaved trees; the silver beech trees have the spreading crowns of pioneers. The second terrace carries forest of red and silver beech, with many of the latter still having the pioneer form. There are also poles less than 20 cm d.b.h. of rimu and miro. The highest terrace supports tall forest of red and silver beech, with large rimu and miro, which appear to be first-generation. The respective soils are recent, a podzolised yellow-brown earth with an iron pan at 9 cm, and a podzol with an iron pan at 16 cm. All are gleyed under an annual rainfall estimated at 6000 mm. Except on the youngest terrace and in the organic horizon of the older soils, bases and nitrogen are already at very low concentrations, comparable to those in the oldest soil in the Reefton sequence. Despite this, forest composition, including the predominance of red beech on the oldest terrace, indicates a fertile status throughout, presumably maintained through organic cycling and continuing release of phosphorus through weathering.

Poorly drained areas on the middle terrace carry thickets of filiramulate shrubs and saplings of kahikatea and matai. Comparable sites on the upper terrace, some of them peaty, are dominated by slender kaikawaka, mountain totara, kahikatea, *Phyllocladus alpinus*, silver pine and rimu.

Below the Fettes Glacier in the Landsborough valley, silver beech seedlings appear on moraine gravels at 450 m within 10 years, and surfaces formed in the mid or late nineteenth century support open and dense pole stands. On early seventeenth-century moraines, pure silver beech forest of near-climax appearance is growing on an incipient podzol, with a leached A horizon at 0–5 cm and a thin iron pan at 17 cm (Wardle 1980*a*).

Fiordland

Fig. 15.3 illustrates successions leading to silver beech/kamahi/*Blechnum discolor* forest on slips of three ages. Slopes are 30–35°, the altitude 300 m, the aspect north, and the parent material gneiss. On the slip face manuka was initially dominant, but seedlings of climax species were well represented in the earliest stage. The debris fan, on the other hand, was occupied by small broad-leaved trees, especially wineberry, and the climax trees were only sparsely represented. *Blechnum 'capense'* and *Polystichum vestitum* prevailed in the respective understoreys. Re-examination of the 15-year-old slip after a further 24 years showed that manuka had grown from 2 to 8 m, and seedlings of tall trees had become saplings. On the debris fan, the wineberry canopy was disintegrating and fuchsia and *Pseudowintera* had increased (Mark *et al.* 1989).

North-west Nelson

Seral vegetation that has colonised slopes devastated by the Murchison earthquake of 1929 is analysed in Table 15.1. On relatively fertile mudstones, the characteristic shrub is *Coprosma propinqua*, and saplings of red beech are up to 5 m tall. This leads to forest communities in which red and silver beech, mountain totara and kamahi are

SLIP-FACE   CLIMAX   DEBRIS-FAN

Age of surface (yr)   15 —— 49 —— 75 —— FOREST —— 49 — 15

Leptospermum scoparium

Metrosideros umbellata

Nothofagus solandri var. cliffortioides

Weinmannia racemosa

Nothofagus menziesii

Griselinia littoralis

Coprosma foetidissima

Coprosma colensoi

Pseudowintera colorata

Coprosma rhamnoides

Carpodetus serratus

Myrsine divaricata

Pseudopanax colensoi

Fuchsia excorticata

Hoheria glabrata

Aristotelia serrata

Trees ■ 10%   Shrubs ▨ 10%   Herbs ☐ 10%   Present |

Astelia nervosa

Lycopodium scariosum

Blechnum "capense"

Blechnum "black spot"

Blechnum discolor

Polystichum vestitum

Histiopteris incisa

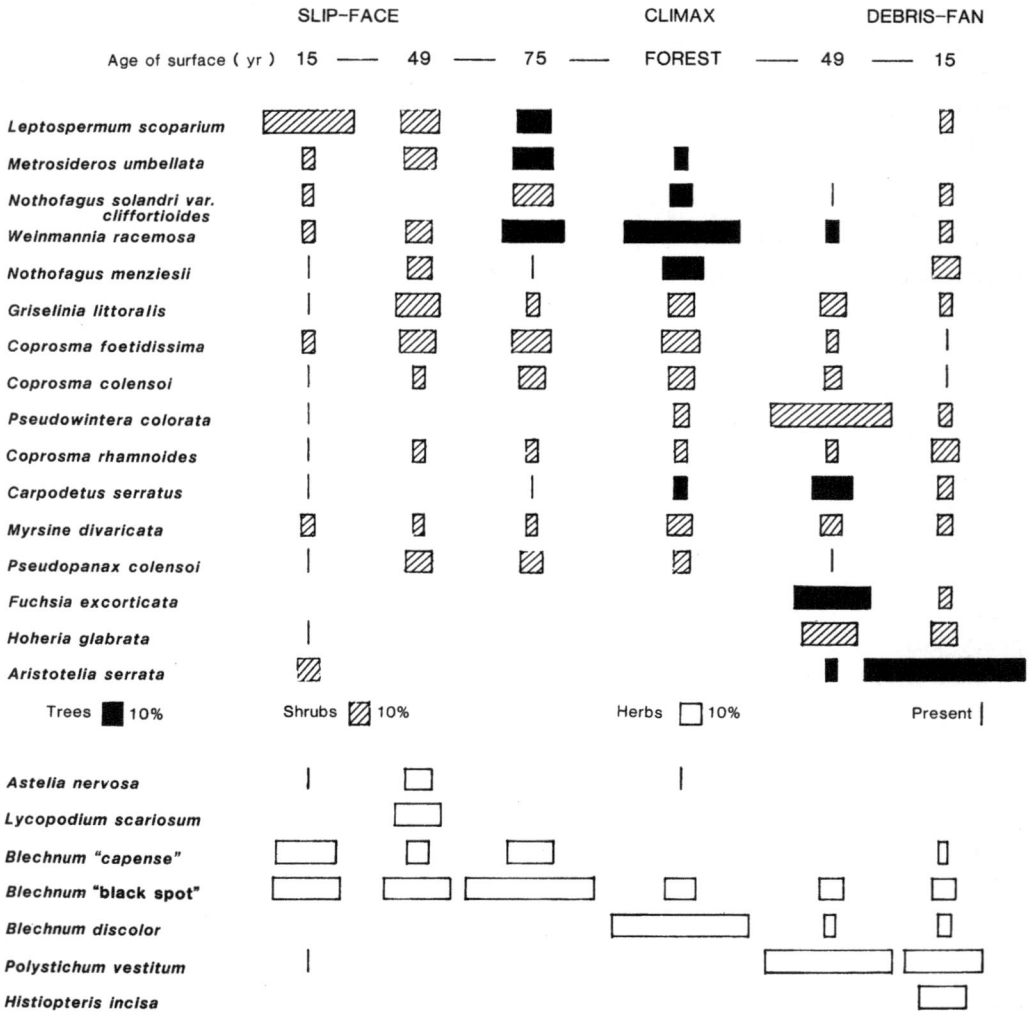

Fig. 15.3. Relative density of main species from successional stages and climax forest above Lake Thomson, Fiordland. Species shown only in their most important tier (from Mark *et al.* 1964).

prominent. On less fertile sandstone slips, manuka is prominent, the other most frequent species being kamahi, rata, *Coprosma foetidissima* and silver beech; the corresponding forests are described as either red–silver–hard beech/kamahi, or beech–southern rata.

Sequences below tree-limit in the eastern South Island

The early stages

The broad river flats and fans east of the South Island Divide provide many opportunities for plant successions, which seldom reach climax except in the most remote valleys. New gravel surfaces close to beech seed-trees can be densely colonised

Table 15.1. *Percentage frequency of species in two seral associations on landslide scars in the Matiri Valley, Western Nelson*

	'fertile' shrubland (16 samples)	'infertile' shrubland (6 samples)
Species more frequent in 'fertile' shrubland		
Coprosma propinqua	100	50
Mycelis muralis	75	0
Hebe salicifolia	75	17
Olearia avicenniifolia	69	17
Uncinia uncinata	63	0
Coprosma rugosa	56	0
Podocarpus hallii	56	0
Helichrysum bellidioides	56	33
Acaena sp.	50	0
Aristotelia serrata	50	0
Corybas trilobus	50	0
Nothofagus fusca	50	0
Carpodetus serratus	50	17
Myrsine divaricata	50	17
Species more frequent in 'infertile' shrubland		
Leptospermum scoparium	13	67
Weinmannia racemosa	13	67
Metrosideros umbellata	19	67
Coprosma foetidissima	38	67
Nothofagus menziesii	44	67
Cyathodes fasciculata	0	50
Gaultheria crassa	0	50
G. antipoda	19	50
Nothofagus soland. cliff.	25	50
N. truncata	0	33

Source: Rose 1985

by beech seedlings within a few years. In inland Canterbury at 1100 m, mountain beech seedlings began to establish in the first mast year, four years after deposition (Wardle 1972) and at 700 m, even-aged beech stands 3 m tall had developed within 40 years (Orwin 1970).

On frost-flats, on surfaces beyond the effective dispersal range of beech seed, or where young beech has not yet formed closed stands, succession is through grasses, forbs and shrubs. Early stages are described by Calder (1961), Wardle (1972) and Orwin (1970). The most favoured position is at margins of sandy depressions, where seedlings are sheltered by stony rises. Although a number of species appear within five years (Fig. 15.4), willow-herbs are by far the most numerous plants. *Raoulia* seedlings appear early and expand by marginal growth to become the main cover, the rapidly growing *R. tenuicaulis* being mainly confined to damp hollows, whereas *R. hookeri* and *R. haastii* grow on higher surfaces. *Muehlenbeckia axillaris* also expands rapidly through substrates too stony for raoulias. Seedlings of matagouri (*Discaria toumatou*),

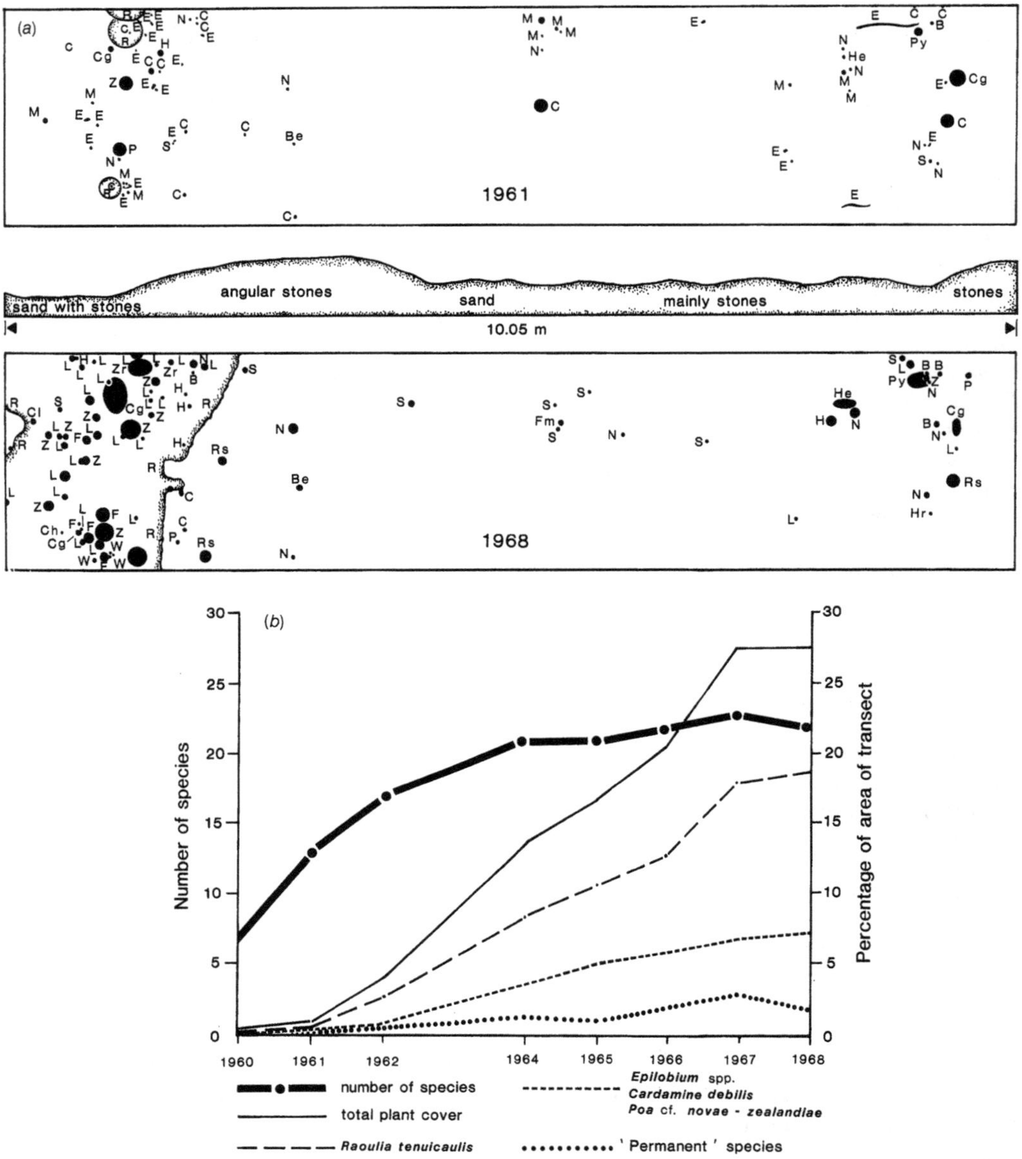

Fig. 15.4. Colonisation of gravel surface deposited by a flood in 1957; 1100 m, Broken River, Puketeraki ER (Wardle 1972).

(a) Charted quadrats; *Epilobium* not shown for 1968. Abbreviations: B, *Blechnum penna-marina*; Be, *Brachyglottis bellidioides*; C, *Cardamine debilis*; Cg, *Carex goyenii*; Ch, *Cerastium fontanum*; Cl, *Celmisia lyallii*; E, *Epilobium brunnescens*; F, *Festuca rubra*; Fm, *Festuca matthewsii*; H, *Hieracium* sp.; He, *Hebe epacridea*; Hr, *Hypochoeris radicata*; L, *Lachnagrostis filiformis* var. *semiglabra*; M, *Epilobium melanocaulon*; N, *Nothofagus solandri* var. *cliffortioides*; Rs, *Rytidosperma setifolium*; P, *Poa ?novae-zelandiae*; Py, *Polytrichum* sp.; R, *Raoulia tenuicaulis*; S, unidentified seedling; W, *Wahlenbergia albomarginata*; Z, *Luzula banksiana* var. *migrata*; Zr, *Luzula rufa*.

(b) Percentage of total area covered by different categories of species, and numbers of species.

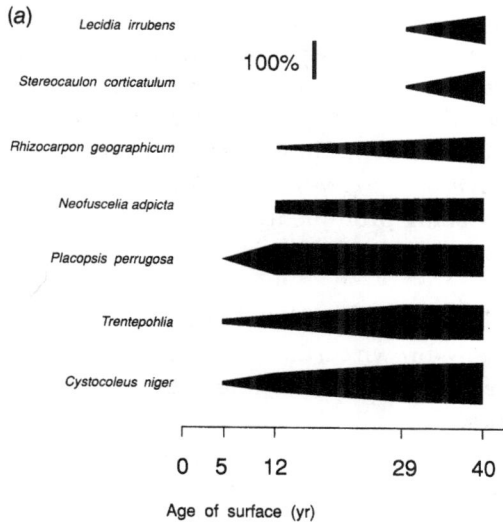

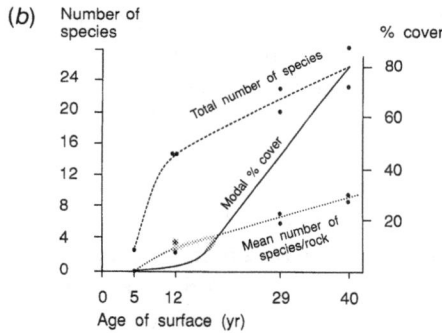

Fig. 15.5. Four stages in colonisation of stones by crustose lichens and *Trentepohlia*, in the upper Waimakariri catchment, Puketeraki ER (from Orwin 1970). (*a*) Changes in percentage of stones supporting seven important species; (*b*) changes in cover and diversity.

Cassinia and filiramulate coprosmas appear within 10–12 years, and the number of species also steadily increases.

Stony herb- and moss-field (p. 367) can persist at least 85 years (Scott 1963). *Poa cita* grassland on fine substrates and matagouri thickets also develop, both benefiting from calcium-bound phosphorus released by the weathering alluvium. Wetland successions ensue in abandoned channels.

Patches of boulders may support little but crustose lichens. These increase in size and number of species with time, and in sheltered valleys at 600–750 m completely covered stones within 40 years (Fig. 15.5). However, large stones are more favourable than small stones, and south aspects better than north aspects. Boulders on the exposed moraines of the Mueller Glacier require 300 years to develop a complete lichen cover. There are also species differences; *Placopsis parellina* prefers south aspects as, to some extent, do *Pertusaria graphica* and the fruticose *Sterecaulon* spp. *Placopsis* spp. generally prefer damper, more sheltered environments, and *Rhizocarpon* is restricted to harsher environments (Orwin 1972).

In mountain valleys that are higher, cooler, wetter or more sheltered, seral vegetation is more luxuriant, with filiramulate and broad-leaved shrubs, and mesic herbs such as *Poa cockayneana, Elymus laevis, Mentha cunninghamii, Parahebe decora* and the larger celmisias being prominent.

Older terraces

Loess blown from dry river beds on to higher terraces forms loamy topsoil; weathering and leaching lead to yellow-brown earths and, in drier areas, yellow-grey earths. Hard tussock (*Festuca novae-zelandiae*), browntop (*Agrostis capillaris*) and sweet vernal (*Anthoxanthum odoratum*) typically dominate, except where severe depletion favours species characteristic of earlier stages (p. 367), and in cooler or wetter areas where *Chionochloa* grasslands prevail. Wetland can occupy depressions and under leaching conditions may support oligotrophic plants such as *Oreobolus pectinatus* and *Sphagnum cristatum*.

Older terraces usually have less scrub than younger, stonier ones, although remaining matagouri may attain small-tree stature. Probably, the closer grass cover restricts establishment of shrub seedlings, but fire is also a factor. Lichens often provide most of the cover where the original stony or gravelly surface remains exposed (p. 256).

Successions to conifer/broad-leaved forest

In the absence of beech, successions once led to mixed forests except in the driest and frostiest localities, but this rarely happens today. An undated sequence on alluvial terraces near Kaikoura (Dobson 1979) would have been typical of low altitudes, but adventive herbs are more prominent as pioneers than native willow-herbs and raoulias. Manuka and kanuka (*Kunzea ericoides*) can establish on bare ground or follow *Coriaria arborea*. The longer-lived kanuka eventually dominates over other pioneers, which also include *Cassinia*, coprosmas and summer-green species of *Coriaria*.

The kanuka and manuka tree-heaths shelter seedlings of broad-leaved trees and shrubs, including *Coprosma robusta*, mahoe (*Melicytus ramiflorus*), *Dodonaea viscosa*, *Pittosporum* spp. and *Myrsine australis*, as well as ferns, *Astelia fragrans* and uncinias, but on drier ground only small-leaved shrubs (*Coprosma* spp., *Cyathodes juniperina* and *Corokia cotoneaster*) may enter. Depending on proximity of seed trees, there is eventually invasion by podocarps and broad-leaved trees such as hinau (*Elaeocarpus dentatus*), *Hedycarya, Griselinia littoralis* and *Pennantia corymbosa*.

Surviving fragments of successions to high-altitude mixed-forest communities seem similar to those described for Westland, but interpretation is clouded by burning and uncertain dating. At 760–920 m a.s.l. on the terminal moraines of the Mueller Glacier near Mt Cook, pioneer herbs are followed by open shrubland, mainly of filiramulate species and *Gaultheria crassa* but also including larger-leaved olearias and hebes and occasional *Phyllocladus alpinus*. According to Gellatly's (1984) dating of surfaces, this shrub phase has endured 1500 years, which suggests prolonging by fire.

Unburnt patches of *Phyllocladus* and mountain totara trees occupy moraine slopes dated to 2500 years and beyond, that are sheltered from föhn winds descending the glacier. By 3350 years, soils are yellow-brown earths, and bog pine (*Halocarpus bidwillii*) is present on small moraines further down-valley that are dated to 4200 and 7200 years.

Kaikawaka is present in some valleys. *Phyllocladus* and *Podocarpus nivalis* form dense scrub on frosty valley floors and above the limit of tree totaras; they appear early in successions, and although slow-growing, can achieve dominance by 500 years through vegetative spread and longevity. By 1000 years *Dracophyllum longifolium* and *D. uniflorum* become co-dominant with *Phyllocladus* (Burrows & Heine 1979).

Successions on alluvial flats dominated by adventive plants

Most lowland flood plains are now colonised almost entirely by willows and adventive herbs and shrubs (p. 365). Other common adventive plants, such as blackberry (*Rubus fruticosus* aggr.), briar (*Rosa rubiginosa*) and elder (*Sambucus nigra*), may invade later. Native forest species can invade flood-plain scrub, but in many districts only propagules of adventives such as pines, sycamore (*Acer pseudoplatanus*), *Clematis vitalba* and *Dryopteris filix-mas* are at hand. Few native plants establish under willows, and their ability to displace rank grasses and forbs is also limited.

Dune sequences

On coastal dunes additional factors in the progressive development of vegetation are stabilisation of mobile sand, increasing shelter from salt-laden winds, and decreasing salinity of ground water in the dune slacks. Because most coastal sand country has been frequented since the beginning of human settlement and is amenable to pastoral farming and exotic forestry, uninterrupted successions to native forest are rare. The best are in southern districts of the South Island. An excellent sequence at Big Bay (Olivine ER) is illustrated in Fig. 15.6.

(1)   *Austrofestuca* and *Desmoschoenus* occupy the foredune.

(2)   These are joined by *Calystegia soldanella*, *Hypochoeris radicata* and *Trifolium dubium*, with *Coprosma propinqua* and *Phormium tenax* appearing further back.

(3)   Dense *Phormium*, *C. propinqua* and *Coriaria sarmentosa*, with 'lanes' of *Coprosma acerosa*, *Hypochoeris*, *Cyathodes fraseri* and *Poa pusilla*.

(4–6) Bush of increasing height and age, with *Ascarina, Griselinia littoralis*, other broad-leaved trees and tree-ferns. There are also podocarp saplings, some of which seem older than the broad-leaved matrix, probably having survived salt storms that kill the latter.

(7)   Kamahi and rata with an understorey of small broad-leaved trees.

(8)   Rata up to 2 m diameter. Radiocarbon dating of a tuatua (*Paphies subtriangulata*) shell midden under the forest shows that the dune stabilised about 650 years ago (R.H. Hooker, personal communication).

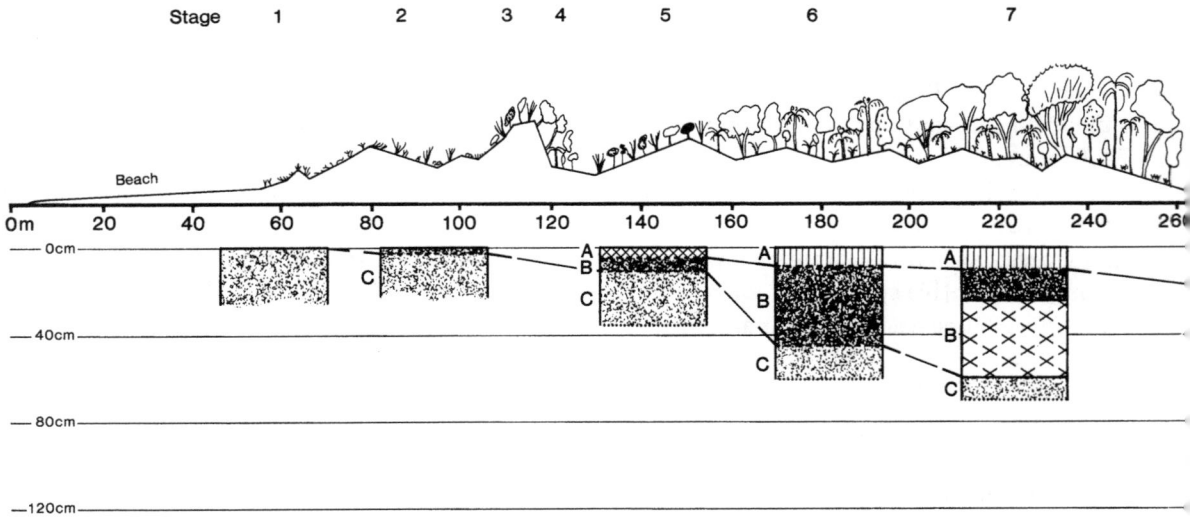

Fig. 15.6. Profiles of vegetation and soil along a sequence of sand dunes at Big Bay, Olivine ER (from drawing by P.N. Johnson). Stage numbers as in text; vertical scales for vegetation approximately 1:500 for stages 1–7, 1:2000 for stages 8–10.

(9)   Rimu and some miro over a kamahi sub-canopy, with *Blechnum discolor* abundant in the understorey.

(10)   Forest on the podzolised innermost dune, similar to 'terrace rimu' stands (p. 128) in that kamahi trees are slenderer than on younger dunes, and rimu and miro trees are accompanied by frequent saplings.

The youngest dune hollows are dry, but those between stages 5 and 9 hold swamps with mainly *Carex virgata* and *C. coriacea*, and peat and peaty silt increasing to a depth of 130 cm. Among the dunes of stage 9 there are peat swamps with thickets of *Gahnia xanthocarpa* and *G. rigida*, wetter areas occupied by *Baumea* spp. and *Lepidosperma australe*, and an open tier of silver pine, rata, kamahi, *Phyllocladus alpinus* and manuka. Narrow swamps are overgrown by masses of *Freycinetia baueriana*. The oldest dune (10) impounds Waiuna lagoon and the wetland communities illustrated on p. 329.

There are many variations on this sequence in South Westland. According to depth, flow and salinity of the water, hollows separating younger dunes can contain dense *Phormium*, groves of young kahikatea or kowhai (*Sophora microphylla*), tidal marshes, and fresh or saline water bodies; infertile swamps and wet tree-heaths prevail between much older crests.

In Catlins ER, under a rainfall of 1200 mm/yr, three dune crests have been dated by tree rings to 130, 440 and 1000 years (Smith *et al.* 1985). Marram (*Ammophila arenaria*) dominates to the crest of the foredune, southern rata is prominent among small broad-leaved trees on its back slope and on the middle crest, and on the third crest rimu forms an overstorey to large, apparently first-generation rata trees. Rata

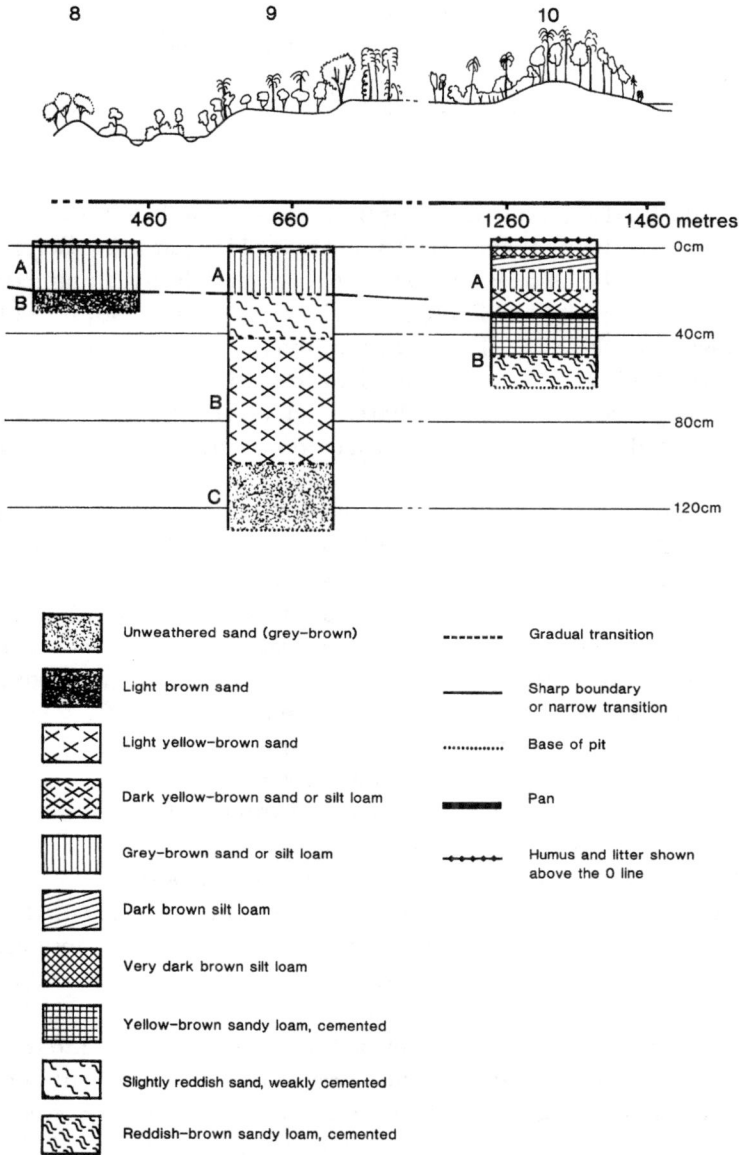

with *Blechnum discolor* beneath occupies the first dune hollow, the second is subject to flooding and mainly contains *Ripogonum* thickets (corresponding to the *Freycinetia* thickets at Big Bay) and the wet hollow behind the third dune supports a rimu/kamahi stand with *Blechnum discolor* and extensive bryophyte cover beneath. Soils show increasing horizon differentiation up to the third crest, except that there is a buried profile in the first hollow; survival of burial may explain the early appearance of rata and *Blechnum discolor* in the sequence. Surface organic matter also increases, with the third hollow containing 50 cm of peat. Shell fragments in the youngest sand result in a surface pH of 7.8, but by the second crest calcium carbonate has disappeared and pH has fallen to 5.4; it is 4.3 in the final hollow. Nutrient levels decline correspondingly.

Successions on volcanic parent materials

Tarawera eruption

On the Volcanic Plateau, explosive eruptions have periodically denuded extensive areas, the most recent being that of Mt Tarawera (1111 m) in 1886. Before this, mixed forest clothed the lower slopes almost to 610 m and fire-maintained shrublands extended to the upper slopes. The exposed summit plateau supported low, open, mossy shrubland (Kirk 1872). The eruption, mainly of basalt lapilli but with mud being ejected from adjacent Lake Rotomahana, destroyed nearly all vegetation on the summit and north-western slopes, whereas on the south-eastern slopes survival graded from isolated trees to complete forest communities, according to distance from the crater.

Within 12 years, *Cortaderia fulvida*, bracken (*Pteridium esculentum*) and tree tutu (*Coriaria arborea*) were vigorously colonising fine materials at the lower altitudes (Turner 1928). In 1915 Aston described diverse cover on denuded slopes between 320 m and 470 m. Shoots of *Litsea calicaris*, tawa (*Beilschmiedia tawa*), and mahoe (and presumably also of pohutukawa (*Metrosideros excelsa*), kamahi and fuchsia) had sprouted from seemingly dead stumps. Sheltered ravines supported dense young bush with pohutukawa, kamahi, *Pittosporum tenuifolium*, *Hebe stricta*, *Cyathea dealbata*, bracken and other ferns. Tall tutu thickets occupied sides of broad valleys, whereas dry slopes carried manuka scrub. Scoria flats, however, supported only sparse, stunted tutu, bracken, *Hebe stricta*, *Olearia furfuracea*, *Coprosma robusta*, kamahi, manuka and *Pimelea prostrata*. *Raoulia* mats on water courses on these flats were colonised by seedlings of tutu and manuka that grew up to shelter wineberry and fuchsia seedlings. Above 500 m, there was only low, open vegetation with *Gaultheria*, *Cyathodes fasciculata*, *Dracophyllum subulatum*, *Muehlenbeckia axillaris*, raoulias, and a few stunted kamahi. Only *Raoulia* patches were recorded from the summit plateau above 900 m.

By the 1960s dense bush generally clothed the slopes to about 850 m. Seedlings of rimu, *Phyllocladus glaucus*, matai and mountain totara occurred only near surviving seed trees, whereas seedlings of kahikatea, miro and lowland totara were found up to 4.0, 4.8 and 4.8 km respectively from survivors (Burke 1974).

The most recent accounts are by Timmins (1983) and Clarkson & Clarkson (1983). Tawa dominates where the forest had been least damaged, but its upper limit is still 150–200 m below its regional limit of 820 m. These tawa stands grade upwards through co-dominance of kamahi and tawa to closed kamahi pole stands that define a still-rising 'bush line'. There are also areas where kamahi is invading and overtopping manuka. Kanuka is prominent on north-western slopes. Scrub of tree tutu, *Olearia furfuracea* and *Coprosma robusta* extends above the 'bush line', especially in sheltered gullies, and contains young broad-leaved trees, mostly kamahi, but its height, density and rate of development decrease with increasing altitude.

Vegetation still covers less than half of the summit plateau, and consists mainly of moss and raoulias (p. 368). Patches of denser vegetation contain shrubs, especially *Dracophyllum subulatum*, tutu, and *Gaultheria* spp. Circular patches of *Muehlenbeckia axillaris*, often growing on otherwise unstable surfaces, are densely colonised by herbs and mosses. Shallow ponds support aquatic plants that are probably dispersed and manured by gulls. Young plants of kamahi and other broad-leaved trees are confined to rhyolite boulders, but sapling pines establish vigorously from seeds blown in from plantations, and will alter the indigenous character of the vegetation unless removed.

Taupo eruption

About 20 000 km² was devastated by the rhyolytic eruption from Lake Taupo in AD 177 (J. Palmer, unpublished). Pyroclastic flows across the lower terrain left sheets of lightly welded pumice, while loose air-fall lapilli blanketed slopes and summits. Concentric zones of forest types were thought to represent stages of succession, with age of vegetation increasing with distance from the eruptive centre (p. 120), but a more complex picture is emerging, in which local survival, differential dispersal, variations in climate, topography and parent material that result in soil and microclimatic differences, and secondary successions initiated by Maori fires, all play a part.

A pollen spectrum from an upland mire (Clarkson *et al.* 1986) shows mainly bracken, *Haloragis*, grasses, composites, manuka and *Pseudopanax* immediately after the eruption. These gave way to bog pine, which was followed by *Phyllocladus*. This probably indicates rapid colonisation of shallow, unconsolidated tephra on the slopes, followed by colonisation of the pumice flats by the slow-growing, shrubby podocarps. In the mire, pollen dominance passed from sedges to sphagnum and *Gleichenia*. Within 450 years, podocarps had reasserted regional dominance, with totara and matai being most prominent at first and rimu increasing later.

Today, in areas that have escaped fire, the lowest depressions and terraces still support tree- or shrub-heath and infertile swamps with bog pine and *Dracophyllum subulatum* (p. 193). Intermediate terraces support *Phyllocladus alpinus*, upland broad-leaved trees such as *Griselinia littoralis* and *Elaeocarpus hookerianus* and, in the southern part of the area, silver pine, kaikawaka and local mountain beech. These communities, especially those with bog pine, seem correlated with intense temperature inversion and porous, droughty parent material of inherently low fertility, in which root penetration and drainage are impeded by welded horizons and iron–humus pans. Phosphorus-occluding allophane clays also form rapidly; these constitute 9, 13–17 and 20–33% of the soil mass in Tarawera, Taupo and older Holocene tephras respectively (Birrell & Fieldes 1968).

The tall, dense podocarp forests for which the region was renowned lie mostly on still higher ground, where frost is less intense and air-fall tephra has buried pre-eruptive top soils to depths usually less than 50 cm. On these sites, succession was probably similar to that now occurring on the lower slopes of Mt Tarawera. However, tall podocarp stands also occur on deep pumice, and may have invaded preceding tree- and shrub-heath as suggested by the pollen record and in the way that matai, especially, invades similar vegetation in frosty basins at present. The lower tiers of podocarp forest on pumice, being composed of species such as mahoe, *Schefflera*, *Dicksonia fibrosa* and *Polystichum vestitum*, indicate higher fertility than heath

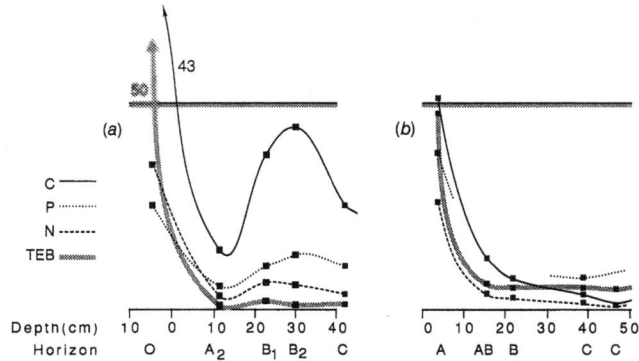

Fig. 15.7. Concentrations of carbon, nitrogen, phosphorus and bases in two profiles in Taupo tephra. The lower horizontal line shows zero concentrations, and the upper shows the following concentrations: C 10%, N 2%, P 20 mg/100 g (citric acid extractable), TEB (total exchangeable bases) 10 meq/100 g (from Vucetich & Wells 1978). (a) Under rimu forest; (b) under 'tussock, cocksfoot, bracken, Leucopogon' (= Cyathodes).

communities growing on comparable soils. This is probably because the trees bring up nutrients from deeper levels and cycle them more vigorously. Fig. 15.7 shows the large cation pool held in the organic horizon under forest.

A stand consisting mainly of rimu trees 400–500 years old (Fig. 16.31) included matai trees 100–200 years older. This confirms the gradual change in composition shown by the pollen record, and presumably is due to slow loss of fertility. A possible further stage is demonstrated in a swampy basin at Pureora, now surrounded by forest of rimu, totara, matai and kahikatea. A preceding forest, flattened and buried during the Taupo eruption, was dominated by tanekaha (*Phyllocladus trichomanoides*) and rimu which, like existing tanekaha stands in the district, was growing on older tephra soils (Clarkson *et al.* 1988).

Stands of mountain, black, red and silver beech are widely scattered through the area affected by the Taupo eruption. Those on leeward slopes, including the large beech forests on the south-western slopes of Mt Ruapehu, survived the blasts, but others result from post-eruption dispersal, further spread having been inhibited once mixed forest had reoccupied the surfaces. Polynesian burning may have begun within 800 years of the eruption, and by the time of European settlement had induced *Dracophyllum subulatum* heath and silver tussock (*Poa cita*) grassland over most of the pumice plains and depressions, grey scrub in frosty gullies, and bracken fernland over higher and steeper ground.

Rangitoto Island (Auckland ER)

The dark, porous scoria of Rangitoto, a basaltic cone in Waitemata Harbour, is colonised almost solely by pohutukawa. Although many forest plants grow up in its shelter, the island is still incompletely vegetated. Most of the eruptions occurred between AD 1200 and 1500, but there was activity as late as AD 1800 and growth rings show that no trees are more than 150–200 years old (Robertson 1986).

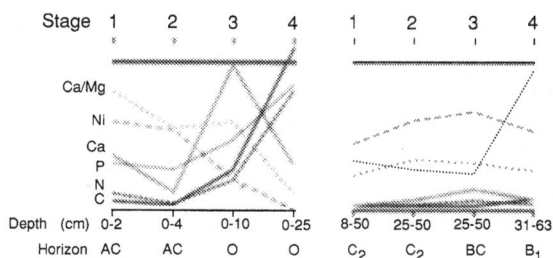

Fig. 15.8. Chemical characteristics of upper (left) and lower (right) horizons in a sequence of soils on ultramafic debris, Little Red Hills, Olivine ER (from Lee & Hewitt 1982). The lower horizontal line shows zero concentrations, and the upper shows the following concentrations: Ni 300 ppm, Ca 5 meq/100 g, P 50 mg/100 g, N 2%, C 50%, Ca:Mg ratio 25%.

Ultramafic surfaces

A sequence at 700–760 m in the Red Hills of South Westland is related to soil changes over ultramafic debris (Lee & Hewitt 1982). Less than 5% of the surface is vegetated, and boulders usually remain bare. The pioneers are *Poa colensoi*, *Rytidosperma setifolium*, *Lachnagrostis lyallii*, *Luzula crinita* var. *petriana*, *Wahlenbergia albomarginata*, *Forstera sedifolia*, *Neopaxia australasica*, *Helichrysum bellidioides*, and the dwarf shrubs *Myrsine nummularia*, *Pimelea oreophila* and *Muehlenbeckia axillaris*. *Racomitrium lanuginosum* forms carpets up to 0.1 m thick.

In a second stage, *Cyathodes juniperina*, *Dracophyllum uniflorum*, *Melicytus alpinus*, *Hebe odora* and *Coprosma* aff. *parviflora* form a canopy over the pioneer species and the ferns *Blechnum procerum*, *B. penna-marina* and *Polystichum vestitum*. The third stage is discontinuous tree-heath of southern rata, *Dracophyllum longifolium* and *Phyllocladus alpinus*, above a tier of the preceding shrubs together with *Pseudopanax colensoi* and manuka. In the tallest patches (> 10 m) mountain beech, *Griselinia littoralis* and pink pine are also present.

Carbon and nitrogen increase as organic material accumulates, mostly as a surface horizon which becomes about 25 cm thick. In contrast to fluvioglacial sequences, phosphorus also increases, possibly because it accumulates in the topsoil faster than it is leached from the total profile. Trends for calcium are erratic, and even at the highest concentrations its ratio to magnesium remains very low (Fig. 15.8). Nickel is the most abundant of the heavier metals but, as with cobalt, copper and manganese, its concentrations are greatly reduced in organic horizons. Slow weathering, and nutrient contents that remain low until an organic fraction with improved ion-exchange and water-conserving qualities builds up, are as likely to account for the stunted vegetation as toxic ions.

Penalpine and alpine successions

In the South Island penalpine belt, primary successions begin with scattered grasses, tufted herbs and, on fine material, *Raoulia tenuicaulis* mats. Eventually *Chionochloa* tussocks dominate, beginning with *C. pallens* or *C.* 'robust', which give way to species such as *C. rigida* or *C.* 'westland' and eventually to *C. macra* or *C. crassiuscula*. This is related to gleying, leaching of nutrients, and a change in phosphorus fractions from apatite, through plant-available iron- and aluminium-bound fractions, to occluded forms (Williams *et al.* 1978*b*). Comparable variations in *Chionochloa* dominance have been described from the Tararua Range in the North Island (p. 243).

Alpine successions run parallel to gradients associated with decreasing exposure or altitude. They are also complicated by other processes active at high altitudes, such as flushing by water from snow banks, that favours peaty top-soil and hygrophilous species such as *Oreobolus pectinatus*; frost-sorting on partly vegetated surfaces, that maintains herbfields of *Marsippospermum gracile* and *Celmisia haastii*; and partial rejuvenation of profiles by erosion, that favours the snow tussock and tall herb communities which ascend northerly aspects. In the North Island, upper slopes of the high volcanoes remain sparsely vegetated, or support patches of vegetation in stages of regeneration and decay (p. 431).

Three South Island examples indicate the time-scales involved.

(1) A sequence on lateral moraines at 1400 m in the Dart valley begins with a sparse alpine herbfield, and grades through successive crests to a prominent moraine (8) that supports a mixture of pioneer and later species on soil with visible profile differentiation (Fig. 15.9). Next, a set of moraines (9–11) supports closed vegetation of *Chionochloa pallens* and *Celmisia lyallii* on yellow-brown earth. A much older moraine (12) has *Chionochloa crassiuscula* with cushion and mat plants, including *Phyllachne colensoi, Drapetes* aff. *lyallii* and *Coprosma perpusilla*. Species diversity increases over the first five crests.

Diameters of *Rhizocarpon geographicum* lichens on rocks indicate that the glacial advance represented by crest 8 occurred about AD 1840. Comparison with Westland sequences suggests that moraines 9–11 may have formed some 2500–5000 years ago, and that 12 is early Holocene. In a parallel sequence at 1100 m, the early pioneers are lower-altitude species such as *Carmichaelia grandiflora, Gingidia montana* and *Acaena caesiiglauca*. Woody plants such as *Podocarpus nivalis, Muehlenbeckia axillaris* and, eventually, *Dracophyllum uniflorum* become increasingly prominent.

(2) Succession through bryophytes to *Chionochloa* on moraines in the Ben Ohau Range (p. 418) is assumed to have taken 2800 years. It is reflected in the inorganic phosphorus fractions (Fig. 15.10), but the final increase of iron- and aluminium-bound phosphorus relative to occluded phosphorus results from gleying, and presumably would not occur in drier soils. Organic-bound phosphorus increases in proportion to the increase of organic carbon. The decrease of total phosphorus seen in longer sequences at low altitudes is not evident. Concentrations (ppm) are indicated by (a) the A-horizon of a recent soil at 1950 m that supports a *Celmisia haastii–Marsippospermum* community and (b) the O-horizon of a gleyed yellow-brown earth at 1750 m that supports a *Chionochloa oreophila* community (Archer & Cutler 1983):

	Ca–P	AlFe–P	Occ. P.	Org. P.	Total P
a	60	105	475	350	990
b	0	73	100	1027	1200

(a) **Moraine Crest**   1 2 3 4 5 6 7 8 9 11

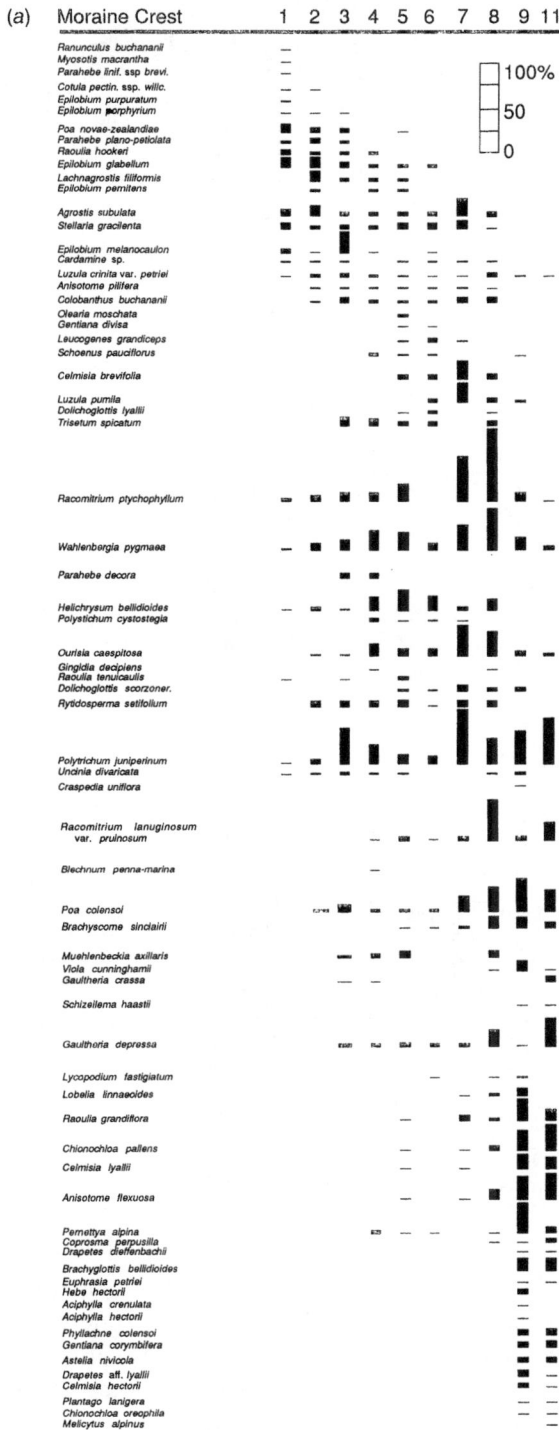

Fig. 15.9 (*a*) (for legend see overleaf)

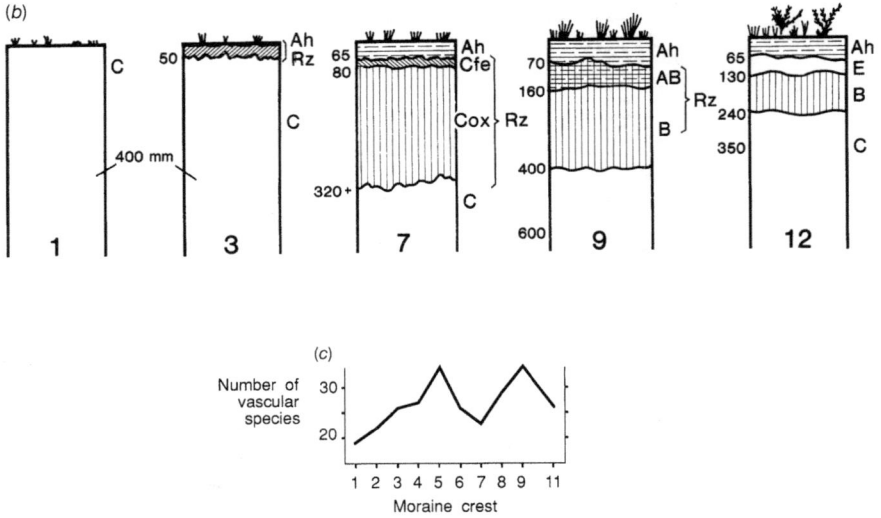

Fig. 15.9. Development of vegetation on a set of moraines at 1400 m; Dart valley, South-eastern Alps (Somerville *et al.* 1982). (*a*) Mean frequencies in 25 × 25 cm quadrats. (*b*) Soil horizons and depths (mm) on five crests: h, humus; fe, iron pan; ox, iron stained; Rz, rooting zone. (*c*) Species diversity.

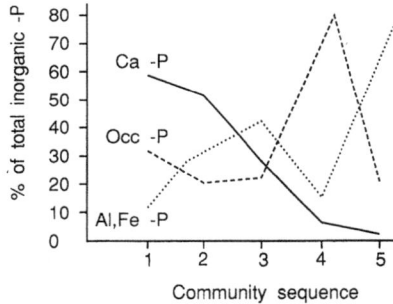

Fig. 15.10. Changes in fractions of inorganic phosphorus along a sequence from (1) bryophyte to (5) *Chionochloa* dominance; Ben Ohau Range, Tasman ER (Archer 1973).

(3) Where an alpine cushion-field and its top soil had been destroyed by road building in Central Otago, the main subsequent changes were an increase of stone pavement and colonisation by *Poa colensoi*. The only other species to achieve more than 1 % cover over 11 years were *Dracophyllum muscoides*, *Epilobium alsinoides*, *Raoulia hectorii*, *Luzula pumila*, *Poa lindsayi* and *Psilopilum australe*. In the undisturbed area, plant cover did not change significantly (Fig. 15.11).

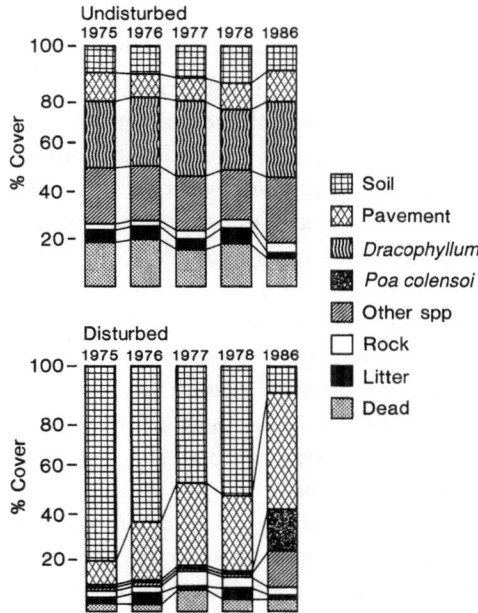

Fig. 15.11. Cover changes on undisturbed cushion field and denuded ground; 1550–1690 m, Old Man Range, Central Otago (Roxburgh *et al.* 1988).

Secondary succession at low altitudes

Some parts of lowland New Zealand that were cleared from forest for pastoral farming 'reverted' to secondary vegetation, because farmers lacked resources to maintain the land in production. Other areas have been allowed to revert for specific reasons, such as catchment protection or rehabilitation of nature reserves.

The course of succession back to forest depends greatly on the extent to which the original cover was destroyed. Selective logging of merchantable trees leaves gaps to be filled by released seedlings and resprouts from stumps or by thickets of wineberry or *Histiopteris incisa* (cf. Baxter & Norton 1989). Localised fires can lead to tree-fern groves, patchy colonisation by kanuka and other seral trees, or even immediate regeneration of 'climax' species, including conifers and beeches. In these cases, there is prompt return to forest cover, although recovery of the original structure and composition may take a long time. Some heavily logged areas are taken over by giant sedge or grass tussocks that suppress forest regeneration, especially *Gahnia xanthocarpa* on wet, podzolised soils and *Cortaderia fulvida* on tephra soils of the Volcanic Plateau. In places, the adventive *C. selloana* has become so dense as to impede silvicultural operations. Most of the following examples of succession have begun on land where forest had been replaced by rough pasture.

Succession on northern off-shore islands

Maori inhabited many of the off-shore islands, and had burnt much of the vegetation and cultivated arable ground. They withdrew soon after European contact and, on some islands, reversion to the original vegetation commenced. Other islands were pastorally farmed, so that grazing-resistant vegetation developed. Several islands have since been destocked in the interests of nature conservation.

On Great Island in the Three Kings group, Maori occupation ceased about 1840, but goats liberated in 1889 induced grassy swards and open kanuka woodland over most of the island. Initial response to eliminating the goats in 1946 was thickening of the sward, increase of kanuka seedlings, and recovery of surviving browsed plants. Other trees and shrubs have increased gradually, especially *Coprosma rhamnoides*, *Geniostoma rupestre*, mahoe and *Cordyline kaspar*. These have small, fleshy fruit that may be dispersed by introduced passerine birds.

Meryta sinclairii and *Coprosma macrocarpa*, which have medium-sized fruit, are spreading slowly. The large-fruited karaka (*Corynocarpus laevigatus*), *Planchonella costata*, *Alectryon grandis*, *Hedycarya arborea* and *Litsea calicaris* have seedlings only close to surviving trees, perhaps because there are no native pigeons (*Hemiphaga novaeseelandiae*) on the island (Cameron *et al.* 1987). Of the two endemic species that had been reduced to a single plant, the liane *Tecomanthe speciosa* is spreading vegetatively, and *Pennantia baylisiana* is a female tree that has yet to reproduce naturally (Wright 1983).

Bush rather than tall forest represents the greatest development of vegetation on most off-shore islands. However, for the interior of Hen Island, Court (1978) suggests that the broad-leaved trees that replace kanuka (i.e. *Vitex lucens*, *Sophora microphylla*, etc.) are eventually replaced by tawa and taraire (*Beilschmiedia tarairi*). With increasing proximity to the shore, seral kanuka and pohutukawa are replaced by karaka, *Pisonia brunoniana*, *Planchonella* and *Streblus banksii*; by karaka and *Meryta*; or by karaka alone.

Little Barrier island is large enough to support conifer/broad-leaved forest and some beech. Hamilton & Atkinson (1961) describe spatial variation in stands generally dominated by kanuka 9–12 m tall, that are about 60 years old.

(1) Presence of grazing animals leads to dense *Coprosma rhamnoides* at 2 m, with *Doodia media*, *Carex dissita*, *C. testacea* and *Uncinia uncinata* beneath.

(2) In sheltered valleys, the shrub layer consists largely of young *Corynocarpus*, *Hedycarya*, *Knightia* and mahoe, together with nikau palms (*Rhopalostylis sapida*).

(3) On lower valley sides, understoreys are dominated either by *Cyathea dealbata* or by *Myrsine australis* with *Coprosma arborea* and *C. rhamnoides*.

(4) On ridges near the coast, pohutukawa is in the canopy with kanuka, and understoreys contain *Pittosporum umbellatum*, *P. tenuifolium*, *Myrsine australis*, five-finger, *Olearia furfuracea*, *Helichrysum lanceolatum*, *Astelia banksii* and *Phymatosorus diversifolius*.

(5) On the leached soils of uneroded remnants of the original volcanic slopes,

manuka is important, and other shrubs are *Olearia furfuracea, Cyathodes juniperina, C. fasciculata, Pittosporum umbellatum* and *Hebe macrocarpa* var. *latisepala*. The herbaceous species, which include *Schizaea fistulosa, Doodia media, Lindsaea linearis, Gahnia lacera, Morelotia affinis, Schoenus tendo* and *Lepidosperma laterale*, are reminiscent of gumland. Among tree seedlings, only kauri (*Agathis australis*) and tanekaha (*Phyllocladus trichomanoides*) are common.

(6) In an older stage, young trees of kauri, hard beech (*Nothofagus truncata*) and occasional northern rata (*Metrosideros robusta*) are overtopping kanuka and manuka.

Succession to lowland forest on the northern mainland

Kauaeranga valley (Coromandel Ecological Region)

The original forest canopy was mainly kauri, lowland podocarps, towai (*Weinmannia silvicola*), tawa, kohekohe (*Dysoxylum spectabile*) and northern rata. Small dicot trees, nikau palms and tree ferns formed a subcanopy, or the main canopy in moist gullies and on slips. This forest was probably intact at the time of European settlement, but was then progressively reduced by logging and fire. Subsequent decades saw pasture being precariously maintained in the face of competition from bracken and invasion by manuka. By 1940 grazing by feral cattle ceased, but pigs are still present, and there have been occasional fires. In 1982 transects were set up on slopes of 25–35° between 150 and 270 m above sea level and ages were determined from growth rings (Fig. 15.12).

In Transect 1, establishment of manuka after fire has been inhibited by bracken clumps and the grassy sward between them, whereas in Transect 2, burnt at the same time, a very large number of manuka plants have grown up. *Leycesteria formosa* has successfully competed with bracken by thrusting long, cane-like stems through the mass of dead and living fronds.

The next four transects show manuka plants decreasing in number as their height increases. Bracken, pasture grasses and herbs become suppressed, and their place taken by the more shade-tolerant *Blechnum* 'black spot', which probably spread from damp gullies. Several heathland plants appear; some of these, notably *Gaultheria antipoda, Drosera peltata, Schoenus tendo* and *Lindsaea linearis*, do not persist in closed forest, whereas others, such as *Morelotia affinis, Dianella nigra, Cyathodes fasciculata* and *Lycopodium volubile*, also thrive in well-lit forest habitats.

Five-finger (*Pseudopanax arboreus*), *Coprosma robusta, Cordyline banksii* and several other species appear as seedlings on the most recent burn. Towai and rewarewa (*Knightia excelsa*) are first recorded on Transect 3, but the largest of these eventual dominants are already taller than the surrounding manuka and show more growth rings, suggesting formation of more than one ring annually. The early appearance of tanekaha and other podocarps is also notable.

Waitakere Range

Mixed stands of manuka and kanuka mostly established on abandoned pasture or after forest fires (Fig. 15.13). As they increase in height, there is continual self-thinning.

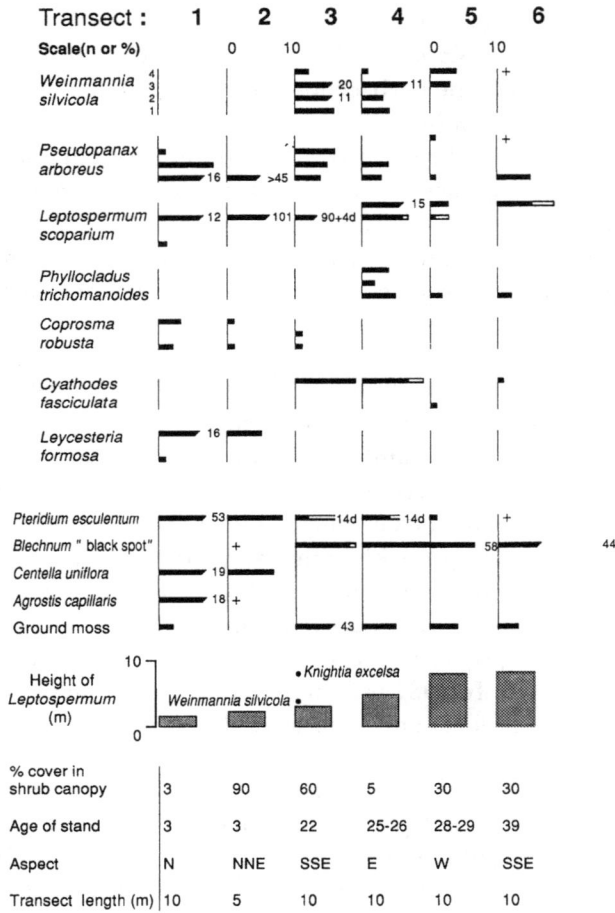

Fig. 15.12. Composition of fire-induced vegetation in the Kauaeranga valley, Coromandel ER. Transects are 1 m wide. Bars show numbers of woody plants in height classes (1 <10 cm, 2 <30 cm, 3 <3 m, 4 ≥3 m), and percentage cover of herbs; each column is 10 units wide; d or open box indicates dead plants.

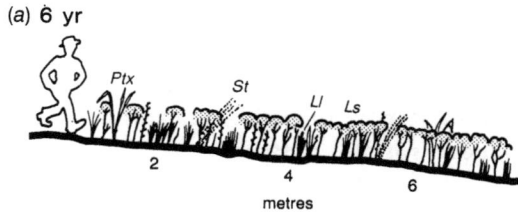

Fig. 15.13 (for legend see p. 536)

(b) **15 yr**

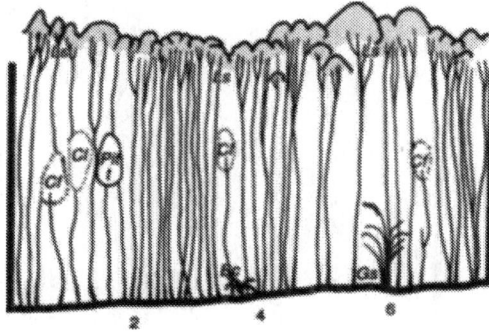

(c) **20 yr**

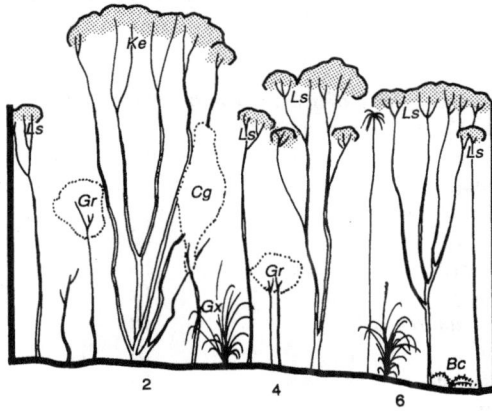

(d) **c.35 yr**

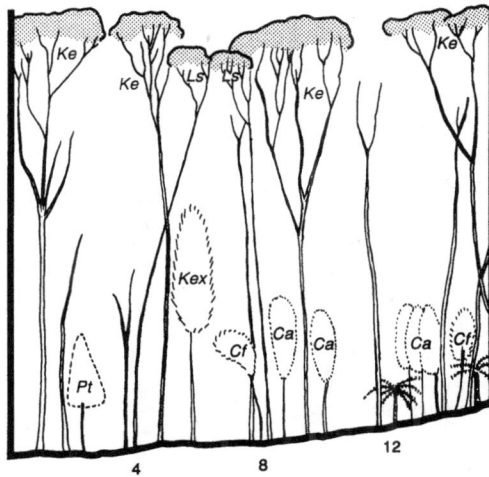

(e) >60 yr

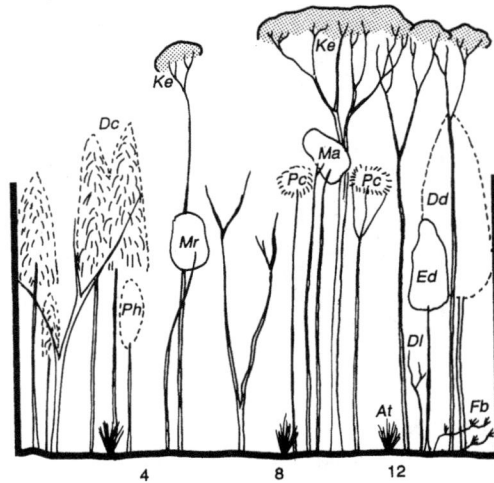

Fig. 15.13. Profile diagrams of manuka and kanuka stands of different ages in the Waitakere Range (Esler & Astridge 1974).

Columns below show abbreviations, average heights (m) of canopy (C) and subcanopy (S), and percentage cover. For understorey species (U) values are percentage frequencies in 20 sub-plots.

Profile	Species	Abbr.	Tier	Ht	%
A	*Leptospermum scoparium*	*Ls*	C	0.8	67
	Lepidosperma laterale	*Ll*	C	0.7	3
	Phormium tenax	*Ptx*			
	Schoenus tendo	*St*			
B	*Leptospermum scoparium*	*Ls*	C	4.1	84
	Others in canopy		C		1
	Geniostoma rupestre		U		19
	Cyathodes fasciculata	*Cf*	U		19
	Pteridium esculentum		U		17
	Gahnia setifolia		U		16
	Blechnum 'capense'	*Bc*			
	Coprosma lucida	*Cl*			
	Gahnia setifolia	*Gs*			
	Pittosporum tenuifolium	*Pit t*			
C	*Leptospermum scoparium*	*Ls*	C	4.5	65
	Kunzea ericoides	*Ke*	C	5.5	15
	Others in canopy		C		5
	Geniostoma rupestre	*Gr*	S	3.1	10
	Coprosma grandifolia	*Cg*	S		
	Others in subcanopy		S		14
	Blechnum 'capense'	*Bc*	U		14
	Coprosma grandifolia		U		14
	Geniostoma rupestre	*Gr*	U		14
	Myrsine australis		U		14
	Uncinia uncinata		U		11
	Coprosma arborea		U		10
	Gahnia xanthocarpa	*Gx*	U		10

Profile	Species	Abbr.	Tier	Ht	%
	Pseudopanax crassifolius		U		10
D	*Kunzea ericoides*	*Ke*	C	14.4	49
	Leptospermum scoparium	*Ls*	C	11.5	22
	Others in canopy		C		2
	Coprosma arborea		S	3.7	12
	Cyathea dealbata		S	3.0	4
	Cyathea medullaris		S	4.5	3
	Knightia excelsa	{ *Kex*	S		
	Others in subcanopy		S		4
	Coprosma arborea	*Ca*	U		19
	Geniostoma rupestre		U		19
	Cyathea dealbata	*Cd*	U		18
	Uncinia uncinata		U		14
	Cyathodes fasciculata	*Cf*	U		12
	Pseudopanax crassifolius		U		12
	Phyllocladus trichomanoides	*Pt*	U		12
	Blechnum 'capense'		U		10
	Hedycarya arborea		U		10
E	*Kunzea ericoides*	*Ke*	C	13.7	61
	Prumnopitys ferruginea		C	14.0	3
	Pseudopanax crassifolius	*Pc*	S	9.8	16
	Quintinia serrata		S	10.0	7
	Elaeocarpus dentatus	*Ed*	S	8.5	5
	Coprosma grandifolia		S	6.6	4
	Melicytus macrophyllus		S	7.4	3
	Dacrydium cupressinum	*Dc*	S	9.0	3
	Dracophyllum latifolium	*Dl*	S		
	Melicytus ramiflorus	*Mr*	S		
	Myrsine australis	*Ma*	S		
	Dacrycarpus dacrydioides	*Dd*	S		
	Podocarpus hallii	*Ph*	S		
	Others in subcanopy		S		17
	Freycinetia baueriana	*Fb*	U		16
	Geniostoma rupestre		U		14
	Coprosma grandifolia		U		13
	Melicytus macrophyllus		U		12
	Alseuosmia macrophylla		U		11
	Blechnum fraseri		U		10
	Astelia trinervia	*At*	U		

Although manuka can be entirely suppressed after 20 years, some persist even in kanuka stands over 35 years old.

Forest plants are vigorously recruited into stands dated at 20 years or older, when the canopy has become uneven and more open. They include seedlings of conifers, notably rimu and tanekaha which are up to 11 m and 12.3 m tall respectively in stands > 65 years old. Of the 117 species listed about 20, including bracken, *Schoenus tendo* and *Phormium tenax*, are confined to the earliest stages. Another 20 are characteristic of the shrubland stage, e.g. *Olearia furfuracea* and *Hakea sericea*. Most of the

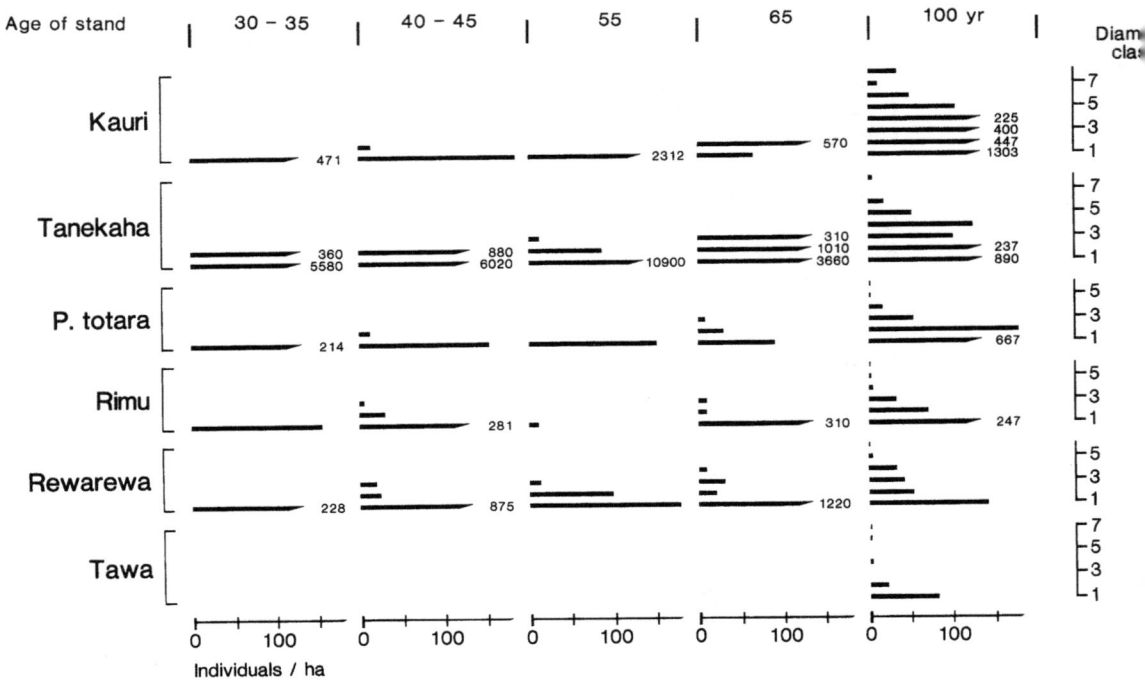

Fig. 15.14. Recruitment of conifers and tall broad-leaved species into kanuka–manuka stands of differing age in the Waitakere Range (Pook 1978). Diameter classes: 1, ≤2.5 cm d.b.h.; 2, >2.5 cm; 3, >5 cm; 4, >10 cm; . . . 8, >30 cm.

remainder would have been present in the original forest, the following being common: *Blechnum* 'capense', *Cyathea dealbata*, lancewood (*Pseudopanax crassifolius*), *Geniostoma, Coprosma grandifolia, C. lucida, C. rhamnoides* and *Gahnia xanthocarpa.*

That recruitment of conifer and rewarewa seedlings begins early and continues vigorously is also shown by Fig. 15.14. Tawa, however, bears out its reputation as a climax species by its late appearance as saplings only. In the oldest stand, kanuka comprises 25%, kauri 28%, tanekaha 14% and *Cyathea dealbata* 15% of stems >10 cm diameter; the remainder includes other podocarps, manuka, lancewood, *Olearia rani, Hakea sericea, Coprosma arborea, Myrsine salicina* and *M. australis.* Conifers have overtopped the kanuka canopy. In stands of young kauri trees, growth ring counts show that accumulation in the seedling classes represents both suppression and continuing recruitment (Ogden 1983).

Forest successions near Wellington

Western side of the Hutt Valley
The original vegetation on soils over greywacke was mixed forest with tawa prominent. Podocarps were logged, and most of the forest burnt and converted to pasture in the later part of the nineteenth century, but by 1949 manuka and *Cassinia*

leptophylla occupied the drier slopes. Elsewhere, there was succession through bracken (Croker 1953). Bracken burnt in 1945 was dense and 1 m tall in 1949, and retained a ground cover of pasture species, including browntop, *Holcus lanatus, Hypochoeris radicata, Plantago lanceolata, Taraxacum officinale, Trifolium repens, T. dubium* and *Hydrocotyle moschata.* There were also young plants of *Geniostoma, Coprosma robusta,* mahoe, *Hebe salicifolia* and *Brachyglottis repanda.* By 1952 the bracken was up to 2 m tall, but discontinuous and equalled in cover by *Coprosma robusta, Geniostoma* and *Leycesteria formosa.*

Where bracken was tall in 1912 and not burnt subsequently, bush 5–6 m tall had developed after 37 years. *Coprosma robusta* or mahoe with 44 to 47 growth rings were dominant. Other species included *Coprosma lucida, C. grandifolia, Brachyglottis,* five-finger, *Myrsine australis, Cyathea* spp. (in gullies), *Uncinia 'australis', Pterostylis* sp., *Lycopodium volubile,* bracken as etiolated fronds, and seedlings of tawa, rewarewa and *Hedycarya.*

Eastern side of the Hutt Valley

At Taita the greywacke is deeply weathered and soils less fertile. Druce (1957*b*) has reconstructed complex successions from historical records and growth rings, in a catchment that extends to 220 m a.s.l. Except in the valley bottom, the original vegetation appears to have been almost pure hard beech forest, with a sparse lower tier of kamahi. This was destroyed by fire around 1850.

Beech forest regenerated over 3% of the area, presumably where seed or seedlings survived the fire. Elsewhere grasses and herbs may have established, but manuka and probably also bracken soon invaded. Development since then has depended on position along a gradient from moist and sheltered to dry and exposed, proximity of seed sources, frequency of fire, and increasingly, entry of adventives, especially *Erica arborea,* gorse and *Pinus radiata.* The various successional paths all lead towards dominance by kamahi. Hard beech seedlings are establishing in the margins between older beech trees and surrounding vegetation, but at greater distance only where successional vegetation has not formed a closed cover.

If there are no more fires, manuka gradually suppresses herbaceous pioneers such as *Rytidosperma* spp., *Deyeuxia* spp., the more shade-tolerant *Microlaena stipoides,* and ground orchids, especially *Thelymitra* spp. After some 20 years, an understorey begins to develop, with *Blechnum 'capense', Gahnia setifolia, Geniostoma* and five-finger. By 40 years, *Knightia* begins to appear in the canopy. After 60–70 years kamahi is dominant, and shelters the beginnings of a forest understorey, including *Blechnum discolor, Phymatosorus diversifolius,* filmy ferns, and the liane *Metrosideros fulgens.*

If scrub already containing kamahi saplings is burnt, the latter resprout to create kamahi stands in which manuka is unimportant. Where manuka grows with gorse or *Erica,* it eventually overtops and outlives them, but the latter increase their hold after each fire; at the time of the study, the further course of succession from these adventive shrubs was not yet apparent. Pine seedlings establish when fires release seed from cones, but increase of pine depends on fires being at long enough intervals to allow young trees to reach reproductive age. Although pine trees suppress surrounding manuka, they will probably be replaced in turn by five-finger and, eventually, kamahi.

In moist gullies, dense bracken can establish after fire; possibly 30 years or more elapse before this is entered by seedlings of five-finger, *Brachyglottis repanda* and *Geniostoma*, which apparently are eventually followed by kamahi. In the valley bottom, three paths lead to mahoe bush: through dense bracken via *Cyathea medullaris*; through broad-leaved shrubs such as *Coriaria arborea* and *Brachyglottis* on stony colluvium; or through kanuka where other pioneer species do not form dense cover. On swampy ground, successions lead to dominance by *Syzygium maire* or *Laurelia novae-zelandiae*.

Secondary succession through legumes in the east of the South Island

On the Port Hills, Banks Peninsula, under a mean annual rainfall of 760–1020 mm, remnants of mixed forest that survived Maori fires were destroyed by European fires. Today there is grassland of native and introduced species on the crests, regenerated bush in gullies, and mainly scrub and bracken on the slopes between. Fig. 15.15 shows that common broom (*Cytisus scoparius*), which has either invaded or re-established in pasture and bracken after fire, is followed in a few years by elder (*Sambucus nigra*). The liane *Muehlenbeckia australis* soon becomes prominent. Mahoe is recorded as seedlings in stage 3, and eventually dominates native bush in which *Myrsine australis, Plagianthus regius, Pittosporum eugenioides, P. tenuifolium, Myoporum laetum* and *Coprosma robusta* are also present. Broom and, to some extent, elder become increasingly suppressed, although the former continues to dominate the soil seedbank for many decades. At intermediate levels of grazing or on unstable soils broom maintains itself without fire.

Gorse can be an effective nurse for native tree seedlings, but in the Dunedin district this may be true only of tall, open gorse scrub with shallow litter and patches of bryophytes or bare soil (Fig. 15.16). Here, native tree species can co-dominate with gorse after 15 years, with the most frequent being *Myrsine australis*, mahoe, *Griselinia littoralis*, kanuka and *Pittosporum tenuifolium*. Other woody plants include *Muehlenbeckia australis*, elder, *Crataegus monogyna*, common broom, *Rubus erythrops* and *Coprosma propinqua*. On 60% of the sites, however, it seems unlikely that native or other introduced species can invade before gorse completes its first generation, which requires 25–30 years; and should invasion fail, there are ample gorse seedlings to perpetuate the stands.

Prevalent stages in secondary succession towards lowland forest

The following summary builds on the examples discussed above, and on schema presented by Silvester (1964), Court (1978) and Esler (1967, 1978b).

(1) Large areas of bracken have become established after forest fires since the beginning of Polynesian settlement; re-burning of secondary growth is especially conducive. Although most of this land has been converted to pasture, bracken can vigorously reinvade where stocking is light and soils well-drained, friable, and of adequate depth and fertility. Spread is usually from moist spots that escape trampling, such as steep banks and gullies, and the rhizomes can penetrate rank grassland that resists invasion by woody plants.

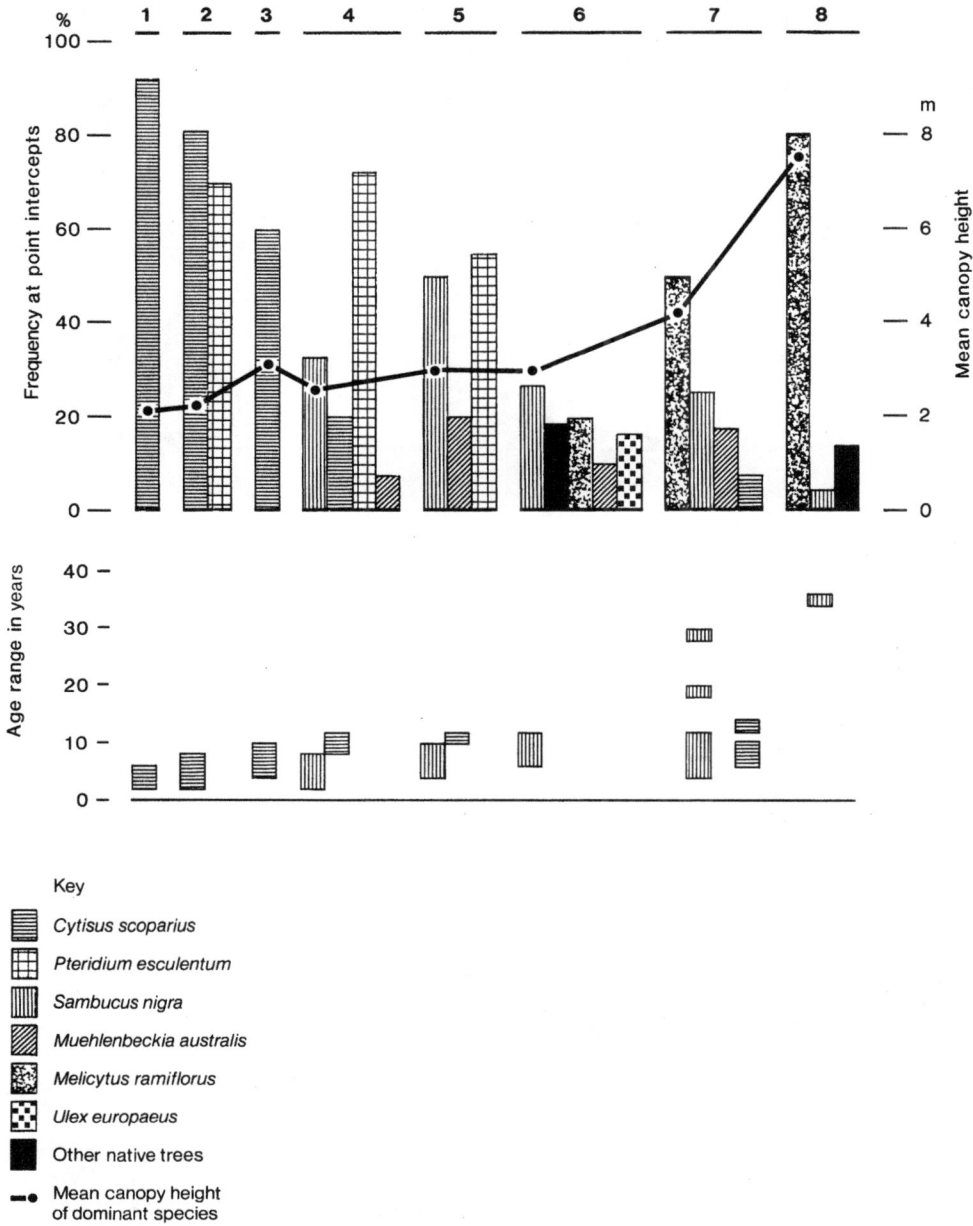

Fig. 15.15. Stages in replacement of common broom by native bush (from Williams 1983).

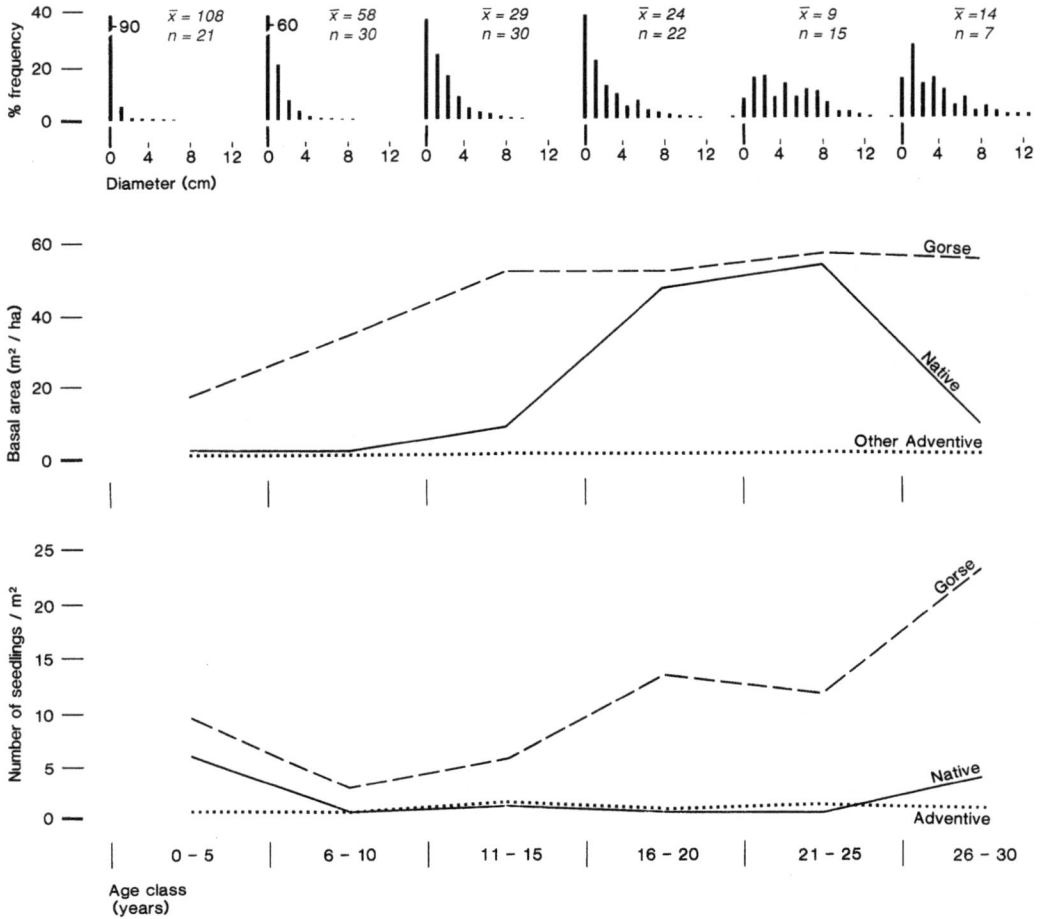

Fig. 15.16. Structure of gorse stands near Dunedin (from Lee *et al.* 1986). Top: frequency of gorse plants of different sizes. $\bar{x}$, Mean no. of stems >5 mm diameter in 2 × 2 m plots; *n*, no. of plots. Middle: basal area of woody plants. Bottom: density of seedlings <5 mm diameter. Age is that of the most mature gorse in each stand.

Where fallen logs remain or in minor erosion slumps, prothalli of *Paesia scaberula* establish on bare, moist, shaded soil and give rise to dense, circular colonies (Moore 1942). Colonies of the tree fern *Dicksonia squarrosa* also arise in this manner. Forms of *Blechnum 'capense'*, mainly 'black spot' persisting from the original forest, develop dense colonies in gullies and on moist, shaded slopes. Other ferns that invade pastures include *Histiopteris incisa* (mainly on moist, sheltered sites), and species of *Adiantum* and *Hypolepis*. *Doodia media* and *Polystichum vestitum* can be abundant in warm and cool districts respectively. *Gleichenia microphylla* and *G. dicarpa* grow on severely leached soils where rainfall is high, notably on gumland and pakihi.

(2) Small-leaved native shrubs invade most readily where bare soil is exposed as in overgrazed pasture on stony or eroding slopes. Dry spurs and ridges are invaded more

slowly. Manuka and kanuka are the most ubiquitous invaders (pp. 195, 197), although the latter does not reach Southland and South Westland. Where both occur, kanuka tends to first establish on comparatively sheltered slopes with better soils, and then to invade open manuka scrub on harsher sites.

Cassinia (p. 211) invades grassland beyond the dispersal range of manuka and kanuka. It seems more tolerant of coastal exposures, high altitudes and competition from grass, but is probably less drought tolerant. Certain filiramulate shrubs are prominent in stony ground of evidently higher fertility, such as the lower parts of debris slopes. The most abundant of these are *Coprosma propinqua*, the liane *Muehlenbeckia complexa*, and, in dry grassland in the east of the South Island, matagouri. The filiramulate juvenile of *Pennantia corymbosa* can also be abundant.

The successional role of native shrubs is being steadily usurped by adventives with attributes that enable them to increase after each successive fire. Seeds of broom and gorse persist in the soil for decades; gorse and *Erica* spp. resprout after fire, and the latter can disperse their seeds widely. *Hakea* spp. have thick serotinous capsules and form dense, exclusive colonies.

(3) Rank herbaceous cover resists invasion by tree and shrub seedlings. The immediate effect of removing livestock from Tiritiri Matangi Island in the Hauraki Gulf was that invasion by shrubs ceased (West 1980). Marginal invasion occurs where grass or fern are overtopped by expanding tree canopies. On dry sites, manuka scrub can suppress grassland along its margin, thereby allowing herbs intolerant of grass competition such as *Centaurium erythraea*, *Veronica plebeia*, *Lagenifera pumila*, *Dichondra repens* and *Microtis unifolia* to persist (Court 1981).

When bracken is burnt, manuka or gorse may usurp dominance before the fern recovers, through mass release of seeds or germination of dormant seeds respectively. In the absence of fire, moderately dense bracken allows steady recruitment of a wide range of broad-leaved trees and shrubs, including *Weinmannia* and other climax species, but the most ubiquitous is *Pittosporum tenuifolium*. Open woodlands of this small tree are prominent on bracken-covered slopes in inland districts of the South Island, especially around the Otago lakes.

Tall, very dense bracken may hold its ground for several decades (Baylis 1958), the only frequent woody species being lianes and weak-stemmed shrubs such as *Leycesteria* that can grow up through the fern despite being pressed down by collapsing fronds. Nevertheless, in the absence of fire its eventual replacement is inevitable, through sporadic entry of woody plants into partial openings provided by stream banks, rocks, logs or animal tracks. Any circumstances that prevent dead fronds from collapsing over a patch of ground and burying it in deep, loose litter may also allow woody seedlings to enter; for example, fronds may be held up by shrubs, or consistently fall away from a central point. The most successful invaders are five-finger and *Coriaria arborea*, the former because it can resume upward growth after being pressed down by fern fronds and the latter because of stout basal shoots that thrust quickly through the bracken cover. Large-leaved coprosmas such as *C. robusta* also invade, but usually do not form enough canopy to suppress the bracken. Tree ferns, especially *Cyathea medullaris*, colonise gullies in fernland.

The filiramulate shrub *Pseudopanax anomalus* invades ungrazed grassland and bracken on Kapiti Island, and persists beneath the small, broad-leaved trees that eventually overtop it (Esler 1967). Lianes, including blackberry, native *Rubus* spp., *Muehlenbeckia australis, M. complexa*, native *Clematis* spp. and *C. vitalba* invade tall fern and rank grass, especially along the forest edge where they can hang as curtains from marginal canopies. They suppress herbaceous growth and create shaded niches for tree seedlings, but are apt to weigh down young trees. The adventive *Berberis* spp., *Eleagnus ×reflexa* and *Rosa rubiginosa* can form dense thickets through vegetative spread.

In some situations where woody plants seem to be invading dense bracken, it is likely that their seedlings established before the bracken had time to thicken after having been kept open by grazing. For example, at Pukatea Bay (North-west Nelson ER) five-finger with lesser amounts of mahoe, kanuka, manuka, *Pittosporum tenuifolium, Myrsine australis* and tree ferns forms a patchy overstorey 5–8 m tall, above bracken with scattered gorse and *Cyathodes fasciculata*. Seedlings of these trees and shrubs, and also of kahikatea, occur only beneath broad-leaved or kanuka canopies.

(4) Open seral vegetation may receive seedlings of climax species at an early stage. Seedlings of *Weinmannia, Metrosideros, Nothofagus*, and less frequently, *Podocarpus totara, P. hallii* and *Libocedrus bidwillii* can colonise bare ground in the open. Open shrubland of young manuka and kanuka is receptive to seedlings of many species; those of rimu, silver pine, *Phyllocladus* spp. and *Knightia* can occur abundantly 200–300 m away from seed sources. Seedlings of trees with succulent fruit are concentrated near isolated trees and shrubs that are taller than surrounding seral vegetation and therefore attract perching birds.

Commonly, however, seral shrubs form very dense stands that not only suppress herbaceous vegetation, but resist entry of further species. Only when growth and self-thinning lead to a higher canopy and wider spacing of stems are shade-tolerant seedlings and understorey plants favoured (Fig. 15.17). Manuka may shelter an understorey of broad-leaved seedlings by the time it reaches a height of 9 m, but manuka itself becomes overtopped by kanuka where the two grow together, and the latter can live 160 years and form a canopy up to 20 m high over broad-leaved bush.

Manuka and kanuka eventually outgrow and suppress gorse, *Erica* or *Cassinia* plants that established at the same time. Dense gorse collapses as it ages, and is subject to invasion by native plants, including bracken, other ferns, and seedlings of broad-leaved trees. Where seed sources of forest species are remote, where the environment is marginal for forest, or where grazing eliminates tree seedlings, seral shrubs and trees can form self-perpetuating communities.

(5) All tree ferns can resprout after forest is burnt. The rhizomatous *Dicksonia squarrosa* develops large colonies in rough pasture, *Cyathea medullaris* establishes readily in damp, fern-choked gullies, and *C. dealbata* is common beneath tall kanuka. Tree ferns growing in the open attract roosting birds that carry seed from nearby forest, and although fallen fronds deeply blanket the ground beneath dense groves of tree-fern, especially *C. medullaris*, the trunks are colonised by epiphytic seedlings, especially those of *Pseudopanax arboreus, Weinmannia* spp. and *Griselinia* spp. Skirts of dead fronds investing the trunks of several species, especially *Dicksonia fibrosa* and

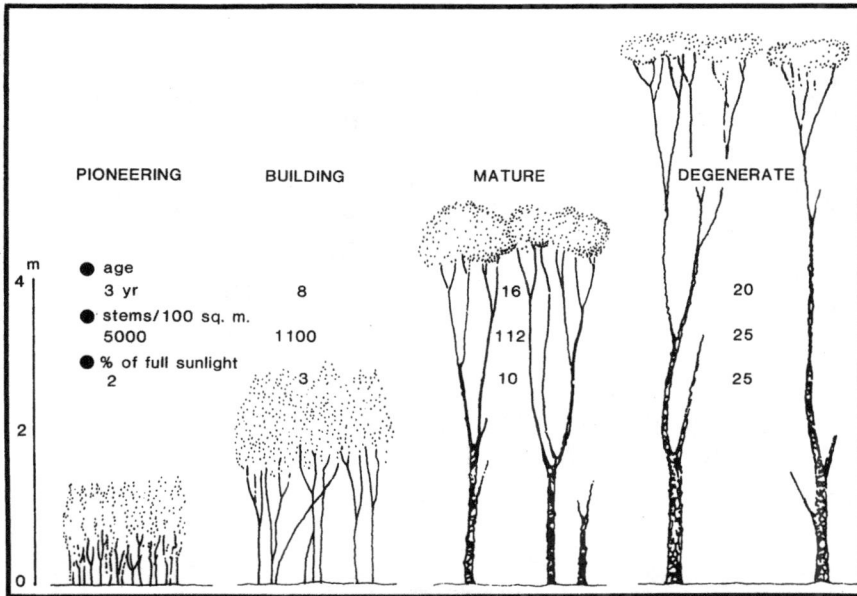

Fig. 15.17. Stages in development of manuka stands in Manawatu ER (Esler 1978a).

most cyatheas, may deter epiphytes and lianes from growing on the upper part of the tree fern and overtopping its crown (Page & Brownsey 1986).

(6) If fire in beech forest coincides with a seed year, beech seedlings may establish abundantly on denuded surfaces near surviving trees, and sporadically at considerable distances, exceptionally over 100 m. Once closed seral vegetation develops, be it grassland, fernland, shrub-heath or bush, beech can recover ground only through slow expansion of margins as seedlings establish beneath beech canopies that have overtopped and suppressed shorter vegetation.

(7) Entry of adventive trees into secondary successions has become increasingly prevalent as plantings increase and native seed sources diminish. Wind carries the seed of *Pinus* spp. many kilometres, and the seedlings establish readily on bare ground and in scrub. *Larix decidua, Eucalyptus* spp., *Racosperma* spp. and *Acer pseudoplatanus* also invade. Smaller adventive trees (pp. 162–3) will probably prove seral to taller native or introduced species.

(8) Forest development is well documented to the stage where small broad-leaved trees form the canopy, or where pole-sized conifers are overtopping kanuka. Less is known about secondary vegetation dating back to the earliest European or pre-European times, largely because most of it has been re-burnt. In the northern zone, tawa and taraire are climax dominants, with tawa alone filling this role on fertile, low-altitude sites in the rest of the North Island and Sounds–Nelson province (Fig. 15.18). On Kapiti Island, tawa stands are growing over decaying trunks of huge *Metrosideros robusta* which, perhaps, began as epiphytes on podocarp trees (Esler 1967). On poorer soils and in the west and south of the South Island, kamahi seems the eventual main

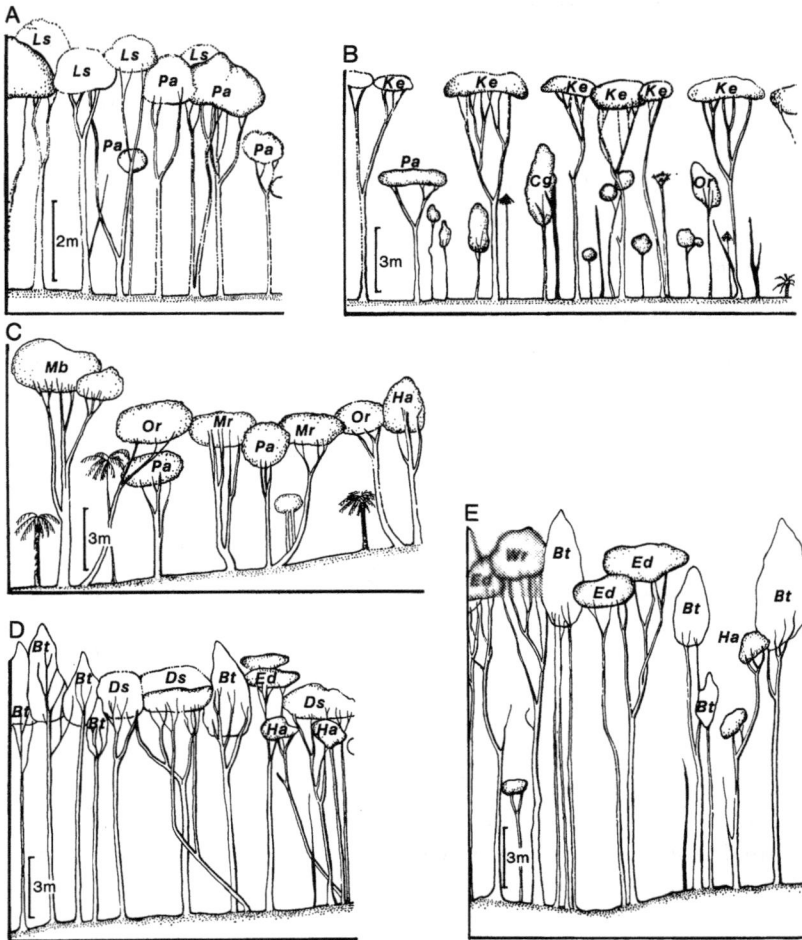

Fig. 15.18. Stages in development from manuka and kanuka heath to broad-leaved forest on Kapiti Island (Esler 1967). Abbreviations: Bt, *Beilschmiedia tawa*; Ds, *Dysoxylum spectabile*; Ha, *Hedycarya arborea*; Mb, *Metrosideros robusta*; Or, *Olearia rani*; Pa, *Pseudopanax arboreus*; Wr, *Weinmannia racemosa*. Other abbreviations as in Fig. 15.13.

broad-leaved dominant. Over much of the eastern side of the South Island, *Griselinia littoralis* can fill the same role because of the regional absence of taller species, especially kamahi and tawa.

Several authors, from Cockayne (1928) onwards, have suggested that although the tall native conifers can persist for centuries, they are not truly part of the climax forest; others argue that inevitable disturbances ensure their continued presence (pp. 602–5). Over much of New Zealand, beech forest is the potential climax at the higher altitudes, in southern districts, and on infertile hill soils. However, as spread of beech into closed vegetation is usually exceedingly slow, tongues of broad-leaved bush or even mixed forest may extend into beech forest at ancient burn margins. Mosaics of beech forest and other vegetation that appear to reflect very much older disturbances are discussed on p. 554.

The increasing extent of both spontaneous and man-made exotic forests presents new paths for succession. *Pinus* plantations in the higher-rainfall regions would be replaced by native forests as they mature, if humans did not intervene. Tree ferns play a major role in this, as they establish more readily than other plants beneath a high *Pinus* canopy (p. 158). In dry districts, closed pine forests have little undergrowth, and pine seedlings may be able to compete when the canopy is opened.

Secondary successions at higher altitudes

In the higher, steeper or more remote parts of New Zealand, there has been both local and large-scale clearance of the original vegetation by fire. In contrast to the lowlands, native species usually dominate throughout the resulting successions, except that short-lived adventive herbs may be abundant during the first few years. On the other hand, return of tall woody vegetation can take much longer than at low altitudes. Often, forest and scrub burnt decades ago has shown little recovery and, at least on drier mountains, fire patterns established centuries ago are still evident, although usually reinforced by later fires.

Mt Cook National Park

On the Liebig Range a remnant of the once-extensive subalpine stands of *Phyllocladus alpinus*, *Podocarpus* spp. and tall dicot shrubs (p. 189) was destroyed by a very hot fire in 1967 (Wilson 1976). After three years, plant cover consisted of the adventive *Senecio sylvaticus* and *Verbascum thapsus* and scattered resprouted *Phormium cookianum*. Matagouri, *Coprosma* spp., *Aristotelia fruticosa*, *Olearia nummulariifolia*, *Pseudopanax colensoi*, and a few trees of *Griselinia littoralis* and *Hoheria lyallii* s.l. showed some regrowth. Most grasses recovered quickly, but slow and incomplete recovery was shown by *Chionochloa pallens* and *C.* 'robust'. Where the fire burnt an area that had carried a fire about 40 years earlier, the fire-tolerant *Phormium*, *Gaultheria crassa* and species of *Celmisia* and *Aciphylla* were growing vigorously. Palatable species, notably resprouting chionochloas, aciphyllas, coprosmas and carmichaelias, were damaged at first by browsing animals. Steep, bare ground has been eroding.

Arthurs Pass, Hawdon Ecological Region

A large area of subalpine vegetation, including scrub, grassland and beech forest margins, was burnt in 1890, and there have been more local fires (Fig. 15.19). The ensuing changes have been followed for nearly a century, at first through L. Cockayne's field notes and later through transects. Table 15.2 illustrates some of the trends. Unlike most mountain localities, grazing has had little impact. The course of succession has been influenced by the degree of importance, in the original vegetation, of perennial herbs and shrubs that resprout after fire. These nearly all belong to the same genera, and mostly to the same species as on the Liebig burn. In tussock-herbfield, fire temporarily weakened *Chionochloa* tussocks, invigorated celmisias and

Fig. 15.19. Area burnt in 1890; 900 m, Arthurs Pass, Hawdon ER, photographed in (*a*) 1931 (Cockayne & Calder 1932) and (*b*) 1963 (Calder & Wardle 1969).

other perennial forbs, and allowed some invasion by hebes and *Cassinia 'vauvilliersii'*, but the pre-fire composition was re-established within 14 years.

Herbaceous members of woody vegetation, especially *Blechnum* 'mountain', *Phormium cookianum* and *Astelia nervosa* increased after fire, to the extent that after a second fire they resisted re-entry of shrubs. Fires in tall, dense scrub, however, left mainly bare ground. *Marchantia* and the fire-weeds *Senecio wairauensis, Epilobium* spp., *Pseudognaphalium luteoalbum*, and the adventive *Rumex acetosella, Stellaria media, Cirsium vulgare, Holcus lanatus* and *Agrostis capillaris* were the first colonists, but mostly disappeared within two years. *Hebe* spp. and *Cassinia* invaded rapidly, but in less than 40 years their demise was well under way. *Brachyglottis buchananii* also germinated in quantity, attained its greatest importance as the hebes declined and thereafter declined very slowly in favour of longer-living species. The slower-growing *Dracophyllum longifolium* and *D. uniflorum* appeared soon after fires, and gradually increased through further recruitment of seedlings and layering of procumbent stems. Adventitious rooting was even more important in the increase of *Phyllocladus alpinus* and *Podocarpus nivalis*. Kaikawaka and pink pine showed sparse seedling establishment and slow growth, but the former does not increase vegetatively, and the latter does so only on open, boggy, sites. Three plants of *Dracophyllum traversii* that appeared soon after the 1890 fire increased their heights from 0.6–1 m to 0.8–2.5 m between 1932 and 1965.

The time required to regain original status and stature is about 75 years for *Dracophyllum longifolium*, longer for *Phyllocladus*, and probably not less than 200 years for *Dracophyllum traversii*, pink pine and kaikawaka. Undisturbed scrub has shown little change among these species over 34 years. Beech forest has recovered far more rapidly, with seedling numbers increasing throughout the study, although only on areas that were formerly occupied by beech trees, and within those, only where scrub has been suppressed by the spreading canopies of beech trees and in enclaves of herbfield.

Tararua Range

McQueen (1951) and Macdonald (1978) describe the vegetation on burns in silver beech forest at 900 m. *Marchantia berteroana* covered most of the surface within two years. Within three years, other plants were becoming important, especially *Polytrichum juniperinum* on drier ridges and *Histiopteris* near the surviving forest edge, and seedlings of woody species were sparingly present. Twelve years after a forest fire, north and west aspects carried herbfield with willow-herbs, *Raoulia glabra, Blechnum procerum, Isolepis* sp., *Astelia nervosa* and *Phormium cookianum*, and there were occasional shrubs of *Coprosma* aff. *parviflora, C. pseudocuneata* and, in the shelter of logs, *Pseudopanax simplex, Carpodetus* and fuchsia. South and east aspects had more woody cover, together with scattered *Blechnum discolor* and *Cyathea smithii* that had survived the fire.

Ninety years after a burn, open scrub of *Dracophyllum filifolium* and kamahi grew

Table 15.2. *Development of vegetation after a fire in 1890, at the margin of beech forest against subalpine scrub, at Arthurs Pass*

species	1897–98	1931–32	1965–66	1965–66 compared with 1931–32
Nothofagus sol. cliff.		+, 1 m	3%*, 4.5 m	increased from 1 to 14 plants
Hebe canterburiensis	seedlings frequent, 1 m	6%, 1.2 m	f, 0.2–0.8 m	plants more numerous and dispersed but smaller
Hebe subalpina	seedlings abundant, 1 m	<1%, 1 m	+, 0.5 m	plants different
Brachyglottis buchan.	seedlings abundant, 1 m	20%, 1.2–1.8 m	20%*, 1.2–3 m	plants mainly same, fragmented
Coprosma aff. parvifl.	locally plentiful, 5 cm	5%, 1.5 m	9%, 1.5–1.8 m	original and new plants
Coprosma pseudocun.	from burnt stumps, scarce	12%, 1.5–1.8 m	35%, 1.5–2.4 m	original and new plants
Olearia ilicifolia	seedlings occasional, 10 cm	11%, 1.8–2 m	9%, 2.4–3.6 m	same 2 plants[a]
Olearia lacunosa		+, 1.8 m	+, 3.3 m	same plant
Olearia ilicif. × lac.		7%, 1–2 m	9%, 3 m	1 plant, now moribund[a]
Dracophyllum longif.		+, 1 m	4%, 1.5 m	1 original and new plants
Phyllocladus alpinus	seedlings rare, small	2%, 1.2–1.5 m	19%, 2.4–2.7 m	11 original and about 8 new plants
Hoheria glabrata	1 from burnt stump, 0.3 m	+, 1.2–2 m	1%, 2–3.3 m	plants mostly different and suppressed
Aristotelia fruticosa	seedlings frequent, 0.6 m	1%, 1 m	+, 0.6–1 m	plants suppressed
Cassinia 'vauvillier.'	seedlings frequent, 0.3 m		1%, 1.5 m	2 plants in openings
Pseudopanax colensoi var. ternatus	grown from burnt stump, 0.3 m	+, 1–1.2 m	+, 0.5–2 m	tallest plant the same, others different
Podocarpus nivalis	a few plants	<1%	8%	same colony, mainly in opening
Coprosma depressa	large patches probably survived fire	not recorded	f	at many points
Myrsine nummularia	in quantity, probably survived fire	not recorded	f	at many points
Phormium cookianum	in quantity	b 2%	b 2%	original plants mainly small and fragmented
Astelia nervosa		b 1%	+	plants smaller and more dispersed
Chionochloa rubra		not recorded	+	1 large tussock in opening
Polystichum vestitum		b <1%	b 1%	
Hypolepis millefolium		+	+	
Blechnum 'capense'	main feature	not recorded	12%	rather localised
total species	19	24	56	21 spp. present in 1932 and 1966

Notes:

Information for 1897–98 from L. Cockayne, unpublished. Data for subsequent years averaged from transects III and IV in Calder & Wardle (1969). Total area of transects 96 m².

Figures are percentage canopy cover and plant height, except for those prefixed 'b', which are percentage basal area.

* Refers to plants rooted inside transect. When overhanging canopies are included, average values become 8% for *Nothofagus* and 27% for *Brachyglottis*.

° These plants resprouted after the fire.

+, Present; f, frequent.

Fig. 15.20. Area of burnt subalpine mixed forest, now supporting tussock-herbfield; 900 m, Croesus Track, Paparoa Range, North Westland.

on a north-west aspect, and had scarcely gained on the 1 m height reached 30 years earlier. On the more sheltered north-east aspect *Dracophyllum* and kamahi were dense on steep slopes, whereas gentle slopes carried herbfield. Substantial regeneration of beech was restricted to a steep south aspect, and presumably commenced soon after the fire. During the past 30 years, browsing by deer has prevented establishment of faster-growing, more palatable woody plants, greatly reduced *Chionochloa conspicua* and *Astelia nervosa*, and encouraged spread of unpalatable species in the herbfield, especially *Lycopodium fastigiatum*.

Other localities

Burning of tall subalpine woody vegetation has often led to prolonged replacement by low shrubs, tussock grasses and large herbs, with much of the cover being provided by fire-resistant species that were present in the original understoreys. For instance, the tussock-herbfield shown in Fig. 15.20, consisting mainly of *Chionochloa conspicua*, *Phormium cookianum* and *Astelia nervosa* with *Schoenus pauciflorus* and *Blechnum* 'mountain' beneath, has replaced southern rata (kaikawaka–pink pine–*Dracophyllum traversii*) forest burnt about 1930. Scattered shrubs of *Brachyglottis buchananii*, *Hebe salicifolia*, *Dracophyllum longifolium* and *Coprosma foetidissima* suggest the eventual development of tall scrub, but this has been set back by later fires.

Burnt pink pine/*Olearia colensoi* scrub on Maungapohatu (Huiarau Range) has

Fig. 15.21. Effects of different management on Maungatua, Lammerlaw ER. There are tussocks of *Festuca novae-zelandiae* in a short sward in the foreground, dense scrub of *Cassinia* and *Hebe odora* on the upper left, and *Chionochloa rigida* grassland on the upper right (Mark 1955).

been replaced by *Coprosma foetidissima–C. pseudocuneata/C. depressa* scrub on slopes, and by herbfield on flatter ground (Cranwell & Moore 1931). In Westland National Park, tall subalpine scrub burnt in the 1890s was replaced by *Chionochloa* grassland, and shrubs are returning only very slowly.

On large areas of mid-altitude country in the east of the South Island, much of which was cleared of forest and tall scrub by early Maori fires, there are vegetation mosaics determined by pastoral use. At 660–870 m on Maungatua in Lammerlaw ER, *Dracophyllum longifolium* heath represents the vegetation that replaced subalpine silver-beech forest after Maori fires. It burns readily, to be temporarily replaced by *Cassinia, Hebe odora* and *Chionochloa rigida. Dracophyllum* eventually returns in the absence of further fires, but otherwise, pure *Chionochloa* grassland may be created and maintained under a regime of extensive early-spring fires followed by light grazing. Heavier grazing may lead to *Festuca novae-zelandiae* tussocks increasing at the expense of *Chionochloa*. Concentrated grazing, such as occurs when small areas are burnt, is likely to kill tussocks through removing all new leaves; at first, small grasses such as *Rytidosperma* spp., sweet vernal and browntop may increase, but *Cassinia* and *Hebe* then invade. These form dense fire-resistant scrub, as *Cassinia* resprouts after fire and *Hebe* does not carry fire (Fig. 15.21). Fire and grazing in the penalpine belt also reduce or eliminate *Chionochloa* tussocks, leading to communities dominated by shorter grasses, especially *Poa colensoi*, or celmisias which are most prevalent on stony or eroded ground.

Invasion

Adventive plants have formed many new communities in areas modified by humans and their agents (Healy 1968). They have also entered open vegetation, especially seral communities, with little or no human assistance, often ousting native pioneers and deflecting successions to paths that will only slowly, if ever, lead back to indigenous vegetation. Examples include *Spartina* on estuaries, willow, gorse and common broom on flood plains, *Elodea canadensis* in lakes, marram on dunes, and thickets of self-seeded pines and hawkweed (*Hieracium*) mats in tussock country.

Such adventives may out-compete native plants by pre-empting resources through greater height, more vigorous growth or reproduction, or greater dispersal ability. They may exploit resources that are not tapped by resident natives, examples being vernal adventives such as *Veronica serpyllifolia* and species of *Bromus* and *Aira* that find niches in dry native grassland, or *Cirsium arvense* that has deep, fast-growing rhizomes that allow it to occupy alluvial silts before native seedlings can establish on the loose, droughty surface. Some, such as cold-resistant conifers, have environmental tolerances not matched by native plants. Hawkweeds produce allelopathic compounds but their displacement of tussock grasses is more likely to be due to competition for water and nutrients (Makepeace *et al.* 1985).

Another probable factor in rapid spread of introduced plants is the absence of predators and parasites that control them in their native countries. For instance, in Britain six species of insects parasitise the seeds or pods of common broom and drastically reduce seed production (Waloff 1968), whereas in New Zealand such parasites are unknown (P.A. Williams, personal communication). Restoring the balance is the main thrust of biological control, the best success in New Zealand having been reduction of St John's wort (*Hypericum perforatum*) by the beetle *Chrysolina hyperici*. Some adventive species have increased when an animal predator has fallen to low numbers, examples being briar (*Rosa rubiginosa*) (p. 212) and hawkweeds (p. 249) in tussock grasslands after rabbits were reduced in the 1950s. In many nature reserves, exclusion of grazing has led to rank growth of pasture grasses that inhibits succession towards forest and overwhelms small native plants that thrive in grazed swards.

Small remnants of native vegetation are especially vulnerable to invasion by foreign plants. In part, this is due to edge effects, such as increased light and exposure, and drift of fertilisers and herbicides from surrounding agricultural land. The sheer abundance of exotic seed is also a factor, that may partly account for the gradual demise of roadside remnants of native grassland. In forest remnants, both soil seed bank and current seed influx can be dominated by adventive species (p. 64), so that seedlings of plants such as elder, privets (*Ligustrum* spp.) and *Prunus* spp. can greatly outnumber those of native trees. Herbaceous adventives such as *Tradescantia fluminensis* can take over the floor (p. 164) and lianes such as *Clematis vitalba* can damage the canopy (p. 163). There is a danger that once such plants have established niches within small areas of native vegetation, they may be positioned to invade more extensive tracts.

Native plants also spread into new localities, often from human introductions, as in the case of taupata (*Coprosma repens*) thickets on islands in Foveaux Strait (p. 396).

Possibly, human introductions account for karaka in the Chatham and Kermadec Islands and *Olearia lyallii* in the Auckland Islands (pp. 434, 436, 447).

Throughout the Quaternary era, massive geological and climatic disturbances have influenced the distribution of species. The content of vegetation developing on disturbed surfaces is limited to species within dispersal distance. Species absent from a region but with potential niches therein can enter only through spread and invasion of existing vegetation; and just as closed native vegetation resists invasion by exotics, so it resists invasion by other native species, even those that seem well suited to the locality.

Present distribution of beeches in the South Island is seen as emanating from small refugia, to which they had been confined during periods of glacial severity. Because beeches usually form sharply bounded stands that differ in structure from adjoining vegetation, invasion patterns are obvious, especially in silver beech. The evidence for invasion is of three kinds: first, the fossil record shows spread and increase of beech during the Holocene (Fig. 2.5); second, boundaries between beech forest and adjoining vegetation often fail to conform with evident environmental gradients and boundaries, so that beech is absent from apparently suitable terrain; and third, zones of young plants usually surround beech forest margins and isolated beech groves.

Although some of these irregularities in beech distribution may be related to undetected (McGlone 1985) or subtle environmental gradients (Haase 1989), in most instances it seems likely that rates of invasion, rather than present positions of boundaries, are environmentally determined. This might explain, for example, why the central South Island 'beech gap' (Fig. 7.25) is absolute for 160 km in Westland, but only partial in the drier regions east of the Southern Alps.

In South Westland, boundaries between beech and adjoining forest and heath communities extend from sea level to the upper forest limit (Fig. 15.22), and lie across soils of widely different fertility. Spread of beech is at an average rate of 6 m per century, and depends on canopies of marginal beech trees creating a zone in which adjoining vegetation is suppressed. However, chance dispersal of seeds, presumably on to open ground, enables outliers of beech to establish at distances of up to 6 km from the main front. Similar patterns are described from North Westland (S.R. June, unpublished) and Catlins ER (Allen 1987). On the Volcanic Plateau, the distribution of beech seems related both to refugia sheltered from volcanic blasts and to post-eruption dispersal (p. 526). Other environmentally discordant, advancing beech forest margins relate to slow recovery of ground lost to fire during the human millennium.

Other species may also be expanding their range. In South Westland, for example, scattered trees of *Quintinia acutifolia* and hinau (*Elaeocarpus dentatus*) exceed the main southern limits by 40–50 km. Doubtless, there were earlier episodes of expansion and contraction and some isolated occurrences may be relicts from these. In particular, the pockets of hard beech in Northland may prove to be remnants of forests that were extensive under colder climates.

Conclusions

Although successions in New Zealand are endlessly varied, some patterns emerge. Central Westland provides primary sequences of unsurpassed length and complete-

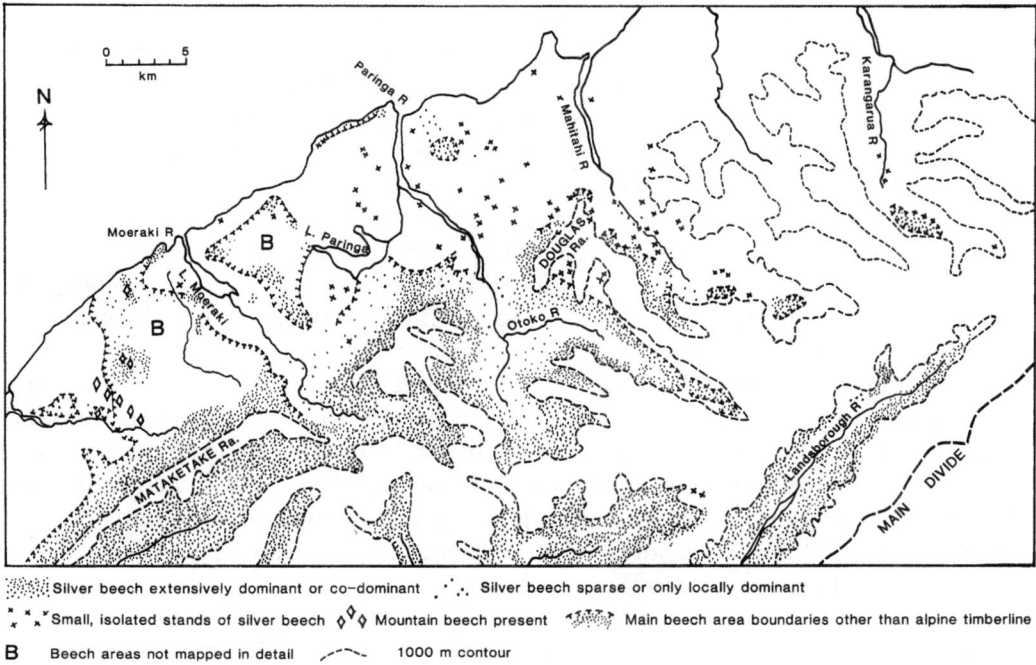

Fig. 15.22. Distribution of silver beech in Paringa ED and surrounding areas, South Westland, showing major area of occurrence, and scattered occurrences in the conifer/broad-leaved region to the north-east (Wardle 1980*b*).

ness, notably on fluvioglacial surfaces amenable to dating. Dense woody vegetation can establish within a decade on fine-textured, well-drained substrates, especially those derived from schist breccia near the Alpine Fault despite rapid leaching of nutrients under rainfall that exceeds 10 m/yr.

At low altitudes, nitrogen-fixing *Carmichaelia, Coriaria* and *Gunnera* species are abundant during early stages, but are rather sporadic above 300 m. Lowland forest communities still have some seral characteristics at 1000 years, but at 5000 years moraines carry climax stands on yellow-brown earths that are locally gleyed and podzolised. Species indicating retrogression appear by 14000 years, and on older surfaces heathland vegetation is widespread.

Alluvial terraces support different sequences of species, and develop iron pans and impeded drainage even where initially well-drained. Dense rimu stands and wet heath on gley podzols overlain by structureless peat can occupy Holocene terraces. Mobile coastal dunes, unlike unstable sediments in river beds, have a specialised florula. Once fixed, dunes and beach ridges on prograding coasts support successions similar to those on alluvium, except that the alteration of well-drained rises and troughs close to the water table lead to sequences of alternating dryland and wetland stages.

In the mountains, yellow-brown earths develop by 500–1000 years and gley podzols by 2000 years, presumably as a result of high precipitation, low evaporation and low rates of plant growth and organic cycling. Accordingly climax forest, grading upwards into subalpine bush, gives way to heathy vegetation. With increasing altitude herbaceous successions occupy increasing areas, especially on flat ground and gentle slopes. There are also successions from fen to bog. However, as landscape rejuvenation is the dominant process in the mountains, only tiny areas attain advanced stages of succession and soil development.

Sheltered lowland boulders and bed-rock can become covered by mosses within nine years, whereas exposed subalpine boulders may support little vegetation other than crustose lichens after several centuries. Vegetation also develops slowly on ultramafic terrain, with patches of tree- and shrub-heath being well established long before vegetation has become continuous.

Understorey species are as characteristic of each successional stage as canopy dominants. However, species do not necessarily become prominent in the same order as that in which they enter the succession. On lowland moraines, for example, southern rata seedlings enter about the same time as those of *Olearia avicenniifolia* but, as trees, form the canopy some 150 years later; and *Phyllocladus alpinus* can be abundant in the understorey of tall forest on surfaces several thousands of years younger than those where it contributes to the canopy of tree-heath.

In beech forest regions west of the Main Divide, beech seedlings are among the early pioneers. The vegetation quickly gains biomss, but floristic diversity and complexity may well decrease, especially at higher altitudes as grassy and shrubby communities are suppressed by pure beech stands. By 300–400 years, silver-beech forest can be essentially climax on podzolised yellow-brown earths. At low altitudes complex (conifer)/beech/broad-leaved stands are more likely to develop.

Successions below tree-limit in South Island valleys east of the Main Divide seem much slower than in the west, reflecting lower rainfall and humidity, and greywacke parent material instead of schist. Disruption by fire is also frequent, which may explain why vegetation on moraine sequences can seem unduly young in comparison with ages of surfaces. Drought-tolerant plants, especially matagouri, are prominent in seral stages, and relatively xeric mountain beech or mountain totara forests can be climax. In the lowlands of the North Island and east of the South Island, most primary surfaces are initially dominated by adventive plants, and few long, unbroken sequences are available.

Large areas of successional vegetation were induced by volcanic eruption in the North Island, especially on the Volcanic Plateau, although it is hard to find dateable sequences of equivalent surfaces in proximity. Unburnt vegetation on the rhyolitic landscape derived from the Taupo eruption includes podocarp forests similar to those on recent alluvial terraces in Westland, and heathlands with more in common with fluvioglacial surfaces that are tens of thousands of years old. Clearly, successions are complicated by special factors, including depth to buried soils, climatic effects (especially cold-air drainage), and a high content of glass that weathers to phosphorus-occluding allophane clays.

On the slopes of the volcanic mountains, primary surfaces vary in chemistry, texture, and form of emplacement, i.e. as lava, pyroclastic flow, lahar, debris flow, or tephra. Lava and scoria are the least amenable for colonisation. For example, Rangitoto Island is still only partly wooded after 200–400 years, despite its sheltered location, warm-temperate climate and basaltic composition. Rates of succession diminish sharply with increasing altitude, and are also influenced by distance to seed sources and extent of survival of pre-eruptive vegetation.

In the penalpine belt, the snow tussocks (*Chionochloa*) form series related to the nutrient status of the soils, especially in respect of phosphorus fractions. These can be related to successions, with species such as *C. crassiuscula* representing retrogressive stages. Although complete sequences can be found on Holocene moraines, dating is provisional.

Secondary successions can be superimposed on any stage of primary succession. Accordingly, pioneering species can differ even where climate and topography are similar. For example, wineberry is likely to appear after climax forest has been burnt, and manuka after tree-heath.

Very large areas are clothed by plant communities maintained by fire or grazing. Within these communities, there can be distinct phases dependent on periodic return of the disturbance factor. Examples involving fire include temporary dominance of *Gleichenia* and rush-like sedges on gumlands and pakihi, alternations of bracken and manuka, and the temporary removal of matagouri and other shrubs from short-tussock grasslands. Burning, especially if accompanied by grazing, allows celmisias to increase at the expense of snow tussocks, and drought or outbreaks of grass-grub allow weeds to invade pastures.

Even if burning or grazing cease, succession can be slow and uneven, because of inadequate seed sources and competition from the resident vegetation. Sometimes later successional stages are enduring, as with the broad-leaved bush that replaced beech forest after Maori fires. Again, recovery is slowest at high altitudes and in dry regions, and early stages at low altitudes are increasingly dominated by adventive plants. Although there is ample evidence that, given time, native forest indistinguishable from the original can re-establish, the outcome is becoming less predictable where the original vegetation was shrubby or herbaceous.

Succession, in its broad sense, also involves invasion of sites by species that have expanded their ranges to within effective dispersal distance. Invasion, whether by adventive or native species, takes place most readily into open, seral vegetation or during advanced stages of retrogression. Climax vegetation, especially forest, appears to resist invasion, but this may be only on the brief time-scale of ecological observation in New Zealand. Over millennia, there has obviously been massive replacement of sets of species by other sets through the impetus of climatic change; to what extent were these replacements expedited by local or large-scale disturbances by wind, fire and earth movements?

The same quetion could be asked of the future. Will the rise to dominance of adventive plants be mainly restricted to open and disturbed habitats as at present, or will replacement become universal, especially if culturally induced atmospheric changes affect regional climates?

Some successional concepts in the New Zealand context

Early pioneers may be regarded as *facilitators* of succession, in that they improve the habitat for species that enter later, through providing shelter and adding organic matter to the soil. Nitrogen-fixing plants are especially important in this role.

Facilitation occurs even late in succession. For example, it is unusual for podocarp seedlings to establish in primary successions before broad-leaved trees or shrubs have formed a canopy. Epiphytic seedlings obviously depend on prior establishment of host plants, as instanced by kamahi seedlings that occupy tree fern trunks.

However, *inhibition* of succession by resident species is surely the rule once vegetation has formed a closed canopy or occupied all the rooting space. Further changes are likely to be driven by onset of conditions that are no longer tolerated by resident species, which can be regarded as facilitators only in so far as they contribute to their own demise, for example through producing leachates that hasten podzolisation. Strong inhibition is seen where grassland pre-empts alluvial flats at low altitudes. Once this occurs, tree seedlings are excluded by a thick sod, by grazing animals that the grassland attracts, and by sharp frosts that affect grassy openings. Forest invasion is mainly along margins where tree canopies provide shelter and suppress the grassland.

Plants have been classed as *r-selected*, if adapted to seral stages through traits such as copious production of small, readily dispersed seeds, intolerance of competition, rapid growth and short lives, and *K-selected*, where heavy seeds, shade-tolerance and long lives enable them to thrive in climax vegetation. Willow-herbs, raoulias, manuka, wineberry, most hebes and a host of adventive weeds are among plants with an array of *r*-selected characteristics. None of the larger native trees is strongly *r*-selected, which partly explains why adventive trees, especially pines and willows, are often more prominent as early pioneers.

At the other end of the spectrum, few New Zealand plants fully meet the criteria of *K*-selection. Instead they show various intermediate strategies, that have much bearing on regeneration processes in forests and will be discussed in Chapter 16.

Bray (1989) uses redevelopment of forest after clearance to illustrate facilitation, inhibition, and the validity of alternative successional models. Through concluding that recurrence times of disturbances tend to be shorter than the life spans of most species present, he proposes that traditional concepts of succession and climax should give way to concepts in which disturbance is recognised as central to all vegetation development and change, including regeneration. Nevertheless, this emphasis on disturbance would seem compatible with traditional views, providing one accepts that succession and disturbance occur in infinite variety and at many scales, and that climax is not interpreted to mean static and unvarying.

16

DISTURBANCE, REGENERATION AND TRENDS IN NATIVE FOREST

Study of forest processes is an important but difficult field in which knowledge rests largely on proxy observations, as most trees live far longer than their observers; indeed, many large trees far pre-date human presence in New Zealand. Seed production varies greatly among species and years; modes and distances of effective dispersal are no less varied. Seedlings must cope with an environment very different from that exploited by mature trees, but vegetative renewal may lessen dependence on seed and seedlings. As young trees grow towards the canopy, they must compete with their contemporaries, older cohorts of their species, and other species in the tiers they pass through.

Effective regeneration may depend on canopy gaps, and formation of gaps in turn usually depends on disturbance at some scale; but the return period of disturbance may exceed the length of the historical record, especially when this is short as in New Zealand. Massive disturbances can initiate successional trends in forest composition. Proportions of species may be changing in response to secular climatic changes, or there may be delayed response to earlier change, as was suggested in the previous chapter to explain invasion by *Nothofagus*. The faunal changes that have accompanied human colonisation have profound impacts on regeneration, but the direction is not always apparent.

This chapter begins by considering the types of stand structure that should be associated with species with different life histories and requirements. Next, tolerances of seedlings and the role of vegetative reproduction are discussed. Disturbance is discussed under the headings of die-back in the canopy, depletion of understoreys by animals, and collapse of forest structure through multiple causes. The final sections deal with population structure and regeneration in beech forests, broad-leaved stands other than beech, and conifer/broad-leaved forests, including those dominated by *Libocedrus* and *Agathis*.

Stand structure in relation to life history of species

Forest dynamics may be described in terms of interplay among canopy-forming species with r- and K-selected characteristics (p. 558). The former would be expected to prevail during seral stages and in forests subject to severe disturbance; the latter should prevail if canopies remain substantially intact in the absence of significant disturbance. Among K-selected species, regeneration should be continuous, so that population structures conform to the power or exponential curves described as 'reverse-J'. Regeneration of r-selected species should occur within canopy gaps, giving rise to discrete cohorts, each approximating to a bell-shaped (unimodal) diameter-distribution curve. As cohorts age, modal diameters shift to larger diameter classes. This might lead to cohort senescence and thereby to further episodes of synchronised regeneration, even in the absence of renewed disturbance. When patches that individually conform to bell-shaped curves are integrated over increasingly large areas, they also should increasingly approach the reverse-J distribution. If they do not, it must be suspected either that episodic factors regionally synchronise regeneration or deplete particular size classes, or that ontogenetic variations in growth rates lead to peaks and troughs in the curve.

Few species fully fit the criteria of r- and K-selection, and there are complex interactions between canopy and understorey species. Since correlations between age and diameter vary from satisfactory to poor there is a large element of uncertainty in deducing regeneration patterns from diameter measurements alone. Even-aged cohorts often range widely in diameter, because some trees are suppressed and others released. Conversely, densely stocked stands of trees of similar size may differ widely in age, because young trees growing up in gaps have overtaken older, slow-growing trees. Analysis of population structure through growth ring measurements is more accurate, although these too are subject to uncertainty because of multiple and missing rings, the unknown time taken by seedlings to grow to the height of coring, or failure to reach the chronological centres of trees with heart rot, eccentric growth, or large diameters. Most studies in New Zealand have relied on diameter measurements that usually are backed by some age determinations.

Requirements of native tree seedlings

Field observations, with some experimental support, suggest that most New Zealand trees have low light compensation points. Values include 1.9–3.9% of incident light for red beech (*Nothofagus fusca*) (June & Ogden 1975), 2% for kauri (*Agathis australis*) (Bieleski 1959) and about 1% for the species in Fig. 16.1. Differences among species lie rather in their responses to increased light. The more vigorous response by the two broad-leaved species in Fig. 16.1, as compared with the conifers, seems true of these groups as a whole. Beeches also respond well to increased light up to an optimal level; under stronger sunlight assimilation is likely to be inhibited by drought or frost. For red and mountain beech (*N. solandri* var. *cliffortioides*) the optimal value is about 35% of incident light (Wardle 1970; June & Ogden 1975). Silver beech (*N. menziesii*) is reputedly more shade-tolerant.

Tall canopy-forming species that are truly shade tolerant, in that they can grow up under a closed canopy, include tawa (*Beilschmiedia tawa*), kohekohe (*Dysoxylum spectabile*) (Court & Mitchell 1988) and, probably, taraire (*B. tarairi*) and pukatea (*Laurelia novae-zelandiae*). None of these reach southern New Zealand. There are no

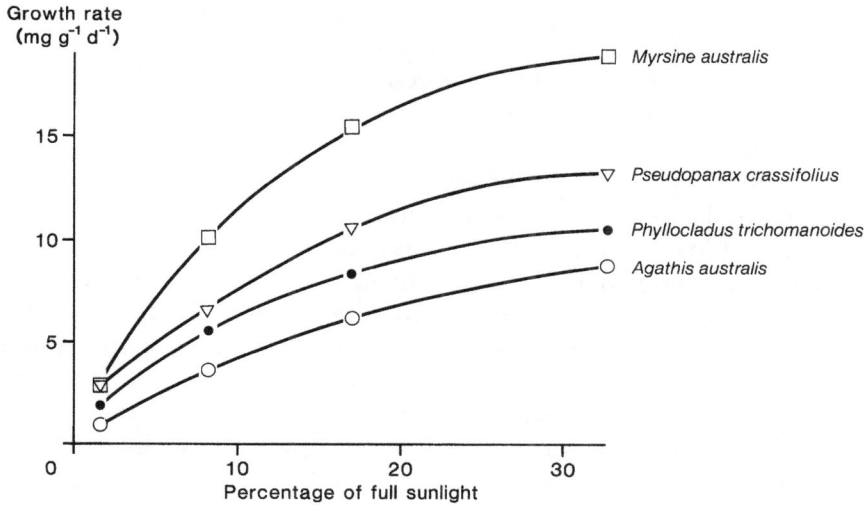

Fig. 16.1. Effect of shade on relative growth rates of seedlings (Pook 1979).

tall, fully light-demanding species that establish in gaps and grow rapidly to the high canopy, such as are important in tropical forests, but northern (*Metrosideros robusta*) and southern rata (*M. umbellata*) are light-demanders that contribute to the canopy through beginning life as epiphytic seedlings, as well as forming seral stands. *Weinmannia* species also have these abilities, but tolerate more shade.

Small trees exhibit a wide range of shade tolerance. *Hedycarya* and species of *Pseudowintera* can reach maturity beneath a tall canopy, whereas *Myrsine* spp. and lancewood (*Pseudopanax crassifolius*) persist in shade as seedlings but require gaps to become trees. Other *Pseudopanax* species and *Griselinia littoralis* often have epiphytic beginnings. Even the light-demanding wineberry (*Aristotelia serrata*) finds a niche in climax forest because of effective dispersal and rapid growth in gaps, especially where tree-falls have turned up mineral soil.

Water relations are at least as important as light for survival of seedlings. The vigorous root systems of shaded tawa seedlings enable them to survive drought better than rimu (*Dacrydium cupressinum*) seedlings (Fig. 3.5). Strongly light-demanding seedlings, such as those of kowhai (*Sophora* spp.) and pioneering shrubs, can retain descending tap-roots (Fig. 4.4).

Deep, loose litter precludes most tree seedlings, because their roots cannot reach underlying soil with adequate water-holding capacity. Conditions are ideal where thin humus or moss cover moist mineral soil, but mossy rotting logs and stumps are also favoured. However, a cover of readily decomposed litter favours seedlings of kohekohe (Court & Mitchell 1988), and probably those of other large-seeded trees. There is general intolerance of dense growth of tall ferns, grasses, sedges or fast-growing broad-leaved shrubs and saplings; thick litter under such vegetation is, perhaps, more inhibiting than shade *per se*. Beeches and, infrequently, conifer seedlings can establish on mineral soils in the open, although partial shade is best.

Whereas suppressed seedlings of beeches and many broad-leaved species, including kamahi (*Weinmannia racemosa*) and *Quintinia*, respond vigorously to the formation of canopy gaps, slow-growing conifer seedlings are likely to be overwhelmed by growth of other plants. The best conditions for their establishment are where a tree canopy suppresses the understorey, without itself providing too much shade, competition, or litter. On the infertile gley podzols of Westland, the herbaceous tier can be poorly developed even under quite open canopies, and regeneration of conifers tolerant of low fertility is relatively unrestrained. On fertile soils, however, opportunities for conifer regeneration are limited in both space and time.

Although allelopathy has rarely been considered in New Zealand forests, V.A. Froude (unpublished) found that water-soluble leaf extracts from 36 species of native forest and scrub mostly inhibited germination and radicle growth of test species; *Hymenophyllum demissum*, *H. sanguinolentum*, manuka (*Leptospermum scoparium*), totara (*Podocarpus totara*) and *Schlefflera digitata* were the most inhibitory. In the forest, patches of the two filmy ferns contained only 34 and 37%, respectively, of the number of dicot seedlings present in control plots. Molloy *et al.* (1978) reported that extracts of podocarp leaves and litter are toxic for kahikatea (*Dacrycarpus dacrydioides*) seedlings.

The role of vegetative reproduction

Many broad-leaved forest trees coppice when the crown loses vigour or dies. In tawa, seedlings coppice from lignotubers if the original stem dies (Cameron 1963), and tree stumps resprout after logging (Smale & Kimberley 1983). Kamahi saplings coppice and weak young stems of *Quintinia* and tawa layer, so that clusters of trees can arise from a single seedling. In heaths on gley podzols, manuka and the small podocarps exist as colonies of procumbent, rooting stems that give rise to erect, dominant stems in canopy gaps. Even kamahi, southern rata and mountain totara (*Podocarpus hallii*) become reduced to slender, layering shrubs in such vegetation.

At the upper forest limit, the normally erect, single-stemmed beeches can form layered colonies (p. 21). Root suckering is a mode of reproduction in silver pine (*Lagarostrobos colensoi*) (Moar 1955), *Olearia ilicifolia* (Burrows 1963) and its hybrids, *Myrsine chathamica* and, probably, other species.

In subalpine heath and bush, the lower branches of shrubs and small trees lie on the ground and take root, especially on the downhill side on steep slopes, probably because of snow pressure and soil creep. This is a major form of reproduction among the canopy plants, including podocarps, *Coprosma pseudocuneata*, *Olearia colensoi*, and *Dracophyllum* species other than *D. traversii*. Established seedlings of canopy species are rare in closed stands, but those of *Dracophyllum* spp. including *D. traversii*, *Hoheria glabrata* and *Olearia ilicifolia* (Haase 1986c–f, 1987) and *O. colensoi* can be frequent on open areas such as slips and burns.

Damage to forest canopies and understoreys

When scattered trees in a forest die of old age, young trees and lateral growth from neighbouring crowns quickly repair the canopy, but when numerous canopy trees die

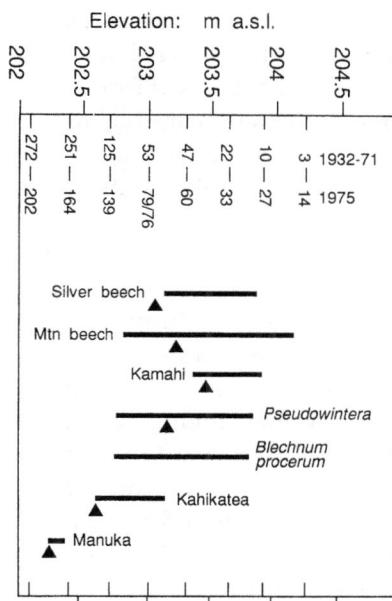

Fig. 16.2. Ranges of elevation for dead plants of seven species on the Te Anau lake shore (Mark *et al*. 1977). Columns on the left show the duration of the largest flood (days > each level) in 1975 and during the 40 year period before the lake was controlled in 1971. Triangles indicate lowest elevations for surviving plants.

within a short period over considerable areas, the visual and ecological impact is profound. The term die-back refers to death that begins with the uppermost twigs and proceeds downwards. Trees such as *Weinmannia* and *Metrosideros* tend to produce epicormic shoots from limbs and trunks while dying gradually; these shoots sometimes lead to recovery. Beeches and most native conifers, on the other hand, usually die rapidly, producing few or no epicormic shoots.

Mass die-back in native forests has been widely reported, especially since the 1950s. Although many apparent causes have been identified, it is often difficult to ascertain which of these are primary, since moribund or unthrifty trees attract a wide spectrum of secondary parasites and pathogens that can inflict the *coup de grâce*. Moreover, catastrophe-initiated regeneration of forests leads to stands of one or more even-aged cohorts (Stewart & Veblen 1982) that may grow senescent together, and become increasingly vulnerable to life-shortening events leading to simultaneous die-back. The following account begins with die-back and other kinds of canopy and understorey damage that have readily identifiable physical or biological causes, and concludes with examples of drastic collapse of forest structure that demand complex explanations.

Die-back attributable to damaged root systems

Drowning of root systems by raised water tables often causes die-back along lake shores, especially those dammed artificially or by landslides. Fig. 16.2 shows the susceptibility of several species to high lake levels, although eventual death may be due to pathogens and sluggish drainage after floods have receded.

Burial of roots by silt can kill large patches of forest on flood plains. If the silt is not too deep or wet, many trees recover through producing adventitious roots from buried boles. In this way, kahikatea can survive burial to depths of 60 cm (P. Wardle 1974) and totara to depths of over 1 m (Foweraker 1929). Trees partly buried by tephra during the Tarawera eruption resprouted.

Although the root pathogen *Phytophthora cinnamomii* is blamed for catastrophic die-back in Australian forests, in New Zealand wild native plants generally seem to tolerate it (Robertson 1970). However, local unthriftiness in native shrub communities has been attributed to *P. cinnamomii*, and it caused severe die-back in kauri and tanekaha (*Phyllocladus trichomanoides*) on a flat ridge top in the Waitakere Range, possibly having been induced by water-logging during a very wet summer (Podger & Newhook 1971). *Phytophthora heveae* may cause unthriftiness of kauri seedlings on wet soil (Gadgil 1974). *Armillaria mellea*, which causes root- and heart-rot, is ubiquitous but of low pathogenicity in native forests; however, it can severely damage exotic trees planted in logged native forest (Gilmour 1966).

Trampling by ungulates locally destroys the humus horizon which contains much of the fine root system and can cut deeply into mineral horizons, especially on well-tracked ridges and spurs, but has not been reported as a cause of die-back. Damage to roots by machinery is said to kill rimu trees in selectively logged stands in Westland.

Die-back attributable to pathogens attacking stems

Many canker-inducing, bracket-forming and wood-rotting fungi and gall-forming, sap-sucking and wood-boring insects are described from New Zealand trees, but large-scale die-back from these is reported only from *Nothofagus* forests (Wardle 1984). Much damage in beech forest is attributed to the pin-hole beetle *Platypus*, through its role in introducing the pathogenic fungus *Sporothrix* to inner sapwood (Milligan 1974). The amount of damage depends on the concentration of beetles, which is likely to be greatest where forests have been damaged by wind, snow or logging, because recently dead wood is favoured for brood-rearing. Vulnerable trees include those stressed by events such as drought and waterlogging. Fast-growing trees in pole stands can also be heavily attacked, which is a serious impediment for beech management. *Platypus* also attacks other hardwoods, including kamahi (Payton 1989b).

Sooty moulds are conspicuous in native woody vegetation, especially low-altitude stands of fusca-group beeches. They depend on phloem sap exuded by coccids, which appear to affect the vitality only of trees already under stress. A notable exception is the widespread death of manuka through attack by *Eriococcus orariensis* (p. 197).

Agents of defoliation

Frosting of foliage is normally followed by recovery during the following growing season, but has led to severe die-back in tawa, and is involved in die-back of beech at the subalpine forest limit (p. 504). Salt-laden gales can scorch canopies several kilometres inland, with glabrous, fast-growing species such as mahoe being most susceptible.

Table 16.1. *Occurrence of species in woody diet of possums, preference ratio, and mean basal area of dead stems between 250 m and the forest limit, Taramakau Valley, North Westland*

species	occurrence in diet[1]	preference ratio[2]	dead basal area[3]
Rubus spp.[4]	0.6		
Fuchsia excorticata	0.1	55.0	—
Coprosma foetidissima	3.6	45.4	—
Pseudopanax spp.	4.3	39.2	—
Pseudopanax simplex	2.3	11.9	—
Podocarpus hallii	0.9	4.3	0.8
Melicytus ramiflorus	13.2	4.0	—
Metrosideros perforata[4]	1.5		
Metrosideros umbellata	27.5	2.6	1.8
Schefflera digitata	0.1	1.4	—
Aristotelia serrata	2.2	1.0	—
Coprosma spp.	1.5	0.9	—
Weinmannia racemosa	37.7	0.8	1.7
Myrsine divaricata	0.2	0.8	—
Libocedrus bidwillii	2.2	0.7	1.3
Coprosma rotundifolia	0.2	0.3	—
Dracophyllum traversii	0.2	0.2	—
Pennantia corymbosa	0.1	0.2	—
Pittosporum eugenioides	0.1	0.1	—
Prumnopitys ferruginea	0.1	0.1	—
Griselinia littoralis	0.2	0.0	—
Quintinia acutifolia	0.2	0.0	—

Notes:
1 Percentage of total width of cuticle fragments in guts.
2 Percentage occurrence in diet/percentage total basal area.
3 Dead stems (m²/ha), mean from 304 plots; —, <0.2.
4 Lianes; basal area very small.
Source: Coleman *et al.* 1985

Many insects, mainly lepidopteran larvae, chew leaves or destroy buds. Epidemics of defoliation occur in fusca-group beeches (p. 567), *Brachyglottis buchananii* and the deciduous *Hoheria glabrata*, but severe mortality seems to require predisposing events. None of the bacterial, viral and many fungal pathogens that attack foliage are known to significantly damage native forests. However, the needle-cast fungus *Dothistroma pini*, which appeared in 1962, causes die-back in pine forests.

Native pigeons (*Hemiphaga novaeseelandiae*) strip leaves from several tree species, including kowhai, fuchsia (*Fuchsia excorticata*), hoherias and *Plagianthus regius*, and the now-rare kokako (*Callaeus cinerea*) feed on a wide variety of foliage and fruit (Clout & Hay 1989). However, in today's forests the main browser of tree foliage is the brushtail possum (*Trichosurus vulpecula*). These marsupials often completely and repeatedly defoliate their preferred species (Table 16.1), to the point that whole stands

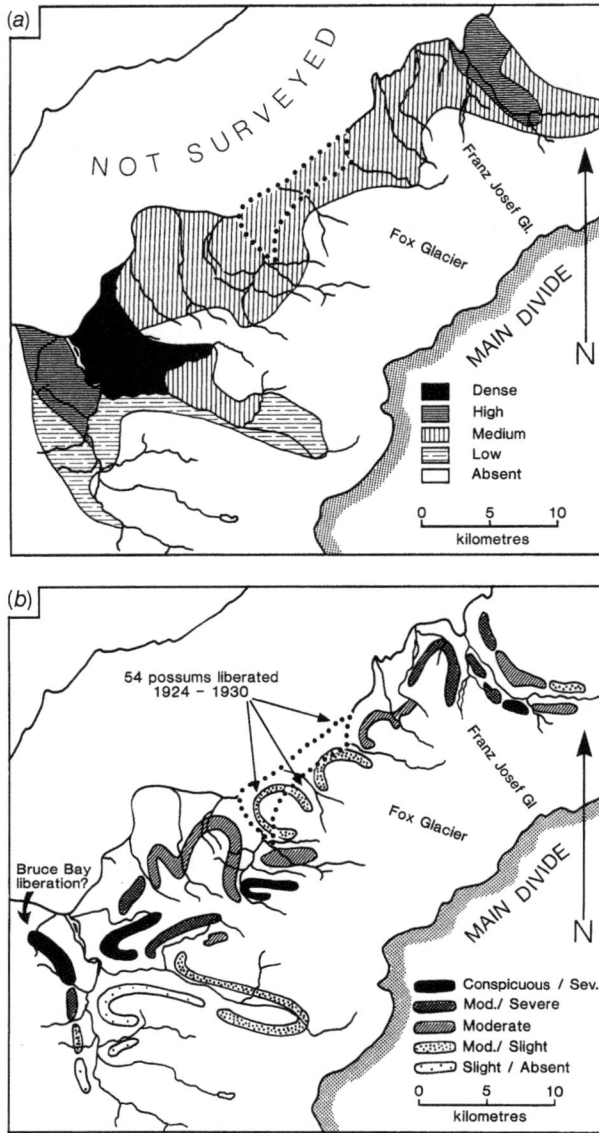

Fig. 16.3. Possums and canopy defoliation in montane mixed forest, Westland National Park (from Pekelharing & Reynolds 1983). (a) Densities of possums; (b) Severity of canopy defoliation, as recorded in 300 m wide swathes at 950–1200 m a.s.l., along the flight path of a helicopter. The area enclosed by dots indicates where possums were dense in 1950, and canopy defoliation then conspicuous.

are killed (p. 582). They increase to very high numbers after colonising an area, and this is followed by death of susceptible trees (Fig. 16.3). Over recent years, spread of possums up the Northland peninsula has been flagged by die-back of northern rata, pohutukawa (*Metrosideros excelsa*) and other trees. Barriers such as deep ravines can, for a time, separate healthy from devastated forest, and severely damaged trees have recovered when fitted with metal sheaths that deny possums access (Meads 1976).

Possums have a minor influence on beech forests, other than exterminating the showy *Peraxilla* and *Alepis* mistletoes (Wilson 1984). Today, mistletoes brighten the beech forests only in areas where possums are scarce, and at forest margins east of the South Island Divide where, under drier, sunnier climates, the plants seem more resilient.

Die-back related to drought

Tawa, mahoe (*Melicytus ramiflorus*), rimu and kahikatea trees died within a year after a summer drought in 1969–70 in the Manawatu district (Atkinson & Greenwood 1972), when rainfall was half the average over an eight-month period. Mountain-beech trees died on deep pumice soils in the Kaimanawa Range after drought in 1945–46 (Grant 1984), and on well-drained sites near Arthurs Pass after the dry summer of 1970–71 (P. Croft, unpublished).

After a spring (September–November) drought in Fiordland, when rainfall was 40% below average, significant numbers of mountain beech trees died when attacked by the defoliating moth *Proteoides carnifex*. In the Maruia Valley (Spenser ER) the years 1976–78 were drier than usual, and included a spring when rainfall was 40% below normal. During 1976 and 1977 the leaf and twig scale *Inglisia fagi* severely attacked red beech and 50–100% of trees in transects lost leaves in October 1978. During 1979 these trees were severely attacked by *Platypus* beetles, and by February 1980 75% were dead. Among surviving trees of red beech and, especially, silver beech (which was little affected by the epidemic), diameter growth dramatically increased (Fig. 16.4).

After a severe summer drought on the Mamaku Plateau (Volcanic Plateau ER) in 1972–73, diameter growth of hard beech (*Nothofagus truncata*) sharply declined in 1973–74. On a dry site, the trees then showed die-back and death over the following decade, attributed to the leaf-mining weevil *Neomyceta palicaris* and the teneid moth *Heliostibes vibratrix* destroying young leaves (Fig. 16.5). These trees were only 100–150 years old, suggesting a history of periodic mortality. For mountain beech in the Kaweka Range, Hosking & Hutcheson (1988) found that die-back and death, although initiated by events such as drought, mainly involved old stands. Drought also seems ultimately responsible for beech die-back in other localities, including the Ruahine Range (p. 580) and Ruapehu (Skipworth 1983).

Canopy destruction by wind and snow

Snapping of trunks or overturning of whole trees occurs during violent storms associated with deep depressions, when gusts can reach 100–200 km/h. In northern and central New Zealand, these are usually tropical cyclones bringing torrential rain that causes floods and landslides (Figs 16.6 and 16.7). East of the South Island Divide, most damage is caused by föhn gales.

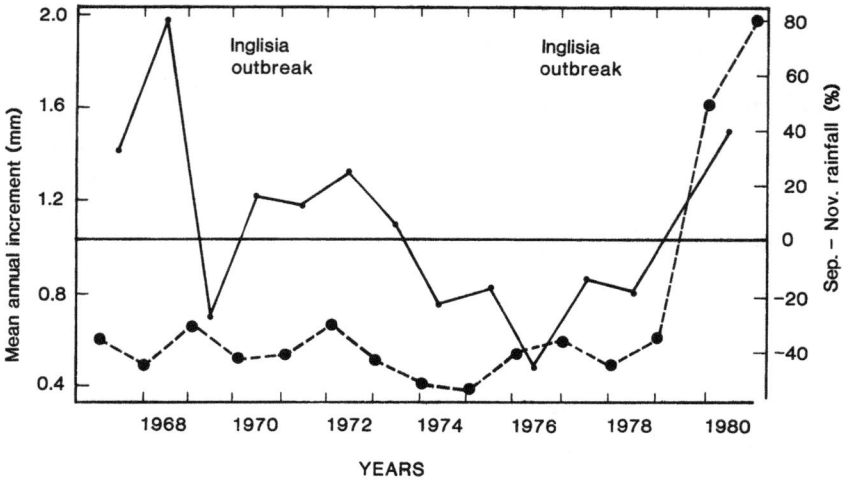

Fig. 16.4. Departure from mean September–November rainfall (solid line) and radial increment in silver beech (dashed line), showing response to die-back of red beech after the second drought (from Hosking & Kershaw 1985).

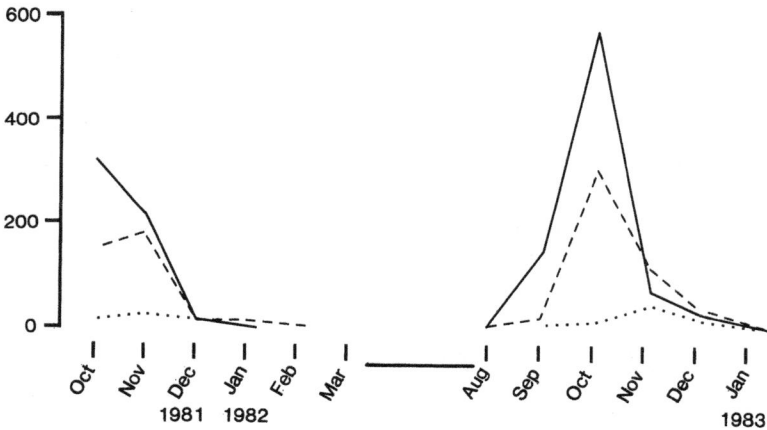

Fig. 16.5. Number of newly flushed leaves/m² lost each week to insect attack in hard beech on drought-affected (solid line), intermediate (dashed line) and unaffected (dotted line) sites on Mamaku Plateau (Hosking & Hutcheson 1986).

The most vulnerable beech forests are on lower lee slopes and consist of trees over 18 m tall with large diameters; trees constantly exposed to prevailing winds may be subjected to minor breakage and wind-training but are usually wind-firm, as are those rooted in bed-rock (Fig. 16.8). Single blow-downs can cover many hectares, and the pattern can be repeated over hundreds of kilometres. Damage in mixed forest also is caused mainly by gusts striking stands that are normally sheltered, and effects vary among species. Kauri are blown over frequently whereas large rimu are relatively

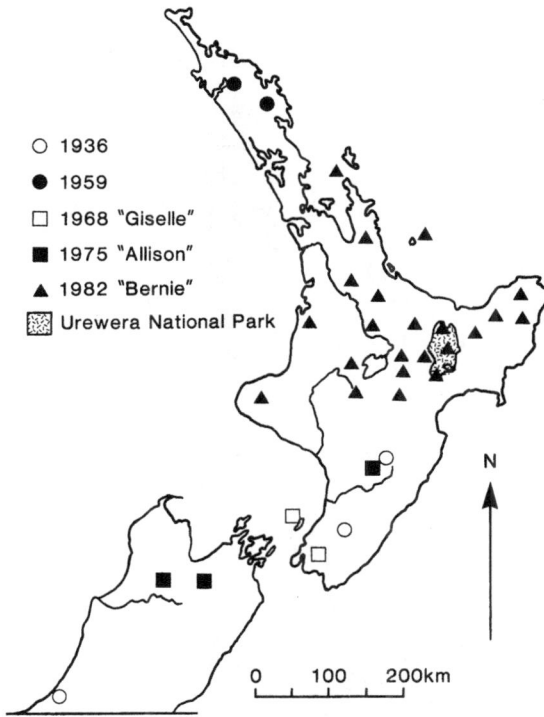

Fig. 16.6. Known locations of indigenous forest damaged by tropical cyclones during the period 1936–82 (Shaw 1983).

wind-firm (references in Shaw 1983). However, wind can cut swathes through shallow-rooted rimu stands on Westland terraces.

Wind-throw has major influences on regeneration, not least through providing varied microsites for seedlings. Decaying logs support seedlings that cannot compete with the undergrowth on the forest floor. Overturned root plates provide mounds of topsoil, and hollows that can extend to the C horizon. Wind-throw also performs natural ploughing that disrupts incipient podzols and maintains yellow-brown earth profiles (Campbell & Mew 1986). In subalpine mountain-beech forests of inland Canterbury, concentration of litter and run-off in wind-throw hollows promotes strong leaching and development of podzolised lenses (Fig. 16.9). Over expanses of now grassy landscape, hummocks and hollows imprinted by wind-throw are evidence of former forests. Where topsoil drift has smoothed these features, cuttings may reveal charcoal trapped in former hollows and the podzol lenses beneath them.

Damage from glaze storms has never been reported in New Zealand forests, but heavy snow can break limbs or snap or uproot whole trees, damage being most severe at moderate altitudes where heavy falls are exceptional. In beech forests, severe damage by wind or snow may be followed by death of many of the remaining trees from attack by insects or pathogens (Wardle & Allen 1983).

Snow avalanches follow predictable routes, but vary in frequency, mass and extent.

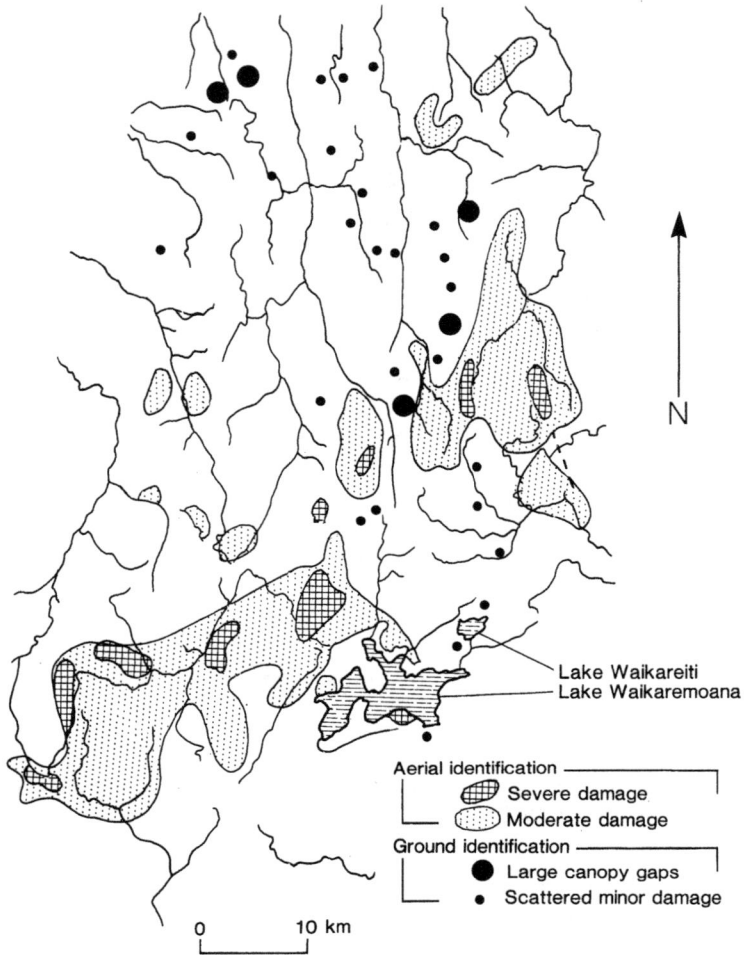

Fig. 16.7. Damage caused by cyclone Bernie (1982) in Urewera National Park (Shaw 1983).

In forests struck on rare occasions, trees may be broken and uprooted. Severe avalanches strip vegetation and soil from slopes and initiate primary successions. Areas swept regularly can maintain tolerant vegetation, including grassland, herbfield and scrub of resilient plants such as *Podocarpus nivalis, Phyllocladus alpinus* and coprosmas (Fig. 16.10).

In beech forest, there are usually enough seedlings to ensure vigorous regeneration after catastrophe (Fig. 16.11). Indeed, Jane (1986) suggested that this response allows beech forest to resist encroachment by more shade-tolerant species. On the other hand, there are areas in the Canterbury foot-hills where mountain beech has largely disappeared from the canopy and shows little tendency to regenerate through dense growth of small broad-leaved trees, shrubs and ferns.

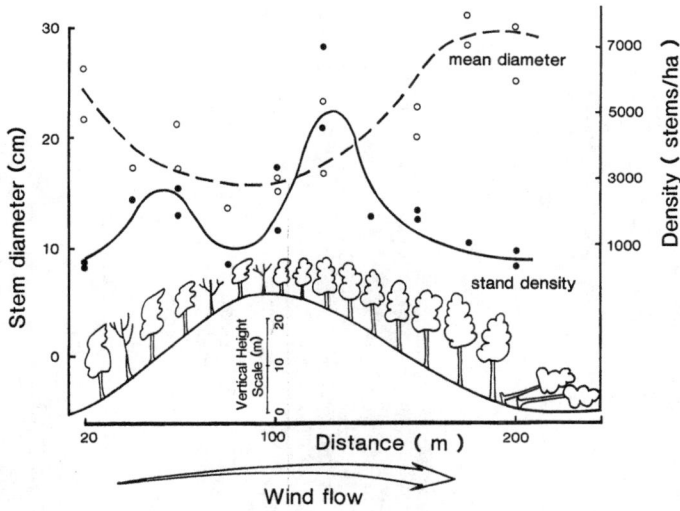

Fig. 16.8. Incidence of wind-pruning and wind-throw in mountain beech forest in inland Canterbury, in relation to wind flow, stand density and tree diameters (Jane 1986).

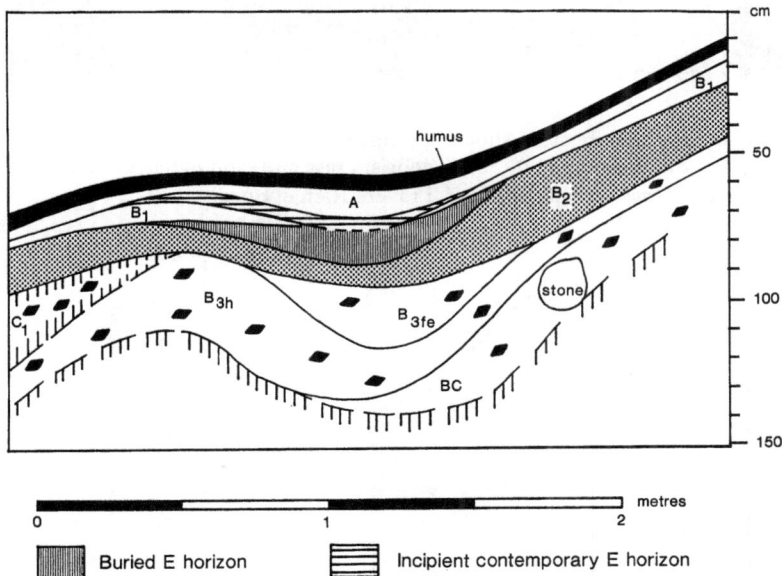

Fig. 16.9. Profile through a wind-throw hollow in mountain beech forest, showing buried and newly forming E horizons, and inversion of the profile in the mound on the downhill side; 1130 m, Craigieburn Range, Puketeraki ER (Ives *et al.* 1972). fe, h: Iron- and humus-cemented portions of the B horizon.

Fig. 16.10. Avalanche track through subalpine mountain beech forest, beginning in scree and bluffs above and occupied by rocky debris, grassland, and patches of shrubs, mainly *Phyllocladus alpinus*. Western side of Travers Range, Spenser ER.

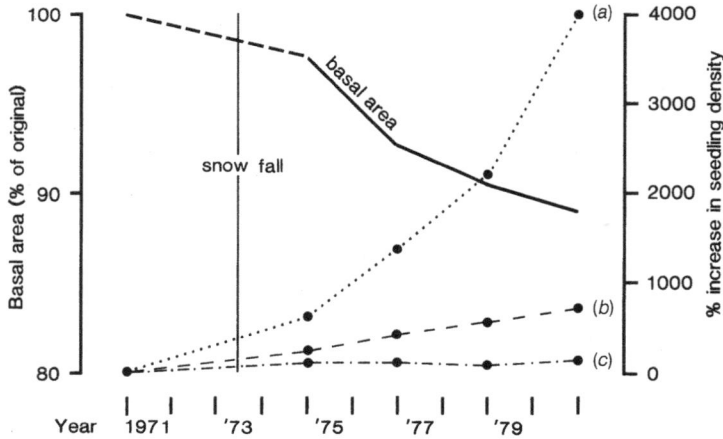

Fig. 16.11. Reduction of basal area of a mountain beech forest after a damaging snow-fall in inland Canterbury, with concurrent release of seedlings (Wardle 1984). Seedling height (cm): (a) 76–135; (b) 46–75; (c) 15–45.

Modification of lower tiers by introduced mammals

Response to disturbance of the canopy depends on the composition and condition of the understorey, especially its content of young plants of dominant species; a prevailing theme of ecological writing in New Zealand has been decimation of understoreys by introduced animals. The widespread red deer (*Cervus elaphus*) and locally concentrated goats are the most important, but other species of deer, pigs, cattle, wallabies (*Macropus rufogrisea*) and even the usually arboreal possum are all significant.

Unless controlled, animals build up rapidly when they colonise a new area. Numbers and vigour then decline as palatable species are exhausted, but enough animals remain to keep pressure on the vegetation (Clarke 1976). The spectrum of damage is similar among the various animals, although there are idiosyncrasies. For instance, goats chew the bark of more species than deer, but the latter damage some unpalatable plants, such as podocarp saplings, through antler-rubbing.

Table 16.2 ranks susceptibility to deer browsing in a Westland valley, through comparing the abundance of species between the heights of 0.3 and 1.8 m, relative to their abundances at 1.8–4.6 m where they are out of reach to deer, and below 0.3 m where they tend to be ignored. The order reflects both browsing intensity and ability to recover. Highly ranked species tend to have largish, soft, glabrous leaves and high nutrient contents, whereas low palatability is conferred by harsh foliage with low nutrient content as in epacrids and most podocarps, or distasteful ingredients as in *Pseudowintera*. Some species occupy positions that reflect inherent population structure rather than influence of browsing. For example, miro (*Prumnopitys ferruginea*) is ranked too high because a large number of small seedlings is typical of the species. *Chionochloa* and *Pseudopanax colensoi* are ranked too low, the former because most plants are tussocks in the 0.3–1.8 m range, and the latter because trees are killed by bark-chewing.

Nevertheless, the list is widely valid, although many species could be added for richer forests at lower altitudes and latitudes. Further palatables include *Geniostoma rupestre*, *Macropiper excelsum*, five-finger (*Pseudopanax arboreus*), large-leaved coprosmas, *Ripogonum scandens* and *Phymatosorus diversifolius*. Further unpalatable plants include *Myrsine salicina*, kauri, manuka and kanuka (*Kunzea ericoides*).

Degree of damage is influenced by availability of alternative food; thus, beech seedlings become vulnerable when more palatable species have been eliminated. It may also be influenced by environment; *Polystichum vestitum* is susceptible when growing in the understorey, but plants in open glades may show little damage. The ability of fuchsia to withstand possum browsing at forest margins, as compared with complete demise in the forest interior, is probably because growth is more vigorous in the better-lighted environment, and pasture grasses and clover provide alternative food.

Information about modification of understoreys can be gained by comparing modified areas with similar areas, such as small islands or exclosures, that are not frequented by animals. Fig. 16.12 shows how palatable species remained abundant in the understorey on an island in Lake Waikareiti whereas unpalatable species increased

Table 16.2. *Browsing susceptibility ratios of forest species in the headwaters of the Grey River, North Westland*

shrubs and herbs			trees		
species		SR	species		SR
Asplenium bulbiferum		40.00	Fuchsia excorticata		13.00
Polystichum vestitum		13.13	Griselinia littoralis		9.07
Leptopteris superba		9.28	Hoheria glabrata		4.21
Cyathea colensoi		3.25	Aristotelia serrata		4.17
Coprosma foetidissima		1.72	Prumnopitys ferruginea	↓	3.75
Olearia ilicifolia	↓	1.70	Brachyglottis buchananii		3.38
Coprosma ciliata		1.62	Melicytus ramiflorus		3.00
Coprosma colensoi		1.43	Schefflera digitata		2.69
Astelia nervosa	↑	1.26	Pseudopanax colensoi	↑	2.60
Olearia lacunosa		1.13	Pseudopanax crassifolius		2.19
Cyathodes fasciculata		1.13	Carpodetus serratus		2.06
Podocarpus nivalis	↓	1.07	Pseudopanax linearis		1.93
Blechnum discolor	↓	1.05	Pseudopanax simplex		1.92
Coprosma banksii		1.05	Weinmannia racemosa		1.81
Coprosma aff. parviflora		0.87	Nothofagus fusca		1.39
Coprosma pseudocuneata		0.80	Nothofagus truncata		1.35
Neomyrtus pedunculata		0.78	Quintinia acutifolia		1.25
Olearia colensoi		0.75	Nothofagus solandri var.		
Myrsine divaricata		0.69	cliffortioides		1.20
Phyllocladus alpinus		0.67	Nothofagus menziesii		0.98
Chionochloa 'robust'	↑	0.65	Metrosideros umbellata	↑	0.98
Pseudowintera colorata		0.64	Dicksonia squarrosa	↑	0.96
Archeria traversii		0.54	Dracophyllum traversii		0.89
Coprosma rhamnoides		0.52	Elaeocarpus dentatus		0.87
Chionochloa pallens	↑	0.50	Libocedrus bidwillii		0.87
Dracophyllum uniflorum	↑	0.49	Cyathea smithii	↑	0.79
Phormium cookianum		0.47	Podocarpus hallii		0.75
Gahnia procera		0.43	Halocarpus biformis		0.69
Dracophyllum longifolium		0.39	Dacrydium cupressinum		0.60
Chionochloa conspicua	↑	0.36			
Pittosporum divaricatum		0.29			

Notes:
SR, susceptibility ratio; see text for derivation.
↑↓ Arrows indicate that true susceptibility is probably greater or less than the value indicated.
Source: J.A. Wardle 1974

in deer-frequented forest on the adjacent mainland. Fig. 16.13 shows differing proportions of saplings to larger plants among palatable trees, according to presence or absence of deer in coastal forest of Stewart Island. Table 16.3, from the same locality, indicates that seedlings and low herbs actually increased in the presence of deer, because destruction of the shrub storey increased the amount of light reaching the forest floor; in most regions *Cardamine debilis, Mycelis muralis* and small poas and uncinias could be added to the list.

MAINLAND   ISLAND

Phyllocladus glaucus
Dracophyllum traversii
Quintinia acutifolia
Nothofagus menziesii
Podocarpus species
Nothofagus fusca
Pseudowintera colorata
Cyathodes fasciculata
Eleocarpus hookerianus
Myrsine divaricata
Griselinia littoralis
Weinmannia racemosa
Ixerba brexioides
Small leaved coprosmas
Large leaved coprosmas
Pseudopanax species
Dicksonia lanata
Blechnum discolor
Asplenium polyodon
Phymatosorus diversifolius
Uncinia species
Astelia nervosa

146 *

0.3 – 0.45 m tall
0.45 – 1.5 m tall
1.5m – 10 cm d.b.h.

80 70 60 50 40 30 20 10   10 20 30 40 50

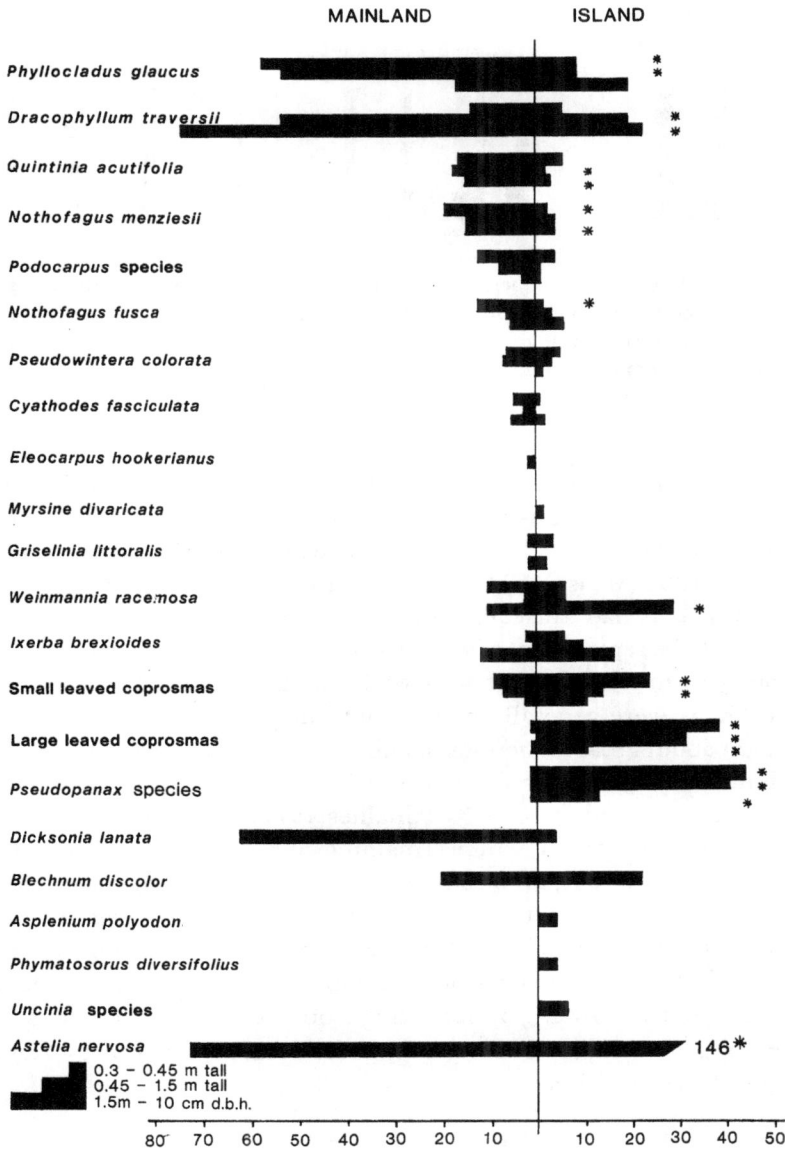

Fig. 16.12. Numbers of woody and herbaceous plants in samples on the mainland accessible to deer, and on an island in Lake Waikareiti, Urewera ER (James & Wallis 1969). Asterisks indicate significant difference ($p \leq 0.05$).

Exclosures in close-canopied, even-aged beech forests may hardly change over many years because lower tiers are sparse even in the absence of herbivores, whereas exclosures in open-canopied lowland forests on relatively fertile soils may contain dense thickets of fast-growing, palatable broad-leaved saplings. Fig. 16.14 shows that in Urewera forests, young plants of the highly palatable *Geniostoma* and mahoe are practically confined to exclosures, whereas young *Pseudowintera* are abundant on plots

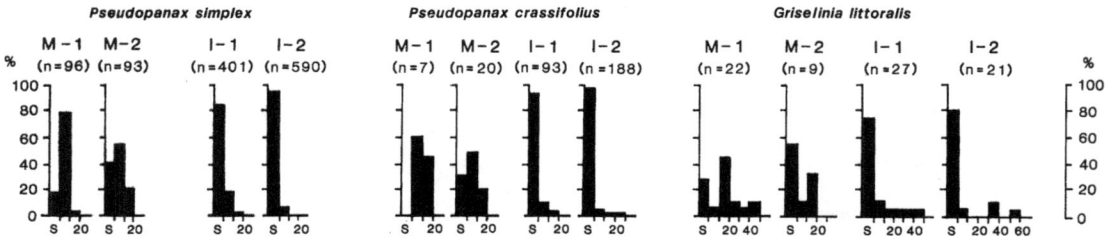

Fig. 16.13. Percentage frequency distributions in size-classes of three palatable subcanopy tree species on the Stewart Island mainland (stands M-1, M-2) and on an on-shore island inaccessible to deer (I-1, I-2) (Veblen & Stewart 1980). S = saplings < 5 cm d.b.h. and ≥ 1.4 m tall. Following classes are 5–10, and 10–20, 20–30 cm d.b.h., etc.

accessible to animals. Kamahi saplings are scarce both inside and outside exclosures, since this species establishes as dense, even-aged stands after disturbance.

Concern about the effects of deer was being expressed only 30 years after the first introduction in 1861 (Walsh 1892). In 1934, Moore and Cranwell reported that feral cattle, pigs and goats had caused once-dense lower tiers in tawa-towai (*Weinmannia silvicola*) forest to be replaced by *Microlaena avenacea* swards. By 1950, animal control programmes were in place through most of the mountainlands, to combat depletion of vegetation. Gains were generally modest until deployment of helicopters from the 1960s brought about great reductions in animal numbers and corresponding recovery of vegetation. In the Arawata valley in South Westland, for example, the reduction in deer herds was estimated at 80–95% (Challies 1977). Recovery has been greater in short vegetation than in forest, where remaining animals find refuge from helicopters.

As Jane (1983) points out, even complete removal of browsing mammals is unlikely to lead quickly to return of the original structure and composition, especially in complex forests. The composition of understoreys will have changed, and palatable tree species selectively removed from the canopy and regeneration tiers. Instead, species unaffected by browsing or benefiting from increased light will have increased their share of regeneration, and may dominate for decades or centuries. Kanuka stands developing in Urewera ER are an example (p. 202).

Other animal influences

The role of mammalian herbivores in primeval New Zealand was at least partly filled by flightless birds. Two that survive, but very precariously, are the takahe (*Notornis mantellii*), which depends on tiller bases of *Chionochloa pallens* and rhizomes of *Hypolepis millefolium* (Mills *et al.* 1980), and the kakapo parrot (*Strigops habroptilus*), which grazes various shrubs and herbs, some being of seemingly low palatability (Best 1984). Competition from deer is a factor in the decline of takahe, and probably kakapo as well.

However, moa were the most notable. Preserved crops from two eastern South Island localities contain seeds from 26 species of trees, shrubs, and herbs (Burrows 1980). Leaves and twigs, on the other hand, come only from matai (*Prumnopitys*

Table 16.3. *Percentage frequency of common species in 0.75 m² plots on the forest floor on the Stewart Island mainland (S), where deer are present, and Bench Island (B) where deer are absent*

species	height ≤ 15 cm		height > 15 cm	
	B	S	B	S
Dacrydium cupressinum	—	16.5*	—	0.5
Prumnopitys ferruginea	—	42.5*	—	1.5
Griselinia littoralis	10.9	28.0*	—	1.0
Pseudopanax simplex	10.9	68.5*	6.3	0.5*
Myrsine divaricata	1.1	26.5*	0.6	1.5
Coprosma foetidissima	24.1	82.5*	4.0	3.0
Cyathodes juniperina	—	10.0*	—	9.5*
Ripogonum scandens	1.1	7.0*	1.1	—
Dicksonia squarrosa	5.2	19.5*	10.9	5.0
Blechnum discolor	—	2.0	—	3.5
Blechnum 'capense'	0.6	3.0	24.1	—*
Asplenium bulbiferum	7.5	—*	4.0	—
Phymatosorus diversifolius	37.9	2.0*	71.3	1.0*
Rumohra adiantiformis	1.1	5.0	—	2.0
Grammitis billardierei	0.6	14.0*	—	—
Hymenophyllum demissum	12.6	74.0*	—	—
Hymenophyllum ferrugineum	0.6	7.0*	—	8.5*
Hymenophyllum multifidum	—	8.5*	—	—
Tmesipteris tannensis	0.6	8.0*	—	—
Nertera aff. *dichondrifolia*	10.9	9.0	0.6	—
Luzuriaga parviflora	2.3	27.0*	—	—
Chiloglottis cornuta	—	8.5*	—	—
Caladenia catenata	—	9.5*	—	—
Corybas oblongus	0.6	6.5*	—	—

Note:
*, Probability of difference <0.01. Total number of species on plots: Bench I. 32, Stewart I. 44.
Source: Veblen & Stewart 1980

taxifolia), kahikatea, *Podocarpus* cf. *hallii, Phyllocladus alpinus, Plagianthus regius, Rubus* spp., *Pseudopanax* sp., manuka, *Myrsine divaricata, Olearia virgata, Phormium* and sedges. The absence of plants with larger, softer leaves may reflect differential destruction in the gizzard.

Certain growth forms, especially the filiramulate shrub and juvenile tree, might have evolved as a protection from moa grazing (p. 32); apparently anomalous population structures of long-lived conifers may have a basis in the former abundance and eventual extinction of moa and other herbivores (p. 605).

Seed is destroyed by native parrots and parakeets, finches and other introduced birds, rodents (Campbell 1978), pigs and possums. Weta (p. 7) are important

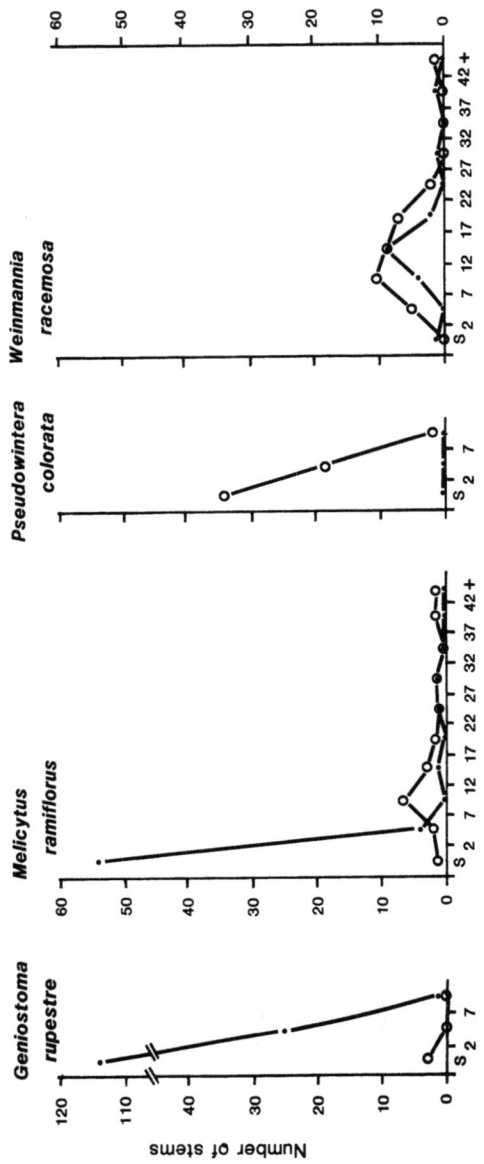

Fig. 16.14. Numbers of stems in exclosure (filled circles) and control (open circles) plots, for four woody species in Urewera National Park. The exclosures had been set up 12–19 years (Allen *et al.* 1984). S, saplings. Subsequent figures show boundaries of 5 cm diameter classes.

predators of fallen seeds of kauri (Mirams 1957) and podocarps (Beveridge 1964). Seedlings are vulnerable to ground-feeding herbivores, both vertebrate and invertebrate. Seed years in beeches and podocarps lead to eruptions of mice and rats and in turn to increases in number of stoats (*Mustela erminea*), which were originally introduced to control rabbits (King 1983). During the period between the eventual population crash of rodents and that of stoats, native birds are heavily preyed upon, and this may be a factor in the continuing decline of the less resilient species.

Animals congregating in large numbers have drastic effects on vegetation, including forest. Die-back is common around sea-bird colonies, because solutes from guano are absorbed via roots or deposited on foliage. Relatively resistant plants include succulents, *Coprosma repens* and pohutukawa (Gillham 1960*a*). At Peel Forest (Pareora ER), roosting pigeons (*Columba livia*) are killing large totara trees. Burrowing petrels, terns and shearwaters create areas of unstable, heavily fertilised ground. Seals and other phocids also modify coastal vegetation. Today, these effects are most apparent on remote islands, where sea-bird and seal colonies are now concentrated.

Forest collapse through multiple causes

Mass die-back in beech forest through physical damage interacting with insect predation has already been mentioned. In more complex forests, especially at high altitudes, catastrophic collapse is widespread; both the causes and the eventual outcome are often unclear. A north-to-south sequence of examples is presented.

Kaimai Range

Upper slopes and crests between 600 and 900 m support summit-warm-temperate forest in which tawa, *Ixerba* and silver beech dominate in close altitudinal sequence. All canopy species, including silver beech, *Ixerba, Quintinia*, toro (*Myrsine salicina*), kamahi, kaikawaka (*Libocedrus bidwillii*) and pink pine (*Halocarpus biformis*), are subject to die-back, and the forest shows degrees of collapse up to complete replacement of trees by a sward of *Microlaena avenacea, Hierochloe redolens* and *Uncinia distans* (Figs 16.15 and 16.16). Partial canopy destruction has led to shrubby communities dominated by *Pseudowintera axillaris, Hedycarya, Coprosma grandifolia* or *Quintinia*.

According to Jane & Green (1986), the vulnerability of the forest results from the cool, foggy climate. The growing season is shortened, and evapotranspiration and carbon uptake are reduced. Soils are infertile, low in available phosphorus because of high allophane content, and usually waterlogged. During clear, dry weather, trees are prone to drought stress, probably because of inadequate root systems (p. 43).

Growth-ring analyses indicate that many canopy trees died after droughts in 1914 and 1946. This set further deterioration in train, through increased exposure to wind and higher water tables that drowned roots (Jane & Green 1983). Although goats and possums are locally abundant, their effect is mostly to delay recovery of lower-altitude forests.

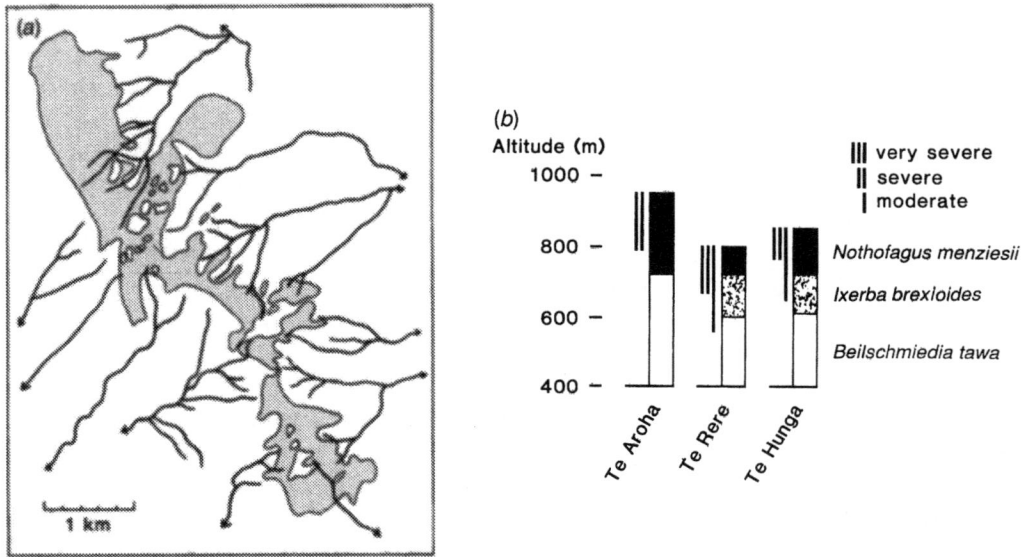

Fig. 16.15. Forest die-back on the Kaimai Range, Coromandel ER (Jane & Green 1983). (a) Areas of die-back (shaded) on the crest of Te Rere. (b) Altitudinal distribution of forest types defined by dominants, and severity of die-back on three mountains.

Ruahine Range

Through much of this range canopies have been decimated, herb and shrub communities have replaced trees, and gullies and slips are greatly in evidence. Montane kamahi forest in the south seems the most degenerate, but beech forest is also affected. Cunningham (1979) linked the changes to the influx, first of sheep and cattle, and then of red deer, goats and possums. By the 1930s understoreys had been destroyed in many places, with palatable subcanopy trees such as *Pseudopanax* spp. reduced to skeletons and the fern *Polystichum vestitum* to apparently dead stumps. On the other hand, unpalatable or browse-tolerant plants such as *Pseudowintera colorata*, toro, *Histiopteris, Dicksonia lanata* and *Blechnum discolor* increased. Die-back of northern rata and kamahi trees was very evident by the 1950s, and attributed to possums; tawa, mountain totara, *Cyathea medullaris* and other trees were also dying. Since the 1960s, the more open canopies and fewer deer have led to local increase of palatable species, such as mahoe, *Hedycarya* and *Carpodetus*.

A run of dry years, culminating in 1914–15 in the driest summer on record, was followed by death of a large proportion of red and mountain beech trees and by local depression of the upper forest limit by as much as 90 m. Other episodes of beech death were also preceded by drought (Grant 1984). Snow breakage in 1961 also mainly affected beech. Cyclonic storms impacting on steep, fault-weakened greywacke terrain rejuvenate old slips and release new ones, leading to bare upper slopes and deep aggradation along water courses (Moseley 1978).

Kaikawaka stands consist largely of dead and dying trees, and the rarity of young trees, saplings and seedlings, can only be partly attributed to bark-rubbing by deer.

Fig. 16.16. Area at 600 m, Kaimai Range, with complete canopy loss and replacement by grasses with scattered *Pseudowintera colorata* (photo G.T. Jane).

However, kaikawaka has regenerated well on some old burns, and in at least one place has replaced beech forest destroyed by a Maori fire. Stands on limestone are also vigorous (Elder 1965).

Tararua Range

Partial destruction of subalpine silver-beech forest on steep slopes in the Tararua Range, presumably by the gale of 2 February 1936, led to an understorey of *Chionochloa conspicua*, browsing by deer apparently preventing either regeneration of beech or development of the *Olearia colensoi* understorey that is characteristic of other tall, open silver-beech stands in the range (Wardle 1962). Browse-tolerant turf occupies many glades, with *Hymenophyllum multifidum* extending beneath canopies.

In the high-altitude mixed forests most mountain totara trees have died, leaving kamahi dominant.

In 1958, much *Olearia colensoi* scrub in the range was dead. As in other localities where similar death has occurred, shrubs on steep ground were healthy, and damage increased towards gullies and flatter areas. This suggests effects of either drainage or accessibility to deer, but the only obvious damage was due to insects, especially the bud-destroying moth *Agriophara coricopa* (Wardle *et al.* 1971).

South Island rata–kamahi stands

On the steep, western flanks of the Southern Alps, die-back is severe in southern rata–kamahi forests. Although possums are the primary agent of destruction there is also cohort-senescence, in so far as old trees are more sensitive to defoliation (Payton 1983), and badly-affected stands are mostly > 300 years old (Stewart & Veblen 1982). The gnarled, often hollow trunks of old rata and kamahi trees may also provide lairs for possums. Both species carry a heavy parasite load, especially of coccids (Hoy 1958), but this seems lethal only in trees damaged through other causes.

Since the soils are mainly young, the forests contain fast-growing, palatable understorey species that, until depleted, help to support large possum populations. The long-term effect of possums may be to shorten the time from colonisation of slips to the death of rata and kamahi trees. Where deer are also present, however, the succession is likely to be deflected, through encouragement of browse-tolerant plants at the expense of rata and kamahi seedlings.

The condition of these forests on unstable slopes contrasts with much healthier rata and kamahi on stable granite country, where continuous replacement seems the prevailing regeneration mode and lower fertility leads to lesser abundance of fast-growing, palatable species and therefore fewer introduced mammals (Reif & Allen 1988). Landscape stability possibly also explains the healthy forests in Catlins ER, where possums have been established since the 1890s.

Northern rata on the coastal ranges of Western Nelson is also generally healthy, although possums have been present for many years. This contrasts with much of the North Island, where huge northern-rata trees have been reduced to dead spars.

Kaikawaka and totara stands in the South Island

Dead kaikawaka trees were a feature of upper montane forests in Central Westland in the 1890s (Holloway 1957) and are also evident near Dunedin (Wardle & Mark 1956). Although these could result from either unusual mortality or durable stems standing a very long time, it is known that every adult kaikawaka tree but one on Banks Peninsula died between 1941 and 1961; low rainfall during the years 1947–49 may be implicated (Wardle 1978a).

Mature totara, both mountain and lowland, often have sparse crowns with slowly growing shoots that support dense *Usnea*. Buds and young foliage are also attacked by tortricid larvae. Although seemingly moribund trees can survive many years, there have been regional episodes of fatal die-back. In Westland, mountain totara trees have died over the same areas affected by rata and kamahi die-back, often leaving only juveniles; although totara is not a major food source for possums, browsing and

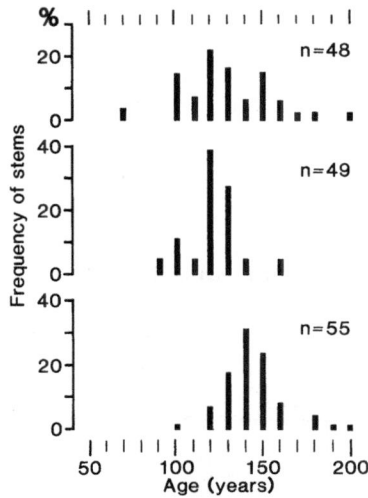

Fig. 16.17. Age class distribution in three mountain beech stands; Andrews catchment, Hawdon ER (Jane 1986).

scratching of bark may suffice to kill trees. At Wainakarua (Kakanui ER) many trees of both species were dead or unhealthy in September 1982 (P.N. Johnson & R.B. Allen, unpublished), probably because of a 20 month drought in a district near the lower rainfall limits for native forest.

Population structure and regeneration

This section opens with the comparatively simple beech forests, which have been studied most in relation to regeneration because they have seemed to offer the best prospects for management for timber. More complex mixed forests are considered first in respect of their broad-leaved element, wherein montane rata–kamahi forests of more or less seral status contrast with northern lowland stands containing the shade-tolerant tawa and kohekohe. The section concludes with the native conifers; regeneration patterns among these, the longest-living organisms in New Zealand, present the greatest problems in interpretation.

Beech forests

Analysis of pure mountain-beech forest in inland Canterbury reveals several regeneration peaks, each resulting from release after wind-throw (Fig. 16.17). However, when all diameter measurements for mountain beech from plots distributed widely through the South Island are pooled, the resulting histogram resembles a reverse-J curve (Fig. 16.18). Histograms for red and silver beeches are similar, although reflecting differences in seedling density, final girth, longevity, and in silver beech, a more continuous mode of regeneration (Allen 1988). The dip in the youngest step of the mountain beech histogram is possibly due to deer depleting its seedlings

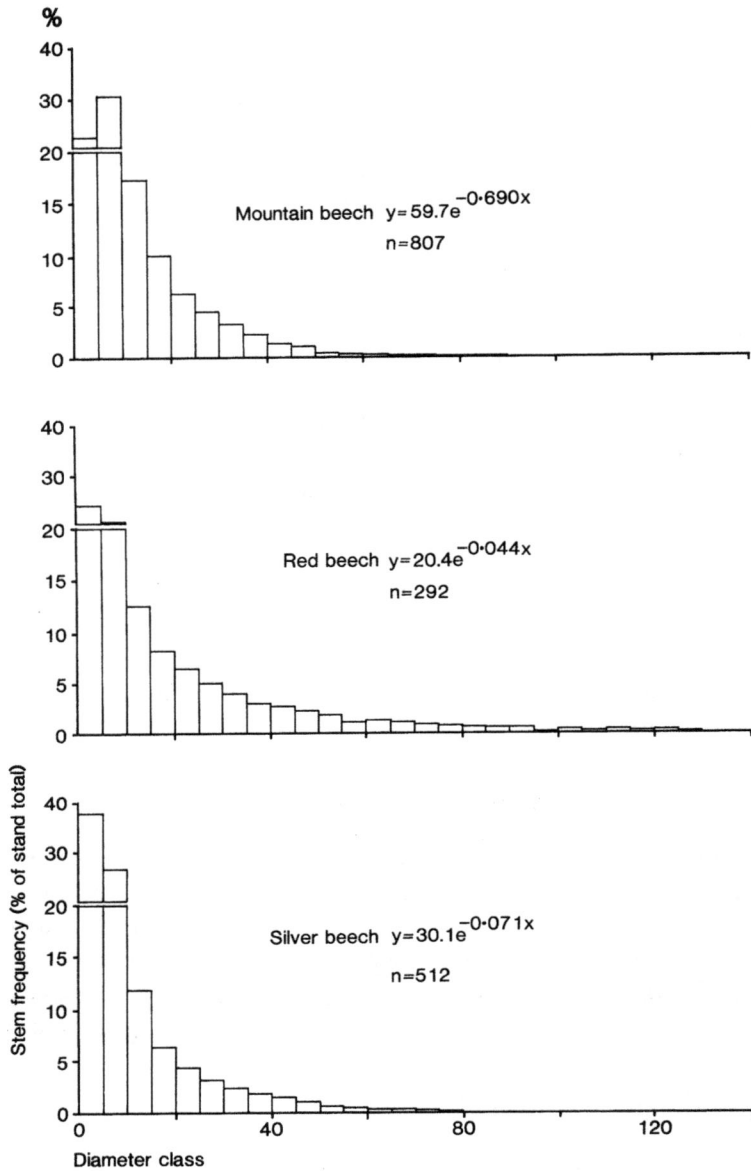

Fig. 16.18. Frequency of stems (percentage of total) in 5 cm d.b.h. classes; data pooled from 400 m² plots distributed over the ranges of three beech species (Wardle 1984); n, number of plots.

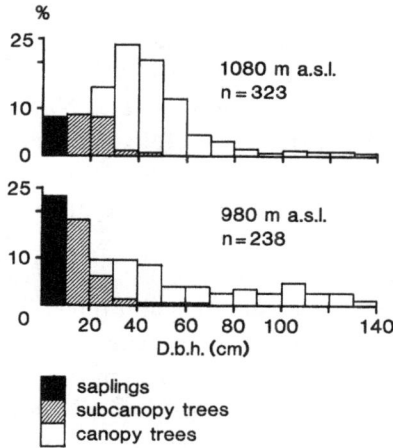

Fig. 16.19. Percentage frequency of red beech stems in diameter classes at two sites on Mt Colenso, Ruahine Range (June & Ogden 1978).

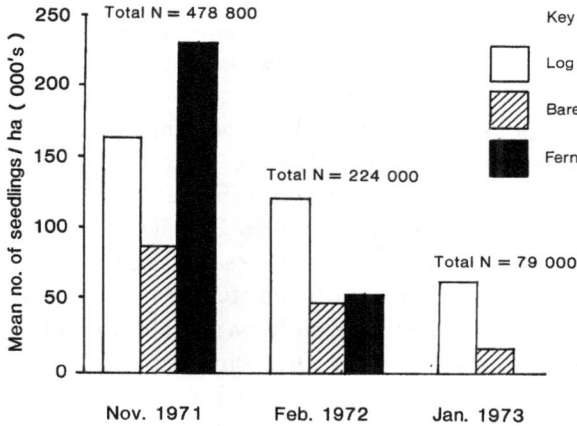

Fig. 16.20. Recruitment and survival of first-year red beech seedlings on three kinds of microsite; 950–1080 m a.s.l., Mt Colenso, Ruahine Range (June & Ogden 1975).

more than those of other beeches, which are often associated with more palatable plants (Table 16.2).

Fig. 16.19 illustrates red-beech forest in the Ruahine Range, that contains thickets of broad-leaved shrubs, *Rubus* tangles, and dense fern, mostly *Dicksonia lanata*. The upper stand seems predominantly even-aged and probably arose after a climatic catastrophe, whereas the lower stand is multi-aged, with trees having established in smaller, more frequently created gaps. Seedlings are numerous after a good seed year, but decrease rapidly. The best sites for survival are decaying logs; bare ground is less favourable, and virtually no seedlings survive under fern (Fig. 16.20). Even-aged red-beech forest should gradually become multi-aged unless renewed by disturbance. As logs decompose within 80–300 years, they are likely to disappear from undisturbed stands before canopy gaps develop, thereby limiting the sites available to seedlings and leaving stands vulnerable to invasion by other species.

Mixtures of red and silver beech at Station Creek, at 400 m a.s.l. in the Maruia Valley, are typical of large areas in inland parts of North Westland. Other vascular plants constitute only a sparse undergrowth, but some 60% of the floor is occupied by mosses, mainly *Dicranoloma* spp. Red beech generally forms the canopy, with trees up to 1.5 m in diameter. Most silver beech trees are subcanopy and less than 60 cm in diameter. Beech seedlings are well distributed over the forest floor, although concentrated on rotting logs. Gaps formed by the death or wind-throw of one to several trees are filled by release of saplings and poles <15 cm d.b.h. and by lateral growth of existing canopies. Although the two species grow at similar rates at first, red beech eventually outgrows and suppresses most of the silver beech. Gaps larger than 400 m² show a much greater ratio of red beech to silver beech seedlings than smaller gaps, leading to almost pure stands of red beech on sites of massive wind-throw (G.H. Stewart, personal communication; Smale *et al.* 1987).

In silver-beech forest, apparently mixed-aged structure seems more prevalent than in other beech forests, perhaps because it is less prone to canopy loss; the reputed shade-tolerance of silver beech may also have a bearing. At high altitudes, on well drained spurs, or on infertile soils in South Westland, most silver beech seedlings are rooted on the ground, whether humus- or moss-covered or bare. On moister, lower sites, with increasing competition from other tree and understorey species, silver beech forms less of the canopy, and its seedlings become largely restricted to elevated sites such as logs and tree bases (Wardle 1980*b*).

There are three regeneration modes among silver beech and its broad-leaved competitors in a valley-floor forest in northern Fiordland (Fig. 16.21). In all-aged stands (a), young silver beeches are confined to tree-fall gaps, and seedlings of this and most other tree species grow primarily on elevated sites such as logs, because of the dense understorey of ferns and shrubs. Seedlings and older plants of silver beech and kamahi therefore cluster in small patches. On a wind-thrown plot (b), silver beech and kamahi are mostly in relatively even-aged groups, but silver beech has attained greater diameters through faster growth. Silver beech on a former landslide (c) is nearly even-aged, randomly distributed, and apparently established directly on a bared surface.

The outliers of hard beech in South Westland (p. 150) occupy gley podzols developed in a veneer of glacial gravel over granite hills. On Nisson Hill it co-dominates with rimu and southern rata, whereas on MacFarlane Mound hard and mountain beech are a minor component of forest dominated by rimu, rata and kamahi. On both hills, hard beech includes a range of diameters, and there is a significant regression between diameter and age (Fig. 16.22).

Forests dominated by broad-leaved trees

The forests of southern rata and kamahi that clothe steep western flanks of the Southern Alps grade from dense, even-aged stands on landslide scars to 'old-growth' stands of greater age range and floristic content that grow on podzolised soils (pp. 511–12). In Fig. 16.23 three of the canopy-forming species in Stand A appear to show intermittent regeneration; only *Quintinia* seems to have been steadily recruited

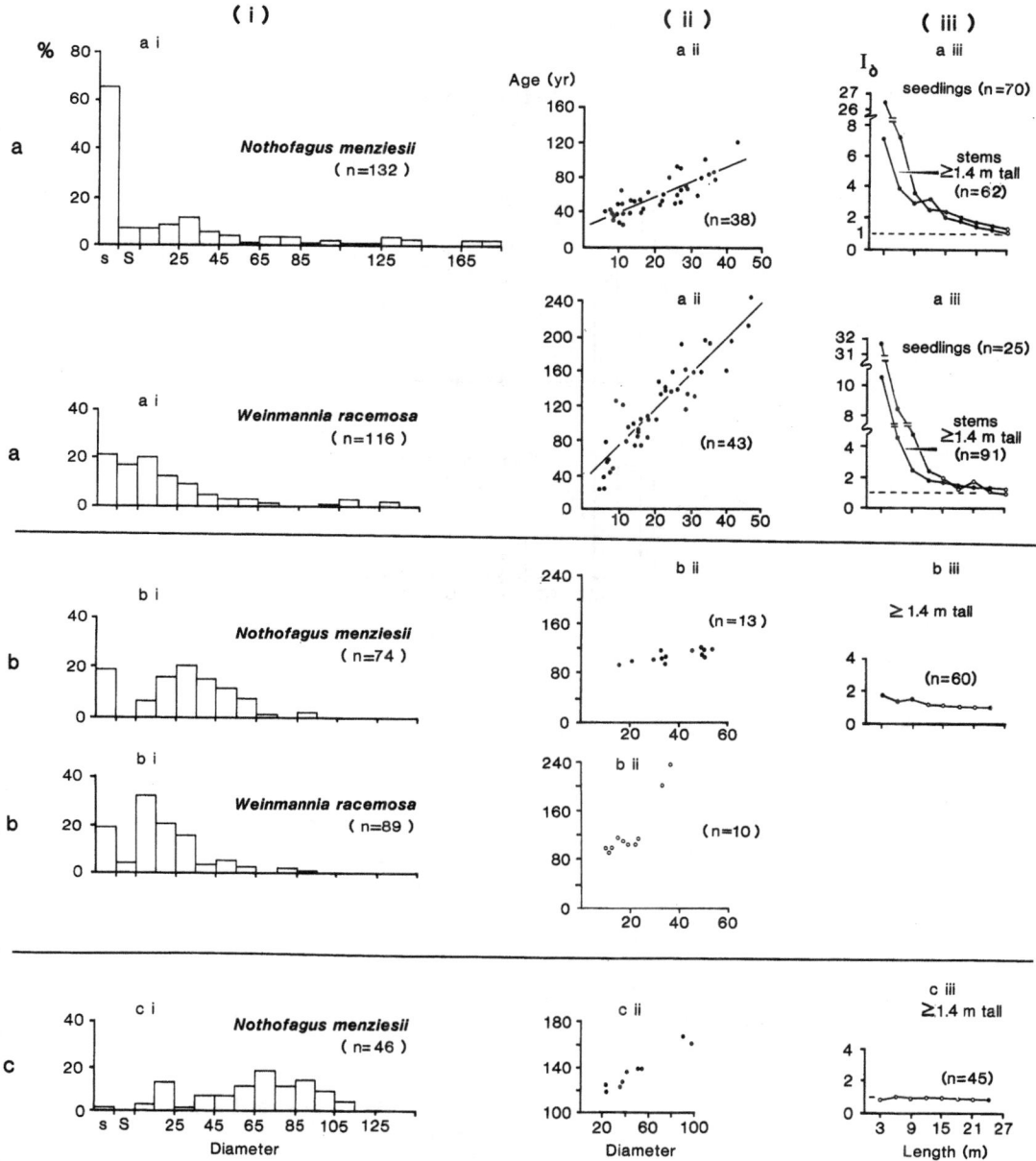

Fig. 16.21. Regeneration of silver beech and kamahi near Lake Hankinson, northern Fiordland (Stewart 1986). (a) Frequently disturbed forest; (b) wind-thrown forest; (c) forest established on a landslide. (i) Frequency in size classes: s, seedlings; S, saplings; subsequent classes as cm d.b.h.; n, no. of plants ≥0.3 m tall. (ii) Relation of age at 1 m to d.b.h. (iii) Values of Morisita's index I_δ at different quadrat sizes. The value 1.0 (dashed line) represents random patterns. Filled symbols indicates I_δ significantly ($P < 0.05$) greater than 1.0, i.e. significant clustering. Values for quadrat size represent the length of the side of a square.

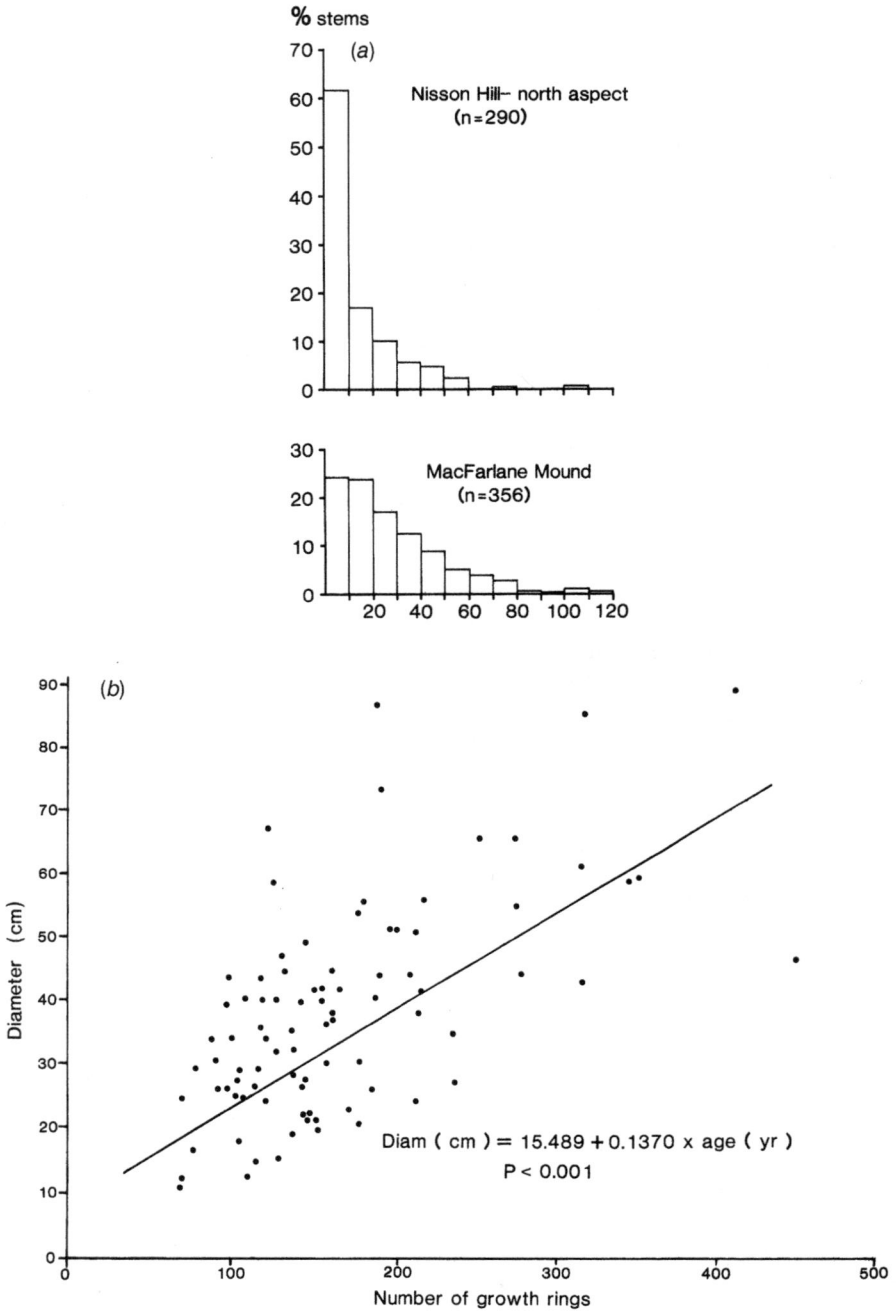

Fig. 16.22. Percentage frequency of hard beech stems > 3 m tall in 10 cm d.b.h. classes on two hills in South Westland (*a*) and relation between diameter and age at MacFarlane Mound (*b*) (from Mark & Lee 1985).

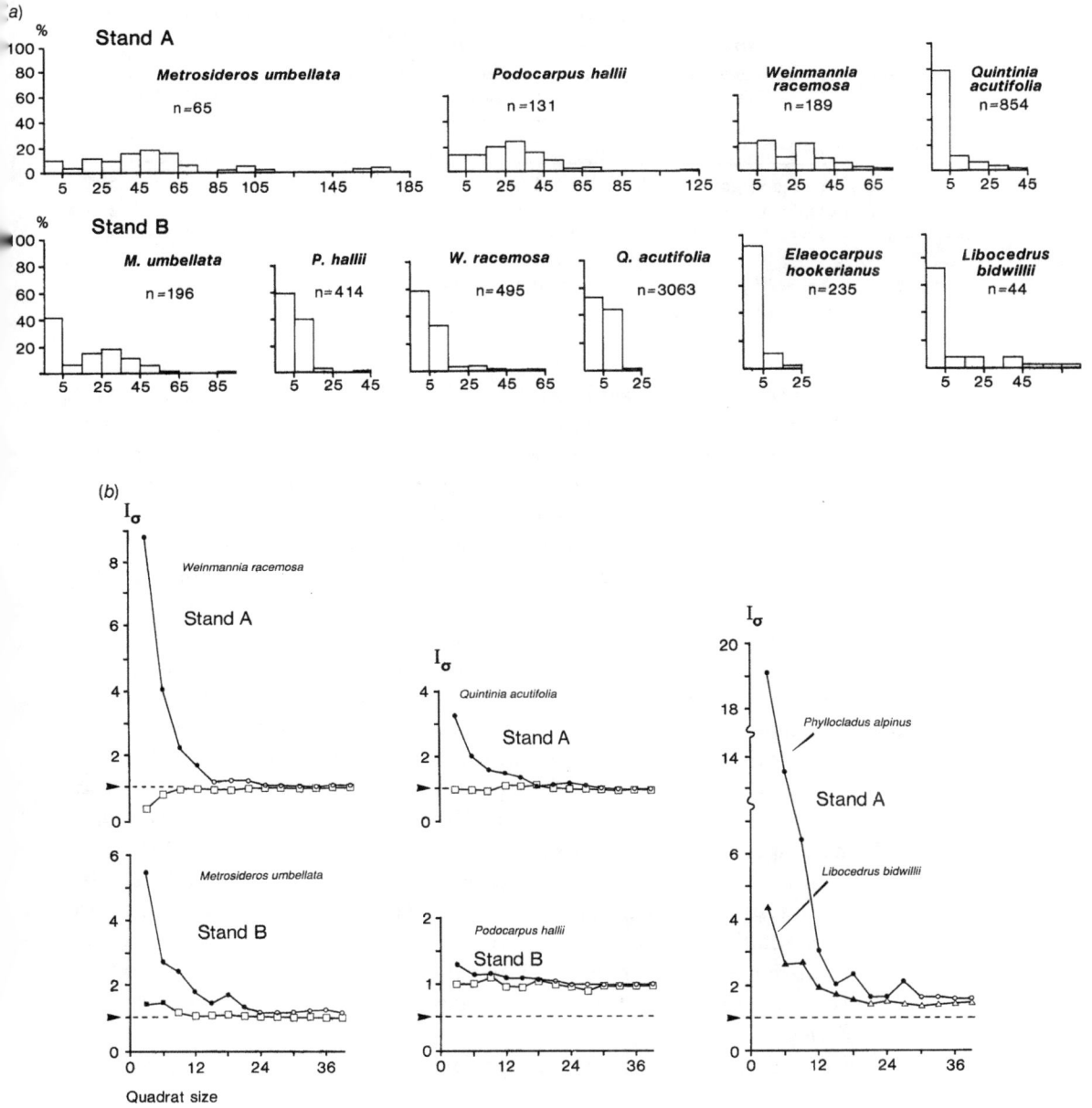

Fig. 16.23. Structure of two 'old-growth' stands of southern rata – kamahi forest at 600 m, Kellys Creek, Central Westland (Stewart & Veblen 1982). (*a*) Percentage frequencies of stems in diameter classes (cm d.b.h.). S, saplings < 5 cm d.b.h. and ≥ 1.4 m tall. (*b*) Values of Morisita's index (see Fig. 16.21 iii). Circles or triangles represent stems ≤ 10 cm diameter; squares represent stems > 10 cm diameter.

Table 16.4. *Ratio of tawa juveniles to kohekohe juveniles in a warm-temperate broad-leaved forest; Northern Volcanic Plateau ER*

	Importance Value		Relative Density	
Gap-maker or canopy species:	T	K	T	K
ratio T : K of juveniles in gaps	1.0	7.4**	1.0	7.5**
ratio T : K of juveniles under canopy	0.1	0.4 NS	0.1	0.2 NS

Notes:
T, tawa; K, kohekohe; **, differences significant at $p < 0.01$; NS, not significant.
See text for derivation of importance value and relative density.
Source: Smale & Kimberley 1983

from an abundant population of saplings. Stand B has numerous logs and stumps of totara, rata and kaikawaka, and dense thickets of saplings and small trees of all canopy species, although most of the 'young' rata appear to have layered from fallen trees. The small trees *Phyllocladus alpinus, Pseudopanax simplex* and *Griselinia littoralis* are represented by numerous stems in both A and B, although there are fewer saplings of the last two species, presumably because of browsing by deer. All species cluster at small quadrat sizes, indicating regeneration in gaps; clusters resulting from vegetative multiplication in kamahi, *Quintinia* and *Phyllocladus* were allowed for by recording each as a single individual.

Southern rata, kamahi and rimu are the main canopy trees in forests on the eastern coast of Stewart Island and on Bench Island lying 5 km offshore (Veblen & Stewart 1980). Again, kamahi saplings are clustered because they regenerate in large gaps, and rata is clustered because procumbent trees layer and seedlings establish on logs.

On tephra soils at Rotoehu forests in the Bay of Plenty, tawa dominates in two study areas at 250–350 m, with kohekohe, rewarewa (*Knightia excelsa*) and mangeao (*Litsea calicaris*) being important; kamahi disappeared through die-back after 1957, and podocarps had been logged before this. In gaps and closed canopy (mature) phases, young plants of canopy-forming species were assessed by various criteria, of which the most useful were relative density (i.e. the proportions of all stems made up by each species) and importance (the average of relative density and proportion of total basal area). Table 16.4 shows that tawa juveniles are more important in gaps left by death of kohekohe and vice versa. Both species are shade-tolerant; the reciprocal relation also holds in shaded understoreys but is not statistically significant. Canopy replacement appears to depend largely on seedlings that predate and survive gap formation; tawa achieves this better than kohekohe, which is near its limit of frost-tolerance. Mangeao and rewarewa established beneath the canopy in tawa forest seldom survive to pole size unless gaps form. Rewarewa seedlings especially also colonise gaps. Fig. 16.24 shows that shade- and exposure-tolerance among species differ most in Area 1, which includes ridges where drought and disturbance are more severe than elsewhere.

Kaikawaka

In Westland, kaikawaka stands often contain numerous dead trees (p. 582), and living trees and young plants tend to form clusters of similar-sized plants, indicating

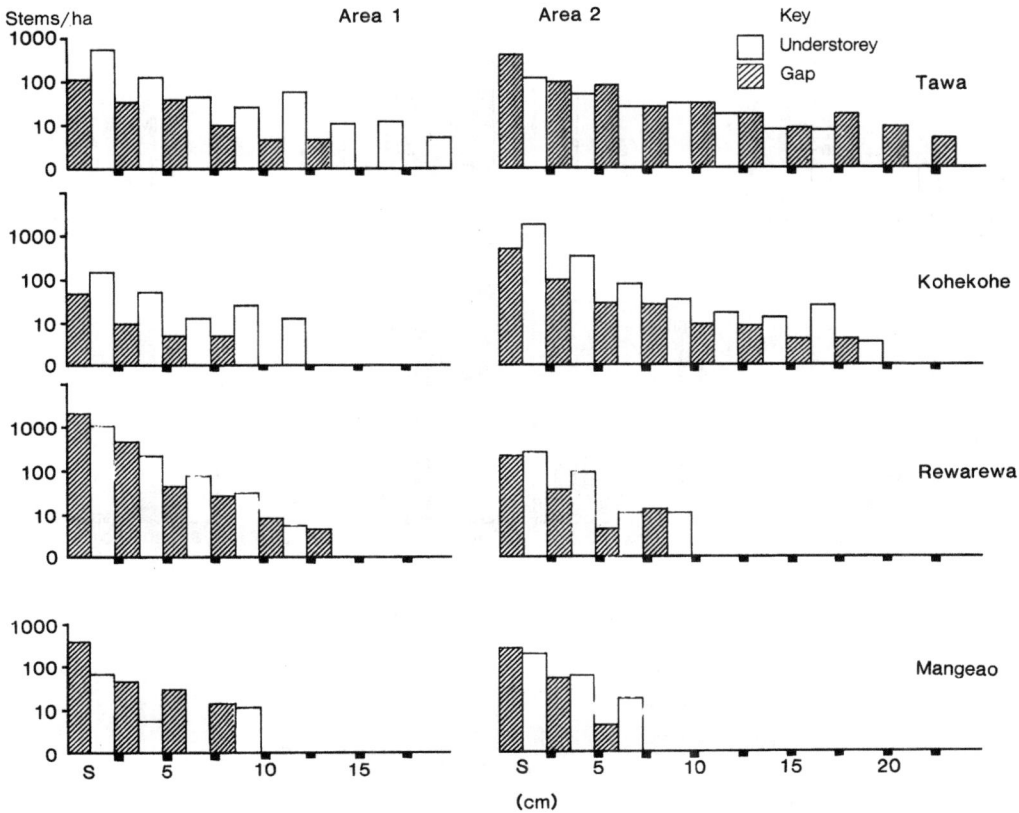

Fig. 16.24. Density of stems in diameter classes for the main regenerating species in gap and understorey plots in tawa-dominated stands at Rotoehu Forest, Northern Volcanic Plateau (Smale & Kimberley 1983). S, saplings < 2.5 cm d.b.h. and > 1 m tall.

establishment in canopy gaps (Veblen & Stewart 1982). Norton (1983) attributes such clustering to regeneration initiated through disturbances, that include wind-throw and slips. Haase (1986b) and others also mention fire.

High-altitude populations, mainly occupying spur crests, show a pronounced deficit of stems in the 0–30 cm diameter range (Fig. 16.25a). Since growth–diameter relations are quite good in these very sparse stands (Fig. 16.26), the deficit is reasonably interpreted as a decline in regeneration from around AD 1600. Semi-fertile debris cones in the upper reaches of a valley in the Kaimata Range support high-altitude bush in various stages of development, with kaikawaka present only as large trees and scattered seedlings. With increasing soil maturity the deficit in intermediate sizes becomes progressively less evident (Fig. 16.25b). However, even on gley-podzols few seedlings have established during the past 200 years, perhaps because of less frequent windthrow (Stewart & Rose 1989).

Most analyses for Westland reveal the trough centred on the 0–30 cm diameter range, although the vigour of recent regeneration and the position of size-class peaks among older trees both vary (Fig. 16.25c–f). The exceptions, where clusters of trees of

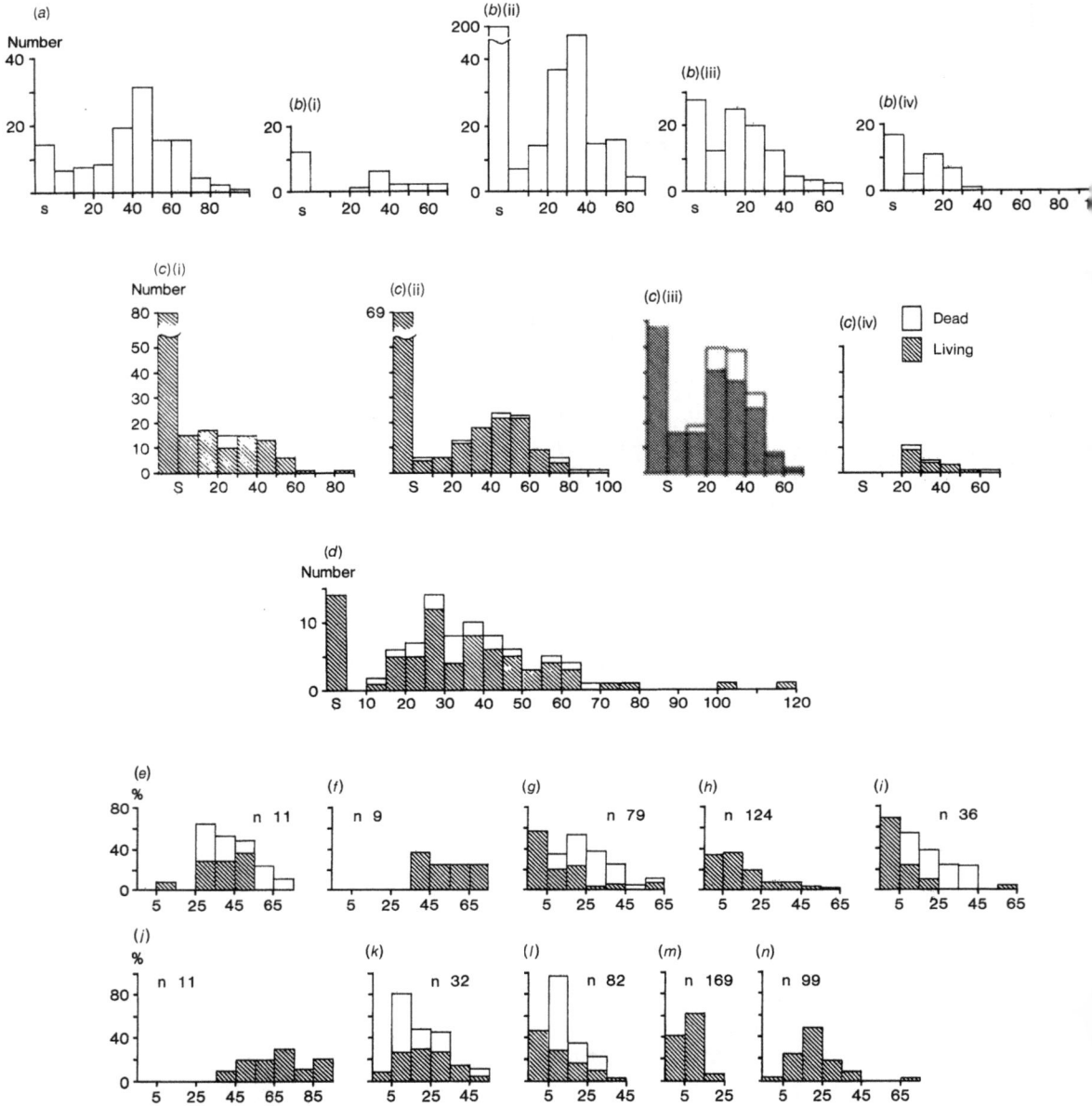

Fig. 16.25. Frequencies of kaikawaka stems, expressed either as number or percent of total in successive diameter classes (cm d.b.h.). S, either < 5 cm d.b.h. or < 1 – 1.2 m tall; n, number of living trees. (*a*) Total for seven sites at 760–1040 m a.s.l., Westland National Park (Wardle 1978*a*). (*b*) At 870–1070 m, Camp Stream, Kaimata Range, North Westland (P. Wardle, unpublished). (i) Associated with *Hoheria glabrata* on raw debris. (ii) All semi-fertile sites including (i). (iii) All infertile sites including (iv). (iv) Associated with pink pine on gley podzols. (*c*) Four sites at 770–1000 m, Cropp River, Central Westland (Norton 1983). (*d*) At 850–900 m, Otira River, Central Westland (Haase 1986*b*). (*e*) In rata–kamahi forest on podzolised yellow-brown earth at 720 m,

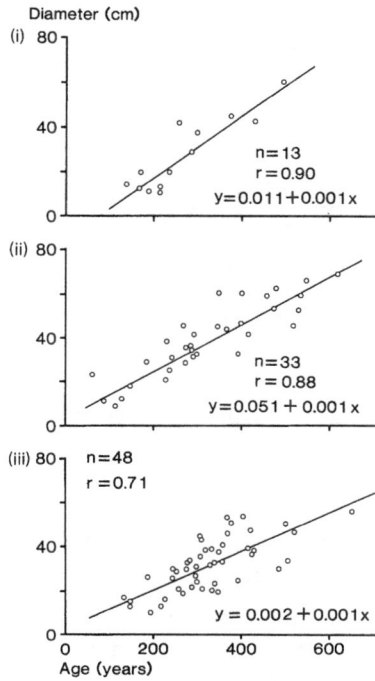

Fig. 16.26. Relation between age and diameter in kaikawaka for Sites (C i–iii) in Fig. 16.25.

similar size add up to a fairly continuous regeneration mode, are all from sites with gley-podzol soils, where species such as silver pine, pink pine (*Halocarpus biformis*), *Phyllocladus alpinus* or sphagnum are prominent. In stands (*g–l*) kaikawaka has established in gaps on waterlogged ground; the large numbers of dead trees may represent self-thinning (*l*) or sensitivity to changes in drainage.

East of the Main Divide, in a tributary of the Wilberforce River, kaikawaka appears to have established with *Phyllocladus* after episodes of deposition of debris on fans (*m–n*). Co-existence of species with such divergent fertility needs as *Hoheria glabrata* and pink pine suggests complex patterns of soil and vegetation development.

On Mts Pirongia and Te Aroha (Auckland province) kaikawaka stands can be assigned to separate age groups (Fig. 16.27), which appear to date back to wind-throw events and landslides. The slopes are either gentle and presumably poorly drained, or

Kelly Range, Central Westland. (*f*) Kaikawaka trees scattered in mountain and silver beech forest on debris fan at 720 m, Rahu Saddle, North Westland. (*g*) Locality as (*e*), but rata–kamahi forest on gley podzol, and containing pink pine. (*h*) Locality as (*f*), but poorly drained area with *Sphagnum* under beech forest. (*i*) Poorly drained terrace with gley podzol soil and pink pine at 450 m, Toaroha River, Central Westland. (*j*) Locality as (*i*), but soil from recent alluvium. (*k*) Mature phase of kaikawaka population in rimu forest on poorly drained fluvioglacial terrace at 250 m, Lake Hochstetter, North Westland. (*l*) As (*k*), but kaikawaka in regenerating phase. (*m–n*) On debris fan at 900 m, Unknown River, D'Archiac ER. (Parts (*e–n*) from Veblen & Stewart 1982.)

| Youngest | Middle | Oldest

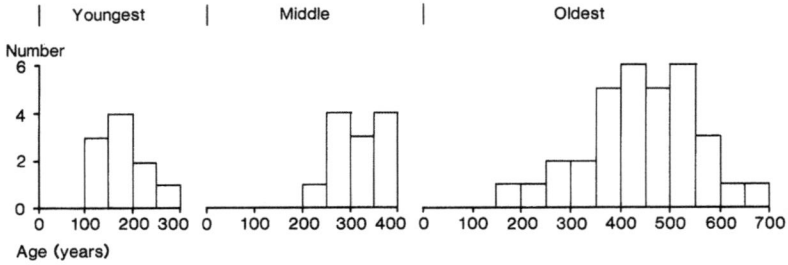

Fig. 16.27. Frequency of stems (excluding seedlings) in age classes, for kaikawaka stands at 850–960 m, Mt Pirongia, Tainui ER (Clayton-Greene 1977).

very steep with soil < 20 cm deep. Associated trees include mountain totara, *Ixerba brexioides, Quintinia serrata*, kamahi, toro and, on Mt Te Aroha, *Phyllocladus glaucus* and *P. alpinus*.

Tall podocarps in the west of the South Island

In rimu stands on the infertile gley podzols of western fluvioglacial terraces, Hutchinson (1932) recognised a mixed age structure wherein rimu seedlings grow up in gaps left by death of old trees, and even-aged patches where young trees establish in gaps formed by wind-throw. Poole (1937) suggested that a cycle is involved, whereby kamahi and *Quintinia* fill gaps in the rimu canopy, and are eventually replaced by the next generation of rimu.

A refinement of these concepts recognises five phases that ensue in response to formation of canopy gaps that may be as small as 0.2 ha and involve only 5–10 trees or be wind-thrown areas as large as 20 ha. Fig. 16.28a illustrates three of these phases. The first has young rimu and silver pines growing up through a broad-leaved canopy, and scattered veteran rimu. The third phase shows a complete canopy of strongly competing podocarps with broad-leaved trees relegated to the subcanopy. In the fifth phase, the podocarp canopy is beginning to break up, and broad-leaved trees are reasserting dominance. Fig. 16.28(b–c) demonstrates that cohorts of podocarps move through as separate peaks, whereas other species tend to maintain a reverse-J population structure.

As conditions diverge from those on the wet, leached terraces, representation of the smaller size classes of rimu decreases (Fig. 16.29). On dissected hill country in North Westland, June (1983) found that regeneration was 'at a constant and relatively high rate from 780–1480 AD, steadily declined until 1780–1880 AD, and partially recovered after 1880 AD.'. J.L. Bathgate (unpublished) found a similar pattern on the coastal side of the Longwood Range in Southland, and less intensive surveys indicate that it prevails widely through the wetter parts of the South Island. The regeneration gap is widest on semi-fertile sites with dense understoreys, where podocarps are commonly represented only by widely scattered very large trees, small saplings and seedlings. It seems much less apparent on Stewart Island, and towards the upper altitudinal limits of rimu.

In ecology and distribution, miro closely parallels rimu, except that usually far

Fig. 16.28. Structure of rimu forest on fluvioglacial terraces; Saltwater Forest, Central Westland (Six Dijkstra *et al.* 1985). (*a*) Profiles and plans of phases 1, 3 and 5. Abbreviations: C, *Coprosma lucida*; D, *Dicksonia squarrosa*; Dc, *Dacrydium cupressinum*; E, *Elaeocarpus hookerianus*; G, *Griselinia littoralis*; H, *Hedycarya arborea*; Lc, *Lagarostrobus colensoi*; M, *Myrsine australis*; P, *Phyllocladus alpinus*; Pf, *Prumnopitys ferruginea*; Ph, *Podocarpus hallii*; P2, *Pseudopanax crassifolius*; Q, *Quintinia acutifolia*; W, *Weinmannia racemosa*. (*b*) Ages of trees at stump height in the five phases: ●, rimu; ▲, silver pine; *, miro; ■, mountain totara; +, minimum age. (*c*) Densities of trees > 1 m tall in diameter classes in the five phases. (Figure continues on pp. 596–8.)

more seedlings persist, probably because of reserves derived from the large seeds (p. 62). Patches of miro trees may have arisen through release of seedlings after windthrow. Kahikatea, matai and lowland totara are confined to deep or relatively fertile soils, and do not occur on the infertile, wet soils that allow continuous regeneration of rimu. All three enter primary successions on alluvial flats, with kahikatea being most abundant on the wettest ground and totara on the driest. For Central Westland, P. Wardle (1974) has described (1) first-generation kahikatea developing through seral broad-leaved bush; (2) stands composed of separate cohorts, representing damage

(a) Phase 3

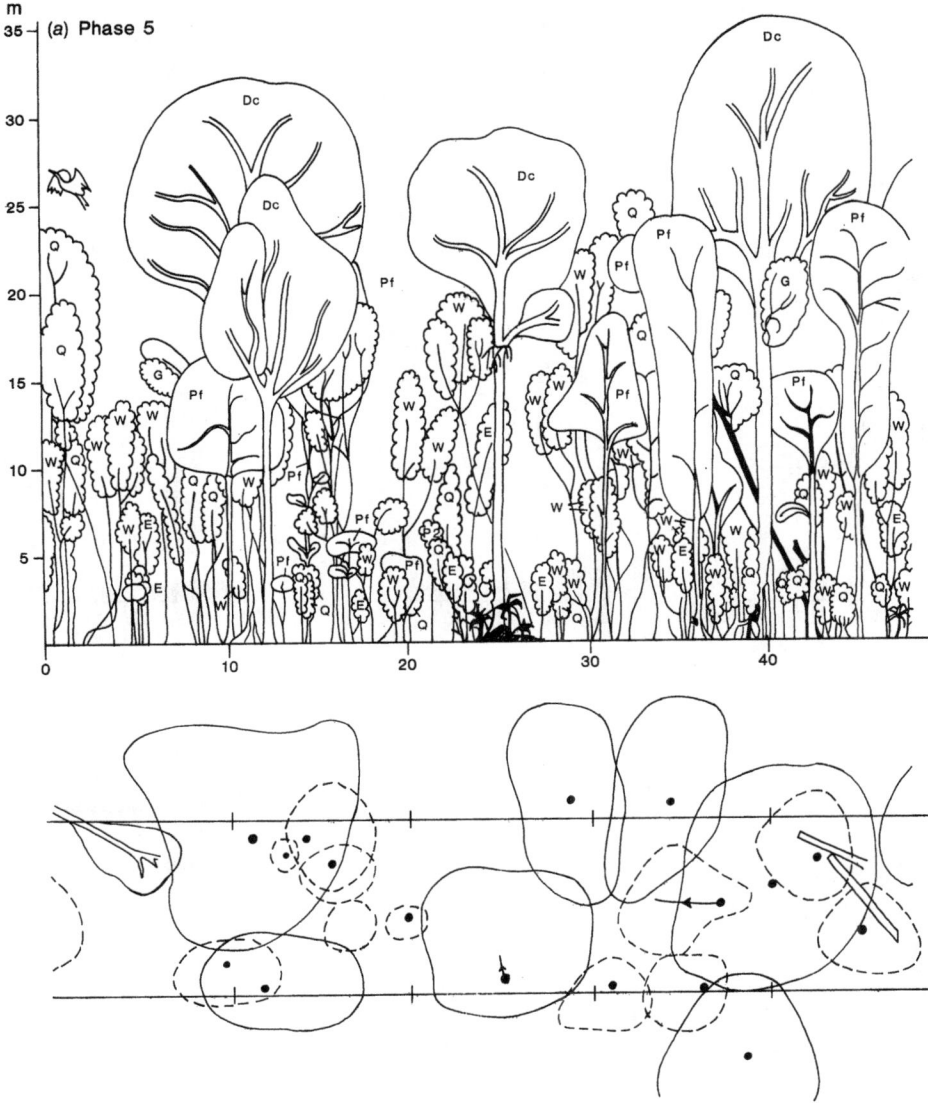

Fig. 16.28(*a*) (*cont.*)

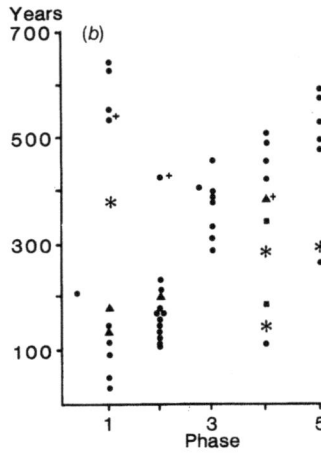

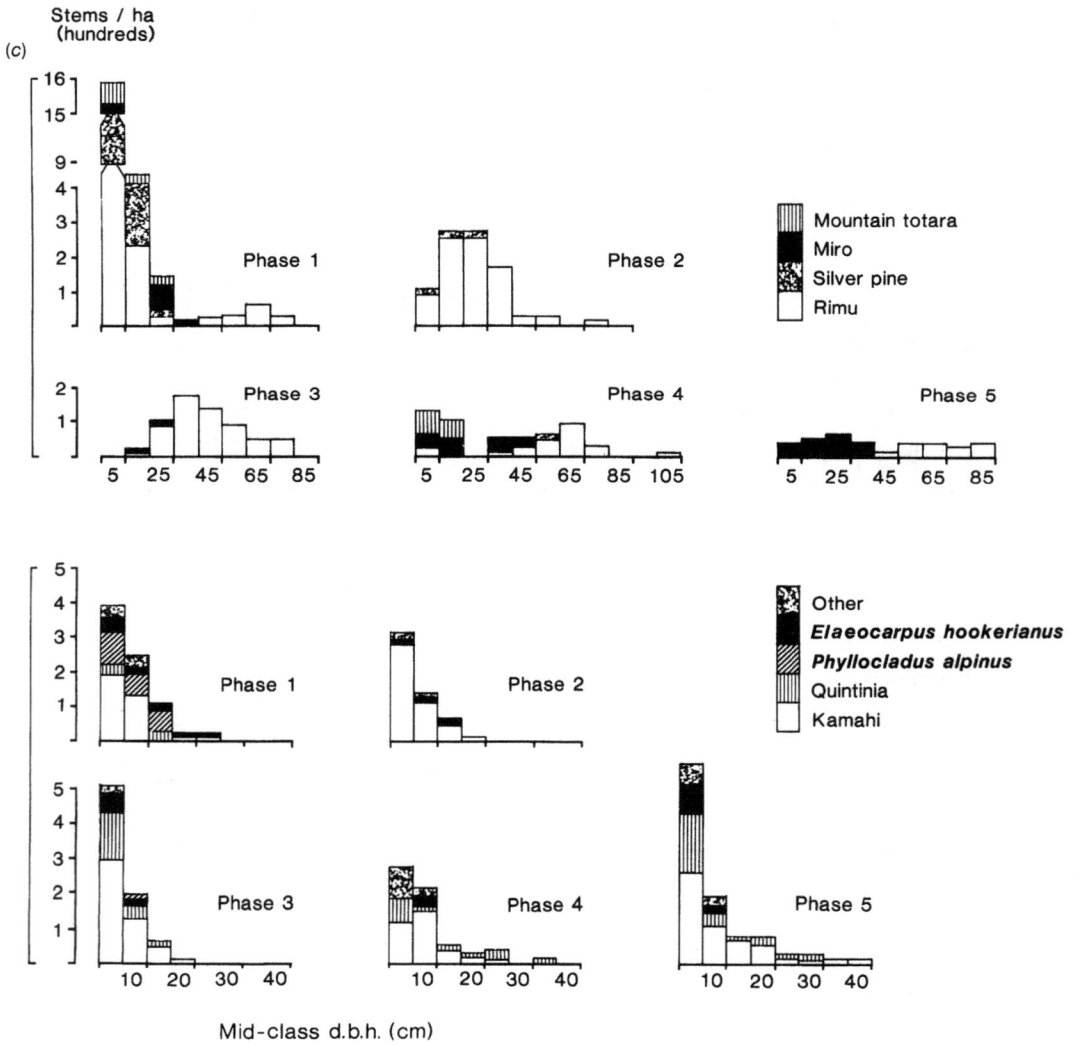

Fig. 16.28(b–c) (for legend see p. 595)

Number

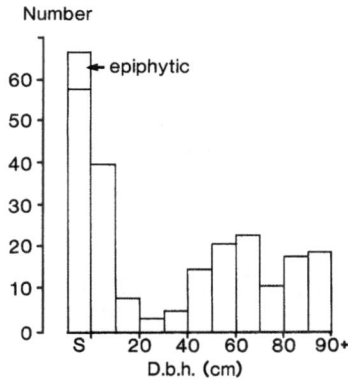

Fig. 16.29. Frequency of rimu stems in diameter classes on slopes in Westland National Park (Wardle 1978a). S, < 1.2 m tall.

through silting up after floods and subsequent recruitment of kahikatea seedlings; (3) mature stands, including admixtures with rimu, in which young plants are usually poorly represented; (4) stunted stands transitional to swamp, in which small size-classes are well-represented.

In the same region, forest of *Podocarpus totara* var. *waihoensis* and matai on stony, well-drained flats tends to be even-sized and also even-aged, as there is good regression between age and diameter (Fig. 16.30). Occasional much larger trees, especially of totara, have probably survived aggradation through production of adventitious roots.

With mountain totara, it is adult trees rather than seedlings and saplings that are often sparsely represented. In mixed forest, this is at least partly related to die-back (p. 582). Many beech forests on either side of the Main Divide contain abundant mountain totara seedlings that are often sprawling and suppressed, but mature trees are rare except where rocky knolls and gorges interrupt the beech canopy.

Tall podocarps in the eastern South Island and North Island

In the Longwood Range (Te Wae Wae ER) the abundance of matai, totara and kahikatea relative to rimu, and the proportion of mature podocarps to young trees both increase along the gradient from the coast to the inland flanks, which are colder in winter and drier (J.L. Bathgate, unpublished). Similar gradients in relative abundance occur throughout the eastern South Island and the North Island, and the paucity of young podocarps has often been noted (Cockayne 1928; Holloway 1954; Nicholls 1956). Around the Volcanic Plateau, forests in which podocarps occur mainly as a mature overstorey grade into dense podocarp forests on Taupo tephra (p. 120). At Pureora on the western side of the Plateau, Beveridge (1973) describes a regeneration cycle in stands where rimu and other large podocarps form a sparse overstorey above tawa. It begins as follows:

(1) Wind-throw of an aged podocarp.
(2) Development of dense tree ferns, especially the rhizomatous *Dicksonia squarrosa*, which cast dense litter that inhibits all tree seedlings.

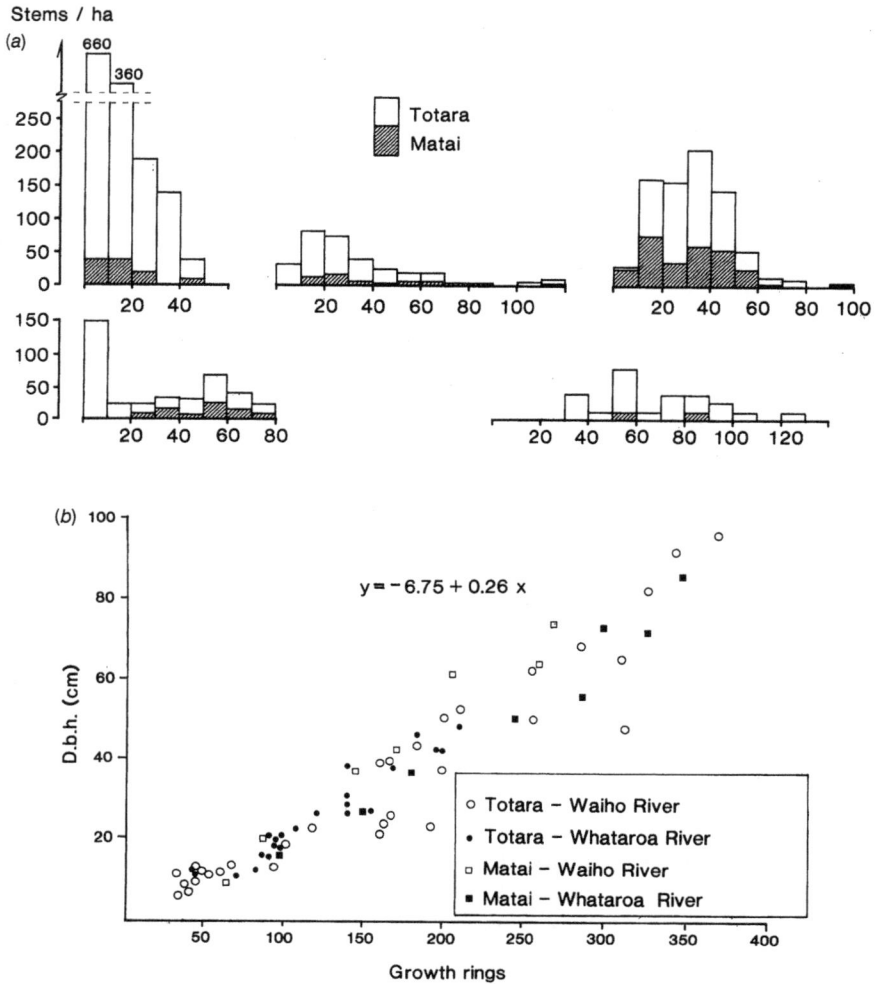

Fig. 16.30. Structure in five totara–matai remnants on alluvial flats in Central Westland (McSweeney 1982). (a) Density of stems >2 cm d.b.h. in diameter classes. (b) Diameter–age relations.

(3) Establishment of broad-leaved tree seedlings, especially kamahi, on tree-fern trunks. Kamahi also establishes on upturned root plates.

(4) Suppression of tree ferns by broad-leaved trees; kamahi grows large and becomes a suitable perching tree, especially for native pigeons.

(5) Recruitment of podocarp seedlings, mainly from bird-dispersed seed.

(6) Development of a podocarp sapling group as the kamahi crown thins and dies.

Stages 1–6 probably take 200–300 years, and entry of podocarps depends largely on the prior success of kamahi; as kamahi is rare in this forest type, young podocarps between 10 and 30 cm in diameter are also rare.

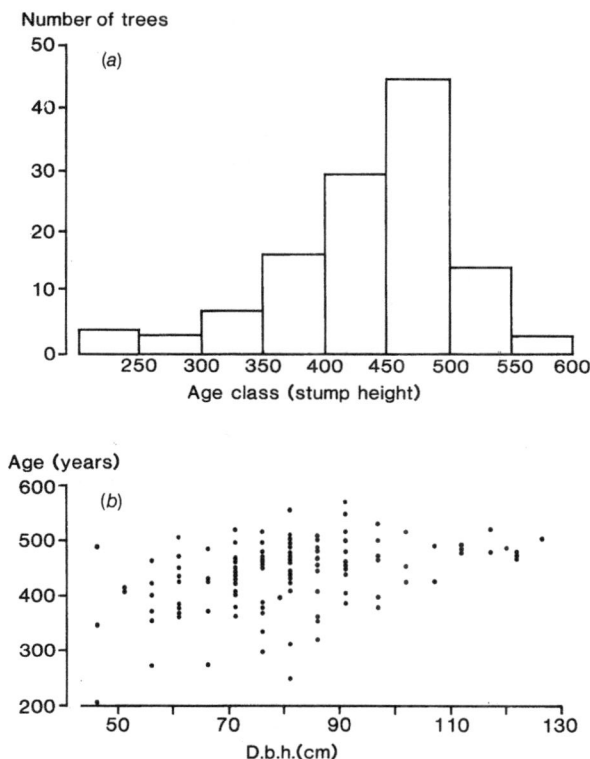

Fig. 16.31. Age class distributions (a) and diameter–age relation (b) in a sample of rimu trees from a dense podocarp stand at Tihoi, Western Volcanic Plateau (Herbert 1980).

Throughout the Volcanic Plateau, dense podocarp stands tend to be even-sized and although the correlation between age and diameter is poor it still translates to relative paucity of younger age classes (Fig. 16.31; Table 7.2). In contrast to rimu and matai, tawa and miro may show reverse-J curves (Fig. 16.32), and at altitudes below 700 m tawa seems likely to gradually replace podocarps, from both undisturbed forests and forests thinned by selective logging and subsequent wind-throw.

Kauri

Profuse regeneration of kauri in secondary succession is illustrated in data for ricker stands by Table 16.5, which also shows that seedlings germinate in considerable numbers in mature stands and persist through tolerance of shade. Population structures range from reverse-J distributions, mainly in young, dense stands, through stands of generally intermediate age and density that peak at various diameters, to sparsely stocked stands that include trees of a great spread of diameters, some very large (Fig. 16.33a). There is a fair linear relation between age and diameter (b), although dense young stands show mutual suppression (c). According to the authors, dense young stands result from disturbance through fire or cyclones, and subsequent cohorts are released in tree-fall gaps. However, as only 5–15% of such gaps are

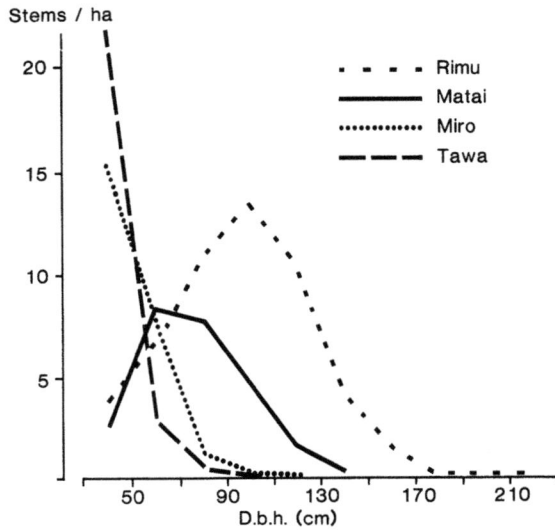

Fig. 16.32. Distribution in diameter classes of major species in dense podocarp forest at Whirinaki, Eastern Volcanic Plateau (Smale *et al.* 1985).

reoccupied by kauri, the remainder being lost to broad-leaved trees or miro, successive cohorts will contain fewer kauri trees unless there is further massive disturbance. This analysis discounts claims that mature kauri stands tend to lack younger stages, but does not explore the possibility that kauri communities in different habitats may differ in life history.

Conclusions

In most mature beech and broad-leaved stands, sequential stages of regeneration can be recognised, that usually involve patchiness on various scales. Discontinuities in population structure can usually be readily attributed to disturbance, succession, or browsing by animals. Poor representation of young red beech near the upper limits of the species in the Ruahine Range (Ogden 1971) and the Taramakau catchment (Haase 1989) has been attributed to sensitivity to climatic trends; perhaps the predominantly very large trees of silver beech that grow with mixed forest at low altitudes in the southern zone (Simpson & Thomson 1928) require a similar explanation.

Native conifers can enter seral stages in large numbers, especially while a broad-leaved canopy provides just enough shade to suppress the herbaceous understorey. Conifers tolerating low fertility also regenerate well in mature forest where conditions inhibit herbaceous understoreys; this may involve regeneration cycles with alternating conifer and broad-leaved phases. In other forests, conifers are represented by disproportionate numbers of mature trees; young trees, and often also saplings and seedlings, seem curiously scarce.

Recent authors, including Odgen (1985), suggest that this 'conifer regeneration

Table 16.5. *Kauri population as mean no. of individuals/ha in mature and ricker communities at various sites; brackets show standard deviation as percentage of mean*

	cotyledonary (1 yr) seedlings	seedlings <2 cm dbh	saplings 2–10 cm	trees ≥10 cm dbh	basal area*
tree-fall gaps†	1940 (105)	8149 (71)	256 (53)	84 (97)	7 (10)
ricker stands	798 (108)	10549 (43)	1669 (67)	672 (34)	33 (33)
mature stands	1302 (57)	1071 (109)	239 (88)	109 (61)	49 (51)

Notes:
* Area in m²/ha, kauri only; † gap size 0.03–0.05 ha.
Source: Ogden *et al.* 1987

gap' can be explained in terms of gap-phase regeneration following disturbances. This implies that the return period of disturbances is very long, or that their intensity or frequency have undergone New Zealand-wide secular changes, or that the claimed synchroneity of the gap results from sampling bias that causes younger stages to be missed. There is, indeed, an overt bias in that young conifer stands originating in fire-initiated successions have been omitted from consideration, on the basis that most forests showing the conifer regeneration gap have no proven fire history. Robbins (1962) interpreted the conifers as an ancient element being gradually ousted by the more recent and vigorous broad-leaved trees, while allowing that stands can be rejuvenated by catastrophic events, including fires and volcanic eruptions. However, the time scales implied seem irrelevant to current ecological processes.

Holloway (1954) developed a hypothesis, which dominated forest ecology for the next 20 years, to the effect that the paucity of young conifers results from a change to colder and, in rain-shadow districts, drier climates since the fourteenth century. Wardle (1978a) suggested that although the regeneration gap varied in duration between localities and sites, it is centred on the period AD 1600–1800, since when there has been some increase in regeneration. He drew a parallel with the record of glacial advances, although it is difficult to conceive of a climatic cause that could impact so widely over New Zealand. Furthermore, recent information indicates a more complex glacial chronology (Gellatly *et al.* 1985).

The paucity of young conifers is most apparent on relatively fertile soils where mature conifers form an open overstorey to a broad-leaved or beech canopy. These appear to be the sites where gap-phase regeneration is least likely to succeed, as disturbances lead to vigorous growth of ferns, shrubs, and broad-leaved saplings that exclude conifer seedlings. However, these are also the sites that are most attractive to browsing mammals, and conifer regeneration can be encouraged where animals have destroyed competing understorey species. There is a fine balance, as heavy browsing and trampling destroy conifer seedlings (McSweeney 1982) and antler-rubbing and bark-chewing can kill saplings. The best conditions for regeneration, given good seed years (p. 57), may be where animals maintain only a shifting impact on the understorey, or immediately after animals have been removed but before understoreys have recovered.

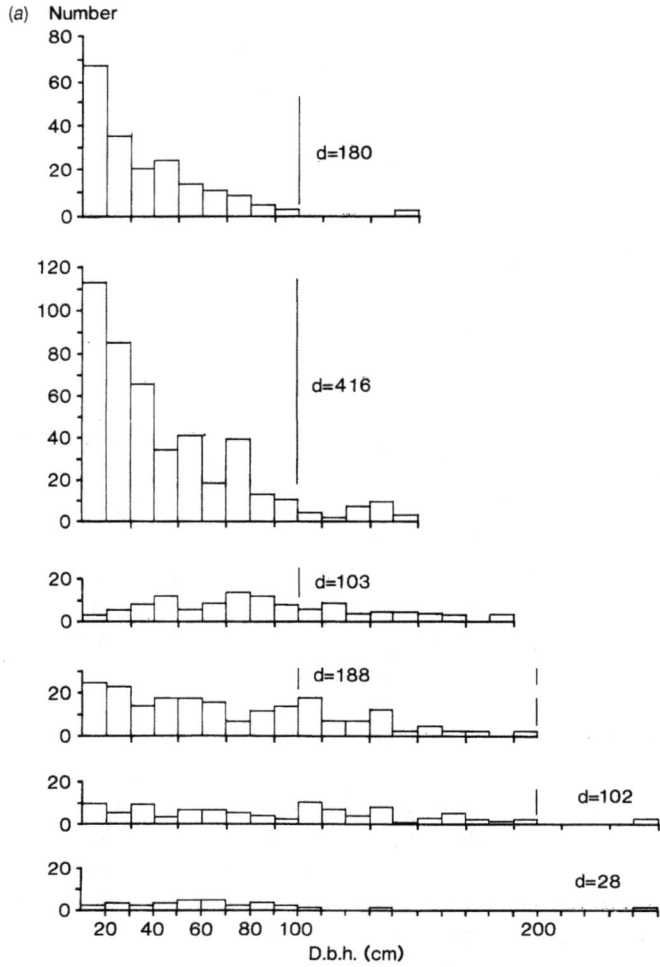

(a) Number

d=180

d=416

d=103

d=188

d=102

d=28

D.b.h. (cm)

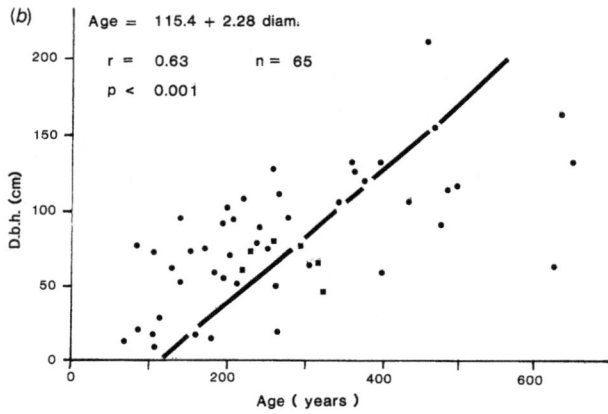

(b)   Age =   115.4 + 2.28 diam.

r =   0.63   n = 65

p <   0.001

D.b.h. (cm)

Age (years)

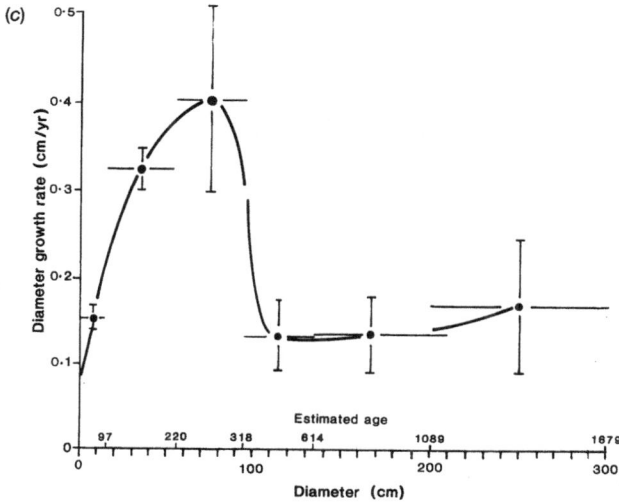

Fig. 16.33. Structure and growth of kauri stands (Ahmed & Ogden 1987). (*a*) Frequency of stems > 10 cm d.b.h. in diameter classes. Vertical lines mark 1 m and 2 m class boundaries; d, no. of stems > 10 cm d.b.h./ha. (*b*) Diameter–age relations derived from 2–3 cores per tree (circles) and cross sections (squares). (*c*) Diameter growth at different diameters. Horizontal lines, periods over which each point is calculated; Vertical lines, 95% confidence limits.

Primeval New Zealand forests also contained abundant herbivores. Moa concentrated in relatively fertile habitats (Anderson 1982) and perhaps reduced understoreys sufficiently to favour conifer seedlings. The widespread decline of conifer regeneration might well be related to the extinction of such animals, with irruptions of the seed-destroying *Rattus exulans* and the decline of seed-dispersing birds and reptiles as contributing factors.

The perceived deterioration of forest canopies and understoreys and collapse of forest structure has been a matter of increasing concern. Until the 1970s the irruption of introduced herbivores was seen as a sufficient explanation, that justified intensive slaughter of wild animals with the object of preventing erosion that could threaten land and livelihood in lowlands downstream. Gradually, the role of droughts, storms and, especially, the tectonic instability of New Zealand steeplands became appreciated, as have the probably significant effects of the herbivores that died out out after Polynesian colonisation. It is also realised that vigorous growth following removal of animals is largely of *r*-selected species filling canopy gaps. Response beneath closed canopies or by seedlings of dominant trees is often lacking. There is now more effort towards understanding the balance among the various factors, which should lead to animal control being concentrated where it is most necessary.

EPILOGUE

Up to this point, I have described the present-day vegetation of New Zealand in terms of the environmental factors and historical events that have shaped it. In these final pages I discuss the values of our vegetation for people, its significance to science, and how environmental, biological and economic forces may affect its future.

A little more than 1000 years ago, New Zealand was uninhabited and supported the last substantial area of truly primitive vegetation in the world. The Maori were equipped with Neolithic technology and accompanied only by a few crop plants and weeds, a species of rat, and a dog. Yet, within a few centuries of their arrival, great tracts of forest had been destroyed, and an avian 'megafauna' and many smaller animals had become extinct; we can only speculate as to how the faunal changes affected vegetation.

In neither extent nor intensity were these changes as far-reaching as those that have taken place over the 150 years of European settlement. In contrast to some other lands recently colonised by Europeans, such as Australia and North America, where native plants remain conspicuous even in closely settled areas, there has been near-total removal of native vegetation from wide areas, and much of what remains has been profoundly modified through fire and grazing. Today only some far-southern islands and a few off-shore stacks are truly unmodified. At first, as the original plants lost ground, their place was taken mainly by native plants such as bracken, manuka and danthonias, but adventive plants have increasingly dominated the landscape. However, many adventives are scarce and local, and many belong to communities, such as 'improved' pastures, that are sustained by intensive human effort.

Mixtures of native and introduced plants occur mostly where human impact is extensive or intermittent; the vast area of low-producing pastures, including short-tussock grasslands, is the major example. Adventives that have entered intact native forest are few and mostly insignificant. Although considerable areas of heath, grassland and wetland are still wholly indigenous, such vegetation is more receptive than forest to adventive plants, including trees that completely alter the physiognomy. Adventive lianes are a particular threat at bush margins. Some exotic trees, especially

606

Pinus contorta, can spread great distances and to as high as the alpine belt; otherwise, few adventive species occur at high altitudes. In contrast, open lowland habitats such as flood-plains, sand dunes and depleted grasslands are extremely vulnerable to invasion, and native plants dominate only where aggressive adventives have yet to arrive.

Although spread of adventive plants in response to changes wrought by humans is a world-wide phenomenon, the extent of their success in New Zealand is because they possess pioneering attributes in greater measure than the native flora. For example, more effective dispersal, faster growth and greater height allow adventive trees and shrubs to invade and suppress shorter native vegetation, especially if they also tolerate exposure, drought or cold better than native trees or shrubs. Many adventive plants can pre-empt frequently disturbed habitats because of short life-cycles or, as in woody legumes, prolonged seed dormancy and ability to fix nitrogen.

Adventive species also seem better adapted to co-exist with the introduced fauna. More plants have prickles or thorns that discourage browsing, and more flowers are brightly coloured, which may make them more attractive to introduced insects, especially bees. Tolerance of introduced grasses to intense grazing, efficient nitrogen fixation by introduced legumes, and the stimulus of added fertilisers have created rapidly cycling pasture systems that few native plants enter.

The limited competitiveness of native plants is particularly apparent in open, disturbed or harsh environments that probably appeared only at the end of the Tertiary, long after New Zealand had separated from other land masses. The new habitats could be claimed only by species with wide ecological tolerances or able to overcome the barriers of distance, or by herbs and shrubs that could evolve quickly enough by virtue of an early reproductive age. European colonisation opened the flood-gates to a host of plants that were pre-adapted to thrive here, especially after human activity had created bridgeheads.

The superior adaptation of adventive plants can be exaggerated. Cabbage tree (*Cordyline australis*), pohutukawa (*Metrosideros excelsa*), ngaio (*Myoporum laetum*) and *Olearia traversii* are as tolerant of coastal exposure as any trees in the world, and the adaptation of plants such as *Haastia* and *Raoulia* 'vegetable sheep' is so thorough as to cast some doubt on the recency of the alpine environment. The same applies to competitiveness; as Cockayne (1928) saw, native forest could ultimately reclothe the moist, temperate regions of New Zealand, were humans and their livestock to leave the scene.

The reality, however, for many of our native plant communities and species is that if we wish to retain them, we need to control invasion and competition by adventive plants. One may well ask, why intervene, when introduced plants can meet our needs for food, fodder, shelter and timber, cover bare, eroding ground better than natives, and provide more varied hues of flower and foliage? The answers lie in the uniquely New Zealand quality of the natural landscape and the native plants and animals that contribute to our awareness of being New Zealanders – as exemplified by the silver fern and kiwi. Most of us cherish this indigenous quality, despite depending economically on introduced plants and animals and enjoying them aesthetically.

To scientists, the value and interest of the native flora and vegetation rest in their

Gondwanan origins, development in isolation, adaptations to a wondrously varied landscape, responses to ceaseless change, and interdependence with a unique fauna. The adventive flora, for its part, teaches a great deal about the phenomena of naturalisation and invasion.

The pressures to maintain the extent and integrity of native vegetation in the face of economic and biological forces come from a strong conservation movement, and the implementation rests mainly with official organisations. The country's first national park (Tongariro) was gifted by Hepi Te Heuheu of the Tuwharetoa tribe in 1887, and the largest conservation body (Royal Forest and Bird Protection Society) was constituted in 1923. In 1987, most government responsibility for conservation was vested in a new Department of Conservation, which directly administers 30% of New Zealand's land area. Nevertheless, there are numerous areas of high conservation value under other jurisdiction. This includes local authorities and, not least, private landowners, many of whom have already formally protected features on their land.

Idealistic goals would be to protect all remaining native vegetation, and totally remove introduced plants and animals from protected lands. Although two Acts of Parliament embody these goals, there are few places, mainly small islands, where 'purity' still exists in large measure or could be restored. Much larger areas, especially in the wetter mountains and larger forest tracts, are free of ecologically significant adventive plants, and could be kept that way. No part of the New Zealand mainland is without introduced vertebrates, let alone insects; and for most species and most areas, eradication is no longer logistically possible, or even politically acceptable in the case of the large 'game' animals. However, grazing animals, rats and cats have been removed from several outlying and offshore islands, with benefit to native vegetation and fauna. Barriers to eradication on other uninhabited islands are more political than logistic.

Levels of legal protection notwithstanding, there will always be pressures for economic use of the indigenous estate, be it grazing, mining, timber, access to fishing grounds, or tourism. Conservationist opposition to economic pressures has had much success, leading to accusations of 'locking up resources'. Unquestionably, there are areas of such high conservation value that no compromise should be countenanced, but in other areas, the conservation ethos may endure more securely where it seeks to co-exist with compatible economic use. For instance, it is not necessarily incompatible with conservation to harvest special-quality timber from native forests, provided harvesting is at a level known to be sustainable without significantly compromising ecological values. Mining might be acceptable provided disturbance scarcely exceeds that engendered by natural events. A precedent – albeit reached from adversary positions – is the control of Lakes Te Anau and Manapouri at levels that can produce nearly 600 megawatts, while allowing survival of valued shoreline vegetation.

The greater part of our native vegetation, including the ranges of many locally endemic species, is in districts where introduced plants and animals are abundant or predominant. Here, any conservation initiative has to take account of economic and proprietary claims on the land. Increasingly, this means taking account of Maori views on conservation and use of resources. Nor will protection of native biota in mountain grasslands that have been extensively grazed for over 140 years be achieved without acknowledging the rights of graziers, although representative examples should be totally protected if only to provide bench-marks. And where biotic communities have become reduced to unique remnants through attrition of the natural estate, their conservation and scientific values should be held to outweigh those accruing from other uses.

Until recently, the main objective of conservation in populated areas has been to salvage remnants of native forest and restore them to their supposed condition in AD 1840, as though 1000 years of Maori presence had no impact. A far better objective is to retain the range of habitats that provides the greatest security to indigenous biota. This places value, not only on pristine forest, but on vegetation that has been induced through exploitative use, such as northern gumlands, Central Otago scabweed (*Raoulia*) deserts and, not least, the vast areas of tussock grassland that resulted from ancient natural and Maori fires. It also accepts that disturbances, such as fire, wind-throw, floods and landslides, are necessary to create niches for many or most indigenous species; disturbance can be accepted with greater equanimity, however, in large reserves than in tiny remnants.

Adventive plants and animals must be tolerated in many indigenous reserves, not only because it would be too disruptive or costly to remove them, but because the purpose of reservation can be achieved only with their continued presence. Controlled grazing is needed to remove rank growth that would otherwise overwhelm most lower-altitude native grasslands; the same may apply in the understorey of some forest remnants. As a more speculative example, mammalian herbivores possibly enhance regeneration of podocarps through substituting for the extinct moa (but this cannot be an excuse for allowing feral herbivores to regain the numbers that severely depleted native forests and mountainland vegetation up to the 1960s). Although spread of gorse into new areas should be prevented, destroying existing gorse scrub can set back succession to native forest. However, whether exotics should be specially planted to provide a nurse for native vegetation is a moot point.

There are times when the conservation option must be vigorously promoted in opposition to other sectional interests. Recent examples where it should have been the preferred option have been subsidised farm development that created uniformity at the expense of biological diversity, and contributed to ruinous soil erosion and flooding; burning native forests after logging, to replace them with exotic plantations; and proliferation of inadequately restrained goat herds.

Conservation of indigenous flora and fauna is too narrow an aim, if pursued in isolation from other environmental concerns, especially those of global relevance. For instance, tourism can help conservation by raising awareness of natural values, and providing an alternative to rural employment based on logging and mining; but how do these advantages weigh against the degradation of natural and cultural values and waste of resources that the tourist industry engenders at other points in New Zealand and overseas? How realistic is conservation planning, if it does not have contingencies for changes in local and global climate brought about by ozone depletion or enhanced carbon dioxide levels? In the worst scenarios the niches of many endangered species would no longer exist within the areas set aside for their perpetuation.

While conservation management calls for a large measure of pragmatism, it must also be guided by underlying principles. The environmental sciences are uniquely qualified to formulate these principles; and at a more detailed level, they can provide inventories of taxa, describe natural systems and how they work, assess values, recognise problems and discover solutions. This '*Vegetation of New Zealand*' attempts to carry understanding forward from the point reached by its predecessor (Cockayne 1928). More important than this are the questions posed and left partly or wholly unanswered. Only through continually seeking scientific knowledge will we be able to fully understand, enjoy and protect the unique life that has evolved on these islands.

APPENDIX 1

Abbreviations

aff.   unnamed taxon with affinity to . . .

aggr.   aggregate of 'microspecies'

a.s.l.   above sea level

cf.   unnamed taxon similar to . . .

d.b.h.   diameter over bark at breast height

ED   Ecological District

ER   Ecological Region

s.l.   sensu lato

sp., spp.   species, singular or plural

s.s.   sensu stricto

Units

d   day

dm   decimetre

h   hour

ha   hectare

MPa   megapascals

t   tonne

yr   year

611

APPENDIX 2

Common names and Latin equivalents

(Common names are listed only where used in isolation from the corresponding scientific name.)

Beech	*Nothofagus* spp.
black	*N. solandri* var. *solandri*
hard	*N. truncata*
mountain	*N. solandri* var. *cliffortioides*
red	*N. fusca*
silver	*N. menziesii*
Bellis daisy	*Bellis perennis*
Blackberry	*Rubus fruticosus* aggr.
Blue tussock	*Poa colensoi*
Bog pine	*Halocarpus bidwillii*
Bracken	*Pteridium esculentum*
Briar	*Rosa rubiginosa*
Bristle tussock	*Rytidosperma setifolium*
Broom (common)	*Cytisus scoparius*
Browntop	*Agrostis capillaris*
Brush wattle	*Paraserianthus lophantha*
Buffalo grass	*Stenotaphrum secundatum*
Cabbage tree	*Cordyline australis*
Catsear	*Hypochoeris radicata*
Carpet grass	*Chionochloa australis*
Cocksfoot	*Dactylis glomerata*
Dandelion	*Taraxacum officinale*
Danthonia	*Rytidosperma* spp.

Filmy fern	Hymenophyllaceae
Five-finger	*Pseudopanax arboreus*
Fuchsia	*Fuchsia excorticata*
Glasswort	*Sarcocornia quinqueflora*
Gorse	*Ulex europaeus*
Hard tussock	*Festuca novae-zelandiae*
Hawkweed	*Hieracium* spp.
Hinau	*Elaeocarpus dentatus*
Kahikatea	*Dacrycarpus dacrydioides*
Kaikawaka	*Libocedrus bidwillii*
Kaikomako	*Pennantia corymbosa*
Kamahi	*Weinmannia racemosa*
Kanuka	*Kunzea ericoides*
Karaka	*Corynocarpus laevigatus*
Kauri	*Agathis australis*
Kikuyu grass	*Pennisetum clandestinum*
Kiekie	*Freycinetia baueriana*
Kohekohe	*Dysoxylum spectabile*
Kowhai	*Sophora* spp., esp. *microphylla*
Lancewood	*Pseudopanax crassifolius*
Lupin (tree)	*Lupinus arboreus*
Mahoe	*Melicytus ramiflorus*
Mangeao	*Litsea calicaris*
Mangrove	*Avicennia resinifera*
Manoao	*Halocarpus kirkii*
Manuka	*Leptospermum scoparium*
Marram	*Ammophila arenaria*
Matagouri	*Discaria toumatou*
Matai	*Prumnopitys taxifolia*
Miro	*Prumnopitys ferruginea*
Nettle	*Urtica* spp.
Ngaio	*Myoporum laetum*
Nikau	*Rhopalostylis* spp.
Pine	*Pinus* spp.
Pingao	*Desmoschoenus spiralis*
Pink pine	*Halocarpus biformis*
Pohutukawa	*Metrosideros excelsa*
Pokaka	*Elaeocarpus hookerianus*
Pukatea	*Laurelia novae-zelandiae*

Puriri	*Vitex lucens*
Rangiora	*Brachyglottis repanda*
Rata	*Metrosideros* spp.
northern	*M. robusta*
southern	*M. umbellata*
Red tussock	*Chionochloa rubra*
Rewarewa	*Knightia excelsa*
Rimu	*Dacrydium cupressinum*
Ryegrass	*Lolium* spp.
Scabweed	herbaceous *Raoulia* spp.
Silver fern	*Cyathea dealbata*
Silver pine	*Lagarostrobos colensoi*
Silver tussock	*Poa cita*
Snow tussock	*Chionochloa* spp.
Sorrel (sheep's)	*Rumex acetosella*
Sweet vernal	*Anthoxanthum odoratum*
Tall fescue	*Festuca arundinacea*
Tanekaha	*Phyllocladus trichomanoides*
Tarahinau	*Dracophyllum arboreum*
Taraire	*Beilschmiedia tarairi*
Tauhinu	*Cassinia leptophylla* s.l.
Taupata	*Coprosma repens*
Tawa	*Beilschmiedia tawa*
Titoki	*Alectryon excelsus*
Toetoe	*Cortaderia* spp. (native)
Toro	*Myrsine salicina*
Totara	*Podocarpus* spp.
lowland	*P. totara*
mountain	*P. hallii*
snow	*P. nivalis*
Westland	*P. totara* var. *waihoensis*
Towai	*Weinmannia silvicola*
Tutu	*Coriaria* spp.
tree	*C. arborea*
White clover	*Trifolium repens*
Willow	*Salix* spp.
Willow-herb	*Epilobium* spp.
Wineberry	*Aristotelia serrata*
Yellow-silver pine	*Lepidothamnus intermedius*
Yorkshire fog	*Holcus lanatus*

REFERENCES

Adams, J.A. (1976). Nutrient requirements of four *Nothofagus* species as shown by foliar analysis. *New Zealand Journal of Botany* **14**: 211–13.

Adams, J.A. & Walker, T.W. (1975). Nutrient relationships of radiata pine in Tasman Forest, Nelson. *New Zealand Journal of Forestry Science* **5**: 18–32.

Ahmed, M. & Ogden, J. (1987). Population dynamics of the emergent conifer *Agathis australis* (D. Don) Lindl. (kauri) in New Zealand. 1. Population structures and tree growth rates in mature stands. *New Zealand Journal of Botany* **25**: 217–29.

Aldridge, R. (1968). Throughfall under gorse (*Ulex europaeus*) at Taita, New Zealand. *New Zealand Journal of Science* **11**: 447–51.

Allan, H.H. (1926). The surface roots of an individual matai. *New Zealand Journal of Science and Technology* **8**: 233–4.

Allan, H.H. (1927). The vegetation of Mt Peel, Canterbury, New Zealand. Part 2. The grasslands and other herbaceous communities. *Transactions and Proceedings of the New Zealand Institute* **57**: 73–89.

Allan, H.H. (1937). A consideration of the 'Biological Spectra' of New Zealand. *Journal of Ecology* **25**: 116–52.

Allen, R.B. (1987). Ecology of *Nothofagus menziesii* in the Catlins Ecological Region, South-east Otago, New Zealand. (I) Seed production, viability and dispersal. *New Zealand Journal of Botany* **25**: 5–10.

Allen, R.B. (1988). Ecology of *Nothofagus menziesii* in the Catlins Ecological Region, South-east Otago, New Zealand. (III) Growth and population structure. *New Zealand Journal of Botany* **26**: 281–95.

Allen, R.B., Payton, I.J. & Knowlton, J.E. (1984). Effects of ungulates on structure and species composition in the Urewera forests as shown by exclosures. *New Zealand Journal of Ecology* **7**: 119–30.

Anderson, A.J. (1982). Habitat preferences of moa in Central Otago AD 1000–1500, according to palaeobotanical and archaeological evidence. *Journal of the Royal Society of New Zealand* **12**: 321–36.

Archer, A.C. (1973). Plant succession in relation to a sequence of hydromorphic soils formed on glacio-fluvial sediments in the alpine zone of the Ben Ohau Range, New Zealand. *New Zealand Journal of Botany* **11**: 331–48.

Archer, A.C. & Cutler, E.J.B. (1983). Pedogenesis and vegetation trends in the alpine and upper subalpine zones of the northeast Ben Ohau Range, New Zealand. 2. Plant communities and plant succession. *New Zealand Journal of Science* **26**: 151–71.

Archer, A.C., Simpson, M.J.A. & Macmillan, B.H. (1973). Soils and vegetation of the lateral moraine at Malte Brun, Mount Cook region, New Zealand. *New Zealand Journal of Botany* **11**: 23–48.

Ashton, D.H. & Gill, A.M. (1965). Pattern and process in a Macquarie Island feldmark. *Proceedings of the Royal Society of Victoria* **79**: 235–45.

Aston, B.C. (1915). Vegetation of the Tarawera mountains, New Zealand. *Transactions of the New Zealand Institute* **48**: 304–14.

Aston, B.C. (1933). The Napier-Ahuriri lagoon lands. A survey of soils and their plant associations. *New Zealand Journal of Agriculture* **46**: 69–77.

Atkinson, I.A.E. (1981). *Vegetation Map of Tongariro National Park, North Island, New Zealand, Scale 1:50 000.* Department of Scientific and Industrial Research, Wellington.

Atkinson, I.A.E. (1985). Derivation of vegetation mapping units for an ecological survey of Tongariro National Park, North Island, New Zealand. *New Zealand Journal of Botany* **23**: 361–78.

Atkinson, I.A.E. & Greenwood, R.M. (1972). Effects of the 1969–70 drought on two remnants of indigenous lowland forest in the Manawatu district. *Proceedings of the New Zealand Ecological Society* **19**: 34–42.

Atkinson, I.A.E. & Greenwood, R.M. (1989). Relationships between moas and plants. In *Moas, Mammals and Climate in the Ecological History of New Zealand* (ed. M.J. Rudge), pp. 67–96. (*New Zealand Journal of Ecology* **12** (supplement).)

Bagnall, R.G. (1972). The dry weight and calorific value of litter fall in a New Zealand *Nothofagus* forest. *New Zealand Journal of Botany* **10**: 27–36.

Bagnall, R.G. (1975). Vegetation of the raised beaches at Cape Turakirae, Wellington, New Zealand. *New Zealand Journal of Botany* **13**: 367–424.

Bagnall, R.G. & Ogle, C.C. (1981). The changing vegetation structure and composition of a lowland mire at Plimmerton, North Island, New Zealand. *New Zealand Journal of Botany* **19**: 371–87.

Ball, P.R. (1982). Nitrogen balances in intensively managed pasture systems. In *Nitrogen Balances in New Zealand Ecosystems* (ed. P.W. Gandar), pp. 47–66. Plant Physiology Division, Department of Scientific and Industrial Research, Palmerston North.

Bannister, P. (1984a). Winter frost resistance of leaves of some plants from the Three Kings Islands, grown outdoors in Dunedin, New Zealand. *New Zealand Journal of Botany* **22**: 303–6.

Bannister, P. (1984b). The seasonal course of frost resistance in some New Zealand pteridophytes. *New Zealand Journal of Botany* **22**: 557–63.

Bannister, P. (1986a). Observations on water potential and drought resistance of trees and shrubs after a period of summer drought around Dunedin, New Zealand. *New Zealand Journal of Botany* **24**: 387–92.

Bannister, P. (1986b). Winter frost resistance of leaves of some plants growing in Dunedin, New Zealand, in winter 1985. *New Zealand Journal of Botany* **24**: 505–7.

Bannister, P. & Fagan, B. (1989). The frost resistance of fronds of *Blechnum penna-marina* in relation to season, altitude and short-term hardening and dehardening. *New Zealand Journal of Botany* **27**: 471–6.

Bannister, P. & Lee, W.G. (1989). The frost resistance of fruits and leaves of some *Coprosma* species in relation to altitude and habitat. *New Zealand Journal of Botany* **27**: 477–9.

Bannister, P. & Smith, J.M. (1983). The heat resistance of some New Zealand plants. *Flora* **173**: 399–414.

Barker, A.P. (1953). *An Ecological Study of Tussock Grassland, Hunters Hills, South Canterbury.* (DSIR Bulletin 107.) Department of Scientific and Industrial Research. Wellington.

Barton, I.R. (1983). Growth and regeneration of kauri. In *Lowland Forests in New Zealand* (ed. K. Thompson, A.P.H. Hodder & A.S. Irving), pp. 121–34. University of Waikato.

Basher, L.R., Tonkin, P.J. & Daly, G.T. (1985). Pedogenesis, erosion and revegetation in a mountainous, high-rainfall area – Cropp River, central Westland. In: *Proceedings of the Soil Dynamics and Land Use Seminar, Blenheim, May 1985* (ed. I.B. Campbell), pp. 49–64. New Zealand Society of Soil Science and New Zealand Soil Conservators Association.

Bate, G.C. & Smith, V.R. (1983). Photosynthesis and respiration in the subantarctic grass *Poa cookii*. *New Phytologist* **95**: 533–43.

Baxter, W.A. & Norton, D.A. (1989). Forest recovery after logging in lowland dense rimu forest, Westland, New Zealand. *New Zealand Journal of Botany* **27**: 391–9.

Baylis, G.T.S. (1958). A botanical survey of the small islands of the Three Kings Group. *Records of the Auckland Institute and Museum* **5**: 1–12.

Baylis, G.T.S. (1959). An example of winter injury to silver beech at moderate altitude. *Proceedings of the New Zealand Ecological Society* **6**: 21–2.

Baylis, G.T.S. (1967). Experiments on the ecological significance of phycomycetous mycorrhizas. *New Phytologist* **66**: 231–43.

Baylis, G.T.S. (1969). Mycorrhizal nodules and growth of podocarps in nitrogen-poor soil. *Nature* **223**: 1385–6.

Baylis, G.T.S. (1975). The magnolioid mycorrhiza and mycotrophy in root systems derived from it. In *Endomycorrhizas: Proceedings of a Symposium held at the University of Leeds, 22–25 July 1974* (ed. F.E. Sanders, B. Mosse & P.B. Tinker), pp. 373–90. Academic Press, London.

Baylis, G.T.S. (1980). Mycorrhizas and the spread of beech. *New Zealand Journal of Ecology* **3**: 151–3.

Bayly, I.A.E., Edwards, J.S. & Chambers, T.C. (1956). The crater lakes of Mayor Island. *Tane* **7**: 36–46.

Beets, P.N. (1980). Amount and distribution of dry matter in a mature beech/podocarp community. *New Zealand Journal of Forestry Science* **10**: 395–418.

Beever, R.E. & Beever, J. (1983). Frost damage to native plants 1982. *Auckland Botanical Society Newsletter* **38**: 1–3.

Bell, C.J.E. (1973). Mountain soils and vegetation in Owen Range, Nelson. 2. The vegetation. *New Zealand Journal of Botany* **11**: 73–102.

Benecke, U. & Evans, G.C. (1987). Growth and water use in *Nothofagus truncata* (hard beech) in temperate hill country, Nelson, New Zealand. In *Temperate Forest Ecosystems* (ITE Symposium No. 20) (ed. Yang Hanxi, Wang Zhan, J.N.R. Jeffers & P.A. Ward), pp. 131–40. Institute of Terrestrial Ecology, Grange over Sands, Cumbria.

Benecke, U. & Havranek, W.M. (1980). Phenological growth characteristics of trees with increasing altitude, Craigieburn Range, New Zealand. In *Mountain Environments and Subalpine Tree Growth* (Proceedings of IUFRO Workshop, November 1979, Christchurch, New Zealand) (ed. U. Benecke & M.R. Davis), pp. 155–74. New Zealand Forest Service, Wellington.

Benecke, U. & Nordmeyer, A.H. (1982). Carbon uptake and allocation by *Nothofagus solandri* var. *cliffortioides* (Hook. f.) Poole and *Pinus contorta* Douglas ex Loudon ssp. *contorta* at montane and subalpine altitudes. In *Carbon Uptake and Allocation in Subalpine Ecosystems as a Key to Management* (Proceedings of an IUFRO Workshop) (ed. R.H. Waring), pp. 9–21. Forest Research Laboratory, Oregon State University.

Best, H.A. (1984). The foods of kakapo on Stewart Island as determined from their feeding sign. *New Zealand Journal of Ecology* **7**: 71–83.

Beuzenberg, E.J. & Hair, J.B. (1984). Contributions to a chromosome atlas of the New Zealand flora – 27. Compositae. *New Zealand Journal of Botany* **22**: 353–6.

Beveridge, A.E. (1964). Dispersal and destruction of seed in central North Island podocarp forests. *Proceedings of the New Zealand Ecological Society* **11**: 48–55.

Beveridge, A.E. (1973). Regeneration of podocarps in a central North Island forest. *New Zealand Journal of Forestry* **18**: 23–35.

Bieleski, R.L. (1959). Factors affecting growth and distribution of kauri (*Agathis australis* Salisb.). I. Effect of light on the establishment of kauri and of *Phyllocladus trichomanoides* D. Don. II. Effect of light on seedling growth. III. Effect of temperature and soil conditions. *Australian Journal of Botany* **7**: 252–94.

Birrell, K.S. & Fieldes, M. (1968). *Soils of New Zealand*, part 2. (Soil Bureau Bulletin 26.) Department of Scientific and Industrial Research, Wellington.

Bliss, L.C. & Mark, A.F. (1974). High-alpine environments and primary production on the Rock and Pillar Range, Central Otago, New Zealand. *New Zealand Journal of Botany* **12**: 445–83.

Boyce, W.R. & Newhook, F.J. (1953). Investigations into yellow-leaf disease of *Phormium*. *New Zealand Journal of Science and Technology* **34A** (Supplement 1), 1–11.

Bray, J.R. (1989). The use of historical vegetation dynamics in interpreting prehistorical vegetation change. *Journal of the Royal Society of New Zealand* **19**: 151–60.

Brockie, R.E. (1986). Periodic flowering of New Zealand flax (*Phormium*, Agavaceae). *New Zealand Journal of Botany* **24**: 381–6.

Brockie, R.E. & Moeed, A. (1986). Animal biomass in a New Zealand forest compared with other parts of the world. *Oecologia* **70**: 24–34.

Brown, J.M.A. (1975). Ecology of macrophytes. In *New Zealand Lakes* (ed. V.H. Jolly & J.M.A. Brown), pp. 244–62. Auckland University Press and Oxford University Press.

Brownsey, P.J., Given, D.R. & Lovis, J.D. (1985). A revised classification of New Zealand pteridophytes with a synonymic checklist of species. *New Zealand Journal of Botany* **23**: 431–89.

Brumley, C.F., Stirling, M.W. & Manning, M.S. (1986). *Old Man Ecological District*. (New Zealand Protected Natural Areas Programme 3.) Department of Lands and Survey, Wellington.

Bulloch, B.T. (1973). A low altitude snow tussock reserve at Black Rock, Eastern Otago. *Proceedings of the New Zealand Ecological Society* **20**: 41–7.

Burke, W.D. (1974). Regeneration of podocarps on Mt Tarawera, Rotorua. *New Zealand Journal of Botany* **12**: 219–26.

Burns, B.R. & Ogden, J. (1985). The demography of the temperate mangrove [*Avicennia marina* (Forst.) Vierh.] at its southern limit in New Zealand. *Australian Journal of Ecology* **10**: 125–33.

Burrell, J. (1965). Ecology of *Leptospermum* in Otago. *New Zealand Journal of Botany* **3**: 3–16.

Burrows, C.J. (1963). The root habit of some New Zealand plants. *Tuatara* **11**: 78–80.

Burrows, C.J. (1972). The flora and vegetation of Open Bay Islands. *Journal of the Royal Society of New Zealand (Botany)* **2**: 15–42.

Burrows, C.J. (1977). Alpine grasslands and snow in the Arthur's Pass and Lewis Pass regions, South Island, New Zealand. *New Zealand Journal of Botany* **15**: 665–86.

Burrows, C.J. (1980). Some empirical information on the diet of moas. *New Zealand Journal of Ecology* **3**: 125–30.

Burrows, C.J. & Dobson, A.T. (1972). Mires of the Manapouri-Te Anau lowlands. *Proceedings of the New Zealand Ecological Society* **19**: 75–99.

Burrows, C.J. & Heine, M. (1979). The older moraines of the Stocking Glacier, Mount Cook region. *Journal of the Royal Society of New Zealand* **9**: 5–12.

Burstall, S.W. & Sale, E.V. (1984). *Great Trees of New Zealand*. A.H. & A.W. Reed, Wellington.

Calder, D.M. (1961). Plant ecology of subalpine shingle river-beds in Canterbury, New Zealand. *Journal of Ecology* **49**: 581–94.

Calder, J.W. & Wardle, P. (1969). Succession in subalpine vegetation at Arthur's Pass, New Zealand. *Proceedings of the New Zealand Ecological Society* **16**: 36–47.

Cameron, E.K., Baylis, G.T.S. & Wright, A.E. (1987). Vegetation quadrats 1982–83 and broad regeneration patterns on Great Island, Three Kings Islands, northern New Zealand. *Records of the Auckland Institute and Museum* **24**: 163–85.

Cameron, R.J. (1960). Natural regeneration of podocarps in the forests of the Whirinaki River Valley. *New Zealand Journal of Forestry* **8**: 337–54.

Cameron, R.J. (1963). A study of the rooting habits of rimu and tawa in pumice soils. *New Zealand Journal of Forestry* **8**: 771–85.

Campbell, A.D. (1981). Flowering records for *Chionochloa*, *Aciphylla*, and *Celmisia* species in the Craigieburn Range, South Island, New Zealand. *New Zealand Journal of Botany* **19**: 97–103.

Campbell, D.J. (1978). The effects of rats on vegetation. In *The Ecology and Control of Rodents in New Zealand Nature Reserves* (ed. P.R. Dingwall, I.A.E. Atkinson & C. Hay), pp. 99–120. Department of Lands and Survey, Wellington.

Campbell, D.J. (1984). The vascular flora of the DSIR study area, lower Orongorongo Valley, Wellington, New Zealand. *New Zealand Journal of Botany* **22**: 223–70.

Campbell, E.O. (1962). The mycorrhiza of *Gastrodia cunninghamii* Hook.f. *Transactions of the Royal Society of New Zealand (Botany)* **1**: 289–96.

Campbell, E.O. (1963). *Gastrodia minor*, an epiparasite of manuka. *Transactions of the Royal Society of New Zealand (Botany)* **2**: 73–81.

Campbell, E.O. (1964). The restiad peat bogs at Motumaoho and Moanatuatua. *Transactions of the Royal Society of New Zealand (Botany)* **2**: 219–27.

Campbell, I.B. & Mew, G. (1986). Soils under beech forest in an experimental catchment area near Nelson, New Zealand. *Journal of the Royal Society of New Zealand* **16**: 193–223.

Challies, C.N. (1977). Effect of commercial hunting on red deer densities in the Arawata valley, South Westland, 1972–76. *New Zealand Journal of Forestry Science* **7**: 263–73.

Challis, G.A. (1971). *Chemical Analyses of New Zealand Rocks and Minerals with C.I.P.W. Norms and Petrographic Descriptions 1917–57. Part 1: Igneous and Pyroclastic Rocks.* (Geological Survey Bulletin 84.) Department of Scientific and Industrial Research, Wellington.

Chambers, T.C. (1958). Pattern and process in a New Zealand subalpine plant community. *Vegetatio* **8**: 209–14.

Chapman, V.J. & Ronaldson, J.W. (1958). *The Mangrove and Salt-marsh Flats of the Auckland Isthmus, New Zealand.* DSIR Bulletin 125. Department of Scientific and Industrial Research, Wellington.

Clarke, C.M.H. (1976). Eruption, deterioration and decline of the Nelson red deer herd. *New Zealand Journal of Forestry Science* **5**: 235–49.

Clarkson, B.D. (1985). The vegetation of the Kaitake Range, Egmont National Park. *New Zealand Journal of Botany* **23**: 15–31.

Clarkson, B.D. (1986). *Vegetation of Mount Egmont National Park, New Zealand.* (National Parks Scientific Series 5.) National Parks Authority, Wellington.

Clarkson, B.D., Clarkson, B.R. & McGlone, M.S. (1986). Vegetation history of some west Taupo mires. In *Ecological Research in the Central North Island Volcanic Plateau Region* (ed. B. Veale and J. Innes), pp. 34–7. Forest Research Institute, New Zealand Forest Service, Rotorua.

Clarkson, B.R. (1984). Vegetation of three mountain mires, west Taupo, New Zealand. *New Zealand Journal of Botany* **22**: 361–75.

Clarkson, B.R. & Clarkson, B.D. (1983). Mt Tarawera: 2. Rates of change in the vegetation and flora of the high domes. *New Zealand Journal of Ecology* **6**: 107–19.

Clarkson, B.R., Patel, R.N. & Clarkson, B.D. (1988). Composition and structure of forest overwhelmed at Pureora, central North Island, New Zealand, during the Taupo eruption (c. A.D. 130). *Journal of the Royal Society of New Zealand* **18**: 417–36.

Clayton-Greene, K.A. (1977). Structure and origin of *Libocedrus bidwillii* stands in Waikato district. *New Zealand Journal of Botany* **15**: 19–28.

Clayton-Green, K.A. & Wilson, J.B. (1985). The vegetation of Mt Karioi, North Island, New Zealand. *New Zealand Journal of Botany* **23**: 533–48.

Clout, M.N. & Hay, J.R. (1989). The importance of birds as browsers, pollinators and seed dispersers in New Zealand forests. In *Moas, Mammals and Climate in the Ecological History of New Zealand* (ed. M.R. Rudge), pp. 27–33. (*New Zealand Journal of Ecology* **12** (Supplement).)

Cockayne, L. (1909). The ecological botany of the subantarctic islands of New Zealand. In *The Subantarctic Islands of New Zealand*, vol. 1(ed. C. Chilton), pp. 182–235. Government Printer, Wellington.

Cockayne, L. (1928). *The Vegetation of New Zealand.* 2nd edition. Engelmann Press, Leipzig.

Cockayne, L. & Calder, J.W. (1932). The present vegetation of Arthur's Pass (New Zealand) as compared with that of thirty-four years ago. *Journal of Ecology* **20**: 270-83.

Coleman, J.D., Gillman, A. & Green, W.Q. (1980). Forest patterns and possum densities within podocarp/mixed hardwood forests on Mt Bryan O'Lynn, Westland. *New Zealand Journal of Ecology* **3**: 69–84.

Coleman, J.D., Green, W.Q. & Polson, J.G. (1985). Diet of brushtail possums over a pasture-alpine gradient in Westland, New Zealand. *New Zealand Journal of Ecology* **8**: 21–35.

Connor, H.E. (1964). Tussock grassland communities in the Mackenzie country, South Canterbury, New Zealand. *New Zealand Journal of Botany* **2**: 325–51.

Connor, H.E. (1965). Tussock grasslands in the middle Rakaia valley, Canterbury, New Zealand. *New Zealand Journal of Botany* **3**: 261–76.

Connor, H.E. (1966). Breeding systems in New Zealand grasses VII. Periodic flowering of snow tussock, *Chionochloa rigida*. *New Zealand Journal of Botany* **4**: 392–7.

Connor, H.E. (1967). Interspecific hybrids in *Chionochloa* (Gramineae). *New Zealand Journal of Botany* **5**: 3–16.

Connor, H.E. & Edgar, E. (1987). Name changes in the indigenous New Zealand flora, 1960–1986 and Nomina Nova IV, 1983–1986. *New Zealand Journal of Botany* **25**: 115–70.

Connor, H.E. & MacRae, A.H. (1969). Montane and subalpine tussock grasslands in Canterbury. In *The Natural History of Canterbury* (ed. G.A. Knox), pp. 167–204. A.H. & A.W. Reed, Wellington.

Cook, J.M., Mark, A.F. & Shore, B.F. (1980). Responses of *Leptospermum scoparium* and *L. ericoides* (Myrtaceae) to waterlogging. *New Zealand Journal of Botany* **18**: 233–46.

Cooper, K.M. (1976). A field study of mycorrhizas in New Zealand ferns. *New Zealand Journal of Botany* **14**: 169–81.

Court, A.J. & Mitchell, N.D. (1988). The germination ecology of *Dysoxylum spectabile* (Meliaceae). *New Zealand Journal of Botany* **26**: 1–6.

Court, D.J. (1978). Forest regeneration on Hen Island. *Tane* **24**: 103–18.

Court, D.J. (1981). The spread of tea-tree (*Leptospermum* spp.) scrub on Motukawanui Island, Cavalli Group, northern New Zealand. *Tane* **27**: 135–51.

Craig, J.L. & Stewart, A.M. (1988). Reproductive biology of *Phormium tenax*: a honeyeater-pollinated species. *New Zealand Journal of Botany* **26**: 453–63.

Cranwell, L.M. & Moore, L.B. (1931). The vegetation of Maungapohatu. *Records of the Auckland Institute and Museum* **1**: 71–80.

Cranwell, L.M. & Moore, L.B. (1936). The occurrence of kauri in montane forest on Te Moehau. *New Zealand Journal of Science and Technology* **18**: 531–43.

Croker, B.H. (1953). Forest regeneration of the western Hutt Hills, Wellington. *Transactions of the Royal Society of New Zealand* **81**: 11–21.

Croker, B.H. (1955). Comments on the shingle vegetation of the Horokiwi stream. *Transactions of the Royal Society of New Zealand* **83**: 333–43.

Crush, J.R. (1973). Significance of endomycorrhizas in tussock grassland in Otago, New Zealand. *New Zealand Journal of Botany* **11**: 645–60.

Cunningham, A. (1979). A century of change in the forests of the Ruahine Range, 1870–1970. *New Zealand Journal of Ecology* **2**: 11–21.

Daly, G.T. (1964). Leaf-surface wax in *Poa colensoi*. *Journal of Experimental Botany* **15**: 160–5.

Daly, G.T. (1969). The biology of matagouri. *Proceedings of the New Zealand Weed and Pest Control Conference* **22**: 195–200.

Daniel, M.J. & Adams, J.A. (1984). Nutrient return by litterfall in evergreen podocarp-hardwood forest in New Zealand. *New Zealand Journal of Botany* **22**: 271–83.

Dawson, J.W. (1954). Trio and Stephens Islands: home of the Tuatara. *Bulletin of the Wellington Botanical Society* **27**: 2–7.

Dawson, J.W. (1988). *Forest Vines to Snow Tussocks: the Story of New Zealand Plants*. Victoria University Press.

Dawson, J.W. & Sneddon, B.V. (1969). The New Zealand rainforest: a comparison with tropical rainforest. *Pacific Science* **23**: 131–47.

Delph, L.F. & Lively, C.M. (1989). The evolution of floral color change: pollination attraction versus physiological restraints in *Fuchsia excorticata*. *Evolution* **43**: 1252–62.

Department Of Lands and Survey (1982). *Franz Josef – Mount Cook Vegetation*. New Zealand Land Inventory 1:100000, NZMS 290 Sheet H34/35/36. Department of Lands and Survey, Wellington.

Dobson, A.T. (1975). Vegetation of a Canterbury subalpine mire complex. *Proceedings of the New Zealand Ecological Society* **22**: 67–75.

Dobson, A.T. (1977). Mire vegetation. In *Cass: History and Science in the Cass District, Canterbury, New Zealand* (ed. C.J. Burrows), pp. 259–70. University of Canterbury.

Dobson, A.T. (1979). Vegetation. In *Ecology of Kowhai Bush, Kaikoura* (ed. D.M. Hunt & B.J. Gill), pp. 11–15. *Mauri Ora* (Special Publication 2).

Dodson, J.R., Enright, N.J. & McLean, R.F. (1988). A late Quaternary vegetation history for far northern New Zealand. *Journal of Biogeography* **15**: 647–56.

Dromgoole, F.I. (1988). Carbon dioxide fixation in aerial roots of the New Zealand mangrove *Avicennia marina* var. *resinifera* (Note). *New Zealand Journal of Marine and Freshwater Research* **22**: 617-19.

Druce, A.P. (1957a). The vegetation of Mt Kaiparoro. *Bulletin of the Wellington Botanical Society* **29**: 7-13.

Druce, A.P. (1957b). *Botanical Survey of an Experimental Catchment, Taita, New Zealand.* (DSIR Bulletin 124.) Department of Scientific and Industrial Research, Wellington.

Druce, A.P. (1966). Tree-ring dating of recent volcanic ash and lapilli, Mt Egmont. *New Zealand Journal of Botany* **4**: 3-41.

Druce, A.P. (1971). The flora of the Aorangi Range, southern Wairarapa with notes on the vegetation. *Bulletin of the Wellington Botanical Society* **37**: 4-29.

Druce, A.P. (1989). *Coprosma waima* (Rubiaceae) – a new species from northern New Zealand. *New Zealand Journal of Botany* **27**: 119-28.

Druce, A.P. & Atkinson, I.A.E. (1958). Forest variation in the Hutt Catchment. *Proceedings of the New Zealand Ecological Society* **6**: 41-5.

Druce, A.P., Bartlett, J.K. & Gardner, R.O. (1979). Indigenous vascular plants of the serpentine area of Surville Cliffs and adjacent cliff tops. *Tane* **25**: 187-206.

Druce, A.P. & Williams, P.A. (1989). Vegetation and flora of the Ben More-Chalk Range area of southern Marlborough, South Island. *New Zealand Journal of Botany* **27**: 167-99.

Druce, A.P., Williams, P.A. & Heine, J.C. (1987). Vegetation and flora of Tertiary calcareous rocks in the mountains of western Nelson, New Zealand. *New Zealand Journal of Botany* **25**: 41-78.

Dunwiddie, P.W. (1979). Dendrochronological studies of indigenous New Zealand trees. *New Zealand Journal of Botany* **17**: 251-66.

Eagle, A. (1982). *Eagle's Trees and Shrubs of New Zealand.* Second series. Collins, Auckland.

Ecroyd, C.E. (1982). Biological flora of New Zealand 8. *Agathis australis* (D. Don) Lindl. (Araucariaceae). Kauri. *New Zealand Journal of Botany* **20**: 17-36.

Egunjobi, J.K. (1969). Dry matter and nitrogen accumulation in secondary successions involving gorse (*Ulex europaeus* L.) and associated shrubs and trees. *New Zealand Journal of Science* **12**: 175-93.

Elder, N.L. (1962). Vegetation of the Kaimanawa Ranges. *Transactions of the Royal Society of New Zealand (Botany)* **2**: 1-37.

Elder, N.L. (1965). Vegetation of the Ruahine Range: an introduction. *Transactions of the Royal Society of New Zealand (Botany)* **3**: 13-66.

Enright, N.J. (1985). Age, reproduction and biomass allocation in *Rhopalostylis sapida* (nikau palm). *Australian Journal of Ecology* **10**: 461-7.

Enright, N.J. & Cameron, E.K. (1988). The soil seed bank of a kauri (*Agathis australis*) forest remnant near Auckland, New Zealand. *New Zealand Journal of Botany* **26**: 223-36.

Esler, A.E. (1962). Botanical features of Abel Tasman National Park. *Transactions of the Royal Society of New Zealand (Botany)* **1**: 297-311.

Esler, A.E. (1967). The vegetation of Kapiti Island. *New Zealand Journal of Botany* **5**: 353-93.

Esler, A.E. (1974). Vegetation of the sand country bordering the Waitakere Range, Auckland: the southern beaches. *New Zealand Journal of Ecology* **21**: 72-7.

Esler, A.E. (1975). Vegetation of the sand country bordering the Waitakere Range, Auckland: Piha Beach. *Proceedings of the New Zealand Ecological Society* **22**: 52-6.

Esler, A.E. (1978a). *Botany of the Manawatu District.* (DSIR Information Series 127.) Department of Scientific and Industrial Research, Wellington.

Esler, A.E. (1978b). Botanical features of Tiritiri Island, Hauraki Gulf, New Zealand. *New Zealand Journal of Botany* **16**: 207-26.

Esler, A.E. (1980). Botanical features of Motutapu, Motuihe, and Motukorea, Hauraki Gulf, New Zealand. *New Zealand Journal of Botany* **18**: 15-36.

Esler, A.E. (1988a). The naturalisation of plants in urban Auckland, New Zealand. 4. The nature of the naturalised species. *New Zealand Journal of Botany* **26**: 345-85.

Esler, A.E. (1988b). The naturalisation of plants in urban Auckland, New Zealand. 6. Alien plants as weeds. *New Zealand Journal of Botany* **26**: 585-618.

Esler, A.E. & Astridge, S.J. (1974). Tea tree (*Leptospermum*) communities of the Waitakere Range, Auckland, New Zealand. *New Zealand Journal of Botany* **12**: 485–501.

Esler, A.E. & Rumball, P.J. (1975). Gumland vegetation at Kaikohe, Northland, New Zealand. *New Zealand Journal of Botany* **13**: 425–36.

Evans, G.R. (1980). Phytomass, litter and nutrients in montane and alpine grasslands, Craigieburn Range, New Zealand. In *Mountain Environments and Subalpine Tree Growth*. (Proceedings of IUFRO Workshop November 1979. Christchurch, New Zealand) (ed. U. Benecke & M.R. Davis), pp. 95–110. New Zealand Forest Service, Wellington.

Evans, G.R. & Kelland, C.M. (1982). Nitrogen balance studies in tussock grassland. In *Nitrogen Balances in New Zealand Ecosystems* (ed. P.W. Gandar), pp. 41–6. Plant Physiology Division, Department of Scientific and Industrial Research, Palmerston North.

Evans, L.T. (1953). The ecology of the halophytic vegetation at Lake Ellesmere, New Zealand. *Journal of Ecology* **41**: 106–22.

Fineran, B.A. (1962). Studies on the root parasitism of *Exocarpus bidwillii* Hook.f. I. Ecology and root structure of the parasite. *Phytomorphology* **12**: 339–55.

Fineran, B.A. (1964). An outline of the vegetation of the Snares Islands. *Transactions of the Royal Society of New Zealand (Botany)* **2**: 229–36.

Fineran, B.A. (1966). The vegetation and flora of Bird Island, Foveaux Strait. *New Zealand Journal of Botany* **4**: 133–46.

Fineran, B.A. (1969). The flora of the Snares Islands, New Zealand. *Transactions of the Royal Society of New Zealand (Botany)* **3**: 237–70.

Fisher, F.J.F. (1952). Observations on the vegetation of screes in Canterbury, New Zealand. *Journal of Ecology* **40**: 156–67.

Foggo, M.N. & Meurk, C.D. (1981). Notes on a visit to Jacquemart Island in the Campbell Island group. *New Zealand Journal of Ecology* **4**: 29–32.

Foggo, M.N. & Meurk, C.D. (1983). A bioassay of some Campbell Island soils. *New Zealand Journal of Ecology* **6**: 121–4.

Foweraker, C.E. (1929). The podocarp rain forests of Westland, New Zealand. 2. Kahikatea and totara forests, and their relationship to silting. *Te Kura Ngahere* **2**: 6–12.

Franklin, D.A. (1968). Biological flora of New Zealand 3. *Dacrydium cupressinum* Lamb. (Podocarpaceae). Rimu. *New Zealand Journal of Botany* **6**: 493–513.

Franklin, D.A. (1973). Growth rates in South Westland terrace rimu forest. I. *New Zealand Journal of Forestry Science* **3**: 304–12.

Gadgil, P.D. (1974). *Phytophthora heveae*; a pathogen of kauri. *New Zealand Journal of Forestry Science* **4**: 59–63.

Gadgil, R.L. (1982). Biological nitrogen inputs to exotic forests. In *Nitrogen Balances in New Zealand Ecosystems* (ed. P.W. Gandar), pp. 123–30. Plant Physiology Division, Department of Scientific and Industrial Research, Palmerston North.

Galloway, D.J. (1985). *Flora of New Zealand Lichens*. Government Printer, Wellington.

Garnock-Jones, P.J. & Johnson, P.N. (1987). *Iti lacustris* (Brassicaceae), a new genus and species from southern New Zealand. *New Zealand Journal of Botany* **25**: 603–10.

Gellatly, A.F. (1984). The use of rock weathering-rind thickness to redate moraines in the Mount Cook National Park, New Zealand. *Arctic and Alpine Research* **16**: 225–32.

Gellatly, A.F., Rothlisberger, F. & Geyr, M.A. (1985). Holocene glacier variations in New Zealand (South Island). *Zeitschrift für Gletscherkunde und Glazialgeologie* **21**: 265–73.

Gibbs, G.W. (1973). Cycles of macrophytes and phytoplankton in Pukepuke Lagoon following a severe drought. *Proceedings of the New Zealand Ecological Society* **20**: 13–20.

Gibson, N. & Kirkpatrick, J.B. (1985). A comparison of the cushion plant communities of New Zealand and Tasmania. *New Zealand Journal of Botany* **23**: 549–66.

Gillham, M.E. (1960a). Vegetation of New Zealand shag colonies. *Transactions of the Royal Society of New Zealand* **88**: 363–80.

Gillham, M.E. (1960b). Vegetation of Little Brother Island, Cook Strait, in relation to spray-bearing winds, soil salinity and pH. *Transactions of the Royal Society of New Zealand* **88**: 405–24.

Gilmour, J.W. (1966). *The Pathology of Forest Trees in New Zealand. The Fungal, Bacterial and Algal Pathogens.* (Technical Paper 48.) New Zealand Forest Service, Wellington.

Given, D.R. (1971). Montane-subalpine vegetation near Lake Shirley, Fiordland. *New Zealand Journal of Botany* 9: 3–26.

Given, D.R. (1980). Vegetation on heated soils at Karapiti, central North Island, New Zealand and its relation to ground temperature. *New Zealand Journal of Botany* 18: 1–13.

Godley, E.J. (1965). Notes on the vegetation of the Auckland Islands. *Proceedings of the New Zealand Ecological Society* 12: 57–63.

Godley, E.J. (1968). Transoceanic dispersal in *Sophora* and other genera. *Nature* 218: 495–6.

Godley, E.J. (1979). Flower biology in New Zealand. *New Zealand Journal of Botany* 17: 441–66.

Godley, E.J. (1982). Breeding systems in New Zealand plants 6. *Gentiana antarctica* and *G. antipoda*. *New Zealand Journal of Botany* 20: 405–20.

Godley, E.J. (1985). Paths to maturity. *New Zealand Journal of Botany* 23: 687–706.

Godley, E.J. (1989). The flora of Antipodes Island. *New Zealand Journal of Botany* 27: 531–63.

Goh, K.M., Molloy, B.P.J. & Rafter, T.A. (1977). Radiocarbon dating of Quaternary loess deposits, Banks Peninsula, Canterbury, New Zealand. *Quaternary Research* 7: 177–96.

Grant, D.A. (1967). Factors affecting the establishment of manuka *(Leptospermum scoparium)*. *Proceedings of the New Zealand Weed and Pest Control Conference* 20: 129–34.

Grant, P.J. (1984). Drought effect on high-altitude forests, Ruahine Range, North Island, New Zealand. *New Zealand Journal of Botany* 22: 15–27.

Green, T.G.A. (1982). Biological nitrogen inputs and turnover in native forests. In *Nitrogen Balances in New Zealand Ecosystems* (ed. P.W. Gandar), pp. 151–6. Plant Physiology Division, Department of Scientific and Industrial Research, Palmerston North.

Green, T.G.A., Horstmann, J., Bonnett, H., Wilkins, A. & Silvester, W.B. (1980). Nitrogen fixation by members of the Stictaceae (Lichenes) of New Zealand. *New Phytologist* 84: 339–48.

Greenwood, R.M. (1978). Rhizobia associated with indigenous legumes of New Zealand and Lord Howe Island. In *Microbial Ecology* (ed. M.W. Loutit & J.A.R. Miles), pp. 402–3. Springer-Verlag, New York.

Greer, D.H. (1984). Seasonal changes in photosynthetic activity of snow tussocks (*Chionochloa* spp.) along an altitudinal gradient in Otago, New Zealand. *Oecologia* 63: 271–4.

Guthrie-Smith, W.H. (1953). *Tutira; the Story of a New Zealand Sheep Station.* 3rd Edition. London, Blackwood.

Haase, P. (1986*a*). Flowering records of some subalpine trees and shrubs at Arthur's Pass, New Zealand. *New Zealand Journal of Ecology* 9: 19–23.

Haase, P. (1986*b*). A study of a *Libocedrus bidwillii* population at Pegleg Flat, Arthur's Pass, New Zealand. *New Zealand Journal of Ecology* 9: 153–6.

Haase, P. (1986*c*). An ecological study of the subalpine tree *Dracophyllum traversii* (Epacridaceae) at Arthur's Pass, South Island, New Zealand. *New Zealand Journal of Botany* 24: 69–78.

Haase, P. (1986*d*). Continuation of wood increment in *Olearia ilicifolia* during the winter of 1984. *New Zealand Journal of Botany* 24: 179–82.

Haase, P. (1986*e*). An ecological study of the subalpine shrub *Senecio bennettii* (Compositae) at Arthur's Pass, South Island, New Zealand. *New Zealand Journal of Botany* 24: 247–62.

Haase, P. (1986*f*). Phenology and productivity of *Olearia ilicifolia* (Compositae) at Arthur's Pass, South Island, New Zealand. *New Zealand Journal of Botany* 24: 369–79.

Haase, P. (1987). Ecological studies on *Hoheria glabrata* (Malvaceae) at Arthur's Pass, South Island, New Zealand. *New Zealand Journal of Botany* 25: 401–9.

Haase, P. (1989). Ecology and distribution of *Nothofagus* in Deception Valley, Arthur's Pass National Park, New Zealand. *New Zealand Journal of Botany* 27: 59–70.

Hall, I.R. (1975). Endomycorrhizas of *Metrosideros umbellata* and *Weinmannia racemosa*. *New Zealand Journal of Botany* 13: 463–72.

Hamilton, W.M. & Atkinson, I.A.E. (1961). Plant associations. In *Little Barrier Island (Hauturu)* (DSIR Bulletin 137, 2nd edition) (compiled by W.M. Hamilton), pp. 87–121. Department of Scientific and Industrial Research, Wellington.

Harris, W. (1970a). The distribution of ryegrass cultivars and other herbage species in respect to environmental heterogeneity within fields. *New Zealand Journal of Agricultural Research* **13**: 862–8.

Harris, W. (1970b). Yield and habit of New Zealand populations of *Rumex acetosella* at three altitudes in Canterbury. *New Zealand Journal of Botany* **8**: 114–31.

Hawkins, B.J. & Sweet, G.B. (1989). Evolutionary interpretation of a high temperature growth response in five New Zealand forest trees. *New Zealand Journal of Botany* **27**: 101–7.

Hayward, B.W. & Hayward, G.C. (1974). Botany of the Shoe and Slipper Island Group – Coromandel Peninsula, Part III: Lichens. *Tane* **20**: 72–85.

Healy, A.J. (1959). Contributions to a knowledge of the adventive flora of New Zealand. 8. The succulent element. *Transactions of the Royal Society of New Zealand* **87**: 229–34.

Healy, A.J. (1961). The interactions of native and adventive plant species. *Proceedings of the New Zealand Ecological Society* **9**: 39–43.

Healy, A.J. (1968). The adventive flora in Canterbury. In *The Natural History of Canterbury* (ed. G.A. Knox), pp. 261–333, A.H. & A.W. Reed, Wellington.

Herbert, J. (1973). Growth of silver beech in northern Fiordland. *New Zealand Journal of Forestry Science* **3**: 137–51.

Herbert, J. (1980). Structure and growth of dense podocarp forest at Tihoi and impact of selective logging. *New Zealand Journal of Forestry* **25**: 44–57.

Hill, R.D. (1963). The vegetation of the Wairarapa in mid-nineteenth century. *Tuatara* **11**: 83–9.

Hinds, H.V. & Reid, J.S. (1957). *Forest Trees and Timbers of New Zealand*. New Zealand Forest Service Bulletin 12.

Hoglund, J.H. & Brock, J.L. (1982). Biological nitrogen inputs in pastures. In *Nitrogen Balances in New Zealand Ecosystems* (ed. P.W. Gandar), pp. 67–78. Plant Physiology Division, Department of Scientific and Industrial Research, Palmerston North.

Holdaway, R.N. (1989). New Zealand's pre-human avifauna and its vulnerability. In *Moas, Mammals and Climate in the Ecological History of New Zealand* (ed. M.R. Rudge), pp. 11–25. (*New Zealand Journal of Ecology* **12** (supplement).)

Hollinger, D.Y. (1987). Photosynthesis and stomatal conductance patterns of two ferns from different forest understoreys. *Journal of Ecology* **75**: 925–35.

Holloway, J.T. (1954). Forests and climates in the South Island of New Zealand. *Transactions of the Royal Society of New Zealand* **82**: 329–410.

Holloway, J.T. (1957). Charles Douglas – Observer extraordinary. *New Zealand Journal of Forestry* **7**: 35–40.

Hosking, G.P. & Kershaw, D.J. (1985). Red beech death in Maruia Valley, South Island, New Zealand. *New Zealand Journal of Botany* **23**: 201–11.

Hosking, G.P. & Hutcheson, J.A. (1986). Hard beech (*Nothofagus truncata*) decline on the Mamaku Plateau, North Island, New Zealand. *New Zealand Journal of Botany* **24**: 263–9.

Hosking, G.P. & Hutcheson, J.A. (1988). Mountain beech (*Nothofagus solandri* var. *cliffortioides*) decline in the Kaweka Range, North Island, New Zealand. *New Zealand Journal of Botany* **26**: 393–400.

Hoy, J.M. (1958). Coccids associated with rata and kamahi. *New Zealand Journal of Science* **1**: 179–200.

Hoy, J.M. (1961). Eriococcus orariensis Hoy and Other Coccoidea (Homoptera) associated with Leptospermum *Forst. Species in New Zealand*. (DSIR Bulletin 141.) Department of Scientific and Industrial Research, Wellington.

Hubbard, J.C.E. & Wilson, J.B. (1988). A survey of the lowland vegetation of the Upper Clutha district of Otago, New Zealand. *New Zealand Journal of Botany* **26**: 21–35.

Huiskes, A.H.L. (1979). Biological flora of the British Isles. *Ammophila arenaria* (L.) Link. *Journal of Ecology* **67**: 363–82.

Hutchinson, F.E. (1932). The life history of the Westland rimu stands. *Te Kura Ngahere* **3**: 54–61.

Irving, R., Skinner, M. & Thompson, K. (1984). *Kopuatai Peat Dome*. (Crown Land Series 12.) University of Waikato and Department of Lands and Survey, Hamilton.

Ives, D. (1973). Nature and distribution of loess in Canterbury, New Zealand. *New Zealand Journal of Geology and Geophysics* **16**: 587–610.

Ives, D., Webb, T.H., Jarman, S.M. & Wardle, P. (1972). The nature and origin of 'wind-throw podzols' under beech forest in the Lower Craigieburn Range, Canterbury. *New Zealand Soil News* **20**: 161–77.

James, I.L. & Wallis, F.P. (1969). A comparative study of the effects of introduced mammals on *Nothofagus* forest at Lake Waikareiti. *Proceedings of the New Zealand Ecological Society* **16**: 1–6.

Jane, G.T. (1983). The impact of introduced herbivores on lowland forests of the North Island. In *Lowland Forests of New Zealand* (ed. K. Thompson, A.P.H. Hodder & A.S. Irving), pp. 135–52. University of Waikato, Hamilton.

Jane, G.T. (1986). Wind damage as an ecological process in mountain beech forests of Canterbury, New Zealand. *New Zealand Journal of Ecology* **9**: 25–39.

Jane, G.T. & Green, T.G.A. (1983). Episodic forest mortality in the Kaimai Ranges, North Island, New Zealand. *New Zealand Journal of Botany* **21**: 21–31.

Jane, G.T. & Green, T.G.A. (1985). Patterns of stomatal conductance in evergreen tree species from a New Zealand cloud forest. *Botanical Gazette* **146**: 413–20.

Jane, G.T. & Green, T.G.A. (1986). Etiology of forest dieback areas within the Kaimai Range, North Island, New Zealand. *New Zealand Journal of Botany* **24**: 513–27.

Jenkin, J.F. (1975). Macquarie Island. In *Structure and Function of Tundra Ecosystems* (Ecological Bulletin 20) (ed. T. Rosswell & O.W. Heal), pp. 375–97. Swedish Natural Sciences Research Council, Stockholm.

Jenkin, J.F. & Ashton, D.H. (1979). Pattern in *Pleurophyllum* herbfields on Macquarie Island (Subantarctic). *Australian Journal of Ecology* **4**: 47–66.

Johnson, P.N. (1972). Applied ecological studies of shoreline vegetation at Lakes Manapouri and Te Anau, Fiordland. Part 2: The lake edge flora habitats and relation to lake levels. *Proceedings of the New Zealand Ecological Society* **19**: 120–42.

Johnson, P.N. (1975). Vegetation and flora of the Solander Islands, southern New Zealand. *New Zealand Journal of Botany* **13**: 189–213.

Johnson, P.N. (1976a). Vegetation associated with kakapo in Sinbad Gully. *New Zealand Journal of Botany* **14**: 151–9.

Johnson, P.N. (1976b). Effect of soil phosphate level and shade on plant growth and mycorrhizas. *New Zealand Journal of Botany* **14**: 333–40.

Johnston, J.A. (1981). The New Zealand Bush: early assessments of vegetation. *New Zealand Geographer* **37**: 19–24.

Johnston, W.B. (1961). Locating the vegetation of early Canterbury: a map and the sources. *Transactions of the Royal Society of New Zealand (Botany)* **1**: 5–15.

June, S.R. (1983). Rimu regeneration in a north Westland podocarp forest. *New Zealand Journal of Ecology* **6**: 144–5.

June, S.R. & Ogden, J. (1975). Studies on the vegetation of Mount Colenso, New Zealand. 3. The population dynamics of red beech seedlings. *Proceedings of the New Zealand Ecological Society* **22**: 61–6.

June, S.R. & Ogden, J. (1978). Studies on the vegetation of Mount Colenso, New Zealand. 4. An assessment of the processes of canopy maintenance and regeneration strategy in a red beech (*Nothofagus fusca*) forest. *New Zealand Journal of Ecology* **1**: 7–15.

Kelly, D. (1987). Slow recovery of *Beilschmiedia tawa* after severe frosts in inland Taranaki, New Zealand. *New Zealand Journal of Ecology* **10**: 137–40.

Kelly, D. & Skipworth, J.P. (1984). *Tradescantia fluminensis* in a Manawatu (New Zealand) forest. 1. Growth and effects on regeneration. *New Zealand Journal of Botany* **22**: 393–7.

Kelly, G.C. (1968). Waituna lagoon, Foveaux Strait. *Bulletin of the Wellington Botanical Society* **35**: 9–19.

Kennedy, P.C. (1978). Vegetation and soils of North Island, Foveaux Strait, New Zealand. *New Zealand Journal of Botany* **16**: 419–34.

King, C.M. (1983). The relationships between beech (*Nothofagus* spp.) seedfall and populations of mice (*Mus musculus*), and the demographic and dietary responses of stoats (*Mustela erminea*), in three New Zealand forests. *Journal of Animal Ecology* **52**: 141–66.

Kirk, T. (1872). Notes on the flora of the Lake District of the North Island. *Transactions and Proceedings of the New Zealand Institute* **5**: 322–45.

Kissel, R.M., Wilson, J.B., Bannister, P. & Mark, A.F. (1987). Water relations of some native and exotic shrubs of New Zealand. *New Phytologist* **107**: 29–37.

Knowles, B. & Beveridge, A.E. (1982). Biological flora of New Zealand. 9. *Beilschmiedia tawa* (A. Cunn.) Benth. et Hook. F. ex Kirk (Lauraceae). Tawa. *New Zealand Journal of Botany* **20**: 37–54.

Leach, F., Anderson, A., Sutton, D., Bird, R., Duerden, P. & Clayton, E. (1986). Obsidian. *New Zealand Journal of Archeology* **8**: 143–70.

Leathwick, J. (1987). *Waipapa Ecological Area: a Study of Vegetation Pattern in a Scientific Reserve.* (FRI Bulletin 130.) Forestry Research Institute, Rotorua.

Leathwick, J.R., Wallace, S.W. & Williams, D.S. (1988). Vegetation of the Pureora Mountain Ecological Area, west Taupo, New Zealand. *New Zealand Journal of Botany* **26**: 259–80.

Lee, W.G., Allen, R.B. & Johnson, P.N. (1986). Succession and dynamics of gorse (*Ulex europaeus*) communities in the Dunedin Ecological District, South Island, New Zealand. *New Zealand Journal of Botany* **24**: 279–92.

Lee, W.G. & Hewitt, A.H. (1982). Soil changes associated with development of vegetation on an ultramafic scree, northwest Otago, New Zealand. *Journal of the Royal Society of New Zealand* **12**: 229–42.

Lee, W.G. & Johnson, P.N. (1984). Mineral element concentration in foliage of divaricate and non-divaricate *Coprosma* species. *New Zealand Journal of Ecology* **7**: 169–74.

Lee, W.G., Kennedy, P.C. & Wilson, J.B. (1983a). The ecology and distribution of *Olearia lyallii* on the subantarctic Auckland Islands. *New Zealand Journal of Ecology* **6**: 150.

Lee, W.G., Mark, A.F. & Wilson, J.B. (1983b). Ecotypic differentiation in the ultramafic flora of the South Island, New Zealand. *New Zealand Journal of Botany* **21**: 141–56.

Lee, W.G. & Partridge, T.R. (1983). Rates of spread of *Spartina anglica* and sediment accretion in the New River Estuary, Invercargill, New Zealand. *New Zealand Journal of Botany* **21**: 231–6.

Lee, W.G., Wilson, J.B. & Johnson, P.N. (1988). Fruit colour in relation to the ecology and habit of *Coprosma* (Rubiaceae) species in New Zealand. *Oikos* **53**: 325–31.

Levett, M.P., Adams, J.A. & Walker, T.W. (1985a). Nutrient returns in litterfall in two indigenous and two radiata pine forests, Westland, New Zealand. *New Zealand Journal of Botany* **23**: 55–64.

Levett, M.P., Adams, J.A., Walker, T.W. & Wilson, E.R.L. (1985b). Weight and nutrient content of above-ground biomass and litter of a podocarp-hardwood forest in Westland, New Zealand. *New Zealand Journal of Forestry Science* **15**: 23–35.

Levy, E.B. (1970). *Grasslands of New Zealand.* 3rd edition. Government Printer, Wellington.

Line, M.A. & Loutit, M.W. (1973). Studies on non-symbiotic nitrogen fixation in New Zealand tussock-grassland soils. *New Zealand Journal of Agricultural Research* **16**: 87–94.

Lloyd, D.G. (1972). A revision of the New Zealand, Subantarctic, and South American species of *Cotula*, section *Leptinella*. *New Zealand Journal of Botany* **10**: 277–372.

Lloyd, D.G. (1985). Progress in understanding the natural history of New Zealand plants. *New Zealand Journal of Botany* **23**: 707–22.

Lloyd, D.G., Webb, C.J. & Primack, R.B. (1980). Sexual strategies in plants II. Data on the temporal regulation of maternal investment. *New Phytologist* **86**: 81–92.

Lough, T.J., Wilson, J.B., Mark, A.F. & Evans, A.C. (1987). Succession in a New Zealand alpine cushion-plant community: a Markovian model. *Vegetatio* **71**: 129–38.

Lyon, G.L., Brooks, R.R., Peterson, P.J. & Butler, G.W. (1971). Calcium, magnesium and trace elements in a New Zealand serpentine flora. *Journal of Ecology* **59**: 421–9.

Macdonald, M.C. (1978). Mt Reeves. *Bulletin of the Wellington Botanical Society* **40**: 31–3.

MacEwen, W.M. (1987). *Ecological Regions and Districts of New Zealand.* 3rd edition. Department of Conservation, Wellington.

Macmillan, B.H. (1972). Biological flora of New Zealand. 7. *Ripogonum scandens* J.R. et G. Forst. (Smilacaceae). Supplejack, Kareao. *New Zealand Journal of Botany* **10**: 641–72.

Macmillan, B.H. (1976). Biological reserves of New Zealand 2. Bryophytes of Bankside reserve, Canterbury Plains. *New Zealand Journal of Botany* **14**: 131–3.

Madden, E.A. & Healy, A.J. (1959). The adventive flora of the Chatham Islands. *Transactions and Proceedings of the Royal Society of New Zealand* **87**: 221–8.

Madgwick, H.A.I., Jackson, D.S. & Knight, P.J. (1977). Above-ground dry matter, energy and nutrient contents of trees in an age series of *Pinus radiata* plantations. *New Zealand Journal of Forestry Science* **7**: 445–68.

Madgwick, H.A.I., Oliver, G. & Halten-Anderson, P. (1982). Above-ground biomass, nutrients and energy contents of trees in a second-growth stand of *Agathis australis*. *New Zealand Journal of Forestry Science* **12**: 3–6.

Makepeace, W., Dobson, A.T. & Scott, D. (1985). Interference phenomena due to mouse-ear and king devil hawkweed. *New Zealand Journal of Botany* **23**: 79–90.

Manley, B.R. & Knowles, R.L. (1980). Growth of radiata pine under the direct saw log regime. *New Zealand Journal of Forestry* **25**: 15–34.

Mark, A.F. (1955). Grassland and shrubland on Maungatua, Otago. *New Zealand Journal of Science and Technology* **37A**: 349–66.

Mark, A.F. (1970). Floral initiation and development in New Zealand alpine plants. *New Zealand Journal of Botany* **8**: 67–75.

Mark, A.F. (1977). *Vegetation of Mount Aspiring National Park, New Zealand*. (National Park Scientific Series 2.) National Parks Authority, Wellington.

Mark, A.F. & Adams, N.M. (1973). *New Zealand Alpine Plants*. A.H. & A.W. Reed, Wellington.

Mark, A.F. & Baylis, G.T.S. (1963). Vegetation Studies on Secretary Island. Part 6: The subalpine vegetation. *New Zealand Journal of Botany* **1**: 215–20.

Mark, A.F. & Bliss, L.C. (1970). The high alpine vegetation of Central Otago, New Zealand. *New Zealand Journal of Botany* **8**: 381–451.

Mark, A.F., Dickinson, K.M. & Fife, A.J. (1989). Forest succession on landslides in the Fiord Ecological Region. *New Zealand Journal of Botany* **27**: 369–90.

Mark, A.F., Johnson, P.N. & Wilson, J.B. (1977). Factors involved in the recent mortality of plants from forest and scrub along the Lake Te Anau shoreline, Fiordland. *Proceedings of the New Zealand Ecological Society* **24**: 34–42.

Mark, A.F. & Lee, W.G. (1985). Ecology of hard beech (*Nothofagus truncata*) in southern outlier stands in the Haast Ecological District, South Westland, New Zealand. *New Zealand Journal of Ecology* **8**: 97–121.

Mark, A.F. & McSweeney, G.D. (1987). Eyre-Cainard, biological treasure trove. *Forest and Bird* **18(2)**: 10–2.

Mark, A.F., Rawson, G. & Wilson, J.B. (1979). Vegetation patterns of a lowland raised mire in eastern Fiordland, New Zealand. *New Zealand Journal of Ecology* **2**: 1–10.

Mark, A.F. & Sanderson, F.R. (1962). The altitudinal gradient in forest composition, structure and regeneration in the Hollyford Valley, Fiordland. *Proceedings of the New Zealand Ecological Society* **9**: 17–26.

Mark, A.F., Scott, G.A.M., Sanderson, F.R. & James, P.W. (1964). Forest succession on landslides above Lake Thomson, Fiordland. *New Zealand Journal of Botany* **2**: 60–89.

Mark, A.F. & Smith, P.M.F. (1975). A lowland vegetation sequence in South Westland: pakihi bog to mixed beech-podocarp forest. Part 1: The principal strata. *Proceedings of the New Zealand Ecological Society* **22**: 76–92.

Mason, R. (1969). The vegetation of the coast. In *The Natural History of Canterbury* (ed. G.A. Knox), pp. 95–105. A.H. & A.W. Reed, Wellington.

Masters, S.E., Holloway, J.T. & McKelvey, P.J. (1957). *The National Forest Survey of New Zealand*. Government Printer, Wellington.

McCracken, I.J. (1980). Mountain climate in the Craigieburn Range, New Zealand. In *Mountain Environments and Subalpine Tree Growth* (Proceedings of IUFRO Workshop November 1979, Christchurch, New Zealand) (ed. U. Benecke & M.R. Davis), pp. 41–59. New Zealand Forest Service, Wellington.

McCracken, I.J., Wardle, P., Benecke, U. & Buxton, R.P. (1985). Winter water relations of tree foliage at timberline in New Zealand and Switzerland. In *Establishment and Tending of Subalpine Forest: Research and Management* (Proceedings of IUFRO Workshop September 1984, Riederalp, Switzerland) (ed. H. Turner & W. Tranquillini), pp. 85–93. Swiss Federal Institute of Forestry Research, Birmensdorf.

McGlone, M.S. (1983). Polynesian deforestation of New Zealand: a preliminary synthesis. *Archeologia Oceania* **18**: 11–25.

McGlone, M.S. (1985). Plant biogeography and the late Cenozoic history of New Zealand. *New Zealand Journal of Botany* **23**: 723–49.

McGlone, M.S. (1988). New Zealand. In *Handbook of Vegetation Science 7. Vegetation History* (ed. B. Huntley & T. Webb III), pp. 557–99. Kluwer Academic Publishers, Dordrecht.

McGlone, M.S. & Bathgate, J.L. (1983). Vegetation and climate history of the Longwood Range, 12000 B.P. to present, South Island, New Zealand. *New Zealand Journal of Botany* **21**: 293–315.

McGlone, M.S. & Webb, C.J. (1981). Selective forces influencing the evolution of divaricating plants. *New Zealand Journal of Ecology* **4**: 20–8.

McIndoe, K.G. (1932). An ecological study of the vegetation of the Cromwell District, with special reference to root habit. *Transactions and Proceedings of the New Zealand Institute* **62**: 230–66.

McIntosh, P.D. & Lee, W.G. (1986). Soil-vegetation relationships on the Dun Mountain Ophiolite Belt at West Dome, Southland, New Zealand. *Journal of the Royal Society of New Zealand* **16**: 363–79.

McKelvey, P.J. (1963). *The Synecology of the West Taupo Indigenous Forest*. New Zealand Forest Service Bulletin 14. New Zealand Forest Service, Wellington.

McKelvey, P.J. (1973). The pattern of the Urewera forests. *New Zealand Forest Service Technical Paper* **59**: 1–48.

McKelvey, P.J. (1984). Provisional classification of South Island virgin indigenous forests. *New Zealand Journal of Forestry Science* **14**: 151–78.

McKelvey, P.J. & Nicholls, J.L. (1959). The indigenous forest types of North Auckland. *New Zealand Journal of Forestry* **8**: 29–45.

McQueen, D.R. (1951). Succession after forest fires in the Southern Tararua Mountains. *Bulletin of the Wellington Botanical Society* **24**: 10–19.

McSweeney, G.D. (1982). Matai/totara flood-plain forests in South Westland. *New Zealand Journal of Ecology* **5**: 121–8.

Meads, M.J. (1976). Effects of opossum browsing on northern rata trees in the Orongorongo Valley, Wellington, New Zealand. *New Zealand Journal of Zoology* **3**: 127–39.

Mejstrick, V. (1972). The classification and relative frequency of mycorrhizae in *Nothofagus solandri* var. *cliffortioides*. *New Zealand Journal of Botany* **10**: 243–53.

Melville, J. & Sears, P.D. (1953). Pasture growth and soil fertility. II. The influence of red and white clovers, superphosphate, lime, dung and urine on the chemical composition of pasture. *New Zealand Journal of Science and Technology* **35A** (*Supplement* 1): 30–41.

Meurk, C.D. (1978). Alpine phytomass and primary productivity in Central Otago, New Zealand. *New Zealand Journal of Ecology* **1**: 27–50.

Meurk, C.D. (1980). Plant ecology of Campbell Island. In *Preliminary Report of the Campbell Island Expedition 1975–6*, pp. 90–6. Department of Lands and Survey, Wellington.

Meurk, C.D. (1982). Regeneration of subantarctic plants on Campbell Island following exclusion of sheep. *New Zealand Journal of Ecology* **5**: 51–8.

Meurk, C.D. & Foggo, M.N. (1988). Vegetation response to nutrients, climate and animals in New Zealand's 'subantarctic' islands, and general management implications. In *Diversity and Pattern in Plant Communities* (ed. H.J. During, M.J.A. Werger, & J.H. Willems), pp. 47–57. SPB Academic Publishing, The Hague.

Mew, G. (1983). Application of the term 'pakihi' in New Zealand – a review. *Journal of the Royal Society of New Zealand* **13**: 175–80.

Mildenhall, D.C. (1980). New Zealand late Cretaceous and Cenozoic plant biogeography – a contribution. *Palaeogeography, Palaeoclimatology, Palaeoecology* **31**: 197–234.

Millener, L.H. (1947). A study of *Entelea arborescens*. *Transactions of the Royal Society of New Zealand* **76**: 267–88.

Miller, R.B. (1963). Plant nutrients in hard beech. *New Zealand Journal of Science* **6**: 365–413.

Milligan, R.H. (1974). Insects damaging beech (*Nothofagus*) forests. *Proceedings of the New Zealand Ecological Society* **21**: 32–40.

Mills, J.A., Lee, W.G., Mark, A.F. & Lavers, R.B. (1980). Winter use by takahe (*Notornis mantelli*) of the summer-green fern (*Hypolepis millefolium*) in relation to its annual cycle of carbohydrates and minerals. *New Zealand Journal of Ecology* **3**: 131–7.

Mirams, R.V. (1957). Aspects of the natural regeneration of the kauri (*Agathis australis* Salisb.). *Transactions of the Royal Society of New Zealand* **84**: 661–80.

Mitchell, R.J., Grace, N.D. & Fordham, R.A. (1987). The nitrogen and mineral content of seven native plant species preferred by feral goats (*Capra hircus* L.) in lowland rimu-rata-kamahi forest on eastern Mt Taranaki (Mt Egmont). *New Zealand Journal of Zoology* **14**: 193–6.

Moar, N.T. (1955). Adventitious root-shoots of *Dacrydium colensoi* in Westland. *New Zealand Journal of Science and Technology* **37A**: 207–13.

Moar, N.T. (1958). Notes on the botany of the Auckland Islands. *New Zealand Journal of Science* **1**: 466–79.

Moar, N.T. & Suggate, R.P. (1979). Contributions to the Quaternary history of the New Zealand flora 8. Interglacial and glacial vegetation in the Westport district. *New Zealand Journal of Botany* **17**: 361–88.

Molloy, B.P.J. (1976). An analysis of sweet briar on Molesworth. In *The Changing Vegetation of Molesworth Station, New Zealand, 1944 to 1971*. (DSIR Bulletin 217) (ed. L.B. Moore), pp. 90–110. Department of Scientific and Industrial Research, Wellington.

Molloy, B.P.J. & Ives, D.W. (1972). Biological reserves of New Zealand 1. Eyrewell Scientific Reserve, Canterbury. *New Zealand Journal of Botany* **10**: 673–700.

Molloy, B.P.J., Ferguson, J.D. & Fletcher, P.J. (1978). The allelopathic potential of kahikatea. *New Zealand Journal of Ecology* **1**: 183–4.

Molloy, L. (1988). *Soils in the New Zealand Landscape. The Living Mantle*. Mallinson Rendel Publishers Ltd, Wellington.

Moore, L.B. (1942). Significance of spores in hard-fern infestations. *New Zealand Journal of Science and Technology* **23B**: 113–25.

Moore, L.B. (1954). Some *Rumex acetosella* communities in New Zealand. *Vegetatio* **5–6**: 268–78.

Moore, L.B. (ed.) (1976). *The Changing Vegetation of Molesworth Station, New Zealand, 1944 to 1971*. (DSIR Bulletin 217.) Department of Scientific and Industrial Research, Wellington.

Moore, L.B. & Cranwell, L.M. (1934). Induced dominance of *Microlaena avenacea* (Raoul) Hook.f., in a New Zealand rain-forest area. *Records of the Auckland Institute and Museum* **1**: 219–38.

Morrison, T.M. & English, D.A. (1967). The significance of mycorrhizal nodules of *Agathis australis*. *New Phytologist* **66**: 245–50.

Moseley, M.P. (1978). Erosion in the south-eastern Ruahine Range: its implications for downstream river control. *New Zealand Journal of Forestry* **23**: 21–48.

Newnham, R.M., Lowe, D.J. & Green, J.D. (1989). Palynology, vegetation and climate of the Waikato lowlands, North Island, New Zealand, since 18,000 years ago. *Journal of the Royal Society of New Zealand* **19**: 127–50.

Newsome, P. (1987). *The Vegetative Cover of New Zealand*. Ministry of Works and Development, Wellington.

New Zealand Meteorological Service (1982). *Meteorological Observations for 1980*. (Miscellaneous Publication 109.) New Zealand Meteorological Service, Wellington.

Nicholls, J.L. (1956). The historical ecology of the indigenous forest of the Taranaki Upland. *New Zealand Journal of Forestry* **7**: 17–34.

Nordmeyer, A.H. (1980). Phytomass in different tree stands near timberline. In *Mountain Environments and Subalpine Tree Growth*. (Proceedings of IUFRO workshop, November 1979, Christchurch, New Zealand) (ed. U. Benecke & M.R. Davis), pp. 111–24. New Zealand Forest Service, Wellington.

Nordmeyer, A.H. & Kelland, C.M. (1982). Nitrogen balance studies in protection forests. In *Nitrogen Balances in New Zealand Ecosystems* (ed. P.W. Gandar), pp. 143–50. Plant Physiology Division, Department of Scientific and Industrial Research, Palmerston North.

Norton, D.A. (1983). Population dynamics of subalpine *Libocedrus bidwillii* forests in the Cropp River Valley, Westland, New Zealand. *New Zealand Journal of Botany* 21: 127–34.

Norton, D.A., Herbert, J.W. & Beveridge, A.E. (1988). The ecology of *Dacrydium cupressinum*: a review. *New Zealand Journal of Botany* 26: 37–62.

Norton, D.A. & Ogden, J. (1987). Dendrochronology: A review with emphasis on New Zealand applications. *New Zealand Journal of Ecology* 10: 77–95.

Norton, D.A. & Schoenenberger, W. (1984). The growth forms and ecology of *Nothofagus solandri* at the alpine timberline, Craigieburn Range, New Zealand. *Arctic and Alpine Research* 16: 361–70.

O'Connor, K.F. (1982). The implications of past exploitation and current developments to the conservation of South Island tussock grasslands. *New Zealand Journal of Ecology* 5: 97–107.

Ogden, J. (1971). Studies on the vegetation of Mount Colenso, New Zealand. 2. The population dynamics of red beech. *Proceedings of the New Zealand Ecological Society* 18: 66–75.

Ogden, J. (1978). On the diameter growth rates of red beech (*Nothofagus fusca*) in different parts of New Zealand. *New Zealand Journal of Ecology* 1: 16–8.

Ogden, J. (1983). The scientific reserves of Auckland University. II. Quantitative vegetation studies. *Tane* 29: 163–80.

Ogden, J. (1985). An introduction to plant demography with special reference to New Zealand trees. *New Zealand Journal of Botany* 23: 751–72.

Ogden, J. & Caithness, T.A. (1982). The history and present vegetation of the macrophyte swamp at Pukepuke Lagoon. *New Zealand Journal of Ecology* 5: 108–20.

Ogden, J., Wardle, G.M. & Ahmed, M. (1987). Population dynamics of the emergent conifer *Agathis australis* (D. Don) Lindl. (kauri) in New Zealand. II. Seedling population sizes and gap-phase regeneration. *New Zealand Journal of Botany* 25: 231–42.

Ogle, C.C. (1987). The retreat of Cook's scurvy grass. *Forest and Bird* 18(1): 26.

Ogle, C.C. & Bartlett, J.K. (1981). Whangamarino swamp resources study. *Waikato Valley Authority Technical Publication* 20: 35–46.

Orwin, J. (1970). Lichen succession on recently deposited rock surfaces. *New Zealand Journal of Botany* 8: 452–77.

Orwin, J. (1972). The effect of environment on assemblages of lichens growing on rock surfaces. *New Zealand Journal of Botany* 10: 37–47.

Page, C.N. & Brownsey, P.J. (1986). Tree-fern skirts: a defence against climbers and large epiphytes. *Journal of Ecology* 74: 787–96.

Park, G.N. (1967). Vegetation and flora of Castlepoint and Cape Turnagain. *Bulletin of the Wellington Botanical Society* 34: 7–18.

Park, G.N. (1968). The vegetation and flora of Mt Stokes. *Bulletin of the Wellington Botanical Society* 35: 42–3.

Park, G.N. (1972). Variation in soil water and soil air contents associated with a vegetation-soil sequence in the Tararua mountains, New Zealand. *Proceedings of the New Zealand Ecological Society* 19: 57–64.

Parkes, J.P. (1984). Feral goats on Raoul Island. II. Diet and notes on the flora. *New Zealand Journal of Ecology* 7: 95–101.

Partridge, T.R. (1989). Soil seed banks of secondary vegetation on the Port Hills and Banks Peninsula, Canterbury, New Zealand, and their role in succession. *New Zealand Journal of Botany* 27: 421–36.

Partridge, T.R. & Wilson, J.B. (1987). Salt tolerance of salt marsh plants of Otago, New Zealand. *New Zealand Journal of Botany* 25: 559–66.

Partridge, T.R. & Wilson, J.B. (1988). Vegetation patterns in salt marshes of Otago, New Zealand. *New Zealand Journal of Botany* 26: 497–510.

Partridge, T.R. & Wilson, J.B. (1989). Methods for investigating vegetation/environment relations – a test using the salt marsh vegetation of Otago, New Zealand. *New Zealand Journal of Botany* 27: 35–47.

Patel, B.K.C., Jasperse-Herst, P.M., Morgan, H.W. & Daniel. R.M. (1986). Isolation of anaerobic, extremely thermophilic, sulphur metabolising archaebacteria from New Zealand hot springs. *New Zealand Journal of Marine and Freshwater Research* **20**: 439–45.

Payton, I.J. (1983). Defoliation as a means of assessing browsing tolerance in southern rata (*Metrosideros umbellata* Cav.). *Pacific Science* **37**: 443–52.

Payton, I.J. (1989*a*). Seasonal growth patterns of southern rata (*Metrosideros umbellata*), Camp Creek, Westland, New Zealand. *New Zealand Journal of Botany* **27**: 13–26.

Payton, I.J. (1989*b*). Fungal (*Sporothrix*) induced mortality of kamahi (*Weinmannia racemosa*) after attack by pinhole borer (*Platypus* spp.). *New Zealand Journal of Botany* **27**: 359–68.

Payton, I.J., Allen, R.B. & Knowlton, J.E. (1984). A post-fire succession in the northern Urewera forests North Island, New Zealand. *New Zealand Journal of Botany* **22**: 207–22.

Pearce, A.J., Rowe, L.K. & O'Loughlin, C.L. (1982). Hydrologic regime of undisturbed mixed evergreen forests, south Nelson, New Zealand. *Journal of Hydrology (New Zealand)* **21**: 98–116.

Pekeharing, C.J. & Reynolds, R.N. (1983). Distribution and abundance of browsing mammals in Westland National Park in 1978, and some observations on their impact on vegetation. *New Zealand Journal of Forestry Science* **13**: 247–65.

Podger, F.D. & Newhook, F.J. (1971). *Phytophthora cinnamomi* in indigenous plant communities in New Zealand. *New Zealand Journal of Botany* **9**: 625–38.

Pole, M. (1989). Early Miocene floras from Central Otago, New Zealand. *Journal of the Royal Society of New Zealand* **19**: 121–5.

Pook, E.W. (1978). Population and growth characteristics of tanekaha in *Leptospermum* scrub and secondary forest of the Waitakere Ranges, Auckland, New Zealand. *New Zealand Journal of Botany* **16**: 227–34.

Pook, E.W. (1979). Seedling growth in tanekaha; effects of shade and other seedling species. *New Zealand Journal of Forestry Science* **9**: 193–200.

Poole, A.L. (1937). A brief ecological survey of the Pukekura State Forest, South Westland. *New Zealand Journal of Forestry* **4**: 78–85.

Poole, A.L. (1988). *New Zealand Beeches*. (DSIR Information Series 162.) Department of Scientific and Industrial Research, Wellington.

Powell, C.L. (1975). Rushes and sedges are non-mycotrophic. *Plant and Soil* **42**: 481–4.

Powlesland, M.H., Philipp, M. & Lloyd, D.G. (1985). Flowering and fruiting patterns of three species of *Melicytus* (Violaceae) in New Zealand. *New Zealand Journal of Botany* **23**: 581–96.

Prendergast, H.D.V. & Hattersley, P.W. (1987). Australian C4 grasses (Poaceae): Leaf blade anatomical features in relation to C4 acid decarboxylation types. *Australian Journal of Botany* **35**: 355–82.

Radcliffe, J.E. (1968). Soil conditions on tracked hillside pastures. *New Zealand Journal of Agricultural Research* **11**: 359–70.

Radcliffe, J.E. & Baars, J.A. (1987). The productivity of temperate grasslands. In *Managed Grasslands, B. Analytical Studies* (ed. R.W. Snaydon), pp. 7–17. Elsevier, Amsterdam.

Raunkiaer, C. (1934). *The Life Forms of Plants and Statistical Plant Geography*. Clarendon Press, Oxford.

Raven, P.H. (1973). Evolution of subalpine and alpine plant groups in New Zealand. *New Zealand Journal of Botany* **11**: 177–200.

Reif, A. & Allen, R.B. (1988). Plant communities of the steepland conifer-broadleaved hardwood forests of central Westland, South Island, New Zealand. *Phytocoenologia* **16**: 145–224.

Rigg, H.H. (1962). The pakihi bogs of Westport, New Zealand. *Transactions of the Royal Society of New Zealand (Botany)* **1**: 91–108.

Robbins, R.G. (1962). The podocarp-broadleaf forests of New Zealand. *Transactions of the Royal Society of New Zealand (Botany)* **1**: 33–75.

Robertson, D.J. (1986). A paleomagnetic study of Rangitoto Island, Auckland, New Zealand. *New Zealand Journal of Geology and Geophysics* **29**: 405–11.

Robertson, G.I. (1970). Susceptibility of exotic and indigenous trees and shrubs to *Phytophthora cinnamomi* Rands. *New Zealand Journal of Agricultural Research* **13**: 297–307.

Rogers, G.M. (1989). The nature of the lower North Island floristic gap. *New Zealand Journal of*

Botany **27**: 221–41.

Rooney, D. (1986). Pine barrens – or boons? *New Zealand Botanical Society Newsletter* **16**: 13–15.

Rose, A.B. (1983). *Succession in fescue* (Festuca novae-zelandiae) *grasslands of the Harper-Avoca Catchments, Canterbury, New Zealand.* (FRI Bulletin 16.) Forest Research Institute, Christchurch.

Rose, A.B. (1985). 5. The forests. 6. The high-altitude grasslands. In *Report on a Survey of the Proposed Wapiti Area, West Nelson* (FRI Bulletin 84) (ed. M.R. Davis & J. Orwin), pp. 68–124. Forest Research Institute, New Zealand Forest Service, Christchurch.

Rose, A.B., Harrison, J.B.J. & Platt, K.H. (1988). Alpine tussockland communities and vegetation-landform-soil relationships, Wapiti Lake, Fiordland. *New Zealand Journal of Botany* **26**: 525–40.

Ross, C.W., Mew, G. & Searle, P.L. (1977). Soil sequences on two terrace systems in the North Westland area. *New Zealand Journal of Science* **20**: 231–44.

Rowley, J. (1970). Lysimeter and interception studies in narrow-leaved snow tussock grassland. *New Zealand Journal of Botany* **8**: 478–93.

Roxburgh, S.H., Wilson, J.B. & Mark, A.F. (1988). Succession after disturbance of a New Zealand high-alpine cushionfield. *Arctic and Alpine Research* **20**: 230–6.

Rudge, M.R. & Campbell, D.J. (1977). The history and present status of goats on the Auckland Islands (New Zealand subantarctic) in relation to vegetation changes induced by man. *New Zealand Journal of Botany* **15**: 221–53.

Rumball, P.J. & Esler, A.E. (1968). Pasture pattern on grazed slopes. *New Zealand Journal of Agricultural Research* **11**: 575–88.

Sakai, A., Paton, D.M. & Wardle, P. (1981). Freezing resistance of trees of the South Temperate Zone, especially subalpine species of Australasia. *Ecology* **62**: 563–70.

Sakai, A. & Wardle, P. (1978). Freezing resistance of New Zealand trees and shrubs. *New Zealand Journal of Ecology* **1**: 51–61.

Schoenenberger, W. (1984). Above ground biomass of mountain beech (*Nothofagus solandri* (Hook.f) Oerst. var. *cliffortioides* (Hook.f.) Poole in different stand types near timberline in New Zealand. *Forestry* **57**: 59–73.

Scott, D. (1963). Erosional effects of recent and past cloudbursts in the Godley Valley, Lake Tekapo. *Proceedings of the New Zealand Ecological Society* **10**: 19–20.

Scott, D. (1970). Relative growth rates under controlled temperatures of some New Zealand indigenous and introduced grasses. *New Zealand Journal of Botany* **8**: 76–81.

Scott, D. (1975). Some germination requirements of *Celmisia* species. *New Zealand Journal of Botany* **13**: 653–64.

Scott, D. (1977). Plant ecology above timberline on Mt Ruapehu, North Island, New Zealand. I. Site factors and plant frequency. *New Zealand Journal of Botany* **15**: 255–94.

Scott, D. (1984). Hawkweeds in run country. *Review; Journal of the Tussock Grasslands and Mountainlands Institute* **42**: 33–48.

Scott, D., Dick, R.D. & Hunter, G.G. (1988). Changes in the tussock grasslands in the central Waimakariri River basin, Canterbury, New Zealand, 1947–1981. *New Zealand Journal of Botany* **26**: 197–222.

Scott, D., Menalda, P.H. & Rowley, J.A. (1970). CO_2 exchange of plants 1. Technique and response of seven species to light intensity. *New Zealand Journal of Botany* **8**: 82–90.

Scott, G.A.M. & Rowley, J.A. (1975). A lowland forest sequence in South Westland: pakihi bog to mixed lowland forest. Part 2: Ground and epiphytic vegetation. *Proceedings of the New Zealand Ecological Society* **22**: 93–108.

Shaw, W.R. (1983). Tropical cyclones: determinants of pattern and structure in New Zealand's indigenous forests. *Pacific Science* **37**: 405–14.

Silvester, W.B. (1964). Forest regeneration problems in the Hunua Range, Auckland. *Proceedings of the New Zealand Ecological Society* **11**: 1–5.

Silvester, W.B. (1978). Nitrogen fixation and mineralisation in kauri (*Agathis australis*) forest in New Zealand. In *Microbial Ecology* (ed. M.W. Loutit & J.A.R. Miles), pp. 138–43. Springer-Verlag, New York.

Silvester, W.B. & Smith, D.R. (1969). Nitrogen fixation by *Gunnera–Nostoc* symbiosis. *Nature* **224**: 1231.

Simpson, G. & Thomson, J.S. (1928). On the occurrence of silver beech in the neighbourhood of Dunedin. *Transactions of the Royal Society of New Zealand* **59**: 326–42.

Simpson, M.J.A. (1976). Seeds, seed ripening, germination and viability in some species of *Hebe*. *Proceedings of the New Zealand Ecological Society* **23**: 99–108.

Simpson, M.J.A. & Burrows, C.J. (1978). Fruits of *Myriophyllum elatinoides* (Haloragaceae). *New Zealand Journal of Botany* **16**: 163–5.

Simpson, M.J.A. & Webb, C.J. (1980). Germination in some New Zealand species of *Gentiana*: a preliminary report. *New Zealand Journal of Botany* **18**: 495–501.

Six Dijkstra, H.G., Mead, D.J. & James, I.L. (1985). Stand structure in terrace rimu forest of Saltwater Forest, South Westland, and its implications for management. *New Zealand Journal of Forestry Science* **15**: 3–22.

Skipworth, J.P. (1983). Canopy dieback in a New Zealand mountain beech forest. *Pacific Science* **37**: 391–5.

Smale, M.C. (1984). White Pine Bush – an alluvial kahikatea (*Dacrycarpus dacrydioides*) forest remnant, eastern Bay of Plenty, New Zealand. *New Zealand Journal of Botany* **22**: 201–6.

Smale, M.C., Bergin, D.O., Gordon, A.D., Pardy, G.F. & Steward, G.A. (1985). Selective logging of dense podocarp forest at Whirinaki: Early effects. *New Zealand Journal of Forestry Science* **15**: 36–58.

Smale, M.C. & Kimberley, M.O. (1983). Regeneration patterns in *Beilschmiedia tawa*-dominant forest at Rotoehu. *New Zealand Journal of Forestry Science* **13**: 58–71.

Smale, M.C., van Oeveren, H., Gleason, C.D. & Kimberley, M.D. (1987). Dynamics of even-aged *Nothofagus truncata* and *N. fusca* stands in north Westland, New Zealand. *New Zealand Journal of Forestry Science* **17**: 12–28.

Smith, S.M., Allen, R.B. & Daly, B.T. (1985). Soil-vegetation relationships on a sequence of sand dunes, Tautuku Beach, Southeast Otago, New Zealand. *Journal of the Royal Society of New Zealand* **15**: 295–312.

Smith, S.M. & Lee, W.G. (1984). Vegetation and soil development on a Holocene river terrace sequence, Arawata Valley, South Westland, New Zealand. *New Zealand Journal of Science* **27**: 187–96.

Sommerville, P., Mark, A.F. & Wilson, J.B. (1982). Plant succession on moraines of the upper Dart Valley, southern South Island, New Zealand. *New Zealand Journal of Botany* **20**: 227–44.

Stark, J.D., Fordyce, R.E. & Winterbourn, M.J. (1976). An ecological survey of the hot springs area, Hurunui River, Canterbury, New Zealand. *Mauri Ora* **4**: 35–52.

Stevens, G.R. (1980). *New Zealand Adrift: the Theory of Continental Drift in a New Zealand Setting*. A.H. & A.W. Reed, Wellington.

Stewart, G.H. (1986). Forest dynamics and disturbance in a beech/hardwood forest, Fiordland, New Zealand. *Vegetatio* **68**: 115–26.

Stewart, G.H. & Harrison, J.B.J. (1987). Plant communities, landforms, and soils of a geomorphically active drainage basin, Southern Alps, New Zealand. *New Zealand Journal of Botany* **25**: 385–99.

Stewart, G.H. & Rose, A.B. (1989). Conifer regeneration in New Zealand: dynamics of montane *Libocedrus bidwillii* stands. *Vegetatio* **79**: 41–9.

Stewart, G.H. & Veblen, T.T. (1982). Regeneration patterns in southern rata (*Metrosideros umbellata*) – kamahi (*Weinmannia racemosa*) forest in central Westland, New Zealand. *New Zealand Journal of Botany* **20**: 55–72.

Sykes, W.R. (1977). *Kermadec Islands Flora*. (DSIR Bulletin 219.) Department of Scientific and Industrial Research, Wellington.

Taylor, B.W. (1955). The flora, vegetation and soils of Macquarie Island. *Australian National Antarctic Research Reports (Botany) Series B* **2**: 1–192.

Taylor, R.H. (1971). Influence of man on vegetation and wildlife of Enderby and Rose Islands, Auckland Islands. *New Zealand Journal of Botany* **9**: 225–68.

Thomson, J.S. & Simpson, G. (1936). Notes on hydrogen-ion concentration of forest soils in the vicinity of Dunedin, New Zealand. *Transactions and Proceedings of the Royal Society of New Zealand* **66**: 192–200.

Thomson, R.C., Rodgers, K.A. & Braggins, J.E. (1974). The relationship of serpentine and related floras to laterite and bedrock type at North Cape. *New Zealand Journal of Botany* **12**: 275–82.

Timmins, S.M. (1983). Mt Tarawera 1. Vegetation types and successional trends. *New Zealand Journal of Ecology* **6**: 99–105.

Tomlinson, P.B. & Esler, A.E. (1973). Establishment growth in woody monocotyledons native to New Zealand. *New Zealand Journal of Botany* **11**: 627–44.

Torrey, J.G. (1978). Nitrogen fixation by actinomycete–nodulated angiosperms. *Bioscience* **28**: 586–91.

Troughton, J.H. & Card, K.A. (1974). Leaf anatomy of *Atriplex buchananii*. *New Zealand Journal of Botany* **12**: 167–77.

Turner, E.P. (1928). A brief account of the re-establishment of vegetation on Tarawera Mountain since the eruption of 1886. *Transactions and Proceedings of the New Zealand Institute* **59**: 60–6.

Turton, S.M. (1985). The relative distribution of photosynthetically active radiation within four tree canopies, Craigieburn Range, New Zealand. *Australian Journal of Forest Research* **15**: 393–4.

Veblen, T.T. & Stewart, G.H. (1980). Comparison of forest structure and regeneration on Bench and Stewart Islands, New Zealand. *New Zealand Journal of Ecology* **3**: 50–68.

Veblen, T.T. & Stewart, G.H. (1982). On the conifer regeneration gap in New Zealand: the dynamics of *Libocedrus bidwillii* stands on South Island. *Journal of Ecology* **70**: 413–36.

Vucetich, C.G. & Wells, N. (1978). *Soils, Agriculture, and Forestry of Waiotapu Region, Central North Island, New Zealand.* (Soil Bureau Bulletin 31.) Department of Scientific and Industrial Research, Wellington.

Walker, T.W. & Syers, J.K. (1976). The fate of phosphorus during pedogenesis. *Geoderma* **15**: 1–19.

Wallace, S. (1984). The vegetation of a mire complex in the Waipapa Ecological Area, Pureora. *Rotorua Botanical Society Newsletter* **3**: 18–21.

Waloff, N. (1968). Studies on the insect fauna on scotch broom *Sarothamnus scoparius* (L.) Wimmer. *Advances in Ecological Research* **5**: 87–208.

Walsh, P. (1892). The effect of deer on the New Zealand bush: A plea for the protection of our forest reserves. *Transactions of the Royal Society of New Zealand* **25**: 435–9.

Wardle, J.A. (1970). The ecology of *Nothofagus solandri*. *New Zealand Journal of Botany* **8**: 494–646.

Wardle, J.A. (1974). Influence of introduced mammals on the forest and shrublands of the Grey River headwaters. *New Zealand Journal of Forestry Science.* **4**: 459–86.

Wardle, J.A. (1984). *The New Zealand Beeches: Ecology, Utilisation and Management.* New Zealand Forest Service, Wellington.

Wardle, J.A. & Allen, R.B. (1983). Dieback in New Zealand *Nothofagus* forests. *Pacific Science* **37**: 397–404.

Wardle, J.A. & Guest, R. (1977). Forests of the Waitaki and Lake Hawea catchments. *New Zealand Journal of Forestry Science* **7**: 44–67.

Wardle, J.A., Hayward, J. & Herbert, J. (1973). Influence of ungulates on the forests and scrublands of South Westland. *New Zealand Journal of Forestry Science* **3**: 3–36.

Wardle, P. (1962). Subalpine forest and scrub in the Tararua Range. *Transactions of the Royal Society of New Zealand (Botany)* **1**: 77–89.

Wardle, P. (1963). Growth habits of New Zealand subalpine shrubs and trees. *New Zealand Journal of Botany* **1**: 18–47.

Wardle, P. (1969). Biological flora of New Zealand 4. *Phyllocladus alpinus* Hook. f. (Podocarpaceae); mountain toatoa, celery pine. *New Zealand Journal of Botany* **7**: 76–95.

Wardle, P. (1971). Biological flora of New Zealand. 6. *Metrosideros umbellata* Cav. [Syn. *M. lucida* (Forst. f.) A. Rich.] (Myrtaceae); Southern Rata. *New Zealand Journal of Botany* **9**: 645–71.

Wardle, P. (1972). Plant succession on greywacke gravel and scree in the subalpine belt in Canterbury, New Zealand. *New Zealand Journal of Botany* **10**: 387–98.

Wardle, P. (1974). The kahikatea (*Dacrycarpus dacrydioides*) forest of South Westland. *Proceedings of the New Zealand Ecological Society* **21**: 62–71.

Wardle, P. (1977). Plant communities of Westland National Park (New Zealand) and neighbouring lowland and coastal areas. *New Zealand Journal of Botany* **15**: 323–98.

Wardle, P. (1978*a*). Regeneration status of some New Zealand conifers, with particular reference to *Libocedrus bidwillii* in Westland National Park. *New Zealand Journal of Botany* **16**: 471–7.

Wardle, P. (1978*b*). Origin of the New Zealand mountain flora, with special reference to trans-Tasman relationships. *New Zealand Journal of Botany* **16**: 535–50.

Wardle, P. (1978*c*). Seasonality in New Zealand plants. *The New Zealand Entomologist* **6**: 344–9.

Wardle, P. (1980*a*). Primary succession in Westland National Park and its vicinity, New Zealand. *New Zealand Journal of Botany* **18**: 221–32.

Wardle, P. (1980*b*). Ecology and distribution of silver beech (*Nothofagus menziesii*) in the Paringa district, South Westland, New Zealand. *New Zealand Journal of Ecology* **3**: 23–36.

Wardle, P. (1985). New Zealand timberlines. 3. A synthesis. *New Zealand Journal of Botany* **23**: 263–71.

Wardle, P. & Campbell, A.D. (1976). Seasonal cycle of tolerance to low temperatures in three native woody plants, in relation to their ecology and post-glacial history. *Proceedings of the New Zealand Ecological Society* **23**: 85–91.

Wardle, P., Field, T.R.O. & Spain, A.V. (1971). Biological flora of New Zealand 5. *Olearia colensoi* Hook. f. (Compositae). Leatherwood, Tupari. *New Zealand Journal of Botany* **9**: 186–214.

Wardle, P. & MacRae, A.H. (1966). Biological flora of New Zealand. 1. *Weinmannia racemosa* Linn. f. (Cunoniaceae). Kamahi. *New Zealand Journal of Botany* **4**: 114–31.

Wardle, P. & Mark, A.F. (1956). Vegetation and climate in the Dunedin district. *Transactions of the Royal Society of New Zealand* **84**: 33–4.

Wardle, P., Mark, A.F. & Baylis, G.T.S. (1970). Vegetation studies on Secretary Island, Fiordland. Part 9: Additions to Parts 1, 2, 4, and 6. *New Zealand Journal of Botany* **8**: 3–21.

Wardle, P., Mark, A.F. & Baylis, G.T.S. (1973). Vegetation and landscape of the West Cape district, Fiordland, New Zealand. *New Zealand Journal of Botany* **11**: 599–626.

Wards, I. (1976). *New Zealand Atlas*. 2nd edition. Government Printer, Wellington.

Webb, C.J. (1986). Breeding systems and relationships in *Gingidia* and related Australasian Apiaceae. In *Flora and Fauna of Alpine Australasia. Ages and Origins* (ed. B.A. Barlow), pp. 383–400. CSIRO, Australia.

Webb, C.J., Johnson, P.N. & Sykes, W.R. (1990). *Flowering plants of New Zealand*. DSIR Botany, Christchurch.

Webb, C.J., Sykes, W.R. & Garnock-Jones, P.J. (1988). *Flora of New Zealand*, volume IV. *Naturalised Pteridophytes, Gymnosperms, Dicotyledons*. Botany Division, Department of Scientific and Industrial Research, Christchurch.

Webb, D.R., Rattray, M.R. & Brown, J.M.A. (1988). A preliminary survey for crassulacean acid metabolism (CAM) in submerged aquatic macrophytes in New Zealand. *New Zealand Journal of Marine and Freshwater Research* **22**: 231–35.

West, C.J. (1980). Regeneration studies on Tiritiri Matangi Island. *New Zealand Journal of Ecology* **3**: 158.

Whitaker, A.H. (1987). The role of lizards in New Zealand plant reproductive strategies. *New Zealand Journal of Botany* **25**: 315–28.

White, E.G. (1975*a*). A survey and assessment of grasshoppers as herbivores in the South Island alpine tussock grasslands of New Zealand. *New Zealand Journal of Agricultural Research* **18**: 73–85.

White, E.G. (1975*b*). An investigation and survey of insect damage affecting *Chionochloa* seed production in some alpine tussock grasslands. *New Zealand Journal of Agricultural Research* **18**: 163–78.

Whitehouse, I.E., McSaveney, M.J. & Chinn, T.J. (1980). Dating your scree. *Review; Journal of the Tussock Grassland and Mountain Lands Institute* **39**: 15–24.

Wilcox, M.D. & Ledgard, N.J. (1983). Provenance variation in the New Zealand species of *Nothofagus*. *New Zealand Journal of Ecology* **6**: 19–31.

Will, G.M. (1959). Nutrient return in litter and rainfall under some exotic conifer stands in New Zealand. *New Zealand Journal of Agricultural Research* **2**: 719–34.

Williams, P.A. (1975). Studies of the tall-tussock (*Chionochloa*) vegetation/soil systems of the southern Tararua Range, New Zealand. 2. The vegetation/soil relationships. *New Zealand Journal of Botany* **13**: 269–303.

Williams, P.A. (1977). Growth, biomass, and net productivity of tall-tussock (*Chionochloa*) grasslands, Canterbury, New Zealand. *New Zealand Journal of Botany* **15**: 399–442.

Williams, P.A. (1980). *Vittadinia triloba* and *Rumex acetosella* communities in the semi-arid regions of the South Island. *New Zealand Journal of Ecology* **3**: 13–22.

Williams, P.A. (1983). Secondary vegetation succession on the Port Hills, Banks Peninsula, Canterbury, New Zealand. *New Zealand Journal of Botany* **21**: 237–47.

Williams, P.A. (1984). Flowering currant (*Ribes sanguineum*) shrublands in the lower Waitaki Valley, South Canterbury. *New Zealand Journal of Agricultural Research* **27**: 473–8.

Williams, P.A. (1989). Vegetation of the Inland Kaikoura Range. *New Zealand Journal of Botany* **27**: 201–20.

Williams, P.A. & Buxton, R.P. (1986). Hawthorn (*Crataegus monogyna*) populations in mid-Canterbury. *New Zealand Journal of Ecology* **9**: 11–17.

Williams, P.A., Grigg, J.L., Nes, P. & O'Connor, K.F. (1978a). Macro-element composition of *Chionochloa pallens* and *C. flavescens* shoots, and soil properties in the North Island, New Zealand. *New Zealand Journal of Botany* **16**: 235–46.

Williams, P.A., Mugambi, S., Nes, P. & O'Connor, K.F. (1978b). Macro-element composition of tall-tussocks (*Chionochloa*) in the South Island, New Zealand, and their relationship with soil chemical properties. *New Zealand Journal of Botany* **16**: 479–98.

Williams, P.A., Nes, P. & O'Connor, K.F. (1977). Macro-element pools and fluxes in tall-tussock (*Chionochloa*) grasslands, Canterbury, New Zealand. *New Zealand Journal of Botany* **15**: 443–76.

Williamson, J.H. (1939). *The Geology of the Naseby Subdivision, Central Otago.* (Geological Survey Bulletin 39.) Department of Scientific and Industrial Research, Wellington.

Wilson, C.M. & Given, D.R. (1989). *Threatened Plants of New Zealand.* (DSIR Field Guide Series 2.) Department of Scientific and Industrial Research, Wellington.

Wilson, H.D. (1976). *Vegetation of Mount Cook National Park, New Zealand.* (National Parks Authority National Parks Scientific Series 1.) National Parks Authority, Wellington.

Wilson, H.D. (1987). *Vegetation of Stewart Island, New Zealand.* New Zealand Journal of Botany 27 (supplement).

Wilson, P.R. (1984). The effects of possums on mistletoe on Mt Misery, Nelson Lakes National Park. *Proceedings of the 15th Pacific Science Congress*: 53–60.

Winterbourn, M.J. (1973). Ecology of the Copland River warm springs, South Island, New Zealand. *Proceedings of the New Zealand Ecological Society* **20**: 72–8.

Wraight, M.J. (1965). Growth rates and potential for spread of alpine carpet grass, *Chionochloa (Danthonia) australis*. *New Zealand Journal of Botany* **3**: 171–9.

Wright, A.E. (1976). The vegetation of Great Mercury Island. *Tane* **22**: 23–49.

Wright, A.E. (1977). Vegetation and flora of the Moturoa Island Group, Northland, New Zealand. *Tane* **23**: 11–29.

Wright, A.E. (1983). Conservation status of the Three Kings Islands endemic flora in 1982. *Records of the Auckland Institute and Museum* **20**: 175–84.

Yates, T.E., Brooks, R.R. & Boswell, C.R. (1974). Factor analysis in botanical methods of exploration. *Journal of Applied Ecology* **11**: 563–74.

Yin, R., Mark, A.F. & Wilson, J.B. (1984). Aspects of the ecology of the indigenous shrub *Leptospermum scoparium* (Myrtaceae) in New Zealand. *New Zealand Journal of Botany* **22**: 483–507.

Zotov, V.D. (1965). Grasses of the subantarctic islands of the New Zealand region. *Records of the Dominion Museum* **5**: 101–46.

INDEX OF PLANT GENERA AND SPECIES

The index lists all plant genera and species mentioned in the text that occur naturally in New Zealand. Binomials are mostly according to Volume 4 (C.J. Webb *et al.* 1988) and earlier volumes of the *Flora of New Zealand*, and Connor & Edgar (1987). Ferns follow Brownsey *et al.* (1985) and lichens Galloway (1985). Grass names have been checked by Drs H.E. Connor and E. Edgar, and bryophytes and algae are according to checklists held at Land Resources Division, DSIR.

The entry for each vascular genus is followed by:

(1) Any other generic name referring to *all* the New Zealand representatives, that occurs in the ecological literature from Cockayne (1928) onwards [square brackets].

(2) The numbers of endemic (•), native but not endemic (°), and adventive or naturalised (*) species, and species of doubtful nativity (†). Some species cover more than one category, and many of the numbers are provisional.

(3) The family to which the genus belongs, abbreviated by deletion of the ending '-aceae'.

Endemic genera are prefixed by •. The entry for each vascular species is prefixed by •°*† as above, and followed by:

(1) Any other specific names or further generic names that occur in the ecological literature from Cockayne (1928) onwards [square brackets]. These include legitimate synonyms as well as illegitimate usages, and may apply to only part of the species or include other species as presently understood.

(2) Taxonomic comment, particularly suggestions as to how future revisions may affect present entities (round brackets). This is mainly based on unpublished checklists compiled by A.P. Druce.

Quotation marks indicate informal names and obsolete names from the literature that cannot be confidently assigned to a particular currently recognised taxon. Infraspecific taxa are listed only where New Zealand plants belong to a named portion of a non-endemic species.

Genera of non-vascular plants are assigned to broad categories (moss, foliose lichen, etc.), and only the total number of species in New Zealand is given, excluding those that are fully marine; nearly all the non-vascular species listed are considered native.

Abrotanella 8˙ Aster.
˙*caespitosa* 341, 423
˙*inconspicua* 109, 411, 420–3, 425
˙*linearis* 229, 514
˙*muscosa* 396
˙*spathulata* 455
Acaena 12˙ 1° 1† 2* Ros. 38, 58–9, 266, 362, 374, 517
˙agnipila [ovina] 249, 359
˙*anserinifolia [pusilla, sanguisorbae, viridior]* 29, 38, 228, 249, 252, 257, 262, 266, 274–5, 372, 392, 444, 512
˙*buchananii* 255
˙*caesiiglauca [sanguisorbae]* 237, 249, 253, 261, 370, 528
˙*glabra* 29, 373–4, 429
˙*inermis [microphylla]* 29
°*magellanica [adscendens]* 463
˙*microphylla* 252
˙*minor [sanguisorbae]* 454, 460, 462, 464–5
†*novae-zelandiae* 266, 268, 270, 275, 446
˙*profundeincisa [hirsutula, sanguisorbae]* 343
˙*saccaticupula* 429
Acarospora 4 Fungus 417
Acer 2* Acer.
˙pseudoplatanus 160, 521, 545
Achillea 4* Aster.
˙millefolium 262, 267, 273–4, 359
Achrophyllum [Pterygophyllum] 2 Moss
quadrifarium 325, 328
Acianthus 3° Orchid.
˙*fornicatus* var. ˙*sinclairii [sinclairii]* 199
Aciphylla c.40˙ Api. 5, 13, 29, 38, 41–2, 54, 57, 78, 190, 220–1, 234, 245, 248, 255, 340, 547
˙*aurea [colensoi]* 47, 61, 220, 236, 239, 248, 251, 388
˙*colensoi* 220, 225, 243
˙*congesta* 29, 109, 230, 412–13
˙*crenulata* 221, 228, 529
˙*crosby-smithii* 103
˙*dieffenbachii [Coxella]* 445
˙*dissecta [Anisotome]* 221
˙*divisa* 221, 412
˙*dobsonii* 109, 241, 412, 420
˙*ferox* 221, 225, 390–1, 430
˙*glaucescens* 221, 234, 390, 429
˙*hectorii* 109, 421–2, 529
˙*hookeri [townsonii]* 100, 221
˙*horrida* 42, 192, 221, 241
˙*kirkii* 109
˙*lecomtei* 108, 420
˙*lyallii* 220, 231–2
˙*monroi* 38, 221, 237, 428, 430
˙*multisecta* 221, 410, 412
˙*pinnatifida* 29, 103
˙*scott-thomsonii [maxima]* (= *colensoi* var.?) 38, 54, 192, 221–2, 241
˙*similis* 420
˙*simplex* 412, 420
˙*spedenii* 108
˙*squarrosa* 264, 385
˙*stannensis* 110
˙*subflabellata [squarrosa]* 220, 234, 238, 248, 261
˙*takahea* 103, 230
˙*traillii* 110

˙*traversii* 442
Ackama 1˙ Cunoni. 9
˙*rosifolia [Caldecluvia]* 79, 92, 119, 140
Acrocladium 2 Moss 454
chlamydophyllum [auriculatum] 271, 453
Acromastigium 7 Leafy liverwort 182
Actinidia 1* Actinidi.
˙deliciosa 95
Actinotus 1˙ Api.
˙*novae-zelandiae [Hemiphues suffocata]* 321, 334, 335, 337, 340
Adenochilus 1˙ Orchid.
˙*gracilis* 39
Adiantum 3˙ 4° 2* Pterid. 436, 542
˙*cunninghamii [affine]* 384–5, 392
°*hispidulum [pubescens]* 397
Aeonium 4* Crassul.
˙arboreum 396
* × *velutinum [Sempervivum* sp.] 396
Agapanthus 1* Lili.
˙orientalis 396
Agaricus 15 Fungus
˙subperonatus 493
Agathis 1˙ Araucari. 5, 8
˙*australis* 20, 37, 45, 58, 61–2, 64, 70, 71, 77, 79, 92, 111–12, 117, 468–9, 477–8, 481, 486, 502, 533, 559–61
Ageratina [Eupatorium] 2* Aster. 202
˙adenophora 267
˙riparia 267
Agrostis c.10˙ 1˙ 4* Po. 248, 320
˙capillaris [tenuis] 50–1, 90, 235, 247, 251, 261, 263, 268, 271, 273–4, 276–7, 279, 342, 369, 385, 394, 444, 454, 457, 460, 484, 489, 520, 534, 548
˙*dyeri* 218, 264, 394
°*magellanica* 411, 419, 458, 465
˙*muscosa* 34, 250, 257, 307, 368
˙*pallescens* 342
˙stolonifera 29, 265, 275–6, 279, 289–90, 295–7, 299, 302, 312, 314, 317, 342, 353, 360, 372
˙*subulata* 235, 259, 395, 411, 416–18, 420–4, 426, 428, 455, 460, 529
'*canina*' 309
'*perennans*' *[parviflora]* 264
Aira 4* Po. 198, 248, 265, 364, 384, 445, 553
˙caryophyllea 235, 251, 270, 275, 366, 369
Alectoria 1 Fruticose lichen
nigricans 395, 417, 421, 425
Alectryon 2˙ Sapind. 9
˙*excelsus* 37, 105, 120, 167, 388
˙*grandis [excelsus]* (= *excelsus* var.?) 92, 532
˙*Alepis [Elytranthe]* 1˙ Loranth.
˙*flavida* 36, 143, 159, 567
Aleurites 1† 1* Euphorbi.
†*moluccana* 434
Alisma 2* Alismat.
˙plantago-aquatica 302, 320
Allium 4* Lili.
˙triquetrum 30
Alocasia 1* Ar.
˙macrorrhiza 435
Alopecurus 3* Po.
˙geniculatus 275–6, 316–17
˙*Alseuosmia* 4˙ Alseuosmi. or Caprifoli. 24, 53, 58, 188

*banksii [atriplicifolia, ligustrifolia, linariifolia] 25
*macrophylla 37, 116–18, 130, 139, 537
*pusilla 128, 132
Alternanthera 1† 2* Amaranth.
*philoxeroides 267
†sessilis 16
Amanita 14 Fungus
*muscaria 493
Ammophila 1* Po.
*arenaria 272, 299, 352, 356, 446, 522
Amphibromus 1° Po.
°fluitans 311
Amphidium 1 Moss
cyathicarpum 382
Anabaena 16 Blue-green alga 496
Anacystis 1-3 Blue-green alga
cyanea 303
Anagallis 2* Primul.
*arvensis 267, 371, 384, 387
Anaphalis [Gnaphalium] 4° Aster. 13
*keriensis 380, 384, 392
*rupestris [G. hookeri] 380
*subrigida 380, 384
*trinervis [G. lyallii] 366, 380, 391–2
Anarthropteris [Polypodium] 1° Grammitid.
*lanceolata [P. dictyopteris] 35, 39, 382
Andreaea 5 Moss 301, 395–6, 415–18, 431, 460
acutifolia 423, 465
australis 417
°Anemanthele [Oryzopsis, Stipa] 1° Po.
*lessoniana [O. lessoniana, rigida; S. arundinacea]
133
Anemone 1° 1* Ranuncul.
*tenuicaulis 109, 221
Aneura 15 Thallose liverwort 182, 301, 325
?palmata 325
Anisotome 15° Api. 13, 53–4, 57
*antipoda 449, 455, 462
*aromatica 29, 34, 38, 221, 224, 238, 240, 242,
244, 247, 341, 430–1, 513
*capillifolia 109, 412
*filifolia 237, 253, 262, 374, 388
*flexuosa [aromatica] 34, 221, 230, 237, 395, 398,
411, 416–19, 422–3, 425–6, 529
*haastii 29, 151, 175, 190, 192, 221, 229, 231, 233,
394, 396
*imbricata 34, 411, 418, 420–2, 426–8, 430
*lanuginosa 107
*latifolia 29, 38, 449, 453, 456, 459, 461–3
*lyallii 281, 374, 393, 396
*pilifera 29, 392, 395, 412, 419, 427–9, 529
Anthoceros 12 Hornwort 383
Anthoxanthum 1* Po.
*odoratum 214, 235, 240, 247, 250, 261, 263, 268,
270, 273–4, 276–7, 279, 317, 342, 359, 369, 372,
444, 493, 520
Aphanes 3* Ros.
*arvensis 254, 258, 281, 371
Apium 1° 2* Api.
*prostratum [australe, filiforme] 278, 289–90, 292–
3, 299, 354, 359, 374, 383–4, 386–7, 396, 445–6,
461
Aponogeton 1* Aponogeton.
*distachyus 301
Aporostylis [Caladenia] 1° Orchid.

*bifolia 339
Aptenia 1* Aizo.
*cordifolia 395–6
Archeria 2° Epacrid. 59, 154, 378
*racemosa 26, 140, 154, 188
*traversii 26, 61, 135, 146, 158, 176–9, 513–14,
574
Arctotheca 1* Aster.
*calendula [Cryptostemma calendulacea] 267
Arenaria 1* Caryophyll.
*serpyllifolia 254, 269, 371, 388, 394
Argyranthemum [Chrysanthemum] 1* Aster
*frutescens 395
Aristotelia 2° Elaeocarp. 9, 39, 54, 58
*fruticosa 25, 136, 151, 173, 195–6, 207, 209–11,
214, 224, 243, 262, 388, 392, 547, 550
*serrata 23, 25, 37, 65, 70, 77, 121, 124, 151, 158,
162, 173, 207, 214, 352, 366, 475–6, 478, 481,
489–90, 500, 502, 506, 508, 511, 516–17, 561,
565, 574
Armillaria 3 Fungus
mellea 36, 564
Arrhenatherum 1* Po.
*elatius 276, 279
Arthropodium 2° Asphodel. 9
*candidum 382
*cirratum 38–9, 382–3, 387, 397
Arthopyrenia 8 Crustose lichen
sublitoralis [Verrucaria] 383
Ascarina 1° Chloranth. 8, 52, 58
*lucida [lanceolata] (var. lanceolata merits species
rank?) 15, 23, 37, 128, 132, 140, 389, 435, 499,
502, 521
Asparagus 4* Lili.
*scandens 164
Asplenium 7° 9° Aspleni. 31, 436
°bulbiferum 60, 113, 128, 135, 151, 158, 164, 168,
170, 173, 486, 493, 509, 574, 577
*chathamense 446
°flabellifolium 172, 359, 382, 388
°flaccidum 35, 117, 135, 151, 159, 166, 171, 173,
175, 187, 313, 383
°hookerianum 382
*lyallii [anomodum] 382, 388
*oblongifolium [lucidum] 39, 117, 166, 382, 387,
389
°obtusatum 168, 191, 281, 383, 396, 436, 446, 451,
455, 461–2
°polyodon [adiantoides, falcatum] 35, 39, 117, 119,
128, 170, 441, 575
*richardii 172, 388, 393–4
°terrestre 382, 385, 388, 439
°trichomanes 375, 382, 388
Astelia 13° Asphodel. 5, 13, 29, 39, 41, 52–3, 58, 78,
112, 151, 312
*banksii 383, 397, 532
*chathamica [nervosa] 439
*fragrans [nervosa] 39, 58, 128, 138, 168, 188, 328,
520
*graminea 397
*grandis [nervosa] 129, 173, 315–16
*linearis 39, 58, 174, 224–5, 229, 232, 243, 335,
338–40, 396, 410, 454
*nervosa [cockaynei] (species aggregate) 134–5,
158, 173, 176–8, 186, 188, 206, 222, 229, 243,

Astelia 13˙ Asphodel. (*cont.*)
   340, 389, 391, 516, 548–51, 574–5
   ˙*nivicola* (var. *moriceae* merits species rank?) 222,
      228, 529
   ˙*petriei* 222, 229
   ˙*skottsbergii* 100, 391
   ˙*solandri* [*cunninghamii*] 35, 117, 119, 128, 140,
      383, 389
   ˙*subulata* 101, 338, 340, 454, 459
   ˙*trinervia* 116, 118, 120, 130, 139, 184, 389, 537
Aster 4* Aster.
   subulatus 291, 293–4, 312
Atriplex 1˙ 1† 5* Chenopodi. 28
   ˙*buchananii* 299, 379, 485
   †*prostrata* [*hastata, novae-zelandiae*] 289, 291,
      295, 299, 353, 359
Australina 1˙ Urtic.
   ˚*pusilla* 38, 112, 439
Austrofestuca [*Festuca, Poa*] 1˙ Po.
   ˚*littoralis* [*P. triodioides*] 353, 356–7, 521
Avena 6* Po.
   barbata 384
Avicennia 1˙ Avicenni.
   ˚*resinifera* [*marina, officinalis*] 43, 55, 60, 79, 80,
      93, 209, 291, 293–5, 500, 502
Axonopus 1* Po.
   affinis 265, 277, 279, 355
Azolla 1˙ 1* Salvini. 496
   ˚*filiculoides* [*rubra*] 301, 305, 313, 333
   pinnata 313
Azorella 1˙ Api.
   ˚*selago* 463, 465

Barbula 6 Moss
   calycina [*Tortella*] 205
Bartramia 5 Moss 382, 419
   papillata 417
Baumea [*Cladium*] 2˙ 5˚ Cyper. 12, 72, 130, 313,
      318, 327, 329, 336–7, 522
   ˚*articulata* 310, 312–14, 323
   ˙*complanata* 199, 402
   ˚*huttonii* 313, 321, 323, 325, 334
   ˚*juncea* 289, 293–4, 310, 312, 323
   ˚*rubiginosa* [*C. glomeratum*] 198, 201, 312, 314,
      316, 321, 323–4, 326, 330–2, 335
   ˙*tenax* [*C. gunnii*] 313, 325, 327–8, 331–2, 443
   ˚*teretifolia* 198–9, 313, 321, 323–6, 328–30, 335
Bazzania 9 Leafy liverwort 147, 182
   novae-zelandiae 328
Beilschmiedia 3˙ Laur. 5, 8, 21, 52, 58, 64, 78
   ˙*tarairi* 37, 43, 77, 79, 93, 112–13, 117, 119, 166,
      481, 502, 532, 560
   ˙*tawa* 21, 37, 44, 55, 57, 61–3, 70, 77, 79, 99, 112–
      13, 117, 119, 121, 124, 167, 387, 469–70, 478,
      481, 500, 524, 546, 560, 580
   ˙*tawaroa* (= *tawa* syn.?) 93, 166
Bellis 1* Aster.
   perennis 267, 274–5, 277
Berberis 5* Berberid. 544
   darwinii 212
Beta 1* Chenopodi.
   vulgaris 28
Bidens 1† 2* Aster.
   frondosa 313–14
   †*pilosa* 16
Blechnum 11˙ 7˚ Blechn. 8, 159, 320

   ˙*banksii* 392, 446
   ˚*chambersii* [*lanceolatum*] 173, 393
   ˚*colensoi* [*patersonii*] 39, 382
   ˙*discolor* 31, 39, 74, 77, 80, 113, 118–19, 128, 134,
      140, 144–5, 147, 149, 151, 159, 168, 172, 304,
      328, 483, 485, 509, 515–16, 522–3, 539, 549,
      574, 577, 580
   ˙*durum* (= *banksii* syn.?) 191, 281, 396, 446, 451,
      455–6, 461–2
   ˙*filiforme* 35, 112, 116, 119–20, 126
   ˚*fluviatile* 124, 138, 144, 151, 159, 164, 173, 389,
      392
   ˚*fraseri* 31, 39, 117, 537
   ˚*minus* [*capense, procerum*] 159, 311–12, 314–16,
      329–30, 443
   ˙*nigrum* 39
   ˚*penna-marina* 31, 159, 223, 238, 259, 262, 320,
      374, 388, 392, 394, 416–17, 429, 454, 503, 505,
      512–13, 518, 527, 529
   ˚*procerum* [*capense, minus*] 113, 128, 138, 182, 328,
      442, 527, 549, 563
   ˚*vulcanicum* 382, 391
   'black spot' [*capense*] 39, 113, 116, 128, 156, 215,
      311, 315–16, 328, 382, 384, 391–2, 441–2, 511,
      516, 533–4, 542
   '*capense*' [includes several spp.] 31, 119, 124, 139,
      144, 151, 158, 164, 172, 186, 198, 202, 215,
      311–13, 320, 325, 327, 333, 335, 339, 382, 385,
      443–4, 515–16, 536–9, 542, 550, 577
   'green bay' [*capense*] 382, 384–5
   'mountain' [*capense*] 135, 149, 176, 178, 185, 187,
      189, 192, 215, 243, 382, 388, 391, 451–2, 457–8,
      462, 514, 548, 551
Boehmeria 1˙ Urtic.
   ˚*australis* var. ˙*dealbata* [*dealbata*] 434–5
Bolboschoenus [*Scirpus*] 3˚ Cyper. 39, 301
   ˚*caldwellii* [*S. maritimus*] 289, 297
   ˚*fluviatilis* 289, 320
   ˚*medianus* [*S. maritimus*] 289, 293, 313
Botriochloa 2* Po.
   maxima [*Dicanthium annulatum*] 265
Brachyglottis [*Senecio*] 30˙ Aster. 5, 13, 41, 53, 72,
      191, 220, 249, 379
   ˙*adamsii* 98, 101, 375, 388
   ˙*bellidioides* (species aggregate) 38, 205, 220, 224,
      237, 240, 261, 345, 347–8, 396, 406, 518, 529
   ˙*bidwillii* 77, 109, 192, 247, 390
   ˙*bifistulosa* 14, 376
   ˙*buchananii* [*bennettii, elaeagnifolia*] (= *rotundifolia*
      vars?) 24, 37, 57, 61–3, 144, 147, 158, 178, 189,
      478, 481, 502, 548, 550–1, 565, 574
   ˙*cassinioides* 27, 189, 212
   ˙*compacta* (= *monroi* var?) 98, 384
   ˙*elaeagnifolia* (= *rotundifolia* var.?) 24, 174, 184,
      186, 472
   ˙*greyi* 376, 385
   ˙*haastii* 212, 236, 253, 368
   ˙*hectorii* 23, 39, 41, 52, 100, 376, 389
   ˙*huntii* 438, 441
   ˙*kirkii* [*Urostemon*] 23, 35, 41, 52, 116, 139–40
   ˙*lagopus* (incl. *saxifragoides*) 238, 363, 379, 385
   ˙*laxifolia* (= *greyi* var.?) 376, 389
   ˙*monroi* 189, 376, 385, 388
   ˙*myrianthos* (includes *pentacopa* as var.?) 93
   ˙*perdicioides* 96
   ˙*repanda* (var. *arborescens* merits species rank?) 23,

37, 41, 92, 120, 124, 162, 203, 387, 389, 500, 539–40
revoluta 101, 109, 241, 375
rotundifolia [reinoldii] 24, 169, 174, 191, 396
sciadophila 25, 34, 163
stewartiae 26, 110, 191
turneri 97, 376
Brachyscome 5˙ 1* Aster. 109, 249
linearis 28, 108
sinclairii [pinnata] 236, 252, 257, 416, 418, 426, 529
Brachythecium 8 Moss 312, 419
albicans 266, 272–4, 278
rutabulum 364
salebrosum 465
Breutelia 3 Moss 182, 260, 262, 364, 382
affinis 205, 249
elongata 340, 382, 454, 459–60, 465
pendula 249, 275, 312, 320, 325, 340, 343, 345–6, 382, 454, 458, 465
Briza 2* Po. 265
major 385
minor 270, 277
Bromus 16* Po. 248, 265, 320, 359, 364, 553
diandrus 270, 272, 355, 384
hordeaceus [mollis] 251, 269–70, 273, 277, 369, 372, 436
sterilis 369
tectorum 251, 279, 368–9
willdenowii [catharticus, uniloides] 265, 271, 276, 279
Bryum ?20 Moss 60, 266, 312, 343, 364, 382, 419
billardierei [truncorum] 205, 461
blandum 382, 415
laevigatum 341, 382, 415
pseudotriquetrum 325
Buddleja 5* Buddlej.
davidii 366
Buellia 18 Moss 383, 395, 423, 425, 461
Bulbinella [Chrysobactron] 6˙ Asphodel. 30, 40, 53–4, 59, 248, 338
angustifolia [hookeri] 223, 234, 237, 248, 253
gibbsii 223, 230, 243, 340
hookeri 223, 244
modesta 335
rossii 450, 452–6, 458–61
Bulbophyllum 2˙ Orchid. 29, 35, 485
pygmaeum 38, 117, 140, 199

Caesalpinia 1* Fab.
decapetala 435
Cakile 2* Brassic.
edentula 353
maritima 353
Caladenia 2˙ Orchid.
catenata [carnea, exigua] 199, 577
lyallii 339
Callitriche 2˙ 2˙ 1† 2* Callitrich.
antarctica 456, 459, 465
petriei 301, 304, 311, 330
†*stagnalis [verna]* 301, 305, 312–13, 320, 333, 372
Calluna 1* Eric.
vulgaris 193, 244
Calochilus 3˙ Orchid.
paludosus 335
Caloplaca 21 Crustose lichen

holocarpa 383
Calystegia 3˙ 2* Convolvul. 34
†*sepium* 17, 163, 306, 312, 374
silvatica 163
°*soldanella* 29, 351, 353, 356–7, 359–60, 521
°*tuguriorum* 163, 196, 215, 351, 359, 374, 385
Camptochaete 6 Moss
angustata 312
arbuscula 147
Campylium 2 Moss
polygamum 307
stellatum 341
Campylopus 10 Moss 322, 333, 337, 339–40, 355, 364, 396, 455
acuminatus [kirkii] 324–5, 331
clavatus 262, 336, 364, 385, 403 ˙
holomitrium 401, 403
introflexus 325, 335, 403, 443
pallidus 460
pyriformis [torquatus] 403
Canavalia 1˙ Fab.
°*rosea [obtusifolia]* 435, 496
Candelaria 1? Crustose lichen 463
Candelariella 3 Crustose lichen
vitellina 461
Capsella 1* Brassic.
bursa-pastoris 267, 273, 371
Cardamine 5˙ 3* Brassic. 13, 54, 230, 413, 419, 428–9, 529
corymbosa 464–5
debilis [heterophylla] (species aggregate) 124, 330, 359, 385, 397, 413, 518, 574
depressa [stellata] 455, 461
hirsuta 28, 59
subcarnosa [glacialis] 455, 460
Carduus 4* Aster.
nutans 266, 273
pycnocephalus 266, 271
tenuiflorus 266, 275, 370
Carex 61˙ 12˙ 22* Cyper. 5, 6, 30, 39, 40, 60, 73, 231, 310, 313, 319, 334, 374
albula 368
°*appressa* 448, 453, 459, 462
berggrenii 303–4
°*breviculmis* 235, 251, 261, 268, 309, 369
chathamica 444
cockayneana 388, 512, 514, 522
colensoi 235, 251, 262
comans 266, 275
coriacea [ternaria] 39, 193, 238, 244, 262, 266, 302, 310, 315–17, 320, 337–8, 342, 346, 348, 365, 384, 493–4, 514, 522
demissa 304
devia 397–8
°*diandra* 310, 318–20, 332
dipsacea [Vignea] 337
dissita 310, 330, 332, 532
divisa 293
°*echinata [stellulata]* 244, 262, 321, 332, 335, 338, 342–3, 348
edgariae 338
elingamita 92
fascicularis [pseudo-cyperus] 313
flagellifera [lucida] 266, 289, 299, 310
°*flaviformis [oederi]* 298
°*gaudichaudiana* 302–4, 310, 315, 321, 326, 330,

Carex 61˙ 12° 22* Cyper. (*cont.*)
335, 338, 341–6, 348
˙*geminata* [*ternaria*] 315, 330
˙*goyenii* 518
˙*hectorii* 423
inyx 266
˙*kermadecensis* 436
˙*kirkii* 423
°*lachenalii* subsp. ˙*parkeri* [*Vignea*] 109, 338, 341, 423
˙*lessoniana* 310, 314, 325
˙*libera* 338
˙*litorosa* 289
longibrachiata 266
˙*maorica* [*fascicularis*] 310, 316–17, 330, 332
˙*muelleri* 368
ovalis 310, 342
˙*pleiostachys* 354
˙*pterocarpa* 414
°*pumila* 30, 353, 355–7
pyrenaica var. °*cephalotes* 12, 230, 414, 416–19, 422–3, 425–7, 429, 431
˙*resectans* 251, 368
˙*rubicunda* 337
scoparia 313
˙*secta* 30, 266, 302, 305–6, 310, 312, 314–18, 320, 324, 329–32, 344, 346, 402
˙*sectoides* [*secta*] 30, 444, 461–2
˙*sinclairii* 224–5, 310, 315–17, 319–20, 329–30, 332, 338, 342–3
˙*solandri* 168
˙*spinirostris* 397
˙*ternaria* [*darwinii*] 444, 448, 462–3
˙*testacea* 60, 262, 266, 354, 532
˙*trachycarpa* 338, 398
°*trifida* 191, 281, 310, 396, 446, 448, 453, 464
˙*virgata* 238, 266, 270, 310, 312–13, 316, 325, 522
˙*wakatipu* 235, 240, 251, 428–9
Carmichaelia 15–38˙ Fab. 5, 13, 53, 59, 72, 207, 234, 245, 251, 372, 432, 555
˙*aligera* [*australis*] (= *arborea* var.?) 383
˙*appressa* (= *arborea* var.? 356
˙*arborea* 193, 196, 208, 329, 385
˙*astonii* 385
˙*compacta* 107, 499, 501
˙*corrugata* 364
˙*enysii* 249, 368
˙*flagelliformis* (= *arborea* var.?) 384
˙*grandiflora* (= *odorata* var.?) 178, 196, 363, 366–7, 391, 509, 511–14, 528
˙*kirkii* [*gracilis*] 207
˙*monroi* 236, 251, 363, 369
˙*nigrans* (= *prona* syn.?) 363, 366
˙*ovata* [*subulata*] (= *arborea* var.?) 189, 268, 385
˙*petriei* (= *arborea* var.?) 42, 48, 196, 207, 235, 255,/373
˙*robusta* (= *arborea* var.?) 196, 205, 255, 271
˙*uniflora* 363
˙*violacea* (= *robusta* syn.?) 359
˙*williamsii* 41, 53, 383
Carpha ?2° Cyper. 12
°*alpina* 140, 224–5, 229, 243, 321, 326, 334–6, 339–43, 345–9, 391, 396, 400, 455, 513
aff. *schoenoides* 335
Carpobrotus 2* Aizo. 395

edulis 353, 356, 360, 396
Carpodetus 1˙ Escalloni.
˙*serratus* 23, 25, 32, 37, 52, 58, 61, 77, 119, 126, 138, 144, 151, 158, 162, 169–72, 202–3, 316, 516–17, 549, 574, 580
Cassinia 1˙ 1* Aster. 52, 206, 210–12, 216, 519–20, 543–4, 552
˙*leptophylla* 211–12, 351, 355, 359, 385, 387–8, 538
'*amoena*' (part of *leptophylla*) 200, 397
'*fulvida*' (part of *leptophylla*) 46, 196, 205, 211, 235, 262, 357, 502
'*retorta*' (part of *leptophylla*) 211, 355–6, 383
'*vauvilliersii*' [*albida*] (part of *leptophylla*) 186, 188–9, 211, 234, 243, 337, 339–40, 388, 397, 451–2, 454–5, 548, 550
Cassytha 1° 1* Laur. 41
°*paniculata* [*pubescens*] 36, 79, 92, 198, 200, 397
Catapodium [*Desmazeria*] 1* Po.
rigidum 388
Celmisia ?50˙ Aster. 5, 13, 28, 38, 40–1, 52, 54, 57, 60, 62, 78, 216–17, 219–20, 242, 248, 340, 391, 487, 547
˙*adamsii* (= *gracilenta* var.?) 379, 383
˙*allanii* 242, 428–9
˙*alpina* 335, 339
˙*angustifolia* 109, 236, 395, 416, 418, 426
˙*argentea* 338, 341–2, 411, 420–1
˙*armstrongii* 176, 219, 228–9, 513
˙*bellidioides* 389, 391, 395
˙*bonplandii* 109, 394
˙*brevifolia* 420, 529
˙*clavata* (part of *argentea*?) 340
˙*cordatifolia* 99
˙*dallii* 100, 337
˙*densiflora* 237
˙*discolor* [*intermedia*] (= *incana* syn.?) 219, 225, 513
˙*du-rietzii* (= *allanii* syn.?) 219, 232, 418
˙*glandulosa* 29, 38, 229, 233, 339, 343, 345, 431, 514
˙*gracilenta* (species aggregate) 38, 193, 205, 220, 237, 240, 245, 248, 253, 261, 320, 322, 339, 343, 345, 348–9, 418, 513
˙*graminifolia* (species aggregate) 231, 234, 322, 324, 335, 339
˙*haastii* 230, 412, 416–18, 420, 422–3, 425, 528
˙*hectorii* 109, 230, 418, 425, 529
˙*hieraciifolia* 219
˙*holosericea* 394
˙*hookeri* 379
˙*inaccessa* 103, 394
˙*incana* 219, 242, 247, 349
˙*laricifolia* 109, 224–5, 420–2, 426, 429
˙*lateralis* 430
˙*lindsayi* 379
˙*lyallii* 41–2, 219–20, 234, 236, 241–2, 395, 416, 418, 426, 487, 518, 528–9
˙*mackaui* 106, 379
˙*macmahonii* 99
˙*major* (= *gracilenta* var.?) 244, 379, 383, 431
˙*markii* 109
˙*monroi* 259, 379, 385, 388, 390, 428–9
˙*morganii* 379
˙*parva* 336

*petriei 41–2, 101, 109, 219, 231
*philocremma 14, 108
*polyvena 110
*prorepens (= densiflora syn.?) 241, 423, 425
*ramulosa 29
*rigida (= verbascifolia var.?) 396
*rutlandii 99
*semicordata [coriacea] (= monroi var.?) 29, 38, 190, 219, 224–5, 228–9, 234, 241, 379, 391–2, 429
*sessiliflora 220, 228, 241, 340, 411, 416, 418, 422, 427–9
*similis 336–7
*spectabilis 29, 57, 61, 219–20, 224–5, 234, 236, 238, 242, 247–8, 251, 257, 259, 262, 349, 385, 397–9, 428–9, 483–4, 488
*spedenii 108, 400–1
*thomsonii 54, 108
*traversii 219
*verbascifolia [incl. petiolata, rigida] 219, 225, 228, 231, 379, 391, 394, 418
*vespertina 109, 228, 392, 418
*viscosa 38, 57, 109, 219, 242, 259, 412, 420–7, 469
*walkeri 176, 219, 228, 230
Centaurium 2* Gentian.
  *erythraea [umbellatum] 271, 307, 384, 394, 543
Centella 1* Api.
  *uniflora [asiatica] 198, 275, 298, 304, 314, 327, 444–5, 534
Centranthus 1* Valerian.
  *ruber 396
Centrolepis [Gaimardia] 2* 1† 1* Centrolepid.
  *ciliata 321, 324, 327, 329–31, 334–5, 338–42, 344, 348, 454–5, 459
  *pallida 233, 303–4, 338, 342, 455, 460
Cephaloziella ?2 Leafy liverwort 322
Cerastium 5* Caryophyll. 254, 258, 262, 267, 274–5, 277, 388, 444–5
  *fontanum [triviale, vulgatum] 237, 269, 320, 359, 371, 429, 454, 518
  *glomeratum 271, 371
Ceratocephalus 1* Ranuncul.
  *pungens [Ranunculus falcatus] 368
Ceratodon 1 Moss
  purpureus 364, 417
Ceratophyllum 1* Ceratophyll.
  *demersum 301, 314
Cetraria 2 Fruticose lichen
  islandica subsp. antarctica [ericetorum] 416, 421–5
Chamaecyparis 1* Cupress. 156
Chamaecytisus 1* Fab.
  *palmensis 36, 213, 395, 500
Chandonanthus 1 Leafy liverwort
  squarrosus 453
Chara 5 Green alga 300, 305, 308
  corallina [australis] 303
  fibrosa 303
Cheesemania [Nasturtium] 5* Brassic. 13, 413
  *enysii 413, 417
  *fastigiata 28, 413, 428
  *gibbsii 430
  *latisiliqua 429
  *wallii 108
Cheilanthes [Notolaena] 2* Pterid. 382

°distans 384
°sieberi 269, 371, 384, 403
Cheiranthus 1* Brassic.
  *cheirii 395
Chenopodium 2* 1° 11* Chenopodi. 28
  *ambrosiides 353
  glaucum var. °ambiguum 291, 298, 353, 445–6
  *pumilio 267
Chiloglottis 1* 1° Orchid.
  °cornuta 199, 325, 577
Chiloscyphus 29 Leafy liverwort 182, 303
  billardierei 328
  compactus 325
Chionochloa [Danthonia] ?22* Po. 5, 13, 40, 50, 56–7, 62, 70, 73, 75–6, 78, 106–7, 151, 172, 192, 206, 216–17, 223, 245, 255–6, 338, 341, 344, 363, 395, 414, 418, 467, 471, 475, 496–7, 504, 514, 520, 557, 573
  *acicularis 30, 182–3, 217, 230–3, 336, 346, 394
  *antarctica 447, 453–5, 457–61
  *australis 99, 108, 218, 224–7, 337, 391, 427, 429, 468
  *beddiei 98, 264, 381, 385
  *bromoides 93, 263, 381, 383
  *cheesemanii 143
  *conspicua [cunninghamii] 39, 154, 178, 185, 217, 225, 243, 363, 372, 391–2, 551, 574, 581
  *crassiuscula 218–19, 226–33, 241–2, 279, 337, 416, 427, 487–8, 528, 557
  *defracta [rubra] 217, 242–3, 398–9
  *flavescens [raoulii] 90, 217, 218, 241, 243, 348, 486, 488
  *flavicans 96, 381, 384
  *juncea 217, 226, 336
  *lanea 218, 234, 396
  *macra 50, 61, 218, 222, 235, 238, 241–2, 259, 418, 420, 422–3, 425, 427, 468, 473, 476, 480, 482–3, 487–91, 493–4, 497, 499, 528
  *oreophila 218, 227, 229–30, 233, 241–2, 259, 414, 416–19, 425–7, 528–9
  *ovata 103, 218, 232, 394
  *pallens 39, 61, 74, 98, 217, 224–30, 233, 241–3, 259, 390, 392, 416, 418, 427–9, 468, 486–8, 491, 499, 512–14, 527–9, 547, 574, 576
  *pungens 218, 234, 340
  *rigida [flavescens] 50–1, 84, 109, 217–18, 232, 234–5, 238–42, 250, 255, 279, 340, 342, 468, 473, 476, 480, 482–4, 487, 488–91, 493–4, 528, 552
  *rubra [raoulii] 42, 186, 193, 217–18, 224–6, 235, 238, 241, 247, 250, 255, 261, 284, 336, 340, 342, 345–7, 400, 472, 482–4, 486–8, 511–13, 550
  *spiralis 218
  *teretifolia 218, 232
  'fiord' [flavescens] 218, 230–4, 396
  'robust' [flavescens; sp(a) in Druce et al. 1987] 90, 192, 217–18, 224–5, 227–9, 235, 238, 241–2, 250, 255, 279, 381, 388, 390–1, 430, 487–8, 527, 547, 574
  'westland' [aff. rigida] 176, 218, 226, 228–9, 345–6, 400, 512, 528
Chionohebe [Pygmea, Veronica] 3* 2° Scrophulari. 13, 52, 411, 417
  *armstrongii (= densiflora × ciliolata?) 109, 415
  °ciliolata 13, 41, 109, 419, 427, 430

Chionohebe [*Pygmea, Veronica*] 3˙ 2° (*cont.*)
  °*densifolia* [*Hebe dasaphylla*; *P.* or *V. tetragona*]
    415
  ˙*myosotoides* 107
  ˙*pulvinaris* 109, 416, 425–7, 429
  ˙*thomsonii* 109, 420–1
Chlamydomonas 15 Green alga 417, 463
Chodatella 5 Green alga
  brevispina 417
Chondropsis 1 Foliose lichen
  semiviridis 368
˙*Chorodospartium* 2˙ Fab. 13
  ˙*muritai* 41
  ˙*stevensonii* 41, 190
Christella [*Dryopteris, Thelypteris*] 2° Thelypterid.
  °*dentata* 400
Chrysanthemoides 1* Aster.
  monilifera [*Osteospermum*] 395
Chrysoblastella 1 Moss
  chilensis 364
Cirsium [*Cnicus*] 4* Aster.
  arvense 215, 262, 267, 273, 275, 278, 354, 359,
    365, 368, 370, 553
  palustre 266
  vulgare [*lanceolatus*] 254, 266, 269, 271, 275, 277,
    307, 354, 359, 368, 370, 388, 394, 441, 548
Cladia [*Cladonia*] 6 Fruticose lichen
  aggregata 203–4, 337, 343, 403
  retipora 193, 204–5, 234, 256–7, 337, 340, 403,
    444, 459
  sullivanii 333
Cladina [*Cladonia*] Fruticose lichen 193
  leptoclada [*Cladonia alpestroides*] (*confusus*) 193,
    203, 205, 240, 257, 331, 333, 343, 403
  mitis 343
Cladomnion 1 Moss
  ericoides 140, 187
Cladonia 41 Fruticose lichen 193, 256, 340, 343, 355,
    416, 424–5
  campbelliana 460
  capitellata 403
  solida 403
Cladophora 29 Green alga 305
Clematis 8˙ 5* Ranuncul. 34, 53–4, 58, 207, 544
  ˙*afoliata* 41, 196, 385
  ˙*cunninghamii* [*parviflora*] 163
  ˙*foetida* 25, 163, 385
  ˙*forsteri* [*incl. australis, colensoi, hexasepala,*
    hookeriana, petriei] 163, 170, 359, 385
  ˙*marata* 25, 196
  ˙*marmoraria* 100
  ˙*paniculata* [*indivisa*] 52, 163–4, 388
  ˙*quadribracteolata* 196
  ˙*vitalba* 163, 396, 521, 544, 553
  '*australis* var. *rutifolia*' (part of *forsteri* complex)
    397
Clianthus 1˙ Fab.
  ˙*puniceus* 53, 96, 377
Climacium 1 Moss
  dendroides 312
Coelocaulon 1 Fruticose lichen
  aculeata [*Cornicularia*] 421
Collospermum [*Astelia*] 2˙ Asphodel. 35, 52–3, 58
  ˙*hastatum* 78, 117, 119, 126, 130, 133, 170
  ˙*microspermum* 136

Colobanthus 11˙ 2° Caryophyll. 10, 34, 41, 252, 342,
    362, 398, 400, 411
  ˙*acicularis* 429
  °*affinis* 411, 428, 431
  °*apetalus* [*crassifolius*] 395, 411, 454, 461–2, 465
  ˙*brevisepalus* 236, 370
  ˙*buchananii* 395, 416–17, 423, 426, 429, 529
  ˙*canaliculatus* 109, 419, 423, 429
  ˙*hookeri* 455
  ˙*masoniae* (= *wallii* syn.?) 429
  ˙*monticola* 419
  ˙*muelleri* 354, 396
  ˙*muscoides* 455, 461, 465
  ˙*strictus* [*muelleri*] 371, 385
  ˙*wallii* [*quitensis*] 397
Colocasia 1* Ar.
  esculenta 16
Conium 1* Api.
  maculatum 28, 165
Conostomum 3 Moss 417
Convolvulus 2˙ 2* Convolvul.
  ˙*fractosaxosa* 374, 429
  ˙*verecundus* [*erubescens*] 252, 257, 268, 270, 370,
    385
Conyza [*Erigeron*] 5* Aster. 267, 365, 384, 388
  albida [*bonariensis, floribunda*] 355
  bonariensis 403, 436
Coprosma c.48˙ 2° Rubi. 5, 9, 13, 16, 27–8, 52, 59,
    63, 78, 119–20, 145, 148–9, 162, 202, 207, 211,
    242, 312, 326, 363, 372–3, 397, 458, 486, 488,
    520, 547, 565, 573, 575
  ˙*acerosa* 351, 355–7, 385, 446, 521
  ˙*acutifolia* 435
  ˙*arborea* 25, 166, 389, 532, 536–8
  ˙*areolata* 133, 325, 487
  ˙*atropurpurea* [*petriei*] 34, 249, 262, 368
  ˙*banksii* (part of *colensoi*) 574
  ˙*brunnea* (= *acerosa* var.?) 264, 351, 363
  ˙*chathamica* 437, 439–40, 442, 444
  ˙*cheesemanii* 33, 192, 224, 228–9, 231–2, 236, 240,
    247, 398, 513
  ˙*ciliata* [*myrtillifolia*] 65, 119, 128, 135, 151, 158,
    172, 184, 189, 196, 209, 211, 392, 454, 457,
    461–2, 574
  ˙*colensoi* [incl. *banksii*] 128, 151, 158, 176–7, 181,
    328, 516, 574
  ˙*crassifolia* 133, 209, 351, 359, 385, 487
  ˙*crenulata* 231, 513
  ˙*cuneata* [incl. *astonii*] 135, 151–2, 158, 181, 452,
    454–5, 457, 461
  ˙*depressa* [*ramulosa*] 119, 147, 159, 178, 189, 228,
    236, 243–4, 550, 552
  ˙*dodonaeifolia* [*lucida*] 188
  ˙*foetidissima* 77, 119, 128, 135, 139, 144, 147, 151,
    158, 181, 184, 187, 304, 328, 330, 389, 391,
    451–2, 454–5, 516–17, 551–2, 565, 574, 577
  ˙*grandifolia* [*australis*] 23, 37, 70, 116, 119, 126,
    139–40, 215, 389, 486–7, 500, 536, 537–9, 579
  ˙*intertexta* 196, 205, 210
  ˙*linariifolia* 80, 146, 171, 389
  ˙*lucida* 23, 56, 58, 116, 119, 126, 128, 139, 151,
    159, 162, 168, 188, 200, 203–4, 487–8, 502, 538–
    9, 595
  ˙*macrocarpa* 58, 92, 166, 502, 532
  ˙*microcarpa* 144–6, 152

°*niphophila* 411
obconica 388, 397
°*perpusilla* [*pumila*] 33, 58–9, 228–9, 241, 338, 342–3, 345, 411, 416, 418, 420, 422–6, 428, 431, 454–5, 458–60, 462–5, 528–9
petiolata 435–6
petriei 33, 236, 251, 261, 368
propinqua 25, 32, 50, 58, 61, 170, 193–6, 205, 207–10, 225, 234, 243, 261, 272, 291, 298, 304, 306, 315–16, 320, 325, 329–30, 346, 351, 359–60, 385, 387–9, 392, 394, 396, 398, 400, 438, 442, 487, 498, 501, 513, 517, 521, 540, 543
pseudocuneata [*egmontiana*] 27, 119, 135, 144, 147, 151, 158, 176–8, 184, 186, 189, 192, 236, 395, 513–14, 549–50, 552, 562, 574
repens [*baueri, retusa*] 23, 166–7, 383, 386–7, 553, 579
rhamnoides [*polymorpha*] 58, 61, 128, 145, 151, 158, 170, 173, 201, 205, 209, 275, 389, 397, 487, 502, 516, 532, 574
rigida 25, 196, 325, 351, 359
robusta 23, 25, 32, 37, 58, 61, 156, 162, 166, 306, 314, 316, 360, 384, 403, 481, 487, 493–6, 498, 500, 520, 524, 533–4, 539–40, 543
rotundifolia 37, 58, 126, 159, 173, 487, 565
rubra 133, 487
rugosa [incl. *antipoda*] 178, 196, 209, 228, 367, 388, 391–2, 461–2, 511–12, 517
serrulata 28, 176, 229, 391–2
spathulata 25, 397
talbrockiei 14, 336
tenuicaulis 210, 316, 324–5
tenuifolia 119, 138, 186
virescens 133
waima 93
wallii 196
aff. *intertexta* [sp.(f) in Eagle 1982] 322, 327, 330, 332, 345
aff. *parviflora* [*myrtillifolia, parviflora,* sp.(t) in Eagle 1982] 145, 147, 151, 158, 172, 184, 186–7, 193, 196, 202, 206, 209, 214, 304, 315–16, 327, 329–30, 388, 398, 527, 549–50, 574
'*parviflora* var. *dumosa*' [sp.(p) in Eagle 1982] 244, 502
'penalpine' [sp.(a) in Eagle 1982] 192, 390
Corallospartium 1 * Fab. 13
crassicaule [*racemosa*] 41, 189, 196
Cordyline 5 * 1* Asphodel. 9, 40, 52, 54, 56, 58, 72
australis 24, 61, 64, 68, 105, 162, 234, 306, 325, 355, 393, 481, 607
banksii 24, 389, 533
indivisa 24, 37, 188
kaspar 502, 532
pumilio 24, 37
terminalis 16, 432
Coriaria 7 * Coriari. 9, 40, 52, 58, 209, 264, 372, 496, 506, 520, 555
angustissima 372
arborea 24, 37, 61, 162, 170, 215, 329, 366, 368, 372, 383–4, 388–9, 391–2, 394, 475–6, 479, 481, 489–90, 497, 506, 509, 511, 520, 524, 540, 543
kingiana [*lurida*] 372, 388
plumosa [*lurida*] 228, 244, 372, 391–2
pottsiana 96, 372
pteridoides 372

sarmentosa [*ruscifolia*] 245, 372, 388, 521
Corokia 3 * Corn. 53–4, 58, 63
buddleioides 25, 139, 188
cotoneaster 25, 193, 196, 207, 385, 394, 397–8, 400, 520
macrocarpa 58, 61–2, 440–1
Coronopus 2* Brassic.
didymus 267
Cortaderia [*Arundo*] 5 * 2* Po. 39, 263, 311, 363, 372, 403
fulvida [*A. conspicua*] 156, 354, 368, 373, 524, 531
richardii [*A. conspicua*] 245, 316, 318, 320, 329, 354, 358, 388–9
selloana 354–5, 531
splendens 311, 354, 383, 397
toetoe 198, 314, 337, 355
turbaria 444
Corybas [*Corysanthes*] 6 * 2° Orchid. 39
°*aconitiflorus* 199
macranthus 199, 392
oblongus 199, 577
rivularis [*orbiculatus*] 325
trilobus 517
Corynocarpus 1 * Corynocarp.
laevigatus 16, 22, 37, 58, 61–2, 64, 77, 79, 80, 117–18, 162, 167, 384, 434, 532
Cotula 2† 2* Aster.
†*australis* 268, 271, 355
†*coronopifolia* 17, 289, 291, 293, 295–6, 299, 372, 445
Cotyledon 1* Crassul.
orbiculata 395–6
Craspedia 6 * Aster. 220, 237, 249, 253, 349, 362, 379, 388, 412
incana 41, 412
lanata [var. *elongata* = sp.(a) in Druce & Williams 1987] 237, 429
major (= *minor* syn.?) 389, 392
minor [*maritima*] 238
robusta 379, 394
uniflora (species aggregate) 342, 379, 385, 460, 529
viscosa 238
Crassula [*Tillaea*] 7 * 6° 6* Crassul. 38, 288, 485
°*helmsii* 288
°*moschata* 379, 396, 446, 454–5, 461, 465
multicaulis 339
sieberiana [records include *tetramera*] 42, 268, 270, 370, 379, 384
sinclairii 303, 307
tetragona 396
Crataegus 1* Ros.
monogyna 212, 479, 500, 540
Cratoneuropsis 1 Moss
relaxa 382
Crepis 4* Aster.
capillaris 214, 238, 249, 253, 257, 262–3, 266, 269, 271, 273, 364, 370, 394
Crocosmia 2* Irid.
* ×*crocosmiiflora* 30
Cryptostylis 1° Orchid.
°*subulata* 12, 324
Ctenopteris [*Grammitis, Polypodium*] 1° Grammitid.
°*heterophylla* [*P. grammitidis*] 35, 39, 128, 385, 442
Cuscuta 4* Convolvul.
epithymum 36, 366

Cyathea 5˚ 2° Cyathe. 8, 118, 124, 539
˚colensoi [Alsophila] 31, 39, 77, 119, 135–6, 144, 151, 172, 389, 574
°cunninghamii 114, 119, 132, 439, 441
°dealbata [tricolor] 114, 116–17, 119, 124–6, 133, 146, 156, 169, 203, 439, 524, 532, 537–8, 544
˚kermadecensis 435
°medullaris 80, 114, 116–17, 126, 132, 156, 168, 441, 481, 500, 537, 540, 543–4, 580
˚milnei 435
˚smithii [Hemitelia] 77, 80, 82, 114, 119, 124–5, 128, 135, 137–8, 140, 144, 151, 158, 170–3, 439, 451, 549, 574
Cyathodes [Styphelia] 6˚ 2° Epacrid. 27, 42, 58, 145, 526
˚colensoi [Leucopogon colensoi, S. °suaveolens] 195, 205, 224, 236, 240, 251, 257, 430
˚empetrifolia [S. taxifolia] 26, 61, 174, 181, 187, 247, 322, 324, 332, 335, 339, 349, 454–5
˚fasciculata [Leucopogon] 26, 61, 70, 116–18, 120, 124, 144, 146–7, 152, 198–9, 202–4, 324, 383, 389, 398, 517, 524, 533–4, 536–7, 544, 574–5
˚fraseri [Leucopogon fraseri, S. nesophila] (var. muscosa merits species rank?) 28, 61, 193, 195, 198, 200, 204, 236, 240, 251, 257, 261, 268, 324, 354, 368–9, 383, 397, 418, 430, 521
°juniperina [acerosa, oxycedrus] 26, 37, 119, 144, 146–7, 170, 176, 195, 200, 202–5, 330, 339, 394, 397–8, 520, 527, 533, 577
°parviflora [Leucopogon] 200, 397, 446
˚pumila [C. °dealbata, S. minuta] 28, 229, 338, 340–1, 430
˚robusta 443
Cyclosorus [Dryopteris, Thelypteris] 1° Thelypterid.
°interruptus [D. or T. gongylodes] 312, 323
Cynodon 1* Po.
*dactylon 265, 353, 355
Cynosurus 2* Po.
*cristatus 265, 270, 272–3, 275–6, 279, 316–17
*echinatus 271
Cyperus 1˚ 13* Cyper. 310
*brevifolius 435
*congestus 266
*eragrostis 266, 320
*polystachyos 266
*tenellus 266
˚ustulatus [Mariscus] 266, 289, 293, 306, 310, 312, 354, 360, 435–6
Cystocoleus 1 Filamentous lichen
niger 519
Cystopteris 1° 1* Dryopterid.
°tasmanica [fragilis, novae-zelandiae] 375, 382, 429
Cytisus 2* Fab.
*multiflorus 213
*scoparius 59, 157, 198, 314, 352, 395, 500–1, 540–1
Cyttaria 3 Fungus 143

Dacrycarpus [Podocarpus] 1˚ Podocarp. 6, 38
˚dacrydioides 16, 20, 57, 61, 63, 65, 102, 112, 119, 121, 126–7, 154, 284, 304, 328, 332, 469–70, 502, 510, 537, 562
Dacrydium 1˚ Podocarp. 6, 8, 38
˚cupressinum 14, 16, 20, 40, 44, 55, 57, 61, 63, 65, 70, 74, 77, 79, 112, 117, 119, 121, 124, 126–7,

131, 145, 151, 154, 159, 168, 203, 286, 327–8, 400, 468–70, 472, 475, 477–8, 481, 488, 502, 509, 537, 561, 574, 577, 595
˚Dactylanthus 1˚ Balanophor.
˚taylorii 36
Dactylis 1* Po.
*glomerata 90, 165, 248, 250, 261, 268, 270, 273–4, 277, 279, 320, 355, 444, 484
˚Damnamenia [Celmisia] 1˚ Aster. 13, 54
˚vernicosa [campbellensis] 449, 455–6, 459
Daucus 1° 1*
*carota 267
°glochidiatus 268
Dawsonia 1 Moss
superba 112
Dendrobium 1˚ Orchid.
˚cunninghamii 29, 35, 117, 128, 140, 204
Dendroligotrichum 1 Moss
dendroides 135
Deparia 1˚ 1° Dryopterid.
°petersenii [Athyrium or Diplazium japonicum] 164
Deschampsia 4˚ 1° Po. 248, 381
°caespitosa 12, 298, 311, 338, 344, 465
˚chapmanii 307, 454, 459
˚tenella 307, 460
˚Desmoschoenus 1˚ Cyper.
˚spiralis [Scirpus frondosus] 39, 352, 356, 446, 521
Deyeuxia 4˚ 2° Po. 248, 264, 539
˚aucklandica 396
˚avenoides 187, 205, 235, 240, 250, 262–3
°billardierei 383, 397
°quadriseta 264
Dianella 1° Phormi. 58
°nigra [intermedia] 156, 198–9, 204, 324–5, 335, 389, 403, 533
Dianthus 4* Caryophyll.
*armeria 254, 371
Dichelachne 3˚ 2* Po. 12, 263–4
°crinita 235, 248, 250, 263, 268, 367, 384–5, 394, 403
Dichondra 1˚ 1° 1* Convolvul. 29, 281
˚brevifolia 266, 445
°repens 252, 266, 268, 270, 275, 337, 370, 383, 444, 543
Dicksonia 3˚ Dicksoni. 8
˚fibrosa 114, 123–4, 126, 156, 173, 439, 441, 525, 544
˚lanata 31, 39, 116, 120, 128, 144, 187, 575, 580, 585
˚squarrosa 31, 114, 116, 119, 124, 128, 135, 140, 151, 156, 158, 170–1, 173, 203, 328, 366, 439, 441–2, 481, 542, 544, 574, 577, 595, 599
Dicnemon 3 Moss
calycinum 140
Dicranoloma 9 Moss 144, 146, 148, 176, 182, 187–8, 231, 339–40, 442, 458, 586
billardierei 312, 325, 328, 343
plurisetum 147
robustum 147, 187, 202, 205, 223, 322, 326, 331, 333, 337, 340, 348, 398, 442, 454, 459–60, 465, 513
Dicranopteris 1° Gleicheni.
°linearis [Gleichenia] 400, 403
Dicranoweisia 2 Moss
antarctica 395, 417, 461

Digitalis 1* Scrophulari.
 *purpurea 165, 372
Digitaria 4* Po.
 *sanguinalis 265, 403
Discaria 1˙ Rhamn. 6, 12, 41–2
 ˙*toumatou* 25, 39, 50, 54, 59, 62, 72, 101, 172, 196, 205, 217, 235, 251, 261–2, 268, 363, 369, 479, 496–7, 506, 517
Disphyma [*Mesembryanthemum*] 2˙ 1* Aizo.
 ˙*australe* 353, 378, 383–4, 386–7, 396, 408, 436
 ˙*papillatum* [*australe*] 445–6
Distichium 1 Moss
 capillaceum [difficile] 382
Ditrichum 13 Moss
 strictum [*Blindia maxwellii*] 465
Dodonaea 1˙ Sapind.
 ˚*viscosa* 15, 23, 37, 52, 58, 61, 79, 132, 166–7, 352, 356–7, 385, 388–9, 520
˙*Dolichoglottis* [*Senecio*] 2˙ Aster. 13, 38, 40, 379
 ˙*lyallii* 53, 220, 340, 380, 389, 529
 ˙*scorzoneroides* 38, 52, 220–1, 231–2, 380, 391–2, 419, 529
Donatia 1˚ Donati. 12
 ˚*novae-zelandiae* 224–5, 243, 334–6, 338–43, 345–8, 400, 513–14
Doodia 2˙ 3˚ Blechn. 39
 media subsp. ˚*australis* 166, 202, 397, 532–3, 542
Dothistroma Fungus
 *pini 565
Dracophyllum c.30˙ Epacrid. 5, 13, 26–7, 40, 59, 72, 74, 78, 81, 174, 190, 206, 209–11, 216, 323, 458–9, 562
 ˙*acerosum* [*acicularifolium*] 46, 189, 236, 503
 ˙*adamsii* 188
 ˙*arboreum* 438, 442, 444
 ˙*filifolium* (= *longifolium* var.?) 98, 185–6, 189, 193, 389, 472, 549, 551
 ˙*fiordense* 27, 37, 178, 180, 378, 391, 394
 ˙*kirkii* 378, 391, 395, 410, 415–18
 ˙*latifolium* (incl. *matthewsii*) 27, 116, 154, 537
 ˙*lessonianum* 26, 198–9, 324–5
 ˙*longifolium* (incl. *oliveri*) 26, 38, 61, 70, 77, 83, 158, 171, 176–9, 182, 184–5, 189, 191–2, 196, 225, 229, 231, 234, 244, 328–30, 332, 334, 339–40, 344, 346, 388, 391, 393, 397–8, 447, 451–5, 457, 502, 512–14, 521, 527, 548–52, 574
 ˙*menziesii* 27, 37, 101, 180, 182, 231
 ˙*muscoides* 38, 109, 411, 420–5, 469, 530–1
 ˙*palustre* 194, 327, 335, 337
 ˙*pearsonii* 339
 ˙*politum* 109, 338–40, 378, 396, 415
 ˙*pronum* 61, 109, 236, 242, 398–9, 415–16, 418, 425–7, 430
 ˙*prostratum* (part of *muscoides*?) 338, 340, 342, 344
 ˙*pubescens* 378, 390
 ˙*recurvum* 27, 77, 98, 186, 192, 243, 247, 339, 349, 415, 430–1
 ˙*scoparium* [*paludosum, subantarcticum*] 442–4, 457, 459
 ˙*strictum* 27, 312, 378
 ˙*subulatum* 26, 93, 193–4, 196, 208, 259, 326, 337, 368, 524–6
 ˙*townsonii* 27, 101, 184
 ˙*traversii* (incl. *pyramidale*) 27, 57, 61–3, 70, 77, 101, 140, 144–5, 154, 158, 175–6, 178, 180, 184,

186, 188, 389, 391, 478, 481, 512–14, 548–9, 551, 562, 565, 574–5
 ˙*trimorphum* (= *pubescens* var.?) 378
 ˙*uniflorum* 26, 39, 77, 83, 98, 178, 189, 192, 206, 224–5, 229, 231–2, 235, 243, 262, 339–40, 388, 390–1, 415, 430, 478, 512–14, 521, 527–8, 548, 574
Drapetes [*Kelleria*] 5˙ Thymelae. 10, 398
 ˙*dieffenbachii* 231, 236, 428, 529
 ˙*lyallii* (South Island plants merit separate species rank) 396, 411, 414, 416–20, 422–6, 528–9
 ˙*villosus* 422
Drepanocladus 5–6 Moss 312, 333, 345
 aduncus 465
 fluitans 301, 307, 339
 fontinaliopsis 303, 305
Drosanthemum 1* Aster.
 *floribundum 395
Drosera 1˙ 5˚ Droser. 36
 ˚*arcturi* 224–5, 321, 340, 342–3
 ˙*binata* 312, 321, 323–5, 327, 330–31, 334, 442, 444
 peltata subsp. ˚*auriculata* [*auriculata*] 35, 198, 533
 ˚*pygmaea* 198, 321
 ˚*spathulata* 321, 324–5, 331, 334–5, 340
 ˙*stenopetala* 174, 321, 339, 454
Drymoanthus [*Sarcochilus*] 1˙ Orchid.
 ˙*adversus* 35
Dryopteris 3* Dryopterid.
 *filix-mas 164, 500, 505, 521
Dysoxylum 1˙ Mel. 8
 ˙*spectabile* 22, 37, 54, 58, 77, 112, 117–19, 132, 162, 167, 533, 546, 560

Earina 2˙ Orchid. 29, 35, 117, 128
 ˙*autumnalis* 151, 204
 ˙*mucronata* 204, 389, 441
Echinochloa 3* Po.
 *crus-gallii 265
Echinopogon 1˚ Po.
 ˚*ovatus* 265, 271
Echium 4* Boragin.
 *candicans 395
 *vulgare 365, 368, 370, 388, 429
Egeria 1* Hydrocharit.
 *densa 301, 320
Eichhornia 1* Pontederi.
 *crassipes 301
Einadia 2˙ 1˚* 1* Chenopodi. 35, 379
 ˙*allanii* [*Chenopodium*] 360, 383
 *nutans [*Rhagodia*] 396
 ˙*triandra* [*Rhagodia nutans*] 58, 167, 383–4, 386–7
Elaeagnus 1* Elaeagn.
 * ×*reflexa* 163, 544
Elaeocarpus 2˙ Elaeocarp. 8, 21, 58, 64, 121
 ˙*dentatus* 25, 37, 43, 61–2, 65, 69, 77, 79–80, 112, 116–17, 119, 121, 124, 127, 169, 472, 481, 502, 520, 537, 554, 574
 ˙*hookerianus* 21, 25, 37, 69, 112, 121, 124, 127–8, 144, 151, 154, 178, 304, 316, 328, 525, 575, 589, 595, 598
Elatine 1˙ Elatin. 303, 307
 ˙*gratioloides* 288, 303, 316
Elatostema 1˙ Urtic.
 ˙*rugosum* 38, 79, 98, 119, 380

Eleocharis [*Heleocharis*] 1˙ 4° Cyper.
  °*acuta* 270, 272, 302–4, 312–13, 315–17, 320, 329–
    30, 332, 342, 344, 346, 355, 444–5
  °*gracilis* [*cunninghamii*] 275, 298, 310, 314, 316
  ˙*neozelandica* 355–6
  °*sphacelata* 302, 312, 324, 331, 349
˙*Elingamita* 1˙ Myrsin. 11
  ˙*johnsonii* 92, 502
Elodea 1* Hydrocharit.
  *canadensis 301, 303, 305–7, 320, 468, 553
Elymus 5˙ 1˙† 1* Po. 248, 263–4
  ˙*apricus* [*Agropyron scabrum*] 107, 245, 258, 279
  ˙*laevis* [*Asperella, Cockaynea*] 520
  ˙*multiflorus* [*Agropyron kirkii*] 264, 397
  ˙*narduroides* [*Asperella* or *Cockaynea gracilis*] 259,
    264, 390
  *pycnanthus [*Agropyron junceiforme*] 289
  ˙*trectisetus* [*Agropyron scabrum*] 215, 235, 248,
    250, 259, 261, 263–4, 268–70, 279, 367, 369,
    386–8, 390, 394
  ˙*tenuis* [*Agropyron scabrum*] 60, 259
Elytrigia 2* Po.
  *pungens [*Agropyron, Elymus*] 290, 299
  *repens [*Agropyron*] 353
˙*Embergeria* 1˙ Aster. 11
  ˙*grandifolia* [*Sonchus*] 38, 445–6
Empodisma 1° Restion. 30
  °*minus* [*Calorophus minor, Hypolaena lateriflora*]
    50, 130, 176, 193, 198, 224–6, 230, 234, 242,
    247, 313, 321, 323–8, 330–2, 334–8, 341, 343,
    345–9, 511
˙*Entelea* 1˙ Tili. 11, 59
  ˙*arborescens* 23, 37, 61, 70, 162, 166, 478, 481, 506
Enteromorpha 10 Green alga 288
  nana 303
Epacris 2˙ 1* Epacrid.
  ˙*alpina* 26, 247
  ˙*pauciflora* 26, 61, 188, 199, 323–5
Epilobium c.34˙ 6° 5* Onagr. 5, 10, 12, 13, 53–4, 58,
    249, 367, 370, 441, 462, 465, 548
  ˙*alsinoides* [*elegans, findlayi, novae-zelandiae*]
    (*cockayneanum*, subsp. *atriplicifolium*, subsp.
    tenuipes merit species rank?) 230, 237, 253, 261,
    343, 462, 530
  ˙*angustum* [*nummularifolium*] 309
  °*billardiereanum* [*cinereum, junceum, tetragonum*]
    53
  ˙*brevipes* 388
  °*brunnescens* [*pedunculare*] 13, 28, 38, 344, 362,
    366, 394, 411, 426, 458, 511, 518
  ˙*chionanthum* 316, 344
  ˙*chlorifolium* 214, 237
  ˙*cockayneanum* (part of *alsinoides* complex) 428
  ˙*confertifolium* 451, 454, 460
  ˙*crassum* 373, 395, 412
  ˙*forbesii* 407, 429
  ˙*glabellum* [*erubescens*] (*rubromarginatum,
    vernicosum* merit species rank?) 38, 237, 253,
    259, 362, 366, 380, 392–3, 395, 406, 412, 416,
    419, 429–31, 511, 529
  °*gunnianum* 12
  ˙*hectorii* [*simulans*] 240, 253
  ˙*insulare* 311, 316
  ˙*komarovianum* [*nerteroides*] 303, 445
  ˙*macropus* 339, 395, 429

  ˙*matthewsii* 380
  ˙*melanocaulon* 28, 47, 362, 518, 529
  ˙*microphyllum* 362, 366
  ˙*nerteroides* [*pedunculare*] 330, 362, 392
  ˙*nummularifolium* 372
  °*pallidiflorum* 311–13, 316
  ˙*pedunculare* [*linnaeoides*] 316, 392, 454, 458
  ˙*pernitens* 529
  ˙*porphyrium* 259, 395, 412, 416–17, 428, 430, 529
  ˙*purpuratum* 109, 529
  ˙*pycnostachyum* 109, 259, 395, 407–8, 417, 425,
    429–30
  ˙*rostratum* 253
  ˙*rubromarginatum* (part of *glabellum* complex) 417,
    429
  °*tasmanicum* 13, 109, 411, 417, 426, 428, 451
  ˙*wilsonii* 53, 380, 388
Equisetum 1* Equiset.
  *arvense 365
Eragrostis 4* Po. 403
  *brownii [*elongata*] 265, 275, 355
Erica 6* Eric. 212, 543–4
  *arborea 197, 498, 539
  *lusitanica 196–7, 204
Erodium 4* Gerani. 267
  *cicutarium 269, 271, 275, 371
  *moschatum 271
Erophila 1* Brassic.
  *verna [*Draba*] 370
Eryngium 1° 3* Api. 42
  °*vesiculosum* 29, 40, 298, 354
Eschscholzia 1* Papaver.
  *californica 365
Eucalyptus 20* Myrt. 6, 9, 12, 156, 545
Euphorbia 1˙ 14* Euphorbi. 6
  ˙*glauca* 353, 446
  *peplus 371, 388
Euphrasia 15˙ 1* Scrophulari. 10, 12, 36, 52, 59,
    221, 413
  ˙*cockayneana* 53
  ˙*cuneata* [*tricolor*] 247
  ˙*disperma* [*Siphonidium longiflorum*] 54, 64, 95,
    327, 335
  ˙*dyeri* 340, 342
  ˙*monroi* 397
  ˙*petriei* 413, 418–19, 529
  ˙*repens* 298
  ˙*revoluta* 413, 418
  ˙*townsonii* 413
  ˙*zelandica* [*antarcticum*] 221, 237, 253
Eurhynchium 6 Moss
  austrinum 301
Ewartia 1˙ Aster.
  ˙*sinclairii* 379
Exocarpus 1˙ Santal. 41
  ˙*bidwillii* 36, 175, 390

Festuca 4˙ 1° 1˙* c.5* Po. 30, 73, 248, 429, 497
  *arundinacea 265, 279, 289–90, 293, 299, 360, 365
  °*contracta* 463, 465
  ˙*coxii* 445–6
  ˙*matthewsii* 138, 228, 235, 238–9, 259, 264, 367,
    390, 392, 512–13, 518
  ˙*multinodis* 235, 265, 388
  ˙*novae-zelandiae* 39, 50–1, 234–5, 238, 250, 258,

261, 263, 265, 268, 279, 306, 354, 369, 468, 483–4, 489, 491, 494, 497, 499, 520, 552
 *rubra 12, 259–61, 264–5, 268, 272, 274–6, 279, 281, 320, 360, 364, 374, 394, 518
 *tenuifolia 260–1
 'petriei' 265, 271
Fissidens 20 Moss
 asplenioides 303
 rigidulus 301, 382
Foeniculum 1* Api.
 *vulgare 267, 354, 396
Forstera 4* Stylidi. 10, 13, 222, 340
 *bidwillii 431
 *mackayi 391
 *sedifolia (var. oculata merits species rank?) 229, 231–2, 418, 527
 *tenella 237, 391, 394
Fragaria 1* Ros.
 *vesca 394
Frankia (= Plasmodiophora syn.?) Actinomycete 496
Freycinetia 1° Pandan. 8, 170, 173
 °baueriana var. *banksii [banksii] 34–5, 37, 53, 68, 77–9, 93, 112, 119, 128, 132, 140, 168, 522–3, 537
Frullania 24 Leafy liverwort 182, 452
Fuchsia 3* 2* Onagr. 13, 53, 58
 *excorticata 23, 37, 39–40, 53–4, 61, 121, 124, 151, 158, 162, 173, 372, 451, 493, 500, 502, 516, 565, 574
 *perscandens 34, 360
 *procumbens 27, 58
Fumaria 5* Fumari.
 *muralis 28
Furcraea 1* Agav.
 *foetida 435

Gahnia 5* 1° Cyper. 60, 112, 335
 *lacera 533
 *pauciflora 117, 170, 204, 389
 *procera 134, 175–7, 180, 182, 186, 188, 339, 349, 398, 574
 *rigida [robusta] 321, 324, 327–30, 335, 522
 *setifolia 139, 198, 204, 536, 539
 °xanthocarpa 39, 116, 118, 120, 130, 132, 188, 310, 316, 329–30, 522, 531, 536
Gaimardia 1° Centrolepid. 12
 °setacea 335, 338, 340–2
Galium 2* 1° 9* Rubi.
 *aparine 215, 267, 359, 394
 *palustre 267, 306, 311
 *perpusillum [Asperula] 236, 264, 275, 309, 370, 373
 °propinquum 307
 *trilobum [tenuicaule] 388
 *uliginosum 277
Gastrodia 2* 1° Orchid.
 *cunninghamii 36
 *minor 36
 °sesamoides 36
Gaultheria 7* 1° 1* Eric. 13, 27, 58, 524–5
 *antipoda 27, 188, 517, 533
 *colensoi 192, 247, 430
 *crassa [rupestris] 37, 61, 158, 186, 189, 192, 236, 378, 388, 391–2, 395, 517, 520, 529, 547
 °depressa (var. novae-zelandiae merits species

rank?) 28, 61, 224, 236, 251, 261, 392, 411, 416, 418, 431, 529
 *oppositifolia 378
 *paniculata 378
 *rupestris 27, 209, 231, 378, 391
 *subcorymbosa (= rupestris syn.?) 378
Gazania 2* Aster. 395
Geniostoma 1° Logani. 9, 52, 59
 °rupestre var. *ligustrifolium [ligustrifolium] 37, 61, 64, 116, 119, 123, 166, 203, 397, 532, 536–40, 573, 575, 578
Gentiana [Gentianella] c.25* Gentian. 5, 10, 12, 52–4, 59–60, 62–3, 242, 245, 248, 253, 366, 397
 *antarctica 457, 459–60
 *antipoda 461–2
 *astonii 385
 *bellidifolia [flaccida] 229, 343–4, 398, 418, 421, 428, 430
 *cerina 450, 455
 *chathamica 442
 *concinna 450, 454
 *corymbifera 28, 38, 236, 248, 341, 416, 418, 529
 *divisa 28, 392, 395, 413, 419, 529
 *filipes 429
 *grisebachii 28, 304, 339
 *lineata 340
 *matthewsii 339, 342
 *montana 231, 429
 *patula (= montana syn.?) 388, 418
 *saxosa 354, 396
 *spenceri 335
 *tenuifolia 397
 *townsonii 335
Geranium 4* 3° 8* Gerani. 38, 53, 249
 *microphyllum 230, 237, 244, 253, 257, 360, 458
 *molle 267, 271–2, 278, 372
 °retrorsum [pilosum] 396
 *robertianum 372
 *sessiliflorum 29, 38, 237, 240, 253, 257, 261, 268, 270, 354, 357, 362, 370, 429
 *traversii 445–6
Geum 4* 1° 1† 1* Ros. 221
 *leiospermum 237
 °parviflorum [albiflorum] 38, 222, 391, 394–5, 455
 *uniflorum 109, 238
Gingidia 5* 1° Api.
 *decipiens [Angelica] 109, 237, 394, 429, 529
 *flabellata [Anisotome] 396
 °montana [Angelica] 13, 38, 58, 245, 264, 379, 385, 388–9, 391, 395, 528
Glaucium 2* Papaver.
 *flavum 354, 388
Gleichenia 2° Gleicheni. 8, 72–3, 200, 274, 284, 313, 327, 334, 525, 557
 °dicarpa [alpina, circinnata] 181, 193, 198–9, 201, 215, 226, 230, 234, 247, 312, 322, 324–8, 333–7, 339, 343, 348–9, 442–3, 511, 542
 °microphylla [circinnata] 176, 215, 322, 324–5, 542
Glossostigma 1* 1° Scrophulari. 288, 307, 309
 °elatinoides 298, 303, 316
 *submersum 303
Glyceria 4* Po. 276, 302, 314
 *declinata 279, 312–13, 320
 *fluitans 316, 372
 *maxima 314, 320

Gnaphalium 7˙ 7° 8* Aster. 33, 202, 214, 362
  ˙audax [collinum] 38, 262, 444
  *coarctatum [purpureum, subspicatum] 267, 274,
    365
  °delicatum [collinum] 277, 337
  °limosum [collinum] 316
  ˙mackayi 307, 411, 423
  ˙paludosum 339, 345, 419
  ˙ruahinicum [audax, collinum] 512
  °sphaericum [involucratum, japonicum] 403
  *subfalcatum 403
  °traversii [mackayi] 234, 339
  'collinum' [incl. several spp.] 236, 252, 257, 268,
    314
Goebelobryum 1 Liverwort 415
  unguiculatum 325
Gonocarpus [Haloragis] 3˙ 1° Halorag.
  ˙aggregatus [H. depressa] 234, 238, 262, 278, 298,
    304, 392
  micranthus subsp. °micranthus 264, 275, 335, 337
  ˙montanus [H. procumbens] 198-9, 237
Gracilaria c.13 Red alga 288
Grammitis [Polypodium] 5˙ 4° Grammitid. 35, 39
  °billardierei 35, 128, 159, 182, 393, 577
  ˙givenii 393
  °patagonica 393
  °poeppigiana [armstrongii, pumilum] 39, 392, 394-
    5, 415, 417, 455, 465
  ˙rigida [crassa] 396
Gratiola 2° Scrophulari. 301
  °nana 304
  °sexdentata [peruviana] 316
Grimmia 8 Moss 382, 395, 415-17, 431
  trichophylla 423
Griselinia 2˙ Corn. or Griselini. 9, 58, 64, 544
  ˙littoralis 22, 35, 37, 61, 65, 77, 82, 84, 119, 121,
    127-8, 133-40, 144, 147, 151, 158, 162, 169-74,
    178, 184-7, 190, 196, 203, 205, 214, 304, 316,
    330, 385, 388-9, 391, 393-4, 398, 470, 472-6,
    493-4, 498, 500, 502, 511-12, 516, 520-1, 525,
    527, 540, 546-7, 561, 565, 574-7, 590, 595
  ˙lucida 22, 35, 77, 79, 117, 119, 126, 128, 132,
    167, 387, 502
Gunnera 5˙ 1* Gunner. or Halorag. 8, 29, 38, 58,
    496-7, 555
  ˙dentata (incl. arenaria?) 29, 303-4, 362, 392
  ˙hamiltonii 354
  ˙monoica (incl. albocarpa?) 59, 240, 275, 362, 384,
    391-2
  ˙prorepens 311, 324, 345
Gymnomitrion 3 Leafy liverwort 415, 418
Gypsophila 1† 1* Caryophyll.
  †australis [tabulaeformis, tubulosa] 254, 258, 368,
    371

˙Haastia 3˙ Aster. 12, 108, 607
  ˙pulvinaris (includes 2 spp.?) 41, 409, 412, 429
  ˙recurva 41, 109, 412, 425, 429
  ˙sinclairii 109, 412, 417, 420, 429-30
Hainardia 1* Po.
  *cylindrica 290, 296
Hakea 4* Prote. 72, 197, 212, 543
  *gibbosa [pubescens] 197
  *salicifolia [saligna] 198, 204
  *sericea [acicularis] 197-9, 204, 324, 537-8

˙Halocarpus [Dacrydium] 3˙ Podocarp. 12, 38, 40,
    64
  ˙bidwillii 26, 69, 186, 196, 234, 326, 332, 503-4,
    521
  ˙biformis 26, 40, 70, 77, 112, 127, 135, 154, 176-7,
    234, 327, 345, 391, 398, 478-9, 503, 512-13,
    574, 579, 593
  ˙kirkii 20, 61, 79, 112, 188
Haloragis 1˙ 1* Halorag. 525
  ˙erecta [cartilaginea, colensoi] 28, 351, 359, 383-5,
    389, 397
Hebe [Veronica] c.78˙ 2° Scrophulari. 5, 10, 13, 27,
    38, 40, 52-5, 59, 62, 178, 189, 206, 209-11, 216,
    372, 377, 383-4, 397, 548
  ˙albicans 378, 389, 391, 502
  ˙amplexicaulis [allanii] 378
  ˙armstrongii [Leonohebe] 195
  ˙barkeri [gigantea] 437, 441, 445
  ˙benthamii [Leonohebe] 53, 447, 455
  ˙bollonsii 383
  ˙brachysiphon [traversii] (= venustula syn.?) 196,
    209, 211, 503
  ˙breviramosa 435
  ˙buchananii 395, 415
  ˙canterburiensis [vernicosa] 550
  ˙carnosula [pinguifolia] 398
  ˙chathamica 445-6
  ˙cheesemanii [Leonohebe] 415, 429
  ˙ciliolata [gillesiana, Leonohebe] 229, 392, 395,
    415, 419, 427, 429
  ˙coarctata (= tetragona var.?) 222, 224
  ˙cockayneana 378
  ˙colensoi 378
  ˙cupressoides [Leonohebe] 14, 69, 207
  ˙decumbens 388, 415
  ˙dieffenbachii [dorrien-smithii] 445-6
  °elliptica 59, 191, 378, 386-7, 392-3, 396, 452-3
  ˙epacridea [Leonohebe] 109, 373, 395, 415, 420,
    425, 429, 518
  ˙evenosa 98
  ˙fruticeti 109
  ˙gibbsii 99, 378
  ˙glaucophylla 211
  ˙gracillima 102, 316
  ˙haastii [macrocalyx, Leonohebe] 109, 415, 417,
    427-9, 447
  ˙hectorii 222, 529
  ˙hulkeana 53, 378, 385
  ˙insularis 92
  ˙lavaudiana 106, 378
  ˙ligustrifolia 200, 397
  ˙lycopodioides [Leonohebe] 69, 222, 242, 418
  ˙macrantha 229
  ˙macrocarpa (var. brevifolia merits species rank?)
    200, 397, 533
  ˙odora [buxifolia, menziesii?; Leonohebe] (species
    aggregate) 37, 151, 186, 211, 231, 234, 240, 244,
    247, 339-40, 342, 344-6, 454, 478, 527, 552
  ˙pareora 106, 378
  ˙parviflora [leiophylla] (consists of vars.
    angustifolia and arborea, each probably a
    distinct species) 24, 170, 243, 385
  ˙pauciflora [Leonohebe] 103
  ˙pauciramosa [odora, Leonohebe] 225, 241, 339,
    341

*petriei [Leonohebe] 109, 415
*pimeleoides (var. rupestris merits species rank?) 249, 378
*pinguifolia 414
*propinqua [Leonohebe] 341
*rakaiensis [scott-thomsonii] 196, 211, 261
*ramosissima (= haastii var.?) 415, 429
*raoulii 378
*rigidula 378
*rupicola 378
°salicifolia 24, 37, 55, 59, 61, 170, 178, 316, 366, 378, 481, 500, 511–12, 517, 539, 551
*speciosa 53, 383, 387
*stricta [cookiana, macroura, salicifolia] 24, 166, 186–8, 193, 196, 314, 324–5, 377, 383–4, 524
*strictissima 106
*subalpina [montana] 158, 178, 211, 378, 388, 512–13, 550
*tetragona [Leonohebe] 192, 222, 247
*tetrasticha [Leonohebe] 415, 427
*topiaria [cockayneana] 211, 224–5, 390
*townsonii 378, 389
*traversii 189, 211, 385
*treadwellii 229, 415
*tumida [Leonohebe] 415
*urvilleana 397
*venustula [laevis] 211, 385, 397
*vernicosa 398
/Hectorella 1 Hectorell. or Portulac. 12
*caespitosa 411, 416, 418–19, 422, 427
Hedera 2* Arali.
*helix 396
Hedwigia 2 Moss 382
ciliata 205
Hedycarya 1* Monimi. 52
*arborea 23, 37, 58, 61–2, 65, 77, 79, 116, 119–20, 124, 126, 128, 132, 158, 166–71, 520, 532, 537, 539, 546, 561, 579–80, 595
Hedychium 2* Zingiber.
*gardnerianum 164
Helichrysum 10* 2* Aster. 12–13, 27, 38, 41
*aggregatum [glomeratum] (part of lanceolatum?) 385, 394
*bellidioides [prostratum] (= Anaphalis?) 33, 225, 228, 230, 232, 237, 240, 259, 380, 392, 395, 398, 454–5, 517, 527, 529
*coralloides 377
*depressum 363, 367
*dimorphum 106, 377
*filicaule 33, 236, 240, 252, 262, 264, 268, 275, 354, 444
*intermedium [selago] 377, 388, 390, 415, 429
*lanceolatum 397, 532
*parvifolium [microphyllum] 377
*plumeum 106
'alpinum' (similar to bellidioides but 2n = 84, cf. 28) 244, 430–1
Herpolirion 1* ? Asphodel.
°novae-zelandiae 234, 240, 307, 334–5, 337, 349
Hibiscus 1† 2* Malv.
†diversifolius 16, 42
Hieracium 9* Aster. 247, 518, 553
*caespitosum [pratense; Pilosella] 249
*lepidulum [lachenalii] 249, 253, 260, 262–3
*pilosella [Pilosella officinarum] 29, 33, 193, 237–8,

249, 253, 256–7, 260, 262–3, 269, 309, 370, 489
*praealtum [Pilosella] 238, 249, 253, 262, 370
Hierochloe 6* 1° Po. 218, 264
*brunonis 448, 460
*novae-zelandiae [fraseri] 218, 235
°redolens [Anthoxanthum] 188, 191, 265, 279, 311, 320, 337, 448, 453, 483–4, 579
Histiopteris 1° Dennstaedti.
°incisa 31, 40, 137, 144, 151, 159, 215, 313, 325, 374, 403, 441–2, 451, 457, 462, 493, 508, 516, 531, 542, 549, 580
Hoheria 5 Malv. 6, 58, 174
*angustifolia 25, 37, 61, 133, 171, 500
*glabrata (= lyallii var.?) 25–6, 39, 58, 65, 70, 77, 112, 135, 145, 154, 158, 170, 173, 178–9, 184–6, 190, 196, 209, 282, 392, 478, 481, 503, 511, 516, 550, 562, 565, 574, 592–3
*lyallii 25–6, 39, 58, 100, 171, 190, 209, 547
*populnea 25, 61, 166, 481
*sexstylosa (= populnea several vars.?) 25–6, 389
'tararua' [sexstylosa; sp.(a) in Eagle 1982] 184
Holcus 2* Po.
*lanatus 90, 215, 235, 248, 251, 261, 263, 265, 268, 270, 272, 274–7, 279, 299, 312–13, 316–17, 320, 359–60, 364, 369, 372, 394, 444, 539, 548
Holomitrium 1 Moss
perichaetiale 140
Homalanthus 1* 1* Euphorbi.
*polyandrus 434–5
Hordeum [Critesium] 8* Po. 275
*hystrix 299
*jubatum 299
*marinum 290, 295–7
*murinum 248, 265, 270, 273, 275, 278–9, 369
Hydatella 1* Centrolepid.
*inconspicua 301
Hydrocleys 1* Butom.
*nymphoides 301
Hydrocotyle 8* 1° 1* Api. 185, 266, 275, 281, 362, 374, 441, 465
*heteromeria [americana] 354
*hydrophila [tripartita] 304, 309
*moschata 314, 384, 539
*novae-zelandiae (an aggregate of ≥ 6 taxa) 178, 214, 236, 252, 262, 268, 275, 282, 298, 313, 316, 367, 373, 392, 429
*sulcata [tripartita] 320, 330, 344
Hymenophyllum 14* 7° Hymenophyll. 8, 159, 202
*armstrongii 39
°bivalve 31
*demissum 31, 441–2, 562, 577
°ferrugineum 577
°flabellatum 159, 391
*malingii 36
*minimum 39, 382
*multifidum 31, 135, 140, 159, 187, 304, 385, 391, 441–2, 455, 458, 462, 511, 513, 577, 581
*pulcherrimum 35
*sanguinolentum 562
*scabrum 328
*villosum 460
Hypericum 2° 10* Clusi.
*androsaemum 165
°gramineum 252
*perforatum 237, 368, 553

Hypnodendron 8 Moss 137, 182
    marginatum 312
    menziesii [*Sciadocladus*] 328
    sect. *Mniodendron* [*Mniodendron comosum*] 328
Hypnum 3 Moss
    cupressiforme 147, 205, 223, 249, 260, 262, 266,
        271, 364, 452, 458
Hypochoeris 3* Aster. 29
    glabra 253, 266, 271, 364, 370, 384
    radicata 193, 198, 224, 237, 240, 249, 253, 262–3,
        266, 269, 271, 273–4, 277, 299, 355–7, 359–60,
        364, 370, 387–8, 403, 444–6, 512, 518, 521, 539
Hypolepis 5* 2° Dennstaedti. 31, 312, 542
    ambigua [*tenuifolia*] 164, 325
    °*dicksonioides* [*tenuifolia*] 403
    °*distans* 325
    millefolium 40, 145, 147, 159, 178, 215, 223, 228,
        232, 373, 550, 576
    rufobarbata [*rugosula*] 441
Hypopterygium c.3 Moss 182, 312
    rotulatum [*novae-seelandiae*] 328
Hypoxis 1† Hypoxid.
    †*hookeri* 30, 40, 356
Hypsela 1* Lobeli.
    rivalis 301, 304

Ileostylus 1* Loranth. 9, 52
    micranthus [*Loranthus*] 36, 500
Imperata 1* 1* Po.
    cheesemanii 435
Iphigenia 1* ?Asphodel.
    novae-zelandiae 30, 40
Ipomoea 2° 4* Convolvul.
    batatas 16
    pes-caprae var. °*brasiliensis* 435
Isachne 1° Po.
    °*globosa* [*australis*] 279, 311–13
Ischnocarpus [*Sisymbrium*] 1* Brassic.
    novae-zelandiae 249
Isoetes 2* Isoet. 301, 468
    alpinus (part of *kirkii*?) 304, 307
    kirkii 303, 485
Isolepis [*Scirpus*] 6* 9° 4* Cyper. 316, 549
    °*aucklandica* 302, 304, 307, 309, 340, 342, 344,
        400, 419, 429, 454–5, 459–60, 465
    basilaris 298
    °*cernua* 289, 293, 296, 298–9, 356, 383, 396,
        454–5
    distigmatosa [*sulcatus* var.] 302
    °*habra* 178, 444, 458
    °*inumdata* 312, 325, 330, 444
    °*nodosa* [*Scirpoides*] 266, 275, 296, 299, 353, 355–
        6, 359–60, 383–5, 392, 396, 436, 443, 446
    praetextata 396, 455
    °*prolifer* 311–13, 316
    reticularis 311, 330
    sepulcralis [*S. antipodus, chlorostachyus*] 312–13
Isotachis 4 Leafy liverwort
    lyallii 382, 391
    montana 340
Iti 1* Brassic.
    lacustris 108
Ixerba 1* Escalloni. 9, 11, 52
    brexioides 21, 37, 43, 79, 96, 112, 116–18, 120,
        139–40, 155, 188, 501, 575, 579–80, 594

Jamesoniella 3 Leafy liverwort
    colorata 340
Jovellana 2* Scrophulari.
    repens 380
    sinclairii 96, 380, 384
Juncus 2* 14° 31* Junc. 6, 30, 54, 311
    °*antarcticus* 262, 338, 342, 429
    articulatus 29, 266, 275–6, 298, 302, 311–17, 320,
        324, 329–30, 332, 338, 365, 372, 384, 392, 394,
        445, 494
    °*australis* 266, 275, 277
    bufonius 266, 277, 303, 317, 372
    bulbosus 301, 316, 394
    °*caespiticius* 317, 389
    canadensis 311, 316
    °*distegus* 248, 266, 273, 317
    effusus 248, 262, 266, 275, 278, 311, 316–17, 338,
        342, 365, 372
    filicaulis 270
    gerardii 290
    °*gregiflorus* [*polyanthemus*] 248, 262, 266, 270,
        274–5, 277, 303–4, 311, 314–15, 317, 320, 330,
        332, 338
    maritimus var. °*australiensis* 38, 289–99, 359, 396
    novae-zelandiae 338, 343, 392, 431, 494
    °*pallidus* 266, 303, 311, 445
    °*planifolius* 324, 493–4
    °*pusillus* 303, 338
    °*sarophorus* 266, 275, 277, 314
    °*scheuchzerioides* 448, 454, 459–60, 465
    tenuis 266, 274, 277
    °*usitatus* 436
Jungermannia 3 Leafy liverwort 382

Kirkianella [*Crepis*] 1* Aster. 28
    novae-zelandiae 237, 249, 387
Knightia 1* Prote. 11, 58
    excelsa 21, 37, 50, 53, 61, 79, 82, 112, 116–17,
        121, 124, 139, 167, 203, 481, 502, 532–3, 537,
        539, 544, 590
Koeleria [*Trisetum*] 2* Po. 248, 250
    cheesemanii 414
    novo-zelandica [*kurtzii, superba*] 235
Korthalsella 3* Visc.
    clavata [*lindsayi*] 36
    lindsayi 36
    salicornioides 36, 198
Kunzea [*Leptospermum*] 2* Myrt. 6, 36, 52, 59
    ericoides [*lineatum*] 27, 34, 64, 70, 72, 79, 106,
        112, 146, 157, 162, 203, 275, 324, 368, 403–4,
        481, 490, 500–2, 520, 536–7, 573
    sinclairii [*ericoides*] 27
Kurzia 8 Leafy liverwort
    hippuroides 328

Lachnagrostis [*Agrostis*] 5* Po. 58, 248, 250, 263,
        337
    filiformis [*A. avenacea, Deyeuxia filiformis,
        forsteri*] (vars. *littoralis, semiglabra* merit species
        rank?) 235, 289, 299, 362, 383, 436, 518, 529
    leptostachys 448, 458
    lyallii 264, 290, 298, 362, 366, 512, 527
    richardii [*A. pilosa*] 384, 391
    striata 298
Lagarosiphon 1* Hydrocharit.

*major 301, 303, 305–6
Lagarostrobos [Dacrydium] 1˙ Podocarp. 38
˙colensoi 11, 20, 61, 70, 112, 127, 154, 176, 196,
   327–8, 393, 502, 511, 562, 595
Lagenifera [Lagenophora] 7˙ 1† Aster. 38, 59, 159,
   202, 249, 252
˙cuneata 236, 314
˙petiolata [barkeri, purpurea] (var. multidentata
   merits species rank?) 392, 454
˙pumila 389, 543
˙strangulata [petiolata] 143
Lagurus 1* Po.
*ovatus 272, 290, 355–6
Larix 1* Pin. 43, 156
*decidua 157, 545
Lastreopsis 4˙ 1° Dryopterid.
˙glabella [Ctenitis, Dryopteris] 31, 168, 170
°hispida [Polystichum, Rumohra] 31, 128, 151, 159,
   170
˙velutina [Ctenitis, Dryopteris] 164, 170
Lathyrus 8* Fab.
*tingitanus 395
Laurelia 1˙ Laur. 5, 9, 52, 58
˙novae-zelandiae 21, 61, 77, 79, 93, 112–13, 117,
   119, 127, 469, 481, 540, 560
Lavatera 5* Malv.
*assurgentifolia 396
Lecidia 46 Crustose lichen
   irrubens 519
Lemna 1° Lemn. 301, 305
°minor 313, 333
Leontodon 2* Aster.
*autumnalis 296, 357, 359
*taraxacoides 266, 272, 275–6, 355, 436
Lepicolea 2 Leafy liverwort 182
   scolopendra 140, 328
Lepidium 5˙ 2° 12* Brassic. 13, 54
*desvauxii 381
°flexicaule 381
˙kirkii 107, 299
°oleraceum 91, 381, 383, 387, 396, 461
˙sisymbrioides [kawarau, matau] 29, 49, 249, 368,
   381
Lepidolaena 6 Leafy liverwort 182, 452, 455
Lepidosperma 1˙ 2° Cyper. 72
˙australe [Cladium vauthiera] 198–9, 204, 304,
   324, 326, 329–30, 332, 335–7, 349, 444, 522
°filiforme 321, 336
°laterale 198, 321, 324, 533, 536
Lepidothamnus [Dacrydium] 2˙ Podocarp. 12, 38
˙intermedius 26, 61, 112, 154, 174, 176, 330–1,
   400, 510
˙laxifolius 26, 28, 61, 174, 186, 247, 322, 331, 333,
   335, 337–8, 343, 345, 348–9, 503, 513
Lepidozia 12 Leafy liverwort 182, 382
   glaucophylla 328, 403
   kirkii 328
   microphylla 328
   pendulina 328
Lepilaena [Althenia] 1° Zannichelli.
°bilocularis 288, 301
Leptinella [Cotula] c.24˙ 1° Aster. 5, 13, 38, 54, 281
˙atrata 109, 408, 425
˙dendyi [atrata] 109, 408, 429
˙dioica 289, 291, 296, 298–9, 354

˙featherstonii [renwickii] 29, 445–6
˙goyenii 411, 420–3
˙lanata 451, 454–5, 461
˙maniototo 368
˙minor [haastii] 106
˙pectinata [linearifolia, villosa, willcoxii] 238, 255,
   370, 395, 417–18, 529
°plumosa 29, 451, 453–5, 460–1, 464–5
˙potentillina 445–6
˙pusilla [angustata, perpusilla] 205, 236, 247, 252,
   303–4, 360, 368
˙pyrethrifolia 388, 400, 411, 426, 428
˙squalida 38, 230, 255, 264, 282, 298, 344, 371,
   392, 445, 512–13
Leptocarpus 1˙ Restion. 30
˙similis [simplex] 290–1, 293–6, 298–300, 303–4,
   321, 323, 326, 329–32, 355–7, 374, 383, 444–5
Leptolepia 1˙ Dennstaedti.
˙novae-zelandiae 173
Leptopteris [Todea] 2˙ Osmund.
˙hymenophylloides 133
˙superba 31, 39, 128, 135, 137–8, 151–2, 159, 574
Leptospermum 1° 1* Myrt. 6, 12, 52, 59
°scoparium 27, 36, 46, 53–4, 60–3, 72, 106, 127,
   144, 171, 196, 199, 216, 247, 271, 293, 303–4,
   325, 327–8, 332, 336, 351, 476, 481, 490, 493–5,
   500, 502–3, 511, 516–17, 534, 536–7, 562
Leptotheca 1 Moss
   gaudichaudii 147, 202
Lethocolea 1 Leafy liverwort
   squamata 325
Leucanthemum 2* Aster.
*vulgare [Chrysanthemum leucanthemum] 372, 394
Leucobryum 1 Moss
   candidum 144, 146
˙Leucogenes 2˙ Aster. 13, 410
˙grandiceps 379, 392, 395–6, 406, 417, 419, 429–
   30, 529
˙leontopodium 379
Leycesteria 1* Caprifoli.
*formosa 165, 214, 533–4, 539, 543
Leymus 1* Po.
*racemosus [Elymus giganteus] 356
Libertia 3˙ 1° Irid. 30, 39, 52, 59, 389
˙grandiflora 385
˙ixioides 353, 388
˙peregrinans 30, 353, 444
°pulchella [Sisyrinchium] 128, 139–40, 389
Libocedrus 2˙ Cupress. 5, 21, 38, 43, 58, 111
˙bidwillii 21, 36, 63, 70, 77, 93, 113, 119, 127–8,
   154, 174, 177, 316, 398, 469–70, 472, 478, 503,
   512–13, 544, 559, 565, 574, 579, 589
˙plumosa 61, 113, 132, 502
Lichina 1 Fruticose lichen
   confinis 383
˙Lignocarpa [Antisotome] 2˙ Api. 13
˙carnosula 407–8, 425
˙diversifolia 408–9, 429
Ligustrum 4* Ole. 553
*sinense 162, 313
Lilaeopsis 2˙ Api. 38, 295, 297–8, 309, 344, 445
˙novae-zelandiae [incl. lacustris, orbiculatus] 288–
   9, 303–4, 307, 485
Limosella 1˙ 1° Scrophulari. 303, 307, 355
°lineata [aquatica, tenuifolia] 298, 301, 356, 445

Lindsaea 1˙ 2° Dryopterid.
  °*linearis* 102, 198, 200, 204, 444, 533
  °*trichomanoides [cuneata]* 128
Linum 1˙ 4* Lin.
  *bienne [marginale] 269, 388
  *catharticum 28, 238, 254, 262, 275, 320, 371
  ˙*monogynum* 380, 383–5, 396
  *trigynum [gallicum] 267, 384
Liparophyllum 1° Gentian. or Menyanth. 12
  °*gunnii* 321, 335, 340
Litsea 1˙ Laur.
  ˙*calicaris* 21, 43, 112, 116, 166, 524, 532, 590
Lobelia 2˙ 1° 1* Lobeli.
  °*anceps* 298, 383–4, 392, 436, 444, 446
  ˙*linnaeoides* 222, 237, 529
  ˙*roughii* 408, 425, 429–30
Logania 1˙ Logani.
  ˙*depressa* 6
Logfia [Filago] 2* Aster.
  *minima 370
Lolium 3* Po. 276, 473
  *perenne 215, 265, 268, 270, 273–4, 276–7, 279,
    289, 359, 387, 444, 482, 484, 493
Lonicera 3* Caprifoli.
  *japonica 314, 351
Lophocolea 28 Leafy liverwort 182, 223, 382, 392,
    417
  cf. *amplectens* 460–61
  cf. *novae-zelandiae* 460
  semiteres 325
˙*Lophomyrtus [Myrtus]* 2˙ Myrt. 58
  ˙*bullata* 25
  ˙*obcordata* 25, 61, 133
Lotus 6* Fab.
  *angustissimus 266
  *pedunculatus [major] 214, 266, 272, 274–5, 277,
    313, 315–18, 360, 365
  *suaveolens [subbiflorus] 266, 278, 355, 360
˙*Loxsoma* 1˙ Loxomat.
  ˙*cunninghamii* 6
Ludwigia 2* Onagr.
  *palustris 301, 303
  *peploides 313
Lupinus 5* Fab.
  *arboreus 213, 352, 358, 365, 497, 501
  *polyphyllus 365
Luzula 12˙ 4* Junc. 12, 30, 231, 251, 309, 360, 400,
    414, 444
  ˙*banksiana [campestris]* (includes 3 spp.?) 388,
    396, 518
  ˙*celata* 368
  ˙*colensoi* 34, 411, 417–18
  *congesta 266, 277
  ˙*crenulata* 107
  ˙*crinita* (var. *petrieana* merits species rank?) 400,
    419, 427, 455, 458, 460–2, 465, 527, 529
  ˙*leptophylla* 342
  ˙*picta* (includes 2 spp.?) 159, 392
  ˙*pumila [cheesemanii, triandra]* 34, 411, 416–18,
    420–3, 426, 428–9, 529–30
  ˙*rufa* (includes 3 spp.?) 230, 235, 240, 248, 255,
    261, 369, 421, 423, 427–9, 518
  ˙*traversii* (includes 2 spp.?) 426, 429
  ˙*ulophylla* 34, 368
Luzuriaga 1˙ Philesi.

˙*parviflora [Enargea]* 29, 128, 187, 394, 577
Lycium 2* Solan.
  *ferocissimum 212, 395
Lycopodium 1˙ 9° Lycopodi. 6, 159
  °*australianum [selago]* 415, 460
  °*cernuum* 198–9, 401, 403
  °*deuterodensum [densum]* 198, 324
  °*fastigiatum* 31, 193, 223, 228, 230–1, 234, 238,
    240–1, 343, 416, 418, 431, 463, 529, 551
  °*laterale* 198–9, 322, 324–5
  ˙*ramulosum* (= *laterale* syn.?) 322, 333–5, 337,
    339, 443, 454
  °*scariosum* 31, 176, 202, 204, 463, 511, 516
  °*serpentinum* 199, 322–3, 325
  °*varium [billardierei, novae-zelandicum]* 31, 35,
    117, 128, 135, 140, 175, 382, 389, 443, 462
  °*volubile* 31, 35, 202, 204, 391, 533, 539
Lygodium 1˙ Schizae.
  ˙*articulatum* 35, 116
Lyperanthus 1˙ Orchid.
  ˙*antarcticus* 339

Machaerina [Cladium] 1° Cyper.
  °*sinclairii* 381, 384
Macromitrium 15 Moss 461
  longipes 140
Macropiper 1˙° Piper. 9
  ˙*excelsum* (var. °*psittacorum* also on Lord Howe
    and Norfolk Is) 37, 58, 92, 106, 126, 132, 166,
    168, 215, 385, 387, 389, 439, 573
Malva 6* Malv. 267
Marattia 1° Maratti.
  °*salicina [fraxinea]* 31, 39
Marchantia 5 Thallose liverwort 60, 339, 341, 364,
    383, 441, 444, 462, 548
  berteroana 316, 322, 324, 333, 454, 460, 549
Marrubium 1* Lami.
  *vulgare 267, 271, 280, 371
Marsippospermum 1˙ Junc.
  ˙*gracile* 219, 230, 233, 414, 416, 418–19, 426, 455,
    460, 528
Marsupidium 5 Leafy liverwort 182
Mastigophora 1 Leafy liverwort
  flagellifera 187
Matthiola 2* Brassic.
  *incana 384
Mazus 1˙ 1° Scrophulari.
  ˙*radicans* 311, 392
Medicago 8* Fab.
  *arabica 266
  ˙*lupulina* 266, 272, 355
  *nigra [polymorpha] 266
Melicope 2˙ Rut. 52
  ˙*simplex* 25, 133, 397
  ˙*ternata* 25, 132, 166–7
Melicytus 8˙ 2° Viol. 9, 13, 40, 52, 54, 58
  ˙*alpinus [Hymenanthera dentata]* (also in
    Australia?) 25, 42, 49, 59, 196, 205, 207, 209–
    11, 235, 243, 245, 248, 251, 261, 351, 359, 367–
    9, 373, 378, 397–8, 400, 415, 527, 529
  °*angustifolius [Hymenanthera dentata]* (N.Z. plants
    probably an endemic species) 193, 196
  ˙*chathamicus [Hymenanthera]* 64, 437, 440–1
  ˙*crassifolius [Hymenanthera]* 25, 351, 359–60, 378,
    385

˙*lanceolatus* 23, 187
˙*macrophyllus* 23, 116, 140, 537
˙*micranthus* 25, 61, 133, 397
˙*novae-zelandiae [Hymenanthera]* 23,166–7
˚*obovatus [Hymenanthera]* 23, 378, 387, 389
˚*ramiflorus* subsp. ˙*ramiflorus* 23, 25, 37, 52, 55,
  61, 77, 86, 118–19, 151, 158, 162, 167, 171, 203,
  352, 435, 481, 486, 488, 498, 500, 520, 537, 541,
  565, 567, 574, 578
Melilotus 3* Fab.
  indicus 272, 355
Mentha 1˙ 5* Lami.
  ˙*cunninghamii* 252, 257, 320, 354, 392, 520
  * × *piperita* 317, 320
  **pulegium* 267, 275, 278, 312
Meryta 1˙ Arali.
  ˙*sinclairii* 23, 37, 92, 166, 502, 532
Metrosideros 12˙ Myrt. 5, 8, 34–5, 37, 52–3, 59, 80,
  112, 168, 374, 544, 563
  ˙*albiflora* 116, 139
  ˙*bartlettii* 92
  ˙*carminea* 97
  ˙*colensoi* 79, 120, 130, 132, 378
  ˙*diffusa [hypericifolia]* 117, 128, 132, 135, 159,
    170, 173
  ˙*excelsa [tomentosa]* 21, 41, 79–80, 96, 153, 166,
    352, 500, 502, 524, 567, 607
  ˙*fulgens [scandens]* 55, 77, 79, 117, 126, 128, 132,
    135, 170, 539
  ˙*kermadecensis [villosa]* 434–5
  ˙*parkinsonii* 376
  ˙*perforata* 116, 119, 128, 132, 168, 170, 378, 565
  ˙*robusta* 21, 35, 61, 77, 79–80, 112, 116–17, 119,
    121, 124, 127, 167, 376, 472, 502, 533, 545–6,
    561
  ˙*umbellata [lucida]* 21, 35, 55–7, 60, 62, 70, 77, 82,
    112–13, 119, 127, 154, 158, 170, 329–30, 352,
    451–2, 455, 470, 475, 478, 493, 495, 502, 509,
    516–17, 561, 565, 574, 589
Metzgeria 10 Thallose liverwort 452, 455
  furcata 460–1
Microlaena [Ehrharta] 4˙ 1˚ Po.
  ˙*avenacea [E. diplax]* (incl. *carsei*) 119–20, 128,
    135, 138, 140, 144, 151–2, 159, 170, 173, 202,
    208, 328, 392, 576, 579
  ˙*colensoi [Petriella]* 218, 228, 230, 259, 392, 419
  ˙*polynoda* 29, 35
  ˚*stipoides* 202, 205, 263–4, 269–70, 272, 275, 278–
    9, 355, 435, 444, 539
  ˙*thomsonii [Petriella]* 101, 338–9
Microseris 1˚ Aster. 28
  ˚*scapigera [forsteri]* 224–5, 237, 253, 388, 429
Microtis 1˙ 2˚ Orchid. 39
  ˚*parviflora* 324
  ˚*unifolia* 52, 199, 249, 270, 278, 316–17, 543
Mida 1˙ Santal.
  ˙*salicifolia* 36, 79, 98, 119
Mimulus 1˙ 3* Scrophulari. 302, 318
  **guttatus* 311, 316–17, 372
  **moschatus* 311, 316, 320
  ˙*repens* 288–9, 295, 297–9
Miscanthus 2* Po.
  **nepalensis* 373
Mitrasacme 1˙ 1˚ Logani.
  ˚*montana* var. ˙*helmsii* 101, 336

˙*novae-zelandiae* 335, 340, 345, 347
Modiola 1* Malv.
  **caroliniana* 267, 278
Monoclea 1 Thallose liverwort
  forsteri 112, 383
Montia 1˚ * Portulac.
  fontana subsp. **chrondrosperma* 267, 278
  fontana subsp. ˚*fontana* 301, 330, 339, 342, 344,
    454, 456, 465
Morelotia 1˙ Cyper.
  ˙*affinis [Gahnia gahniaeformis]* 198, 321, 324, 397,
    533
Muehlenbeckia 3˙ 2˚ Polygon. 34, 39, 52, 58, 163, 374
  ˙*astonii* 25, 207, 351
  ˚*australis* 25, 159, 163, 173, 196, 325, 351, 360,
    385, 439–40, 540–1, 544
  ˚*axillaris* 28, 211, 236, 251, 262, 264, 304, 360,
    369, 373, 388, 390, 429, 512, 517, 524–5, 527–9
  ˙*complexa* 25, 163, 167, 196, 207, 251, 268, 270,
    272, 293, 306, 351, 355–6, 359–60, 369, 373,
    384–5, 387, 395, 543–4
  ˙*ephedroides* 28, 41, 351, 359
Muelleriella 3 Moss
  angustifolia 461
Mycelis 1* Aster.
  **muralis* 124, 164, 172, 214, 224, 517, 574
Myoporum 2˙ 1† 1* Myopor. 58
  †*debile* 12
  **insulare* 352
  ˙*kermadecense [obscurum]* 434–5
  ˙*laetum* 22, 37, 61, 79, 162, 352, 387, 437, 498,
    500, 540, 607
˙*Myosotidium* 1˙ Boragin. 11, 53
  ˙*hortensia* 29, 38, 442, 445–6
Myosotis ?30˙ 2˚ 6* Boragin. 5, 10, 12, 53, 103, 381,
    394, 413
  ˙*albosericea* 107
  ˙*angustata* 429
  ˚*antarctica* 457, 460
  ˙*arnoldii* 385
  ˚*australis* (species aggregate, incl. endemic forms)
    53, 237
  ˙*capitata* 451, 455
  ˙*concinna* 429
  **discolor [versicolor]* 275, 371
  ˙*laeta* 397
  **laxa* subsp. *caespitosa* 301, 303, 314, 316–17, 320,
    344
  ˙*lyallii* 394, 419
  ˙*macrantha* 28, 53, 412, 529
  ˙*monroi* 397
  ˙*pulvinaris* 109, 411
  ˙*pygmaea* (var. *drucei*=sp.(c) in Druce & Williams
    1987, may merit species rank) 354, 357, 414,
    419, 428–9, 457
  ˙*rakiura [albida]* 110, 396
  ˙*suavis* 417, 419
  ˙*tenericaulis* 381
  ˙*traversii* 408, 417, 429
  ˙*uniflora* 362, 368
Myosurus 1˚ Ranuncul.
  ˚*minimus* subsp. ˙*novae-zelandiae [novae-zelandiae]*
    28, 368
Myriangium Fungus
  thwaitesii 197

Myriophyllum 3˙ 2° 1* Halorag. 63, 468
  aquaticum [brasiliense] 301, 314, 320, 324
  °*pedunculatum* subsp. ˙*novae-zelandiae* 288, 298, 309, 444
  °*propinquum* 33, 288, 303–5, 307, 313, 339, 344, 442
  ˙*robustum* 301
  ˙*triphyllum [elatinoides]* 301, 303–5, 307, 309, 330, 344, 465
  ˙*votschii* 288, 355–6
Myrsine [Suttonia] 8˙ Myrsin. 8, 54, 561
  ˙*australis* 23, 58, 116, 119, 128, 151, 158, 162, 166–7, 171, 187, 202–3, 304, 312, 387, 397, 500, 520, 532, 536–40, 544, 561, 595
  ˙*chathamica* 168, 437, 441, 562
  ˙*coxii* 438, 442
  ˙*divaricata* 25, 33, 61, 65, 119, 128, 151, 158, 171, 173, 178, 181, 184, 186, 188, 193, 196, 208, 304, 315–16, 328–30, 346, 389, 391, 398, 451–5, 457, 513, 516–17, 565, 574–5, 577
  ˙*kermadecensis* 434–5
  ˙*nummularia* 174, 189, 192, 229–31, 340, 391, 513, 527, 550
  ˙*oliveri* 92
  ˙*salicina* 23, 25, 37, 43, 58, 61–2, 77, 116, 124, 501, 538, 573, 579
  ˙aff. *divaricata* 389

Nassella [Stipa] 1* Po.
  trichotoma 156, 265, 268
Navarretia 1* Polemoni.
  squarrosa 360, 371
Neofuscelia [Parmelia] 20 Foliose lichen 417
  adpicta 519
˙*Neomyrtus [Myrtus]* 1˙ Myrt.
  ˙*pedunculata* 25, 58, 124, 128, 151, 159, 187, 304, 328, 330, 509, 574
Neopaxia 1° Portulac.
  °*australasica [Claytonia, Montia, M. calycina]* 307, 309, 392, 400, 406, 411, 419, 423, 428, 430–1, 527
Nephrolepis 2° 1* Davalli. 9, 39
  °aff. *cordifolia* 400, 402–3
Nertera 5˙ 2° Rubi. 28, 58–9, 112, 144, 362
  ˙*balfouriana* 334, 339, 343
  ˙*ciliata* 366
  °*depressa* 38, 128, 159, 202, 275, 316, 330, 384, 392–4, 439, 442, 454
  ˙*scapanioides* 321, 324–5, 334
  ˙*setulosa* 236, 252, 266, 275
  ˙aff. *dichondrifolia [dichondraefolia]* 38, 128, 159, 170, 173, 202, 577
Nestegis [Gymnelaea, Olea] 3˙ 1° Ole. 9, 21, 52, 121, 132, 155
  °*apetala* 162, 166
  ˙*cunninghamii* 37, 58, 79, 104, 112, 116–17, 472
  ˙*lanceolata* 58, 117, 119, 167
  ˙*montana* 58, 117
Neuropogon 5 Fruticose lichen 416
  ciliatus 417, 426
Nitella 8 Green alga 300, 303, 307
  hookeri 303
  pseudoflabellata [gracilis] 303
Nostoc 19 Blue-green alga 383, 496
Nothofagus 4˙ Fag. 5, 8, 9, 17, 21, 36, 43, 45, 54,
56–8, 60–2, 64, 71, 109, 111, 492, 544, 559, 564
  ˙*fusca* 16, 21, 25, 37, 39, 56, 61, 70, 77–8, 93, 112, 141, 151, 154, 158, 208, 469–70, 477–8, 481, 483, 486, 500, 502, 514, 517, 560, 574–5
  ˙*menziesii* 11, 16, 21, 37, 43, 56, 58, 61, 63, 70, 77–8, 82–3, 93, 127, 141, 143, 145, 151, 154, 158, 174, 208, 231, 328, 346, 389, 467, 470, 476–8, 481, 483, 486, 495, 514, 516–17, 560, 574–5, 580, 587
  ˙*solandri* [incl. *cliffortioides*] 21, 25, 37, 56, 61, 63, 65, 70, 74, 78, 81, 97, 99, 112, 141, 145, 154, 158, 174, 204, 227, 262, 304, 328, 331–2, 388, 467–8, 470, 473, 476–8, 481,–2, 486, 503–4, 514, 516–18, 550, 560, 574
  ˙*truncata* 37, 47, 61, 70, 74, 97, 112, 118, 141, 145, 154, 184, 468, 470, 473, 476, 480, 482, 486, 490, 502, 514, 517, 533, 567, 574
Notospartium 3˙ Fab. 13, 377
  ˙*carmichaeliae* (inc. *glabrescens*?) 189, 385
  ˙*torulosum* 377
˙*Notothlaspi* 2˙ Brassic. 10, 13, 413
  ˙*australe* 413, 429
  ˙*rosulatum* 28, 109, 407–8, 425, 429
  ˙aff. *australe* 398, 400
Nymphaea 2* Nymphae.
  alba 301

Ochrolechia 4 Crustose lichen
  parella 461
Oenanthe 3* Api.
  pimpinelloides 267, 277
Oenothera 7* Onagr.
  stricta 355
Olea 1* Ole.
  europaea subsp. *africana* 434
Olearia 32˙ Aster. 5, 12–13, 39, 41, 52–3, 72, 162, 207, 209, 376
  ˙*albida [angulata]* 397
  ˙*arborescens* 23–5, 37, 178, 187, 366, 389, 391, 511
  ˙*avicenniifolia* 23–4, 168, 196, 329, 352, 360, 366, 388–91, 394, 502, 511–12, 517, 556
  ˙*capillaris* 25
  ˙*chathamica* (= *oporina* var.?) 442, 446
  ˙*colensoi* 24, 37, 57, 63, 70, 77, 81, 98, 135, 144, 154, 158, 174, 176–7, 179, 182, 184–6, 188, 192, 231, 234, 243, 339, 391, 478, 512–14, 551, 562, 574, 581–2
  ˙*coriacea [Shawia]* 385
  ˙*crosby-smithiana* 42, 180
  ˙*cymbifolia [nummulariifolia* var.] 27, 38, 42, 189
  ˙*fragrantissima* 23, 25
  ˙*furfuracea* 23, 524, 532–3, 537
  ˙*hectorii* 25
  ˙*ilicifolia* 24, 61–3, 70, 77, 135, 145, 158, 170–1, 173, 178, 184, 186–7, 190, 196, 478, 481, 512, 514, 550, 562, 574
  ˙*lacunosa* 24, 37, 42, 70, 77, 145, 176, 178, 180, 184, 513–14, 550, 574
  ˙*lineata [virgata]* 25, 196
  ˙*lyallii* 24, 70, 191, 447, 452–3, 554
  ˙*moschata* 24, 37, 228, 259, 375, 512, 529
  ˙*nummulariifolia* 27, 38, 42, 186, 189, 244, 247, 512, 547
  ˙*odorata* 25, 196, 210, 260
  ˙*oporina* (incl. *angustifolia*) 24, 110, 179, 191–2,

393, 396
˙*paniculata* [*Shawia*] 23, 167, 169, 385, 387
˙*rani* 23, 37, 41, 116, 118–20, 167, 170, 538, 546
˙*semidentata* 442
˙*serpentina* [*virgata*] 397
˙*solandri* 27, 212, 306, 352, 355, 359
˙*traversii* 25, 37, 437, 439, 444, 446, 607
˙*virgata* [*divaricata, laxiflora*] (species aggregate)
   25, 193–4, 196, 208, 262, 327, 339, 577
Opegrapha 6 Crustose lichen
   diaphorhiza 461
Ophioglossum 2° Ophiogloss. 39
   °*coriaceum* 238, 254, 262
Oplismenus 2° Po.
   °*hirtellus* [*imbecillis, undulatifolius*] 166, 202, 265
Oreobolus 3˙ Cyper. 12, 39, 340, 343, 514
   ˙*impar* 229, 232, 339, 396
   ˙*pectinatus* 187–8, 225, 243–4, 247, 307, 321, 323–
      4, 331–2, 334–5, 337–42, 344–5, 347–9, 400,
      431, 454–5, 459, 512, 520, 528
   ˙*strictus* 321, 332, 337, 345, 513
Oreomyrrhis 3˙ Api. 253
   ˙*colensoi* 224, 230, 240, 428, 445
   ˙*rigida* [*andicola*] 236, 261, 388
Oreoporanthera 1˙ Euphorbi.
   ˙*alpina* [*Poranthera*] 411, 429
Oreostylidium 1˙ Stylidi. 10, 12
   ˙*subulatum* 38, 335, 340
Orobanche 1* Orobanch.
   minor 36
Orthoceras 1° Orchid.
   °*strictum* 199
Othonna 1* Aster.
   **capensis* 395
Ottelia 1* Hydrocharit.
   **ovalifolia* 301
Ourisia 14˙ Scrophulari. 12, 52–3, 216, 221, 511
   ˙*caespitosa* 28, 221, 366, 373, 392, 394–5, 411,
      417, 419, 423, 426, 529
   ˙*macrocarpa* 221, 231, 391–2
   ˙*macrophylla* 38, 188, 221, 244, 389, 394
   ˙*sessilifolia* 411, 419
   ˙*vulcanica* 98, 247
Oxalis 3° 15* Oxalid. 6, 38
   °*exilis* [*corniculata*] (records probably include
      °*rubens*) 236, 252, 266, 268, 270, 275, 278, 281,
      354, 360, 370, 388, 429
   °*magellanica* [*lactea*] 230, 238, 366, 384, 392

˙*Pachycladon* 2˙ Brassic. 13
   ˙*crenata* 103, 413
   ˙*novae-zelandiae* 109, 413
˙*Pachystegia* 3˙ Aster. 13, 41, 52, 376
   ˙*insignis* 377, 385, 388
Paesia 1˙ Dennstaedti.
   ˙*scaberula* 156, 202, 215, 316, 374, 430, 542
Pannoparmelia 2 Foliose lichen
   angustata 431
Papillaria 4 Moss 135
   crocea 435
Parahebe [*Veronica*] 11˙ Scrophulari. 12–13, 28,
   52–4
   ˙*birleyi* 413, 417, 419–20
   ˙*canescens* 309
   ˙*catarractae* [*diffusa, irrigans, lanceolata*] (subsp.

martinii merits species rank?) 380, 384–5, 394
   ˙*cheesemanii* 108, 408, 427
   ˙*decora* [*bidwillii*] 520, 529
   ˙*hookeriana* 380, 430
   ˙*linifolia* 380, 392, 395, 429, 529
   ˙*lyallii* 380, 391–2
   ˙*planopetiolata* 109, 413, 529
   ˙*spathulata* 98, 408, 430
   ˙*trifida* 107, 413
Parapholis 2* Po.
   **incurva* 290, 296
Paraserianthes 1* Fab.
   **lopantha* [*Albizia*] 163, 395
Parentucellia 2* Scrophulari.
   **viscosa* 36, 267, 278, 316, 366
Parietaria 1° 2* Urtic.
   °*debilis* 28, 165, 383, 440
Parmelia 9 Foliose lichen 383, 395
Parmotrema 12 Foliose-lobate lichen
   perlatum 403
   reticulatum 403
Parsonsia 2˙ Apocyn. 34, 58, 163
   ˙*capsularis* 25, 53, 133, 196, 207, 373, 397
   ˙*heterophylla* 173
Paspalum 1° 4* Po.
   **dilatatum* 265, 279–80, 293, 403
Passiflora 1˙ 6* Passiflor. 34, 54
   **mollissima* 163
   ˙*tetrandra* [*Tetrapathaea*] 79, 132, 502
Pelargonium 1˙ 5* Gerani.
   * ×*hortorum* 395
   †*inodorum* 267
   **peltatum* 395
Pellaea 1˙ 2° Pterid. 39, 382
   °*falcata* 436
   °*rotundifolia* 133
Peltigera 7 Foliose lichen
   dolichorhiza 203
Pennantia 2˙ Icacin. 9, 52, 58
   ˙*baylisiana* [*Plectomirtha*] 25, 92, 502, 532
   ˙*corymbosa* 25, 37, 61, 69, 126, 162, 207, 359, 500,
      520, 543, 565
Pennisetum 5* Po.
   **clandestinum* 29, 265, 277, 279, 353, 356, 360
Pentachondra 1° Epacrid. 58
   °*pumila* 28, 61, 186, 224–5, 228–9, 234, 236, 240,
      242–3, 247, 333–4, 337–41, 345, 347, 349, 396,
      513
Peperomia 1˙ 2° Piper. 35, 485
   ˙*urvilleana* 379, 383, 387, 435
˙*Peraxilla* [*Elytranthe*] 2˙ Loranth. 36, 53, 567
   ˙*colensoi* 143
   ˙*tetrapetala* 143, 159
Pernettya 3° 1* Eric. 13, 27–8, 58
   ˙*alpina* 411, 418, 423, 426, 529
   ˙*macrostigma* 234, 240, 264, 340
   ˙*nana* 234, 249, 334
Pertusaria 23 Crustose lichen 383, 395–6, 463
   dactylina 417
   graphica 461, 520
Petroselinum 1* Api.
   **crispum* 396
Phalaris 4* Po.
   **aquatica* [*tuberosa*] 279
   **arundinacea* 320

Phebalium 1˙ Rut.
  ˙*nudum* 117
Philonotis 3 Moss 364, 382, 395
  pyriformis [australis] 316, 415
  tenuis 417
Phleum 1* Po.
  pratense 265, 275–6, 279
Phormium 1˙ 1° Phormi. 9, 17, 29, 39, 40, 54, 56–9,
    73, 285, 432, 577
  ˙*cookianum [colensoi]* 39, 56, 151, 159, 175, 177–8,
    182, 185, 189, 191–2, 206, 222, 224–5, 229, 231,
    381, 384–5, 387–94, 397–8, 502, 513, 547–51,
    574
  °*tenax* 39, 53–4, 56, 130–1, 168, 193, 245, 270,
    272, 293, 300, 306, 310, 312–20, 325, 327, 329–
    32, 334, 344, 346, 349, 354, 358–60, 381, 383,
    392–3, 397, 402, 444, 446, 502, 521–2, 536–7
Phyllachne 2˙ 1° Stylidi. 13, 420
  ˙*clavigera* (= *colensoi* syn.?) 454–5, 458–60
  °*colensoi* 34, 229, 242–3, 338, 340, 342, 344, 347,
    396, 411, 416, 418, 421–3, 425–7, 429, 528–9
  ˙*rubra* 340, 411, 421, 423
Phyllocladus 3˙ Podocarp. or Phylloclad. 5, 9, 37,
    41, 525, 544
  ˙*alpinus [aspleniifolius]* 11, 26, 62–5, 70, 83–4, 93,
    113, 127–8, 130, 134, 136, 138, 140, 144, 146–7,
    151, 158, 171, 176, 178–9, 185–9, 193–6, 206,
    208–9, 225, 243, 247, 326, 328, 330–1, 349, 389,
    397–8, 478, 503–4, 509, 512–15, 520–2, 525,
    527, 547–50, 556, 570, 572, 574, 577, 589–90,
    593–5, 598
  ˙*glaucus* 21, 70, 79, 96, 113, 139, 155, 188, 524,
    575, 594
  ˙*trichomanoides* 21, 61, 79, 112–13, 117, 121, 155,
    202, 397, 478, 502, 526, 533–4, 537, 561, 564
Phylloglossum 1° Lycopodi. 6
  °*drummondii* 31, 199
Phymatosorus [Phymatodes, Polypodium] 1˙ 2°
    Polypodi. 35, 39
  °*diversifolius* 117, 151, 159, 169–71, 173, 313, 374,
    382, 441, 451–2, 505, 532, 539, 573, 575, 577
  °*scandens [pustulatum]* 126, 173, 382
Physcia 7 Foliose lichen 395
Phytolacca 3* Phytolacc.
  octandra 267, 354
Phytophthora ?5 Fungus
  cinnamomii 564
  heveae 564
Pilularia 1˙ Marsile.
  ˙*novae-zelandiae* 301, 303
Pimelea 12˙ Thymelae. 10, 27, 52, 205
  ˙*arenaria* 351, 355–6, 385, 446
  ˙*aridula* 368
  ˙*concinna* 255
  ˙*gnidia* 38, 179, 204, 385
  ˙*longifolia* 38, 385, 389
  ˙*lyallii* 351, 357
  ˙*oreophila [pseudolyallii]* (= *suteri* var.?) 34, 236,
    240, 251, 262, 418, 527
  ˙*prostrata [urvilleana]* 58, 193, 198, 249, 255, 264,
    268, 324, 345, 351, 356, 384, 388–9, 396–7, 524
  ˙*pulvinaris* 34, 251, 368
  ˙*sericeovillosa* 255
  ˙*suteri* 397
  ˙*tomentosa* 196

˙*traversii* 189, 196, 236, 261
Pinus 13* Pin. 16–17, 105, 403, 545, 547
  contorta [murrayana] 156, 160, 193, 244, 467–8,
    471, 473, 477, 480, 482–3, 486, 497, 607
  nigra 156, 158, 160
  patula 156
  pinaster 204
  ponderosa 156
  ˙*radiata [insignis]* 43, 156, 158, 201, 204, 352, 452,
    468–9, 473, 475–7, 479, 481, 483, 485–6, 488–
    90, 497, 506, 539
  strobus 156
Pisonia [Heimerliodendron] 1° Nyctagin. 37
  °*brunoniana* 22, 58, 64, 77, 162, 166, 435, 532
Pittosporum 26˙ 1† Pittospor. 40, 53, 58, 162, 207,
    387, 520
  ˙*colensoi (tenuifolium* var.?) 23
  ˙*cornifolium* 35, 117, 130
  ˙*crassicaule (rigidum* var.?) 176–7, 398
  ˙*crassifolium* 23, 41, 166, 397
  ˙*dallii* 53, 143, 378
  ˙*divaricatum* 25, 574
  ˙*eugenioides* 23, 37, 53, 59, 61, 79, 133, 171, 500,
    502, 540, 565
  ˙*kirkii* 35, 53, 140
  ˙*obcordatum* 6
  ˙*patulum* 25, 143
  ˙*pimelioides* var. *major [michiei]* 397
  ˙*rigidum* 25
  ˙*tenuifolium* 23, 37, 61, 171, 202–3, 205, 389, 394,
    500, 502, 524, 532, 536, 540, 543–4
  ˙*turneri* 24–5, 193, 196
  ˙*umbellatum* 532–3
  ˙*virgatum* 25
Placopsis 13 Crustose lichen 395, 520
  parellina 519
  perugosa 519
  trachyderma 425
Plagianthus 2˙ Malv. 59
  ˙*divaricatus* 25, 209, 289, 291, 293, 295–6, 298–9,
    359
  ˙*regius [betulinus, chathamicus]* 22, 25, 37, 39, 61,
    112, 162, 171–3, 196, 207, 437, 439–40, 500,
    502, 510, 540, 565, 577
Plagiochila 25 Leafy liverwort 182, 187, 223, 382
  gigantea 328
  stephensonii 112
Planchonella [Sideroxylon] 1° Sapot.
  °*costata [novo-zelandica]* 22, 37, 58, 77, 79, 162,
    166, 502, 532
Plantago 6˙ 2° 7* Plantagin. 253, 264, 307, 457
  ˙*aucklandica* 451, 455
  australis [hirtella] 316, 389, 392
  °*coronopus* 289, 291, 293, 296–7, 299, 353, 359
  lanceolata 266, 273–5, 277, 359, 394, 444, 539
  °*lanigera [incl. novae-zelandiae]* 29, 237, 240, 411,
    423, 431, 529
  major 266, 274–5, 278, 394
  ˙*raoulii* 38, 354, 392
  ˙*spathulata* 238, 388
  ˙*triandra [hamiltonii]* (subsp. *masoniae* merits
    species rank?) 262, 298, 304
  °*triantha [brownii, carnosa, subantarctica]* 451, 455
  ˙*uniflora* 340, 396
Pleurophascum 1 Moss

grandiglobum 335
Pleurophyllum 3˙ Aster. 13, 53, 57
  ˙*criniferum* 448–9, 453, 459, 462–3
  ˙*hookeri* 448–50, 455–6, 459, 464–5, 468, 473
  ˙*speciosum* 38, 449, 453–5, 461
Pleurosorus [*Gymnogramma*] 1° Aspleni.
  °*rutifolius* 382, 385
Pneumatopteris [*Dryopteris, Thelypteris*] 1°
  Thelypterid.
  °*pennigera* 124, 215
Poa 34˙ 1° 11* Po. 5, 12, 30, 73, 191, 218, 298, 359,
  381, 394, 574
  ˙*acicularifolia* 385, 400
  ˙*anceps* 263–5, 314, 373, 383–5, 436, 447
  ˙*antipoda* 454
  annua 265, 273–5, 277, 279, 281, 372, 451, 454,
    458, 464–5
  ˙*astonii* 283, 393, 396
  ˙*aucklandica* 448, 460
  ˙*breviglumis* [*imbecilla*] 145, 178, 374, 388, 454,
    458
  ˙*buchananii* [*sclerophylla*] 414, 425, 429
  ˙*chathamica* 446–7
  ˙*cita* [*caespitosa, laevis*] 30, 50–1, 90, 105, 187,
    193–4, 234, 250, 260–1, 263–4, 268, 270, 273,
    279, 289–90, 297–9, 354, 366–7, 369, 374, 384–
    8, 394, 396, 429, 444, 447, 483–4, 494, 519, 526
  ˙*cockayneana* 30, 225, 228, 259, 264, 363, 366,
    444, 447, 512, 520
  ˙*colensoi* 39, 217–18, 225–6, 228–31, 233, 235,
    238–41, 243, 245, 247–8, 250, 258–9, 261, 263,
    279, 342–3, 346, 367, 369, 373, 385, 391–2, 400,
    414, 416, 418–26, 428–31, 468, 472–3, 475–6,
    493–4, 513, 527, 529–31, 552
  °*cookii* 463–4, 480, 482
  ˙*dipsacea* 279, 414, 429
  ˙*foliosa* 281, 283, 447, 452–5, 460–2, 464, 467–8,
    473, 480
  ˙*hesperia* [*colensoi*] 414
  ˙*incrassata* 423
  ˙*kirkii* 218
  ˙*lindsayi* 235, 250, 257, 309, 369, 530
  ˙*litorosa* 30, 447, 452–5, 457–62
  ˙*maniototo* 34, 250, 257, 369
  ˙*novae-zelandiae* 259, 363, 366, 391–2, 395, 406,
    414, 416–17, 419, 429, 431, 511, 518, 529
  pratensis 224–5, 248, 251, 259–61, 265, 273, 275–
    6, 279, 342, 353, 359–60, 369, 444
  ˙*pusilla* 353, 360, 363, 392, 521
  ˙*pygmaea* 34, 107, 411, 420, 423
  ˙*ramosissima* 448, 456, 458, 460–1
  ˙*subvestita* [*novae-zelandiae*] 391–2
  ˙*tennantiana* 281, 283, 452
  trivialis 265, 270, 275–6, 278–9, 312
Podocarpus 4˙ Podocarp. 5, 8, 16, 37, 470, 547, 575
  ˙*acutifolius* 26, 128
  ˙*hallii* 20, 40, 76–7, 84, 112, 117, 119, 121, 127,
    137–8, 151, 154, 158, 170–2, 178, 189–90, 304,
    326, 328, 389, 398, 472, 500, 503, 509, 513, 517,
    537, 544, 562, 565, 574, 577, 589, 595
  ˙*nivalis* 26, 78, 83, 138, 147, 151, 158, 171, 174,
    176, 178, 186, 189–90, 192, 206, 209, 242, 247,
    373, 388–90, 395, 430–1, 503, 513–14, 521, 528,
    548, 550, 570, 574
  ˙*totara* 20, 40, 84, 112, 121, 127–9, 154, 168, 173,

202, 388, 478, 481, 488, 493, 510, 514, 538, 544,
  562, 599
Pogonatum [*Polytrichastrum, Polytrichum*] 2 Moss
  alpinum 417, 465
Pohlia 8 Moss 395
  cruda 417
Polycarpon 1* Caryophyll.
  **tetraphyllum* 267, 269, 271, 360, 384, 436
Polygala 5* Polygal.
  **myrtifolia* 213
Polygonum 1° 1† 11* Polygon. 311–13
  **hydropiper* 267, 305, 313, 317–18, 320, 372
  †*prostratum* 360
  °*salicifolium* [*decipiens*] 267, 305, 311–12, 320
Polypogon 3° Po.
  **monspeliensis* 289–90, 436
Polystichum 5˙ 3* Dryopterid. 31
  ˙*cystostegia* 40, 392, 415, 419, 429, 455, 460, 529
  ˙*richardii* [*aristatum*] 133, 385
  ˙*vestitum* 31, 39, 135, 144, 147, 149, 151–2, 159,
    163, 170, 172–4, 178, 185, 187, 189, 191, 208,
    215, 223, 232, 243, 320, 339, 392, 415, 439,
    451–3, 457–62, 464, 505, 512, 515–16, 525, 527,
    542, 550, 573–4, 580
Polytrichadelphus 1 Moss
  magellanicus 382, 391
Polytrichum 4 Moss 260, 262, 325, 417, 518
  commune 234, 312, 320, 325, 339
  formosum 343
  juniperinum 205, 223, 240, 249, 343, 364, 367–8,
    417, 422–3, 529, 549
  longisetum [*gracile*] 343
Pomaderris 3˙ 4° 1* Rhamn. 59, 198, 324
  °*apetala* 12, 16
  ˙*kumeraho* [*elliptica*] 198–9
  °*oraria* var. ˙*novae-zelandiae* (merits species
    rank?) 397
  phylicifolia var. °*ericifolia* [*ericifolia*] (species rank
    should be restored?) 27, 198–9, 205
  ˙*prunifolia* var. *edgerleyi* (merits species rank?)
    200, 397
Populus 6* Salic.
  **nigra* 160
Poranthera 1° Euphorb.
  °*microphylla* 380
Portulaca 2* Portulac.
  **oleracea* 403, 436
Potamogeton 1° 3° 1* Potamogeton. 307–8
  °*cheesemanii* 301, 303–5, 307, 313, 330, 344
  **crispus* 301, 303, 305, 313, 320
  °*ochreatus* 301, 303, 305, 309
  °*pectinatus* 288, 301, 305
  ˙*suboblongus* [*polygonifolius*] 301, 316, 324, 330,
    333
Potentilla 1˙ 6* Ros.
  **anglica* 275
  ˙*anserinoides* [*anserina*] 38, 303, 317, 444
Prasiola 5 Green alga 463
Prasophyllum 2˙ 2° Orchid. 39
  ˙*colensoi* 52, 199, 222, 237, 249, 253, 261, 339
  ˙*pumilum* 199
Pratia 4˙ 1* Lobeli. 362
  ˙*angulata* (species aggregate) 159, 224, 228, 243,
    264, 266, 304, 384, 392, 513
  ˙*arenaria* 444–5

Pratia 4˙ 1*. (*cont.*)
˙*macrodon* 230, 411, 417, 419
˙*physaloides* [*Colensoa*] 28, 53
Prumnopitys [*Podocarpus*] 2˙ Podocarp. 5, 37, 58, 63
˙*ferruginea* 16, 20, 25, 58, 61–3, 65, 77, 112, 117, 119, 121, 126–7, 151, 154, 158, 168, 328, 389, 470, 472, 502, 509, 537, 565, 573–4, 577, 595
˙*taxifolia* [*Podocarpus spicatus*] 14, 20, 25, 58, 61, 70, 84, 112–13, 119, 121, 124, 127, 196, 203, 388, 470, 475, 478, 514, 576
Prunella 2* Scrophulari.
vulgaris 254, 262, 267, 274, 277, 316, 384, 394
Prunus 11* Ros. 553
avium 163
Pseudephebe 2 Foliose lichen
miniscula [*Alectoria*] 417
Pseudocyphellaria 42 Foliose lichen
coronata [*Sticta*] 112
Pseudognaphalium 1° Aster.
°*luteoalbum* [*Gnaphalium*] (species aggregate) 41, 356, 366, 388, 394, 396, 436, 445, 548
Pseudopanax 13˙ Arali. 37, 57–8, 120, 162, 525, 561, 565, 575, 577, 580
˙*anomalus* [*Neopanax*, *Nothopanax*] 25, 151, 173, 193, 196, 330, 543
˙*arboreus* [*Neopanax*, *Nothopanax*] (var. *kermadecensis* merits species rank?) 23, 35, 37, 55, 61, 116, 119, 140, 166, 204, 312, 387–8, 435, 501, 533–4, 544, 546, 573
˙*chathamicus* 437, 440–2
˙*colensoi* [*Neopanax*, *Nothopanax*] (var. *ternatus* merits species rank?) 23, 26, 35, 128, 135, 138–9, 144, 151, 158, 171, 173, 178, 181, 186–8, 192–3, 203, 231, 328, 330, 385, 390–1, 394, 516, 527, 547, 550, 573–4
˙*crassifolius* 23, 37, 64–5, 69, 77, 116, 119, 128, 145, 151, 158, 162, 196, 304, 316, 328–9, 389, 537–8, 561, 574, 576, 595
˙*discolor* 162, 188
˙*edgerleyi* 23, 128, 137, 140, 151
˙*ferox* 23, 37, 388
˙*laetus* [*Neopanax*, *Nothopanax*] 23, 93, 140
˙*lessonii* 23, 166, 397
˙*linearis* [*Nothopanax*] 26, 37, 135, 147, 151, 158, 176, 574
˙*macintyrei* [*Nothopanax*] 378, 389
˙*simplex* [*Neopanax*, *Nothopanax*; *sinclairii*] 23, 25, 64–5, 69, 119, 128, 135, 138–40, 144, 151, 158, 176, 178, 184, 186–7, 451–2, 455, 549, 565, 574, 576–7, 590
Pseudoparmelia 7 Foliose lichen 57–8
pseudosorediosa [*Parmelia caperata*] 461
Pseudotsuga 1* Pin.
menziesii 156
˙*Pseudowintera* [*Drimys*, *Wintera*] 3˙ Winter. 8, 9, 52, 54, 58, 561, 573
˙*axillaris* 23, 37, 120, 124, 139, 144, 579
˙*colorata* 23, 61, 119, 123–4, 128, 135, 137–8, 144, 151, 158, 170, 173, 184, 187, 196, 210, 328, 471, 511, 515–16, 563, 574–5, 578, 580–1
˙*traversii* 26, 100, 184
Psilopilum 3 Moss
australe 223, 421–2, 530
Psilotum 1° Psilot. 6, 31
°*nudum* [*triquetrum*] 382–3, 401

Psoralea 1* Fab.
pinnata 213
Psoroma ?25 Foliose lichen
buchananii 421, 425
microphyllizans 461
Psychrophila [*Caltha*] 2˙ Ranuncul. 10
˙*novae-zelandiae* 241, 345, 411, 418
˙*obtusa* 55, 411, 420, 422–3, 425
Pteridium 1° Dennstaedti.
°*esculentum* [*aquilinum*] 16, 31, 40, 72, 93, 156, 163, 213, 245, 254, 269, 278, 316, 320, 325, 353, 403, 443, 483, 485, 493, 524, 534, 536, 541
Pteris 2˙ 2° 1* Pterid.
˙*macilenta* 164, 168, 389
Pterostylis 11˙ 8° Orchid. 52, 249, 539
˙*alobula* 199
˙*banksii* 199
˙*graminea* 199
˙*humilis* 397
°*mutica* 429
°*nana* [*puberula*] 199
°*plumosa* [*barbata*] 199
˙*trullifolia* 199
Ptychomitrium 2 Moss
australe [*Orthotrichum hurunui*] 394
Ptychomnion 2 Moss
aciculare 147, 182, 202, 271, 312, 325, 328, 442, 458, 465
Puccinellia [*Atropis*] 5˙ 2° 2* Po. 290
˙*antipoda* 461
˙*chathamica* 445–6, 461
°*fasciculata* 289, 299
˙*macquariensis* 463, 465
˙*novae-zelandiae* 289, 299
°*stricta* 289–90, 295–7, 299, 384
Pulchrinodus 1 Moss
inflatus [*Eucamptodon*] 322, 331, 333, 335
˙*Pyrrhanthera* [*Triodia*] 1˙ Po.
˙*exigua* 30, 250, 257, 363, 368
Pyrrhobryum [*Rhizogonium*] 3 Moss
bifarium 147
mnioides 147
Pyrrosia [*Cyclosorus*] 1° Polypodi.
°*serpens* (= *elaeagnifolia* syn.) 35, 39, 313, 374, 382, 385, 436

Quintinia 1–3˙ Escalloni. 21, 59, 124–5, 137, 155, 188, 562, 579
˙*acutifolia* (= *serrata* var.?) 43, 61, 65, 112, 119, 128, 132, 144, 154, 184, 391, 468, 470, 502, 509–10, 512, 554, 565, 574–5, 586, 589–90, 595, 598
˙*serrata* (incl. *elliptica*) 77, 116, 120, 127, 139, 326, 501, 537, 594

Racomitrium 5 Moss 367, 418
crispulum 205, 366, 382, 395, 415–17, 422, 460, 465, 511
lanuginosum 187, 193, 205, 223, 234, 247, 260, 262, 340, 343, 349, 364, 367–8, 373, 382, 398, 415–16, 418, 426, 430–1, 527, 529
ptychophyllum 367, 415, 529
Racosperma [*Acacia*] 13* Fab. 9, 12, 496, 545
dealbata 163
melanoxylon 36, 156

Ranunculus 36˙ 4° 13* Ranuncul. 5, 10, 12, 40, 53–4, 59, 78, 216, 224, 293, 330
  °*acaulis* 354–6
  ˙acris 311, 316
  °*amphitrichus* [*rivularis*] 298, 301, 303, 313, 316
  °*biternatus* 463, 465
  ˙buchananii 52, 109, 412, 420, 529
  ˙crithmifolius [*chordorhizos, paucifolius*] 408, 420
  ˙enysii [*berggrenii, novae-zelandiae*] 238
  ˙flammula 303, 311, 313, 315
  ˙foliosus 343, 429
  °*glabrifolius* 234
  ˙godleyanus 109, 412, 417, 419
  ˙gracilipes [*sinclairii*] 38, 109, 237, 341, 345, 348, 422, 429
  ˙grahamii 109, 412, 417
  ˙haastii [*scott-thomsonii*] 407–8, 420, 425
  ˙insignis [*lobulatus, monroi*] 190, 225, 385, 388, 390, 412, 429–30
  ˙limosella 301
  ˙lyallii 29, 38, 52, 175, 190, 221, 223, 228, 380, 391–2, 394, 410
  ˙maculatus 345
  ˙macropus 301, 345
  ˙membranifolius 29
  ˙multiscapus [*lappaceus*] 29, 237, 249, 253, 262, 269
  ˙nivicola 244, 412, 430
  ˙pachyrrhizus 55, 412, 422–3, 425
  parviflorus 267, 277, 355
  ˙pinguis 451, 455, 461
  ˙recens 95, 304
  ˙reflexus [*hirtus*] 159, 178, 185, 282, 392
  repens 29, 267, 275, 277, 311–14, 316, 318, 372, 384
  sceleratus 305, 311
  ˙scrithalis 108, 408
  ˙sericophyllus 55, 230, 412, 416–17, 419
  ˙subscaposus [incl. *aucklandicus, subantarcticus*] 451, 454, 460
  trichophyllus [*fluitans*] 301, 303, 305, 308
  ˙verticillatus [*geraniifolius*] 242, 412
˙*Raoulia* c.24˙ Aster. 6, 13, 41, 52, 73, 109, 259, 367, 427, 510, 524, 607, 609
  ˙albosericea [*hookeri*] 96, 187, 368, 410, 430
  ˙apicinigra [*hookeri*] 193, 236, 252, 368, 410, 429
  ˙australis [*lutescens*] 33–4, 48, 205, 252, 257, 354, 356, 362, 367–8, 370
  ˙bryoides 410, 428
  ˙buchananii 109, 410
  ˙cinerea 410, 429
  ˙eximia 41, 108, 395, 409–10, 417, 420, 425
  ˙glabra 228, 354, 360, 366, 368, 549
  ˙goyenii 340, 396, 410
  ˙grandiflora 307, 395, 410, 416, 418–19, 422–3, 426–7, 429, 529
  ˙haastii 362, 517
  ˙hectorii 109, 410, 420–4, 530
  ˙hookeri [*australis*] (species aggregate; s.l. includes albosericea, apicinigra) 236, 252, 354, 357, 360–2, 366, 368, 370, 517, 529
  ˙mammillaris 109, 410, 420–1, 425–6
  ˙monroi 33, 238, 252, 362, 368
  ˙parkii 362, 368
  ˙petriensis 410, 420

  ˙rubra (= *eximia* var.?) 98, 410, 430
  ˙subsericea 47, 236, 249, 252, 262, 368, 370
  ˙subulata 410, 417, 419, 423, 425
  ˙tenuicaulis 36, 47, 362, 366, 512, 517–18, 527, 529
  ˙youngii 109, 410, 416–17, 420–1
Reseda 4* Resed.
  luteola 370, 394
˙*Rhabdothamnus* 1˙ Gesneri. 11
  ˙solandri 25, 32, 52, 166
Rhacocarpus 1 Moss
  purpurascens 337, 339–41, 382, 396, 415, 455
Rhizobium Bacterium 496
Rhizocarpon 8 Crustose lichen 395, 417, 431, 520
  geographicum 417, 519, 528
Rhizophagus Fungus
  tenuis 495
Rhopalostylis 1˙ 1° Palm. 8, 24, 37, 58, 432
  °*baueri* var. *˙cheesemanii* 434
  ˙sapida 61–2, 68, 72, 77–9, 93, 111, 119, 162, 389, 437, 475, 481, 533
Ribes 5* Grossulari.
  sanguineum 207, 500
  uva-crispa 207
Riccardia 27 Thallose liverwort 182, 325
  alcicornis 328
  cochleata 340
  oppositifolia 328
  ?striolata 328
Riccia 1 Thallose liverwort
  fluitans 301
Ricciocarpus 1 Thallose liverwort
  natans 301
Ripogonum 1˙ Smilac.
  ˙scandens 34, 58–9, 61–2, 68, 77, 80, 116–17, 123–4, 126, 128, 133, 140, 168, 170–1, 173, 439, 441–2, 488, 523, 573, 577
Rorippa [*Nasturtium*] 1˙ 1† 4* Brassic.
  nasturtium-aquaticum [*N. officinale*] 302, 305, 312–13, 498
  †*palustris* [*islandica*] 445
Rosa 8* Ros. 6
  rubiginosa [*eglanteria*] 196, 207, 210, 251, 260–1, 367, 369, 489, 501, 521, 544, 553
Rostkovia 1° Junc.
  °*magellanica* 338, 341, 343, 448, 455
Rubus 5˙ 5* Ros. 6, 34, 42, 54, 58, 80, 374, 544, 565, 577, 585
  ˙australis [*schmidelioides*] 117, 119, 171, 173
  ˙cissoides [*australis*] 42, 77, 128, 135–8, 144, 159, 163, 170, 172, 186–7, 196, 210
  erythrops (part of *fruticosus* aggregate) 540
  fruticosus (species aggregate) 165, 213, 267, 278, 313, 325, 351, 440, 521
  laciniatus (part of *fruticosus* aggregate) 213
  ˙parvus 27–8, 58
  ˙schmidelioides [*cissoides, subpauperatus*] 163, 173, 196, 207, 374
  ˙squarrosus [*cissoides*] 41, 163, 351, 359
Rumex 2˙ 11* Polygon. 10, 311, 354, 365
  acetosella 237–8, 254, 262, 269, 273, 275, 309, 354, 356, 360, 368, 371, 548
  brownii 271
  conglomeratus 275, 313, 320
  crispus 267, 317

Rumex 2˙ 11*. (cont.)
˙flexuosus 392
˙neglectus 354, 445, 454
*obtusifolius 267, 273–4, 372
Rumohra [Polystichum] 1° Dryopterid.
°adiantiformis 35, 128, 151, 159, 441–2, 577
Ruppia 2˙ Rupp. 60, 295–6
˙megacarpa [maritima] 288, 298
˙polycarpa [maritima] 288, 301, 305
Rytidosperma 15˙ 3° 9* Po. 5, 12, 30, 39, 73, 90, 157,
    202, 248, 250, 261, 263–4, 270, 297, 321, 360,
    369, 385, 397, 444, 539, 552
°australe [Erythranthera, Triodia] 342
˙biannulare [Notodanthonia] 198
˙buchananii [Danthonia, Notodanthonia] 394
*caespitosum 278
˙clavatum [Notodanthonia] 264, 267–8, 271, 275,
    279
°gracile [Danthonia, Notodanthonia] 205, 235, 240,
    264, 275, 335
˙nigricans [Danthonia, Notodanthonia] 337–8, 343
°pumilum [Erythranthera, Triodia] 235, 241, 248,
    250, 420, 422–3, 426, 428
*racemosum [Danthonia, Notodanthonia] 264, 435
˙setifolium [Danthonia, Notodanthonia] 96, 187,
    217, 228, 230–1, 233, 235, 241, 247, 258–9, 263,
    279, 283, 363, 367, 373, 381, 388–96, 398, 400,
    414, 416, 419, 428–30, 512, 518, 527, 529
˙thomsonii [Danthonia, Notodanthonia] 248, 369
˙unarede [Notodanthonia; Danthonia
    semiannularis) 264, 356, 359, 384
˙viride 372

Sagina 3* Caryophyll.
*procumbens 267, 275, 277, 316, 383, 392, 394, 454
Salix 11* Salic. 6, 366
*alba 312
*cinerea 296, 311, 313–14, 318, 325
*fragilis 160, 302, 313, 318, 368
Salsola 1* Chenopodi.
*kali 354
Salvia 7* Lami.
*verbenaca 371
Salvinia 1* Salvini.
*molesta 301
Sambucus 2* Caprifoli.
*nigra 163, 352, 521, 540–1
Samolus 1° Primul.
°repens 289, 291–4, 298–9, 359, 374, 384, 388, 392,
    396, 408, 436, 445–6
Sarcocornia 1° Salicorni. [Chenopodi.]
quinqueflora subsp. °quinqueflora [Salicornia
    australis] 28, 38, 289–91, 293–9, 359, 383, 386–
    7, 396, 408, 445–6
. Scaevola 1˙ Goodeni.
gracilis 435
˙Scandia [Angelica] 2˙ Api.
˙geniculata 25, 34, 207, 264, 396
˙rosifolia 25, 378, 383–4
Schefflera 1˙ Arali.
˙digitata 23, 37, 58, 61, 70, 118–20, 128, 135, 138,
    140, 158, 162, 170, 173–4, 488, 509, 525, 562,
    565, 574
Schistidium 2 Moss 382
apocarpum 354

Schistochila s.s. 15 Leafy liverwort 182, 382
glaucescens 328
nobilis 328
Schizaea 4° Schizae.
°bifida 322
°dichotoma 117
°fistulosa (small plants probably referable to
    australis) 181, 199, 312, 322, 325, 334–5, 339,
    349, 454, 533
Schizeilema 11˙ Api.
˙cockaynei 298, 303
˙haastii (var. cyanopetalum merits species rank?)
    230, 392, 396, 412, 419, 426, 430, 529
˙hydrocotyloides 109, 513
˙reniforme 455
˙roughii 388
Schoenoplectus 2° Cyper. 30, 39, 301
°pungens [Scirpus americanus] 288–9, 293, 295,
    297–9
°validus [Scirpus lacustris] 288, 297–8, 312–13
Schoenus 3˙ 5° Cyper. 30, 39
°apogon 277
°brevifolius 198–200, 313, 321, 324–5
°carsei 310
°maschalinus 29, 311, 316, 330
˙nitens (var. concinnus merits species rank?) 289,
    298–9, 314, 355
˙pauciflorus 176, 219, 224–5, 228–9, 232–5, 241,
    244, 247, 251, 304, 310, 317, 319–20, 334, 338,
    341, 343, 345–6, 348–9, 381, 388–9, 391–2, 395,
    400, 444, 513, 529, 551
˙tendo 198–9, 321, 533, 536–7
Scirpus 1† 1* Cyper.
†polystachys 12
Scleranthus 2˙ 1° 2* Caryophyll. 255, 262, 354, 362
°biflorus 354
˙brockiei 238
˙uniflorus 236, 252, 268, 309, 370
Scutellaria 1˙ Lami.
˙novae-zelandiae 380
Sedum 12* Crassul.
*acre 42, 371, 379, 394, 396
*album 395–6
*praealtum 395–6
*reflexum 395–6
Selaginella 3* Selaginell.
*kraussiana 164
Selliera 1° Gooden.
°radicans (inland plants separable as S.
    microphylla?) 289, 291–4, 296–9, 303–4, 309,
    355–6, 359, 383, 396, 408, 444–6
Sematophyllum 7 Moss
amoenum 325
contiguum 325
Senecio 13˙ 5° 16* Aster. 6, 13, 26, 38, 53
*angulatus 163, 395
˙banksii [colensoi] 379, 384
*bipinnatisectus [Erechtites atkinsoniae] 267
*cinerea 395
*elegans 354, 356
˙glaucophyllus (species aggregate; var. discoideus
    merits species rank?) 237, 379, 385, 388, 390,
    407–8, 425, 429
°glomeratus [Erechtites arguta] 356
˙hauwai 388

*jacobaea 202, 262, 267
°lautus 53, 379, 383–5, 387, 396, 446
*mikanioides 34, 163, 395
°minimus [Erechtites] 38, 316
°quadridentatus [Erechtites] 360, 388
˙radiolatus [antipodus] 445–6, 461–2
˙rufiglandulosus [latifolius] 379
˙sterquilinus [lautus] 379, 386
*sylvaticus 547
*vulgaris 365
˙wairauensis [Erechtites] 548
Setaria 7* Po. 265
Sherardia 1* Rubi.
*arvensis 267, 273, 275, 277
Sicyos 1° Curcurbit.
°australis [angulata] 34, 167, 374, 384
Silene 12* Caryophyll.
*gallica [anglica] 267, 269, 271, 370, 397
Silybum 1* Aster.
*marianum 266, 271, 354, 359
˙Simplicia 2˙ Po.
˙buchananii 381
˙laxa 381
Siphula 7 Lobate lichen 340, 396, 425
decumbens 421
Sisymbrium 5* Brassic.
*officinale 267, 278
Sisyrinchium 4* Irid. 31
Solanum 2° 1† 19* Solan. 6
†americanum [nodiflorum] 16, 64, 164, 435
°aviculare 28, 58, 70, 165, 351, 385, 440, 481
*chenopodioides [gracile] 165
*dulcamara 306
°laciniatum 28, 58, 165, 351, 440, 493, 500
*linnaeanum [sodomeum] 267, 355
*mauritianum [auriculatum] 162
*nigrum 28, 164, 313, 493
*pseudocapsicum 313
Soliva 2* Aster.
*sessilis 281
Solorina 2 Foliose lichen
crocea 425
Sonchus 1˙ 3* Aster. 354, 359–60, 385, 392, 396
*asper 379, 394, 462
˙kirkii [littoralis] 379, 392, 436
*oleraceus 271, 379, 387, 436, 446
Sophora [Edwardsia] 2˙ 1° Fab. 53, 63–4, 561
°microphylla 13, 25, 39, 50, 55, 60–1, 64–5, 133,
    162, 196, 207, 304, 385, 394, 437, 500, 522, 532
˙prostrata 25, 196, 207, 210, 269, 270
˙tetraptera (incl. howinsula of Norfolk Island?) 25,
    162
Sorbus 1* Ros.
*aucuparia 163
Sparganium 1° Spargani.
°subglobosum 311, 324, 330
Spartina 2* Po. 287, 291, 485, 553
*alterniflora [townsendii] 290
*anglica [townsendii] 289–90, 298–9
Spartium 1* Fab.
*junceum 213
Spergula 1* Caryophyll.
*arvensis 371–2
Spergularia 1° 3* Caryophyll.
°media [marginata, maritima] 28, 289, 291, 295–7,

299, 383–4, 388
Sphaerophorus 11 Fruticose lichen
tener 182
Sphagnum 8 Moss 128, 176, 193, 198, 244, 287, 302,
    312, 315, 319, 322–3, 326, 331, 334–5, 337, 339–
    40, 344–6, 348, 443, 459, 593
australe [antarcticum] 333
cristatum 262, 316, 319–20, 322–3, 325–6, 328,
    331, 333, 340–1, 343–5, 403, 520
falcatulum 301, 320, 322, 324–5, 328, 331, 333,
    341, 465
subsecundum 325
Spinifex 1° Po.
°sericeus [hirsutus] 30, 58, 352, 356–7, 485
Spiranthes 1° Orchid.
sinensis subsp. °australis 311, 324, 335
Spirodela 1* Lemn.
*punctata 301, 305
Spirogyra 20 Green alga 303, 305
˙Sporadanthus 1˙ Restion. 12
˙traversii 30, 321, 324–5, 442–4
Sporobolus 2* Po.
*africanus 265, 277, 279, 355, 435
Sprengelia 1° Epacrid. 59
°incarnata 12, 26, 230, 336
Stackhousia 1˙ Stackhousi.
˙minima 252, 337, 368
Stellaria 4˙ 1° 3* Caryophyll.
*alsine 311, 316, 342
˙decipiens 458, 460, 462, 464
˙gracilenta 41, 237, 253, 309, 362, 370, 428, 529
*graminea 311
*media 372, 387, 451, 454, 548
°parviflora (= decipiens syn.?) 214, 375, 388
˙roughii 40, 108, 408, 425, 429
Stenotaphrum 1* Po.
*secundatum 265, 279, 353, 435
Stereocaulon 10 Fruticose lichen 205, 395, 431, 460,
    511, 520
caespitosum 417
corticulatum 519
ramulosum 463
Sticherus [Gleichenia] 1˙ 1° Gleicheni.
˙cunninghamii 113, 128, 135, 139, 176–7, 186, 509
°flabellatus 101
Sticta 13 Foliose lichen 425
˙Stilbocarpa 3˙ Arali. 38
˙lyallii [Kirkophytum] 169, 191, 396
˙polaris 38, 449–50, 453, 455–6, 462–5
˙robusta [Kirkophytum] 110, 191
Stipa 1˙ 1° 9* Po. 265, 269, 276, 297
*bigeniculata 270
*nodosa [variabilis] 269, 270, 356, 360
˙petriei 107
°stipoides [teretifolia] 290, 293–4, 397
Stokesiella 1 Moss
praelonga [Eurhynchium] 266, 271, 274, 278, 364
Streblus [Paratrophis] 3˙ Mor. 9, 58
˙banksii [P. opaca] 25, 166, 532
˙heterophyllus [P. microphylla] 25, 37, 119, 133, 173
˙smithii 92, 502
Stuartina 1* Aster.
*muelleri 271
Suaeda 1˙ Chenopodi.
˙novae-zelandiae [maritima] 288–91, 299, 386, 408

Suillus [*Boletus*] 8* Fungus
  luteus 493
Swainsona 1˙ Fab.
  ˙*novae-zelandiae* 374, 496
Syzygium 1˙ 1* Myrt. 9
  ˙*maire* [*Eugenia*] 43, 58, 112, 118, 140, 540

Tanacetum 2* Aster.
  **parthenium* [*Chrysanthemum*] 372, 394
Taraxacum 1˙ 1* Aster. 28, 53
  °*magellanicum* 249, 429
  **officinale* 262, 266, 271, 273–4, 364, 370, 539
Tecomanthe 1˙ Bignoni. 11
  ˙*speciosa* 92, 502, 532
Telaranea 19 Leafy liverwort
  gottscheana 325, 328
  herzogii 325
  tetradactyla 325
Teline [*Cytisus*] 3* Fab.
  **monspessulana* [*C. candicans*] 213, 395, 500
  **stenopetala* 395
Tetrachondra 1˙ Tetrachondr. 12
  ˙*hamiltonii* 95, 108, 301
Tetracymbiella 2 Leafy liverwort
  decipiens 328
Tetragonia 1˙ 1° Aizo. 351
  °*tetragonioides* [*expansa*] 28, 60
  ˙*trigyna* 34, 58, 167, 191, 356, 374, 379, 384, 395–6, 446
Tetraria [*Cladium*] 1° Cyper.
  °*capillaris* [*capillaceum*] 198–9, 321, 325, 335
˙*Teucridium* 1˙ Verben. 11
  ˙*parvifolium* 25
Thamnolia 1 Tubular lichen
  vermicularis 340, 395, 416, 421–2, 424–5
Theleophyton [*Atriplex*] 1° Chenopodi.
  °*billardierei* [*A. chrystallina*] 353, 445, 485
Thelymitra 7˙ 5° Orchid. 335, 539
  °*carnea* [*imberbis*] 199
  °*longifolia* 199, 237, 249, 253, 385
  ˙*pulchella* 199
  °*venosa* [*unifolia*] 324–5, 335, 337, 339
Thelypteris 1˙ Thelypterid.
  °*confluens* [*palustris* var. *squamigera*] 312, 323
Thinopyrum 1* Po.
  **junceiforme* [*Agropyron*] 356
Thismia 1° Burmanni.
  °*rodwayi* [*Bagnisia hillii*] 36
Thuidium 4 Moss
  furfurosum 249, 266, 275, 278, 312, 325, 458, 465
Thymus 2* Lami.
  **vulgaris* 368–9, 498, 501
Tmesipteris 1˙ 3° Psilot. 6, 9, 31, 35–6, 117, 140
  °*tannensis* 128, 182, 577
Torilis 3* Api. 165, 275
  **arvensis* 267
˙*Toronia* [*Persoonia*] 1˙ Prote.
  ˙*toru* 37, 58, 188
Tortula 12 Moss 364
  princeps [*tenella*] 355
  robusta [*rubra*] 415
Tradescantia 1* Commelin.
  **fluminensis* 164, 473–4, 553
Tragopogon 3* Aster.
  **porrifolius* 267

˙*Traversia* 1˙ Aster. 52
  ˙*baccharoides* 26, 376
Trentepohlia 10 Green alga 140, 383, 519
Trichocolea 5 Leafy liverwort 147, 182
  mollissima 328
Trichomanes 3˙ 3° Hymenophyll.
  °*elongatum* 382
  ˙*reniforme* [*Cardiomanes*] 35, 39, 117, 128, 140, 382, 441–2
  ˙*strictum* [*rigidum*] 330, 382, 391
  °*venosum* 36
Tridontium 1 Moss
  tasmanicum 303, 382
Trifolium 25* Fab. 6
  **arvense* 254, 266, 269, 272, 365, 368, 370
  **dubium* [*minus*] 237, 254, 262, 266, 269, 271–3, 275, 277, 365, 370, 372, 387, 394, 521, 539
  **fragiferum* 317
  **glomeratum* 271, 275
  **hybridum* 249, 254, 320
  **pratense* 54, 262, 266, 273, 278
  **repens* 215, 224–5, 237, 249, 254, 262, 269, 271–4, 276, 277, 313, 317, 359, 372, 394, 444, 473, 539
  **striatum* 266, 275
  **subterraneum* 266, 269, 271, 275
Triglochin 2° Juncagin.
  °*striatum* 288–9, 292–3, 295–9, 330, 359, 445
˙*Trilepidea* [*Elytranthe*] 1˙ Loranth.
  ˙*adamsii* 6, 36, 93
Triquetrella 1 Moss
  papillata 205, 266, 271, 355, 364
Trisetum ?2˙ 1° Po. 264
  ˙*antarcticum* 248, 259
  °*spicatum* [*subspicatum*] 12, 414, 426, 429, 461, 529
  ˙*youngii* 248
Tropaeolum 3* Tropaeol.
  **majus* 395
˙*Tupeia* 1˙ Loranth.
  ˙*antarctica* 36
Tussilago 1* Aster.
  **farfara* 365
Tylimanthus 3 Leafy liverwort
  saccatus 328
Typha 1° Typh. 30, 73
  °*orientalis* [*angustifolia, muelleri*] 40, 293, 297, 305–6, 312–14, 319

Ulex 2* Fab.
  **europaeus* 59, 84, 157, 168, 267, 274, 277, 313, 352, 444, 497, 541
Ulothrix 16 Green alga 300, 305
  subtilis 303
Ulva 16 Green alga 288
Umbilicaria 11 Foliose-lobate lichen 395, 417
  cylindrica 420
  hypoborea 417
  vellea 417
Uncinia 30˙ 2° Cyper. 5, 58–9, 112, 144–5, 151, 170, 173, 219, 374, 390, 574–5
  ˙*aucklandica* 448
  ˙*caespitosa* 243
  ˙*distans* 579
  ˙*divaricata* 228, 230, 366, 395, 414, 419, 494, 529
  ˙*drucei* 414
  ˙*egmontiana* 173, 208

°*filiformis* 159
fuscovaginata 421
gracilenta 159
hookeri [*riparia*] 448, 458, 460–2, 465
purpurata 240
rubra 240
rupestris 137, 159, 187, 328, 442
°*uncinata* [*pedicellata*] 117, 144, 159, 168, 170,
172–3, 202, 389, 392, 517, 532, 536–7
'*australis*' (= *uncinata* and related species) 539
Urtica 4˙ 1° 2* Urtic. 91
australis [*aucklandica*] 28, 445–6, 455, 462
ferox 39, 165, 170, 359, 374
incisa 38, 165
linearifolia 311
urens 280
Usnea 15 Fruticose lichen 135, 396, 403, 415, 582
Utricularia 1˙ 3° 1* Lentibulari. 36
australis [*mairii, protusa*] 303, 322
°*lateriflora* [*delicatula*] 322, 325
°*monanthos* (part of *novae-zelandiae*?) 298, 321,
324–5, 327, 331, 334–5, 341–2
novae-zelandiae [*colensoi*] 304, 321

Verbascum 5* Scrophulari. 28, 374
thapsus 41, 165, 254, 269, 365, 367–8, 370, 394,
547
virgatum 365, 368, 370
Veronica 16* Scrophulari. 6
anagallis-aquatica 301, 305, 308, 311
arvensis 267, 274, 278
plebeia 543
serpyllifolia 275, 553
verna 371
Verrucaria 10 Crustose lichen 396, 461
maura 383, 463
Vicia 10* Fab. 360
hirsuta 215, 271
sativa 215, 267, 269, 271, 313
Vinca 2* Apocyn.
major 351
Viola 2˙ 1° 5* Viol.
°*cunninghamii* 230, 237, 240, 249, 253, 262, 307,
320, 341, 370, 429, 529
filicaulis 185
lyallii 304, 315
Vitex 1˙ Verben.
lucens 21, 58, 63, 77, 79–80, 96, 112–13, 117, 119,
166, 532
Vittadinia 1˙ 4* Aster.
australis 237, 253, 268, 369, 385, 388
gracilis [*triloba*] 368–9, 371
Vulpia 2* Po. 248, 265, 270, 275, 278, 281, 364, 369,
384

bromoides [*dertonensis*] 250, 268, 272, 435
myuros 272, 366

Wahlenbergia 9˙ 1°* Campanul. 53, 387
albomarginata 29, 237, 240, 249, 253, 259, 261,
309, 363, 370, 373, 418, 426, 518, 527
cartilaginea 408, 429
congesta (= *albomarginata* var.?) 354
gracilis (adventive plants probably include other
species, e.g. *marginata*) 29, 215, 249, 252, 269–
70, 385, 397
matthewsii 385
pygmaea (= *albomarginata* var.?) 247, 367, 392,
529
Watsonia 4* Irid.
bulbillifera 31
Weinmannia 2˙ Cunoni. 5, 8, 21, 52, 55, 58, 543–4,
561, 563
racemosa 37, 43–4, 61, 63, 65, 70, 77–9, 112–13,
119–21, 124, 127, 145, 151, 154, 158, 168, 171,
203, 304, 312, 328, 366, 468, 470, 472, 478, 486,
488, 493, 495, 501–3, 509, 516–17, 546, 562,
565, 574–5, 578, 587, 589, 595
silvicola (= *racemosa* vars?) 113, 117, 119, 140,
188, 200, 533–4, 576
Weissia 4 Moss
controversa 364
Weymouthia 2 Moss 135, 187
Wolffia 1° Lemn.
°*australiana* [*arrhiza*] 301, 305, 316

Xanthoparmelia 16 Foliose lichen 355
Xanthoria 4 Foliose lichen 383, 396
Xeronema 1˙ ?Phormi. 11
callistemon 53, 92, 381, 383

Yoania 1˙ Orchid.
australis 36, 101

Zannichellia 1° Zannichelli.
°*palustris* 301, 305
Zantedeschia 1* Ar.
aethiopica 164
Zizania 2* Po.
latifolia 302
Zoopsis 4 Leafy liverwort
caledonica 328
Zostera 2° Zoster. 288, 290–1, 293
°*capricorni* 288, 292
°*muelleri* 288, 290, 292, 298
Zoysia 3˙ Po. 353, 485
minima [*pungens*] 356, 385
Zygodon 7° Moss 382

GENERAL INDEX

(Bold page numbers indicate main references)

Abel Tasman National Park 204, 214
Ahipara plateau 199–201
Ahuriri valley 81
akeake 23
allelopathy **562**
alligator weed 267
allophane **86**, 525, 556, 579
alpine 78, Chapter 12
Alpine Fault 14, 101, 178, 555
Alps province **108–9**
altitudinal belts **74–8**
Anglem (Mt) 182
animal biomass 467
aniseed 245
annuals 28, 247, 368–9
Antipodes Islands **461–3**
Aorangi Range 97, 153, 385
apatite 85
aquatic plants **286**, 485
aquatic system **285**
Arawata River 136, 515
area (New Zealand) 1
area (outlying islands) 1, 433
Arthur (Mt) 100, 429–30
Arthur Ecological District 100, 259, 385
Arthurs Pass 108, 179, 229, 428, 547–50
arum lily 164
ash (see tephra)
Auckland (city) xix, 355–6
Auckland (province) **93**
Auckland Islands **446–56**
avalanche 84, 569–70, 572
Awarua Plain 331, 334
Awarua Point 374
axial ranges (North Island) **98**

Banks Peninsula 105–6, 215, 269–72, 582
barberry 212
basalt 105, 463, 526
bats 6, 52
Bay of Plenty 120
beaches (shingle) 351, **359–61**

beech (see also black, hard, mountain, red, silver beech) 15–16, 71, 101–2, 104, 106–8, 125–6, 132, 136, **140–59**, 170–1, 174, 184, 202–3, 214, 469, 474–5, 488, 499, 504, 508, 514, 526, 545–6, 554, 556, 561–4, 567, 570, 580
beech gaps 134, 141, 150, 179, 184, 554
beech (regeneration) **583–6**
bees 54
bell-bird 53
belts (altitudinal) 11, **74–8**
Ben Ohau Range 238, 318, 418, 528
biennials 28, 247
Big Bay 327–31, 521–3
bioassay 457
biological spectra 20
biomass **466–7**, 490, 505–7
bird colonies 386–7, 448, 458, 460, 579
Birdlings Flat 359
birds 6
birds (coastal) 379, 381
birds (dispersal agents) 59–60, 162, 532, 600
birds (effect of guano) 381, 579
black beech 97, 141, 145–7, 153, 155–6, 388, 394
blackberry 213–14, 267, 272, 314, 360, 440–1, 444, 521, 544
blight (manuka) (see also scale insects) 197
Blue Mountains 340
blue tussock 245, 258
blue wheatgrass (see wheatgrass)
bog (see also string bogs) 285, Chapters 10, 13
bog pine 186, 193–5, 208, 234, 243, 326–7, 331, 336, 340, 348, 521, 525
botanical provinces (map) **94**
Bounty Islands 463
boxthorn 212
bracken 16, 72, 93, 156, 163, 199, 204, **213–15**, 245, 258–9, 275, 320, 353, 356, 443, 446, 483, 499, 524, 526, 533, 537, 539–40, 543–4, 606
briar 212, 367, 521, 553
bristle tussock 258–9
Broken River 210, 512
broom (common) 157, 198, 213, 314, 352, 365–6,

377, 496, 540, 553
broom (native) 245, 256, 356, 373
broom (pink) 377
browntop 90, 247–8, 259–60, 265, 272, 275, 394, 520, 539, 552
browsing 32, 170, 203, 377, 547, 573–6, 590, 603
brush wattle 163
bryophyte 112, 140, 182, 187, 205, 340, 364, 367, 403, 415, 452
buds 40
buffalo grass 265, 276
Burnett (Mt) 389
bush **72**, Chapter **8**

cabbage tree 23, 162, 166, 173, 245, 311, 313, 315–16, 355, 388, 607
calcicole 218, 378–9
calcifuge 218, 225
calcium 85, 87, 400
Campbell Island 456–61
candlenut 432
Canterbury Plains 105, 157, 204, 267–8
Canterbury province **104–6**
carbon dioxide 466, 480, 499
Cargill (Mt) 84
carices (i.e. *Carex* spp.) 248, 302, 310, 316, 321, 338, 343
carpet grass 218, 225–6, 242, 337, 343
Cass 318–20
Castle Hill 408
Castlepoint 384
catena **89**, 113, 126
Catlins 107, 522–4, 582
catsear 214, 249, 257, 272, 364, 397
cattle 276, 310, 318, 436, 447, 454, 456, 575
celery pine 26
Central Otago 48, 104, **106–7**, 150, 205, 258–9, 299, 340–1, 368, 419–24
Characeae 288, 300, 305
charcoal 84, 136, 197, 202, 205, 287
Chatham Islands 287, **436–46**
cherry 163
clay (role in ion exchange) 86
climatic change 14, 603
climax **91**, 207, 338, **507–8**, 545–6, 556
cloud 4, 41–2, 108
clover (*see also* red, white clover) 275, 281, 573
coal measure 100, 184, 226
coccids (*see* scale insects)
cocksfoot 90, 214–15, 248, 272, 275–6, 278, 320, 359, 372, 394, 526
cohorts 560, 594–8
coltsfoot 365
comb sedge 338
community interrelations (diagrams) 124, 227, 233, 293, 297, 299, 452, 510
compensation point (light) 560–1
competition effects 201, 607
conservation 608–9
Cook (Mt; *see also* Mt Cook National Park) 1, 108
Cook's scurvy grass 381
coppicing 562
cord grass 290
Coromandel (Ecological Region, Peninsula) 93, 115, 188, 165, 198

crack willow 302, 311
Craigieburn Range 22, 150, 426, 504, 571
Cropp River 512–13
Cupola basin 55
cupressoid (habit) 38
cupressoid hebes (*see also* whipcord hebes) 415
currant (flowering) 207
cushion bog 330–1, 334–5, 454, 459–60, 465
cushion plant **33–4**, 41, 354, 338, 409–11
cyclones 3, 569

dandelion 266, 364
danthonia 90, 248, 256, 259, 264, 269, 272, 275–6, 444, 606
Dart valley 528
debris 89–90, 100, 168–9, 185, 190, 209, 372–4, 385, 388–90, 395, 398, 400
deciduousness 26, **39**
deer 7, 178, 185, 191, 202–3, 217, 247, 282, 573–6, 580–1, 590
depletion 221, 367–8
diameter increment 475–7
die-back 563–4, 568, 579–80, 590
dioecy 54
diorite 86, 109, 182
disjunct distributions 14, 101
dispersal 58–60
divaricating (*see* filiramulate)
dock 267, 365
dodder 36
dolerite 429
dolomite 389
douglas fir 156–7, 160
drainage 89, 285, 287, 310, 318
drought 3, 84, 264, 275, 280, 366–7, 567, 579, 582
dry matter 471–5
dunes 272, **350–9**, 361, 446, 521–4
dwarf shrubs 27

earthquake (Murchison) 515
ecological regions (map) **94**
ecotypes 141, 397
edelweiss 379
eel-grass 288
Egmont (Mt; *see* Taranaki)
elder 163, 352, 521, 540, 553
Ellesmere (lake) 289, 294–7, 316–18, 357–8
endemic species (to regions) 14, 376–82, 385, 406, 432–5, 437–8, 442, 445, 447–51, 455, Chapter 6
epacrid 26–7
epiphytes **35–6**, 80, 112, 116, 135, 140, 187
erosion 374
European settlement 17, 436, 447
eutrophic 285
evolution 13
exclosures 573–6
Eyre Mountains 108

facilitators (in succession) 557
far-southern zone **80**
Farewell Spit 350
fellfield 258, 405, 464, Chapter **12**
fen **285**, 459, 556
fern (filmy) 35, 112
fernland 73, 213–15

fescue (*see also* tall fescue) 265
fescue tussock 217
fibre ('flax') 310
filiramulate 23, 25, **31–32**, 42, 72, 162, 165, 206–10,
    351, 374, 543, 577
Fiordland (Ecological Region, province) 102–3, 168,
    180, 515–16
fiords 376, 393, 456
fire 14–15, 22, 132, 136–7, 170, 175, 188, 190, 193–4,
    198, 200–2, 205–6, 210, 212, 214, 225, 244–5, 258,
    260, 264, 316, 324, 334–6, 389, 434, 442, 456, 507,
    526, 540, 545, 547–9, 554, 557, 581
five-finger 23, 116, 124, 166–7, 169, 171, 532, 533,
    539–40, 543–4, 573
flatweed 249, 266, 281
flooding 129, 563–4
flower colour 52–4, 448–51
flowering (variable) 56–7
flowering time 55
flushes **285**, 339, 414
fluvioglacial 128, 174
fog (*see also* cloud) 84
föhn 3, 84, 256, 258
forest (exotic) 157–60, 547
forest limit (subalpine) 78, 145, 471, 479, 504
Fortrose dunes 357–9
fossil record (*see also* pollen record) 8–11
Fox Glacier 75
foxglove 214, 372, 374
Franz Josef Glacier 509–11
freezing damage 502–5
frost 4–5, 120, 135, 213, 245, 264, 316, 499–500
fruit (adaptions for dispersal) 58
fruiting (variable) 56–57
fuchsia 120, 133, 162, 169–71, 174, 372, 524, 573
fungi 36, 493, 564

gentian 248, 339, 413, 450
geothermal habitats (*see* thermal habitats)
germination 62–4
ginger 164
Gisborne province **95**
glaciation 1, 5, 14, 82, 100, 506
glass (volcanic) 85
glasswort 290, 295, 298–9
gleying **87**, 147
gneiss 102, 393
goats 168, 389, 434–5, 447, 454, 488, 532, 575,
    579–80
Gondwana 8–9, 12
gooseberry 207
Gorge River 180, 344–8, 392
gorse 84, 157, 168, 197–9, 204, **212–13**, 267, 272,
    275, 313–14, 352, 360, 365–6, 392, 394, 444, 496,
    498, 539–40, 542–4, 553
Gouland Downs 336–7
gradients 112, 279, 451–2, 505
granite 86, 101, 102, 109, 179, 223, 391, 393
grass grub 258, 278, 281
grasses ('English') 214, 245, 280
grasshopper 217
grassland types **50**
grazing 164, 172, 225, 247, 258, 281, 367, 443, 457–
    8, 460, 557
grey willow 311

greywacke 1, 86, 88, 95, 101, 107–8, 192, 258, 406,
    556, 580
gumland **72**, 88, **198–201**, 285, 321

Hankinson (Lake) 587
hard beech 47, 80, 89, 118, 126, 141, 145–6, 153,
    156, 184, 480, 483, 490, 533, 539, 554, 567–8, 586,
    588
hard tussock 245, 248, 256, 258–9, 260, 520
Hauraki Gulf 165, 312–13
Hawkes Bay 275, 384
hawkweed 249, 260, 368, 553
hawthorn 212
heath(land) types **72–3**
heather 193–4, 244
heathlands 12, 441–2, Chapter **8**
Hen Island 92, 532
Herangi Range 93, 349
herbfield **73**
herbicide 161, 190, 377
Hikurangi (Mt) 98, 186, 193, 221
hinau 80, 116, 119, 120, 124–5, 130, 132, 520, 554
Hohonu Range 126
Hokitika River 512, 514
Hollyford valley 151
Horokiwi Stream 372
horsetail 365
humus 86–7
Hunter valley 83, 106, 171–2
Hutt Valley 125, 538–40
hybrids 13, 25, 26, 32, 65, 120, 128, 138, 174, 189–
    90, 206, 217, 220, 238, 340, 351, 363, 365, 389,
    410, 412, 429, 550, 562

ice-plant 353, 378
inhibition (in succession) **558**
insectivorous plants 36, 321–2
insects (*see also* grass grub, scale insects) 6, 52, 54,
    564–5
interception (of precipitation) 84
interior zone 80
invasion 508, **553–4**
inversion (temperature) **81–2**, 147, 173, 179, 193,
    525
isolation (geographical) xviii, 506

juvenile forms 24, **64**, 68–9

K-selection **558–60**
kahikatea 20, 102, 112, 118–20, 123, 126, 128–33,
    145, 284, 313, 316, 327, 330, 511, 515, 522, 525,
    544, 564, 567, 577, 595, 599
kaikawaka 43, 113, 128, 135–8, 140, 144, 174, 178,
    186, 188, 512, 514, 521, 525, 548–9, 551, 579–80,
    582, 590–4
Kaikohe 199–200
kaikomako 162, 167, 172
Kaikoura 520
Kaikoura Range (Inland) 171, 427–9
Kaikoura ranges 103, 108, 136, 138, 189–90, 242
Kaimai Range 43, 579
Kaimanawa Range 153, 349
Kaimata Range 171, 177, 511–12, 592
Kaitorete Spit 356
kaka-beak 377

kakapo 153, 576
kamahi 43, 78, 80, 82, 113, 120–1, 124–7, 128–9,
    132–8, 144, 168, 170, 173–4, 181–2, 185, 204, 312,
    330, 366, 389, 391–2, 400, 488, 494–5, 498, 509,
    512, 514, 521–5, 539–40, 545, 562–4, 575, 579–80,
    582–3, 586–7, 589–90, 594, 598, 600
kanuka 84, 157, 162, 166, 169, 171, **195–206**, 324,
    368, 383–4, 397, 400, 490, 520, 524, 531–2, 537,
    540, 543–4, 573, 576
Kapiti Island 99, 387, 544, 546
karaka 16, 22, 80, 118, 132, 162, 166, 169, 384, 434–
    40, 532
Karamea 100, 318
Karangarua valley 173, 178, 343, 391
Karapiti geothermal area 400–2
karst 100, 389
Kauaeranga valley 533–4
kauri 20, 45, 71, 78, 80, 92, 111–12, 114–17, 139–40,
    188, 198, 201, 324, 469, 470, 475, 483, 488, 494,
    533, 538, 560, 564, 573, 579, 601–5
kawaka 112
Kermadec Islands **432–6**
kiekie 78, 120
kikuyu grass 265, 276
kohekohe 22, 118, 120, 123, 126, 162, 166–7, 560–1,
    590–1
kokako 565
Kopuatai peat dome 313, 324
kowhai 162, 171, 173, 394, 437, 439, 522, 561, 565

Lagoon Saddle 319, 341–2
Lammerlaw Range 341–3
lancewood 23, 116, 162, 167, 171, 173, 316, 538, 561
landslides **89–90**, 104, 512, 586–7
lanes (in heathland) 339, 454–5
larch 157
latitudinal zones **78–80**
layering (see also roots (adventitious)) 26, 562, 590
leaf size categories **37–38**
leaf temperature 42, 449
leaf wax 41, 248, 407
legumes 88, 496
Lewis Pass 108, 227
lianes 34–5, 116, 123, 128, 163, 173, 207, 351
lichens 193, 203, 257, 355, 383, 403, 415–17, 461,
    463, 496, 519–20
life span 64, **70**
limestone 86, 103, 209, 375, 382–3, 385, 388–9, 408,
    581
litter **86**, 474, 490–2, 498, 561–2
litter (mor, mull) **86**
Little Barrier Island 55, 385–7, 532–3
lizards 59
loess 361, 520
logging 115, 590
Lord Howe Island 13, 432
lowland totara (see totara)
lupin 360, 366, 497
lycopod 31, 35, 164

Mackenzie (basin, Plain) 104, 234, 255–6, 258, 306–
    7, 309, 318–19, 365, 367
Macquarie Island 80, **463–5**, 480
macrophytes (algal) 286
macrophytes (freshwater) 300, 305

magnesium 84, 527
mahoe 86, 118, 120, 124, 126, 133, 162–3, 167, 168–
    71, 173, 352, 520, 524–5, 532, 539–40, 544, 567,
    580
Manapouri (Lake) 303–4, 306–7, 331
Manawatu (Ecological Region, River) 272, 275, 305–
    6, 308, 320, 355
mangeao 116, 590–1
mangrove 43, 80, 93, 291–3, 499
Maniototo Plain 107
manoao 112, 118
manuka 27, 130, 171, 175, 178, 180–2, 184, 188,
    193–206, 208, 211, 216, 303, 311–13, 323–4, 326–
    7, 329–31, 334–5, 337, 344, 349, 351, 355, 363,
    385, 388, 391, 394, 397, 400, 490, 494, 498, 511,
    515–16, 520, 524–5, 527, 533, 538–9, 543–4, 558,
    562, 564, 573, 577, 606
Manukau Harbour 293
Maori settlement 15–17, 287, 436, 447
maps xix, 2, 4, 9, 10, 15, 17, 18–19, 75–6, 79, 94,
    142, 317, 464, 555, 569–70, 580
marble 100, 223, 225, 393, 429
Marlborough province **103–4**
Marlborough Sounds 126
marram 272, 352–3, 355–7, 359–60, 446, 497, 522
Maruia valley 148, 567, 585
mat plants **33–4**, 249, 338, 354, 409–11
matagouri 72, 101, 172, 206–7, 210–11, 217, 245,
    256, 258, 260, 363, 367, 373, 517, 519, 556
matai 14, 20, 43, 84, 112, 120, 124, 126, 129, 132,
    145, 149, 388, 475, 514, 524–6, 576, 595, 599–600,
    602
Matiri valley 517
Maungatua (Mt) 552
Mavora Lakes 260–2
Mayor Island 312
medick 266
megafauna 32, 606
mice 579
microclimate 42
Milford Sound 104
Mineral Belt 99, 397
mineralisation 88, 507
mire **73**, 285, Chapters **10**, 13
miro 43, 112, 115, 119–20, 126, 134, 139, 170, 509,
    515, 524, 573, 594–8, 602
mistletoe 36, 143, 198, 567
moa 6–7, 576–7, 605
Moawhango (Ecological Region) 95
moisture **83–4**
monocarpic 28
mor (see litter)
moraine 176, 509–10, 512, 528–9
Moriori 436
moss (see bryophyte)
Motu (gorge, River) 373, 384
Motutapu Island 312–13
mountain beech 22, 78, 84, 103, 143, 146–55, 174,
    179, 184, 186, 346, 349, 388, 400, 467, 469, 471,
    475, 479–80, 483, 488, 514, 517, 527, 556, 560,
    563, 567, 580, 583–4
mountain totara (see totara)
Mt Cook National Park 136, 259, 395, 416–18,
    547
mull (see litter)

mycorrhiza 141, **492–5**, 506
Mysore thorn 435

nectar 53
nettle 91, 165, 280, 374, 445
ngaio 22, 162, 167, 169, 171, 352, 356–7, 385, 388, 437, 439, 607
Ngauruhoe (Mt) 96, 157
nikau 80, 93, 101, 111, 116, 119–20, 126, 132, 162, 166, 168, 434–5, 437, 441, 473–4, 532–3
nitrogen 88, 207, 278, 281, 489, **495–6**, 499
nitrogen fixation 88, 352, **496**, 555
nival **78**, 417
Norfolk Island 432
North Cape 92, 200, 397
northern rata 80, 116, 119–20, 123, 125, 130, 140, 170, 184, 376, 533, 561, 567, 580, 582
Northland 92, 95, 114–20, 140, 165, 168, 198–201, 276–8, 323–4
nutrient elements **485–92**
nutrient transfer by livestock 247, 278

occlusion (phosphorus) 86–7, 525
Ohinemaka 302, 327
Okarito 298, 300
Old Man Range 241, 424, 531
oligotrophic **285**
Omahutu Forest 115, 118
orchids 30, 36, 40, 101, 199, 324, 339
Otago (Central; *see* Central Otago)
Otago province **106–7**
Owen (Mt) 100, 223–4, 429–30
oxygen weed 301

pakihi **72**, 88, 174, **285**, 334–6
palatability 573–4
Paparoa Range 100, 184–5, 226–7, 391, 551
parasites (vascular plants) 36, 41, 221
Paringa 41, 102, 315, 392–3, 555
peat 92, 180, 191, 313, 321, 324, 326–7, 331, 334–5, 343, 348, 393, 436, 442–3, 446–8, 456–7, 462–4, 507, 511
peat types **285**
penalpine **78**
pepperwood 23
petrel burrows 386, 460, 462
pH (soil, water) 313, 315, 319, 324, 326, 341, 387, 401, 457, 524
phosphorus **85–9**, 198, 280–1, 488–9, 492–6, 506, 509, 514, 525–8, 557, 579
photosynthesis **480**, **482–5**
photosynthesis (C4, CAM) **485**
phytoplankton 305
pigeon (introduced) 579
pigeon (native) 59, 532, 565, 600
pigs 214, 436, 447, 454, 575
pine (*see* bog, celery, pink, silver, yellow-silver pine)
pines (exotic) **156–60**, 471, 521, 539, 558
pingao 352, 355–7, 359–60
pink pine 135, 137, 176, 181, 184–6, 194, 344, 348, 400, 512, 514, 548–9, 551, 579, 593
piripiri 249
Pisa Range 190, 206
plant cover (map, percentages) **18–19**
plantations 156–60

Plimmerton swamp 314
podocarps 57, 71, 93, 111, 115, 120–32, 134–5, 144, 170, 173, 470, 494, 579
podzol **88–9**, 127, 174, 194, 509, 512, 514–15, 522, 555, 562, 569, 586, 591
pohutukawa 80, 96, 166, 351, 376, 383, 397, 499, 524, 526, 532, 567, 579, 607
pokaka 144, 147
pollen record 14, 16–17, 195, 201, 214, 324, 525
pollen vectors 52–4
poplar 160
poppy (Californian) 356
porina moth 278
poroporo 351
Port Hills 55, 395–6, 540
possum 7, 565–7, 573, 579–80, 582
precipitation 3–4, 84
privet 162, 553
production 466, 471–80, 505–6
profile diagrams 119, 122, 124, 126, 134–5, 145, 153, 167, 171, 173, 177, 183, 200, 329, 346, 455, 522–3, 535–6, 545–6
psilophyte 31
pukatea 93, 113, 118–20, 126, 560
pumice 85, 89, 120, 188, 193, 337, 349, 526
Pureora (Mt) 137, 176, 187, 326
Pureora Forest 123–4, 599–600
puriri 80, 86, 96, 113, 120, 166

r-selection **558**–60, 605
rabbits 7, 212, 245, 247, 367–8, 447, 454, 463
radiocarbon 198, 202, 521
ragwort 365
Rakaia River 104, 109
Rakiura province **109**
rangiora 23, 120, 123, 126, 162, 166
Rangipo Desert 361
Rangitaiki Plain 193–4, 337
Rangitoto Island 93, 526
rata (*see* northern, southern rata)
rata vines 128, 378
rats 7, 436, 579
ratstail 265, 275
raupo 297, 310–11, 316, 320
red beech 21, 78, 80, 93, 103, 141, 145–51, 153, 155–6, 159, 204, 469, 480, 483, 488, 514–15, 560, 567–8, 580, 583–5, 602
red clover 272, 275, 266
Red Hills (Nelson) 398–9
Red Hills (Westland) 400, 527
red tussock 108, 186, 217, 226, 232, 234, 238, 243, 244, 260, 320–1, 334, 336–7, 343, 349, 511, 514
redtop 265
refugia 14, 376
regeneration 91, 344, 440, Chapter **16**
regeneration cycles 431, 459–60, 599–600, 602
regolith **87**, 419, 428, 463
reserves 205, 440, 609
resin (kauri) 198
restiad (i.e. Restionaceae) 12, 50, 284, 290, 442
'reverse-J' curve **560**, 601
rewarewa 116, 119–20, 124, 533, 538, 590–1
rhizomes 29–31, 40, 212–13, 265, 352, 363
ribbonwood (shore) 291
Richmond Range 99, 397

ricker (kauri) 20, 114
rimu 14, 20–1, 43–4, 112, 115, 118–20, 123–5, 126–
9, 132, 134–6, 139–40, 174, 176, 181, 188, 327,
400, 475, 483, 509, 511, 514–15, 522–6, 544, 561,
564, 567, 590, 594–600, 602
Rimutaka Range 153, 169–70, 243
rock (calcareous; *see also* limestone, marble) 90, 223,
225, 376, 381
rock (mafic) 85, 427
rock (ultramafic) 1, 85, 99, 102, 108–9, 173, 179,
200, 242, **396–400**, 427, **527**
root hairs 50, 493–4
root nodules 43, 494, 496
roots (adventitious; *see also* layering, suckering) 21,
24, 26, 29, 34, 35, 42
roots (*see also* tap-roots) **43–50**, 321, 407, 473, 489,
492–6, 563–4
rosette 28
Rotoiti (Lake) 147, 202, 205
Rotorua lakes 303–5
rowan 163
Ruahine Range 137, 153, 185, 249, 348, 579, 585
Ruapehu (Mt) 76, 155, 186–7, 192, 349, 361, 430–1
rush 266, 272, 310, 365, 444
Russell lupin 365
ryegrass 265, 269, 272, 275, 278, 444

salt damage 191, 564
salt marsh, meadow 209, **285**, 293–9, 445
salt pans 107, 299
salt spray 291, 383, 385
salt tolerance **288**
scabweed 362, 367
scale insects 152, 197, 567, 582
schist 1, 86, 88, 108, 223, 391, 419, 556
scree 373–4, 405–8, 417, 425, 429
scree plants **406–8**
scrub (grey) 72, **207–10**, 245, 258, 373, 401
sea hard-grass 290
sea musk 288
sea rocket 353
sea-barley 290
seals 281, 447, 456, 463, 579
Secretary Island 231
seed bank 64, 540
seed mass 60–2
seed predation 577–9
seed viability **62**
seedlings 43, 65–8, 583–600
semi-arid habitat **84**, 107
sheep 245, 436, 456
sheep treads 276, 278
Shirley (Lake) 393–4
shoot growth 477–9
short-tussock grassland 73, 244–63
Siberia valley 76
silver beech 78, 93, 103, 144–56, 159, 174, 179, 184,
346, 389, 392, 469, 475, 480, 483, 495, 514–15,
554–6, 560, 563, 567, 579, 581, 583–4, 586–7, 602
silver pine 20, 26, 43, 112, 118, 128, 130, 176, 178,
184, 186, 188, 193–4, 327, 331, 349, 400, 511,
514–15, 522, 525, 544, 562, 593
silver tussock 90, 194, 215, 245, 248, 256, 258–9,
265, 269, 290, 354, 356–7, 363, 367, 444, 526
silver wattle 163

Snares Islands 80, 109, 191
snow 78, 108, 227, 414, 504, 562, 580, Chapter 12
snow bank 425
snow tussock 217, 228, 243, 245, 337, 381, 471, 480,
528, 557
soil bases 88
soil drift 89, 569
soil fertility classes **90–2**
soil fertility gradients 112, 279
soil horizons **86–7**
soil pans 87–9, 128, 193–4, 335, 348, 509, 525, 530
soil pH (*see* pH)
soil sequences **88–90**
soil weathering **86**
soils (alluvial) 172, 510
Solander Islands 109, 191, 281
solifluction 337, 399, 405, 407, 419
solifluction topography 417–24
sorrel 249, 257, 368
Sounds–Nelson province **98–9**
Southern North Island province **97**
southern rata 113, 127, 132, 134–5, 139–40, 170,
174, 181, 184, 188, 352, 362, 376, 389, 400, 451–3,
475, 483, 494–5, 509, 511–12, 521–2, 527, 551,
561–2, 582–3, 585, 589–90
Southland province **107**
spaniard 245
speargrass 245
spines 42, 207
spinifex 355
St John's wort 553
Stephens Island 387
Stewart Island 109–10, 174, 180–2, 191, 281, 336,
339, 350, 396, 575, 590
Stokes (Mt) 188, 243
stolons 249, 265
stonecrop 379
stone nets 407, 420
string bogs 286, 331
subnival **78**
succession **91**, 197, 200, 205, 507, Chapter **15**
succulents 353, 378–9, 485
suckering 26, 562
sulphur 280
summit (climate) **82**, 187, 579
sundew 198
sunflecks 483
Surville Cliffs 397
swamp types **285**
swamps 73, 130, 193–4, 207, Chapter **10**
sweet vernal 214–15, 247, 259–60, 265, 272, 275–6,
372, 444, 520, 552
sycamore 160, 521

Table Mountain 188
tagasaste 213
Taita 498
takahe 575
Takitimu Mountains 218, 232
tall fescue 265, 276, 290, 360
tanekaha 113, 116, 120, 525, 533, 537, 564
tap-roots 43, 194, 413
tarahinau 437, 439, 441–2
taraire 93, 113, 116, 166, 532, 545, 560
Taramakau River 101, 565

Taranaki (Mt; also known as Mt Egmont) 96, 123, 139, 174, 186, 244, 349, 431, 472, 488
Taranaki province **96**
Tararua Range 98, 125, 184, 243, 349, 549–51, 581
Tarawera (Mt; including eruptions) 93, 362, 368, 524, 564
tarns 306, 309, 319, 400
Tasman Bay 99, 143
tauhinu 23, 166, 211, 396, 553
Taupo (Lake) 93, 95, 120
Taupo eruption 120, 324, 525–6, 556
tawa 21, 44, 93, 99, 113, 116, 119–121, 123–6, 132, 167, 524, 532, 538–9, 545, 560–1, 567, 575, 579, 590–1, 602
Te Anau (Lake) 103, 149, 331, 563
Te Moehau (Mt) 139, 349
temperature (*see also* leaf temperature) 3, 42, 57, 78, 83
tephra **85**, 93, 98, 188, 193–4, 208, 244–5, 259, 324, 361, 368, 525–6
Tertiary era 8–9, 11, 87, 506
thermal habitats 307, 400–4
thistle ('Californian') 267, 368
thistles 266, 275, 354, 364, 374, 444
Thomson (Lake) 516
Three Kings Islands 11, 92, 532
tidal zones 287
Tiritiri Matangi Island 543
titoki 120, 132, 167, 170
toatoa 113, 120
toetoe 245, 311, 316, 354
tomentum 41, 412
toro 116, 120, 130, 136, 138–9
totara (lowland) 84, 112, 120, 126, 128, 132, 145, 173, 388, 514, 524–6, 562, 582, 595, 599–600
totara (mountain) 84, 112, 115, 120, 127, 134–5, 137–9, 170–1, 174, 178, 184, 186–8, 509, 515, 524, 556, 562, 580, 590, 594–9
towai 533, 576
toxic ions 90, 401, 527
trace elements 280, 466
trampling 287, 443–4, 564
transpiration 497–9
traveller's joy 163
Travers Range 388, 572
tree fern 'skirts' 544
tree ferns 24, 31, 39, 111, 114, 116, 120, 123, 128, 135, 156, 162, 202, 438–9, 451, 509, 521, 542–4, 547, 599
tree lupin 213, 352, 355, 357, 359
Turakirae (Cape) 359–60, 385
turf 34, 281–2, 337, 355, 454, 458
Turks Head Range 82, 390
tussock-herbfield **73**
tussocks (*see also* blue, fescue, hard, red, silver, snow tussock) **29–30**, 70, 84, 193, 353–4
tutu 245, 524

ultra-infertile soil **90**
ultramafic rock (*see* rock; ultramafic)
Urewera (Ecological Region, National Park) 202, 575, 578

valley climate (*see* inversion)
vegetable sheep 108, 409–10, 607
vetch 267, 269
viper's bugloss 365
volcanic cliffs 395
volcanic eruptions (*see also* Tarawera, Taupo) 434–6, 526, 564
Volcanic Plateau 93, 120–30, 193, 208, 400, 599–601
volcanoes 1, 93, 96, 137
volume (increment) 475
volume (standing) 467–9

Waiau River 103, 108
Waikaremoana (Lake) 153, 377
Waikato (basin, Ecological Region) 93, 313, 324
Waima gorge 149, 377, 385, 388
Waimakariri (River, gorge) 106, 109, 195, 394, 425–7, 519
Waiomangu thermal area 402, 404
Waipoua Forest 114–16, 201
Wairarapa Plain 264
Waitakere Range 93, 533–8
Waitemata Harbour 292–5
wasps 7
water table 284, 319, 321, 331, 456, 579
waterfalls 381–2
wattle (*see* brush, silver wattle)
weka 245, 436, 463
West Cape 182–3, 192, 230–1
West Dome 400–1
Western Nelson province **99–101**
Westland 127–30, 176–9, 315–16, 326–8, 334–5, 358, 361, 366–7, 391–2, 509–15, 590–8
Westland National Park 177, 179, 418–19, 510, 566, 592
Westland province **101–2**
Westport 335–6
wet heath (*see* heath)
weta 7, 577
wetland **73**, Chapter **10**
Whangamarino swamp 313
wheatgrass (blue) 245
whipcord hebes 109, 195, 222
white clover 249, 259, 266, 269, 272, 275, 278, 313
willow (*see also* crack, grey willow) 160, 366, 553, 558
willow-herb 362, 412, 451, 558
wind (dispersal agent) 60, 195
wind-throw 567–72, 581, 583, 586–7
wineberry 162, 170–1, 174, 352, 372, 388, 508, 511, 515, 524, 531, 558, 561
woolly nightshade 162

xeromorphy **40–2**

yarrow 267
yellow-brown earth **88**, 127, 509, 511–12, 514–15, 520, 555–6, 569
yellow-grey earth **89**, 520
yellow-silver pine 26, 176, 178, 180, 182, 184
yorkshire fog 90, 444

zone (*see* latitudinal zones, moisture)

ERRATA

Page	Lines up or down	
7	4 up	Replace "Vespa" with "Vespula".
12	15 up	Delete "Deschampsia caespitosa".
12	14-15 up	Replace "the native form of" by "a possibly native form of Festuca rubra."
14	2 down	Should read "Celmisia philocremna" not "Celmisia philocremma".
16	Fig.2.4	Key for ferns (5th square from left) should contain horizontal lines.
24	12 down	Should read "talus" instead of "debris".
24	17 text lines down	Replace "Since the vascular cambium is regenerated from stem parenchyma, Cordyline is seldom killed by fire." With "Recovery from fire is usually through new shoots arising from the vertical, taproot-like rhizome (P. Simpson pers. comm.)"
25	Table 3.1	Entry for Muehlenbeckia astonii should be sd, not sf.
31	11 down	Should read "Histiopteris" not "Histopteris".
37	12 down	"≥ 50 cm2" not $\leq$.
50	1st text line	Insert "(Fig. 3.9)" after "semi-arid conditions".
54	10 down	Should read "3 m tall" not "4 m tall".
61	Table 4.1	Delete "woody" from legend.
77	Fig. 5.2 legend	Height of Mt Anglem is 980 m, not 676 m.
79	Fig. 5.3 legend	After "woody plants" add "and the herbaceous Elatostema rugosum".
94	7 up	Spelled "McEwen" not "MacEwan".
107	7 text lines down	Should read "2500 mm" not "2500 cm".
108	11 down	Should read "Celmisia philocremna" not "Celmisia philocremma".
152	6 text lines down	Delete "one of".
177	Fig. 8.10 legend	Coprosma colensoi should be preceded by Cco, not Cc.
208	3 up	Should be "They can be succeeded by forest" not "They can succeed".
238	15 up	Should read "Geum parviflorum" not "Geum uniflorum".
256	1st text line	Last word should be Cladina not Cladonia.
311	21 down	Should read "Deschampsia cespitosa, a mainly north-temperate species" not "...caespitosa, an otherwise north-temperate species".
340	6 up	Phyllachne colensoi not Phyllachne rubra.

389	3rd para heading	Height of Mt Burnett is (641 m) not 886 m.
446	11 down	Should read "north-east" not "north-west".
469	5 down in legend	Reference 8 should be "Heron ER" not "Puketeraki ER".
507	12 text lines down	Should read "sulphur above ground is lost" not "phosphorus above ground is lost..."
512	18 up	Should read "present in tree heath" not "present in the heath".
557	8 up	Spelled "question" not "question".
626	4 up	Spelled "McEwen" not "MacEwen".
643	2 down	Should read "philocremna" not "philocremma".
644	Column 1, 12 down	Replace "Chorodospartium" with "Chordospartium"
644	Column 1, 12 up	Should read "*vitalba" not "· vitalba".
646	Column 1, 5 down	Should read "·dealbata not "° dealbata".
654	Column 2, 14 up	Should read "*lupulina" not "· lupulina".
656	Column 1, 1 and 2 up	Left register Nostoc and Nothofagus.
657	Column 2, 30 down	Should read "Pelargonium 1† 5*" not "1 5*".
658	Column 2, 9 down	Should read "*radiate" not "· radiate"
662	Column 1, 31 down	Should read "semiannularis]" not ")".
662	Column 1, 10 up	Replace "gracilis" with "gracile".
665	Column2, 3 down	Should read "Wahlenbergia 9·1· *" not "9 · 1°*".
668	Column 1	Should read "forest (exotic) 156-60, 547" not "forest (exotic) 157-60, 547".
669	Column 2	There should be an entry for "mixed forest", with page entries 71 (i.e. bold) and 111.
671	Column 2, 8 up	Should read "takahe 576" not "takahe 575".

www.ingramcontent.com/pod-product-compliance
Lightning Source LLC
Chambersburg PA
CBHW062009190326
41458CB00009B/3024